Handbook of Mobile Broadcasting

DVB-H, DMB, ISDB-T, AND MEDIAFLO

INTERNET and COMMUNICATIONS

This new book series presents the latest research and technological developments in the field of Internet and multimedia systems and applications. We remain committed to publishing high-quality reference and technical books written by experts in the field.

If you are interested in writing, editing, or contributing to a volume in this series, or if you have suggestions for needed books, please contact Dr. Borko Furht at the following address:

Borko Furht, Ph.D.
Department Chairman and Professor
Computer Science and Engineering
Florida Atlantic University
777 Glades Road
Boca Raton, FL 33431 U.S.A.

E-mail: borko@cse.fau.edu

Handbook
of Mobile
Broadcasting

DVB-H, DMB, ISDB-T, AND MEDIAFLO

Editors-in-Chief

Borko Furht

Syed Ahson

CRC Press
Taylor & Francis Group
Boca Raton London New York

CRC Press is an imprint of the
Taylor & Francis Group, an **informa** business

AN AUERBACH BOOK

First published 2008 by Auerbach Publications

Published 2019 by CRC Press
Taylor & Francis Group
6000 Broken Sound Parkway NW, Suite 300
Boca Raton, FL 33487-2742

© 2008 by Taylor & Francis Group, LLC
CRC Press is an imprint of Taylor & Francis Group, an Informa business

First issued in paperback 2019

No claim to original U.S. Government works

ISBN 13: 978-0-367-45265-0 (pbk)
ISBN 13: 978-1-4200-5386-9 (hbk)

Library of Congress Cataloging-in-Publication Data

Handbook of mobile broadcasting : DVB-H, DMB, ISDB-T, and MEDIAFLO / Borko Furht and Syed Ahson, eds.
    p. cm. -- (Internet and communications ; 10)
    Includes bibliographical references and index.
    ISBN 978-1-4200-5386-9 (alk. paper)
    1. Webcasting. 2. Internet television. 3. Digital television. 4. Television broadcasting. I. Furht, Borko.
II. Ahson, Syed. III. Title. IV. Series.

TK5105.887H36 2008
006.7'876--dc22

2007040399

Visit the Taylor & Francis Web site at
http://www.taylorandfrancis.com

and the CRC Press Web site at
http://www.crcpress.com

Contents

PART IV: APPLICATIONS

Preface

Consumption of media has grown rapidly in the previous decade due to advances in digital technology. In recent years, the wireless industry has seen explosive growth in device capability. Ever-increasing computing power, memory, and high-end graphic functionalities have accelerated the development of new and exciting wireless services. Personal video recorders, video on demand, multiplication of program offerings, interactivity, mobile telephony, and media streaming have enabled viewers to personalize the content they want to watch and express their preferences to broadcasters. Viewers can now watch television at home or in a vehicle during transit using various kinds of handheld terminals, including mobile phones, laptops computers, and in-car devices. The concept of providing television-like services on a handheld device has generated much enthusiasm. Mobile telecom operators are already providing video-streaming services using their third-generation cellular networks. Simultaneous delivery of large amounts of consumer multimedia content to vast numbers of wireless devices is technically feasible over today's existing networks, such as third-generation (3G) networks.

National governments are encouraging an all-digital broadcast environment. As conventional analog television services end, broadcasters will exploit the capacity and flexibility offered by digital systems. Broadcasters will provide quality improvements, such as high-definition television (HDTV), which offer many more interactive features and permit robust reception to receivers on the move in vehicles and portable handhelds.

Mobile TV systems deliver a rich variety of content choice to consumers while efficiently utilizing spectrum as well as effectively managing capital and operating expenses for the service provider. Mobile TV standards support efficient and economical distribution of the same multimedia content to millions of wireless subscribers simultaneously. Mobile TV standards reduce the cost of delivering multimedia content and enhance the user experience, allowing consumers to surf channels of content on a mobile receiver. Mobile TV standards address key challenges involved in the wireless delivery of multimedia content to mass consumers and offer better performance for mobility and spectral efficiency with minimal power consumption.

Viewers have indicated their interest in watching television from a handheld device. A number of studies conducted globally confirm this. Several pilot projects and commercial launches that provide live television broadcast services are currently under way or in development in many parts of the world. *Digital video broadcasting–handheld* (DVB-H) services have been launched in several countries across the globe. At present, about 40 countries are promoting or planning to introduce the digital audio broadcasting (DAB) service. Integrated Services Digital Broadcasting–Terrestrial (ISDB-T) was adopted for commercial transmissions in Japan, and the government of the Federative Republic of Brazil adopted a terrestrial digital television standard based upon ISDB-T. Argentina and Venezuela are strongly considering ISDB-T to enjoy common market benefits from the regional South American development of the technology. Several organizations are dedicated to fostering the growth of mobile TV services around the globe.

The *Handbook of Mobile Broadcasting* provides technical information about all aspects of mobile broadcasting. The areas covered in the handbook range from basic concepts to research articles, including future directions. The handbook captures the current state of mobile broadcasting technology and serves as a source of comprehensive reference material on this subject. The handbook comprises four sections: **"Standards," "Optimization and Simulation," "Technology,"** and **"Applications."** It has a total of 24 chapters authored by 80 experts from around the world.

Section I, "Standards," consists of five chapters that describe and compare DVB-H, *digital video broadcasting* via satellite services to handheld devices (DVB-SH), DMB-T, and ISDB-T in great detail. **Section II, "Optimization and Simulation"** consists of five chapters that discuss implementation of H.264/AVC video decoder and *analog-to-digital converters for mobile broadcasting.* **Section III, "Technology,"** consists of ten chapters that analyze synchronization techniques, describe XXXX–orthogonal frequency division multiplexing (MIMO-OFDM) systems, detail end-to-end service discovery process and seamless handovers, and present an overview of radio resource management schemes and error control schemes used in broadcasting systems. Chapters in the technology section also cover air interface enhancements, optimization of packet scheduling schemes, finite-state models, and performance analysis. **Section IV, "Applications,"** consists of four chapters that present the basic architecture and components of video streaming and explore periodic broadcast techniques for video-on-demand (VOD) services in a wireless networking environment.

The targeted audiences for the handbook include professionals who are designers and planners for mobile broadcasting systems, researchers from academia (faculty members and industry graduate students), and those who would like to learn about this field.

The handbook is expected to have the following specific salient features:

- To serve as a single comprehensive source of information and as reference material on mobile broadcasting technology
- To deal with an important and timely topic of emerging technology of today, tomorrow, and beyond
- To present accurate, up-to-date information on a broad range of topics related to mobile broadcasting technology
- To present the material authored by the experts in the field
- To present the information in an organized and well-structured manner

Although the handbook is not precisely a textbook, it can certainly be used as a textbook for graduate courses and research-oriented courses that deal with mobile broadcasting. Any comments from the readers will be highly appreciated.

Many people have contributed to this handbook in their unique ways. The first and foremost group that deserve immense gratitude are the highly talented and skilled researchers who have contributed 24 chapters to this handbook. All of them have been extremely cooperative and professional. It has also been a pleasure to work with Rich O'Hanley, Ms. Jessica Vakili, and Ms. Gail Renard of CRC Press, and we are extremely gratified for their support and professionalism. Our families have extended their unconditional love and strong support throughout this project, and they all deserve very special thanks.

Borko Furht
Boca Raton, Florida

Syed Ahson
Plantation, Florida

Editors

Borko Furht is chairman and professor of computer science and engineering at Florida Atlantic University (FAU) in Boca Raton, Florida. He is the founder and director of the Multimedia Laboratory at FAU, funded by the National Science Foundation. Before joining FAU, he was a vice president of research and a senior director of development at Modcomp, a computer company in Fort Lauderdale, Florida, and a professor at the University of Miami in Coral Gables, Florida. He received his Ph.D. degree in electrical and computer engineering from the University of Belgrade, Yugoslavia. His research interests include multimedia systems and applications, video processing, wireless multimedia, multimedia security, video databases, and Internet engineering. He is currently principal investigator or co-principal investigator and leader of several large multiyear projects including "One Pass to Production," funded by Motorola, and "Center for Coastline Security Technologies," funded by the U.S. government as a Federal Earmark project.

Dr. Furht has received research grants from various government agencies, such as NSF and NASA, and from private corporations, including IBM, Hewlett Packard, Racal Datacom, Xerox, Apple, and others. He has published more than 20 books and about 200 scientific and technical papers, and holds two patents. His recent books include *Multimedia Security Handbook* (2005) and *Handbook of Multimedia Databases* (2004), both published by CRC Press. He is a founder and editor-in-chief of the *Journal of Multimedia Tools and Applications* (Kluwer Academic Publishers, now Springer). He is also consulting editor for two book series on Multimedia Systems and Applications (Kluwer/Springer) and Internet and Communications (CRC Press). He has received several technical and publishing awards and has consulted for IBM, Hewlett-Packard, Xerox, General Electric, JPL, NASA, Honeywell, and RCA. He has also served as a consultant to various colleges and universities. He has given many invited talks, keynote lectures, seminars, and tutorials. He has been program chair as well as a member of program committees at many national and international conferences.

Syed Ahson is a senior staff software engineer with Motorola Inc. He has extensive experience with wireless data protocols (TCP/IP, UDP, HTTP, VoIP, SIP, H.323), wireless data applications (Internet browsing, multimedia messaging, wireless email, firmware over-the-air update), and cellular telephony protocols (GSM, CDMA, 3G, UMTS, HSDPA). He has contributed significantly toward the creation of several advanced and exciting cellular phones at Motorola. Prior to joining Motorola, he was a senior software design engineer with NetSpeak Corporation (now part of Net2Phone), a pioneer in VoIP telephony software.

Syed is a coeditor of the *Handbook of Wireless Local Area Networks*: *Applications, Technology, Security, and Standards* (CRC Press, 2005). Syed has authored "Smartphones" (International Engineering Consortium, April 2006), a research report that discusses smartphone markets and technologies. He has published several research articles in peer-reviewed journals and teaches computer engineering courses as an adjunct faculty member at Florida Atlantic University, Boca Raton, where he introduced a course on smartphone technology and applications. Syed received his B.Sc. in electrical engineering in India in 1995 and M.S. in computer engineering in 1998 at Florida Atlantic University.

Contributors

Cristiano Akamine
Mackenzie Presbyterian University
São Paulo, Brazil

I. Andrikopoulos
Space Hellas S.A.
Athens, Greece

Tommi Auranen
Nokia Corporation
Turku, Finland

Gunnar Bedicks, Jr.
Mackenzie Presbyterian University
São Paulo, Brazil

F. Boronat
Universidad Politécnica de Valencia
Valencia, Spain

Yung-Hung Chang
Department of Computer Science
National Tsing Hua University
Hsin-Chu, Taiwan

Ping Chao
Department of Computer Science
National Tsing Hua University
Hsin-Chu, Taiwan

Jian-Wen Chen
Department of Computer Science
National Tsing Hua University
Hsin-Chu, Taiwan

Liang-Gee Chen
National Taiwan University
Taipei, Taiwan

Qingchun Chen
Southwest Jiaotong University
Chengdu, China

Tung-Chien Chen
National Taiwan University
Taipei, Taiwan

Jean-Yves Chouinard
Laval University
Quebec City, Canada

N. Chuberre
Thales Alenia Space
Toulouse, France

M. Cohen
Alcatel-Lucent Mobile Broadcast
Paris, France

David Coquil
University of Passau
Passau, Germany

Américo Correia
IT/ADETTI
Lisbon, Portugal

O. Courseille
Thales Alenia Space
Toulouse, France

Nikos Dimitriou
Institute of Accelerating Systems
    & Applications
National Kapodistrian
University of Athens
Athens, Greece

Hongfei Du
University of Surrey
Guildford, Surrey, United Kingdom

Régis Duval
Alcatel-Lucent Mobile Broadcast
Paris, France

Barry G. Evans
University of Surrey
Guildford, Surrey, United Kingdom

Linghang Fan
University of Surrey
Guildford, Surrey, United Kingdom

J. Farineau
Thales Alenia Space
Toulouse, France

I. Foukarakis
National Technical University
    of Athens
Athens, Greece

P. Fouliras
University of Macedonia
Thessaloniki, Greece

Tiago Gasiba
Digital Fountain
Fremont, California

Y. Guo
Shanghai MicroScience Integrated
    Circuits Co., Ltd.
Shanghai, China

C. Herrero
Helsinki University of Technology
Espoo, Finland

Heidi Himmanen
University of Turku
Turku, Finland

Günther Hölbling
University of Passau
Passau, Germany

Sheng-Tsung Hsu
Department of Computer Science
National Tsing Hua University
Hsin-Chu, Taiwan

Yu-Wen Huang
National Taiwan University
Taipei, Taiwan

Heiko Hübert
Fraunhofer Institut für Nachrichtentechnik
Sankt Augustin, Germany

Wei-Cheng Hung
Department of Computer Science
National Tsing Hua University
Hsin-Chu, Taiwan

Valery Ipatov
University of Turku
Turku, Finland

Kai-Yuan Jan
Department of Computer Science
National Tsing Hua University
Hsin-Chu, Taiwan

Tero Jokela
University of Turku
Turku, Finland

D. I. Kaklamani
National Technical University of Athens
Athens, Greece

D. A. Kateros
National Technical University of Athens
Athens, Greece

C. Katsigiannis
National Technical University of Athens
Athens, Greece

Harald Kosch
University of Passau
Passau, Germany

E. Kosmatos
National Technical University of Athens
Athens, Greece

P. Laine
Alcatel-Lucent Mobile Broadcast
Paris, France

Kam-Yiu Lam
City University of Hong Kong
Hong Kong, China

Chun-Hsin Lee
Department of Computer Science
National Tsing Hua University
Hsin-Chu, Taiwan

J. Li
State Key Lab of ASIC & System
Fudan University
Shanghai, China

Chung-Jr Lian
National Taiwan University
Taipei, Taiwan

Youn-Long Lin
Department of Computer Science
National Tsing Hua University
Hsin-Chu, Taiwan

Yuan-Chun Lin
Department of Computer Science
National Tsing Hua University
Hsin-Chu, Taiwan

R. Llorente
Universidad Politécnica de Valencia
Valencia, Spain

J. Lloret
Universidad Politécnica de Valencia
Valencia, Spain

Tzu-Jen Lo
Department of Computer Science
National Tsing Hua University
Hsin-Chu, Taiwan

Michael Luby
Digital Fountain
Fremont, California

Gunther May
Technical University of Braunschweig
Braunschweig, Germany

Thinh Nguyen
School of EECS
Oregon State University
Corvallis, Oregon

Jarkko Paavola
University of Turku
Turku, Finland

C. Papagianni
National Technical University of Athens
Athens, Greece

Huan-Kai Peng
Department of Computer Science
National Tsing Hua University
Hsin-Chu, Taiwan

Jussi Poikonen
University of Turku
Turku, Finland

G. N. Prezerakos
Technological Education Institute of Piraeus
Athens, Greece

T. Quignon
Thales Alenia Space
Toulouse, France

L. Roullet
Alcatel-Lucent Mobile Broadcast
Paris, France

A. Serment
Alcatel-Lucent Wireless
Velizy, France

Amin Shokrollahi
EPFL and Digital Fountain
Fremont, California

João Carlos Silva
IT/ADETTI
Lisbon, Portugal

Armando Soares
IT/ADETTI
Lisbon, Portugal

Nuno Souto
IT/ADETTI
Lisbon, Portugal

Benno Stabernack
Fraunhofer Institut für Nachrichtentechnik
Sankt Augustin, Germany

Thomas Stockhammer
Digital Fountain
Fremont, California

M. Tatard
Thales Alenia Space
Toulouse, France

Duc A. Tran
Department of Computer Science
University of Dayton
Dayton, Ohio

Jani Väre
Nokia Corporation
Turku, Finland

I. S. Venieris
National Technical University of Athens
Athens, Greece

D. Vicente
TECATEL, S.A.
Beniarjó, Valencia, Spain

P. Vuorimaa
Helsinki University of Technology
Espoo, Finland

Xianbin Wang
Communications Research Centre
Ottawa, Canada

Mark Watson
Digital Fountain
Fremont, California

Kai-Immo Wels
Fraunhofer Institut für Nachrichtentechnik
Sankt Augustin, Germany

Yiyan Wu
Communications Research Centre
Ottawa, Canada

D. A. Zarbouti
National Technical University
  of Athens
Athens, Greece

X. Zeng
State Key Lab of ASIC & System
Fudan University
Shanghai, China

STANDARDS

1

Chapter 1

An Overview of the Emerging Digital Video Broadcasting–Handheld (DVB-H) Technology

David Coquil, Günther Hölbling, and Harald Kosch

Contents

Keywords

digital broadcasting, DVB, DVB-H, handheld device, IP datacast, mobility, forward error correction, time slicing

1.1 Introduction

In recent years, television broadcasting systems all over the world have been gradually switching over from analogue to digital. Meanwhile, the market of handheld mobile devices (mainly mobile phones and smart phones but also personal digital assistants) has seen a tremendous growth, prompting such improvements in the capabilities of the devices that the most technologically advanced are now up to the task of efficiently displaying video content. This aroused interest in making digital television and video programs in general available to these devices, and led to the introduction of a number of relevant commercial proposals and international standards. Among these, the Digital Video Broadcasting (DVB) Group has released the Digital Video Broadcasting–Handheld devices (DVB-H) standard, for which this chapter provides an extensive introductory overview. Before describing the main features of DVB-H, let us first outline general information about the DVB Group and the DVB-H project.

1.1.1 The DVB Group

The DVB Group is a consortium of around 300 public and private organizations that was founded in 1993.[1] The original members were primarily European, but as the project developed, members from other continents joined. As of late 2006, about 35 different nationalities were represented in the project. The members are essentially companies that have an interest in digital video broadcasting: content producers, broadcast providers, network operators, software producers, and digital devices manufacturers.*

---

* Up-to-date information about the DVB Project is published on its website at http://www.dvb.org/.

The project is the driving force of the development of all aspects of digital television in many countries, especially in Europe. It is mostly concerned with the design of specifications for all areas related to digital video broadcasting. Once validated by DVB, these documents are made publicly available as DVB blue books and passed to the Joint Technical Committee (JTC) Broadcast of the European Broadcasting Union (EBU), jointly formed by the European Telecommunications Standards Institute (ETSI) and the Comité Européen de Normalisation ELECtronique (CENELEC). These organizations amend the specifications before publishing them as international (European) standards. In parallel, DVB also provides implementation guidelines for its specifications and conducts technical experiments to assess the practical validity of its technologies. This process has been notably conducted to define the following standards: DVB-S and its follow-up DVB-S2[2] for satellite-based digital video broadcasting, DVB-T for digital terrestrial television broadcasting,[3,4] DVB-C for cable transmission,[5] and the subject of this chapter, DVB-H, for digital video broadcasting to mobile handheld devices.

1.1.2 DVB-H: Motivations

In 1998, the DVB Project started to study the possibility of mobile reception of the signals defined in the up and coming terrestrial DVB-T standard, which had been published one year earlier.[6] The overall conclusion was that it was feasible if a number of updates and extensions were included in the DVB-T standard.[7] Work consequently started toward a new version of the standard. This effort bore fruit starting from 2002 with the wide launch of mobile DVB-T broadcasting services: in Singapore DVB-T displays were implemented in the buses of the public transportation system, and in Germany mobile services were made available as part of the general switch of the country from analogue to DVB-T-based digital television broadcasting. However, during that period of time customer expectations had changed, and the technology of handheld devices (personal digital assistants, cell phones, smart phones) had made such remarkable progress that they were now capable of mobile reception of digital television signals.

Such a service has a number of important differences with respect to the mode of operation of a DVB-T system (either fixed or mobile DVB-T):

- Although typical DVB-T receivers benefit from a continuous power supply, handheld devices run on batteries with limited autonomy that the users would rather not have to recharge too often.
- The handheld devices can only receive signals using a small and not optimally oriented antenna, contrarily to the rooftop or room antennas used for receiving DVB-T signals.
- This service needs to support extreme mobility, as opposed to the rather limited and controlled mobility of the DVB-T services described above (for example, the path along which a bus moves is fixed and known in advance). It also possibly needs to cope with very unfavorable transmission and display conditions characterized by severe mobile multipath channels and a high-level of man-made noise.
- The available size for the display is much smaller and the devices have limited capabilities (less memory, processing power, storage space) compared to those of the standard equipment used for the reception of DVB-T streams.

These constraints prompted the DVB Group to launch a specific project to define such a system in 2002, led by the DVB ad hoc group TM-H (for technical module handheld). The resulting standard was called Digital Video Broadcasting–Handheld devices, DVB-H (it was originally called DVB-M and later also DVB-X). It was published in November 2004 as ETSI standard EN 302 304.[8] This document defines an umbrella standard, in the sense that it explains how other DVB standards are to be combined to produce the DVB-H system.

1.1.3 DVB-H Overview

The DVB-H system can be precisely defined as a transmission system built out of several DVB standards aiming at efficient terrestrial broadcasting of digital multimedia data to handheld devices, the main type of transmitted data being digital television. Within the framework of the Open Systems Interconnection (OSI) reference model, ETSI standard EN 302 304[8] defines a DVB-H system as a combination of technology elements of the physical layer, elements of the data link layer (as defined in the OSI network reference model), and service information.

The main standards composing the system are the following:*

- ETSI EN 300 744,[4] which is the main DVB-T standard. DVB-H is based on DVB-T; in particular, it uses its physical transmission layer.
- ETSI EN 301 192[9] defines the DVB data broadcast standard. It specifies a number of concepts used by several or all DVB standards, including DVB-H.
- ETSI EN 300 468[10] defines the format of the service information (SI) in DVB systems, including DVB-H.

In addition, the implementation guidelines detailed in ETSI TR 102 377[11] provide very complete information on the nature of future DVB-H systems. The development of DVB-H also had an influence on the DVB Single-Frequency Networks Megaframe specification,[12] which describes the synchronization of terrestrial single-frequency networks within the framework of DVB.

At the physical layer level, DVB-H essentially makes use of the corresponding elements of the DVB terrestrial standard (DVB-T), with a number of extensions adapting it for mobile handheld reception. These extensions have either been integrated in the latest DVB-T version or do not directly modify it; DVB-T and DVB-H are thus fully compatible, as evidenced by the possibility of them coexisting within a single multiplex. This compatibility is an important feature of DVB-H, as many DVB-T systems have already been launched all over the world.

At the link layer level, Multi-Protocol Encapsulation (MPE) sections are transmitted. The main technological innovations of DVB-H are implemented there:

- **Time slicing**, a data transmission method enabling lower average power consumption for the battery-powered devices as well as a seamless handover between different services for the mobile DVB-H users.
- **MPE-FEC**, which adds an additional level of forward error correction (FEC) to that already provided by DVB-T, as a means of coping with the possibly extremely difficult reception conditions arising from the mobility of DVB-H devices.

Although the above is strictly speaking sufficient to define a full system according to the definition of ETSI standard EN 302 304,[8] DVB-H has been conceived from the start to be fully

---

* Up-to-date versions of most DVB standards related to DVB-H are available online at http://www.dvb-h.org/. ETSI standards in general are archived at http://www.etsi.org/.

compatible with data networks. This is outlined by the use of MPE at the link layer level, as this method has been chiefly created to provide a means of transporting data network protocols, and in particular the Internet Protocol (IP), in its IPv4[13] and IPv6[14] versions, on top of an MPEG-2 transport stream. DVB-H can serve as the bottom-most layer of a completely IP-based system, on condition that a number of additional components are specified. The DVB Internet Protocol Datacast (DVB-IPDC[42]) specification was introduced to fulfill this requirement. In particular, it can be used to combine a unidirectional network involving a wide transmission area and high data transmission rates (typically DVB-H broadcast channel) with a bidirectional communication network such as the current popular cell networks (General Packet Radio System [GPRS], Universal Mobile Telecommunications System [UMTS], etc.).

The rest of this chapter is organized as follows. In section 1.2, we describe the components of DVB-H corresponding to its physical layer (mostly based on DVB-T with a number of extensions), as well as its service information features. Section 1.3 describes the link layer of DVB-H Multi-Protocol Encapsulation (MPE), on which the data transmission is based at this level, and the main technological innovations of DVB-H: time slicing and MPE-FEC. The DVB-H IP datacast system is then described in section 1.4. Finally, a summary of the chapter can be found in section 1.5.

1.2 DVB-H Physical Layer and Service Information

The physical layer of DVB-H systems essentially conforms to the DVB-T specification, with a limited number of extensions.

DVB-T is the DVB standard for the terrestrial broadcasting of digital television signals in the atmosphere. It primarily covers signal reception using fixed-rooftop aerial antennas or room antennas, but also provides support for specific mobile devices incorporated in cars or buses.[15]

The latest version of DVB-T was adopted by ETSI as a standard in October 2004.[4] As of 2006, DVB-T has been selected for digital terrestrial television by most European countries, and a number of them have launched services (Germany, United Kingdom, France, Spain, etc.).* The standard has also been adopted by India, and trials are taking place in a number of southeastern Asian countries, including China. Services have launched in Australia and Taiwan.

The service information features of DVB-H are also similar to those of DVB-T, in the sense that they both follow the global DVB standard in this matter.[10] The lowest-level parameters of a program are transmitted using DVB-H signaling, which is basically identical to DVB-T signaling with a limited number of extensions.

1.2.1 DVB Networks

Figure 1.1 represents the global hierarchical structure of DVB broadcast networks, as used by DVB-T and DVB-H. The overall principle is based on MPEG-2 transport streams, as defined in the corresponding MPEG-2 standard.[17]

At the lowest level are the elementary streams (ESs). One ES carries one specific type of consistent streaming content, usually output by one MPEG-2 encoder. Video and audio elementary streams are the most common, but other types of data are also possible (for example, textual information for subtitles).

---

* The DVB Group maintains an updated summary of the status of adoption and deployment of DVB-T and DVB standards in general around the world. It is available on its website at http://www.dvb.org.

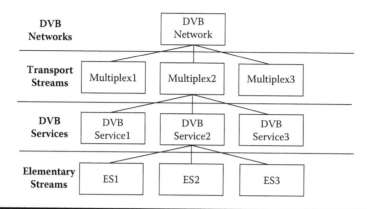

Figure 1.1   General structure of DVB networks (adapted from reference 16).

A set of ESs that constitutes a semantic entity when grouped and synchronized together is called a DVB service (also a program, from the vocabulary used in the MPEG-2 standard). An example is one DVB service carrying a TV channel that comprises one ES for its video content, one ES for its audio content, and one ES for each language in which subtitles are available.

For transmission purposes, several DVB services are multiplexed together in the form of a single signal (a terrestrial radio signal in the case of DVB-T/DVB-H). This is performed to obtain an MPEG-2 transport stream (TS) as defined by the MPEG-2 standard. Note that while one TS carries exactly one multiplex, the same multiplex may be carried by several TSs.

At the highest level of the structure are DVB networks, which comprise several TSs.

1.2.2   MPEG-2 Data Transmission

DVB-T generally follows the data transmission methods for MPEG-2 transport streams defined in the MPEG-2 standard.[17]

ES data is first divided in packetized elementary stream (PES) packets (see figure 1.2). PES packets have a variable size (up to 65,536 bytes). They start by a 6-byte protocol header, which includes a 1-byte stream identifier (SID) of the source of the payload. This fixed-length header may optionally be followed with a variable-length extended header comprising various informational fields: cycle redundancy check code calculated over the payload, rate of encoding of the data, when

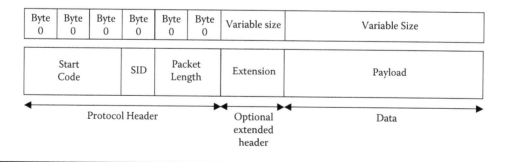

Figure 1.2   PES packet format.

Table 1.1   Structure of TS Packets

Field Name	Size (in bits)	Description
Synchronization byte	8	Value fixed at 0×47
TS error indicator	1	Set by lower-level equipment upon detection of an unrecoverable error
Payload unit start	1	Set if the payload of this packet is the start of a PES packet
Priority	1	
PID (packet identifier)	13	
Scrambling control	2	
Adaptation field control	2	Indicate if there is an adaptation field extending the header information: 01: no adaptation field, payload only 10: adaptation field only, no payload 11: adaptation field followed by payload 00: reserved
Continuity counter	4	Counter incremented for each packet corresponding to one PID; reset to 0 when it overflows its capacity
Adaptation field	0 or more	Depends on the adaptation field control
Data	0 or more	

to decode the frame (decode time stamp [DTS]), when to display the frame (presentation time stamp [PTS]), and so on.

The PES packets are then broken down into fixed-size transport packets (TS packets). Doing so provides a way to combine several ESs into one single TS. The structure of TS packets is detailed in table 1.1.

The size of the TS packets is fixed at 188 bytes each (or 204 if 16 bytes of optional error correction code is added). Each packet starts with a header of 4 bytes. For the most common packets, that is, those without an adaptation field, the header is followed by 184 bytes of payload. The most important fields of the header are the synchronization byte and the packet identifier (PID). The synchronization byte allows the receiver to identify the start of a TS packet. The PID associates the TS packet with one of the elementary streams carried by the transport stream.

1.2.3 Physical Transmission

1.2.3.1 Channel Coding and Modulation

The MPEG-2 TS packets described in the preceding section must undergo a number of transformations before they can be transmitted. This is due to several reasons. First, DVB-T being a digital terrestrial television technology, its signals are to be transmitted on radio frequencies. This implies a high probability of transmission errors, notably burst errors. Given the lack of inherent protection of MPEG-2 transport streams from such errors, specific techniques are necessary: interleaving to reduce the consequences of transmission errors, and forward error correction to allow the receiver to correct them. Another constraint specific to DVB-T is that it is designed to operate within frequencies where analogue transmissions are currently operating, meaning that provisions have to be taken so that the DVB-T signals neither disturb nor are disturbed by these other signals. Finally, a digital signal needs to be modulated before it can be transmitted via radio airwaves.

Taking these constraints into account, the following processes are successively applied to the data stream starting from the base band TV signal:[4]

- Transport multiplex adaptation
- Randomization for energy dispersal
- Outer coding
- Outer interleaving
- Inner coding
- Inner interleaving
- Mapping and modulation
- Orthogonal frequency division multiplexing (OFDM) transmission

Let us now describe the exact purpose and technical aspects of these processes. Their complete specifications can be found in Ladebusch and Liss.[4]

The **transport multiplex adaptation** phase transforms several input MPEG-2 encoded elementary streams (video, audio, or other) into a single stream of multiplexed MPEG-2 TS packets as defined in section 1.2.2. The data is then subjected to a **randomization** program. Its purpose is to avoid periodic bit patterns in the transmitted data. Such patterns would create discrete spectral lines in the modulated signal, possibly causing transmission problems (interferences). The bits of the base band signal are thus combined with a generated pseudorandom binary sequence.

The **outer coding** process consists in computing a Reed–Solomon code to provide error correction capabilities to the receiver. The code is computed using an RS (204, 188) shortened algorithm obtained by adding 51 null bytes of padding to the 188 bytes of a TS packet and passing them to an RS (255, 239) process.

The **outer interleaver** that follows is called Forney convolution interleaving. Its purpose is to scatter potential burst errors by shuffling the data bytes over 12 packets.

Next the data undergoes **inner coding**, an inner error protection code optimized for bit errors correction that is implemented by a convolutional encoder with a basic rate of ½ with 64 states. This generates a high level of redundancy that can be reduced by puncturing, that is, not transferring all the calculated bits to reduce the overhead. This induces a trade-off with the robustness of the resulting code; the available code rates per DVB-T definition are 1/2, 2/3, 3/4, 5/6, and 7/8.

The next process is **inner interleaving**, performed to alleviate the bursty structure of the stream. It consists in a bit-wise interleaving followed by a symbol interleaving.

The data output by the symbol interleaving is then mapped into a signal constellation. This is performed OFDM frame by OFDM frame (OFDM is the transmission mode used by DVB-T; see next section). Several modulation scheme options are available: quadrature phase-shifting keying (QPSK), 16-quadrature amplitude modulation (QAM), 64-QAM, nonuniform 16-QAM, and nonuniform 64-QAM. Two, four, or six consecutive bits is allocated to the carrier using a Gray mapping.

1.2.3.2 OFDM Transmission

For transmission, the system uses orthogonal frequency division multiplexing (OFDM; see Scott[18]). The two main OFDM transmission modes of DVB-T are 2K and 8K—the names corresponding to the number of points of the fast Fourier transform that they use (2,048 or 8,192). The standard

Table 1.2   Parameters for DVB-H OFDM Modes

OFDM parameter	2K	4K	8K
Number of global carriers	2,048	4,096	8,192
Number of modulated carriers	1,705	3,409	6,817
Number of useful carriers	1,512	3,024	6,048
Duration of an OFDM symbol (in μs)	224	448	896
Duration of the guard interval(in μs)	7, 14, 28, 56	14, 28, 56, 112	28, 56, 112, 224
Carriers spacing (in kHz)	4,464	2,232	1,116
Maximal distance coverage of transmitters (in km)	17	33	67

also defines an optional 4K mode; it is designed specifically with the requirements of DVB-H in mind. The detailed parameters of the three DVB-H OFDM modes can be found in table 1.2.

In DVB-T, the signal is transmitted in OFDM frames composed of 68 symbols. A symbol consists of 6,817 carriers in the 8K mode, 3,409 carriers in the 4K mode, and 1,705 carriers in the 2K mode. Each OFDM symbol is composed of a useful part (payload) and of a guard interval meant to protect the transmitted signal from echoes and reflections. Four consecutive frames constitute a superframe.

An OFDM frame also contains other informative parts in addition to the data:

■ Scattered pilot cells and continual pilot carriers that can be used by the receivers for the equalization and synchronization phase
■ Transmission parameter signaling (TPS) carriers (see section 1.2.5)

The characteristics outlined by the summary of OFDM parameters found in table 1.2 summarize the purpose of the introduction of the 4K OFDM mode for DVB-H systems: to provide a compromise between the size of the transmission cells (optimal in the 2K mode) and the resistance to echo effects (best achieved with the 8K mode). More precisely, the trade-off is the following:[6]

■ The DVB-T 8K mode can be used both for single-transmitter operation (multiple-frequency networks [MFNs]) and for small, medium, and large single-frequency networks (SFNs). It provides a Doppler tolerance allowing for high-speed reception.
■ The DVB-T 4K mode can be used both for single-transmitter operation and for small and medium SFNs. It provides a Doppler tolerance allowing for very high-speed reception.
■ The DVB-T 2K mode provides the best mobile reception performance because of its larger intercarrier spacing. However, because of the OFDM symbol and guard interval, it is only suitable for small-size SFNs.

The DVB-T system allows for hierarchical transmission, which means that two different transport streams can be transmitted simultaneously. Typically, this is used to transmit two streams on the same carrier, a low-bit-rate stream and a more error-prone stream of higher quality, allowing the receiver to dynamically switch between them depending on its reception conditions or its own capabilities.

1.2.3.3 Frequency Bands

One of the design constraints of the DVB-T system is that it should be compatible with the VHF and UHF spectrum frequencies currently reserved for analogue transmissions, that is, for Europe

7 MHz wide for the VHF range and 8 MHz wide for the UHF range; 6 MHz was considered as well, because it is used in other parts of the world. The OFDM system described in the previous section is therefore defined for 8, 7, and 6 MHz channel spacing. These bandwidths are supported by DVB-H as well, with the addition of a 5 MHz mode, providing an option for using it outside the classical broadcast bands.

1.2.4 Service Information

In addition to video and audio data, an MPEG-2 transport stream also needs to carry so-called service information. This information is meant to help receivers navigate around the services offered by a DVB system as well as to learn about network and signal parameters required for the technical servicing.

To this end, DVB-H and DVB-T use the program-specific information (PSI) defined in the MPEG standard.[17] The PSI is organized in the form of tables. Like regular data, these tables are carried in MPEG-2 TS packets (as described in section 1.2.2) and have their own PIDs. The tables are the following:

- Program Association Table (PAT): Contains a breakdown of the TS into individual programs with the correspondino PIDs for each of them in the Program Map Table (PMT). The PAT always has PID 0x0000. A PAT identifies a TS with the field "transport_stream_id." These identifiers are not guaranteed to be unique across DVB networks; therefore, the PAT also carries an identifier of its DVB network (original_network_id).
- Program Map Table (PMT): Contains the list of PIDs carrying data from a specific program, as well as metadata about each program. The PMT identifies a DVB service, with its "service_id" field.
- Conditional Access Table (CAT): Contains data that can be used to restrict the access to some programs. It can be used for pay-per-view services, for example. Contrary to the others, this table is optional.
- Network Information Table (NIT): Contains general information relating to the physical organization of the multiplexes/TSs carried via a given network and the characteristics of the network itself. This information is supposed to be sufficient for a receiver to tune in to a TS. The information about other TSs is optional, but the NIT must contain information about the TS it belongs to. An NIT table contains an identifier of a DVB network called "network_id."

Although the MPEG-2 standard specifies these four types of table, it does not define the content of the CAT and the NIT, leaving it up to projects using the standard. For DVB, the definition is to be found in EN 300 468.[10] Moreover, the standard allows for the definition of private tables, whose structure and semantic can be freely defined by the projects using MPEG-2, as the standard only provides a template structure for these tables. The NIT is considered a private table. The DVB Project took advantage of this possibility by introducing the following additional tables:

- Service Description Table (SDT): Complements the PAT. While the PAT provides the correspondence between the PIDs and a program, the SDT adds extended metadata information about the programs it contains. The SDT of a TS should at least include information about all programs of the TS it belongs to, and can also optionally provide information about the programs of other TSs of the same network.

- Event Information Table (EIT): Complements the PMT. While the PMT allows a receiver to locate the TSs currently used by a program, the EIT describes its current and future content: name of the event (e.g., name of a TV show), start time, duration, information about the present event, the next event, and optionally other future planned events.
- Time and Date Table (TDT): Used to update the internal clock of the receivers. It denotes the current date and time.

The following tables are optional:

- Running Status Table (RST): Gives the status of one event; because it is constantly updated, it can be used by a receiver to synchronize with the start of an event.
- Time Offset Table (TOT): Gives the difference between the local time and the universal time.
- Bouquet Association Table (BAT): Provides information about the *bouquet*: bouquet provider, included services. A bouquet is a set of services offered as a single entity to the users, in which the services can possibly span across different networks. For each bouquet, the BAT provides its provider name and the list of its included services.

The tables defined by DVB are called the service information (SI) tables.

Figure 1.3 shows an updated representation of the structure of DVB networks, detailing the table where the information about each type of entity is to be found and the names of the corresponding identifier fields.

Contrary to video and audio data, SI data is not transmitted by means of PES packets but instead within *sections*. These sections may have varying length, but they are limited to 1,024 bytes for the tables introduced in MPEG-2 (PAT, PMT, and CAT) and to 4,096 bytes for the others.

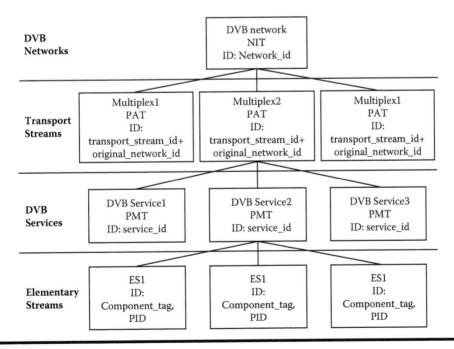

Figure 1.3 DVB network structure.

Each section starts with a 3-byte header, most importantly carrying a 1-byte table_id that identifies the type of table to which the section corresponds. The data of the table is to be found in the body of the section. Each type of table has its own structure and corresponding semantics,* even though the structure used is overall compatible for all tables. The sections are inserted into the payload of TS packets to be transmitted. The longer sections may have to be broken down into several packets; the "payload unit start" field of the TS packets is used in this case.

1.2.5 DVB-H Signaling

The DVB-H receivers need to have an efficient means to discover the services that are available around them. This is accomplished using transmission parameter signaling (TPS).

TPS is inherited from the DVB-T standard.[9] It corresponds to a special channel that is added as part of the OFDM signal. The special TPS channel is used for the transmission of operating parameters (related in particular to channel coding and modulation) that the receivers need to analyze the signal and possibly to synchronize to it if need be. The following information is available: information on modulation, hierarchy information, guard interval, inner code rates, transmission modes (2K or 8K), frame number, and cell identification. The robustness of the TPS signal is increased by transmitting it totally redundant on 17 carrier frequencies (2K), 34 frequencies (4K), or 68 frequencies (8K). Note that the receiver needs to know most of this information beforehand to synchronize, equalize, and decode the signal, and thus to gain access to the information held by the TPS pilots. As such, the TPS data is mostly useful in some specific cases, such as changes in the parameters or fast reinitialization after a signal loss.

DVB-H signaling extends TPS as defined for DVB-T with the following settings and additional parameters: the length indicator is set at 33 bits; 1 bit signals the presence of a time-sliced DVB-H service, and another signals whether MPE-FEC is used. This is implemented in a backward-compatible way with DVB-T on two unused TPS bits. Moreover, the cell identifier of the signal is also always transmitted, whereas it is only an option in the DVB-T standard. For receivers, this makes the discovery of neighboring network cells where the currently selected service is available more efficient, which can prove useful to perform seamless frequency handovers (see also section 1.3.2.1).

1.3 DVB-H Link Layer

1.3.1 Multi-Protocol Encapsulation

At the link layer level of DVB-H systems, Multi-Protocol Encapsulation (MPE) is the standard method for carrying data. MPE has been introduced by the DVB Project (in ETSI EN 301 192[9]) for transporting the network packets from various protocols over DVB broadcast channels, that is, MPEG-2 transport streams at the physical layer level. MPE can be used to carry many different packet-oriented protocols other than IP, using the Logical Link Control/SubNetwork Attachment Point encapsulation method (LLC/SNAP).[19] It has, however, been especially optimized for IP, and is almost exclusively used for this purpose in the context of DVB-H.

The datagrams are carried by MPE sections. These sections are defined as an adaptation of Digital Storage Media—Command and Control (DSM-CC; cf. ISO/IEC 13818[17]) sections, in that

---

* Refer to [DVB-SI][10] for the complete syntax of all types of tables in a DVB system.

they use the same syntax but with an occasionally different semantics. DSM-CC sections are well adapted here, as they were defined to allow the transport of private data over MPEG-2 TSs. Moreover, DVB bases its data carousel* system on that defined in the DSM-CC specification.

As DSM-CC sections are defined based on the MPEG-2 private section format, MPE sections are similar to the SI and PSI information tables described in section 1.2.4. The type of data can be easily distinguished by the receiver using the "table_id" field, which is fixed at the value 0xE3 for the MPE sections.

The full general syntax of an MPE section is detailed in table 1.3. The abbreviations used for the names of the identifiers have the following meanings:

Uismbf: Unsigned integer most significant bit first
Bslbf: Bit string, left bit first
Rpchof: Remainder polynomial coefficients, highest order first

MPE sections used for transmitting IP datagrams (main use for MPE in the DVB-H context) have specific properties. First, they do not use the LLC/SNAP protocol. The "LLC_SNAP_FLAG" field is thus always set to null, and the part of the structure reserved for LLC/SNAP headers is not present. Moreover, the IP datagrams are directly inserted in the payload of the section (the "IP_datagram_data_byte" field of the MPE section header; see table 1.3). The general MPE definition specifies that this payload should be less than 4,086 bytes; as this is also the maximum size for an IP datagram in DVB networks, one MPE section corresponds exactly to one IP datagram. Note that both IPv4 and IPv6 datagrams are supported.

The general structure of an MPE section is therefore the following: a 12-byte header (fixed length), an IP datagram, and a 4-byte data integrity checking part, either a checksum ("checksum" here denotes a 32-bit checksum calculated by treating the data as a series of 32-bit integers, performing an exclusive or (XOR) operation over all of them and taking the complement of the result), or a cyclic redundancy check—32 bits (CRC-32).

1.3.2 Time Slicing

1.3.2.1 Principles

One of the main characteristics of handheld receivers is that they benefit from no constant electrical power supply, but are instead powered by batteries of limited capacity. The users of these devices often do not have immediate access to sources of energy; moreover, they have gotten used to the growing autonomy of handheld devices, and would consider it a nuisance to have to recharge the batteries of their receivers too often. This is a particularly critical issue for DVB-H because of its goal of compatibility with DVB-T and its high-bit-rate streams, whose processing is very costly in terms of energy dissipation. Therefore, the inclusion of specific provisions in the technology itself so as to restrict the power consumption of the devices is required. This is the first purpose of the time-slicing data transmission technique that is implemented in DVB-H. The formal definition of time slicing has been incorporated in the general DVB standard,[9] with extended details in the DVB-H implementation guidelines.[11]

---

* A data carousel is a method for repeatedly delivering data in a continuous cycle. An example of service based on data carousels is the analogue teletext service that was popular in many countries.

Table 1.3   MPE Section Syntax

Syntax	Size in Bits	Unit
Datagram_section{		
table_id = 0xE3	8	Uimsbf
section_syntax indicator	1	Bslbf
private_indicator	1	Bslbf
reserved	2	Bslbf
section_length	12	Uimsbf
MAC_adress_6	8	Uimsbf
MAC_adress_5	8	Uimsbf
reserved	2	Bslbf
payload_scrambling_control	2	Bslbf
address_scrambling_control	2	Bslbf
LLC_SNAP_flag	1	Bslbf
current_next_indicator	1	Bslbf
section_number	8	Uimsbf
last_section_number	8	Uimsbf
MAC_address_4	8	Uimsbf
MAC_address_3	8	Uimsbf
MAC_address_2	8	Uimsbf
MAC_address_1	8	Uimsbf
if (LLC_SNAP == 1) {		
LLC_SNAP()		
} else {		
for (j=0; j<N1; j++) {		
IP_datagram_data_byte	8	Bslbf
}		
}		
if (section_number == last_section_number) {		
for (j=0; j<N2; j++) {		
stuffing_byte	8	Bslbf
}		
}		
if (section_syntax_indicator == "0") {		
checksum	32	Uimsbf
} else {		
CRC_32		
}	32	Rpchof
}		

Source: From reference 9.

A comparison of the reception of time-sliced DVB-H services with that of multiplexed DVB-T can be found in figure 1.4. The principle of time slicing is to transmit data in high-data-rate *bursts* corresponding to a single service rather than continuously multiplexed with other services. Between two consecutive bursts, no data is transmitted for the service. The data rate is deemed high in comparison with the one that would be necessary for the smooth transmission of the same data using traditional continuous streaming techniques. This is feasible because a quality display of a video on a handheld terminal only requires a relatively low transmission bit rate (500 Kbps according to Faria et al.[6])

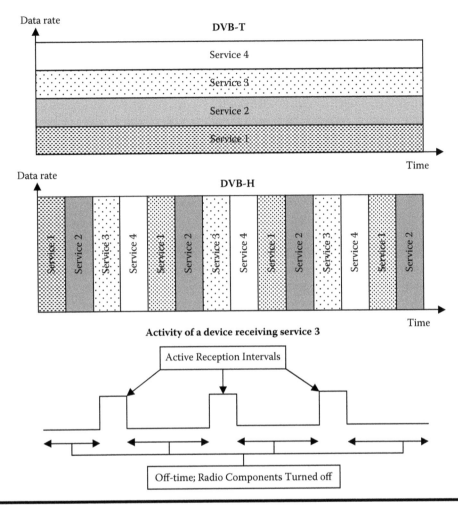

Figure 1.4   Comparison of DVB-T reception with time-sliced DVB-H.

due to the reduced size of the screen, whereas DVB-T channels normally offer bit rates of at least 3 Mbps,[4] leaving ample space for flexibility of the actual transmission bit rate.

A burst is formally defined as a set of MPE sections transmitted on an elementary stream. The burst itself contains the expected starting time of the next burst for the stream. Thus, between the bursts, the receiver can switch off its radio components, whose power consumption is usually very high. Note that on the contrary, the transmitter remains constantly on, always in the process of transmitting one burst at a given time.

It is claimed in the DVB-H literature that time slicing can reduce the power consumption up to 90 percent with respect to a non-time-sliced transmission. In their extensive simulation-based evaluation of the performance of the technique, Yang et al.[20] reached less optimistic conclusions but still found reduced power dissipation of up to 80 and 50 percent on average.

The second purpose of time slicing is to optimize handovers. Indeed, the mobility of the users of DVB-H devices introduces the requirement of cell handovers, that is, the ability to change the frequency and data stream to receive the same content in another radio cell. This is particularly critical in a broadcasting network such as DVB-H, where there is normally no return channel, and thus no way for the infrastructure of the network to know that a user is approaching the border region between two radio cells and to take anticipated action. A service handover may also be required in the case where a neighboring radio cell provides a better signal for the same content than the current cell. To ensure a high quality of service, such handovers should be seamless, that is, have no noticeable influence on the quality of the display on the handheld device.

Time slicing can play a role in the realization of this goal of seamless handovers. While waiting for the next burst, the device can scan the neighboring radio cells, analyze the quality of the signals they provide, compare their quality and possibly decide to hand over to one of them from the current signal. As the receiver finds itself in the "between burst" state most of the time, the handover process can be initialized almost as soon as it becomes necessary. The fact that it is always performed between bursts prevents it from disturbing data reception in any way.

This function can seem somewhat contradictory with the goal of diminished power consumption for which time slicing was introduced, as the device needs to be turned on during the process. However, as noted in ETSI TR 102 377,[11] the power consumption induced by monitoring a single frequency is typically low, due in part to the use of DVB-H signaling that provides a fast way to access signaling, and intelligent methods can be used to reduce the number of signals to check. This ensures that the frequency-checking process only needs to last for a small fraction of the total time separating the transmission of two successive bursts, ensuring a lower bound to the potential loss in power saving.

1.3.2.2 Burst Schedule Information

The burst schedule information is transmitted to the receiver within the current burst using the **delta-t** method. The actual information that is sent consists of the expected length of the interval of time between the current burst and the next one; this is the value that is called delta-t. Defining this information in relative terms removes the need for strict time synchronization throughout the network, especially between transmitters and receivers, and ensures that the method is unaffected by delays in the transmission path. To achieve optimal flexibility, the DVB standard defines no specific values for the characteristics of the bursts.

One potential problem of time slicing lies in the fact that the receiver becomes heavily dependent on its knowledge of the current value of delta-t; upon the loss of this information it cannot know when to expect the next burst and has no other option than to leave its radio components on until it receives the next burst. To prevent this situation, the delta-t value is included in the header of each section of the bursts, so that even in bad reception environments where only one section of a burst is properly received power saving can still be performed. More precisely, with respect to the structure of MPE sections described in section 1.3.1, this information is inserted in the MAC4, MAC3, MAC2, and MAC1 sections of the MPE header. This choice is justified in ETSI TR 102 377[11] by the fact that in the context of DVB-H, these fields corresponding to a part of the MAC address of the receiver that will be either unused or redundant with the information found in the header of the IP datagram encapsulated in the payload of the MPE section.

Two factors can impact the appropriate delta-t setting. The first one is the possibility of jitter within the transmission path of the signal, which might cause a burst to start later than expected. This can be handled by decreasing the delta-t of the burst; according to ETSI TR 102 377,[11] a 10 ms jitter of this kind is acceptable. The second factor is the synchronization time of a receiver, that is, the time it needs to turn itself on, tune to the appropriate signal, and start receiving it. As this

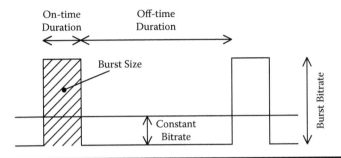

Figure 1.5   Parameters of a time-sliced burst.

delay depends on the implementation of the receiver, it cannot be handled by modifying the delta-t of the burst and should instead be addressed by the device itself.

1.3.2.3 Characteristics of the Bursts

A burst is defined by the following parameters (see figure 1.5):

- **Burst size:** Corresponds to the number of (network layer) bits in one burst. This size should be less than the available RAM in the device, as in the general case the data will be buffered and displayed during the off-time. The section and transport packet headers cause a slight overhead with respect to the size of the transmitted data; this overhead is estimated to 4 percent in ETSI TR 102 377.[11] Apart from the constraint of the memory available in the devices, all burst sizes are possible.
- **Burst bandwidth:** "An appropriate momentary bandwidth used by a time-sliced elementary stream while transmitting a burst."[9]
- **Burst duration:** The length of the interval of time between the beginning and end of a burst. To compensate for the off-times, it is normally greater than the **constant bandwidth**, which corresponds to the bandwidth that would be necessary to transmit the elementary stream if it were not time sliced.
- **Off-time:** Corresponds to the interval of time between the transmissions of two bursts. Its length may not be exactly equal to delta-t because of the jitter and synchronization issues detailed in the previous section.
- **On-time** or **burst duration:** The length of the interval of time during which data is to be transmitted for the burst. A **maximum burst duration** (mbd) parameter is defined for each elementary stream. It is therefore guaranteed that with the end of the current burst as a reference time, the next burst will start in delta-t seconds at the earliest, and stop in delta-t + mbd seconds at the latest. This information can be used by a device to know when a burst has ended in bad reception conditions.

The DVB standards define no specific value for any of these parameters (apart from constraints such as the available memory or the maximum burst duration). Moreover, thanks to the flexibility allowed by the delta-t parameter, all these parameters can freely vary from one burst to another (as well as the duration of the off-time).

Note also that if the bandwidth of one burst is less than the total available bandwidth of the TS, other elementary streams can be transmitted during its on-time, as well as other types of stream, such as DVB-T streams.

1.3.3 MPE-FEC

1.3.3.1 MPE-FEC Frames

MPE-FEC stands for Multi-Protocol Encapsulation–Forward Error Correction. It was introduced in DVB-H to cope with the potential bad reception conditions that are due to the mobility of the device, the limited capacity of their antennas and the impossibility to permanently orientate them toward the signal emitters, and interferences from various sources. To this end, the MPE-FEC method strives to improve the carrier-to-noise (C/N) and Doppler performances in mobile DVB-H channels.

This error correction technique is the second important technological contribution of DVB-H with time slicing. MPE-FEC is fully implemented in the link layer and as such comes as an addition to the error correction mechanisms provided by the DVB-T layer.

MPE-FEC sections are special types of MPE sections as defined in section 1.3.1. The value of the "table_id" field in the headers of these sections is different from that of normal sections, so that the receiver can easily distinguish between the two types of sections. The fixed value of table_id for MPE FEC sections is 0×78.

MPE-FEC is an optional component of DVB-H at the transmitter level as well as at the receiver level. It is implemented in such a way that a device with no MPE-FEC processing module can still receive and display the data carried by an elementary stream using MPE-FEC.

The principle of MPE-FEC is to add Reed–Solomon codes to each time-sliced burst. The parity bits are computed from the datagrams of the bursts. They are encapsulated in special MPE-FEC sections that are transmitted after the last (data) section of a burst in the same elementary stream (so they are also part of the burst). A special MPE-FEC frame is used for the computation of the Reed–Solomon code. The structure of an MPE-FEC frame is presented in figure 1.6. A frame comprises two parts, the **application data table** that contains the datagrams, and the **Reed–Solomon data table** with the parity bits. The data is transmitted column by column and downward each column, starting from the upper left corner of the frame. The size of one element of unique coordinates (row, column) is 1 byte. The number of rows in the frame can be 256, 512, 768, or 1,024.

In the application data table, the datagrams are introduced column by column in the frame. If the size of one datagram is larger than that of a column it is continued in the next column, and so on. Such is the case in figure 1.6 for datagrams 1 and 2.

The length of the application data table is 191 columns. If there is less data than the total capacity of the table, the remaining space is filled by null padding after the last IP datagram.

The size of the RS data table is 64 columns. On each row the content results from the application of the Reed–Solomon code RS (255, 191) to the corresponding row of the application data table. Because an application data table row typically contains data from many different IP datagrams, the interleaving is generally high. One optional feature of MPE-FEC is *puncturing*, which means that a number of columns of the Reed–Solomon code are not actually transmitted to reduce the overhead that they introduce.

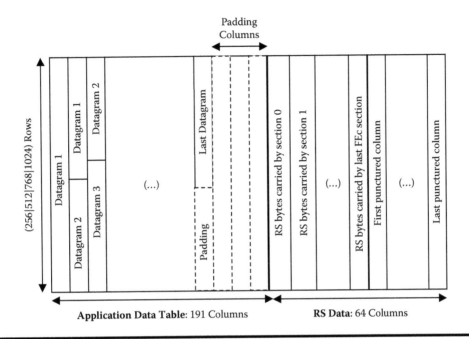

Figure 1.6 Structure of an MPE-FEC frame.

1.3.3.2 Transmission of MPE-FEC Data

As stated previously, MPE-FEC data is transmitted using sections. Although the data's format is slightly different from the MPE section format described in section 1.3.1, it remains sufficiently compatible so that a receiver that is not capable of making use of the FEC data can ignore it and still be able to access the payload of the section as if it were a normal MPE section.

MPE sections start by a header that includes, among other things, the byte (column) number of the IP datagram it carries and the value of delta-t. Sections whose first IP datagram is continued from a previous one bear the byte number of the initial section. With this information, the receiver is able to reconstruct the application data table of a frame.

MPE-FEC sections also have headers with byte numbers so that the RS data table of an MPE-FEC frame can be rebuilt. The headers also bear the number of padding columns in the application data table (if any), as this information must be known to perform the RS decoding.

1.3.3.3 Receiver Decoding Strategy

MPE sections include a data integrity verifier part (checksum or CRC-32; see section 1.3.1). This allows the receiver to check whether a section was correctly received. If all sections pass the check, the receiver can directly turn itself off to save power after it receives the last MPE section (identified by a set "table_boundary" flag in its header), even when MPE-FEC is used, as it has no use for the RS data that follows in the next sections. Non-MPE-FEC-capable receivers also turn themselves off at this point.

While receiving data, the receiver marks sections that failed the check as unreliable. This also stands for RS data columns; the corresponding bytes will not be used for RS data correction. When the whole frame is known, rows are considered. Per the Reed–Solomon definition, the RS data allows the correction of as many errors (data application table bytes marked "unreliable") per row as the number of RS data bytes, that is, 64 here for RS (255, 191)—less if puncturing was used.

1.3.3.4 Performances

A series of experiments about DVB-H were conducted by DVB Group members in 2004. The results are available as an ETSI technical report.[21] Regarding MPE-FEC, the characteristics of DVB-H with MPE-FEC turned on have been compared with those of DVB-T. The tests highlight a significant gain over DVB-T. With the QPSK and 16-QAM modulation schemes with coding rates 1/2 and 2/3, carrier-to-noise ratio gains of 5 to 8 dB for a receiver in a moving vehicle and 3 dB for pedestrian reception were observed. It was also concluded from these experiments that MPE-FEC dramatically increases the speed at which the reception of a DVB-H signal is possible.

1.4 DVB IP Datacast

1.4.1 Principles

DVB IP datacast (DVB-IPDC) is a suite of specifications developed by the DVB Project that defines the components needed to define a complete commercial system on top of an IP interface such as that of DVB-H. It is characterized by the convergence of data networks that have historically evolved independently, the broadcast and mobile communication networks. This should ultimately allow users to consume broadcast programs and to use mobile communication services in a single integrated device.

DVB IP datacast has been defined within the framework of DVB-H and of existing cell phone networks. A DVB IP datacast system can thus be described as a system combining a unidirectional point-to-multipoint DVB(-H) broadcast path with a bidirectional interactivity path, typically a cellular communication network such as General Packet Radio System (GPRS) or Universal Mobile Telecommunications System (UMTS). Both paths are based on the IP protocol, hence the IP datacast name. IP datacast also makes the very large amount of existing IP-based digital content services available with little or no modifications required for broadcasting over DVB-H: video streams, Web pages, music files, software, and so on.

Although the DVB-IPDC specifications have been designed with DVB-H as a physical layer in mind, they are not restricted to it, and are planned to be eventually used with all DVB mobile video systems (such as DVB-SH), and in a larger perspective with all systems offering an IP interface.

The DVB-IPDC specifications were produced by the DVB ad hoc group CBMS (Convergence of Broadcast and Mobile Services), which was formed in March 2001. In 2004, the group released a first set of detailed technical specifications for an "IPDC in DVB-H" system. These form the basis of the current fundamental specifications for IPDC systems. Like DVB-H, the complete DVB-IPDC system is an umbrella standard built upon several standards. Most of them have already been approved and published by ETSI as standards. The relevant documents are the following:

- ETSI TR 102 469:[22] Describes the architecture of the system.
- ETSI TS 102 470:[16] Describes the program-specific information (PSI) and the service information (SI) used in DVB-IPDC.

- ETSI TR 102 471:[23] Describes the electronic service guide (ESG).
- ETSI TS 102 472:[24] Defines the protocols used in a DVB-IPDC system for the delivery of content.
- ETSI TR 102 473:[25] Features detailed examples of IPDC services based on some possible use cases.
- ETSI TS 102 005:[26] Defines the standard content source coding methods for the DVB-IPDC system.
- DVB blue book A100:[27] Describes the mechanisms for service purchase and protection in a DVB-IPDC system.
- DVB blue book A096:[28] Describes the coordination and areas of application of the different standards.
- Set of standards defining DVB-H itself (listed in section 1.1.2).

In addition, the CBMS group has released implementation guideline documents for the ESG[1], for content delivery protocols[2], and for addressing the problem of seamless handover in a DVB-IPDC system.[3]

1.4.2 Architecture and Protocol Stack

The reference architecture for IPDC over DVB-H is specified in ETSI TR 102 469,[22] from which the architectural diagram shown in figure 1.7 is extracted. On the left side of the diagram, the multimedia content to be delivered to the terminal displayed on the right is created.

On the figure, only the components represented by rectangles in bold are within the scope of DVB-IPDC: the content creation process and the architecture of the associated interactive network are excluded. The relevant components are the following:

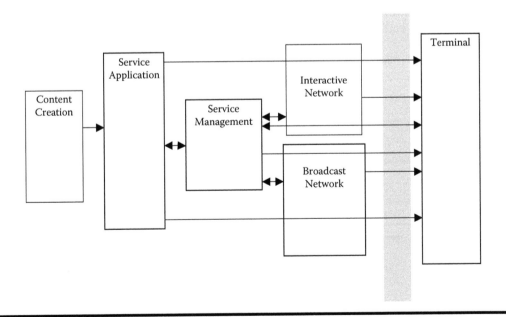

Figure 1.7   DVB-IPDC reference architecture.

■ **Service application:** Provides a direct link between the content provider and the end user, making it possible to deliver the content over different networks. In particular, each content provider provides the user with its electronic service guide (described in section 1.4.4) that lets her or him know the services it offers.

■ **Service management:** Essentially in charge of allocating resources (bandwidth, source/destination IP). It assigns locations (within the broadcast network) and bandwidth to services and schedules them. It also performs some access control on the service applications.

■ **Broadcast network:** Allows for the multiplexing of the service applications at the IP level. It is responsible for the encapsulation of IP datagrams in DVB-H-compatible (i.e., taking time slicing into account) MPE sections.

■ **Terminal:** Corresponds to the DVB-H user device. Within the architecture it represents consumption for content and client of network and service resources. It is also supposed to be capable of interacting with the interactive network.

Content delivery protocols are needed to deliver the services to the users. They are specified in ETSI TS 102 472.[24] Two main cases apply. The first is the delivery of real-time streaming multimedia content, typically digital television programs (audio is possible as well). All other possible DVB-IPDC services fall in the second category of file-based deliveries. Figure 1.8 represents the network protocol stack designed to meet these requirements (with the corresponding components in the interactive network). The first three layers correspond to the DVB-H system as described in sections 1.2 and 1.3, with the Internet Protocol on top of it at the network layer level. The User Datagram Protocol (UDP)[29] is used in the transport layer; Transmission Control Protocol (TCP) cannot be used there as in the interactive network because it needs a return channel, which a system such as DVB-H lacks. The Real-Time Protocol (RTP)[30] is used to take care of the real-time synchronous content. At the same level, file-based operations are conducted using a File Delivery over Unidirectional Transport/Asynchronous Layered Coding (FLUTE/ALC)[31] data carousel. As shown in figure 1.8, applications may be defined on top of these primitives. Note that all protocols that are part of the system are Internet protocols specified by the Internet Engineering Task Force (IETF). This outlines the strong will of the DVB Group to maintain compatibility with the Internet world.

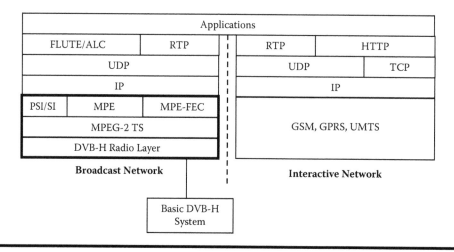

Figure 1.8   Simplified DVB-IPDC protocol stack.

1.4.3 Content Source Coding Methods

The methods of content source coding used in conjunction with delivery via RTP over IP in DVB networks (including but not restricted to DVB-IPDC) have been standardized in ETSI TS 102 005.[26] For video data, the standard specifies two possible coding formats: MPEG-4 H.264/AVC[32] and the Microsoft Windows Media VC-1 codec.[33] In a perspective of compatibility with the Third Generation Partnership Project (3GPP) specifications, H.264/AVC is defined as a "strong recommendation" for DVB-IPDC systems. For audio data, MPEG-4 HE AACv2[34] is the default, with AMR-WB+[35] as an optional second choice.

1.4.4 Metadata Representation

1.4.4.1 DVB-IPDC Service Information

IPDC requires certain extensions to the service information system described in section 1.2.4. These extensions are defined in ETSI TR 102 470.[16] It adds the notion of IP platform to the DVB network, as represented in figure 1.9.

An IP flow corresponds to a set of IP datagrams with the same source and destination IP addresses. An IP platform is a set of IP flows managed by a single entity; IP addresses are unique within a single IP platform. An IP stream maps an IP flow to the broadcast network. It corresponds to a specific elementary stream within a specific transport stream. An ES may actually carry several IP streams at once, in which case the streams are distinguished by the source and destination of the corresponding IP datagrams.

An IP platform is identified by a specific "platform_id" identifier (see below). An IP flow is simply identified by its source and destination address. Identifying a specific IP stream requires identifying its corresponding transport stream (thus, the "transport_stream_id" and "original_network_id" fields; see section 1.2.4), and then its elementary stream (i.e., the PID field; see section 1.2.4).

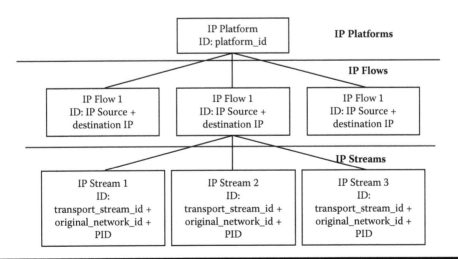

Figure 1.9   IP entities in an IPDC network.

The information relative to an IP platform is conveyed in a table of the same nature as those described in section 1.2.4, the IP/MAC Notification Table (INT). The INT contains an identifier of the platform called "platform_id." There are two ranges of platform_id. The first range is unique across all DVB networks and corresponds to IP platforms that can be spanned across different DVB networks. The second range is used for IP platforms contained within a single DVB network and can be reused in different networks. Apart from this, the INT identifies the IP streams of a TS, and may also announce the IP streams of neighboring TSs. It may contain subtables, one for each IP platform interacting with the TS. The INT contains the full identification and metadata information of each IP stream it announces.

1.4.4.2 Electronic Service Guide

Although the PSI/SI information described above is well adapted to the representation of service information parameters, it is not flexible enough to accurately represent the complex set of services and programs that a commercial system would make available to its subscribers. This is the purpose of the electronic service guide (ESG) defined in ETSI TR 102 471.[23] The ESG can be seen as a central directory of information about the services that are available in a DVB-IPDC system. This allows the user to select the services in which she or he is interested thanks to an ESG browsing application on her or his terminal. The provided information includes both detailed service description information (possibly presented in the form of rich content such as pictures, video extracts, links to other services, etc.) and information that is necessary for the terminal to access the service if selected.

A terminal can access an ESG once it has synchronized with a DVB-H-IPDC transport stream. It can know where to find the ESG data in the stream thanks to the PSI/SI tables.

The ESG is implemented through a layered model. The first layer is the **data model**, which is defined in the form of an eXtensible Markup Language (XML) schema (see figure 1.10). The components of this schema (service, purchase, content, etc.) are encoded in fragments; a fragment is a chunk of XML data extensively describing the component it relates to.

The semantics of the represented fragments are the following:

- **Service:** An IPDC service (for example, a channel).
- **ServiceBundle:** Groups a set of services together to present them as a single entity to the user (typically a set of services that can be accessed through a unique purchase).
- **Content:** Metadata describing specific content that is part of a service (a TV show, for example). It may be instantiated several times.
- **ScheduleEvent:** Defines the broadcast time of an instance of content.

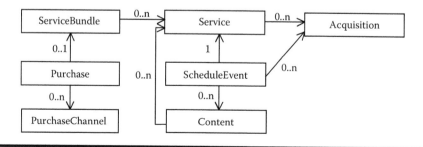

Figure 1.10   ESG data model (from reference 23).

- **Purchase:** Purchase information of a service (bundle).
- **PurchaseChannel:** Interface allowing the user to purchase content.
- **Acquisition:** Information necessary to access a service or content.

The second layer of the ESG is the instantiation of the data model of the first layer to describe the services available in a specific DVB-IPDC network. The result of this operation is the creation of a number of fragments. The third layer is an optional encoding of the fragments: they can be compressed with either GZIP[36] or BiM.[37] The fourth layer encapsulates the fragments in containers (objects that can be delivered by network protocol); in particular, private data that existed only as links in the previous layers (image files, video extracts) is incorporated into the containers. The fifth and last layer corresponds to the delivery of the containers using the FLUTE protocol.

1.4.5 Service Purchase and Protection

The service purchase and protection (SPP) function of the DVB-IPDC system defines the mechanisms implemented to only let authorized users access the content, and how they are related to the service acquisition options, described in the ESG. This is naturally a critical component for a successful commercial deployment of the system. At the time of writing, it is described in the DVB blue book A100.[27]

The SPP in DVB-IPDC is based on a four-layer model (see figure 1.11). At the lowest-layer level, the content and services are encrypted using either IPSEC,[38] ISMACryp,[39] or SRTP,[40] with

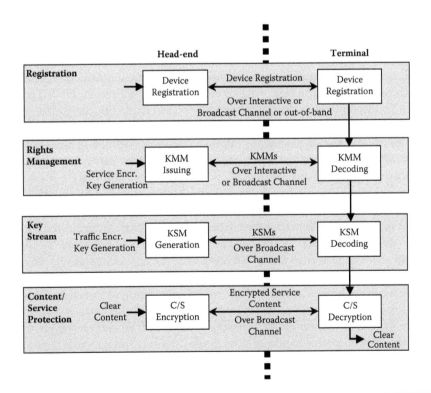

Figure 1.11 DVB-IPDC SPP model (from reference 27).

IPSEC and ISMACryp defined as strong recommendations. The second layer is responsible for the delivery of the service encryption keys. The so-called key stream messages (KSMs) are delivered over the DVB-H channel. The "rights management" layer grants access rights to content/services to the requesting devices. The authorization can, for example, be in the form of a service encryption key that can be used to access the KSMs. These messages, called the key management messages (KMMs), can be delivered via the DVB-H channel or the interactive network.

The highest-level layer in this model is responsible for the registration of the devices. In the SPP model, this may be performed via an out-of-band interface (for example, by buying a prepaid card), via the broadcast channel or the interactive channel.

The IPDC specification defines an overall option that slightly changes the procedures at the key stream, rights management, and registration layers level. They can either use a solution based on the digital rights management (DRM) specifications proposed by the Open Mobile Alliance[41]—this option is referred to as the 18Crypt system in the specifications—or cooperate with an external conditional access provider, in a fashion already used by existing digital TV solutions. This second solution is called the IPDC SPP Open Security Framework. The major difference between the two is that 18Crypt is fully specified with respect to the four-layer model of IPDC SPP, while the OSF considers key stream and rights management to be private and only defines an interface that can be used to interact with any external key management system. In both cases, the encryption at the content/service protection layer remains identical.

1.5 Summary

Digital Video Broadcasting–Handheld (DVB-H) is a standard proposed by the DVB Group for the delivery of digital multimedia content to mobile, handheld devices.

At the physical layer level (in the sense of the OSI network reference model), DVB-H is based on DVB-T (with a number of extensions that were introduced in the standard for DVB-H), the DVB standard for the terrestrial transmission of digital television via radio airwaves. DVB-T itself makes use of the MPEG-2 transport streams defined in the MPEG-2 standard. This compatibility with DVB-T allows DVB-H to use the DVB-T networks that are already widely deployed all over the world.

DVB-H also makes use of the service information format defined in the DVB data broadcasting standard, which defines how metadata about the services, network parameters, and data formats is formatted and transmitted over DVB networks. This information is organized in the form of tables: the program-specific information (PSI) tables specified in the MPEG-2 standard and the service information (SI) tables specified by DVB.

At the link layer level, DVB-H transmits data as Multi-Protocol Encapsulated (MPE) sections, allowing it to transmit the data encapsulated in Internet Protocol (IP) packets. DVB-H implements its main technological elements at this level, with the goal of addressing the specific issues of digital video broadcasting to handheld mobile devices.

The first issue is the limited capacity of the batteries powering the receivers. This requires the implementation of power consumption techniques. In DVB-H this corresponds to time slicing, meaning that instead of constantly multiplexing several services, each service is transmitted in bursts using the whole available bandwidth. As the devices only receive one service at a time, they can turn themselves off between the bursts, resulting in power saving. Time slicing also addresses one issue related to mobility: as the device moves, changing its position with respect to the signal transmitters, one transport stream carrying the same services as the one currently in use may become a better option than the one in use. The device needs to monitor the signals of the adjacent radio cells to determine whether a handover is necessary. With time slicing, this monitoring as well as the handovers can be performed during the off-time between bursts. To ensure that this function does not defeat the main purpose of time slicing (reducing power consumption), special provisions are included in DVB-H to increase the speed of signal scan and handover.

The second main technological element introduced in DVB-H is the Multi-Protocol Encapsulation–Forward Error Correction (MPE-FEC) technique. MPE-FEC adds another level of forward error correction to the MPE sections in addition to the mechanisms defined at the DVB-T physical layer level. The goal is to increase the carrier-to-noise (C/N) and Doppler performances in mobile DVB-H channels to cope with reception conditions that can be very difficult for a mobile device, especially because of typically limited capabilities of the antennas of the handheld devices and of the movement of the terminal during operation. The FEC method consists in adding a Reed–Solomon (RS) code to MPE frames. The calculation of the RS code is organized so that it is applied to chunks of data spanning all the IP datagrams constituting the payload of an MPE frame, ensuring a high interleaving. While time slicing is specified as mandatory in DVB-H, MPE-FEC is optional, and defined so that devices that cannot make use of the error correction possibility that it provides can still process MPE-FEC frames.

The physical and data link layers described above define a DVB-H system. The DVB IP datacast set of specifications (DVB-IPDC) defines the additional components required to build a complete commercial system on top of the IP interface of DVB-H, in a hybrid network combining a unidirectional broadcast network and a bidirectional mobile telecommunications network.

These specifications define a global architecture for the system, which includes both the broadcast and interactive channels together. The expected services are based either on the delivery of real-time content (streaming video or audio) or on file downloads (display of videos on demand, execution of predownloaded applications, etc.). To meet these requirements, the content delivery protocol must be implemented on top of the Internet Protocol. The proposed protocol stack completes DVB-H by using UDP at the transport layer level, RTP for streaming media content delivery, and FLUTE/ALC for file-based downloads.

To format and deliver the different types of metadata needed by the receivers to process the signals, DVB-IPDC defines two complementary systems. The first one is an adaptation of the PSI/SI tables used by DVB-H, which completes it with information about the IP infrastructure. The purpose of the second one, which defines an electronic service guide (ESG), is to represent rich information about the services available in a DVB-IPDC network, allowing the users to select a service and access it.

Finally, the DVB-IPDC specifications define the service purchase and protection system, which specify the methods used in the system to prevent unauthorized users from accessing services, and the process through which users purchasing a service can be granted the access rights. The system is based on encryption, a system of key message exchanges, and digital rights management methods.

Links

1. DVB-H project: http://www.dvb-h.org.
2. DVB Group: http://www.dvb.org.
3. European Telecommunications Standards Institute (ETSI): http://www.etsi.org.
4. The Digital Terrestrial Action Group (DIGITAG): http://www.digitag.org.
5. The IP Datacast Forum: http://www.ipdc-forum.org/.

References

1. Ulrich Reimers, ed. 2005. *DVB: The family of international standards for digital video broadcasting.* 2nd ed. Berlin: Springer-Verlag Berlin.

1a. DVB group. 2007. Electronic Service Guide (ESG) Implementation Guidelines. DVB document A112.

2. ETSI. 2005. *Digital video broadcasting (DVB) user guidelines for the second generation system for broadcasting, interactive services, news gathering and other broadband satellite applications (DVB-S2).* ETSI technical report TR 102 376, V1.1.1.

2a. DVB group. 2007. Content Delivery Protocols (CDP) Implementation Guidelines. DVB document A113.

3. ETSI. 2004. *Digital video broadcasting (DVB); Framing structure, channel coding and modulation for digital terrestrial television.* ETSI standard EN 300 744, V1.5.1.

4. Ladebusch, U., and Liss, C. A. 2006. Terrestrial DVB (DVB-T): A broadcast technology for stationary portable and mobile use. *Proceedings of the IEEE* 94:183–93.

5. ETSI. 1998. *Digital video broadcasting (DVB); Framing structure, channel coding and modulation for cable systems.* ETSI standard EN 300 429, V1.2.1.

6. Faria, G., Henriksson, J. A., Stare, E., and Talmola, P. 2006. DVB-H: Digital broadcast services to handheld devices. *Proceedings of the IEEE* 94:194–209.

7. Burow, R., Fazel, K., Hoeher, P., Klank, O., Kussmann, H., Pogrzeba, P., Robertson, P., and Ruf, M. J. 1998. On the performance of the DVB-T system in mobile environments. In *Proceedings of the IEEE Global Telecommunications Conference (GLOBECOM'98), The Bridge to Global Integration,* 2198–2204. Vol. 4. IEEE Computer Society.

8. ETSI. 2004. *Digital video broadcasting (DVB); Transmission system for handheld terminals.* ETSI standard EN 302 304, V1.1.1.

9. ETSI. 2004. *Digital video broadcasting (DVB); DVB specifications for data broadcasting.* ETSI standard EN 301 192, V1.4.1.

10. ETSI. 2007. *Digital video broadcasting (DVB); Specifications for service information (SI) in DVB systems.* ETSI standard EN 300 468, V1.8.1.

11. ETSI. 2005. *Digital video broadcasting (DVB); DVB-H implementation guidelines.* ETSI technical report TR 102 377, V1.2.1.

12. ETSI. 2004. *Digital video broadcasting (DVB); DVB mega-frame for single frequency network (SFN) synchronization.* ETSI technical specification TS 101 191, V1.4.1.

13. Postel, J. 1981. *Internet Protocol (IP): Specification.* Request for Comments 791.

14. Hinden, R., Deering, S., and Network Working Group. 1998. *IP version 6 addressing architecture.* Request for Comments 2373.

15. Wang, C.-C., Lee, T.-J., Lo, H. K., Lin, S.-P., and Hu, R. 2006. High-sensitivity and high-mobility compact DVB-T receiver for in-car entertainment. *IEEE Transactions on Consumer Electronics* 52:21–25.

16. ETSI. 2006. *Digital video broadcasting (DVB); IP datacast over DVB-H: Program specific information (PSI)/service information (SI).* ETSI technical specification TR 102 470, V1.1.1.

17. ISO/IEC. 2000. *Information technology: Generic coding of moving pictures and associated audio information.* ISO/IEC international standard ISO/IEC 13818.

18. Scott, J. 1998. The how and why of COFDM. EBU technical review 272. http://www.ebu.ch/en/technical/trev/trev_278-stott.pdf (accessed February 21, 2007).

19. ISO/IEC. 1998. *Information technology—Telecommunications and information exchange between systems—Local and metropolitan area networks—Specific requirements—Part 2: Logical link control.* ISO/IEC international standard ISO/IEC 8802-2.

20. Yang, X. D., Song, Y. H., Owens, T. J., Cosmas, J., and Itagaki, T. 2004. Performance analysis of time slicing in DVB-H. In *Proceedings of the joint IST Workshop on Mobile Future and the Symposium on Trends in Communications (SympoTIC'04),* 183–86. IEEE Press.

21. ETSI. 2005. *Digital video broadcasting (DVB); Transmission to handheld terminals (DVB-H); Validation task force report.* ETSI technical report TR 102 401, V1.1.1.
22. ETSI. 2005. *Digital video broadcasting (DVB); IP datacast over DVB-H: Architecture.* ETSI technical report TR 102 469, V1.1.1.
23. ETSI. 2006. *Digital video broadcasting (DVB); IP datacast over DVB-H: Electronic service guide (ESG).* ETSI technical report TR 102 471, V1.2.1.
24. ETSI. 2006. *Digital video broadcasting (DVB); IP datacast over DVB-H: Content delivery protocols.* ETSI technical specification TS 102 472, V1.2.1.
25. ETSI. 2006. *Digital video broadcasting (DVB); IP datacast over DVB-H: Use cases and services.* ETSI technical report TR 102 473, V1.1.1.
26. ETSI. 2007. *Digital video broadcasting (DVB); Specification for the use of video and audio coding in DVB services delivered directly over IP protocols.* ETSI technical specification TS 102 005, V1.3.1.
27. DVB Group. 2005. *IP datacast over DVB-H: Service purchase and protection (SPP).* DVB document A100.
28. DVB Group. 2005. *IP datacast over DVB-H: Set of specifications for phase 1.* DVB document A096.
29. Postel, J. 1980. *User datagram protocol.* Request for Comments 780.
30. Audio-Video Transport Working Group. 1996. *RTP: A transport protocol for real-time applications.* IETF Request for Comments 1889.
31. Paila, T., et al. 2004. *FLUTE: File delivery over unidirectional transport.* IETF Request for Comments 3926.
32. ITU-T. 2005. *Advanced video coding for generic audiovisual services.* ITU-T recommendation H.264 and ISO/IEC 14496-10 (MPEG-4) AVC, V3.
33. Society of Motion Picture and Television Engineers (SMPTE). 2006. *VC-1 compressed video bitstream format and decoding process.* SMPTE 421M.
34. ISO/IEC. *Information technology—Generic coding of moving picture and associated audio information—Part 3: Audio,* including ISO/IEC 14496-3/AMD-1 (2001): *Bandwidth extension*; ISO/IEC 14496-3 (2001) AMD-2 (2004): *Parametric coding for high quality audio.* ISO/IEC 14496-3.
35. ETSI. 2007. *Digital cellular telecommunications system (phase 2+); Universal Mobile Telecommunications System (UMTS); Audio codec processing functions; Extended adaptive multi-rate–wideband (AMR-WB+) codec; Transcoding functions (3GPP TS 26.290 release 7).* ETSI technical specification TS 126 290, V7.0.0.
36. Deutsch, P. 1996. *GZIP file format specification version 4.3.* IETF Request for Comments 1952.
37. ISO/IEC. 2004. *Information technology—Multimedia content description interface—Part 1: Systems.* ISO/IEC International Standard ISO/IEC 15938-1.
38. Kent, S., and Seo, K. 1998. *Security architecture for the Internet protocol.* IETF Request for Comments, 4301.
39. Internet Streaming Media Alliance. *Internet Streaming Media Alliance implementation specification,* V1.0.
40. Baugher, M., McGrew, D., Naslund, M., Carrara, E., and Norrman, K. 2004. *The secure real-time transport protocol (SRTP).* IETF Request for Comments 3711.
41. Open Mobile Alliance. 2006. *Open Mobile Alliance digital rights management specification.* Approved version 2.0.
42. Kornfeld, M. and May, G. 2007. DVB-H and IP datacast: Broadcast to handheld devices. *IEEE Transactions on Broadcasting* 53:161–70.

Chapter 2

An Overview of Digital Video Broadcasting via Satellite Services to Handhelds (DVB-SH) Technology

I. Andrikopoulos (ed.), N. Chuberre, M. Cohen,
O. Courseille, Régis Duval, J. Farineau, P. Laine,
T. Quignon, L. Roullet, A. Serment, and M. Tatard

Contents

Keywords

mobile TV, radio interface, Unlimited Mobile TV

Abstract

Mobile TV is already booming on existing cellular infrastructures in point-to-point mode. But this mode is not optimized to deliver the same content to many users at the same time, preventing mass-market deployment.

For massive access to mainstream TV, channels on mobile, overlay broadcast networks, complementary to existing cellular networks, are necessary. But the design of such broadcast infrastructure shall be geared to meet the following challenges to contribute to the mobile TV market takeoff:

■ A business model based on an attractive monthly fee subscription allowing unlimited usage
■ A service coverage including good in-building performance compatible with Global System for Mobiles (GSM) to address a wide population
■ A guaranteed quality of service, including a good picture and sound quality, a large number of programs and content formats to meet all users' interests, a fast program switching time, an efficient electronic service guide
■ A harmonized technology framework over Europe to prevent market fragmentation and favor a wide terminal choice

To this end, a smart hybrid satellite–terrestrial mobile broadcast concept targeting the IMT2000 frequency band allocated to mobile satellite systems (2.17–2.2 GHz) available worldwide is being developed in Europe. This chapter gives an overview of the system architecture and the radio interface technology, and provides details on the Unlimited Mobile TV system features and performances with respect to a typical dimensioning/deployment scenario.

2.1 System Design Principles

Mobile TV is expected to become a killer application and to generate revenues of several billion euros by 2010 in Europe for the media and mobile industry. Several market surveys have shown that around a third of mobile subscribers are ready to pay a 5–15 euro monthly fee to watch TV on their mobiles. Several third-generation (3G) network operators have already proposed a mobile TV service to their customers in streaming mode. It consists in a multichannel offer made of general and thematic programs to satisfy most user interests. It attracts many subscribers even though access to the service is limited to the current 3G network coverage. Moreover, the service quality is adversely affected by a high cellular network traffic load or a high audience level because the TV programs are delivered on dedicated radio resources to each terminal.

To develop further the mobile TV market in Europe, there is a need to develop new solutions that would overcome these current limitations. To this end, a broadcast/multicast solution is well adapted to deliver popular TV and content to an unlimited audience without any constraints on the usage scenarios. In line with the critical success factors identified by the Universal Mobile Telecommunications System (UMTS) forum for the mobile TV technologies,[1] great care has been taken in the design of this broadcast/multicast solution to:

- Provide seamless coverage with strong indoor requirements because most users will watch mobile TV at home, in the office, or while commuting. Ideally, the service coverage should be compatible with GSMs to minimize a churn rate due to coverage discrepancy.
- Minimize the infrastructure deployment and operational costs. It shall allow a business model based on a monthly fee subscription meeting demand for guaranteed quality of service and unlimited usage. This can be achieved by ensuring that the solution's transmitters can share existing broadcast/cellular sites.
- Address cellular handheld terminals, which are expected to represent the largest market share. To this end, cost and form factor impacts associated with the broadcast function integration shall be minimized. The solution shall also support PDAs and even vehicular terminals. The power consumption of battery-activated terminals shall be minimized to maximize usage duration between two chargings.
- Allow incremental investment associated with the solution to allow the generation of revenues before full deployment.
- Support TV experience comparable to that of fixed TV. This requires the support of a large TV offer, live TV, pay TV, quick and simple access to TV service, fast response time such as off to on, channel switching time, and fair enough viewing quality.
- Ensure interworking with the cellular network at both the user terminal and service platform ends to enrich the service offer with additional TV contents or interactive broadcast applications, potentially creating an additional source of revenue.

Moreover, the solution shall use standardized technology to allow an open framework for application development, a wide range of terminal and network products, and vendors favoring stimulating competition for the benefit of end users.

These are the major requirements considered in the design of the hybrid satellite–terrestrial mobile broadcast system described in this chapter and referred to as the Unlimited Mobile TV solution.

2.2 System Overview

Figure 2.1 provides a high-level view of the Unlimited Mobile TV solution. The overall solution combines a dedicated broadcast system based on a hybrid infrastructure, integrated, at service and application levels, with existing cellular networks to provide end users with a full range of enter-tainment services with interactivity.

This hybrid satellite–terrestrial broadcast system encompasses:

- A space segment made of high-power geostationary satellites for TV broadcast to mobile terminals over nationwide coverage (1 in figure 2.1).
- A network of medium- and low-power repeaters (2 in figure 2.1), co-sited with mobile base stations for TV broadcast to mobile terminals in urban areas. Repeaters in urban areas complement satellite coverage for indoor service quality, which may be weakened by multiple walls and building obstacles. These repeaters can retransmit the satellite signal at the same frequency. Additional capacity can also be offered by the repeaters compared to the satellite for local insertion of programs in urban areas.

The system can interwork at the service level with a cellular network (3 in figure 2.1) to serve mobile terminals with limited-audience TV, video on demand (VoD), and interactive broadcast.

The system supports a high flexibility in frequency plan depending on the service targeted in terms of quality of service (QoS), number of TV programs, and regional content.

The solution operates in the 2,170–2,200 MHz frequency band (S-band), which was allocated to mobile satellite service (MSS) in 1992. This frequency band is adjacent to the frequency bands used by UMTS, which allows a cost-effective integration into cellular networks and terminals. This band is currently available in the whole of Europe, Africa, and Asia. This worldwide avail-ability allows economies of scale for terminals.

This concept system has been validated with experiments carried out within the European MoDiS[2] and MAESTRO[3] R&D projects, and recently during a trial conducted by the French Space Agency CNES in Toulouse. The MoDiS experimental platform demonstrated for the first time an innovative service delivery proposal for mobile users, relying on the synergy of mobile satellite and terrestrial networks. The successful demonstration of representative applications in a particularly challenging propagation environment such as Monaco/Monte Carlo, reinforced the

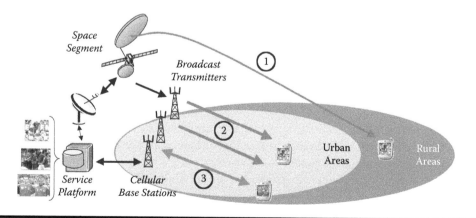

Figure 2.1   System overview.

technical feasibility of the system concept, while allowing real-world experimentation with key system engineering issues. MAESTRO went a step forward by addressing the whole range of issues pertinent to the system concept, from system design and demonstration to regulation and standardization. MAESTRO incrementally validated by both lab and field trials a number of key functions and performances using an experimental platform much closer to the envisaged commercial system and enabled real-world experimentation with more design trade-offs, particularly on the user terminal side, many of which have been made clear during the MoDiS trials.

2.3 Radio Interface Description and Performances

2.3.1 General Description

The purpose of the DVB-SH standard is to provide an efficient transmission system using frequencies below 3 GHz suitable for satellite services to handheld devices, in terms of reception threshold and resistance to mobile satellite channel impairments.

The system relies on a hybrid satellite–terrestrial infrastructure. The signals are broadcast to mobile terminals on two paths:

- A direct path from a broadcast station to the terminals via the satellite.
- An indirect path from a broadcast station to terminals via terrestrial repeaters that form the complementary ground component (CGC) to the satellite. The CGC can be fed through satellite or terrestrial distribution networks.

The system includes two transmission modes:

- OFDM/OFDM mode: The orthogonal frequency division multiplexing (OFDM) signal is based on the DVB-H standard[4,5] with enhancements. It is used on both the direct and indirect paths; the signals are combined in the receiver to strengthen the reception in a single-frequency network (SFN) configuration. This mode is particularly of interest for spectrum-limited systems.
- TDM/OFDM mode: The time division multiplexing (TDM) signal is partly derived from the DVB-S2 standard. Its allows optimizing transmission through satellite toward mobile terminals. It is used on the direct path only. OFDM with the same characteristics as above is used for the indirect path. The system supports code diversity recombination between satellite TDM and terrestrial OFDM signals to increase the robustness of the transmission in relevant areas (mainly suburban). This optional mode may be of interest in power-limited satellite systems.

For equivalent capacity, the TDM/OFDM mode requires a higher spectrum than the OFDM/OFDM mode, and therefore the TDM/OFDM mode is not considered further in this chapter.

Addressing handheld terminals, features already defined within DVB are reused, in particular time slicing for power-saving purposes, handover between frequencies/coverage beams, and Internet Protocol (IP) datacast protocols.

The main specific features are efficient turbo coding, allowing very low coding rates, and extended time interleaving, at the physical layer, for maximum robustness in severe shadowed environments.

The DVB-SH radio interface has been designed to support the application enablers defined by the DVB (Digital Video Broadcast, CBMS group) and in the future by the OMA (Open Mobile

Alliance, BCAST group) forums. No change in these standards will be necessary to support the DVB-SH. The same platform will be able to deliver services via DVB-H/UHF band infrastructure or via DVB-SH/S-band infrastructure.

2.3.2 Physical Layer

The DVB-SH radio interface is based on orthogonal frequency division multiplexing (OFDM) waveform technology well adapted to SFN transmission. It implements a high degree of flexibility in terms of configuration:

- Channel bandwidth: 5, 6, 7, or 8 MHz channel; 5 MHz is the preferred choice in the S-band allocation to align with UMTS canalization selected in the adjacent frequency band.
- FFT size: 2, 4, or 8K. In case of the S-band, 2K is the preferred choice to maximize the Doppler tolerance and hence allow terminal speed as high as 160 km/h with a state-of-the-art synchronization scheme.
- Quadrature phase-shift keying (QPSK) and 16 state quadrature amplitude modulation (16QAM) are digital modulation schemes: The choice results from a trade-off between broadcast capacity and targeted QoS.
- Guard interval can be chosen between 1/4, 1/8, 1/16, and 1/32 depending on the cell range and the SFN requirements.
- Coding rate of the turbo code can be selected down to 1/5 depending on the needed robustness.
- Interleaving length can be tuned up to several seconds. Already, 100 ms offers significant gain in the mobility scenario under terrestrial coverage, while a depth of a few seconds could improve the QoS in mobility conditions under satellite coverage.
- Upper layer FEC can be an added-in option; this FEC scheme is based on the MPE-FEC, with an extension capability of several seconds. This may advantageously replace the extended interleaving at the physical layer.

The radio interface offers a Moving Picture Experts Group–transport stream (MPEG-TS) interface service access point to support all application-enabling features defined in both the Digital Video Broadcast–Convergence of Broadcast and Mobile Services (DVB-CBMS) and the Open Mobile Alliance–Broadcast (OMA-BCAST) standardization work groups. MPEG-TS data is composed of bursts compliant with DVB-H time slicing.[4] Typically a burst transports a given service (or set of services), for example, a TV channel. The size of each burst may vary with time to support variable burst bit rate.

Typical configuration parameters and respective performances in terrestrial propagation are detailed in the table below. The typical urban six-path model for a pedestrian mobility scenario (at 3 km/h) has been selected to be the most representative reception condition.

Radio interface typical configuration		1	2
Channel bandwidth	MHz	5	5
Mode		2K	2K
Modulation		QPSK	16QAM
Coding rate (turbo code)		1/3	1/3
Interleaving depth	ms	125	125
MPE-FEC		None	None
Useful data rate at MPEG-TS level	Mbit/s	For GI 1/4: 2.25	For GI 1/4: 4.5
		For GI 1/8: 2.5	For GI 1/8: 5

Useful data rate at IP level	Mbit/s	For GI 1/4: ~2 For GI 1/8: ~2.25	For GI 1/4: ~4 For GI 1/8: ~4.5
C/N dB @ BER 10-5	dB	4.5 (TU6 at 3 km/h)0.0 (Gaussian)	9.5 (TU6 at 3 km/h)4.5 (Gaussian)

The C/N value corresponds to simulation results to which conservative implementation loss has been added.

Typically, QPSK1/3 is the preferred option for the satellite link to maximize the link margin, while 16QAM1/3 can be used for the terrestrial link to maximize the broadcast capacity. However, this will depend on the service targeted by the system.

In 2K mode, the useful symbol duration is 358.4 µs, and the guard interval duration is respectively 45 and 90 µs, allowing a max distance between two transmission sites of 13.4 and 26.8 km. Both are largely compatible with cellular network site density in urban and suburban environments, leaving sufficient margin to accommodate delay spread associated with multipath even in hilly terrain, as well as synchronization error between terrestrial transmitters and satellite.

Note that a bit error rate (BER) of 10-5 corresponds to a frame error rate of 1.5 percent for an MPEG-TS block size of 188 bytes, which is typically compatible with the error tolerance of a video encoder.

2.3.3 Service Layer

2.3.3.1 Principles

The service layer is compliant with the DVB-IP datacast over DVB-H specifications.[6] It supports two modes of delivery: streaming and download. The streaming mode applies to the delivery of real-time TV and radio programs, whereas the download mode is used to securely broadcast segmented radio and TV contents, music downloads, data files, and rich media contents.

The system targets reception by handset as well as vehicular terminals. QVGA320*240 at 25 frames per second is the typical screen resolution considered for mobile TV services.

The main service layer design challenges are to maximize the number of TV programs the system can broadcast while optimizing the user experience, which mainly depends on the video and audio quality as well as the channel zapping time.

To this end, the use of a state-of-the-art H264 video encoder along with high-efficiency advanced audio codec is recommended.

Video service is by nature variable in terms of bandwidth requirements. For a given video quality, a slow scene with few details will require a low bit rate, whereas a fast scene and complex pictures need a much higher bit rate.

Because, statistically, different video services do not need high bandwidth at the same time, the overall bandwidth requirements can be reduced provided that a statistical multiplexing scheme is used. This scheme, also called variable-bit-rate encoding, applies traffic smoothing while taking into account time-sensitive constraints associated with streaming services as well as some of the download services.

2.3.3.2 Service Capacity

For a video quality characterized by a pixel signal noise ratio of 32 dB, the equivalent bandwidth used by a TV program targeting QVGA320*240 screen resolution at 25 frames per second is

210 kbps, taking into account a 30 percent statistical multiplexing gain due to VBR mode video encoding. Another 24 kbps shall be provisioned for the audio encoded in constant bit rate with HE-AAC (high efficiency AAC) codec for stereo reception.

So the bandwidth requirement at the IP level becomes

- 2,106 kbps for 9 TV channels
- 4,212 kbps for 18 TV channels

This bandwidth requirement is compatible to, respectively, QPSK1/3 and 16QAM1/3 radio interface configurations taking into account an IP overhead of approximately 3.5 percent as well as PSI/SI table signaling. This also leaves some margin to transmit electronic signaling guide tables and encryption keys.

2.3.3.3 Zapping Time

Contributions to zapping time are multiple:

- **Reaction delay of the terminal:** Delay between the user action and the effective channel switching; below 0.1 s.
- **Time slicing:** Zapping action occurs at any time inside the frame so the effective delay between the currently received slice and the reception of the desired slice is variable and can last up to a multiplexing frame, whose duration is typically 1 s (9 TV channels, each one having a slice of 110 ms duration). The average value will be 0.5 s.
- **Video encoding:** We need to recover the next reference frame (I frame) on the new channel; the delay can vary between 0 (there is an I frame inside the first received time slice) to the group of picture size (which can be configured typically to 1 s). Good cooperation between the encoder and the IP encapsulator can ensure the presence of an IDR inside each burst so that the additional delay is null.
- **Reception buffering:** The buffer is set up in relation to the maximum jitter witnessed by IP packets over the network; in our case, jitter is minimized as real-time flows are handled stream-wise, with a minimum queuing delay at each stage; a reasonable buffering delay is 0.3 s.
- **Physical delay:** The DVB-SH physical layers offer an interleaving of duration between 100 ms and several seconds. Although the interleaving scheme is convolutional (meaning we need to wait for the interleaver delay after the end of the burst), extreme configurations provide nonlinear schemes, enabling fast zapping in good conditions. So the additional delay can be limited to 0.2 s.

The overall zapping time is not the addition of each contribution's worst case. Taking into account the probability of each delay value for each contribution, and the way they effectively add to each other, leads to an average zapping time of 1.1 s.

2.3.3.4 Power Saving

The power saving on the reception chain will depend on:

- **Number of TV channels:** selected for reception and the total number of TV channels on the carrier. In case of QPSK1/3, nine TV programs are multiplexed per time-slice period of 1 s; the burst duration is around 110 ms, which leads to a 90 percent theoretical power saving.

- **Chipset acquisition time performance:** The time at which the chipset needs to be powered on before actual service start so that it has synchronized and is able to deliver data; the typical value is in the order of 100 ms.
- **Physical layer interleaving duration:** Because DVB-SH supports a convolutional interleaver, the base-band receiver needs to be kept on for the duration of the interleaving depth after the end of the received burst. The typical interleaving duration is 200 ms.

The following equations can be derived in the convolutional interleaver case:

$$burst\_duration_{ms} = acquisition\_time_{ms} + 1000 * \frac{service\_rate}{channel\_rate} + convolutional\_duration_{ms}$$

$$power\_saving_{ratio} = \left(acquisition\_time_{ms} + 1000 * \frac{service\_rate}{channel\_rate} + convolutional\_duration_{ms} \right) \Big/ 1000$$

Assuming 234 kbit/s average bandwidth per TV programs, 110 ms burst duration, 100 ms acquisition time, and 200 ms interleaving depth, the power-saving ratio can be tuned to 59 percent. The power-saving ratio can be further decreased if several TV programs are simultaneously received or if the burst duration increases.

2.4 System Architecture

The Unlimited Mobile TV solution achieves a global SFN network per carrier. Synchronization between the terrestrial repeaters and the satellite allows the receiver to see the satellite signal as a simple echo of the terrestrial repeater signal. To increase the system's capacity in urban areas, the repeaters can broadcast additional DVB-SH signals over adjacent frequencies.

The Unlimited Mobile TV solution architecture is depicted in figure 2.2.

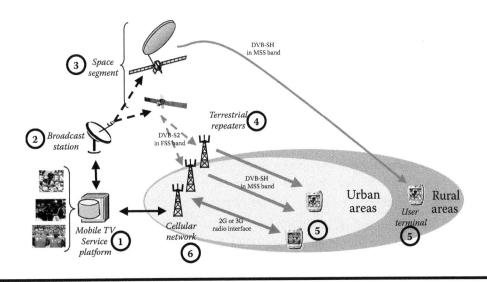

Figure 2.2   Hybrid satellite–terrestrial system architecture.

2.4.1 Overview

This system encompasses:

- A mobile TV service delivery platform that performs content adaptation and aggregates TV programs and rich multimedia services into IP service streams.
- A broadcast station that maps the IP service streams on the satellite and terrestrial repeater radio resources.
- A space segment encompassing dedicated geostationary satellites that amplify and convert the satellite radio signals from the Ka/Ku-band to the targeted IMT2000 frequency band. These signals, based on the DVB-SH radio interface, are received from the broadcast station and are transmitted directly toward the terminals.
- Terrestrial repeaters that receive service bundles from the hub via a satellite backhaul system or via a terrestrial IP network and map them onto the terrestrial carrier resources in DVB-SH radio interface format to the terminals in the targeted IMT2000 frequency band.
- User terminals including 2G/3G handset, PDA or vehicular mounted with the necessary features to receive the DVB-SH radio interface format signals and support mobile TV services.

2.4.1.1 Mobile TV Service Platform

The service platform bundles different types of content (live TV, VoD, podcast, etc.) into IP service streams and selects the transmission bearer, either broadcast (DVB-SH based) or unicast (2G, 3G, etc.), depending on the targeted audience. It implements the following features:

- Live and VoD streaming servers and video encoders.
- An electronic service guide (ESG) to list the available programs and contents. ESG is common to broadcast and unicast modes.
- An interactive services and rich media service node (e.g., to allow the purchase of the MP3 file associated with a clip currently playing on a TV program).
- A security scheme to authorize access to a specific program or specific contents.
- Interactive and personalized advertising.
- An audience measurement engine.
- Content and subscriber management.
- A bearer selection function to send the IP service streams created.

2.4.1.2 Broadcast Station

The broadcast station includes a hub and a mission control center. There is typically at least one broadcast station per dedicated satellite. Several broadcast stations may be colocated.

The hub encompasses:

- A network head end responsible for the mapping of the IP service streams received from the mobile TV service platform into MPEG2-transport streams. It also adds some time-stamp information (Megaframe Information Packet insertion) so that terrestrial repeaters can achieve a single-frequency network with the dedicated satellite in a spot beam.

- A broadcast head end in charge of the transmission of the service streams toward the satellite in DVB-SH format for the direct broadcast link and in DVB-S2 format for the indirect broadcast link. It also monitors and controls the satellite signal transmission.

The mission control center provides tools to manage the spectrum resources. It allocates the frequency carriers to the spot beams of the dedicated satellites. It then transfers the frequency plan to the terrestrial repeater network management system. It also interfaces with the satellite control center that controls and operates the dedicated satellites.

2.4.1.3 Space Segment

The space segment involves high-power, dedicated geostationary satellites (12 to 18 KW power class) with large deployable reflectors (12 m) to accommodate the handset terminal's low G/T performance without any antenna add-ons. European coverage is provided through several beams, each of nationwide size.

These satellites are transparent to the radio interface technology, occupying a 5 MHz channel bandwidth. They can all be colocated at the same orbit location. Typically, a satellite broadcasts one 5 MHz frequency carrier per beam in circular polarization.

Furthermore, one shall note that these dedicated satellites or other standard Ku-band satellites can be used for backhauling the TV programs' multiplex toward the terrestrial repeaters in the Ku-band using the DVB-S2 radio interface format.

2.4.1.4 Terrestrial Repeater Network

The terrestrial repeater network is deployed to provide indoor coverage in urban areas where the satellite signal is insufficient. Each terrestrial repeater can broadcast up to three 5 MHz frequency carriers in the IMT2000 band allocated to mobile satellite systems in linear polarization.

The terrestrial repeaters are designed for smooth integration in existing 2G or 3G cellular sites. The antenna systems can be shared. As well as a clear cost reduction, this integration avoids renegotiation with the site owners because it is a mere upgrade of the existing equipment.

The terrestrial repeaters can be fed either via Ku-band satellite signal using a DVB-S2 radio interface or via terrestrial IP network in case line-of-sight conditions cannot be met for a satellite antenna dish. Feeding the terrestrial repeater by satellite is the most cost-efficient solution to provide common information to a national network. Use of the IP terrestrial network also provides the ability to integrate regional programs on a geographic basis.

The terrestrial repeaters achieve a single-frequency network with the other terrestrial repeaters as well as with the satellite.

2.4.1.5 Terminals

Chipset processing the DVB-H signal may be adapted to take into account the specific parameters of the DVB-SH in the S-band (turbo code, interleaving). The tuners, currently used, may also be adapted to receive S-UMTS band frequencies. As the system is designed to operate at 2.2 GHz, reception diversity can be introduced, allowing significant improvement of the link budget (the signal is received on two antennas separated by a few centimeters, and then recombined). Chipsets currently developed allow indifferent reception of DVB-H in UHF or DVB-SH in the S-band.

Both reception chains can be integrated into a single module interfacing to the base-band host processor via a standard SD/SPI interface. This module includes two multiband tuners able to operate at 2.2 GHz and UHF, a base-band receiver with two OFDM demodulators, and a controller. The module will require less than 50 mm² surface area, with 65 nm CMOS technology in 2007.

This reception module can be embedded in 2.5G or 3G standard mobile phones, smart phones, pocket PCs, and vehicle receivers. The terminal form factor is the same as that for large-screen models from the consumer market.

DVB-CBMS and OMA application-enabling features can be fully reused to support service protection, service, and program guides.

A typical handset configuration would feature a terminal with a G/T of −29 dB/K. This terminal will implement reception diversity with linear polarized antennas.

The technology can be improved in the future to gain a significant margin on the satellite as well as a terrestrial link budget by reducing the noise figure and by using helicoidal or quadrifilar helix antennas in circular polarization.

2.4.2 Terrestrial Repeater Network Dimensioning

To determine the service coverage of the terrestrial repeater, we need to estimate the cell range associated with the terrestrial repeater. Because we consider three-sector antennas, we first assess the cell range in one 120° sector. The cell range is determined with the link budget that corresponds to a system gain. The system gain shall be greater than all possible losses the radio signal may encounter between transmitter and receiver. The loss includes path loss, shadowing loss, building penetration loss, and so on.

2.4.2.1 System Gain

The system gain is the difference in decibels between the transmitter radio frequency (RF) output power and the receiver sensitivity.

■ The receiver sensitivity is defined as the minimum RF input level required to provide an acceptable QoS associated with the broadcast service. It depends on the terminal G/T as well as on the reception bandwidth and system temperature. Here it has been assumed to be a G/T around −29 dB/K.

■ The transmitter RF output power can be computed with the following formula:

RF output power = 10*log(amplifier power/3) − feeder loss + base station antenna gain

We can assume an amplifier for the three signal carriers with a useful transmit power of 20 to 30 W, which means taking into account the necessary back-off from the 1 dB compression point to operate linearly. This transmit power needs to be divided by 3 for the link budget associated with one signal carrier. One shall provide 3 dB for the feeder loss, which corresponds to about 30 m of cable. A typical base station antenna will present a 18 dBi gain at 2.2 GHz.

Assuming a 10 W transmit power for the power amplifier per carrier, the transmitter RF output power will produce an equivalent isotropic radiated power of 55 dBm.

Consequently, the terrestrial link presents an overall 148.1 dB system gain.

2.4.2.2 Average Building Penetration Loss

In case of coverage in indoor environments from an outdoor transmitter, the signal experiences a building penetration loss that depends on:

- The type of building materials (walls, roof, windows)
- The frequency band in which the wireless system operates
- The targeted level of signal penetration (see table below)
- The incidence at which the signal penetrates

Extensive measurement campaigns have shown that the building penetration can be characterized by a lognormal law with an average and a standard deviation. The later component is taken into account in the shadow fading margin computed below.

Targeted Level of Signal Penetration	Reception Conditions
Deep indoor	Reception in a blind room (no windows) or reception behind two walls
Indoor daylight	Reception inside a room with a window
Indoor window	Reception behind a window

To simplify cell range calculations from the link budget we take average values for building penetration loss (BPL) with respect to the different propagation environments at 2200 MHz:

Type of Scattering Environments	Dense Urban Areas (DU)	Urban Areas (U)	Suburban Areas (SU)	Rural Areas (RU)
Average building penetration loss	21 dB	18 dB	14 dB	12 dB
Targeted level of signal penetration	Deep indoor	Deep indoor	Indoor daylight	Indoor daylight

These empiric values comes from propagation measurements at the 2 GHz frequency band.

2.4.2.3 Shadow Fading Margin

It is also called the lognormal or slow fading margin. It is introduced to guarantee a certain coverage probability Pcov in a noise-limited cell. The margin is the difference between the signal level necessary to cover the cell with a probability of coverage F (for the whole cell) and the average signal level at the cell edge.

> n: The slope of the propagation model (in the simplified formula $\alpha + 10 *n* \log (d)$)
>
> *SFM*: The shadow fading margin
>
> σ: The standard deviation of the shadow fading corresponding to the standard deviation of the combined fading effect of the outdoor slow fading and the building penetration loss.

The general formula that links these three parameters is called Jakes' formula. It should be noted that Jakes' calculations are for a single cell and take no account of overlapping cell coverage effects (e.g., macrodiversity; this corresponds to the SFN gain addressed in the following paragraph) that will in fact increase the probability of coverage.

$$P\,\text{cov} = \frac{1}{2}\left[1 - erf(a) + \exp\left(\frac{1-2ab}{b^2}\right) * \left(1 - erf\left(\frac{1-ab}{b}\right)\right)\right]$$

where

$$a = \frac{-SFM}{\sigma\sqrt{2}}$$

and

$$b = \frac{10 * n * \log_{10}(e)}{\sigma\sqrt{2}}$$

and $e = 2.72$.

Therefore, the probability of coverage within the cell will determine the shadow fading margin to be provided in the link budget.

Thanks to propagation measurements carried out in the T-UMTS frequency band for the 3G system, the n value has been characterized to 3.52 and standard deviations to the following:

Target Level of Penetration	Deep Indoor (U)	Indoor Daylight (SU)	Indoor Daylight (RU)
σ	10 dB	8 dB	7 dB

For the DVB-SH case, the shadowing fading margin is then computed with the Jakes' formula:

	DU&U	SU	RU
Shadow fading margin at Pcov = 90%	7.6 dB	5.5 dB	4.4 dB

2.4.2.4 Diversity Gains

Two gains shall be taken into account in the link budget. They can both be added together because they result from two independent phenomena.

2.4.2.4.1 Single-Frequency Network Gain

Because all transmitters transmit the same signal, which is frequency and time synchronized, it contributes to increase the number of signal echoes received or even increase the energies associated with some echoes in those areas covered by at least two adjacent cells. This results in a statistical gain on the received signal power that can range from 3 to 7 dB in the areas where adjacent cells overlap provided that the echoes from the distant cells fall within the guard interval of the OFDM symbol.

Given the cellular network configuration foreseen for DVB-SH technology, a 3 dB gain is conservative.

2.4.2.4.2 Reception Antenna Diversity Gain

We assume that the terminal will implement the maximal ratio combining scheme, which adds the signals received via the two antennas in phase. The diversity gain mainly depends on the correlation coefficient between both signals. The highest gain is achieved with the lowest correlation coefficient, where both signals are uncorrelated in terms of fading behavior. This occurs when the distance between the antennas is greater than the coherent distance of the propagation environments. This distance corresponds to the maximum spatial separation over which the channel response can be assumed constant. It has been observed a coherent distance equivalent to 1/3 of the wavelength (~4.6 cm) in indoor environments within the MAESTRO R&D project. With such minimum spacing between antennas, an antenna diversity gain of more than 6 dB can be experienced in these environments with a de-correlation factor of around 1/2. In outdoor environments, the reception diversity gain has been measured to lie around 2.5 dB.

2.4.2.5 Max Allowable Path Loss (MAPL) and Cell Range

The max allowable path loss is the margin remaining after having subtracted the shadow fading margin, the average building penetration loss from the system gain, but taking into account the diversity gains.

In our case:

Type of Scattering Environments	Dense Urban Areas (DU)	Urban Areas (U)	Suburban Areas (SU)	Rural Areas (RU)
System gain	148.1 dB	148.1 dB	148.1 dB	148.1 dB
Average building penetration loss	21 dB	18 dB	14 dB	12 dB
Shadow fading margin for 90% coverage	7.6 dB (deep indoor)	7.6 dB (deep indoor)	5.5 dB (indoor daylight)	4.4 dB (indoor daylight)
Diversity gains	3 + 6 dB	3 + 6 dB	3 + 6 dB	3 + 6 dB
MAPL	128.5 dB	131.5 dB	137.6 dB	140.7 dB

The cell range can be computed from the MAPL using a propagation model that describes the signal attenuation or path loss with respect to the scattering environments, the distance, and the frequency band.

In this chapter, we consider a model defined within the European COST 231 project. This model is valid for frequencies of 1.5 to 2 GHz. The basic formula for the median propagation loss given by COST 231–Hata is

$$L(dB) = 46.3 + 33.9 \log f_{MHz} - 13.82 \log h_1 - a(h_2) + (44.9 - 6.55 \log h_1) \log d_{km} - K$$

where:

Transmitter height (h_1) = 30 m; same as the base stations because the repeater shares the same antennas.

Receiver height (h_2) = 1.5 m; same as the cellular handset.

$a(h_2)$ and K depend on the environment; see Hata formulae.

Scattering Environments	DU	U	SU	RU
MAPL	128.5 dB	131.5 dB	137.6 dB	140.7 dB
DVB-SH Cell Range for Pcov = 90%	0.65 km	1.1 km	1.9 km	4.4 km

In a cellular topology network, the cell area can be computed with 1.95*(cell range)^2 to take into account overlaps between cells.

Scattering Environments	DU	U	SU	RU
Number of sites per km2	1.2	0.4	0.15	0.03

This dimensioning is comparable to a typical UMTS network dimensioned for 384 kbit/s in downlink and 64 kbit/s in uplink. The terrestrial repeater network can be deployed following a cellular grid compatible to GSM1800 and UMTS, especially in urban and suburban areas:

Scattering Environments	DU	U	SU	RU
MAPL	130.4 dB	131.6 dB	133.7 dB	138.9 dB
DVB-SH cell range for Pcov = 90%	0.72 km	1.08 km	1.4 km	3.8 km

The difference in cell size between DVB-SH and 3G can be used to take into account losses due to possible implementation at the terrestrial repeater side when sharing the same antenna with a base station or other interference effects that may occur in certain areas.

2.4.3 Space Segment Dimensioning

The satellite link budget in decibels can be computed in a simplified manner with the below formula:

$$\frac{C}{N_0} = EIRP - F_{loss} + L_{oss} - A_{loss} + \frac{G}{T} - depol_{loss} + 228.6$$

where:

C/N_0 (dB.Hz) is the carrier power-to-noise density ratio.

$EIRP$ (dBW) corresponds to useful satellite equivalent isotropic radiated power in dBW. In other words, it takes into account the necessary output back-off to prevent nonlinearities on the multicarrier signal. An output back-off of 2.5 dB has been assessed in the MAESTRO project to be sufficient for some satellite payload transponder adapted to S-band transmission. The EIRP depends on the payload RF power associated with the beam. In the case of linguistic beams over Europe and large satellite platforms, the EIRP can reach 74 dBW or more.

F_{loss} (dB) corresponds to the feeder link degradations. It can be minimized by adjusting the earth station EIRP.

L_{oss} (dB) corresponds to the free space loss and is negative according to the definition. It depends on the distance between the terminal and the satellite. The distance relates to the elevation angle at which a geostationary satellite located at 10° E ranges over Europe from 26° (Edimburgh) to 41.5° (Roma). The slant range never exceeds 39,000 km, and hence the free space loss remains above –191 dB at 2.2 GHz using the following definition:

$$L_{oss} = 20 \times Log\left(\frac{\lambda}{4 \times \pi \times R}\right)$$

A_{loss} (dB) corresponds to the atmospheric losses. In the S-band, there is no impact of water or gases in the propagation. The overall atmospheric loss is less than 0.2 dB for an availability greater than 99.99 percent; see ITU 618-7.

G/T (dB/K) qualifies the handset terminal's RF sensitivity. For handsets, –29.1 dB/K is a typical value.

$Depol_{loss}$ (dB) corresponds to the loss created by a polarization mismatch between the circular polarized satellite signal and the linear polarized receiver antenna. It accounts to 3 dB loss.

–228.6 (dBW^{-1}) corresponds to the Boltzmann constant 1.38*10^{-23}.

The satellite channel does not suffer from significant multipaths over 25° elevation angles, which is largely the case with the considered coverage. The reception conditions can be approximated by a Gaussian environment. Shadowing and blocking are the main impairments impacting the satellite signal reception. To simplify, the satellite propagation model is based on a three-state model:[7]

- Line of sight.
- Shadowed with a certain attenuation. This attenuation includes both fast and slow fading phenomena.
- Blocked.

The blocked state depends on the available margin on the satellite link budget. As the signal suffers attenuation (independently of the free space losses), it is necessary to provide link margins to mitigate the attenuation. We can have the following situations (in dB):

- If attenuation < link margin: The terminal experiences reception in line-of-sight or shadowed conditions, and hence the attenuation can be mitigated.
- If attenuation > link margin: The system cannot mitigate the attenuation, and hence the reception is blocked.

Based on this definition, the availability over the covered area can be simply defined as the probability that attenuation ≤ link margin.

Let us compute the link margin, which will allow overcoming the shadowing effect experienced by the terminal as well as the interferences effect:

$$Link\_marg\,in = C/(N_0) - 10 \times Log\,(BW) - C/N$$

where:

C/N (dB) corresponds to the carrier power-to-noise ratio required to ensure the signal reception. For QPSK 1/3 and in Gaussian environments, C/N = 0 dB, including implementation loss.

BW (MHz) is the DVB-SH equivalent signal bandwidth as seen by the receiver. *B* = 4.75 MHz for 5 MHz channel spacing.

Hence the link budget in QPSK1/3 modulation becomes for a typical high-power satellite configuration with 12 KW platform, 12 m diameter reflector, and two linguistic beams per satellite.

General parameters		
Mod code		QPSK1/3
Frequency band	(MHz)	2200
Signal bandwidth	Mhz	4,8
Transmitter (satellite)		
Satellite carrier EIRP	(dBW)	71
Propagation		
Slant range	(km)	39000
Free space propagation	(dB)	−191
Atmospheric loss	(dB)	0
Receiver (handset)		
Terminal G/T	(dB/K)	−29
Polarization mismatch loss	(dB)	3
Signal		
(C/No)	(dB.Hz)	76
C/N	dB	0
Link margin	dB	9

Thanks to the inherent time diversity scheme of the DVB-SH radio interface this link margin is large enough to provide more than 99 percent service availability in a mobile environment. Hence, the above link budget ensures high service availability for outdoor reception.

Moreover, the reception diversity scheme provides additional gain to achieve light indoor penetration in rural areas for rooms oriented southward.

2.5 Conclusion

We have described in this chapter the Unlimited Mobile TV solution, which can be configured to offer a large broadcast capacity for the distribution of high-quality TV and radio programs as rich media content, to satisfy the huge projected demand for mobile TV services in the future. The solution has been designed especially to achieve very good indoor coverage for handset terminals in urban areas relying on low-power terrestrial repeaters deployed in existing cellular sites. It also provides light indoor coverage in rural areas with high-power geostationary satellites. Thus, global coverage is achievable, following a rapid and cost-effective deployment.

This satellite-based system concept offers sufficient flexibility to optimize the performances according to targeted terminal and service types.

Acknowledgments

The authors thank the European Space Agency (ESA), the European Commission (EC/IST), the French Space Agency (CNES), and the Agence de l'Innovation Industrielle (A2I) for their financial and technical support in various R&D projects that contributed to the definition of this solution. Special thanks should be addressed to B. Barani, F. Medeiros, R. di Gaudenzi, J. Rivera, J. M. Casas, F. Louvet, G. Scot, and C. Allemand for their guidance in the design of this solution. Also, all the colleagues at Thales Alenia Space and Alcatel-Lucent Mobile Broadcast should not be forgotten for their contribution to this work.

References

1. Joint Mobile TV Group. 2006. *Mobile TV: The groundbreaking dimension*. White paper, version 2.21. Mobile TV UMTSF/GSMA Joint Work Group.
2. I. Andrikopoulos, A. Pouliakis, I. Mertzanis, et al. 2005. Demonstration with field trials of a satellite-terrestrial synergistic approach for digital multimedia broadcasting to mobile users. *IEEE Wireless Communications Magazine* 12.
3. T. Heyn and C. Wagner. Deliverable D2-2.3 propagation channel characterization. Final report, Fraunhofer Gesellschaft e.V. http://maestro.gfi.fr (accessed February 15, 2006).
4. *Digital video broadcasting (DVB); Transmission system for handheld terminals (DVB-H)*. EN 302 304.
5. G. Faria, J. Hendriksson, E. Stare, and P. Talmola. 2006. Digital broadcast services to handheld devices. *Proceedings of the IEEE* 94.
6. *Digital video broadcasting (DVB); IP datacast over DVB-H: Architecture*. TR 102 469.
7. F. Perez-Fontan, M. A. Vazquez-Castro, S. Buonomo, J. P. Poiares-Baptista, and B. Arbesser-Rastburg. 1998. S-band LMS propagation channel behaviour for different environments, degrees of shadowing and elevation angles. *IEEE Transactions on Broadcasting* 44.

Chapter 3

An Overview of Digital Multimedia Broadcasting for Terrestrial (DMB-T)

C. Katsigiannis, I. Foukarakis, D. I. Kaklamani, and I. S. Venieris

Contents

Keywords

digital multimedia broadcasting, DMB-T, DAB

3.1 Introduction

Mobile TV as analyzed in Samsung[1] is television for people on the move. The mobile viewer is provided with a broad range of television as well as radio programs that can be received by the user on cellular phones or personal digital assistants (PDAs) and displayed on the color screen in very high quality. Technically, there are currently two main ways of delivering mobile TV. The first is via a two-way cellular network, and the second is through a one-way dedicated broadcast network. Using the existing third-generation 3G infrastructure (UMTS) is the fastest and easiest way to get mobile TV off the ground, though there are specific issues like capacity shortage that constrain the transmission of TV program. Specifically, GSM and UMTS can only serve a limited number of viewers supporting point-to-point transmission. Even if multiple viewers watch the same television channel, each of them would have to be served by a dedicated point-to-point transmission, and hence the capacity of the serving radio cell would be quickly exhausted. To deal with these limitations, protocols like Digital Video Broadcasting–Handheld (DVB-H) and Digital Multimedia Broadcasting (DMB) were developed, serving an arbitrary number of users simultaneously.

Mobile TV differentiates from the conventional one-way transmission providing interactive television through feedback channels. The viewer does not passively watch a television program, but makes choices and takes actions, for example, in the context of public opinion polls, prize competitions, home shopping, advertising campaigns, or other interraction patterns. Providers of television programs may also profit from the combination with cellular networks. Audience ratings can be easily determined now, and new payment and receipt models can be established.

In the next sections the key technologies for realizing mobile TV in a DMB broadcast network will be analyzed.

3.2 DMB-T

Digital Multimedia Broadcasting–Terrestrial (DMB-T) is a mobile television service that targets mobile devices such as mobile phones and handheld and portable devices. The first launch of DMB-T services started in December 2005 in Korea. By many people, DMB is considered the first technology available to the market where mobile broadcasting and telecommunications converge.

DMB-T involved broadcasting of audio, video, and data. The appropriate streams are multiplexed and transmitted over a network of transmitters and gap fillers to reach the target devices. The transmitted programs use data rates of 1 ~ 1.5 Mbps, using frequency channels of 1.536 MHz bandwidth. DMB uses the DAB system for transmission, thus being able to be transmitted in a large range of frequencies in the electromagnetic spectrum, from 30 MHz to 3 GHz. However, the most common frequencies used are band III (174–240 MHz) and L-band (1452–1492 MHz).

The broadcast network can be realized by either a single-frequency network (SFN) or a multi frequency network (MFN) (see figure 3.1). SFNs are networks where the same signal is transmitted in the same frequency channel from various transmitters. All transmitters must be synchronized with each other. In MFNs the same signal is transmitted in different frequency channels. While they do not require complex mechanisms for synchronization, MFNs need more spectrum than SFNs to broadcast the same number of channels.

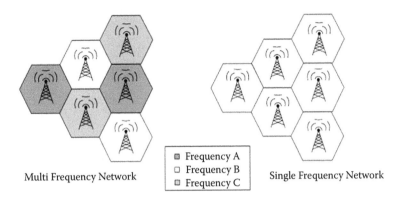

Figure 3.1 MFN and SFN.

DMB is standardized by the European Telecommunications Standards Institute (ETSI)[2] and will be legislated by ITU-R (figure 3.2).

3.3 DMB-T Services

DMB-T is based on Eureka 147 Digital Audio Broadcasting (DAB), standardized by ETSI.[3] It offers audio, video, and data services that are multiplexed in a MPEG-2 transport stream and are transmitted using DAB infrastructure. A detailed view of the services and service components can be seen in figure 3.3.

3.3.1 Audio Service

The goal for the audio services of DMB was to distribute digital radio programs that could replace analog VHF radio. Digital technology offers the means to detect errors in the digital signal due to interferences and to correct them to a certain degree. Compared to analog radio, DAB audio services achieve CD quality by using MUSICAM (MPEG-1, 2 Layer 2), the predecessor of the well-known MP3 format. Radio programs transmitted by this service can be either mono or stereo.

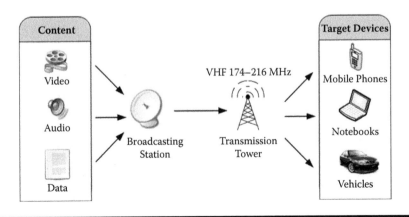

Figure 3.2 DMB-T overview.

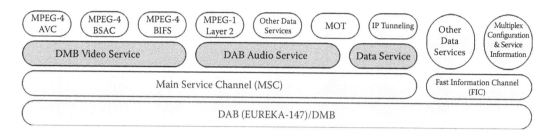

Figure 3.3   DMB services and service components.

Data can also be transmitted at the same time as audio. The data (called program-associated data [PAD]) includes short text messages that can contain information such as song title, program title, artist name, album name, and music genre, or other metadata related to the stream. This data can be synchronized with the audio stream, and is usually displayed on the receiver's screen/display.

3.3.2   Video Service

The DMB video service (figure 3.4) allows delivery of multimedia content to mobile devices through DAB. The video service architecture is divided in to three layers:

- Content compression layer
- Synchronization layer
- Transport layer

The main problem the video service needs to address is the limited bandwidth offered by DMB (maximum of 1.5 Mbps). The content compression layer is responsible for encoding and

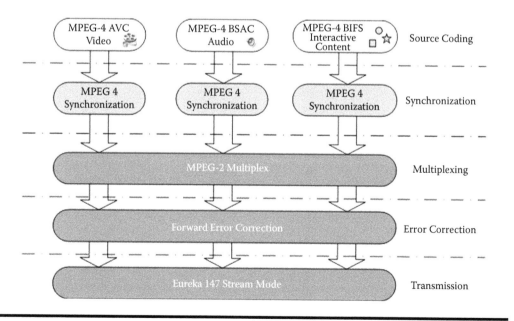

Figure 3.4   DMB video service layers.

compressing the content to be transferred. However, the compression and encoding mechanism need not be complex, as the average mobile device has limited processing and battery power.

Three different encoding schemes are used for video, audio, and interactive data streams. MPEG-4 Advanced Video Coding (MPEG-4 AVC), also known as H.264, is used for video coding. H.264 achieves very high data compression. It has been written by the ITU Video Coding Experts Group and ISO/IEC Moving Picture Experts Group. Its first version was released in 2003. The most common resolution for video is CIF (352 × 288 pixels), with a refresh rate of 30 frames per second. CIF requires a data rate of 250–300 Kbps. Other screen resolutions used include QCIF (176 × 144 pixels), QVGA (320 × 240 pixels), and WDF (384 × 224 pixels), all at a 30 frames per second refresh rate.

Originally audio streams were encoded using MPEG-4 Bit Sliced Arithmetic Coding (MPEG-4 BSAC). However, DMB now supports various versions of MPEG-4 Advanced Audio Coding (MPEG-4 AAC), offering better sound quality (especially if HE-AAC is used).

Interactive content is coded using object-based coding with MPEG-4 Binary Format for Scenes (MPEG-4 BIFS). MPEG-4 BIFS was designed to allow representation of dynamic and interactive data, such as video, text, two dimension and three dimension graphics, images, and audiovisual material. The main concept of BIFS is to allow separate encoding and transmission of the entities comprising a scene. The scene is recomposed when playback starts. For example, a scene may contain the audiovisual material, links to a Web page, active controls such as buttons, and so forth. Interactivity can be achieved in different ways, depending on the nature of the objects. For example, the user is able to hide, move, or show scene objects on demand, play videos, pop up Web pages upon clicking a button or a part of an image, and so on.

The next step after coding the content is to synchronize its elements. There is one data stream for audio, video, and interactive data. Audio and video are synchronized both temporarily and spatially by using MPEG-4 SL. The synchronized streams are then multiplexed to MPEG-2 transport stream (TS) packets.[4] To improve reception capability, a forward error correction scheme that utilizes Reed–Solomon coding is used. Finally, the packets are transmitted through the Eureka 147 stream mode.

3.3.3 Data Service

DMB can also offer a wide variety of data services other than PAD. While PAD is coupled with audio and video streams, this type of data is transferred in fixed-size packets and is called non-program-associated data (NPAD). The most notable applications of these data services are considered the Multimedia Object Transfer (MOT)[5] and IP tunneling.

MOT aims to deliver multimedia content to different types of devices in a unified manner, independently of the application, device manufacturer, and type. As MOT data is transferred over a one-way communication channel, it is possible that the receiver will lose packets because of bad reception conditions, being turned off, and so on. MOT segments data in different levels (objects and packets), and repeats transmission in various patterns, called carousels. When the receiver receives the data, it can replace bad segments or objects with the new ones, thus ensuring good reception of data. If the reception is still not good, the receiver can wait for the next transmission cycle to obtain the correct data.

IP tunneling enables encapsulation of IP packets into the packets of the data service. This way common TCP/IP applications such as the Web, File Transfer Protocol (FTP), and SMTP (e-mail) can be used on top of DMB/DAB service. The tunnel starts from the packet mode encoder on the transmitter and ends at the packet mode decoder at the receiver.

Applications
TCP/UDP
IP
DMB/DAB Data Link Layer

Figure 3.5   Protocol stack for IP tunneling.

Figure 3.5 shows the protocol stack used for IP tunneling.[6] Applications are unaware of the underlying transport layer, and they only use the TCP/UDP packets they receive. Like MOT, IP tunneling is unidirectional. To ensure correct reception of packages, packets at the data link layer are transmitted more than one time. If a correct packet is received, it is forwarded to the IP layer. Duplicate packets are filtered out, as IP cannot handle the same versions of the same IP datagram.

Interactive data services can be developed by combining DMB with a two-way communication channel offered by a mobile cellular network (i.e., Short Message Service offered by GSM). As can be seen in figure 3.6, the communication channel is used not only for sending data from the user's terminal to the content provider, but also for retrieving part of the content. The most common services offered by providers are electronic program guides, news, weather forecasts, traffic information, and so on.

3.4   Receivers

DMB-T aims to deliver multimedia content mostly to mobile terminals. In this section we present the most common cases of such terminals.

3.4.1   Mobile Phones

Mobile phones are one of the main targets of DMB-T technology, mostly because of the high number of such devices in countries where DMB-T has been or is about to be introduced. They have the smallest screen size of all the DMB-T-enabled devices, but they are expected to be the main body of

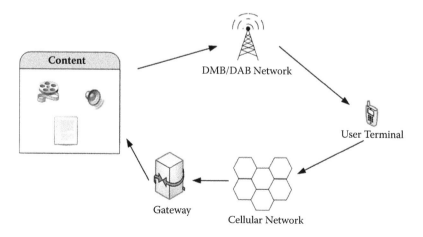

Figure 3.6   Interactive data services in DMB.

mobile terminals the users are going to use. The user is very likely to use the device when he has a few minutes available (i.e., when he is in a train or in a bus). Various manufacturers have presented mobile phones with built-in DMB receivers, both for band III and L-band. To improve the user experience, additional functionality is provided. These functions include PVR (using time shift) so that seamless TV watching can be achieved, TV screen capture, TV-out (allowing the sending of DMB signals to standard TV sets), and digital camera/MP3 player integration. Memory requirements for these applications can be met either by the phone's memory or by using flash cards.

3.4.2 Vehicle Receivers

Vehicle receivers provide a 5- to 7-inch display. They use existing audio systems (i.e., a car's stereo) for sound. Their main benefit is that they have no battery limitation, as they take advantage of the vehicle's plugs (i.e., cigar power jack). Current vehicle receivers support both band III and L-band. They provide various additional types of functionality, varying from navigation software to media players. A combination of navigation software with real-time traffic and travel information and telematics may provide the killer application for these devices. In cases where a display is already available in the vehicle, vehicle set top boxes provide the required functionality for DMB-T reception.

3.4.3 Laptop/USB Receivers

Personal computer users may enjoy DMB-T either by receivers that are built in to their laptops or by using USB or PCMCIA receivers.

3.4.4 Handhelds

Various handheld devices that have a small LCD screen are likely to have DMB-T functionality integrated. Such devices include digital cameras, portable game consoles, PDAs, multimedia players, and so on. Existing devices that support a level of extensibility may support plug-in DMB-T receivers.

3.5 Channel Coding/Multiplexing

Coding and multiplexing are two of the basic functions to be undertaken concerning the transmission of data streams. The main goal of channel coding is to adapt the present form of data streams, resulting from the different services, to the special characteristics of the radio channel. Through channel coding the data streams are prepared in a manner such that errors that may occur during transmission are reliably detected and corrected at the receiver. This is accomplished by calculating redundant data from the data streams. To make the transmission more reliable and to improve the total quality of a service, channel coding increases the total amount of data.

The data streams for all services are encoded by a method known as convolutional coding, which takes n bits from a continuous input data stream and maps them onto m bits (m > n) of an output stream. The output streams are generated by combining the outputs of several linear feedback shift registers in a certain manner. At the receiver the transmission errors can correctly decode the incoming data by a Viterbi decoder.

Data error correction is associated with the code rate, which represents the ratio between the number of input and output data bits (R = n/m). Specifically, the lower the code rate is, the higher the probability of error correction is. But to the contrary, the net data rate that can be achieved at the radio channel is lower. For this purpose, DAB/DMB standards propose dynamic fixes of the appropriate code rate, taking into account the expected degree of interferences, which also depends on the environment where the transmitter operates.

Viterbi decoder can deal only with correction of single bit errors, although the major issue arising in radio transmission is the error bursts. The main characteristic of error burst is the large number of errors in consecutive bits. To deal with error bursts, the output generated by the convolutional coder must be mixed in a next step, and this process is called time interleaving. Data streams are subdivided into code words of fixed length, and in accordance with a certain algorithm, the consecutive bits of a code word are exchanged with the bits of previous and subsequent code words. At the receiver the bit sequence must be reassembled following a de-interleaving procedure, during which error bursts occurring in the transmission are distributed over several code words subdivided into single bit errors that can be corrected by the Viterbi decoder.

On the other hand, while dealing with error bursts, time interleaving, by spreading consecutive bits over several code words, results in increasing the delay. The reassembling of bits to create the original sequence of a code word must be delayed until all bits needed are received. The total delay will be hundreds of milliseconds, but does not cause serious problems for the majority of DAB/DMB services. Instead, it leads to issues concerning the trasmission of large time-sensitive control information. For this purpose, the fast information channel (FIC) was introduced. It is used to carry time-sensitive control information when no interleaving is applied, and it represents a channel that is subject to a convolutional coding more robust than that of other channels.

Moreover, regarding the viewers traveling at high velocities, up to 200 km/h, an additional error correction method has been specified in the standards, concerning data streams originating from the DMB video service. This method is called block coding and is executed before convolutional coding at the transmitter and after convolutional decoding at the receiver. It is also known under the term *outer coding*.

Convolutional coding is performed on a continuous data stream, while the data stream resulting from the MPEG-2 multiplex is subdivided into blocks of 187 bytes in length. The method applied to correct up to eight erroneous bytes per block is called Reed–Solomon code. This code is generated and attached to the respective block as a parity word of 16 bytes in length. After application of the Reed–Solomon code, consecutive blocks and associated parity words are subject to another interleaving process for reducing the probability of error bursts.

Because the data streams resulting from the different services are not suitable for transmission, after being adapted to the special characteristics of the radio channel, through channel coding, they must be multiplexed into a common transport stream prior to transmission.

The time-interleaved data streams coming from the several services are merged into a common transport stream, which is referred to as multiplex and, in the particular case of DMB, ensemble. Furthermore, multiplexing is organized depending on the frame structure, as described in figure 3.7. A transmission frame is subdivided into three fields for carrying data for the synchronization channel (SC), the fast information channel (FIC), and the main service channel (MSC).

■ The SC transfers a standardized, fixed bit pattern, marking the beginning of a frame, and used by the receiver for synchronization with the transmitter.

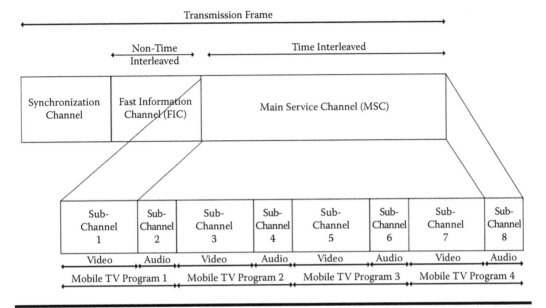

Figure 3.7   Structure of DMB transmission frame.

■ FIC carries the multiplex configuration information (MCI), indicating the structure and organizing the ensemble. Because this information is very time sensitive, this channel is not time interleaved.

■ MSC carries user data from different DMB services and is subdivided into several sub-channels. Through every single subchannel only a single data stream of a particular service is carried, though multiple streams are not allowed to be transferred in the same subchannel. The service to be delivered defines a certain level of data rate, so the size of the fields is dynamically configured. From the receiver's perspective, the information concerning the assignment between services and associated subchannels is obtained through the MCI.

It is obvious that the receiver needs to receive and decode only the data streams that belong to the services actually used. The position of a certain subchannel within the transmission frame is indicated to the receiver through the MCI. To reduce power consumption, the receiver can be activated and decode the specific data streams according to the requested service; thus, during the reception as well as decoding of the rest of the data streams belonging to the same frame, the receiver can be deactivated. More precisely, when receiving an ensemble, which carries four mobile TV programs from which the viewer watches only one, the receiver can be switched off three-quarters of the time. This method is known as time slicing.

3.6   Modulation and Transmission

The transport stream resulting from multiplexing and encoding is broadcasted over a radio channel to the receivers. The parameters of a radio channel are fixed by the carrier and the data rate. The first one is a periodic, sinusoidal electromagnetic wave of a certain frequency, corresponding to the frequency range in the electromagnetic spectrum used by the emiting antenna. The data rate denotes the speed of data transmission and determines the required bandwidth of the radio channel.

Concerning DMB, the bandwidth offered by the radio channel is 1.536 MHz and net data rates are between 1 and 1.5 Mbps, depending on the code rates used for convolutional coding of the various data streams. These data rates are achieved for a large area of coverage, up to 100 m in diameter of a single DMB transmitter. The problem that arises when transmitting at high data rates over long distances is associated with the signal's interference and is called multipath propagation.

The data is transferred over a radio channel in the form of symbols, and each symbol represents one or several bits of the data stream. DMB transmits two bits as a single symbol defining four symbols: 00, 01, 10, and 11. Accordingly, the carrier of the radio channel may adopt four different signal states depending on the next symbol ready for transmission. This process is called modulation or shift keying. The signal generated in this way is subject to phenomena that distort it and make it difficult or impossible to interpret the incoming signal to the receiver. These phenomena are the aforementioned multipath propagation, or attenuation, noise, shadowing, and frequency shifts, which are called Doppler shifts and are caused by the receiver's movement.

The multipath propagation problem, as depicted in the figure 3.8, is associated with the existence of obstacles in the transmission range of the DMB transmitter. Specifically when referring to multipath propagation, reflection, scattering, and diffraction of the transmitted signal, due to buildings, hills, or trees, are implied. The signal is decoupled during transmission and the receiver receives not only the primary impulse of the signal but also several delayed secondary impulses. The sequence of the received impulses introduces a metric called delay spread, which is associated with the intermediate time of arrival of the primary, secondary, and following impulses of the transmitted signal. The size of the delay spread depends on the transmitter's range as well as the density of obstacles in the close surroundings. Specifically, it is in accordance with the density and range, and increases when the range is long and the density is high.

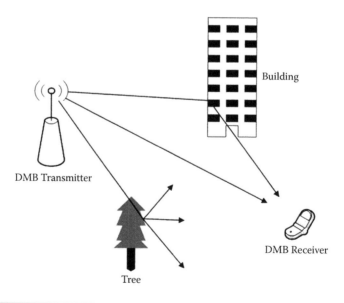

Figure 3.8 Multipath propagation.

If the delay spread is larger than the symbol duration T, used during the transmission, then the multipath propagation may cause heavy interference. The symbol duration denotes the length of time a single symbol is transmitted, and it corresponds to the reciprocal of the symbol rate. There is a possibility that the delayed secondary impulses may destruct the impulses of the subsequent symbols if the symbol duration is much smaller than the delay spread. This effect, which is the main reason for transmission errors, is called intersymbol interference.

DMB operates in a manner very convenient for intersymbol interference. Assuming that the transmission was carried out with a conventional single carrier modulation, then to achieve high data rates the symbol duration should be in the range of a few microseconds. Obviously this is too small compared with the delay spread ranging from 10 to 100 μs, in long-range DMB transmitters operating in urban environments. However, in SFNs the delay spread may be even larger, as receivers also pick up signal impulses from neighboring transmitters far away from the serving one. To avoid intersymbol interference, DMB applies a technique known as multicarrier modulation.

Concerning the multicarrier modulation of DMB, a radio channel of 1.536 MHz bandwidth is subdivided into N subcarriers. Each of the subcarriers is able to transfer data independently of the other. The transport stream is distributed over the N subcarriers, and the symbol duration on each subcarrier can be enlarged by the factor N with respect to the symbol duration that would be used with a single carrier modulation. The intersymbol interferences are avoided, because the symbol duration of each subcarrier is larger than the expected delay spread, assuming that N is chosen to be sufficiently large.

The major problem that may arise in multicarrier modulation is the side lobes, resulting from the out-of-band radiation in the frequency bands below and above each subcarrier. Side lobes do not carry any useful information for the interpretation of the data streams by the user, but they can distort the transmission to the neighboring subcarriers. For this purpose, the main concern, when using multicarrier modulation, is to select an appropriate frequency space between the subcarriers. DMB systems utilize a very well-known and well-used technique called orthogonal frequency division multiplexing (OFDM) where the subcarriers are orthogonal to each other, presented in figure 3.9. The main feature of this method is the mutual neutralization of the side lobes appearing in the radiation pattern of the signal. This is caused by the aforementioned orthogonality of the OFDM carriers.

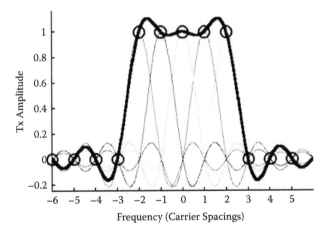

Figure 3.9   Orthogonal frequency division multiplexing.

Table 3.1 Parameters of DMB Transmission Modes

	Mode I	Mode II	Mode III
Network	SFN	MFN	Terrestrial
Frequency range	174–216 MHz	1,452–1,467 MHz	1,452–1,467
Range	96 km	24 km	48 km
Number of subcarriers	1536	384	768
Space between subcarriers	1 kHz	4 kHz	2 kHz
Symbol duration	1 ms	250 μs	123 μs
Guard time	246 μs	62 μs	123 μs
Number of bits/OFDM symbol	3072	768	1,536
Frame duration	96 ms	24 ms	48 ms

In addition, because the subcarriers are placed close together, OFDM is very bandwidth efficient compared to a nonorthogonal multicarrier modulation.

Analyzing the transmission capabilities offered by the DMB system, we could define different modes of transmission, summarized in table 3.1. The differences between them are focused on the number of subcarriers contained in a radio channel of 1.536 MHz bandwidth, on the symbol duration and on the length of the transmission frame. To choose the more efficient transmission mode, several criteria are defined, like the allocated frequency range, the surrounding area (rural, suburban, or urban), and the type of network to be applied (single-frequency network, multifrequency network).

Preferred modes are those with a high number of subcarriers and long symbol duration, for networks consisting of DMB transmitters with long ranges, causing long delay spreads and strong intersymbol interference. This especially holds for transmission in SFNs where the receiver receives several signal impulses from the neighboring transmitters. For the mentioned scenarios, DMB adopts transmission mode I, since it offers a maximum number of 1,536 subcarriers (separated by 1 KHz from one another) and a symbol duration of 1 ms. On the other hand, when referring to networks that request a smaller delay spread, transmission modes with fewer subcarriers and shorter symbol durations may be used. For example, mode II, where 384 subcarriers are applied and the symbol's duration is 125 μs, could be selected. The robustness against intersymbol interference is further improved by including a so-called guard time between consecutive symbols during which no data transmission is allowed.

The frames generated by the multiplexing are finally modulated onto the subcarriers of the OFDM radio channel. According to the symbol that will be transmitted, the DMB system utilizes a modulation scheme where the phase of each subcarrier is shifted. In general, the modulation scheme to be applied is called differential quadrate phase shift keying (DQPSK). The four symbols defined to transmit data streams over a radio channel (00, 01, 10, 11) are assigned to the phase shifts of 0°, 90°, −90°, 180°, respectively. For a symbol to be transmitted, the phase of each subcarrier is changed, taking into account the respective phase shift, and in association with the phase of the previous symbol.

In figure 3.10 the distribution of symbols over the subcarriers is presented, concerning an OFDM radio channel with only N = 8 subcarriers. Transport frames are subdivided into OFDM symbols, each covering exactly the N subcarriers and thus consisting of N single

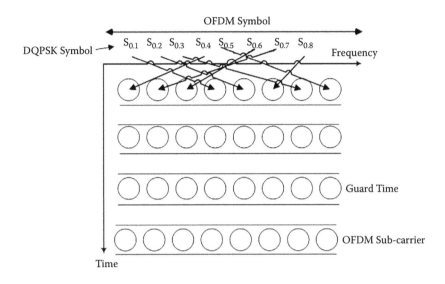

Figure 3.10 Structure of DMB transmission frame.

DQPSK symbols. The DQPSK symbols are mixed according to a certain algorithm in a manner that consecutive symbols are not transmitted in neighboring subcarriers. This procedure is characterized as frequency interleaving. Similarly to time interleaving, frequency interleaving is used to deal with error bursts, which occur if interferences span a number of neighboring subcarriers.

It should be mentioned that depending on the size of the allocated frequency range, several DMB channels can be realized in parallel. Assuming that there is a frequency range, of 8 MHz, which corresponds to the bandwidth of an analog television channel, if this range is used to serve DMB radio channels, then 15 to 20 mobile TV programs can be delivered.

3.7 Advantages

As described in the previous sections, DMB-T is based on the DAB standard for transmitting a multimedia multiplex. By using the existing DAB infrastructure, the cost to introduce DMB-T technology is lower than that required for developing new infrastructure.

Compared to other digital mobile television standards, DMB-T requires lower field strength, thus making it more efficient. Also, it requires only 1.5 MHz of spectrum to transmit one multiplex. While other technologies aim to transmit at the UHF band, DMB-T utilizes band III and L-band, where it is easier to locate available spectrum than the UHF band, which is in use for analog and digital television.

Another important issue is that DMB-T is already active in many countries, while its two main competitors (DVB-H and MediaFLO) are still in development. DMB-T was initially developed for Korea, and the first commercial transmission was aired in December 2005 in Korea. However, a large number of countries are considering the introduction of this technology. The standardization of DMB-T by both ETSI and ITU is one of the reasons that many countries are considering the DMB-T option.

Finally, a large number of DMB-T-enabled devices are already available on the market. These devices vary from mobile phones to USB receivers. As DMB-T is built to require less battery power than other mobile broadcasting technologies, most devices have enough battery for 2 to 3 hours of playback.

3.8 Summary

Mobile TV has introduced a novel set of business models, like composition of new interactive programs. In this chapter we analyzed the mechanisms by which mobile TV can be realized through the DMB system. The DMB system follows the architecture proposed by the DAB system, allowing the encoding of mobile TV programs in different resolutions and frame rates. Furthermore, the bandwidth used by DMB is 1.536 MHz, providing the capability of carrying an ensemble consisting of four or five mobile TV programs.

Transmission in the DMB system is very robust against interference, because a two-step error coding is applied. By incorporating methods of time interleaving and frequency interleaving, error bursts can be subdivided into single bit errors, increasing the reliability of error correction. Furthermore, issues like intersymbol interference, arising from multipath propagation of radio signals, are limited by applying multicarrier modulation in accordance with OFDM.

Finally, the most useful feature of DMB is that it can be implemented by the DAB infrastructure, which has already been developed on a broad basis in many countries. So the cost of realizing a DMB system is affordable, taking into account the advantages coming from this technology, like delivering very high-quality TV programs to viewers traveling at high velocities.

Links

1. WorldDMB: http://www.worlddab.org/.
2. DMB portal: http://eng.t-dmb.org/.

References

1. Samsung. DMB, digital multimedia broadcasting. http://uk.mobiletv.samsungmobile.com/fileadmin/downloads/DMB_en.pdf.
2. ETSI. 2005. Digital audio broadcasting (DAB); DMB video service; User application specification. TS 102 428.
3. ETSI. 2006. Radio broadcasting systems; digital audio broadcasting (DAB) to mobile, portable, and fixed receivers. EN 300 401.
4. ETSI. 2005. Digital audio broadcasting (DAB); Data broadcasting—MPEG-2 TS streaming. TS 102 427.
5. ETSI. 1999. Digital audio broadcasting (DAB); Multimedia object transfer protocol. EN 301 234.
6. ETSI. 2000. Digital audio broadcasting (DAB); Internet protocol (IP) datagram tunnelling. ES 201 735.

Chapter 4

Overview of ISDB-T
One-Segment Reception

Gunnar Bedicks, Jr. and Cristiano Akamine

Contents

Keywords

ISDB-T, ISDTV, Digital TV, BTS OFDM, hierarchical transmission, partial reception

Abstract

The Integrated Services Digital Broadcasting–Terrestrial (ISDB-T) was developed in Japan by the Association of Radio Industries and Businesses (ARIB) in 1998.[1] The ISDB-T provides audio, video, and multimedia services and integrates various kinds of digital contents, each of which may include high-definition TV (HDTV), multiprogram standard-definition TV (SDTV), and low-definition TV for portable reception and data (graphics, text, etc.). In Japan the ISDB-T was launched in 2003 for HDTV and SDTV, and in 2006 for portable reception. Also, in 2006, the Brazilian government adopted the International System for Digital Television (ISDTV), which uses the same modulation as ISDB-T. For both standards, portable reception, also called 1seg, means that the receiver can demodulate 428.57 KHz of the center of the broadcasting channel. This chapter describes the transmission system and the one-segment reception.

4.1 History and Background

The ISDB-T was developed to cover a variety of services in 6 MHz of bandwidth. The system is very flexible, and it is possible to combine many types of services, including HDTV, SDTV, and LDTV. The system uses a modulation called Band Segmented Transmission–Orthogonal Frequency Division Multiplexing (BST-OFDM)[1–4] designed to provide reliable high-quality video, sound, and data broadcasting not only for fixed receivers but also for mobile and portable receivers. In the BST-OFDM each segment has a bandwidth corresponding to 6/14 MHz = 428.57 KHz, and it is possible to combine 13 segments in up to three hierarchical layers called A, B, and C. Each layer has an individual error protection scheme similar to that of digital video broadcasting–terrestrial (DVB-T)[5] (coding rates, 1/2, 2/3, 3/4, 5/6, and 7/8 of inner code; depth of the time interleaving, 0, 100, 200, and 400 ms) and a type of modulation (DQPSK, QPSK, 16-QAM, or 64-QAM). For example, in the most useful configuration one segment is designed to layer A for portable reception (QPSK modulation, 2/3 code rate, and 400 ms of time interleaving), and to layer B 12 segments are used for fixed reception (64-QAM modulation, 3/4 code rate, and 200 ms of time interleaving). In BST-OFDM the common parameters are the inverse fast Fourier transform (IFFT) size (modes 1, 2, and 3) and guard interval ratio (1/4, 1/8, 1/16, and 1/32). The useful bandwidth of the ISDB-T channel is 6/14 × 13/14, corresponding to 5.57 MHz for 6 MHz of broadcasting channel spacing. Also, the ISDB-T showed a very robust modulation for many types of interferences, like impulsive noise, Doppler, and multipath.[6]

4.2 Overview of ISDB-T

We can divide the ISDB-T transmission system in four blocks: audio/video coding, remultiplexing, channel code, and modulation (shown in figure 4.1).

In Japan the ISDB-T system uses MPEG-2 video coding and MPEG-2 advanced audio coding (AAC) for HDTV and H.264 for mobile reception[7] (shown in table 4.1).

In the remultiplex transport stream packets (TSPs) of length 188 bytes are grouped in a multiplex frame. The length of multiplex frame depends on guard interval and IFFT size values. In the output of remultiplex transmission multiplexing configuration control (TMCC) information is added with the configuration of the parameter of the modulator and assigned to each layer. At this point the length of TSPs is 204 bytes and is called the broadcast transport stream (BTS).

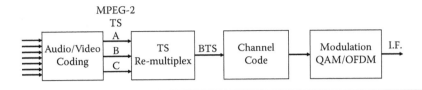

Figure 4.1   Basic block diagram of ISDB-T system.

In the channel coder and modulation, each TSP is assigned to the correct layer and parallel processing occurs. The code rate, type of modulation, and time interleaving length are independent for each layer. After the layers are combined passing to OFDM modulator and guard interval insertion resulting in a BST-OFDM signal of 6 MHz of bandwidth. In the ISDB-T it is possible to use up to three layers, setting different levels of robustness in each layer (shown in figure 4.2).

The parameters of the modulator and association for each segment can be chosen following table 4.2, and the total bit rate for the segment is shown in table 4.3. Table 4.4 shows the desired receiving parameters used in Japan.[8]

4.3 Transmission Scheme of ISDB-T

4.3.1 *Remultiplex*

The ISDB-T system uses a MPEG-2 system[1,9] and can remultiplex up to three layers, called A, B, and C. Furthermore, to achieve an interface between multiple MPEG-2 transport streams in layers A, B, and C to the OFDM transmission system, these transport streams (TSs) are remultiplexed into a single TS called the BTS.

The basic unit of a BTS signal is a TSP with a length of 204 bytes, and the first byte of TSP is 47Hex. A multiplex frame consists of *n* TSPs (shown in table 4.5).[1,9,10] In addition, transmission control information such as channel segment configuration, modulation, channel code, and other transmission parameters is sent to the receiver in the form of a transmission multiplexing configuration control (TMCC) in 8 bytes and also multiplexed as ISDB-T information packets (IIPs).[1] Figure 4.3 shows the outline of the multiplex frame of a BTS signal.[11]

The BTSs establish a fixed transmission clock (512/63 Mbps) by inserting null packets in a multiplex frame. Layer A can be used for partial reception (1 segment) or full transmission (13 segments). The compatibility between OFDM segments and transport stream packets can be ensured by introducing a frame configuration of the OFDM signal and multiplex signal. The packets can be correctly separated and synthesized with respect to their appropriate hierarchical layers by prescribing a packet arrangement in a multiplex frame based on the receiver operation. In partial

Table 4.1   Audio and Video Coding

Video Coding	Audio Coding
HDTV MPEG-2 MP@HL	MPEG-2 AAC
STVD MPEG-2 MP@ML	MPEG-2 AAC
LDTV H.264 BP@L1.2	AAC-SBR

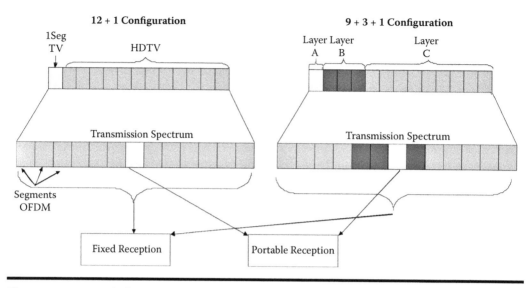

Figure 4.2   Transmission spectrum of ISDB-T.

Table 4.2   Transmission Parameters

Parameters	Mode 1	Mode 2	Mode 3
Number of segments		13	
Segment bandwidth (kHz)		6,000/14 = 428.57	
Number of carriers per segment	108	216	432
Total number of carriers—Nc	1,405	2,809	5,617
Effective symbol length—Tu (µs)	252	504	1,008
Number of symbols per frame		204 OFDM symbols	
Guard interval length		1/4, 1/8, 1/16, or 1/32 (effective symbol length)	
Modulation scheme		QPSK, 16-QAM, 64-QAM, DQPSK	
Outer code		Reed Solomon (204,188)	
Inner code		Convolutional code: 1/2, 2/3, 3/4, 5/6, 7/8	
Time interleaving		0, 0.1, 0.2, or 0.4 s	

		Up to 3 layers (A, B, and C)		
Hierarchical transmission	Examples[a]	Number of segments in layer A	Number of segments in layer B	Number of segments in layer C
	1	13	0	0
	2	1	12	0
	3	1	3	9

[a] In option 1 only one layer (low protection layer) is used for HDTV or multiple SDTV programs.
In option 2 one segment (high protection layer) is used for partial reception and twelve segments (low protection layer) are used for HDTV or multiple SDTV programs.
In option 3 one segment (high protection layer) is used for partial reception, three segments (middle protection layer) for mobile reception in SDTV, and nine segments (low protection layer) for HDTV.

Table 4.3   Bit Rate for Segment

Modulation	Convolutional Code	Number of TSPs Transmitted in One Segment (Mode 1/2/3)	Bit Rate (kbps)			
			Guard Interval			
			1/4	1/8	1/16	1/32
DQPSK QPSK	1/2	12/24/48	280.85	312.06	330.42	340.43
	2/3	16/32/64	374.47	416.08	440.56	453.91
	3/4	18/36/72	421.28	468.09	495.63	510.65
	5/6	20/40/80	468.09	520.10	550.70	567.39
	7/8	21/42/84	491.50	546.11	578.23	595.76
16-QAM	1/2	24/48/96	561.71	624.13	660.84	680.87
	2/3	32/64/128	748.95	832.17	881.12	907.82
	3/4	36/72/144	842.57	936.19	991.26	1021.30
	5/6	40/80/160	936.19	1040.21	1101.40	1134.78
	7/8	42/84/168	983.00	1092.22	1156.47	1191.52
64-QAM	1/2	36/72/144	842.57	936.19	991.26	1021.30
	2/3	48/96/192	1123.43	1248.26	1321.68	1361.74
	3/4	54/108/216	1263.86	1404.29	1486.90	1531.95
	5/6	60/120/240	1404.29	1560.32	1652.11	1702.17
	7/8	63/126/252	1474.50	1638.34	1734.71	1787.28

Table 4.4   Desired Receiving Parameter

	13-segment Receiver	1-segment Receiver
Number of segments	13	1
Mode	2, 3	2, 3
Guard interval ratio	1/4, 1/8, 1/16	1/4, 1/8, 1/16
Interleaving	Mode 3: I = 1, 2, 4	Mode 3: I = 1, 2, 4
	Mode 2: I = 2, 4, 8	Mode 2: I = 2, 4, 8
Modulation and FEC	QPSK: 1/2, 2/3	QPSK: 1/2, 2/3
	16-QAM: 1/2, 2/3	16-QAM: 1/2
	64-QAM: All	

Table 4.5   Number of TSP in a Multiplex Frame

Mode	Guard Interval			
	1/4	1/8	1/16	1/32
1 (2k)	1,280	1,152	1,088	1,056
2 (4k)	2,560	2,304	2,176	2,112
3 (8k)	5,120	4,608	4,352	4,224

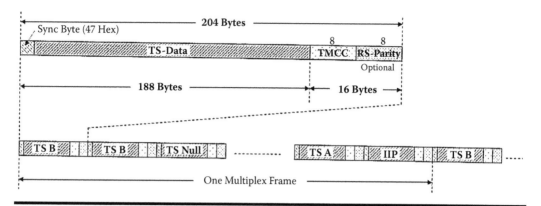

Figure 4.3 Outline of BTS signal.

reception, setting restrictions in the PCR transmission on the transmit side enables simplified reception without special processing, such as PCR correction at the receiver.[10]

4.3.2 Channel Code

A digital signal containing a BTS is first subjected to Reed–Solomon coding. This coder adds 16 bytes and provides a maximum correction capability of 8 random bytes. After this, the signal is divided into hierarchical layers for channel coding in parallel. Figure 4.4 shows an example of a three-layer case. The gray blocks are used in the modulation processes for one-segment transmission.

Each layer has a channel code, and after separation an appropriate randomization ensures the adequate binary transitions and a delay is used to synchronize all layers in an OFDM frame. After this stage, the byte interleaving, also called Forney convolution interleaving, is performed to scatter the errors and improve the outer coding efficiency by spreading the data bytes over 12 packets. The convolutional coder is a 1/2 rate code (G1 = 171_{oct}, G2 = 133_{oct}), and different puncturing patterns' performance can be obtained varying the robustness: R = 1/2, 2/3, 3/4, 5/6, or 7/8.

4.4 Modulation

The complete modulation side is shown in figure 4.5. For each layer it is possible to choose one of four digital modulation schemes: DQPSK, QPSK, 16-QAM, or 64-QAM. DQPSK is a differential type of modulation but should be not used.[8] The others (QPSK, 16-QAM, and

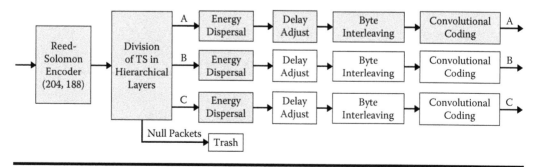

Figure 4.4 Channel code of ISDB-T transmission system.

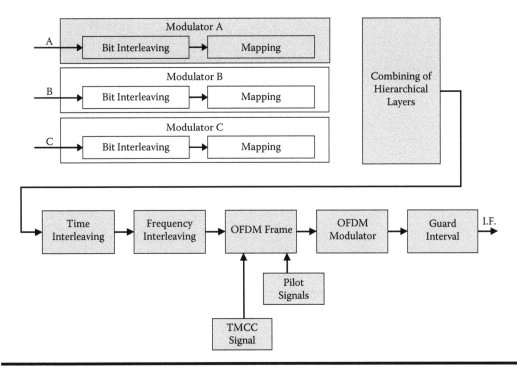

Figure 4.5   Block diagram of modulation side.

64-QAM) are coherent types of modulation. As the number of bits carried by a symbol increases from 2 to 4 and 6 bits, the bit rate increases. At the same time, however, the distance between signal points becomes smaller and the signal becomes less robust to noise and other disturbances.[4] Figures 4.6 and 4.7 show the block diagram of modulations QPSK and 16-QAM, including bit interleaving and the phase diagram that can be used in one-segment transmissions. The bit

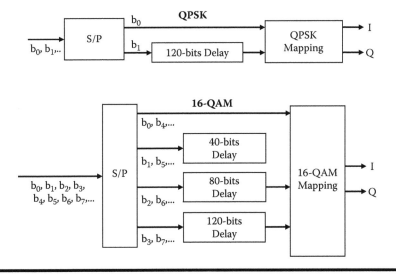

Figure 4.6   Block diagram of bit interleaving and QPSK/16-QAM mapper.

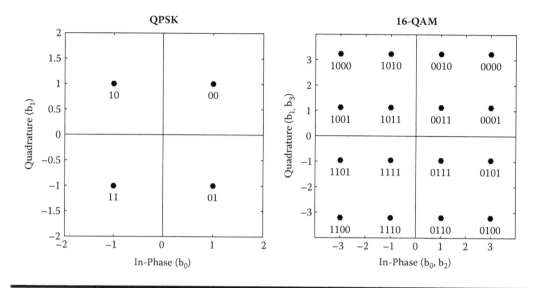

Figure 4.7   QPSK and 16-QAM constellations.

delay after serial/parallel conversion is a form of bit interleaving performed to reduce intercarrier interference.

4.5   Time and Frequency Interleaving

Once signals of different hierarchical layers are combined, they must be time interleaved in units of modulation symbols. The time interleaving circuit is formed by a convolutional interleaving, and the length must be specified for each hierarchical layer, independently of other layers. After this, an appropriate number of symbols are delayed to ensure in the receiver a delay of a multiple number of frames. The time interleaving length can be adjusted for each mode. For example, in mode 3 the number of symbols delayed varied with the parameter I: I = 0 (0 ms), I = 1 (100 ms), I = 2 (200 ms), or (I = 4 400 ms). The scatter plot of time interleaving in mode 3 is shown in figure 4.8.

During segment division, data segment numbers 0 to 12 are assigned sequentially to the partial reception portion and SDTV and HDTV portions. The relationship between the hierarchical layers must be named layers A, B, and C sequentially, in ascending order of data segment number. In the frequency interleaving shown in figure 4.9, the intersegment interleaving must be conducted on two or more data segments when they belong to the same type of modulation portion. Intrasegment carrier rotation and intrasegment carrier randomizing are conducted in all segments and are shown in figures 4.10 and 4.11, respectively.

4.6   OFDM Frame Structure

In mode 3, one segment uses 384 data carriers for transmitting information and the remaining 48 carriers for transmission control. The broadcast OFDM signal features 13-segment frames. Figure 4.12 shows OFDM segment frames for coherent modulation, each having 1 TMCC carrier, 8 AC carriers, and an equivalent of 36 scattered pilots (SPs) arranged in a specific pattern.

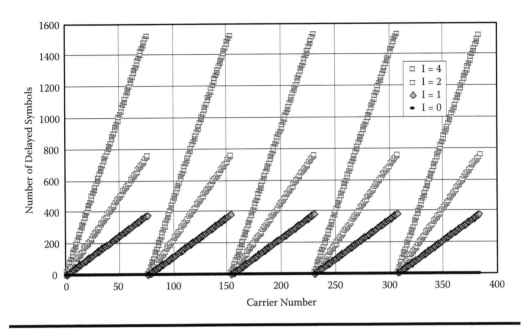

Figure 4.8   Scatter plot of time interleaving in mode 3.

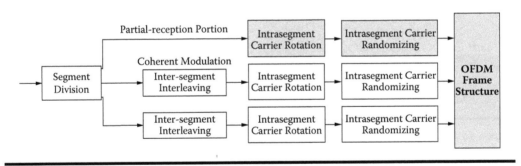

Figure 4.9   Block diagram of frequency interleaving.

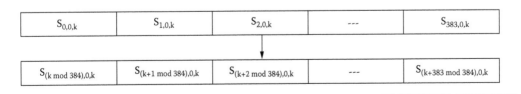

Figure 4.10   Intrasegment carrier rotation for mode 3.

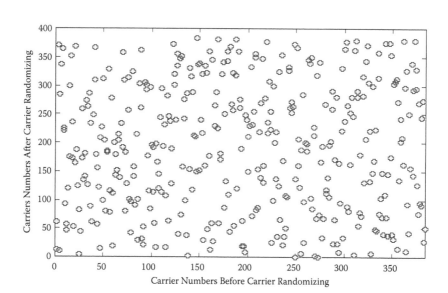

Figure 4.11   Scatter plot of carrier randomizing in mode 3.

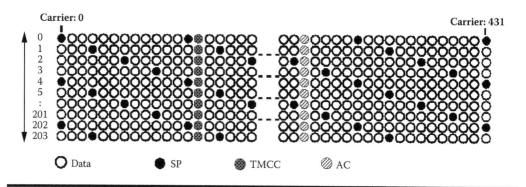

Figure 4.12   OFDM segment frame for mode 3.

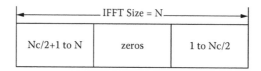

Figure 4.13   QAM arranged for use of IFFT.

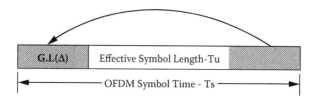

Figure 4.14   OFDM symbol with cyclic extension.

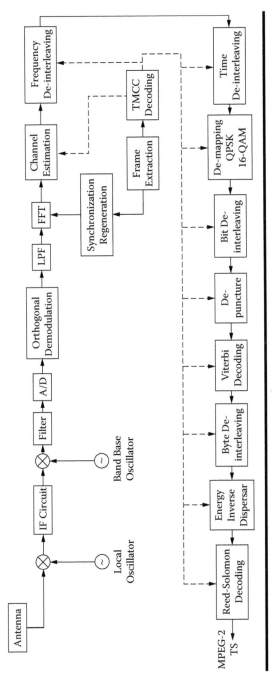

Figure 4.15 One-segment front end.

An SP is inserted once every 12 carriers in the frequency domain and once every 4 symbols in the time domain and is used for synchronization of the receiver.

4.7 OFDM Modulator

The OFDM signal is generated by an inverse fast Fourier transform (IFFT) with size N equal to 2,048, 4,096, and 8,192 for modes 1, 2, and 3, respectively. Because the total number of carriers Nc in each mode are smaller than N, QAM values are padded with zeros to get N input samples that are used to calculate an IFFT. Figure 4.13 shows the QAM arranged to use IFFT.

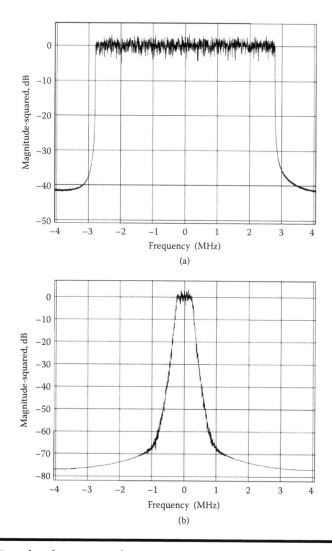

Figure 4.16 (a) Base-band spectrum. (b) LPF output.

4.8 Guard Interval

To eliminate intersymbol interference, each OFDM symbol is cyclically extended in the guard time, as shown in figure 4.14. The values of guard interval (Δ) are proportional (1/4, 1/8, 1/16, or 1/32) to the effective symbol length.

4.9 One-Segment Receiver

The block diagram of a one-segment receiver[12–15] is shown in figure 4.15. The RF signal is tuned and converted to an intermediate frequency (IF) and band limited by a band-pass filter (BPF). After the IF signal is converted to a base band and applied to A/D, it is followed by a quadrature demodulator that converts to the in-phase and quadrature signal (I/Q). After the quadrature demodulator, a low-pass filter (LPF) cuts off frequency at 250 kHz (six-pole Buttenvorth filter) to eliminate interference between the adjacent symbols. Figure 4.15 shows the base-band signal and the signal after the LPF in mode 3.

After this block, automatic frequency control (AFC), symbol synchronization, and clock recovery are performed continuously and applied to FFT. For a one-segment receiver, the FFT sizes are 256, 512, and 1,024 for the modes 1, 2, and 3, respectively. Following the frame synchronization detects the TMCC and a channel estimation estimate the time-varying amplitudes and phases of all subcarriers. Using the values of SP, the channel is equalized and the frequency deinterleaving and time deinterleaving are done; then demapping, bit deinterleaving, and depuncture are done. Then Viterbi decoding and Reed–Solomon (RS) decoding are carried out before the transport stream (TS) packets are finally output.

4.10 Conclusions

In March 2006, NHK announced in the OpenHouse a new way to transmit 13 channels of one-segment signals in one channel of 6 MHz bandwidth to be used for a reception system of a mobile type receiver. This mode is called connected segment.

Links

1. NHK OpenHouse 2007: http://www.nhk.or.jp/strl/open2007/en/tenji/t09.html.

References

1. *Transmission system for digital terrestrial broadcasting.* ARIB Standard STD-B31, ver. 1.6 E2, November 2005.
2. M. Uehara, M. Takada, and T. Kuroda. 1999. Transmission scheme for the terrestrial ISDB system. *IEEE Transactions on Consumer Electronics* 45:101–6.
3. S. Nakahara, M. Okano, M. Takada, and T. Kuroda. 1999. Digital transmission scheme for ISDB-T and reception characteristics of digital terrestrial television broadcasting system in Japan. *Consumer Electronics* 76–77.
4. M. Takada and M. Saito. 2006. Transmission system for ISDB-T. *Proceedings of the IEEE* 94:251–56.

5. Y. Wu, E. Pliszka, B. Caron, P. Bouchard, and G. Chouinard. 2000. Comparison of terrestrial DTV transmission systems: The ATSC 8-VSB, the DVB-T COFDM, and the ISDB-T BST-OFDM. *IEEE Transactions on Broadcasting* 46.

6. G. Bedicks, F. Yamada, F. Sukys, C. Dantas, L. T. M. Raunheitte, and C. Akamine. 2006. Results of the ISDB-T System tests as part of digital TV study carried out in Brazil. *IEEE Transactions on Broadcasting* 52.

7. *Video coding, audio coding, and multiplexing specifications for digital broadcasting.* ARIB standard STD-B32, ver. 1.9, March 2006.

8. *Operational guidelines for digital terrestrial television broadcasting.* ARIB technical report TR-B14, ver. 2.8, May 2006.

9. *Generic coding of moving pictures and associated audio: Systems.* ISO/IEC 13818-1, November 1994.

10. M. Uehara. 2006. Application of MPEG-2 systems to terrestrial ISDB (ISDB-T). *Proceedings of the IEEE* 94:261–68.

11. Methods of measurement for digital terrestrial broadcasting transmission networks. In *JEITA handbook*. Ver. 1.2, October 2006.

12. *Receiver for digital broadcasting.* ARIB standard STD-B21, ver. 4.5, September 2006.

13. M. Okada, T. Masaki, T. Iwasaki, and N. Ueno. 2003. Narrowband OFDM receiver architecture for partial reception of ISDB-T signal. *Consumer Electronics* 17–19:202–3.

14. K. Fukuda, K. Watanabe, M. Ouchi, N. Tokunaga, and T. Kamada. 2003. A 9 mW OFDM demodulator LSI for narrow-bandwidth ISDB-T. *Consumer Electronics* 17–19:90–91.

15. S. Kageyama, I. Kanno, H. Kitamura, T. Kamada, and H. Takami. 2004. Development of an ISDB-T "front-end" module for use in handheld terminals. *Consumer Communications and Networking Conference* 5–8:676–77.

Chapter 5

Comparative Study of Mobile Broadcasting Protocols *DVB-H, T-DMB, MBMS, and MediaFLO*

C. Herrero and P. Vuorimaa

Contents

Keywords

mobile broadcasting, handheld devices, DVB-H, T-DMB, MBMS, MediaFLO

First mobile broadcasting technologies were developed to allow easy portable reception of analog broadcast services transported by radio frequency emissions. The best-known example of mobile broadcasting is probably the successful transmission of frequency-modulated (FM) radio services. It is possible to find FM radio receivers smaller than a matchbox with batteries that last many hours. In addition to FM radios, portable analog TV receivers are also in the market, but their usability is reduced due to the fact that users should be more careful of the antenna positioning and carrying extra batteries.

At this moment, analog audio and video radio transmission technologies are being replaced globally with more advanced digital broadcasting systems, for example, digital audio broadcasting (DAB) for radio services and digital video broadcasting–terrestrial (DVB-T) for TV services. These digital broadcasting technologies have several advantages:[1,2]

- **Bandwidth efficiency.** For example, up to five digital TV channels can be allocated in the bandwidth occupied by a single analog TV channel.
- **Robustness.** Tolerance against radio frequency noise and interference has been improved.
- **Quality of the signal.** Digital audio broadcasting services can have a quality comparable to that of a CD, and digital TV channels can provide a quality of video and audio comparable to that of a DVD.

Moreover, new digital video broadcasting systems allow the transmission of high-definition TV channels and delivery of Java-based interactive applications such as games.[3] Figure 5.1 depicts how a single-frequency channel can allocate different digital services in contrast with the situation when analog TV technology is used.

When transmitting digital radio frequency signals that convey audio or video data, it is possible to select among multiple combinations of modulation and coding schemes. Thus, although only digital audio broadcasting systems are primarily intended for portable reception, by careful selection of the transmission parameters, digital video broadcasting services can be received on

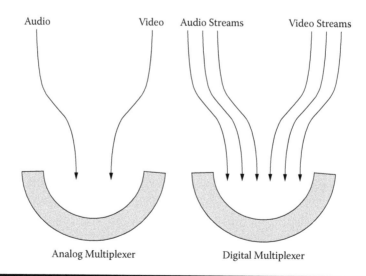

Audio   Video   Audio Streams   Video Streams

Analog Multiplexer   Digital Multiplexer

Figure 5.1   Digital multiplexers can transmit multiple digital services into the same frequency channel.

the move as well.[4] This option, however, has serious limitations, such as that power consumption in the terminal is too high.[5]

First digital audio broadcasting systems cannot be used for transmitting TV channels with good resolution, because the systems were designed to deliver only low- or medium-bit-rate audio services. So, we can see that neither DAB nor DVB-T is a good solution to offer mobile TV services to portable receivers.[6]

Third-generation (3G) mobile phone cellular technologies are optimized to decrease power consumption of the terminals, and the theoretical service bit rate can be as high as 384 Kbps, which is enough for the delivery of a mobile TV channel with good video and audio quality. However, cellular networks are primarily intended to support two-way communications services, for example, phone calls or Internet browsing. Cellular networks make use of the unicast transport mode to send data to the receivers, instead of using the broadcast transport mode. The unicast transport mode means that the number of simultaneous recipients is limited. Thus, the bandwidth is not used efficiently when sending mobile TV channels, making this service very expensive in terms of radio capacity.

To explain the difference between unicast and multicast transmission mode we use one example: the audio guides available at a museum. The number of audio guides is limited, so each new visitor taking an audio guide reduces the possibility of having an available guide for a future visitor. When the museum does not have more guides available, this service is not offered until a user releases his audio guide. This is what happens as well with cellular networks that use unicast mode. The number of audio guides would be the number of bandwidth units, each one needed to transport a single TV channel. If a user takes an audio guide or a bandwidth unit, he reduces the possibility of having available guides or bandwidth units for future users. When the museum or the cellular network does not have more available resources, it must wait until one of these is released to be able to offer the service again. Systems that use broadcast/multicast transmission mode, like DVB-T or DAB, work in a different way. We can explain it by stating that these systems work like a tourist guide using a speaker. It does not matter if there is only one listener or 100, because the only requirement to listen is to be close enough to the speaker. Thus, the tourist guide can accept a new client without caring what number of clients are actually listening to him.

Until now we have seen three different solutions to transmit mobile TV services to portable devices: DAB, DVB-T, and 3G cellular systems. The first solution cannot transmit high-bit-rate services such as mobile TV channels with good quality, because of its bit rate limitations. The second one can transmit high-bit-rate services to multiple users, but the power consumed in the receiver would be too high, emptying the batteries in a short time. Finally, the third one transmits data using unicast mode, so that services can be received only by a very limited number of users in a certain area. Advantages and main drawbacks for those technologies are exposed in table 5.1.

Hence, due to the public interest in having mobile TV services available in portable devices, specially in mobile phones, several more efficient solutions for mobile broadcasting have been

Table 5.1   Digital Transmission Technologies almost Suitable for Mobile Broadcasting

	DVB-T	*DAB*	*3G*
Main advantages	Can transmit any broadcast service already	Designed for portable reception	Excellent market penetration
Drawbacks	Not suitable for battery-powered devices	Not appropriate for high-data-rate services	Limited number of simultaneous users

proposed recently.[7] This chapter introduces some of those technologies and presents a comparative study.

5.1 Summary of Mobile Broadcasting Technologies

This section details the essentials of four modern mobile broadcasting technologies: T-DMB, DVB-H, MBMS, and MediaFLO. The main elements of each technology are introduced to understand how the protocol works and the differences from other, alternative systems.

Based on previous digital transmission standards, DAB and DVB-T, two mobile broadcasting protocols have been proposed: Terrestrial–Digital Media Broadcasting (T-DMB) based on the DAB system, and Digital Video Broadcasting–Handheld (DVB-H), which is a derivative of the DVB-T standard.[8]

The Third Generation Partnership Project (3GPP) suggests an enhancement of current cellular networks to support Multimedia Broadcast/Multicast Services (MBMS). MBMS is a broadcasting service that can be offered via cellular networks using point-to-multipoint links instead of the usual point-to-point links. Because of its operation in broadcast or multicast mode, MBMS can be used for efficient streaming or file delivery to mobile phones. MBMS is the third technology explained in this section.

Finally, we will introduce MediaFLO. MediaFLO is a proprietary technology, developed by Qualcomm, to broadcast data to portable devices. FLO in the name *MediaFLO* stands for "forward link only," which means that data is transmitted in one way. MediaFLO technology is similar in many aspects to T-DMB and DVB-H, as we will show later.

These four technologies are explained in more detail in this section. However, we must point out that in addition to DMB, DVB-H, MBMS, and MediaFLO, other mobile broadcasting protocols have been proposed, for example, DVB-SH and S-DMB protocols that rely on satellite transmissions, and 1seg in Japan, which uses one of the 13 segments available in the Japanese terrestrial digital broadcast system.

5.1.1 T-DMB: Terrestrial–Digital Multimedia Broadcasting

The T-DMB system is based on the Eureka 147 Digital Audio Broadcasting that has been deployed in many countries around the world.[9] A conventional DAB transmission system can be used for T-DMB transmission by adding a multimedia multiplexer and a video encoder to the existing DAB system. The size of a DAB or T-DMB multiplex is 1.5 MHz, so although it uses band III (VHF), which is also in use for terrestrial analog and digital TV, it should be easy to find an available spectrum for a T-DMB multiplex. The deployment of T-DMB systems started in South Korea in 2005, and several other trials are running in other countries as well.

Like many other broadcasting systems, DAB and T-DMB use a form of transmission known as orthogonal frequency division multiplex (OFDM). This transmission technology has proven to offer high data capacity and high resilience to interference, can tolerate multipath effects, and can be used in single-frequency networks (SFNs).

We will now explain the differences between DAB and T-DMB transmission systems:

■ **Modulation.** T-DMB uses DAB's transmission system, that is, OFDM differential modulation with 1,536 subcarriers and QPSK signal constellation. However, it adds an additional layer of forward error correction (FEC) coding, which improves robustness and spectral efficiency.

- **Video and audio compression codecs.** DAB's inefficient MPEG-1 Audio Layer 2 is replaced in T-DMB transmissions by MPEG-4 Part 3, High-Efficiency Advanced Audio Codec (HE-AAC), for audio compression. The old audio codec requires 224 kbps to match FM's audio quality, while HE-AAC provides good audio quality at 64 kbps. For video compression, T-DMB specifies the MPEG-4 Part 10 standard, also known as ITU standard H.264.
- **Number of radio stations per multiplex.** If we keep constant the requirement of carrier-to-noise ratio (C/N), for example, equal to 16 dB, the capacity of a T-DMB multiplex would be 1.416 Mbps using protection level 3B. For the same value of C/N, a DAB multiplex would use protection level 3, which gives a capacity of 1.184 Mbps. The difference of those values is due to the use of an outer layer of FEC coding by the T-DMB system, in addition to DAB's convolutional FEC coding. If the multiplex is used to transmit radio stations with good quality, and applying the bit rates of audio compression codecs, we calculate that the DAB system can transmit 1.184/224 = 5 radio stations, and the T-DMB system 1.416/64 = 22 radio stations. Thus, the T-DMB multiplex is 4.4 times more efficient than DAB for C/N = 16 dB.

We have demonstrated in this section that the transmission cost per service for T-DMB systems is considerably cheaper than for DAB. Moreover, the core of DAB's transmission networks can be used to deliver multiple television services in the same multiplex, due to the use of high-efficiency compression codecs. First, T-DMB services were deployed in South Korea, where television channels have bit rates of 512–544 Kbps, and radio stations 128–160 Kbps. In South Korea, video and audio services are provided for free and broadcasters earn profits from advertisements. However, with the expansion of the consumer base, service operators plan to set up fee-based services. Interactive applications can be delivered as data services, allowing users to vote for quiz shows, download content for the handheld terminal, or purchase products, becoming a new source of revenue for mobile network and service providers.

5.1.2 DVB-H: Digital Video Broadcasting–Handheld

The DVB-H standard is based on the DVB-T system used for terrestrial digital TV broadcasting.[10,11] The new standard brings some features that make it possible to receive TV channels and other high-bit-rate services in handheld devices efficiently. Mobile reception of DVB-T services in handheld devices is also possible, but it is discouraged because it implies a dedicated multiplex to cope with the more demanding requirements of robustness, and because of the huge radio frequency front-end power consumption, which would waste all the energy of battery-powered devices in a few minutes.

DVB-H extends both the link and physical layers of the DVB-T standard, as shown in table 5.2.

The new characteristics added by the DVB-H specification make it possible to receive high-data-rate services in battery-powered devices. To increase power saving, the bit rate required by a single mobile TV channel has to be a small fraction of the total capacity of the multiplex. This can be achieved only if video and audio streams are compressed with newer codecs. Thus, DVB-H specifies MPEG-4 Part 10 for video compression and HE-AAC for audio.

DVB-H trials have been carried out in several countries, and commercial DVB-H services have been launched as well. Most DVB-H receivers nowadays are mobile phones, although DVB-H chipsets were also designed to work attached to personal digital assistants (PDA). For example, the power consumption of the DiBcom DVB-H chipset varies from 20 to 40 mW. Although the real value depends on the size of the data bursts and total bit rate of the multiplex.

Table 5.2   DVB-H Extensions to DVB-T Standard

Link layer	• **Time slicing.** One mobile TV channel has a fairly low bit rate compared to the total bit rate of a multiplex. To reduce drastically the power consumption, the terminal is only turned on to receive data bursts containing streams of a selected TV channel. Thus, the receivers will be turned off most of the time but have to know when to turn on. To do that, each data burst indicates in its header the start time for the following burst of the same TV channel. • **FEC.** Multi-Protocol Encapsulated–Forward Error Correction (MPE-FEC) improves C/N and Doppler performance in mobile channels and tolerance to impulse interferences, but its use is optional.
Physical layer	• **Transmitter parameter signaling (TPS).** Some additional signaling bits have been introduced to indicate whether DVB-H services are present in the multiplex, and if MPE-FEC is in use. • **4K mode.** This additional transmitter mode gives more flexibility for network design. Still, 2K and 8K transmission modes of DVB-T can be used to transmit DVB-H services. • **Symbol interleaver.** The option of an in-depth interleaver is available now for 2K and 4K modes. Thus, bits are interleaved over one, two, or four OFDM symbols, improving robustness and tolerance to impulse noise. • **5 MHz channel bandwidth.** A DVB-H multiplex can be allocated outside traditional broadcasting bands using a frequency channel of 5 MHz.

5.1.3 MBMS: Multimedia Broadcast/Multicast Service

The MBMS standard has been defined by 3GPP to enable efficient point-to-multipoint data distribution services within third-generation cellular networks (Universal Mobile Telecommunications System [UMTS]). The new features are split into MBMS bearer service, which addresses MBMS transmission features below the IP layer, and MBMS user service, for service and application layer protocols.[12,13]

- ■ **MBMS bearer service.** The MBMS bearer service includes two modes—broadcast and multicast. When using the MBMS broadcast mode the network does not receive information about the activity of users in the network. So, it is very similar to other broadcast systems, such as DVB-T or DAB. The multicast mode can be more efficient in using radio access resources because terminals inform the network about their state, passive or active. Therefore, the cellular network can select the most efficient transmission bearer, a point-to-point or a point-to-multipoint link. For core network resources management one new functional entity is added to the UMTS reference architecture—the Broadcast Multicast Service Center (BM-SC). It links external content providers, such as TV broadcasters, with the mobile operator core network. From that point MBMS services share the resources of the mobile operator core network.
- ■ **MBMS user service.** It offers two different delivery methods, streaming and download delivery. The streaming delivery method is intended for continuous media transmission, such as mobile TV or radio services. The download delivery method is more appropriate for "download and play" types of services, such as small video clips. The delivery method does not depend on the selected MBMS transmission mode, that is, broadcast or multicast.

Although the theoretical upper limit of service bit rate for a mobile device is 384 Kbps, the MBMS recommendation sets 256 Kbps as the limit for the channel bit rate. The HE-AAC codec is recommended for audio compression and MPEG-4 Part 10 codec for video.

MBMS services are allocated in the spectrum available for 3G cellular networks, which means that potential users can be found all around the globe. There will be no need to build an MBMS-specific broadcast infrastructure once the 3G cellular network is deployed. Moreover, UMTS terminals can use the same antenna and chipset to receive the new broadcast services. Thus, most of the changes required to make MBMS-compliant mobile operator networks and terminals are SW related. Deployment, better said, adjustments, of the UMTS network would be very cheap, in comparison, for example, with the cost of installing a new network of transmission antennas for DVB-H. Another advantage with respect to unidirectional broadcast systems is that the UMTS uplink channel can be used as an interaction channel, for example, for paying service fees, and mobile operators can continue using the same business models. However, commercial rollout is only scheduled for 2008, and very few trials have been done so far. For this reason, this technology has lost the opportunity to have time-to-market advantage.

5.1.4 MediaFLO: Media Forward Link Only

MediaFLO is Qualcomm's proprietary transmission system to broadcast data to portable devices using unidirectional radio transmissions. The system was designed from scratch without backward compatibility constrains. Thus, while DAB and DVB-T multiplexes can be used to transmit simultaneously other types of broadcast services like T-DMB and DVB-H, a MediaFLO multiplex works only for MediaFLO types of services.[14]

Physical and link layer features are specified within the Telecommunications Industry Association (TIA), while the service layer and protocols are defined in the FLO Forum. The following design requirements were taken into account:

- **Frequency spectrum.** The system works in different bands, for example, UHF and VHF, and transmissions are possible for channel bandwidths of 5, 6, 7, or 8 MHz. This would maximize the possibility of finding available frequency resources to deploy the MediaFLO network.
- **Data rate.** Different modulation and coding schemes can be selected to match the required coverage and C/N. Typical data rates range from 0.47 to 1.87 bits per second per hertz.
- **Power consumption.** Data is transmitted in bursts, and a time-slicing mechanism, similar to the one present in DVB-H, is used. The goal for viewing time with 850 mAh battery-powered devices using 360 Kbps of data rate is 4 hours.

MediaFLO relies on a 3G cellular network, such as UMTS, for providing control functions, to support interactivity and facilitate user authorization to the service, because the services can be received only by subscription. MediaFLO receivers would indeed be integrated within mobile phones, which would require strong collaboration between Qualcomm and other manufacturers. Besides that, there is already a strong competition to get available frequencies; thus, the global success of this technology seems to depend too much on negotiations instead of being a pure technology contest.

5.1.5 Overview of Other Mobile Broadcasting Technologies

We now briefly describe other mobile broadcasting technologies, which are not further analyzed in the rest of the chapter. Satellite DMB (S-DMB) is a new form of satellite broadcasting system for personal portable receivers or automobile receivers to allow reception of TV, radio, and data broadcasts. The services are provided by broadcast centers that transmit the signal to DMB-specialized satellites, which then send the signal directly to the receivers or, in nonsatellite reception areas, using gap fillers. Receivers that allow broadcast services to be viewed are mobile phones, PDAs, automobile systems, or satellite DMB-specific devices. Field trials and commercial development have already started in Korea.[15]

Satellite services to handheld terminals, or DVB-SH, is being developed by DVB Project. DVB-SH systems will be able to deliver IP-based media content and data to handheld terminals like mobile phones and PDAs via satellite. If a line of sight between terminal and satellite does not exist, then terrestrial gap fillers are employed to provide coverage. The DVB-SH system complements the existing DVB-H physical layer standard and uses the DVB IP datacast (IPDC) set of content delivery, electronic service guide, and service purchase and protection standards as well. DVB-SH specifies two operational modes. The SH-A mode specifies the use of COFDM on both satellite and terrestrial links, while the SH-B mode uses a time division multiplex (TDM) on satellite with COFDM on the terrestrial link.

5.2 Comparison of Mobile Broadcasting Technologies

In the previous section we described different mobile broadcast technologies that have been developed to offer high-bit-rate services to mobile devices. However, due to particular technology characteristics, each technology will result in a different user experience. The aim of this section is to present as clear as possible a comparative study of these systems, to see the pros and cons for users and network operators. Thus, besides a series of tables telling about their transmission and reception characteristics, the text tries to clarify what those differences mean. The systems are compared according to different criteria along this section, and a final discussion is presented at the end of the chapter.

5.2.1 Frequency and Channel Bandwidth

Table 5.3 shows the frequency bands where mobile broadcasting systems work, the required bandwidth, and spectral efficiency.

Table 5.3 Comparison of Frequency and Channel Bandwidth

	DVB-H	T-DMB	MBMS	MediaFLO
Frequency band	VHF: 174–240 MHz UHF: 470–862 MHz	VHF	~2 GHz	VHF and UHF
Channel bandwidth	5, 6, 7, 8 MHz	1.536 MHz	5 MHz	5, 6, 7, 8 MHz
Spectral efficiency	0.46–1.86 bps/Hz	0.2–1.2 bps/Hz	0.15–0.3 bps/Hz	0.47–1.87 bps/Hz

As can be seen in table 5.3, DVB-H and MediaFLO systems can achieve better spectral efficiency than their competitors, having more possibilities to find available frequency channels in different broadcasting bands.

If a national mobile broadcasting system is designed to transport 40 radio stations at 64 Kbps bit rate and 10 TV channels at 384 Kbps, the total bit rate transported in the system would be 6.4 Mbps. With extremely good transmission conditions, the system can consist simply of a single DVB-H or MediaFLO multiplex per cell, using a frequency channel of 5 MHz of bandwidth. If the DVB-H or MediaFLO multiplex is allocated in a channel of 8 MHz and the required C/N is 12 dB, a system, which is more robust against noise and interferences, modulation, and coding parameters, can be adjusted to still match enough spectral efficiency of 1.25 bps/Hz. To transmit that amount of radio stations and TV channels, at least four T-DMB multiplexes are needed. However, if protection level 3A is used, to require C/N only equals 12.4 dB, the capacity of a T-DMB multiplex in that case is 1.091 Mbps, and six multiplexes would be needed. Thus, of those technologies that need the installation of new transmission antennas, T-DMB would be the most expensive.

The example of a national system presented above cannot be deployed by a 3G network alone, because the 5 MHz channel can transmit a maximum of 1.5 Mbps of MBMS data traffic. On the other hand, MBMS utilizes the band specified for 3G cellular networks, which has been already granted in each country around the world, and the existent 3G network infrastructure. So, the network can start immediately to transmit MBMS services without waiting for new network licenses or installation of transmission antennas. Indeed, MBMS would be the cheapest technology for deploying a broadcast network of few channels.

5.2.2 Modulation and Coding Parameters

A transmission system that supports a great variety of modulation and coding parameters can be deployed in a larger number of locations. Table 5.4 shows modulation and coding parameters for the technologies of this study.

OFDM modulation is preferred over wideband code division multiple access (WCDMA) for broadcasting systems. The advantage of WCDMA over other radio interfaces is that several users can simultaneously transmit data, but in broadcasting scenarios there is only one transmission antenna that does not share the radio interface. Among the systems that use OFDM, DVB-H presents the highest amount of possible combination for modulation and coding parameters. MediaFLO, however, is the only one that utilizes a more modern and efficient turbo code.

Table 5.4 Comparison of Modulation and Coding Parameters

	DVB-H	*T-DMB*	*MBMS*	*MediaFLO*
Modulation	OFDM	OFDM	WCDMA	OFDM
Constellation	QPSK, 16QAM, 64QAM	QPSK	QPSK	QPSK, 16QAM
Guard interval	1/4, 1/8, 1/16, 1/32	1/4	Not available	1/8
Coding	Convolutional, Reed–Solomon	Convolutional, Reed–Solomon	Turbo	Turbo, Reed–Solomon

Table 5.5 Comparison of Maximum Multiplex and Service Bit Rate

	DVB-H	T-DMB	MBMS	MediaFLO
Maximum multiplex net data rate	23.75 Mbps	1.8 Mbps	1.5 Mbps	24 Mbps
Multiplex data rate in real scenario	Up to 15 Mbps	Up to 1.4 Mbps	Up to 1.5 Mbps	Up to 15 Mbps
Single-service data rate	0–10 Mbps	8 kbps–1.8 Mbps	0–256 kbps	12 kbps–1 Mbps

5.2.3 Multiplex Bit Rate and Service Bit Rate

Previously we have examined the transmission parameters that can be selected by each system to guarantee a certain bit error rate (BER) in channels with different C/N values. In bad transmission scenarios, modulation and coding parameters are configured to offer better protection levels against transmission errors. The cost of using more bits for error correction purposes is to have smaller net capacity of the multiplex. Table 5.5 presents the maximum data bit rate that each system can offer.

From the data shown in table 5.5 we can deduce that MBMS cannot be used to transmit multiple TV channels with excellent quality, even if using MPEG-4 Part 10 for video compression. The DVB-H, T-DMB, and MediaFLO systems can all achieve that. However, in the case of T-DMB, the transmission of high-data-rate services is not recommended from the power consumption point of view.

In all transmission systems that use time slicing, low-bit-rate services (e.g., radio services) consume less battery power than high-bit-rate services (e.g., TV channels). If the same TV channel is transmitted in DVB-H, T-DMB, and MediaFLO with the same bit rate, for example, 570 kbps, the receiver must be switched on for retrieving data at different times. For average multiplex bit rates, a DVB-H receiver should be on 3.8 percent of the time, and the same for a MediaFLO receiver. However, a T-DMB receiver should work 40 percent of the time. It is still a good achievement, but battery power is saved much more efficiently in DVB-H and MediaFLO systems. If the network is intended for transmission of radio stations only, or other low-bit-rate services, then all systems work properly from the power consumption point of view.

5.2.4 Reception of Mobile Broadcasting Services

After the examination of network transmission parameters it is time to consider the implications on the other end of the transmission chain, that is, the receiver. Some useful parameters for comparison are shown in table 5.6.

The main characteristics that a mobile broadcasting system should have are, first, the support for on-the-move reception and, second, the possibility of seamless handover. This is something typical for all modern cellular networks such as 3G cellular networks, and it is valid as well for MBMS. As can be seen in table 5.6, new broadcasting systems are capable of offering those characteristics, and in addition to that, the coverage area is usually larger than for cellular networks. The bigger the coverage area, the lower the number of transmitter points needed to cover a certain extension of land, and the cheaper the implementation cost of the network.

Table 5.6   Reception of Mobile Broadcasting Services

	DVB-H	T-DMB	MBMS	MediaFLO
Maximum distance to transmitter	Up to 40 km	Up to 80 km	500 m–2 km	Up to 25 km
Reception possible at 120 km/h	Yes	Yes	Yes	Yes
Seamless handover	Yes	Yes	Yes	Yes

5.3 Summary

This chapter presented and compared four different mobile broadcasting technologies: T-DMB and DVB-H, which are variants of previous broadcasting technologies, DAB and DVB-T; MediaFLO, which is a proprietary technology developed by Qualcomm; and, finally, MBMS. Three of those technologies, T-DMB, DVB-H, and MediaFLO, share many transmission characteristics, while the fourth one, MBMS, is an extension of current cellular networks to allow point-to-multipoint types of connections. We have shown that DVB-H and MediaFLO possess several advantages if the network is planned to transport high-data-rate services, such as mobile TV channels. In that scenario, DVB-H and MediaFLO can offer better quality for audio and video signals because the total data rate of the network is superior to that of T-DMB or MBMS. Moreover, although all the technologies have been designed to guarantee long watching times between recharges of battery-powered devices, DVB-H and MediaFLO obtain the best results for the time-slicing power-saving mechanism, because the data rate of a single service represents a smaller portion of the total available data rate in the multiplex.

Nevertheless, commercial T-DMB services have been launched already in several countries, and there are multiple T-DMB receivers available, most of them integrated into mobile phones. All this shows that the time to market was shorter for this technology.

Finally, mobile phones are the most popular portable device at this moment, nearing 100 percent market penetration in several countries, and over that figure in others, which means that on average there is more than one mobile phone per person. Thus, the implementation of mobile broadcasting services using MBMS would be very cheap for both transmitters and final users, because there would be no need to find new locations for transmission antennas and to integrate another antenna in the mobile phone for reception of the services.

Although in some countries these technologies will compete fiercely to obtain a free available frequency channel in the future, in other countries they can all be offered to the users, which would create the possibility for collaboration scenarios between them, where the terminals and users could select the technology that best fits their needs at any moment.

Links

1. DVB-H technology: http://www.dvb-h.org/.
2. T-DMB portal: http://eng.t-dmb.org/.
3. MediaFLO technology portal: http://www.qualcomm.com/mediaflo/.
4. Multimedia Broadcast/Multicast Service White Paper, Sonera MediaLab: http://www.medialab.sonera.fi/workspace/MBMSWhitePaper.pdf.
5. Comparison of the DAB, DMB, and DVB-H systems: http://www.digitalradiotech.co.uk/dvb-h_dab_dmb.htm.

References

1. H. Benoit. 1997. *Digital television: MPEG-1, MPEG-2 and principles of the DVB system.* London: Arnold.
2. U. Reimers. 1998. Digital video broadcasting. *IEEE Communications Magazine* XXXVI:104–10.
3. C. Herrero, P. Cesar, and P. Vuorimaa. 2003. Delivering MHP applications into a real DVB-T network, OtaDigi. In *Proceedings of the 6th International Conference on Telecommunications in Modern Satellite, Cable and Broadcasting Services,* 231–34. Vol. I.
4. U. Ladebusch and C. Liss. 2006. Terrestrial DVB (DVB-T): A broadcast technology or stationary portable and mobile use. *Proceedings of the IEEE* 94:183–93.
5. J. Aaltonen, H. Pekonen, T. Auranen, K. Laiho, and P. Talmola. 2002. Power saving considerations in mobile datacasting terminals. Paper presented at Proceedings of IEEE ISCE 2002, Erfurt, Germany.
6. Y. Wu, E. Pliszka, B. Caron, P. Bouchard, and G. Chouinard. 2000. Comparison of terrestrial DTV transmission systems: The ATSC 8-VSB, the DVB-T COFDM, and the ISDB-T BST-OFDM. *IEEE Transactions on Broadcasting* 101–13.
7. C. Herrero and P. Vuorimaa. 2004. Delivery of digital television to handheld devices. In *Proceedings of the 1st International Symposium on Wireless Communications Systems,* 240–44.
8. G. Faria, J. Henriksson, E. Stare, and P. Talmola 2006. DVB H: Digital broadcast services to handheld devices. *Proceedings of IEEE* 94:194–209.
9. S. Cho, G. Lee, B. Bae, K. Tae Yang, C.-H. Ahn, S.-I. Lee, and C. Ahn. 2007. System and services of terrestrial digital multimedia broadcasting (T-DMB). *IEEE Transactions on Broadcasting* 171–78.
10. European Telecommunications Standards Institute. 2004. *Digital video broadcasting (DVB); DVB-H: Transmission system for handheld terminals.* ETSI EN 302 304.
11. European Telecommunications Standards Institute. 2005. *Digital video broadcasting (DVB); DVB-H implementation guidelines.* ETSI TR 102 377.
12. *Multimedia Broadcast/Multicast Service (MBMS); Stage 1.* 3GPP TS 22.146.
13. F. Hartung, U. Horn, J. Huschke, M. Kampmann, T. Lohmar, and M. Lundevall. 2007. Delivery of broadcast services in 3G networks. *IEEE Transactions on Broadcasting* 188–99.
14. Qualcomm MediaFLO. 2007. *Technology comparison: MediaFLO and DVB-H.*
15. S.-J. Lee, S.-W. Lee, K.-W. Kin, and J.-S. Seo. 2007. Personal and mobile satellite DMB services in Korea. *IEEE Transactions on Broadcasting* 179–87.

OPTIMIZATION AND SIMULATION

II

Chapter 6

Hardware and Software Architectures for Mobile Multimedia Signal Processing

Benno Stabernack, Kai-Immo Wels,
and Heiko Hübert

Contents

Keywords

H.264 profiling, SIMD, RISC core coprocessors

6.1 Characteristics of H.264/AVC Video Signal Processing

This section gives an overview of the H.264/AVC video decoder and its implementation as part of a *digital video broadcasting–handheld* (DVB-H) terminal system architecture. The requirements set on implementation by the DVB-H scenario are discussed.

6.1.1 H.264/AVC Video Codec

H.264/AVC is the most recent video compression standard developed by the Joint Video Team (JVT) of ISO/IEC MPEG and ITU-T VCEG.[3] Like its predecessors, H.264/AVC uses a block-based hybrid coding approach, which takes advantage of (motion-compensated temporal and spatial) prediction and transformation of residual data. H.264/AVC adds various new coding features and refinements of existing mechanisms, which lead to a two to three times increased coding efficiency compared to Motion Picture Experts Group (MPEG)-2. However, the computational demands and required data accesses have also increased significantly. Figure 6.1 shows the block diagram of an H.264 decoder, and figure 6.2 the call graph of a typical software implementation.

The decoder consists of five sequential computation steps: bit stream parsing, entropy decoding, prediction, coefficient transformation, and deblocking. The bit stream processing unit parses

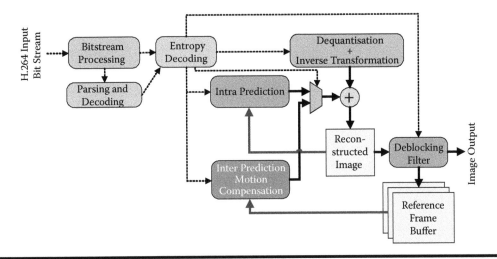

Figure 6.1   Block diagram of an H.264/AVC decoder.

the bit stream for symbols, which are then entropy decoded. H.264/AVC allows two different entropy coding modes, variable length coding (CAVLC) and binary arithmetic coding (CABAC). Both methods are context adaptive (CA), that is, the coding parameters are adapted according to previous data to achieve a high compression. The decoded symbols contain control information, prediction data, and transformed residual data. H.264/AVC provides inter- and intraframe prediction modes to predict image data from previous frames or neighboring blocks, respectively. Interprediction, as depicted in figure 6.3, can be performed on the submacroblock level (down to 4 × 4 blocks), and the motion vector resolution goes down to quarter-pixel precision, requiring interpolation of pixel data.

For intraprediction several modes are defined, for example, horizontal prediction from the left neighboring macroblock or vertical prediction from left and upper neighbors (see figure 6.4). Intraprediction can be performed on either 16 × 16 or 4 × 4 blocks.

The residuals of the prediction are received as transformed and quantized coefficients. After inverse quantization and transformation of the coefficients the residuals are added to the predicted data, which leads to a reconstructed image. The transformation is performed on 4 × 4 blocks (in high profile 8 × 8 is also supported) and is based on integer arithmetic, contrary to the residual transformation in previous video compression standards applying a discrete cosine transformation (DCT). The reconstructed image is postprocessed by a deblocking filter for reducing blocking artifacts at block edges (see figure 6.5). The deblocked image is used for performing interprediction, whereas intraprediction is based on the reconstructed image.

6.1.2   DVB-H Specific Requirements

DVB-H incorporates H.264/AVC as the codec for video compression. The baseline profile has been chosen mainly due to the reduced computational requirements in comparison to the other profiles of the H.264/AVC standard, for example, main profile, taking the limited processing power of mobile devices into account.

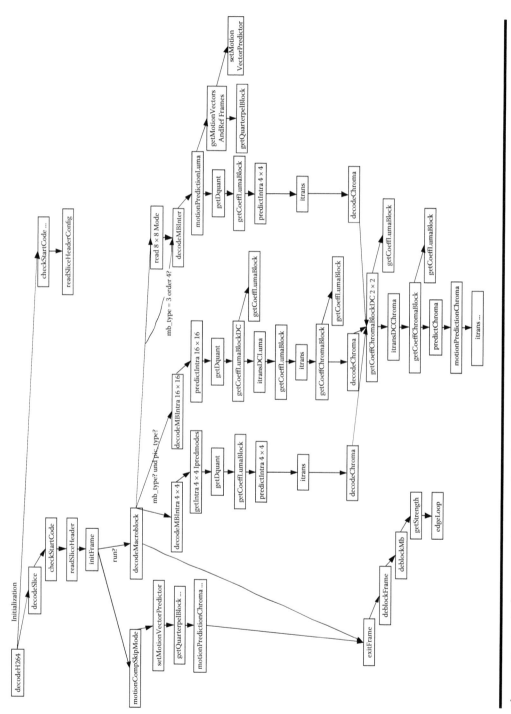

Figure 6.2   Typical software structure of an H.264 video decoder.

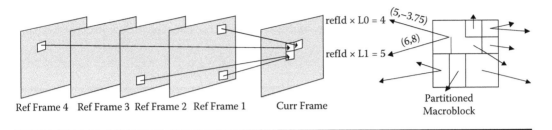

Ref Frame 4   Ref Frame 3   Ref Frame 2   Ref Frame 1   Curr Frame

Partitioned
Macroblock

Figure 6.3   Interprediction: Macroblocks are partitioned and predicted from already decoded frames.

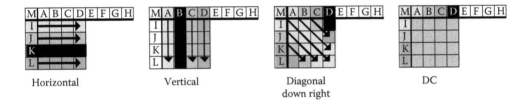

Horizontal   Vertical   Diagonal
down right   DC

Figure 6.4   Intraprediction: Pixels are predicted from neiboring pixels (nine different modes are available for 4 × 4 blocks).

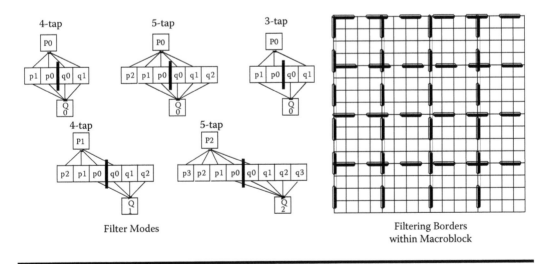

Filter Modes   Filtering Borders
within Macroblock

Figure 6.5   Deblocking: Each 4 × 4 block edge is filtered for reducing blocking artifacts. Several filter modes are available.

Different, so-called capability classes are defined, reflecting the variety of DVB-H terminals regarding picture resolution, frame rate, and used data rates. The table below lists the available classes.

DVB-H Capability Classes

Capability	H.264/AVC Profile—Level	Max Frame Size	Max Frame Rate	Max Bit Rate
A	Baseline—1b	QCIF (176 × 144)	15	128
B	Baseline—1.2	CIF (352 × 288)	15.2	384
C	Baseline—2	CIF (352 × 288)	30	2,000
D	Main—3	625SD (720 × 576)	25	10,000
E	High—4	2 × 1K HD (2048 × 1024)	15	20,000

Further, H.264/AVC-specific requirements for DVB-H terminals are real-time decoding, low latency, low power consumption, and low costs.

6.2 Performance and Memory Profiling for Embedded System Design

The design of embedded hardware/software systems is often underlying strict requirements concerning various aspects, including real-time performance, power consumption, and die area. Especially for data-intensive applications, such as multimedia systems, the number of memory accesses is a dominant factor for these aspects. To meet the requirements and design a well-adapted system, the software parts need to be optimized and an adequate hardware architecture needs to be designed. For complex applications this design space exploration can be rather difficult and requires in-depth analysis of the application and its implementation alternatives. Tools are required that aid the designer in the design, optimization, and scheduling of hardware and software. We present a profiling tool for fast and accurate performance and memory access analysis of embedded systems, and show how it has been applied for an efficient design flow. This concept has been proven in the design of a mixed hardware/software system for H.264/AVC video decoding.

6.2.1 Introduction

The design of an embedded system often starts from a software description of the system in C language, for example, an executable specification developed by the designer or a reference implementation of the application, for example, from standardization organizations or the open-source community. This software code is often not optimized in any manner, because it mainly serves the purpose of functional and conformance testing. Therefore, it has to be transformed into an efficient system, including hardware and software components. The design of the system requires the following steps: system architecture design, hardware/software partitioning, software optimization, design of hardware accelerators, and system scheduling. All these steps require detailed information of the performance requirements of the different parts of the application. Besides the arithmetical demands of the application, memory accesses can have a huge influence on performance and power consumption. This is especially the case for data-intensive applications, such as multimedia systems, due to the huge amount of data to be transferred in these applications. This problem is even increased if the given data bandwidth is not used efficiently.

To reduce the overall data traffic, those parts of the code that require a high amount of data transfer have to be identified and optimized. Because the software of the above-mentioned applications may contain up to 100,000 lines of code, tools are required that help the designer to identify those critical parts. Several analysis tools exist, for example, timing analysis is provided by gprof or VTune. Memory access analysis is part of the ATOMIUM[12] tool suite. However, all these tools provide only approximate results for either timing or memory accesses. A highly accurate memory analysis can be done with a hardware (VHDL) simulator, if a VHDL model of the processor is available, but it implies long simulation times.

To achieve a fast and accurate solution, we developed a specialized profiler, called memtrace,[3] for performance and memory access statistics. This chapter describes the tool with all its features and how such tools can be used for design and optimization of embedded hardware/software systems. As a case study, we show how we used memtrace for the efficient design of a mixed hardware-software system for H.264/AVC video decoding. Starting from a software implementation of an H.264/AVC video decoder, we show how the software can be optimized, an efficient hardware architecture developed, and the system tasks scheduled.

6.2.2 Memtrace: A Performance and Memory Profiler

6.2.2.1 Tool Architecture

Memtrace is a nonintrusive profiler that analyzes the memory accesses and real-time performance of an application, without the need of instrumentation code. The analysis is controlled by information about variables and functions in the user application, which is automatically extracted from the application. Furthermore, the user can specify the system parameters, for example, the processor type and memory system. During the analysis, memtrace utilizes the instruction set simulator ARMulator[5] for executing the application. ARMulator provides memtrace with the information required for the analysis, for example, the program counter, the clock cycle counter, and the memory accesses. Memtrace creates detailed results on memory accesses and timing for each function and variable in the code.

6.2.2.2 Analysis Workflow

The performance analysis with memtrace is carried out in three steps: the initialization, the performance analysis, and the postprocessing of the results.

During initialization memtrace extracts the names of all functions and variables of the application. During this process user variables and functions are separated from standard library functions, such as printf() or malloc(). This is achieved by comparing the symbol table of the executable with the ones of the user library and object files. The results are written to the analysis specification file. The specification file can be edited by the user, for example, for adding user-defined memory areas, such as the stack and heap variables, for additional analysis. Furthermore, the user can define a so-called split function, which instructs memtrace to produce snapshot results each time the split function is called. This can be used in video processing, for example, for generating separate profiling results for each processed frame. Additionally, the user can control if the analysis results (e.g., clock cycles) of a function should include the results of a called function (accumulated), or if they should only reflect the function's own results (self). Typically auxiliary functions, for example, C library or simple arithmetic functions, are accumulated to the calling functions.

In the second step the performance analysis is carried out, based on the analysis specification and the system specification, as shown in figure 6.6. The system specification includes the

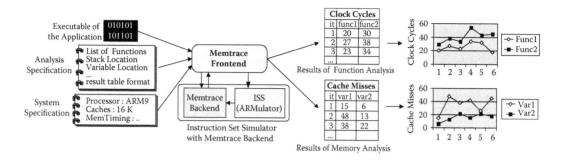

Figure 6.6 Performance analysis tool: Memtrace profiles the performance and memory.

processor, cache, and memory type definitions. The memtrace back end connects to the instruction set simulator for the simulation of the user application and writes the analysis results of the functions and variables to files; see chapter 7 for more details. If a split function has been specified, these files include tables for each call of the split function; table 6.1 shows exemplary results for function profiling. The output files serve as a database for the third step, where user-defined data is extracted from the tables.

In the third step a postprocessing of the results can be performed. Memtrace allows the generation of user-defined tables, which contain specific results of the analysis, for example, the load memory accesses for each function. Furthermore, the results of several functions can be accumulated in groups for comparing the results of entire application modules. The user-defined tables are written to files in a tab-separated format. Thus, they can be further processed, for example, by spreadsheet programs for creating diagrams.

6.2.2.3 Tool Back End: Interface to the ISS

Memtrace communicates with the instruction set simulator (ISS) via its back end, as depicted in figure 6.7. The back end is implemented as a dynamic link library (DLL), which connects to the ISS. Currently only the ARM (Advanced RISC Machines, Ltd.) instruction set simulator ARMulator is supported. The back end is automatically called by the ISS during simulation. During the start-up phase, the back end creates a list of all functions and marks the user and split functions found in the analysis specification file. For each function a data structure is created, which contains the function's start address and variables for collecting the analysis results. Finally, two pointers, called currentFunction and evaluated Function, are initialized. The first pointer indicates the currently executed function, and if this function should not be evaluated, the second pointer indicates the calling function, to which the result of the current function should be added.

Table 6.1 32-Bit Exemplary Result Table for Functions

f	ca	cyl	ls	ld	l8	st	s8	pm	cm	BI	BC	BD
f1	8	215	75	22	7	52	3	42	5	123	92	0
f2	2	295	39	35	3	14	9	17	9	55	153	87
f3	2	432	78	68	4	10	2	31	17	143	289	0

Abbreviations: f, function; ca, calls; cyl, bus (clock) cycles; ls, all load/store accesses from the core; ld, all loads; l8, byte and half-word loads; st, all stores; s8, byte and half-word stores; pm, page misses; cm, cache miss; BI, bus idle cycles; BC, core bus cycles; BD, DMA bus cycles.

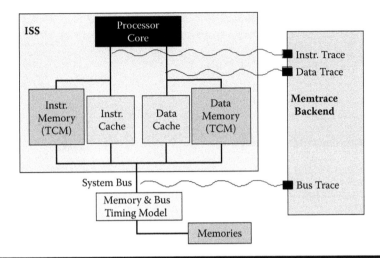

Figure 6.7   Interface between memtrace back end and the ISS.

Each time the program counter changes memtrace checks, if the program execution has changed from one function to another. If so, the cycle count of the evaluatedFunction is recalculated and the call count of the currentFunction is incremented. Finally, the pointers to the currentFunction and evaluatedFunction are updated. If the currentFunction is a split function, the differential results from the last call of this function up to the current call are printed to the result files.

For each access that occurs on the data bus (to the data cache or tightly coupled memory [TCM]), the memory access counters of the evaluatedFunction are incremented. Depending on the information provided by the ARMulator, it is decided if a load or store access was performed and which bit width (8/16 or 32 bit) was used. Furthermore, the ARMulator indicates if a cache miss occurred. Page hits and misses are calculated by comparing the address of the current with the previous memory access and incorporating the page structure of the memory.

For each bus cycle (on the external memory bus) memtrace checks if it was an idle cycle, a core access, or direct memory access (DMA) and increments the appropriate counter of the evaluatedFunction.

At the end of the simulation the results of the last evaluatedFunction are updated and the results of the last call of the split function and the accumulated results are printed to the result files.

6.2.2.4 Memtrace Front End

Memtrace comes with two front ends, a command line interface and a graphical user interface (GUI); a screen shot is given in figure 6.8. The command line interface is very well suited for usage in batch files, for example, for performing a profiling for a set of system configurations or input data. The GUI version allows easy and fast access to all features of the tool. Especially for the quick generation of result diagrams the GUI version is very helpful.

6.2.2.5 Portability to Other Processor Architectures

The current version of memtrace is only targeted at the ARM processor family, as it uses the ISS from ARM (ARMulator). However, the interface of the profiler, as described before, is rather simple

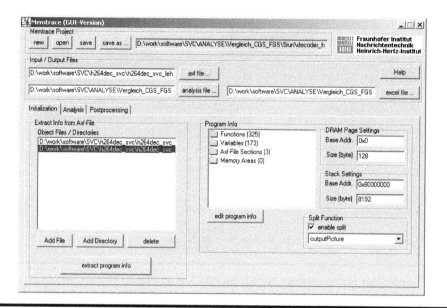

Figure 6.8   Memtrace GUI front end.

and could be ported to other processor architectures if an instruction set simulator is available, which allows debugging access to its memory buses. Our plans for future work include memtrace back ends for other processor architectures.

As long as other back ends are not available, the ARM-based profiling results may function as a rough estimation for the results on other Reduced Instruction Set Computer (RISC) processor architectures. Because all processors of the ARM family can be profiled, a wide variety of architectural features are covered, including variations of pipeline length, instruction bit width, availability of DSP/SIMD instructions, memory management units (MMUs) cache size, and organization, tightly coupled memories, bus width, and detailed memory timing options. For a profiling estimation of a non-ARM processor, an ARM processor with a similar feature set should be chosen. In table 6.2 a list of common embedded processors is given, which have similarities with ARM processors. They have a basic feature set in common, including a 32-bit Harvard architecture with caches, a five- to eight-stage pipeline, and a RISC instruction set. However, it has to be mentioned that some of the processor provides specific features, which may have a significant influence on the performance, for example, the custom instruction extensions of ARC and Tensilica Xtensa processors.

6.2.3 Embedded System Design Flow with Memtrace

This chapter describes how the profiler can be applied for embedded system design. Figure 6.9 shows a typical design flow for such hardware/software systems. Starting from a functionally verified system description in software, this software is profiled with an initial system specification, to measure the performance and see if the (real-time) requirements are met. If not, an iterative cycle of software and hardware partitioning, optimization, and scheduling starts. In this process detailed profiling results are crucial for all steps in the design cycle.

Table 6.2   32-Bit Embedded RISC Processors

Processor	Pipeline	Registers[a]	Instruction/Data Cache, TCM[a]	Special Features
ARM9E	5 stage	16	128k/128k Yes/yes	Coprocessor interface
ARM11	8 stage	16	64k/64k Yes/yes	SIMD Branch pred. 64-bit bus Coprocessor interface
ARC600	5 stage	32 (–60)	32k/32k 512k/16k	Custom instructions Branch pred. extendable register file
ARC700	7 stage	32 (–60)	64k/64k 512k/256k	Custom instructions Branch pred. extendable register file 64-bit bus
Tesilica Xtensa 7	5 stage	64 or >	32k/32k 256k/256k	Custom instructions Windowed registered file Up to 128-bit bus
Ten silica Diamond 232L	5 stage	32	16k/16k—/—	Windowed registered file
Lattice Mico32	6 stage	32	32k/32k—/—	
Altera NIOS II	5–6 stage	32	64k/64k Yes/yes	Direct-mapped cache Custom instructions
Xilinx Micro Blaze v5	5 stage	32	64k/64k Yes/yes	Direct-mapped cache Coprocessor interface
MIPS 4KE	5 stage	32	64k/64k Yes/yes	Coprocessor interface
openRISC OR1200	5 stage	32	64k/64k—/—	Direct-mapped cache Custom instructions
LEON3	7 stage	520	1M/1M Yes/yes	Windowed registered file Coprocessor interface

[a] Many features are customizable; given is the maximum value.

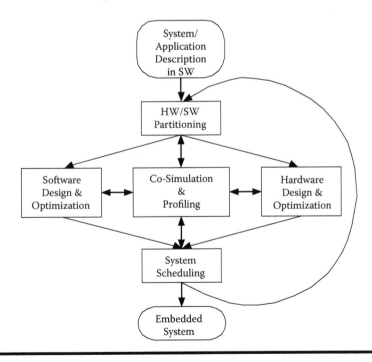

Figure 6.9   Typical embedded system design flow.

6.2.3.1 Hardware/Software Partitioning and Design Space Exploration

For the definition of a starting point of a system architecture an initial design space exploration should be performed. These steps include a variation of the following parameters:

- Processor type
- Cache size and organization
- Tightly coupled memories
- Bus timing
- External memory system and timing (DRAM, SRAM)
- Hardware accelerators, DMA controller

Memtrace can be run in batch mode, and thus different system configurations can be tested and profiled. Therefore, the influence of the system architecture on the performance can be evaluated. This initial profiling also reveals the hot spots of the software. The most time-consuming functions are good candidates for either software optimization or hardware acceleration. Especially computationally intensive functions are well suited for hardware acceleration in a coprocessor. With support of a DMA controller even the burden of data transfers can be taken from the processor. Control-intensive functions are better suited for software implementation, as a hardware implementation would lead to a complex state machine, which requires long design time and often does not allow parallelization. To get a first idea of the influence of hardware acceleration, a (well-educated guessed) factor can be defined for each hardware candidate function. This factor is used by memtrace, to manipulate the original profiling results.

6.2.3.2 Software Profiling and Optimization

After a partitioning in hardware and software is found, the software part can be optimized. Numerous techniques exist that can be applied for optimizing software, such as loop unrolling, loop-invariant code motion, common subexpression elimination, or constant folding and propagation. For computationally intensive parts arithmetic optimizations or SIMD instructions can be applied, if such instructions are available in the processor. If the performance of the code is significantly influenced by memory accesses, as is mainly the case in video applications, the number of accesses has to be reduced or they have to be accelerated. The profiler gives a detailed overview of the memory accesses and allows therewith identifying the influence of the memory access. One optimization mechanism is the conversion of byte (8-bit) to word (32-bit) memory accesses. This can be applied if adjacent bytes in memory are required concurrently or within a short period, for example, pixel data of an image during image processing. A further mechanism is the usage of tightly coupled memories (TCMs) for storing frequently used data. For finding the most frequently accessed data area, the memory access statistics of memtrace can be used. In ITU-T Recommendation H.264[3] these techniques are described in more detail.

6.2.3.3 Hardware/Software Profiling and Scheduling

Besides software profiling and optimization, a system simulation including the hardware accelerators needs to be carried out to evaluate the overall performance. Usually hardware components are developed in a hardware description language (HDL) and tested with an HDL simulator. This task requires long development and simulation times. Therefore, HDL modeling is not suitable for the early design cycles, where exhaustive testing of different design alternatives is important. Furthermore, if the system performance is also data dependent, a huge set of input data should be tested to get reliable profiling results. Therefore, a simulation and profiling environment is required, which allows short modification and simulation time.

For this purpose, we used the instruction set simulator and extended it with simulators for the hardware components of the system. The ARMulator provides an extension interface, which allows the definition of a system bus and peripheral bus components. It already comes with a bus simulator, which reflects the industry standard advanced microcontroller bus architecture (AMBA) bus and a timing model for access times to memory-mapped bus components, such as memories and peripheral modules (see figure 6.10).

6.2.3.3.1 Coprocessors

We supplemented this system with a simple template for coprocessors, including local registers and memories and a cycle-accurate timing. The functionality of the coprocessor can be defined as standard C-code; thus, the software function can be simulated as a hardware accelerator by copying the software code to the coprocessor template. The timing parameter can be used to define the delay of the coprocessor between activation and result availability, that is, the execution time of the task, as it would be in real hardware. This value can be achieved either from reference implementation

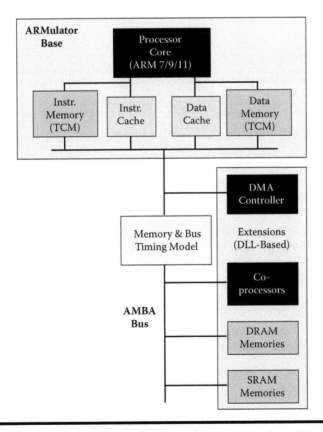

Figure 6.10 **Environment for hardware/software cosimulation and profiling.**

found in literature or by an educated guess of a hardware engineer. Furthermore, often multiple hardware implementations of a task with different execution times (and hardware costs) are possible. In the proposed profiling environment, simply by varying the timing parameter and viewing its influence on the overall performance, a good trade-off between hardware cost and speed-up can be found quickly.

6.2.3.3.2 DMA Controller

For data-intensive applications data transfers have a tremendous influence on the overall performance. To efficiently outsource tasks into hardware accelerators also, the burden of data transfer has to be taken from the CPU. This job can be performed by a DMA controller. The memtrace hardware profiling environment includes a highly efficient DMA controller with the following features:

- Multichannel (parameterizable number of channels)
- One- and two-dimensional transfers
- Activation FIFO (nonblocking transfer, autonomous)
- Internal memory for temporary storage between read and write
- Burst transfer mode

Thus, the designer is enabled to determine the influence of different DMA modes to find an appropriate trade-off between DMA controller complexity and required CPU activity.

6.2.3.3.3 Scheduling

After the software and hardware tasks have been defined, a scheduling of these tasks is required. For increasing the overall performance, a high degree of parallelization should be accomplished between hardware and software tasks. To find an appropriate scheduling for parallel tasks, the following information is required:

- Dependencies between tasks
- The execution time of each task
- Data transfer overhead

Especially for data-intensive application the overhead for data transfers can have a huge influence on the performance. It might even happen that the speed-up of a hardware accelerator is vanished by the overhead for transferring data to and from the accelerator.

The overhead for data transfers to the coprocessors is dependent on the bus usage. Furthermore, side effects on other functions may occur, if a bus congestion occurs or when cache flushing is required to ensure cache coherency. To find these side effects, detailed profiling of the system performance and bus usage is required. Memtrace provides these results; for example, figure 6.11 shows the bus usage for each function depending on the access time of the memory.

6.2.3.3.4 HDL Simulation

In a later design phase, when the hardware/software partitioning is fixed and an appropriate system architecture is found, the hardware component needs to be developed in a hardware description

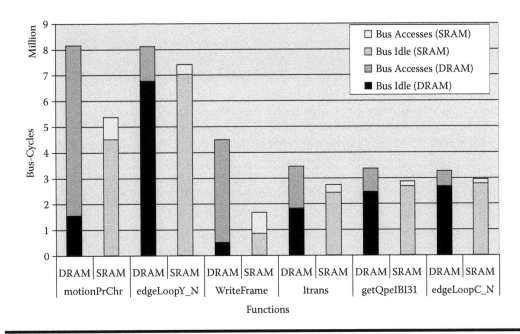

Figure 6.11   Bus usage for each function, depending on the memory type.

language and tested using an HDL simulator, such as Modelsim. Finally, the entire system needs to be verified, including hardware and software components. For this purpose, the instruction set simulator and the HDL simulator have to be connected. The codesign environment PeaCE[13] allows the connection of the Modelsim simulator and the ARMulator.

6.3 Profiling Results and Optimization of the H.264/AVC Video Decoder

The H.264/AVC baseline decoder has been profiled with memtrace using a system specification typical for mobile embedded systems comprising an ARM946E-S processor core, a data and instruction cache (16 kB each), and an external memory. The external bus system is running with half the core clock frequency. Three different external memory types have been simulated:

- SRAM: Fast SRAM memory with 0 wait states (WS)
- DRAM16: Fast DRAM with 1 WS for sequential and 16 WS for nonsequential accesses
- DRAM24: Slow DRAM with 0 WS for sequential and 24 WS for nonsequential accesses

For testing the influence of the encoded video data streams (bit streams) on the decoding performance, numerous video data streams have been evaluated. Also, different bit rates (384 and 256 kBit/s) have been tested. All bit streams have a picture size of 352 × 288 pixels (CIF resolution) and frame rate of 15 frames per second. The I-frame period of all streams is 16, that is, every 16th frame is an I-frame.

6.3.1 Results

6.3.1.1 Overall Analysis

The overall analysis (given in figure 6.12) shows the averaged processor clock frequency needed for real-time decoding of the corresponding data stream, that is, for decoding 15 frames per second. It can be seen that the memory type can influence the performance by almost a factor of 2 (comparing SRAM with DRAM24 version of container_256kb), and the influence of the bit stream goes up to a factor of 1.75 (comparing SRAM versions of container_256kb and tempete_384kb).

In figures 6.13 and 6.14 the comparison of the needed external bus cycles (EBCs; which is running with half of the processor clock frequency) for the different function groups and the used memory architecture according to their picture coding type (I-frame or P-frame) is depicted. The data has been generated using the worst-case pictures of the corresponding bit stream.

6.3.1.2 Cycles per Frame Analysis

The following analysis has been generated to analyze the number of used external bus cycles needed for each frame of a sequence of coded pictures. All three memory architectures have been used for the profiling runs.

The number of external bus cycles has been generated as a sum of all function groups of the decoder and reflects the overall needed real-time performance for the according picture number.

Figures 6.15 and 6.17 show the results for three different bit streams. Once again, the performance difference between the memory types can be observerd. Furthermore, the diagram shows the different clock cycle requirements for I-frames and P-frames. Especially in figure 6.15, the

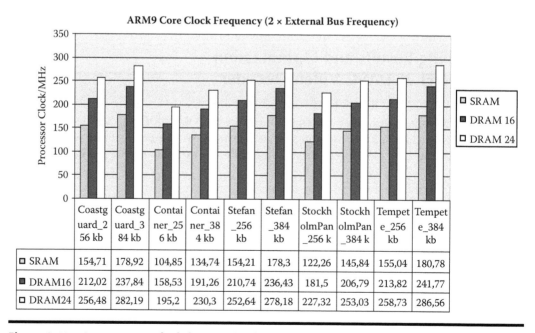

	Coastg uard_2 56 kb	Coastg uard_3 84 kb	Contai ner_25 6 kb	Contai ner_38 4 kb	Stefan _256 kb	Stefan _384 kb	Stockh olmPan _256 k	Stockh olmPan _384 k	Tempet e_256 kb	Tempet e_384 kb
▢ SRAM	154,71	178,92	104,85	134,74	154,21	178,3	122,26	145,84	155,04	180,78
▣ DRAM16	212,02	237,84	158,53	191,26	210,74	236,43	181,5	206,79	213,82	241,77
▢ DRAM24	256,48	282,19	195,2	230,3	252,64	278,18	227,32	253,03	258,73	286,56

Figure 6.12 Average core clock frequencies for different sequences and different memory architectures.

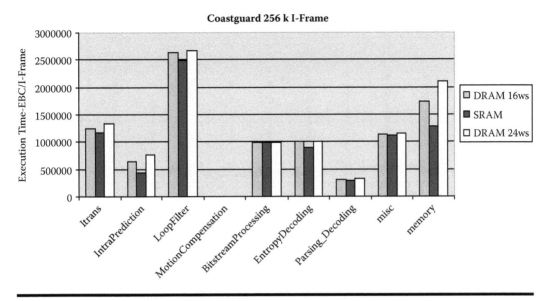

Figure 6.13 **Function group analysis for worst-case I-frame with different memory types for sequence Coastguard 256 k.**

I-frames appear as high peaks, but the results for the other bit streams show that the difference is not always as dominant. The required decoding time for P-frames also varies between the bit streams. Whereas all P-frames in figure 6.15 have a similar clock cycle requirement, the P-frames in figure 6.17 show a high variation.

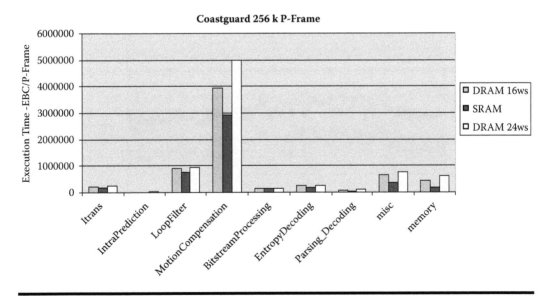

Figure 6.14 **Function group analysis for worst-case P-frame with different memory types for sequence Coastguard 256 k.**

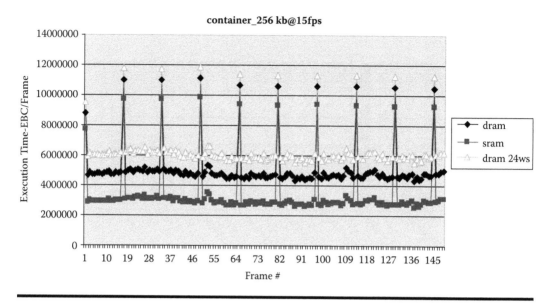

Figure 6.15 **Cycles per frame analysis for different memory types for sequence container 256 k.**

Due to the different memory configurations concerning the parameters for sequential reads for DRAM16 and DRAM24 memory architectures, in some cases the peak performance for DRAM24 can be higher than that for DRAM16. This is the case if the number of sequential read accesses dominates the overall number of accesses, because the DRAM16 is slower in sequential read accesses than the DRAM24.

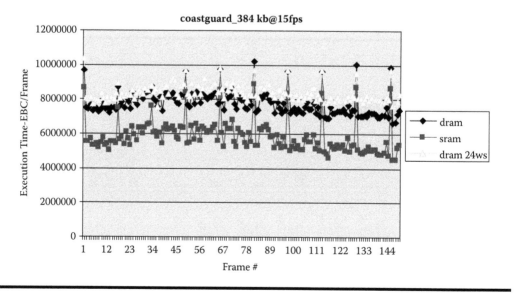

Figure 6.16 **Cycles per frame analysis for different memory types for sequence coastguard 256 k.**

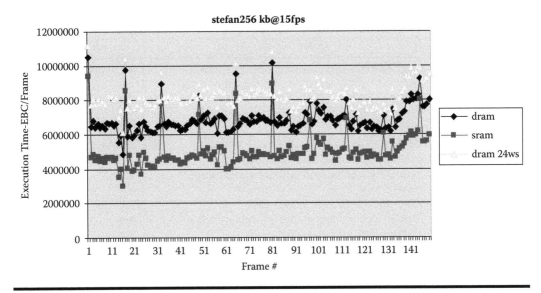

Figure 6.17 Cycles per frame analysis for different memory types for sequence stefan 256 k.

That special case can be observed in figure 6.16. The decoding of the frames (I-frames), where the DRAM16 is slower than the DRAM24 configuration, includes mostly sequential reads accesses.

These results show that the decoding time is highly dependent on the image content, and how it is encoded and varies not only for different memory types and bit streams, but also within a bit stream for each frame. Therefore, an exhaustive testing of numerous bit streams is important for obtaining meaningful profiling results.

6.3.1.3 Memory Access Statistics

The following analysis has been generated for analyzing the number of data memory accesses and data cache read misses. Separate results have been created for different H.264 decoder memory sections. As shown in table 6.3, the decoder memory is split up in the read-only section (ER_RO), read-write section (ER_RW), zero-initialized section (ER_ZI), and the stack and heap variables. The results of this analysis allow an estimation of which data sections use the data cache efficiently.

6.3.1.4 Cache Access Statistics

This analysis distinguishes between memory access and cache misses, as illustrated in figure 6.18. Memory accesses are caused by load and store operations executed by the processor and are served by the data cache. If data is not available in the cache, a load or store operation leads to a cache miss. In case of a read operation, a cache miss leads to a cache fill of a cache line. In case of a store operation, a cache miss leads to writing to the memory via the write buffer.

In the following, memory accesses are counted for both load and store operations, whereas cache misses are only counted for read operations (cache fills). This restriction is caused by the profiling tool.

Table 6.3   H.264 Decoder Memory Sections and Heap Variables

Memory Section	Description	Size
ER_RO	Program code and constant global variables	103,460 bytes
ER_ZI	Global variables, which are initialized to zero	612 bytes
ER_WR	All other global variables (which are not constant and not initialized to zero)	126,920 bytes
stack0	Stack	8,192 bytes
dec_struct	Main memory structure for the decoder (including the current frame, the five reference frames, and additional decoding data)	967,796 bytes
AUBuf	Buffer for the current access unit	16,384 bytes
bs_struct	Structure containing status information about the input bit stream buffer	36 bytes
others	Mainly a 4 kB input bit stream buffer (bs->pi32_buffer)	Approximately 4,096 bytes
sum		~1,17 MByte

Figure 6.19 shows the memory accesses to the sections and heap variables for decoding the sequence Stefan 384 kB. Memory accesses are the number of read or write operations caused by load or store operations. Regardless, if the load or store operation is a byte, half-word, or word access, each one of them is counted as one access. Figure 6.20 shows the read cache misses, which are caused by load operations to the specific memory sections and heap variables.

As can be seen, most load operations access the stack. However, these load operations only cause a few read cache misses. This shows that the stack uses the cache very efficiently.

Further, the diagrams show that for reading the dec_struct, the cache cannot be used as efficiently as for reading the stack. The reason for this is that the dec_struct is large, and the locations (addresses) of accesses to the dec_struct are very random. However, when comparing the total number of accesses to the dec_struct (200 million) with the number of read cache misses (3 million), it can be seen that using the cache still has a significant positive influence on access time to the dec_struct. Therefore, the dec_struct should not be marked as noncacheable.

In the following an overview of the accesses to the main memory (data transfer between caches and DRAM/SRAM) is given. Read accesses to the main memory are caused by read data cache

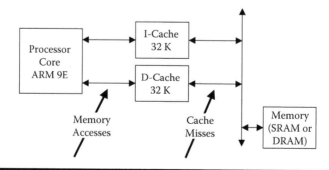

Figure 6.18   Locating data transfers caused by memory accesses and cache misses.

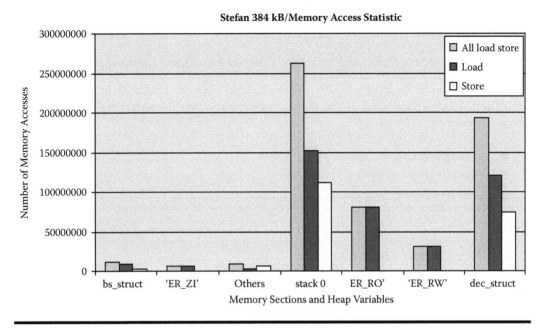

Figure 6.19   Memory accesses to the sections and heap variables for decoding the sequence Stefan 384 kB.

misses and instruction cache misses. Each cache miss leads to a data cache fill of a cache line. When decoding the Stefan 384 kB sequence the following read cache misses occurred:

- Instruction cache misses = 1,548,764
- Data cache read misses = 3,887,992

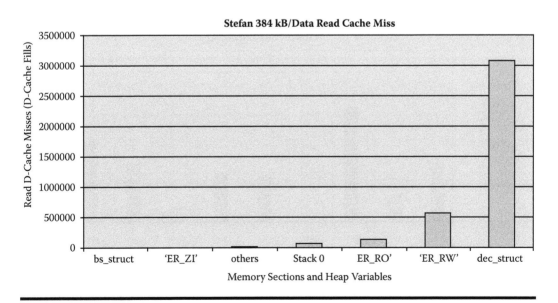

Figure 6.20   Read cache misses for decoding the sequence Stefan 384 kB.

For each cache miss a cache line with a length of eight words (4 bytes each) is read from the main memory. This leads to:

$$(1{,}548{,}764*8) + (3{,}887{,}992*8) = 43{,}494{,}048 \text{ read accesses to the main memory}$$

The write accesses to the main memory were calculated from the results of the memory model simulator. The simulator provides the number of sequential and nonsequential accesses to the memory. These are:

- Nonsequential write accesses = 23,695,488
- Sequential write accesses = 11,168,236

This leads to a total number of 34,863,724 write accesses to the main memory.

6.3.2 Design and Optimizations

Based on the acquired profiling results, several software and hardware architectural optimizations are applied. Our first target is a pure software version of the video decoder for the implementation of a DVB-H terminal on a PDA. In a second step, an embedded hardware/software is developed.

6.3.2.1 Software Implementation and Optimizations

Following Amdahl's law, those parts of the software should be considered for optimization first that take up the most execution time. The execution time for each module of the decoder has been evaluated as depicted in figure 6.21. The results show that the distribution over the modules differs significantly between I- and P-frames. Whereas in I-frames the deblocking has the most influence on the overall performance, in P-frames the motion compensation is the dominant part. Exploring the results of the functions corresponding to the motion compensation,

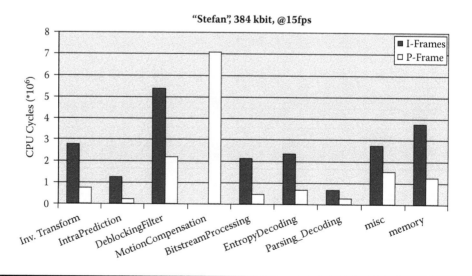

Figure 6.21 Profiling results for the H.264/AVC software decoder.

Table 6.4 32 Clock Cycles and Memory Accesses for Chrominance Motion Compensation (one P-frame)

	Clock Cycles	All Loads	Load 8/16
Before optimization	13,149,109	309,368	104,784
After optimization	9,355,709	196,746	34,584

it can be seen that the function motionCompChroma() requires the most execution time. This function performs the motion compensation for the chrominance pixels, which is mainly based on bilinear interpolation. Focusing on the read memory accesses, which are performed in motionCompChroma(), as given in the second column of table 6.4, more than 30 percent are byte or half-word accesses (third column). This is due to the fact that the pixel values have the size of 1 byte each.

Because the interpolation is applied iteratively on adjacent pixels, the source code can be optimized by reading four adjacent bytes at once. This leads to a reduction of the execution time of the function by almost 30 percent. The speed-up of the function leads to a reduction of the execution time for processing a P-frame by about 5 percent.

Further speed-up of the software could be achieved by applying well-known software optimization techniques and those proposed in Hübert et al.[4] to the functions identified by the profiler. The resulting software decoder has been tested on an Intel PXA270–based PDA within the DVB scenario. The required processor clock frequency for H.264/AVC decoding is about 420 MHz (320×240 pixel resolution, 384 kBit/s).

Considering the dynamic power consumption of CMOS circuits, given in equation (6.1), the rather high system frequency leads to high power consumption. The power consumption is raised even further by the increased core supply voltage, which is required for the high system frequency.

$$P_{dynamic} = \sum_{k=1}^{M} c_k \cdot f_k \cdot V_{DD}^2 \tag{6.1}$$

Thus, for achieving low power consumption, methods need to be applied that allow the reduction of the system frequency. The two steps applied for this purpose are finding an appropriate memory system and the development of hardware accelerators.

6.3.2.2 Memory System

Besides the processing power of the CPU, the memory and bus architecture determine the overall performance of the system. Namely, the cache's size and architecture, the speed and usage of a tightly coupled (on-chip) memory (TCM), the width of the memory bus, the bandwidth of the off-chip memory, and a DMA controller are the most influential factors. Adjusting these factors requires a trade-off among hardware cost, power consumption, and performance. The H.264/AVC decoder has been simulated with different cache sizes to find an appropriate size for the DVB-H terminal scenario (QVGA image resolution). It has been evaluated how the required decoding time changes when either the instruction cache size or the data cache size is increased (see figure 6.22).

The results show that increasing the instruction cache size from 4 kByte to 32 kByte has a minor influence on the overall performance. However, adding a data cache of 4 kByte to the

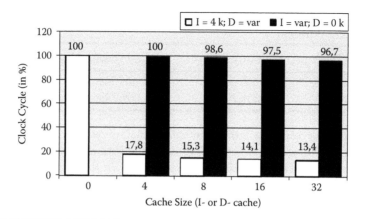

Figure 6.22   Influence of the instruction (I) and data (D) cache sizes on the execution time of the H.264/AVC decoder.

system decreases the decoding time to less than 20 percent. Further increasing the data cache size does not yield a dramatic performance increase. Therefore, data and instruction cache sizes of 4 kByte each are a good trade-off between performance and die area.

The data cache decreases the number of memory accesses, especially for accessing the stack, and the internal memory can be used to speed up memory accesses to frequently accessed data, which cannot be stored efficiently in the cache, such as randomly accessed data areas. As the H.264 decoder requires about 1.1 MByte of data memory (at CIF video solution), only small parts of the data memory (less than 3 percent with 32 kByte of SRAM) can be stored in the internal SRAM. Therefore, it is required to profile the memory accesses to each data area of the decoder to find an optimal partitioning of data areas to SRAM and DRAM. Because a data cache is used, accesses to the memory only occur if data is not available in the cache, that is, cache misses occur. Therefore, the cache misses for each relocatable data area in the decoder need to be profiled, which includes global variables, heap variables, and the stack.

Figure 6.23 shows the profiling results for these data areas. The results are presented as a ratio between the cache misses and the size (in byte) of each data area, which can be considered cache miss density. Thus, the cache miss density reflects a benefit–cost ratio between the expected

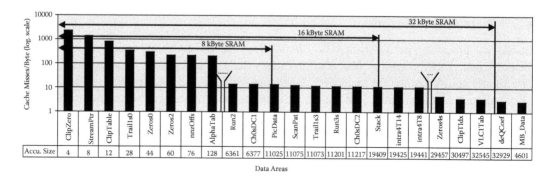

Figure 6.23   Analysis of the cache misses per byte for each relocatable data area in the H.264 decoder.

performance gain due to cache miss reduction and the required SRAM size. The highest cache miss density occurs when accessing the clipping table pointers (clipZero, clipTable) and the variable length code (VLC) tables (ZerosXX, Trail1sXX, RunXX, etc.).

To achieve the optimal partitioning, those data areas with the highest cache miss density should be stored in the SRAM. In figure 6.8 the range of data areas for 8, 16, and 32 kByte of SRAM are marked. When using 32 kByte of internal SRAM with an optimal partitioning, the decoding time is reduced by more than 20 percent.

Using the cache miss density as an indicator for data partitioning is a rather simple approach. More advanced methods have been presented, for example, in Panda et al.[10] and Kandemir et al.,[11] that also support dynamic partitioning. Memtrace could be used to supply these methods with the required analysis results.

6.3.2.3 Hardware/Software Partitioning

To further increase the system efficiency and decrease power consumption and hardware costs, the CPU can be enhanced by coprocessors. Again, the hot spots in the software code should be considered, namely, the loop filter, the motion compensation, and the integer transformation. These are the foremost candidates for hardware implementation. All these components are demanding on an arithmetical level rather than on a control flow level. Therefore, they are well suited for hardware implementation as coprocessors, which can be controlled by the main CPU. To ease the burden of providing the coprocessors with data, a DMA controller can be applied, allowing memory transfers concurrently to the processing of the CPU. The coprocessors should be equipped with local memory for storing input and output data for processing at least one macroblock at a time, preventing fragmented DMA transfers. As the video data is stored in the memory in a two-dimensional fashion, the DMA controller should feature two-dimensional memory transfers. The output of the video data to a display, which is required by a DVB-H terminal, even increases the problem of the high amount of data transfers.

6.3.2.4 Hardware/Software Interconnection and Scheduling

After the software optimization is performed and the hardware accelerators are developed, scheduling of the entire system is required. The scheduling is static and controlled by the software. The hardware accelerators are introduced step by step to the system. Starting from the pure software implementation, at first the software functions are replaced by their hardware counterparts. This also requires the transfer of input data to and output data from the coprocessors. These data transfers are at first executed by load-store operations of the processor, and in a next step replaced by DMA transfers. This also requires flushing the cache or cache lines, which may decrease the performance of other software functions. In a final step, the parallelization of the hardware and software tasks takes place. All decisions made in these steps are based on detailed profiling results.

The following example shows how the hardware accelerator for the deblocking is inserted into the software decoder. The hardware accelerator only includes the filtering process of the deblocking stage; filter strength calculation is performed in software, because it is rather control intensive, and therefore more suitable for software implementation. The filter processes the luminance and chrominance data for one macroblock at a time. It requires the pixel data and filter parameters as an input and provides filtered image data as an output; this sums up to about 340 32-bit words of data transfer. Figure 6.24 shows the results for the pure software implementation when using

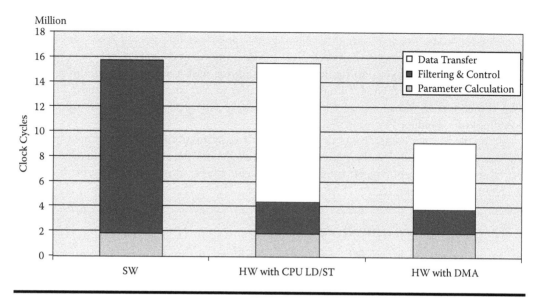

Figure 6.24   Clock cycle comparison of different deblocking implementations.

the filter accelerator with data transfer managed by the processor, and when additionally using the DMA controller. As can be seen, the data needs to be transferred by the processor, the performance gain of the accelerator is vanished by the data transfers, and only in conjunction with the DMA controller can the coprocessor be used efficiently.

6.4   SOC Architecture

Taking account of the various optimization and implementation suggestions made in the previous chapters, a proposal of an H.264 decoder SOC architecture is given as follows.

A mixed hardware-software implementation, including a RISC core and additional coprocessors, is chosen, as it combines a flexible system with moderate implementation effort. Looking at possible hardware optimizations, as discussed in the previous chapter, the loop filter, motion compensation, and integer transformation are suitable functions to be built as dedicated hardware blocks. The system bus connecting the processor core with the main memory and coprocessor components is augmented with a DMA controller, which supports the main processor by performing the memory transfers to the coprocessor units. Furthermore, a video output unit is implemented directly, driving a connected display or video DAC. To avoid a heavy bus load on the mentioned system bus due to transfers from a frame buffer to the video output interface, an extra frame buffer memory and the video output unit are provided by a separate video bus system. The data transfers between both bus systems are also performed by the DMA controller. The suggestions made above lead to the following structure of the SOC architecture.

6.4.1   Bus System

In the following paragraph the bus system of the proposed SOC architecture is described. Facing a reusability of the specific hardware structures combined in the given SOC architecture, an AMBA

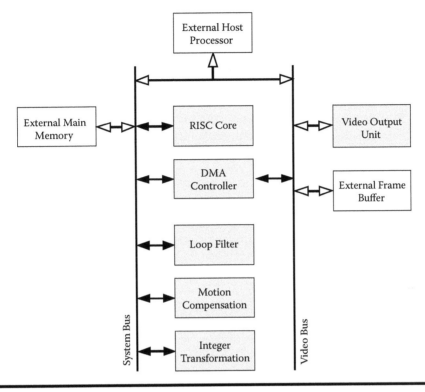

Figure 6.25   H.264/AVC decoder SOC architecture.

AHB bus-compliant implementation of the bus systems is chosen. As the dedicated coprocessors comprise an AHB-conforming interface, they are applicable in other SOC architectures using the same bus model without modifications.

A major bottleneck of many processor architectures implementing a display interface is a single bus system shared for memory accesses of the processor core and data transfers from a frame buffer to the video output interface. As a key feature of the proposed SOC architecture, both types of transfers are separated by implementing a double bus architecture. On the system bus side (see figure 6.25) the processor core is connected to main memory and coprocessor units. Transfers from the frame buffer to a display interface are performed on a separate video bus, leading to a distinct relief of the system bus load. Furthermore, the processor core is supported by a DMA controller performing the data transfers between system and video bus.

On the system bus side the RISC core and DMA controller, and on the video bus the DMA controller and the video output unit operate as bus masters.

6.4.2 RISC Core

The main idea behind the described system architecture is to offload the processing load from an application processor to the proposed companion chip. Therefore, a processor core is needed that is able to run the software decoder accelerated by the dedicated coprocessors. The core runs a version of our H.264/AVC baseline profile decoder optimized using the methods described in the previous section. Besides these optimizations, the decoder provides various software functions to interface the coprocessors, which will be discussed in the following chapters. The main control functionality of the architecture is performed by a 32-bit RISC processor derived from an ARC 625 processor core. The processor features a five-stage pipeline and has been enhanced with special instructions to accelerate functionalities not covered by special hardware blocks, for example, bit stream processing.

Taking the aforementioned cache optimizations into account, 4 kByte of data and instruction cache using a direct-mapped architecture would be sufficient for our implementation. Facing the usage of the SOC for other applications than H.264/AVC decoding, it has been decided to implement 32-kByte-sized caches as a two-way set-associative cache architecture.

Another alternative to the described methods is the usage of special instruction set extensions, which can be applied to configurable processor cores, for example, ARC or Tensilica Xtensa.

6.4.3 SIMD Video Processing for H.264

Historically, traditional embedded processors found in SOC designs for mobile applications mostly lack the ability to perform arithmetic operations on more than one data path in parallel. Typically the ALU performs one operation using the native operand width of the processor, for example, 32 bits. The fact that most operations used in video processing algorithms only use a fraction of the given data word width leads to the split ALU-SIMD concept. In particular, the core component of a split ALU is a functional unit that can perform the same operation on more than one data path in parallel, for example, 4×8 bit add instead of one 32-bit add.

The Intel mobile processor series PXA27x features an architectural implementation (Wireless MMX) similar to the established desktop counterparts. Like the predecessors MPEG-2 and MPEG-4 Part 2, H.264/AVC uses the macroblock concept, dividing images into sections of 16×16 pixels. By applying a 4×4 integer transform on prediction errors (residuals) instead of the previously used 8×8 DCT, the base block size for motion compensation and intraprediction is reduced to 4×4 pixels. The important implication for SIMD approaches is that just 16 pixels are reconstructed in a single video processing transaction of identical parameters, eventually leading to significant function call overhead.

Another consequence results for the operation of the in-loop deblocking filter. Parameters for horizontal and vertical cross-block borders to the filter are derived for a set of four border pixels each, resulting in 32 individual parameter sets and filter operations per macroblock. The resulting requirement to achieve noticeable gains out of SIMD data path parallelism is therefore to maximize the amount of data processed in a single operation. Considering arbitrary predictor locations in memory and different subpixel filtering schemes for reconstructing each 4×4 block in luminance motion compensation, a speed-up approach requires the inclusion of the basic 4×4 block size. By deriving all filtering parameters of a macroblock in a single run, the basic 4-pixel filtering can be extended to include all 16 pixels of a macroblock border.

6.4.3.1 Case Study: Wireless MMX

Verifications of the concepts outlined in the previous section were conducted in the PXA270 architecture by means of assembly code. One finding was that 25 to 33 percent of all executed instructions are needed to rearrange data words in their operand registers to be able to apply massive parallel SIMD operations, which reveals the requirement of efficient data reordering to exploit benefits in data path parallelism.

Further obstacles for successful SIMD operations can be found in the deblocking filter. Adaptive algorithms, including individual thresholds and workflow for each single filtered pixel, set an effective limit on SIMD performance gains. The example below shows the calculation of final clipping values in the regular luminance deblocking filter, depending on the difference of pixels 0–2 left and right of the respective border. Because wireless MMX is 64 bits wide and an H.264 pixel border of the same parameter is 4 pixels wide, the calculations are performed in half-word width. To conserve registers, the no-filter condition was encoded by zero-masked clipping values.

```
wmaxuh wR8, wR0, wR5; // max(R0,R2)
wminuh wR9, wR0, wR5; // min(R0,R2)
wmaxuh wR10, wR2, wR6;// max(L0,L2)
wminuh wR11, wR2, wR6; // min(L0,L2)
// clear constants if !filter
wand   wR7, wR7, wR15;
wand   wR12,wR12,wR15;
wsubh wR8, wR8, wR9; // abs(R0-R2)
wsubh wR10, wR10, wR11; // abs(L0-L2)
// aq: if( beta > wR8 ) 0xff else 0x00
wcmpgtuh wR8, wR7, wR8;
// ap: if( beta > wR10 ) 0xff else 0x00
wcmpgtuh wR10, wR7, wR10;
// calculate CLIP values
// if( aq ) ( C0+1 )
wsubh wR13, wR12, wR8;
// c0_aq if( aq ) C0 else 0
wand   wR8, wR8, wR12;
// if( ap ) ( C0+1 )
wsubh wR13, wR13, wR10;
// c0_ap if( ap ) C0 else 0
wand   wR10, wR10, wR12;
```

6.4.3.2 Custom Instruction Set

The investigations on SIMD performance gains in the previous chapters allowed us to get an impression of required instructions and feasible performance gains in terms of data types, data alignment issues, and reordering overheads. This knowledge enables us to derive a hypothetical instruction set that can be the starting point for the implementation of application-specific processors for SOC environments. Several tools are available at the market, like Lisatek[8] or Tensilica Extensa,[7] which enable the designer to implement very application-specific instruction set processors.

A suitable application-specific instruction set requires a design not only fulfilling the application at hand, but also fitting into implementation and chip complexity constraints. Thus, the basic rule in our approach was to stick to simple operations. Common instructions include logic operations (AND/OR/XOR) and simple arithmetic instructions (ADD/SUB/Shift). To cover conditionally executed filtering operations, two instructions for absolute difference and comparison on 8-bit unsigned operands have been defined.

An example from the loop filter condition and clip constant calculations is shown below. Compared to the wireless MMX implementation demonstrated in the previous section, the instructions are more compact. The shown operations are carried out in 8-bit accuracy per slot. In the 128-bit instruction set implementation example, the conditions can be calculated for all 16 pixels of a macroblock border in a single shot. Therefore, less than one-fourth of the instructions have to get carried out per filtered macroblock, compared with the WMMX version.

```
// calculate ap,aq by aq=(abs(R3-R5)<Beta)
// and ap=(abs(R0-R2)<Beta) for all 16 Pels
ABSDIFF8_128( R8, R3, R5); // ABS( R3-R5 )
// if( R8[n]<Beta) R11[n]=0xff,else 0
SUBCARRY8_128( R11, R8, R7, FLAG_BANKC);
ABSDIFF8_128( R8, R0, R2); // ABS( R0-R2 )
SUBCARRY8_128( R12, R8, R7, FLAG_BANKC);
// C0-0xff == C0+1
USUB8_128( R16, R13, R11 ); // C0+aq
USUB8_128( R16, R16, R12 ); // C0+aq+ap
// calculate clip values for l1 and r1
AND_128( R11, R11, R13 ); // C0 & aq
AND_128( R12, R12, R13 ); // C0 & ap
// mask all slots out where filtering is
// prohibited by (abs()<Alpha...)&&
// (abs()<Beta)&&(strength>0)
AND_128( R16, R16, R10 ); // (C0+aq+ap)&FLAG
```

An observation regarding the available instruction set in desktop-class machines was a rare use of multiply-add operations in favor of low-latency shifts and additions. All necessary scaling coefficients in motion compensation and deblocking filter can be expressed by a single Shift-ADD instruction, so that costly multipliers are not required in our architecture.

6.4.4 Coprocessors

6.4.4.1 Loop Filter

In the following paragraph the structure of an efficient hardware implementation of the deblocking filter is described.

To minimize block artifacts at the borders of decoded 4×4 blocks, an adaptive filtering of these borders is performed in vertical and horizontal directions, whereas deblocking is done in a given order on a macroblock basis. Depending on strength values and weighting factors calculated for a decoded 4×4 block, different filter modes are accomplished. The filtering of one horizontal/vertical edge of a 4×4 block comprises the processing of block data from the adjacent upper/left 4×4 blocks.[9]

Looking at the underlying software model of the deblocking filter functions, the filtering is done on 4×4 block edge basis (horizontal or vertical direction), whereas processing of one block edge is performed as below:

- Select between strong or normal filter mode for one block edge depending on the strength parameter.
- Calculate difference value to determine whether one row of the 4×4 block edge is filtered.
- Perform filtering of one row of a block edge.
- Process next row of the block edge.

As the software model is based on a 4×4 block edge basis, this model is not suitable for a hardware realization because of redundant data transfer due to access on adjacent 4×4 blocks. So a macroblock-wise concept with double buffering of two adjacent macroblocks is chosen.

Thus, redundant data transfers are reduced to the transfer of four 4×4 blocks of the upper adjacent macroblock. Furthermore, the double-buffer concept gives the ability to transfer block data to and from the loop filter coprocessor in parallel to filter processing.

The filter process itself is rearranged by parallelizing the calculation of different filter modes and the difference evaluation, leading to a six-stepped pipeline structure delivering one processed row in one cycle.

As block edges have to be filtered in vertical and horizontal direction, the macroblock-wise processing and desired reuse of the filter arithmetic for both filter directions necessitate the possibility of rotating block data. Therefore, the filter structure is extended by a transpose unit including extra memory to store rotated block data.

Figure 6.26 gives an overview of the complete loop filter structure, whereas only luminance data processing is shown. The final implementation comprises additional components to process luminance and chrominance data in parallel.

An estimation of the calculating effort of the software function (normal filter mode, memory transfers ignored) gives a cost of ~54 arithmetic operations for difference evaluation and calculation of one row of a 4×4 block.

Looking at a macroblock basis, this results in a processing effort of ~6912 operations per macroblock.

Implementing the proposed structure of the loop filter as a dedicated hardware component, the filtering of a whole macroblock is processed in only 232 clock cycles, as described in table 6.5.

6.4.4.2 Motion Compensation

In this paragraph a proposal of a capable hardware structure performing pixel interpolation for motion prediction of luminance samples is given.

Depending on a given motion vector, sample values at quarter-pixel positions are calculated by interpolation of reference frame samples. Thereby, at first a six-tap filter for obtaining sample values at half-pixel and afterwards a two-tap filter for calculating sample values at quarter-pixel positions is used.[6] At a maximal granularity, motion vectors are delivered 4×4 block-wise. According to the use of a six-tap filter for a half-sample calculation, a maximum of 9×9 reference samples are processed, obtaining the prediction values of a 4×4 block.

As motion vector ranges from 0 to 3/4 in the horizontal and vertical directions, there are 16 cases given to be processed in a different way, which could be performed by a variation of horizontal and vertical six-tap filtering (referenced as A and B), horizontal and vertical two-tap filtering (referenced as C and D), and diagonal two-tap filtering (referenced as DIAG).

This leads to a hardware structure comprising three types of filter components—six-tap, two-tap, and two-tap diagonal filter, built as pipelined units processing one reference sample row per cycle in parallel, whereas the six-tap and two-tap units perform filtering in both the horizontal and vertical directions. To rotate block data for filter processing in these two dimensions, additional transpose memories are implemented.

To minimize the number of transposing steps, as they break pipeline processing or give the need of additional data transfer cycles to the transpose memories, respectively, two measures are taken: reordering of the different filter steps and implementing the input memory (containing reference samples) as a transpose unit, too.

Table 6.6 gives an overview of the different filter steps to be performed depending on a given motion vector.

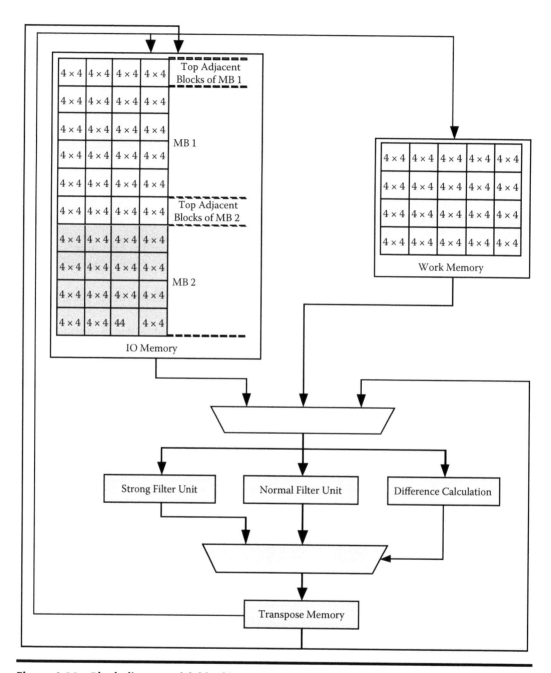

Figure 6.26 Block diagram of deblocking unit.

Figure 6.27 shows the resultant structure of the pixel interpolation unit. Only processing of luminance data is performed by this component, as chroma prediction values are obtained by a simple bilinear interpolation.

To give an estimation of the reached performance gain processing one 4 × 4 block by utilizing the proposed hardware structure, at first a look is taken at the filter functions of the underlying

Table 6.5   Processing Effort of Deblocking Unit

Processing Effort of Filtering, 1 MB		
Edge	*Processing Steps*	*Clock Cycles*
1. Vertical	20 cycles read/store data to transpose memory 6 cycles filtering	26 cycles
2. Vertical	20 cycles read/store data to transpose memory 6 cycles filtering	26 cycles
3. Vertical	20 cycles read/store data to transpose memory 6 cycles filtering 4 cycles transpose/store to work memory	30 cycles
4. Vertical	20 cycles read/store data to transpose memory 6 cycles filtering 12 cycles transpose/store to work memory	38 cycles
1. Horizontal	16 cycles read/store6 cycles filtering 4 cycles store to IO memory	26 cycles
2. Horizontal	16 cycles read/store 6 cycles filtering 4 cycles store to IO memory	26 cycles
3. Horizontal	16 cycles read/store 6 cycles filtering 4 cycles store to IO memory	26 cycles
4. Horizontal	16 cycles read/store6 cycles filtering 12 cycles store to IO memory	34 cycles
		Total: 232 cycles

software model. The number of required arithmetic operations according to a given motion vector is described in table 6.7.

In comparison to the software model, table 6.8 represents the processing effort of the implemented hardware unit in clock cycles.

6.4.4.3 Integer Transformation

Following an efficient hardware structure, performing the inverse integer transformation is given. Compared to earlier coding standards utilizing a discrete cosinus transformation based on an 8 × 8

Table 6.6   Filter Steps in Motion Compensation

Motion Vector	*Filtering Steps*	*Reordered Sequence*
(0 ; 0)	—	—
(1/2 ; 0)	A	A
(1/4 ; 0) (3/4 ; 0)	A, C	A, C
(1/2 ; 1/2)	A, B	B, A
(1/4 ; 1/2) (3/4 ; 1/2)	A, B, C	B, A, C
(1/2 ; 1/4) (1/2 ; 3/4)	A, B, D	A, B, D
(1/4 ; 1/4) (1/4 ; 3/4) (3/4 ; 1/4) (3/4 ; 3/4)	A, B, DIAG	B, A, DIAG
(0 ; 1/2)	B	B
(0 ; 1/4) (0 ; 3/4)	B, D	B, D

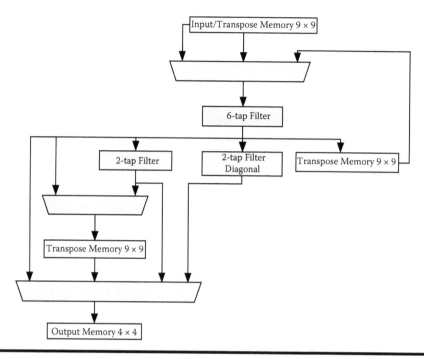

Figure 6.27   Block diagram of pixel interpolation unit.

block size, H.264 implements a 4 × 4 block-wise integer transformation. On the decoder side the calculation is performed by applying a matrix multiplication on a block of coefficients and a given transform matrix. As a character of this transform matrix, the whole transformation is feasible only performing add and shift operations.

Table 6.7   Processing Effort Estimation of Motion Compensation—Software Version

Motion Vector	Estimated Calculation Effort of Software Processing One 4 × 4 Block
(0 ; 0)	—
(0 ; 1⁄4)	224 ops.
(0 ; 1⁄2)	192 ops
(0 ; 3⁄4)	224 ops.
(1⁄4 ; 0)	224 ops.
(1⁄4 ; 1⁄4)	416 ops.
(1⁄4 ; 1⁄2)	416 ops.
(3⁄4 ; 3⁄4)	416 ops.
(1⁄2 ; 0)	192 ops
(1⁄2 ; 1⁄4)	416 ops.
(1⁄2 ; 1⁄2)	224 ops.
(1⁄2 ; 3⁄4)	416 ops.
(3⁄4 ; 0)	224 ops.
(3⁄4 ; 1⁄4)	416 ops.
(3⁄4 ; 1⁄2)	416 ops.
(3⁄4 ; 3⁄4)	416 ops.

Table 6.8 Processing Effort Estimation of Motion Compensation—Hardware Version

Motion Vector	Complete Calculation Effort of Hardware Processing One 4 × 4 Block
(0 ; 0)	0 cycles
(0 ; 1⁄4)	17 cycles
(0 ; 1⁄2)	15 cycles
(0 ; 3⁄4)	17 cycles
(1⁄4 ; 0)	16 cycles
(1⁄4 ; 1⁄4)	30 cycles
(1⁄4 ; 1⁄2)	29 cycles
(3⁄4 ; 3⁄4)	31 cycles
(1⁄2 ; 0)	14 cycles
(1⁄2 ; 1⁄4)	34 cycles
(1⁄2 ; 1⁄2)	27 cycles
(1⁄2 ; 3⁄4)	34 cycles
(3⁄4 ; 0)	16 cycles
(3⁄4 ; 1⁄4)	31 cycles
(3⁄4 ; 1⁄2)	29 cycles
(3⁄4 ; 3⁄4)	31 cycles

Looking at the implementation of a capable hardware transformation unit, the following considerations are taken in account. As the matrix multiplication could be let back on a one-dimensional vector operation performed on rows and columns of the given coefficient matrix, arithmetic components are reusable for calculating results in horizontal and vertical order. Figure 6.28 describes the processing of a row or column, respectively, whereas A, B, C, and D represent a row/column of delivered coefficients, and A', B', C', and D' the according calculation results.

To achieve a high performance gain, the operations mentioned above are performed on all four rows or columns of a 4 × 4 coefficient block in parallel (see figure 6.29). Furthermore, the proposed hardware structure implies the addition of residuals and predicted samples sequencing the transformation step.

To minimize the administration effort utilizing a dedicated hardware unit for integer transformation, an 8 × 8 block-wise processing of coefficients is implemented, giving the advantage of pipelined processing of four 4 × 4 blocks in one step.

The resulting structure of the coprocessor is shown in figure 6.30, comprising input memories for coefficients and sample values, de-multiplexers for supplying the arithmetic unit with 4 × 4 block data in parallel, and an output memory containing result values.

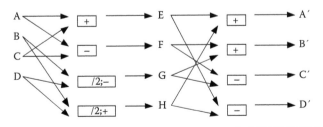

Figure 6.28 Integer transformation.

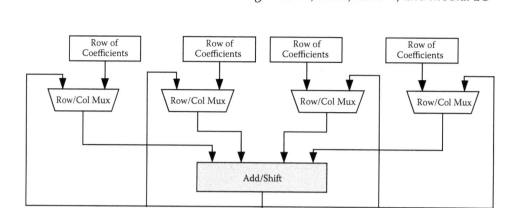

Figure 6.29 Block diagram of integer transformation unit.

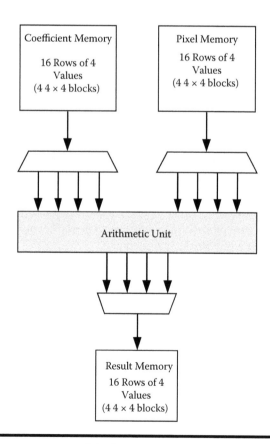

Figure 6.30 Coprocessor structure of the integer transformation unit.

Table 6.9 DMA Controller Functionalities

Functionality	Purpose
16 predefined DMA channels	To minimize parameter transfers to the DMA controller (e.g., source, destination addresses), 16 programmable DMA channels are implemented.
One- and two-dimensional memory transfers	Next to one-dimensional memory transfers, two-dimensional transfers are also supported in face of block transfers (e.g., 4×4, macro blocks) to the dedicated coprocessor structures.
One source to two destination transfers	Gives the ability to transfer results from the loop filter to frame buffer and reference frame buffer, saving redundant read accesses.
Data buffering	Decoupling of read and write transfers facing segmentation of according bus arbitrations; also utilized for one source to two destination transfers.
Postincrement address operations	Allows auto postincrement of source and destination addresses of the programmable DMA channels after a DMA transfer, saving additional address operations and reconfiguration of the according channel; e.g., used if adjacent blocks of a frame are processed consecutively.
Memset mode	Set memory regions to a programmable value; e.g., used for clearing data structures.

Estimating the transformation effort applying the software model, about 320 arithmetic operations are performed, processing four 4×4 blocks (not including address calculations for referencing coefficients and predicted pixels).

In contrast to the software version, the dedicated hardware structure performs the whole computation in 30 clock cycles.

6.4.5 DMA Controller

In this paragraph a functional description of a flexible DMA controller hardware structure is given. As shown in figure 6.26, a DMA controller is added to the H.264 coprocessor design, supporting the implemented RISC core by performing memory transfers between the dedicated coprocessor structures and the external main and video memories. Table 6.9 gives a detailed overview of the implemented DMA controller functions and a description of their purpose.

6.4.6 Video Output Unit

Below a description of the functionality of the implemented video output unit is given. Implementing a SOC architecture for decoding H.264 bit streams, it is desirable to directly drive a connected display. To cover a large amount of different display types, a flexible video output system is demanded. Table 6.10 gives an overview of the main features accomplished by the proposed display video output unit.

The video output unit is responsible for delivering video data to a connected display at any time without delay. For this reason, video data transferred from the frame buffer to the display

Table 6.10 Video Output Unit Functionalities

Free programmable video raster (up to 1024-by-1024 pixel formats, including blank region)
Configurable synchronization signals as vertical/horizontal sync, blank signal, etc.
Color room conversion supporting YUV 4:4:4, 4:2:2, 4:2:0, and RGB 4:4:4 input formats
Supporting variable frame sizes to be displayed independent from the selected video raster
Decoupling of the video output based on an external video clock from system clock
Buffering of frame buffer data to ensure availability of video data permanently

interface is stored in an extra line buffer implemented as a FIFO memory to bridge time gaps due to a busy video bus.

The observation of the fill level of the mentioned FIFO memory and data transfers from the frame buffer to the display unit are performed autonomously by the video output controller without a need of additional DMA transfers.

References

1. *Digital video broadcasting (DVB); Specification for the use of video and audio coding in DVB services delivered directly over IP*. ETSI TS 102 005.
2. B. Stabernack, H. Richter, and E. Müller. Evaluating and implementing H.264 for embedded and mobile streaming applications. *SPIE XXVI Applications of Digital Image Processing*, August 2003.
3. Joint Video Team (JVT) of ISO/IEC MPEG and ITU-T VCEG. 2003. *International standard of joint video specification*. ITU-T Rec. H.264; ISO/IEC 14496-10 AVC; JVT-G050.
4. H. Hübert, B. Stabernack, and H. Richter. 2004. Tool-aided performance analysis and optimization of an H.264 decoder for embedded systems. Paper presented at the Eighth IEEE International Symposium on Consumer Electronics (ISCE 2004), Reading, England.
5. Advanced RISC Machines (ARM) Ltd. 2004. RealView *ARMulator ISS user guide*. Version 1.4, Ref. DUI0207C. http://www.arm.com.
6. T. Wiegand, G. J. Sullivan, G. Bjøntegaard, and A. Luthra. 2003. Overview of the H.264/AVC video coding standard. *IEEE Transactions on Circuits and Systems for Video Technology* 13.
7. R. E. Gonzalez. 2000. Xtensa: A configurable and extensible processor. *IEEE Micro* 20(2):60–70.
8. M. Hohenauer, H. Scharwaechter, et al. 2004. A methodology and tool suite for C compiler generation from ADL processor models. In *Proceedings of the Conference on Design, Automation & Test in Europe (DATE)*.
9. P. List, A. Joch, J. Lainema, G. Bjøntegaard, and M. Karczewicz. 2003. Adaptive deblocking filter. *IEEE Transactions on Circuits and Systems for Video Technology* 13(7): 614–619.
10. P. Panda, N. Dutt, and A. Nicolau. 2000. On-chip vs. off-chip memory: The data partitioning problem in embedded processor-based systems. *ACM Transactions on Design Automation of Electronic Systems* 5:682–704.
11. M. Kandemir, J. Ramanujam, M. J. Irwin, N. Vijaykrishnan, I. Kadayif, and A. Parikh. 2001. Dynamic management of scratch-pad memory space. In *Proceedings of the 38th Design Automation Conference*, Las Vegas, 690–95.
12. Bormans, J., K. Denolf, S. Wuytack, L. Nachtergaele, and I. Bolsens. 1999. Integrating system-level low power methodologies into a real-life design flow. In *PATMOS'99 Ninth International Workshop Power and Timing Modeling, Optimization and Simulation*, Kos, Greece, 19–28.
13. S. Ha, C. Lee, Y. Yi, S. Kwon, and Y.-P. Joo. 2006. Hardware-software codesign of multimedia embedded systems: The PeaCE approach. In *12th IEEE International Conference on Embedded and Real-Time Computing Systems and Applications*, Sydney, 207–14. Vol. 1.

Chapter 7

Development of an H.264/AVC Main Profile Video Decoder Prototype Using a Platform-Based SOC Design Methodology

Huan-Kai Peng, Chun-Hsin Lee, Jian-Wen Chen, Tzu-Jen Lo, Yung-Hung Chang, Sheng-Tsung Hsu, Yuan-Chun Lin, Ping Chao, Wei-Cheng Hung, Kai-Yuan Jan, and Youn-Long Lin

Contents

Keywords

ASIP, AMBA, H.264/AVC, hardware/software codesign, HDTV

Abstract

With the increasing demand of high-quality video on handheld devices, we will see widespread use of the state-of-the-art video compression standard H.264 Advanced Video Coding (H.264/AVC) in cell phones, PDAs, and smart phones in the next few years. Currently, complicated H.264/AVC decoding requires very high processor frequency to meet the real-time requirement, consuming an unacceptable amount of power. Therefore, a low-power implementation is needed. In this chapter, we describe a pure hardwired H.264/AVC main profile video decoder that has been implemented in Verilog RTL, equipped with an AMBA bus interface, and demonstrated in an FPGA prototype.

We present our development process, the SOC platform used, and its associated design methodology as well as lessons learned. Our decoder takes as its input H.264/AVC compressed video bit stream and produces as its output video frames ready for display. We wrap the decoder core with an AMBA-AHB bus interface and integrate it into a multimedia SOC platform. Several architectural innovations such as massive parallelism and data reuse at both IP and system levels are proposed to achieve high performance at a low operating frequency. Running at a 16 MHz FPGA, our decoder can real-time decode D1 video (720 × 480) at 16 frames per second, providing an over 20 times performance improvement over today's mainstream DSP approach. With its low-operating-frequency requirements and thorough consideration of system-level overhead, this H.264/AVC decoding system can significantly improve the power consumption and video quality of real-time mobile video applications.

7.1 Introduction

Advances in semiconductor manufacturing technology and demand in complex consumer applications have made possible implementation of a whole electronics system on a single chip (SOC). With hundreds of millions or even billions of transistors on a chip, we can implement many communication and signal processing functions on the chip. However, it is a great challenge to successfully design such a complex chip. New methodology is needed.

One of the most popular applications is digital video. A series of international standards help bring high-quality video contents to the mass. For example, we have witnessed the success of VCD based on MPEG-1, DVD and DTV based on MPEG-2, and videoconferencing based on MPEG-4. Newer standards achieve a better compression ratio and higher quality at the expense of more computational complexity. To support upcoming high-definition TV, next-generation DVD, and digital video broadcasting (DVB), the standard body has defined H.264 Advanced Video Coding (H.264/AVC).[13]

Like previous video coding standards, H.264/AVC coding divides video frames into 16 × 16 macroblocks (MBs) and codes one MB at a time in three major stages: prediction, transformation, and entropy coding. What differentiates it from the previous ones is that it employs more sophisticated coding tools in every stage.

An MB can be predicted in one of two ways: intraframe prediction in an I-type frame and interframe prediction in a P-type or B-type frame. For intraprediction, an MB is predicted based on the pixels of its neighboring MBs. The H.264/AVC standard defines totally 17 prediction modes, each with a set of prediction functions. The encoder chooses the mode that results in the smallest difference between the prediction and the actual. For interprediction, the encoder finds a motion vector pointing to a region in the reference frame, which results in the smallest prediction errors. Compared with previous standards, H.264/AVC allows variable-size block partitioning, multiple reference frames, and quarter-pixel motion estimation.

The difference between the actual and prediction is called residual error. The data needs to be further processed and transmitted or stored. H.264/AVC employs an integer 4×4 discrete cosine transform. An MB is divided into 16 4×4 blocks, each transformed separately. Discrete cosine transform (DCT) converts data from spatial domain (pixel) to frequency domain (coefficients). It essentially represents a 4×4 image pattern with a weighted sum of 16 basis patterns. Because most of the transformed coefficients are either zero or very small, we can discard them via quantization without suffering much from quality degradation.

Entropy coding takes advantage of probability distribution of symbols to compress data. The most notable approach is the Huffman coding.[10] That is, giving short code to frequent symbols and long codes to less frequent symbols. H.264/AVC employs two entropy encoding methods: context-based adaptive variable-length coding (CAVLC) for baseline profile and context-based adaptive binary arithmetic coding (CABAC) for main profile. CABAC is more effective at the expense of more computation complexity.

Many approaches have been proposed for implementing H.264/AVC encoding, decoding, or codec. They include pure software in a general-purpose CPU[12] or DSP,[16,17,19,22] application-specific instruction set processor (ASIP),[14] and hardware-software codesign[6,15–18] with partial functions accelerated to a pure hardwired approach.[9]

Although system-level analysis is crucial when considering system overhead, most previous research only calculates performance under an ideal system-level environment, making their approaches impractical for real-world applications. Because Iverson et al.[12] and Talla[22] used CPU-based and DSP-based approaches, respectively, their results are closer to actual environments by default. However, Iverson et al.,[12] using Intel's CPU, requires over 1 GHz, while Talla,[22] using TI's DSP, requires 600 MHz to run a real-time H.264/AVC decoding at D1 (720×480) resolution, making power consumption in both cases too high for power-sensitive applications.

In this chapter, we describe our experience in developing a pure hardwired, demonstrable H.264/AVC main profile video decoder prototype. In section 7.2, we present a multimedia SOC platform and its associated design and verification methodology. Section 7.3 describes how we map the decoder onto the platform and major steps during decoding. We then describe our proposed architecture, subfunction IP design and verification, whole decoder integration and verification, and system-level integration and verification in section 7.4. Afterwards, section 7.5 shows that we achieved an FPGA prototype that can real-time decode D1 video (720×480 at 16 frames per second) while running at only 16 MHz. Later, in section 7.6, we present some lessons we have learned over the project course, before we conclude this work in section 7.7.

7.2 Platform Description

The UMVP-2000 multimedia SOC platform from Global UniChip Corp. GUC UMVP-2000[8] provides application-specific support for multimedia system prototyping. It integrates a Reduced Instruction Set Computer (RISC) CPU processor, memory modules, an FPGA holding hardware accelerators, another built-in FPGA containing peripheral IPs, and associated firmware drivers.

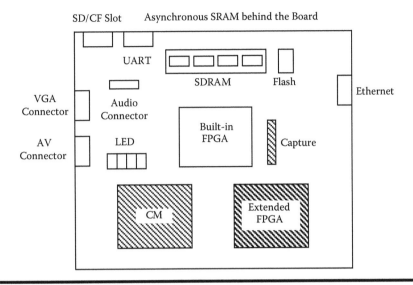

Figure 7.1 Platform layout.

It has been applied to several real-time multimedia applications, including digital still camera, digital photo album, and MPEG-4 codec. In this section, we will introduce the platform layout, bus architecture, and associated development flow.

7.2.1 Platform Layout

Figure 7.1 depicts the platform layout. The core module (CM) is an ARM 926-EJS[3] processor. On-board memory includes 128 MB SDRAM, 1 MB asynchronous SRAM, and 32 MB flash. The built-in FPGA integrates peripheral I/O controllers, while the extended FPGA provides custom logic module (LM) expansion. Its I/O interfaces support most multimedia applications, including a secure digital/compact flash (SD/CF) card, an A/V connector, a 15-pin D-sub connector, a charge-coupled device (CCD) capture module with TV signal, an AC-97 audio extension, a 10/100 Mbps Ethernet local area network (LAN) port, and four 7-segment light-emitting diodes (LEDs).

7.2.2 Platform Architecture

Figure 7.2 shows our platform architecture based on the Advanced Microcontroller Bus Architecture (AMBA),[1] including CPU, memory, and reserved AHB connection ports.

7.2.2.1 AMBA Interconnection System

The Advanced High-performance Bus (AHB) and Advanced Peripheral Bus (APB), consisting of multiplexer, arbiter, and decoder, specify the interconnection between hardware components and the central bus controller. Under this architecture, system designers can quickly integrate new IPs via software visible registers and application drivers.

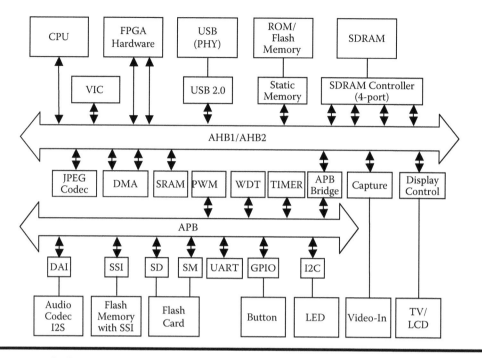

Figure 7.2 Platform architecture.

7.2.2.2 CPU

The ARM926-EJS RISC processor[3] has two execution phases: system boot-up phase and runtime phase. During the boot-up phase, all slaves get their initial data and wait for next instructions from the processor. During runtime, the processor deals with all programmable I/O devices and system exceptions, such as interrupts.

7.2.2.3 Memory

Flash memory, SRAM, and SDRAM are available in the platform, providing memory spaces for multimedia development. Flash memory stores built-in test programs for self-diagnosis. SRAM stores instructions and data for CPU operations. SDRAM supports a large memory space for multimedia application buffers, such as video frame buffer and still-image data.

7.2.2.4 Reserved Masters and Slaves for FPGA Hardware

The platform reserves four master ports and eight slave ports for the AHB1 bus and four additional master ports for the AHB2 bus. Every port can be connected to dedicated hardware inside the extended FPGA. We could utilize both buses to reduce bus contention.

7.2.3 Platform-Based Design Methodology

Figure 7.3 gives an overview of our platform-based design methodology. Given a specification, we use a software reference model to analyze the performance requirement on each major task. Based

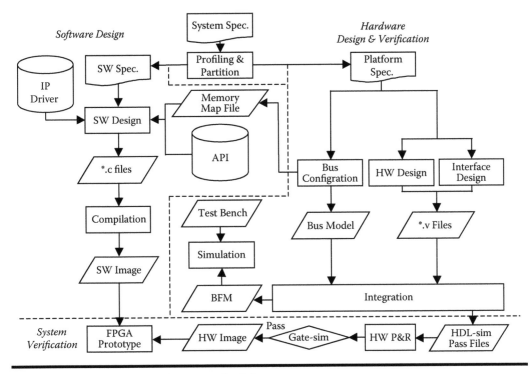

Figure 7.3 Platform-based design methodology.

on the total system performance requirement and the profiling result, we determine whether a task should be implemented in hardware or software. In hardware design and verification, we first design the hardware architecture, which can be divided into several subfunction IP designs. Then, we equip the integrated hardware with an AMBA wrapper and associated bus functional model (BFM) of peripheral IPs. System integration includes device drivers, application programming interface (API), and device memory map development. Finally, we use an FPGA-based platform to verify the correctness of the whole system through prototyping.

7.3 System Description

Figure 7.4(a) depicts the mapping between our H.264/AVC decoder system and the usage of the multimedia platform. Figure 7.4(b) depicts a simplified system diagram.

Table 7.1 lists ten major steps the system performs to decode a video sequence. The amount of data traffic is based on real-time decoding of D1 (720 × 480 at 30 FPS) resolution.

In the beginning, the CPU instructs the DMA to move one block (32 KB in CIF version, 108 KB in D1 version) of compressed video from the SD card to the system SDRAM. Next, the CPU tells the H.264/AVC hardwired decoder to start decoding a frame. The decoder then reads the input buffer and reference frames from the SDRAM, applies decoding methods, and then writes the reconstructed data back to the reference frame buffer and the decoded data to the output buffer in an MB-by-MB order. After receiving an interrupt signal from the decoder, the CPU commands the display controller to read and display the next decoded frame from the SDRAM. Frame by frame, this scheme proceeds until it reaches the end of sequence.

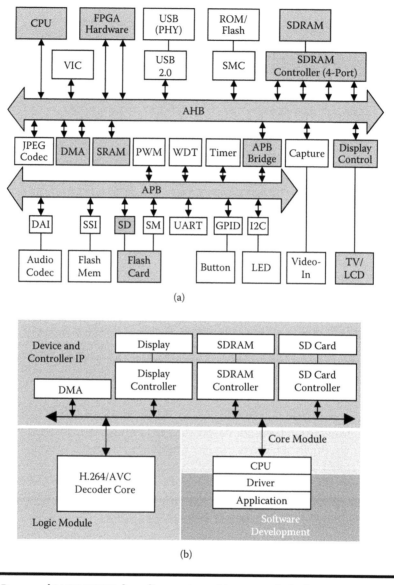

Figure 7.4   Proposed H.264/AVC decoding system: (a) usage of platform components; (b) simplified system diagram.

7.4   Design Flow

To develop the H.264/AVC decoding system, we first study the standard documents and reference software.[20] Then, we design the top-level hardware architecture and all subfunction IPs before we implement, integrate, and verify them in both hardware and system levels. Finally, we analyze our system bottlenecks and employ system-level techniques to improve overall performance.

Table 7.1   Video Decoding Steps

Step	Type	Source	Dist.	Description	Data Traffic
1	Control	CPU	DMA	Command	—
2	Data	SD card	SDRAM	Write of input buffer	1.1 MB/s
3	Control	CPU	Decoder	Starting signal of decoding one frame	—
4	Data	SDRAM	Decoder	Read of input buffer	1.1 MB/s
5	Data	SDRAM	Decoder	Read of reference frame buffer	32.6 MB/s
6	Data	Decoder	SDRAM	Write of reference frame buffer	14.9 MB/s
7	Data	Decoder	SDRAM	Write of output buffer	19.9 MB/s
8	Control	Decoder	CPU	Interrupt	—
9	Control	CPU	Display	Display the decoded frame	—
10	Data	SDRAM	Display	Read of output buffer	19.9 MB/s

7.4.1   Algorithms and Reference Software Study

We started our research by studying the H.264/AVC standard and the Joint Video Team (JVT) meeting reports, which give specification, syntax, and algorithms of decoding tools. Then, we profiled the reference software, called Joint Model (JM), to get the results depicted in figure 7.5. Interprediction consumes most decoding time due to fractional pixel interpolation, variable block size, and multiple reference frames. Both deblocking filter (DF) and context-adaptive binary arithmetic decoder (CABAD) are also time consuming because of the large amount of filtering operations and complex conditional statements, respectively.

7.4.2   Specification Definition

We aim to develop a fully hardwired H.264/AVC main profile decoder IP with standard AMBA bus interface for SOC integration. We intend to real-time demonstrate full D1 resolution on an FPGA-based platform. For higher resolution, we port the system into an application-specific

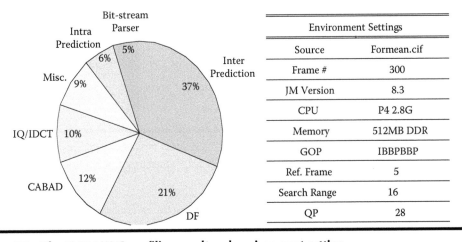

Figure 7.5   The H.264/AVC profiling result and environment setting.

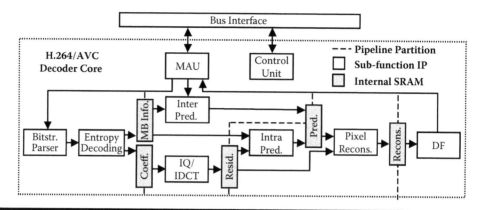

Figure 7.6   H.264/AVC decoder pipelined architecture.

integrated circuit (ASIC) flow to perform real-time decoding at 1,080 HD resolution (1,920 × 1,088 at 30 frames per second). Under all constraints above, we minimize hardware cost, internal/external memory usage and traffic, and required operating frequency.

7.4.3   Architectural Design

The whole decoding data path is divided into a four-stage, MB-level pipelined architecture according to data dependency among subfunctions, with internal memory used as buffer between adjacent stages. Our decoder, as illustrated in figure 7.6, employs a four-stage pipelined architecture and consists of nine components: control unit (CU), memory access unit (MAU), bit stream parser, entropy decoding, inverse quantization and inverse discrete cosine transform (IQ/IDCT), intrapredictor, interpredictor, pixel reconstructor, and deblocking filter (DF). CU interacts with external controller signals and conducts all subfunction IPs to direct the decoding flow, while MAU manages all external memory accesses through bus interface. The bit stream parser takes input bit stream through MAU and decodes all header information. Entropy decoding, including the context-adaptive binary arithmetic decoder (CABAD) and context-adaptive variable-length decoder (CAVLD), decodes slice data to generate a set of quantized coefficients and macroblock (MB) information. IQ/IDCT scales and inversely transforms the coefficients into residuals, while intra- or interpredictor produces prediction data according to MB information. Pixel reconstructor adds the residuals and prediction data together for the deblocking filter, which later writes filtered pixels to external memory through MAU.

7.4.4   Subfunction IP Design

We describe all the architectures of subfunction IPs and show how they operate.

7.4.4.1   Bit Stream Parser

The bit stream parser extracts parameters from input bit streams for other subfunction IPs' use. Though it contributes only 5 percent in profiling (shown in figure 7.5), implementing it via software run on an 150 MHz ARM926EJ-S CPU will on average introduce an 0.08 second overhead for each CIF frame, which is unacceptable in our target decoding system.

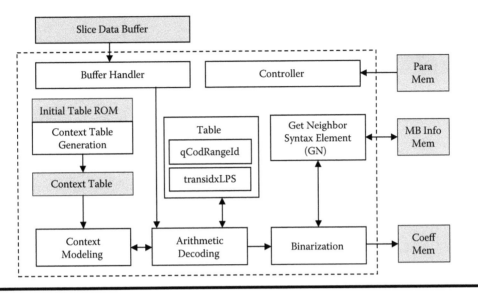

Figure 7.7   CABAD block diagram.

Our hardwired implementation recognizes a full set of syntax in main profile and produces a parameter every cycle. It eliminates the overhead mentioned while greatly simplifying the hardware-software interface at the expense of 20 K logic gates.

7.4.4.2 Entropy Decoding

Our entropy decoding, including CABAD and CAVLD, takes slice data as input and generates MB information and coefficient data.

7.4.4.2.1   Context-Adaptive Binary Arithmetic Decoder (CABAD)

Figure 7.7 depicts our CABAD block diagram. The buffer handler passes slice data to the arithmetic decoder, which looks up the transIdxLPS and qCodlRangeIdx tables. The context table generation module generates a context table for context modeling. The binarization module accesses neighboring syntax elements from the get neighbor module and writes coefficient data to the coefficient memory.

In the CABAD data path, context modeling, arithmetic decoding, and binarization all operate concurrently, resulting in a world-leading CABAD performance of 0.8 bin per cycle.

7.4.4.2.2   Context-Adaptive Variable-Length Decoder (CAVLD)

Figure 7.8 depicts our CAVLD block diagram. The buffer handler passes slice data to both the Exp-Golomb core for decoding MB information and the coefficient decoder for decoding coefficient. The coefficient decoder consists of a level decoder, a priority encoder with a shifter, and combinational tables. The two-stage pipelined level decoder decodes non-T1 (trailing one) and non-zero coefficients by arithmetic calculation. The priority encoder with a shifter calculates the

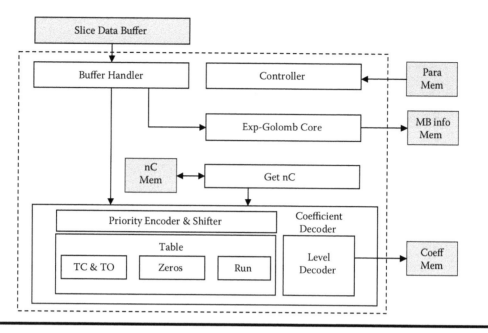

Figure 7.8 CAVLD block diagram.

leading zeros and left bits used as indices to the combinational tables for decoding TotalCoeff, TrailingOnes, totalzeros, and run_befores.

In case of I-frames with QP 28, our CAVLD needs an average of 350 cycles per MB. Besides, the area efficient coefficient decoder uses only 3.4 K logic gates without any table memories.

7.4.4.3 Inverse Quantization and Discrete Cosine Transform

The proposed IQ/IDCT takes the entropy-decoded, quantized coefficient data as input to reconstruct the residual data using all three kinds of inverse transforms: 4×4 Hadamard transform for luma DC block, 2×2 Hadamard transform for chroma DC blocks, and 4×4 core transform for residual blocks. It also supports both skip and direct modes to avoid unnecessary computation.

Figure 7.9(a) depicts our pixel dequantizer, which contains three operators: a scalar, an adder, and a shifter. We employ 16 dequantizers for a 4×4 block. Figure 7.9(b) shows our inverse transform architecture utilizing eight processing elements (PEs) to perform two-dimensional inverse DCT transform. The four PEs in the left-hand side process one row of coefficients at a time, while the four PEs in the right-hand side process one column at a time. The transpose matrix in between enables row-wise input and column-wise output of four pixels simultaneously.

7.4.4.4 Intraframe Prediction

The intraframe prediction module receives the residual data and generates prediction data. According to the prediction mode derived from MB information, a three-layer multiplexer always selects some pixels from neighboring MBs, as shown in figure 7.10, and passes them to the predicted value calculator that calculates the prediction data based on the standard defined equations.

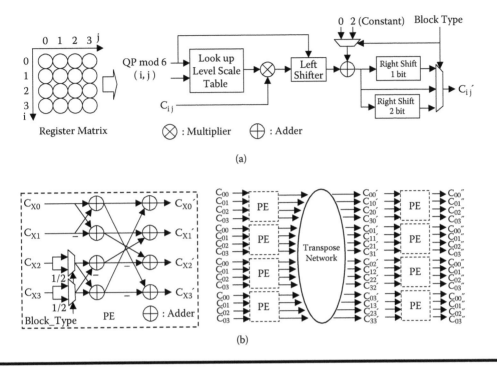

(a)

(b)

Figure 7.9   IQ/IDCT. (a) One-pixel dequantizer. (b) Inverse integer transform.

For area efficiency, we implement our design supporting all nine prediction modes for a 4×4 luma block using a three-layer multiplexer array and a predicted value calculator, leading to 80 percent area reduction compared with a straightforward design. It produces 256 luma and 128 chroma pixels of a single 16×16 MB in just 127 cycles.

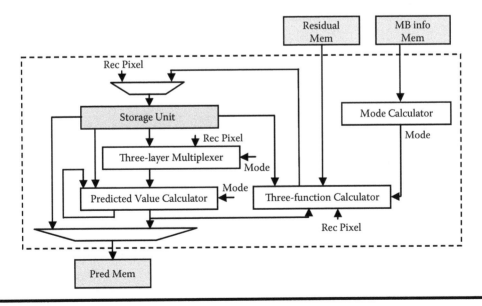

Figure 7.10   Intraframe prediction block diagram.

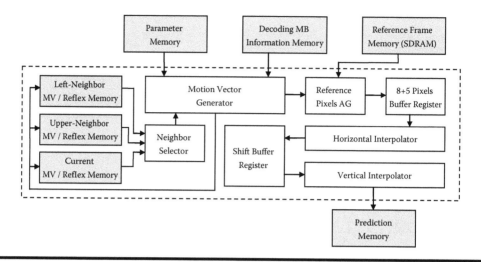

Figure 7.11   Interframe prediction block diagram.

7.4.4.5   Interframe Prediction

The interframe predictor (also called motion compensator) takes a portion of a reference frame according to the motion vector, which is obtained using decoded MB information and spatial-neighboring motion vectors, to produce a prediction block after interpolation. The process involves two functional units, as follows.

7.4.4.5.1   Motion Vector Generator

Motion vectors are obtained by combining motion vector difference (MVD) and motion vector prediction (MVP): MVD is decoded from the bit stream, while MVP is predicted from MVs of neighboring blocks of the block currently being decoded. We use three buffers to store these neighboring MVs, as shown in figure 7.11, to minimize the internal memory requirement and simplify the address generation so that the critical path is shortened.

7.4.4.5.2   Interpolator

A 4 × N interpolator with a shift register[23] is a popular solution to optimize interpolation in a 4 × N (N = 4, 8, 16) block. However, it introduces vertical and horizontal redundancies in reference pixel access when its target decoding system using 4 × 4 block pipelining meets the double z-scan ordering, as shown in figure 7.12(a). These redundancies increase the interpolation latency. Therefore, we propose three schemes to avoid them: MB pipelining, double-width shift register file, and filter-sharing scheme.

Using MB pipelining, instead of 4 × 4 block pipelining, enables us to interpolate a block in an optimal order by ignoring the double z-scan effect caused by entropy decoding, so that we could eliminate the vertical redundancy. However, it still suffers from horizontal redundancy if we still use a 4 × N interpolator, as shown in figure 7.12(b).

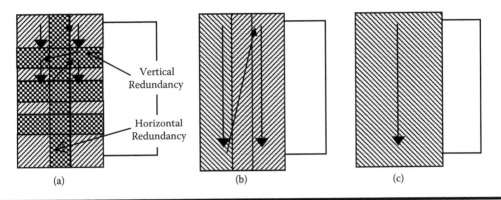

Figure 7.12   Redundancies reduction of MB pipelining and 8 × N interpolator. (a) 4 × 4 block pipelining with 4 × N interpolator. (b) MB pipelining with 4 × N interpolator. (c) MB pipelining with proposed 8 × N interpolator.

It looks reasonable now to use an 8 × N or 16 × N interpolator by a double-width or quad-width shift register file. Nevertheless, it needs two or four times the number of filters. Fortunately, our filter-sharing scheme, which achieves near 100 percent utilization of filters depending on the memory access time of every different block type, helps reduce the number of filters required. Finally, we come up with an 8 × N interpolator using six filters and a double-width shift register file. It can process an 8 × N block directly and reads horizontal-redundant reference pixels only when the block is 16 × 16 or 16 × 8. That makes our interpolator near optimal in any decoder system equipped with a 32-bit wide memory bus.

7.4.4.6   Deblocking Filter

The deblocking filter takes reconstructed frame as its input and generates filtered frame as its output. Figure 7.13 depicts our proposed architecture. The five-stage pipelined edge filter filters boundary edges using neighboring pixels from local mem 0 and local mem 1, and boundary strength from BS/threshold generator. Afterwards, the write-back unit writes the result to external memory through the MAU. Our proposed DF needs 14 cycles to read necessary coding information for the first pair of 4 × 4 blocks, 192 more cycles for the remaining MB, and another 8 cycles to flush the pipeline. Thus, it requires 214 cycles to decode any MB except for 32 extra cycles for the rightmost MB in a row.

7.4.4.7   Memory Access Unit

The memory access unit manages all external memory accesses, such as reference frame, display frame, and input bit stream. For external board-level memory, a high-density, low-cost memory is used. A popular choice is SDRAM, which requires 5 ~ 20 cycles to initialize a sequential access, called a burst initial. Therefore, an efficient external memory management scheme should reduce the number of required burst initials as much as possible. Figure 7.14 shows our proposed MAU architecture composed of four components: bit stream reader, reference pixel reader, reference frame writer, and display writer.

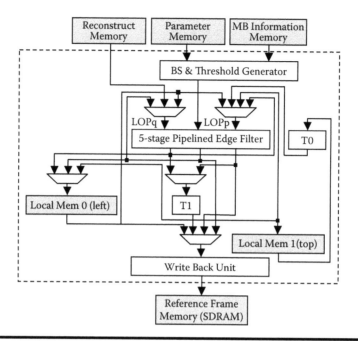

Figure 7.13 **Deblocking filter block diagram.**

7.4.4.7.1   Bit Stream Reader

The bit stream reader sequentially reads in the compressed bit stream from the SDRAM and passes it to the bit stream parser. However, unbalanced production and consumption of input bit stream makes the bit stream parser generally require a burst initial in every entry of external access. We therefore use a 64-entry buffer and allow the bit stream reader to fetch more data only if there are more than 32 empty data entries. This scheme saves 96.9 percent of burst initial cycles.

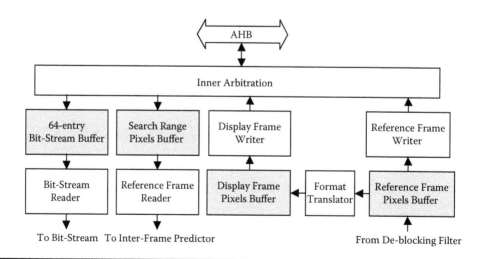

Figure 7.14 **Memory access unit block diagram.**

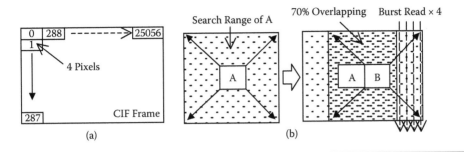

Figure 7.15 **(a) Optimal format of reference frame. (b) The overlapping between two neighbor MBs.**

7.4.4.7.2 Reference Frame Reader

The reference frame reader reads in pixels for the interpolator of the interprediction unit to calculate subpixel predictions. The number of burst initials varies according to different block sizes within the current MB, which means that when 4×4 subblocks appear frequently, the interpolation engine needs to request burst initials more often. We thus use a buffer to save the whole search range of the currently decoding MB to minimize the amount of burst initials. Further, the current MB only needs to load 30 percent of the nonoverlapped search range, as shown as figure 7.15(b).

Rearranging reference frame format also helps. Figure 7.15(a) shows the optimal reference frame format, which leads to only four (16 pixels per MB/4 pixels per 32 bits of data) burst initials needed per search range fetching.

7.4.4.7.3 Display Writer and Reference Frame Writer

The DF writes the decoded results into two different SDRAM locations, one for reference frame and the other for display frame. The reference frame writer writes each filtered MB to the reference frame buffer using 10 burst initials, while the display writer translates the decoded MB format from 4:2:0 to 4:2:2 and writes it out using 16 burst initials. Note that the number of burst initials could be further reduced by more buffers. For example, the display writer requires only 3.2 burst initials, on average, if a buffer of five MBs of pixels is used.

7.4.4.8 Summary of Subfunction IP Innovations

Every subfunction IP in our decoder pipeline has unique characteristics. We design innovative architectures to achieve high performance (low latency) and small area. Table 7.2 summarizes their features.

7.4.5 Hardware Integration and Verification

After designing all subfunction IPs, we integrate them in hardware level. During hardware integration, we eliminate bubble cycles and schedule internal memory accesses carefully to increase throughput. During verification, we ensure the correctness of the integrated decoder by comparing its results with golden patterns generated by the reference software.

Table 7.2   Summary of Subfunction IP Innovation

Module	Features	Results
Bit stream parser	Pure hardwired	Simplified HW/SW interface
CABAD	Parallel context modeling and arithmetic decoding Single context memory by context forwarding	0.8 bin per cycle
CAVLD	Memory-free table lookup scheme	24% gate-count reduction (compared with Wang et al.[23])
Interprediction	Interpolate $8 \times N$, $4 \times N$ blocks directly to avoid redundant filtering 6 luma filters and 2 chroma filters	62% cycle-count reduction (compared with Chang et al.[5])
Intraprediction	Shared common filter operation with shift register	127 cycles per MB
Deblocking filter	5-stage pipelined filter with forwarding path Near-optimal cycles with single filter	214 cycles per MB

7.4.5.1   Hardware Integration

During hardware integration, we adopt an elastic pipelined architecture and a prioritized memory access scheduling to enhance the decoder's throughput. The elastic pipelined architecture allows a new pipeline stage to start right after all subfunction IPs in the previous stage finish their computation. Further, we give higher priority to critical subfunction IPs when they access shared resources, that is, the internal memory. For example, we give entropy decoding higher priority when decoding I-type frames because it generally requires more cycles.

To implement these two techniques above, we use a two-layer FSM architecture in our control unit, as illustrated in Figure 7.16. In the bottom layer, each subfunction FSM handshakes with the main control FSM while managing its own subfunction IP. In the upper layer, the main control

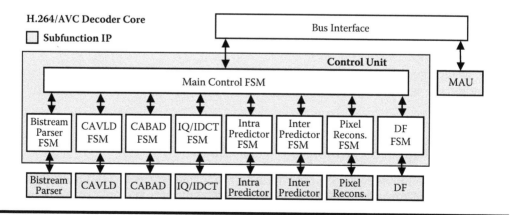

Figure 7.16   Two-layer control unit architecture.

FSM coordinates all bottom-layer FSMs and communicates with the software via the external bus interface.

7.4.5.2 Hardware Verification

During hardware verification, we first construct our software model from the JM reference software and generate all necessary golden patterns corresponding to the buffer memory contents of our decoder. During the verification of subfunction IPs, each of them must have an output identical to that of the golden pattern, given the same input.

During early stages of integration, we compare most memory contents with the golden pattern. It helps us to quickly narrow down the potential area of design errors.

Later on, we concentrate on the outputs of entropy decoding and display frame, to get faster simulation efficiency, while still preserving the information about whether the bugs are located at preentropy or postentropy areas.

In the final stages, we use the iPROVE emulator from Dynalith Systems[11] to run large set of test patterns within a much shorter period.

7.4.6 System Integration and Verification

After we finish designing the H.264/AVC hardwired decoder core, we integrate it into our SOC platform and verify it with a real-time decoding demo.

7.4.6.1 System Integration

Given the verified decoder core, our system integration task consists of two parts: bus protocol wrapping and application software development, as shown in figure 7.17.

For the former, we first connect a master wrapper to the MAU to enable the internally arbitrated data requests to access the system bus. Then another slave wrapper is connected to the CU so that software could directly write its commands to registers inside the decoder core.

We next show the application software's controlling flowchart in figure 7.18.

The application program first initializes the whole system, including all devices and their controlling registers. After getting a new input sequence's file handle, it instructs the DMA to load

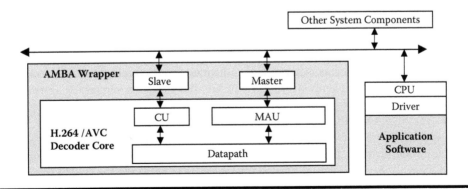

Figure 7.17   System-level integration.

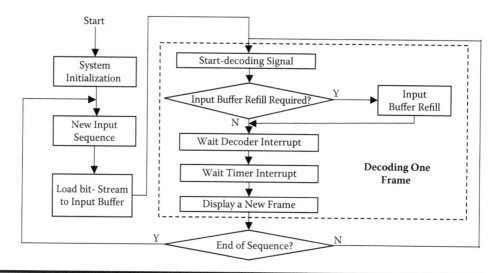

Figure 7.18 Software flowchart

some bit stream from the compressed video file inside the SD card into the input buffer on the SDRAM. Getting ready to enter the main decoding loop, it issues a start signal to the H.264/AVC decoder, checks if the input buffer needs a refill, and then waits for the decoder's interrupt, signaling the completion of a frame's decoding. Afterwards, the application program waits for another interrupt from the system timer, which triggers every 1/30 second to synchronize the decoding/display rate before changing the displaying address to the output buffer containing the newly decoded frame. The program then continues to decode the next frame again and again until it reaches the end of the sequence. Getting another input sequence's file handle, the program proceeds to decode a new sequence.

Note that we put all data-centric controlling steps (bit stream parsing) and massive data transfers into hardware so that we need little CPU intervention along the decoding process. That way the hardware decoder must have a higher degree of integration, rewarded by a higher overall performance.

7.4.6.2 System Verification

We divided the whole system verification into two subprocesses from the beginning, as shown in figure 7.19. These two subprocesses could go in parallel to reduce the overall verification time.

For the hardware side, shown in figure 7.19(a), we build a complete, simulation-based environment composed of HDL behavioral models. These models simulate the interface and overhead of our target system, including bus behavior and contention. For example, the CPU model not only coordinates multiple components in the system, but also keeps on requesting instruction/data memory. The display-controller model always tries to grant the bus to read the output buffer in SDRAM. The SDRAM model simulates the variable-latency property of SDRAM. The input file loading model represents the behavior of the SD card, which dynamically loads the input bit stream with a high-latency and low-bandwidth property. These overheads modeled above could be turned off in the beginning stage of system verification to boost the simulation speed, but they should be turned on later to have a closer simulation environment to the target system. Implementation details are described in section 7.6.3.2.

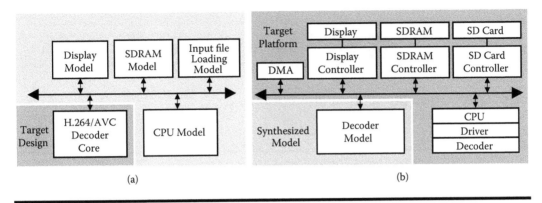

Figure 7.19 **System verification. (a) Hardware side: Simulation-based environment for hardware verification. (b) Software side: Platform-based environment for software verification.**

For the software side, shown as figure 7.19(b), we verify the application software directly on the target platform using a synthesized model. This HDL model is synthesized into the FPGA, with an identical interface as our actual decoder: it takes the starting command from application software, does a lot of external SDRAM accesses, similar to the decoder, and issues an interrupt back to the software. This model enables us to verify the correctness of application software without actually verifying the decoder itself.

When the above two processes are completed, the only steps left are to combine the preverified hardware and software together and to add a big-endian/little-endian translation circuit to the hardware. The translation circuit is necessary because the simulation environment reads the input file as big endian, while the target platform treats it as little endian. Eventually, this stage takes a much shorter time than the two previous processes.

7.4.7 System Bottleneck Analysis and Performance Optimization

The system has two bottlenecks. One is computational latency, and the other is external data traffic. After analyzing and optimizing both, we further utilize some system-level techniques to improve the overall performance.

7.4.7.1 Computation Latency

In our elastic pipelined architecture, the latency of a pipeline stage is decided by the subfunction IP that takes the most cycles to process its current MB, which is often different. We analyze the distribution of pipeline stage dominance for each subfunction IP as shown in figure 7.20. We can see CABAD dominates the most stages in I-type frame decoding, while interpredictor dominates in P-type and B-type frames. An interesting part is that IQ/IDCT and DF also sometimes dominate the latency of pipeline stages.

7.4.7.2 External Data Traffic

Our fast-running decoder needs the assistance of an efficient input bit stream loader, write/read access of intermedium reference frames, and output for display. These external memory accesses

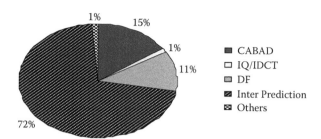

Figure 7.20 **Computation latency distribution: (a) I-frame, (b) P-frame.**

managed by the MAU described in section 7.4.4.7 have a major impact on the system performance. Table 7.3 shows bus cycle reduction contributed by the MAU in different video resolutions. Among them, CIF and D1 versions are obtained through prototype management, and the HD version is projected accordingly.

Figure 7.21 analyzes the distribution of all bandwidth reductions for the D1-resolution version of table 7.3. Note that in table 7.3, the straightforward MAU differs from the improved one in that the later further utilizes the data-reusing technique presented in section 4.4.7.

Observe that among the 89 percent reduction, most are originated from "display buffer write," "reference frame read," and "reference frame write" accesses, which are issued by our decoder and managed by the MAU. Moreover, MAU overlaps (parallelizes) decoding with external data transfer to remove the dependency between them so that the decoder need not stall for requiring external data.

As for "display buffer read" and "SD card to SDRAM," these are fixed factors in this analysis.

7.4.7.3 System-Level Techniques and Their Overall Impact

We have two system-level options: cache enabling and dynamic input bit stream loading. Each of them has a different impact on overall performance.

Enabling cache will reduce the CPU's bus access to instruction and data memory, and thus preserve more bus bandwidth for the hardwired decoder. According to a D1 version implementation,

Table 7.3 **Bus Cycle Requirement in Different Resolution and External Memory Access Management Scheme**

Resolution	External Memory Management Scheme	Bus Cycle Required (M Cycle/s)
CIF 352 × 288	W/O MAU	111.3
	W/ MAU	11.1
D1 720 × 480	W/O MAU	355.0
	W/ MAU	37.8
HD 1920 × 1088 (projection)	W/O MAU	2290.7
	W/ MAU	230.1

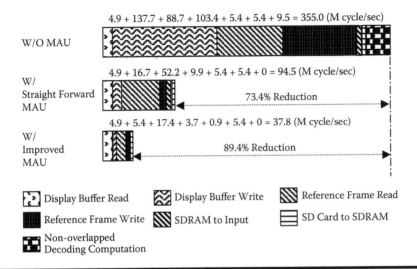

Figure 7.21 Bus cycle consumption analysis in D1 resolution.

cache reduces 89.9 percent of CPU's bus request on average, which contributes to 17.5 percent of system performance improvement, as shown in figure 7.22.

It is necessary to dynamically load input bit stream from the storage device into limited-size external memory, because we cannot always load the whole video sequence file. The nature of this data transfer is high latency, for the bit stream is usually stored in high-latency devices, such as SD cards, but low bandwidth, because the bit stream is already encoded into a smaller size. Fortunately, we can choose how and when it is loaded. For example, we can transfer them by CPU or DMA, and schedule them while decoding or between the decoding of two consecutive frames. Figure 7.23 summarizes different overheads when the dynamic loading is required in different times and by different means. It shows that DMA transfer overlapped with decoding time results in the smallest overhead.

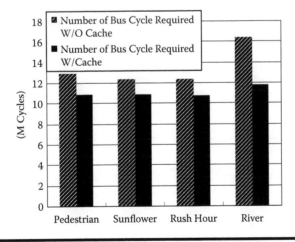

Figure 7.22 Required bus cycles to decode a D1 frame with or without cache.

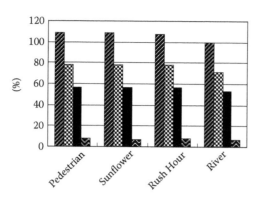

Figure 7.23 Dynamic loading overhead in different schemes.

7.5 Result

In this section, we show the resource usage of our demonstrable decoding system pictured in figure 7.24 and analyze the overall performance in the subfunction IP, HW integrated, and system integrated levels. Finally, we list some records about the whole project development.

7.5.1 Resource Usage

We described our decoder in Verilog HDL and synthesized it toward both TSMC 130 nm CMOS technology and Altera FPGA library. Table 7.4 shows the detailed gate counts of every subfunction IP. To sum up, the whole system costs 238.5 K logic gates in the ASIC version at 166 MHz, and 38,550 lookup tables (LUTs) in the FPGA version at 24 MHz. Figure 7.25 illustrates the internal and external memory requirements for the whole system. Compared with the CIF

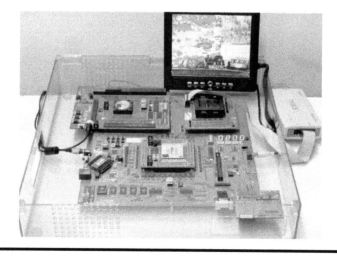

Figure 7.24 H.264/AVC demonstrable decoding system.

TABLE 7.4 ASIC and FPGA Cell Area Result

Frequency	166 MHz	24 MHz	
Technology	TSMC 130 nm	Altera FPGA	
Subfunction IP	*Gate Count*	*No. of LUTs*	*No. of Reg. Bits*
Bit stream parser	20K	4,413	588
CABAD	35.4K	7,335	2,030
CAVLD	17.5K	3,814	637
Intraprediction	20.8K	7,104	575
Interprediction	35K	5,449	2,230
IQ/IDCT	43.5K	5,781	1,188
Pixel reconstruction	2.7K	375	274
Deblocking filter	20.0K	3,637	1,407
Control unit	1.9K	793	93
MAU	41.7K	5,012	1,100
Total	238.5K	38,550	9,339

version, the D1 version consumes more internal and external memory due to non-MB-based memory and buffer for higher resolution, respectively.

7.5.2 *Performance Analysis*

This section describes the performance of our implementation in a bottom-up order, including the performance of subfunction IPs, the individual H.264/AVC hardware decoder, and finally the whole system.

For subfunction IPs, we synthesized them using Synopsis Design Compiler targeting toward the TSMC 130 nm library under the timing constraint of 6 ns, equivalent to 166 MHz in frequency. We simulate each of them using Cadence NC-Verilog with Sun Solaris workstation for five different CIF-resolution QP-28 sequences: carphone, foreman, mobile, news, and tempete. We show the minimum, maximum, and average number of cycles required to process an MB of

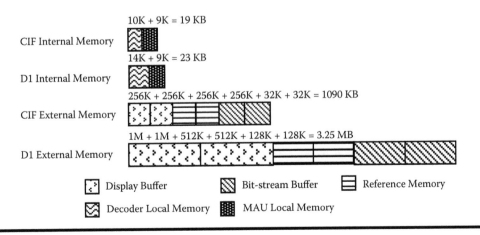

Figure 7.25 Internal memory and external memory usage.

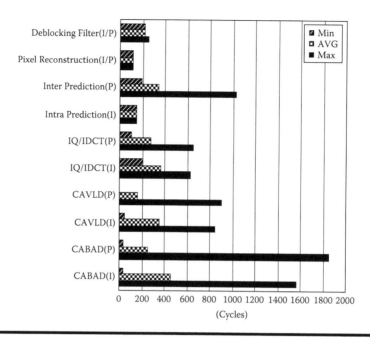

Figure 7.26 Maximum, average, and minimum number of cycles required to process an MB for each subfunction IP.

I-type or P-type frames by different IPs in figure 7.26, in which we can see three kinds of behaviors. First, for CABAD, CAVLD, IQ/IDCT, and interprediction, the required cycle numbers for an MB vary greatly, depending on the MBs' properties, including the number of zero coefficients, skip modes, or block sizes. Second, the deblocking filter usually takes 214 cycles for an MB except when it encounters picture edges; as for intraprediction and pixel reconstruction, they have a fixed number of cycles for processing any MBs. We summarize each individual IP's performance in terms of frame per second (FPS) in CIF resolution in figure 7.27.

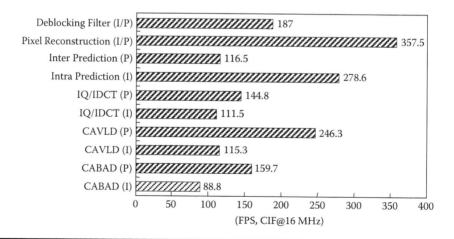

Figure 7.27 Performance of each subfunction IP in FPS for CIF resolution.

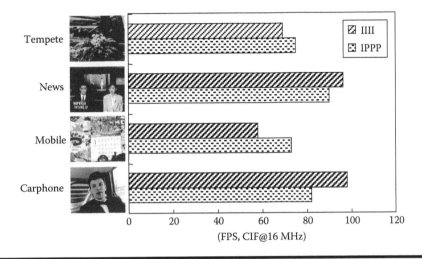

Figure 7.28   Performance of hardwired decoder in FPS for CIF resolution.

At the hardware decoder level, we integrate all subfunction IPs, connect the integrated decoder design to the system model shown in figure 7.19(a), and simulate it with the same machine and sequences as those of subfunction IPs. Figure 7.28 summarizes the performance of the hardwired decoder in terms of FPS under CIF resolution at 16 MHz.

Finally, we integrate the decoder into the 16 MHz target platform, enable the cache, and apply the best scheduling scheme, mentioned in section 7.4.7.2. Figures 7.29 and 7.30 summarize the performance of CIF and D1 versions, respectively, for four different sequences and three different GOP. We can see that although system performance still varies with different sequences, it generally decodes more FPS in I-frames than in P-frames, because in the latter case it needs more bus bandwidth for transferring the data of reference frame.

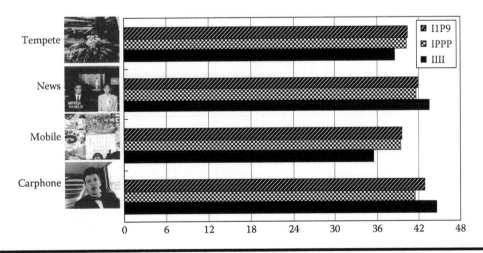

Figure 7.29   Performance of the 16 MHz demonstrable H.264/AVC decoding system in CIF resolution.

River
Rush Hour
Sunflower
Pedestrian
I1P9
IPPP
IIII
0
2
4
6
8
10
12
14
16
18

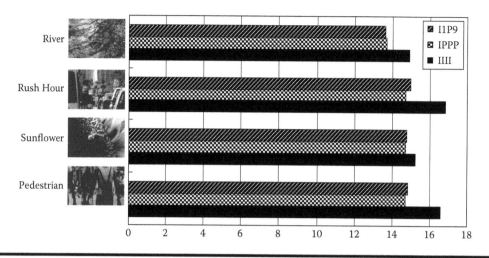

Figure 7.30 Performance of the 16 MHz demonstrable H.264/AVC decoding system in D1 resolution.

7.5.3 Project-Related Records

7.5.3.1 Project Gantt Chart

There are a total of 12 graduate students involved in this project. We use three Sun Solaris workstations for simulation, one emulator, and one platform for system verification. Figure 7.31 shows the Gantt chart of the time allocation of our project.

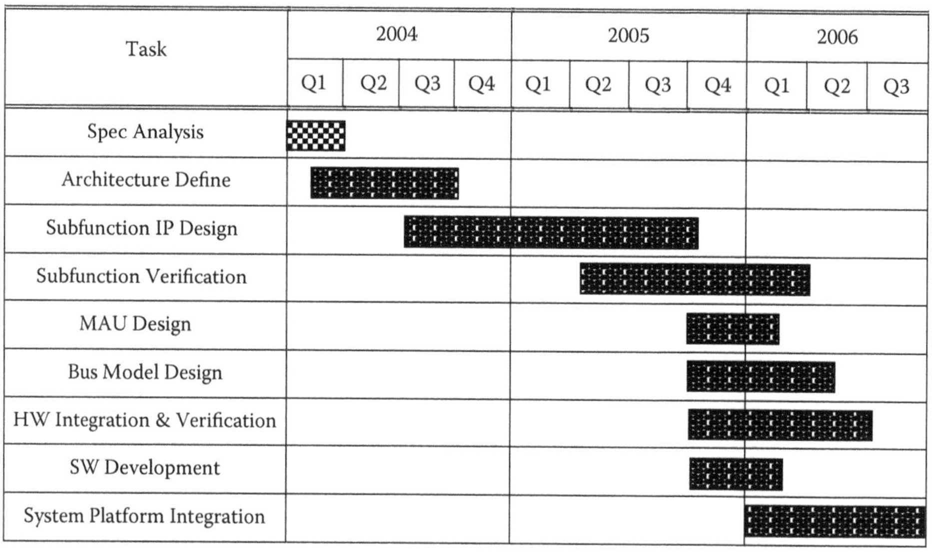

Task
2004
2005
2006
Q1 Q2 Q3 Q4 Q1 Q2 Q3 Q4 Q1 Q2 Q3
Spec Analysis
Architecture Define
Subfunction IP Design
Subfunction Verification
MAU Design
Bus Model Design
HW Integration & Verification
SW Development
System Platform Integration

Figure 7.31 Project Gantt chart.

Table 7.5 Bug Record

Functional bug number in different verification stages	
RTL simulation	16
Gate-level simulation	4
FPGA prototyping	10
Other	2
Functional bug number in different design blocks	
Individual subfunction IP	10
MAU	10
Platform	6
SW	4
Other	2

7.5.3.2 Bugs Found

Although each subfunction IP was verified, we still detect many functional bugs during integration. We trace our verification history by a version control tool for HW integration and verification, and show the bug trace record during different verification stages and design tasks in table 7.5.

7.6 Lessons Learned

In this section, we share our experience on key decisions and design challenges through the whole project. These will cover platform selection, RTL development, verification, and early performance modeling.

7.6.1 Platform Selection

Before designing the system, selecting a suitable platform is crucial and needs a lot of study work. On the one hand, bus bandwidth and storage spaces will impact on design decisions like internal and external memories partition; on the other hand, application API and proven tool chains will enormously reduce the system integration cycle. We share design experiences on several commercial platforms, all of which are constructed from typical AHB or multilayer AHB architectures.

7.6.1.1 Typical AHB and Multilayer AHB

Typical and multilayer AHB are two widely used AMBA architectures. In a typical AHB bus topology, as shown in figure 7.32(a), only one master and one slave can commence with their transfer at a time after the master is granted by the bus arbiter. All slaves share the same address and data buses, making them easily suffer from the bus contention problem. For multilayer topology, there are parallel channels between multiple masters and slaves through the interconnect matrix, as shown in figure 7.32(b). It hence provides more flexible bus bandwidth for applications that require heavy data traffics.

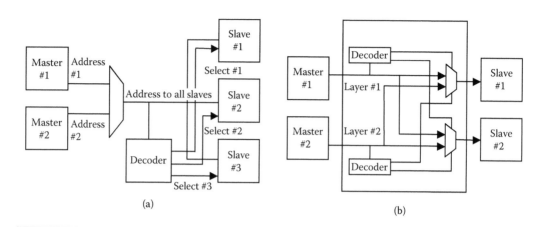

(a)   (b)

Figure 7.32 Typical AHB and multilayer AHB buses.

7.6.1.2 Platform Comparison

In this section we compare the four different platforms we used in the past three years. Table 7.6 lists the major factors that affect system design, including bus architecture, effective recovery transfer, SDRAM access, multimedia application API, and verification supports.

EPXA10[7] is the first Altera FPGA platform supporting the AMBA system. It packages one hard core, named Stripe, and an FPGA into a chip. Stripe included an ARM922T processor and an AMBA backbone with peripheral I/O supports. Moreover, it also has a tool chain that covers system-level integration, hardware design, software design, and cosimulation. However, it produces a lot of bubble cycles because it requires an FPGA-AHB bridge to access the AHB. This is a serious disadvantage for H.264/AVC's high bandwidth requirement.

Both Integrator[2] and Versatile[4] are from ARM. ARM Integrator has an AHB-lite bus architecture that only allows CPU to be the bus master. Therefore, custom hardware with a DMA interface is forbidden, which is inefficient for our application. ARM Versatile is a powerful platform with multilayer bus and rich multimedia I/O interface supports. In its AHB recovery transfer mode, such as split and retry, a master can suspend an unfinished transfer to start another transfer for system utilization balance. In addition, we can observe AHB system signals via built-in AHB monitor to catch the runtime bus situation. However, the I/O controllers and API require a long time to develop, and need to run various applications to firm up the basic infrastructures.

Table 7.6 Platform Comparisons

	Altera EPXA10	ARM Integrator	ARM Versatile	GUC UMVP-2000
Major bus architecture	AHB	AHB-lite	AHB multilayer	Two AHB
Recovery transfer	Support	—	Support	—
SDRAM access	AHB bus	AHB bus	N/A	AHB bus + dedicated FPGA port
Multimedia software API	—	—	—	Support
HW/SW coverification	Support	—	N/A	—
On-board SW verification	Support	Support	Support	Support
On-board HW verification	—	—	AHB monitor circuit	Self-diagnosis circuit

GUC UMVP-2000 has an AMBA backbone with two AHB buses. This architecture allows two AHB buses working at the same time: the ARM processor manipulates I/O operations through AHB1, while custom hardware accesses SDRAM through AHB2. It is helpful to balance the system bandwidth and increase overall system utilization.

Through the above comparisons, we choose the GUC platform as our target multimedia infrastructure. It was evaluated offering enough system bandwidth for our real-time H.264/AVC decoding at D1 resolution and associated preverified multimedia API to speed up our system development cycle.

7.6.2 RTL Development

This section addresses the design consideration of three subfunction IPs.

7.6.2.1 CABAD

CABAD is one of the computation bottlenecks in our decoding system, because it has dependencies not only on input patterns but also on previously decoded results. So the design strategy for CABAD is to analyze the dependency and reveal potential parallelism to reduce the cycle count. Its complex control logic also brings the greatest challenge in subfunction IP verification.

7.6.2.2 Interprediction

After analysis of the whole decoding system, we find that memory bandwidth of interprediction dominates the decoding performance when decoding P-type or B-type frames. Accordingly, the design challenge is fetching reference frame pixels efficiently for different subblock types. Among all interpolators, $16 \times N$ interpolators fully reuse all potential reusable reference frame pixels within a block, at the expense of higher hardware cost. Therefore, it is a good interpolator candidate for decoding of higher resolutions where 16×16 blocks would be most likely to appear. The format of the reference frame itself can also be an important topic. First, it influences the interframe predictor's external memory accessing behavior, which dominates the predictor's processing latency even with an optimum interpolator. Furthermore, DF also accesses the reference frame with another different behavior, so that the individual optimum format for each is not the same. In our case, we choose the format mentioned in figure 7.15(a) as a global optimization, which is the optimum format for interpredictor, while having little overhead for the deblocking filter.

7.6.2.3 Deblocking Filter

The design challenge of the deblocking filter is adjusting filtering order to save filtering cycles as well as local memory usage and memory traffic. It involves analyzing the dependency of the filtering process at both the block level and pixel level. Figure 7.33(a) shows our filtering order of luma block. At the 4×4 block level, instead of filtering all vertical edges before all horizontal edges defined by the standard, our DF filters the edges from left to right to save memory usage. Furthermore, we take advantage of pixel-level dependency to save filtering cycles. Taking filtering steps 5 and 6 shown in figure 7.33 as an example; filtering step 6 requires block LT0 and the transpose of block B0. However, pixels of block B0 are still in the pipeline when the filtering of step 6 begins. Therefore, we add forwarding logic in our five-stage pipeline filter, which enables line 5 to be filtered right after line 4 without any stalls.

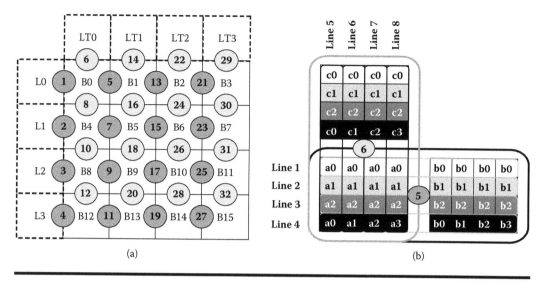

Figure 7.33   Deblocking filter order. (a) Luma filter order. (b) Filtering order in steps 5 and 6.

7.6.3   Verification

This section shares some experience that could help to reveal and resolve bugs in the early verification stage while developing an H.264/AVC decoding system or other complex digital systems.

7.6.3.1   Mismatch between RTL and Gate-Level Simulation

During the verification process, there are times when RTL and gate-level simulation mismatches. This is mainly caused by the different responses when a case statement gets an unknown controlling input signal: it gets into the "else" or "default" branch in RTL simulation and generates an unknown propagation in the gate level. These kinds of controlling signals, in H.264/AVC design, are mainly generated by entropy decoding, that is, CABAD and CAVLD. Therefore, special care should be taken of these signals by assertion-based verification.

7.6.3.2   Simulation-Based Verification Environment and Modeling

An ideal simulation-based verification system, according to former experiences, should provide at least three kinds of modeling to fully reveal system-level problems: interface modeling, bus contention modeling, and program flow modeling. Respectively, we design our environment as shown in figure 7.34.

For interface modeling, we constructed an RTL AMBA backbone with a memory-mapping scenario identical to that of our target system. An SDRAM model is also plugged into the environment to simulate the variable-latency behavior of a real SDRAM. This part helped us locate most bugs about the interface and memory-mapping protocol.

For program flow modeling, we model the program flow by using only four HDL subroutines in the testbench: bus_read, bus_write, file_read, and file_write, where the first two model all bus transactions and the latter two model storage devices' I/Os. This technique helps us to find many

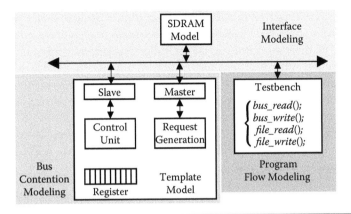

Figure 7.34 Implementation of simulation-based environment by modeling.

potential problems related to controlling signals, such as the start-decoding signal, buffer-refill signal, and interrupts.

For bus contention modeling, we have to model the bus bandwidth consumption of some other system components. We achieve this using a template model by configuring its control unit and request a generation circuit to set its bandwidth consumption according to the component's real requirement. This helped us to locate some critical bugs that only show up in bus contention conditions.

7.6.4 Performance Modeling Challenges

For early system-level analysis, we need to know about the required data bandwidth and the computation latency of the modeling target. In our case, the computation time usually varies greatly for different MBs. Worse still, as figure 7.20 shows, the critical pipeline stage of each MB is not always the same. The latency of these complex decisions done by the H.264/AVC decoder core cannot be easily estimated without architectural details, and thus is difficult to be modeled accurately during early design stages.

One possible approach is to feed back those performance numbers of RTL designs statistically to all subfunction models. For example, depending on different input sequences, we sometimes get an MB with skip mode, in which CABAD requires much fewer cycles to decode it. If we feed back this number to higher-level models, another pipeline stage will interactively take over the latency dominance from CABAD in this stage, and gets a more accurate latency for this stage. That way, we could improve the accuracy of behavioral models while maintaining the fast simulation speed for evaluating different high-level configurations.

7.7 Summary and Future Work

We have presented an H.264/AVC main profile video decoder prototype. Implemented in pure hardware, our system needs no software intervention except signaling the beginning and end of decoding a frame or sequence. Additionally, our elastic four-stage pipelined architecture and optimal design of every stage result in us being able to decode D1 video at 16 FPS running at only 16 MHz on a multimedia SOC platform. This is the lowest operating frequency in open literature.

This project has demonstrated the viability of using an existing industry platform for designing a complex system such as the H.264/AVC decoder in an academic environment. This process should be valuable to those who want to design other complex systems such as next-generation wireless communication systems.

In the future, we will work on extending the prototype to ultra-high-definition resolution (16 times that of present HDTV). Additionally, we will design an encoder system so that we can make a codec for applications such as videoconferencing.

Our H.264/AVC decoding system can reduce power consumption by more than 20 times. Consequently, with the growing demand for high-quality mobile video, the H.264/AVC decoding system has great potential for the mobile video industry.

Acknowledgments

The authors make grateful acknowledgments to C. L. Chen, Mark Liu, Stan Liao, Felix Yang, Kata Yen, and Cica Chen from Global Unichip Corporation. They exert great efforts to help us build our H.264/AVC decoding system. The authors also thank Alice Tsay and Joe Lin for help in editing this paper. Financial support from the National Science Council and Ministry of Economic Affairs of Taiwan, Taiwan Semiconductor Manufacturing Company (TSMC), and Industrial Technology Research Institute (ITRI) are greatly appreciated.

References

1. AMBA. 2006. 2.0 *Specification*. http://www.arm.com/products/solutions/AMBA_Spec.html.
2. ARM9. 2006. *Family*. http://www.arm.com/products/CPUs/ARM920T.html.
3. ARM9E. 2006. http://www.arm.com/products/CPUs/ARM926EJ-S.html.
4. ARM. 2006. *RealView Versatile family*. http://www.arm.com/products/DevTools/VersatileFamily.html.
5. Chang, H. C., Lin, C. C., and Guo, J. I. 2005. A novel low-cost high-performance VLSI architecture for mpeg-4 avc/h.264 cavlc decoding. In *Proceedings of IEEE International Symposium on Circuits and Systems*, 6110–13. Vol. 6.
6. Chen, T. W., Huang, Y. W., et al. 2005. Architecture design of H.264/AVC decoder with hybrid task pipelining for high definition videos. In *Proceedings of IEEE International Symposium on Circuits and Systems*, 2931–34.
7. Excalibur. 2006. http://www.altera.com/literature/lit-exc.jsp.
8. GUC UMVP-2000. 2006. http://www.globalunichip.com/ip-03.html.
9. Hu, Y., Simpson, A., McAdoo, K., and Cush, J. 2004. A high definition H.264/AVC hardware video decoder core for multimedia SoC's. In *Proceedings of IEEE International Symposium on Consumer Electronics*, 385–89.
10. Huffman, D. A. 1952. A method for the construction of minimum redundancy codes. *Proceedings of Institute of Electronics and Radio Engineers* 40:1098–101.
11. iPROVE. 2006. http://www.dynalith.com.
12. Iverson, V., McVeigh, J., and Reese, B. 2004. Real-time H.24-AVC codec on Intel architectures. In *International Conference on Image Processing*, 757–60.
13. Joint Video Team. 2003. Draft ITU-T recommendation and final draft. International standards of Joint Video Specification.
14. Kim, S. D., Lee, J. H., et al. 2006. ASIP approach for implementation of H.264/AVC. In *Proceedings of the 2006 Conference on Asia South Pacific Design Automation*, 758–64.

15. Lee, S. H., Park, J. H., et al. 2006. Implementation of H.264/AVC decoder for mobile video applications. In *Proceedings of the 2006 Conference on Asia South Pacific Design Automation*, 120–21.

16. Lin, C. C., Guo, J. I., et al. 2006. A 160kGate 4.5kB SRAM H.264 video decoder for HDTV applications. In *Proceedings of International Solid-State Circuits Conference*, 660–62.

17. Lin, H. C., Wang, Y. J., et al. 2006. Algorithms and DSP implementation of H.264/AVC. In *Proceedings of the 2006 Conference on Asia South Pacific Design Automation*, 742–49.

18. Lin, T. A., Wang, S. Z., et al. 2005. An H.264/AVC decoder with 4×4 block-level pipeline. In *Proceedings of IEEE International Symposium on Circuits and Systems*, 1810–13.

19. Ramadurai, V., Jinturkar, S., et al. 2005. Implementation of H.264 decoder on Sandblaster DSP. Paper presented at IEEE International Conference on Multimedia & Expo, Amsterdam, the Netherlands.

20. Joint Model. 2006. http://bs.hhi.de/~suehring/tml.

21. Shih, S. Y., Chang, C. R., and Lin, Y. L. 2006. A near optimal deblocking filter for H.264 advanced video coding. In *Proceedings of the 2006 Conference on Asia South Pacific Design Automation*, 170–75.

22. Talla, D. 2006. DaVinci technology for digital video applications. Paper presented at 6th International Symposium on Multiprocessor Systems-on-Chips, Estes Park, CO.

23. Wang, S. Z., Lin, T. A., et al. 2005. A new motion compensation design for H.264/AVC decoder. In *Proceedings of IEEE International Symposium on Circuits and Systems*, 4558–61.

Chapter 8

H.264/AVC Video Codec Design
A Hardwired Approach

Tung-Chien Chen, Chung-Jr Lian, Yu-Wen Huang,
and Liang-Gee Chen

Contents

Keywords

video codec, video coding, compression, hardwired architecture, low power, H.264, AVC, MPEG

Abstract

Video compression is the key enabling technology for multimedia communications. Without compression, digitized raw video data requires too much bandwidth and storage. Usually, the higher the compression ratio we want, the more complex the video coding algorithms that we have to apply. To realize complex video encoding or decoding, we need powerful VLSI circuits to do the computations in real-time.

The Advanced Video Coding (AVC; also known as ITU-T H.264 and MPEG-4 Part 10) standard was developed by the Joint Video Team (JVT), which consists of members of both the ITU-T Video Coding Experts Group (VCEG) and the ISO/IEC Motion Picture Experts Group (MPEG). AVC significantly outperforms previous video coding standards. Its high compression performance comes at a very high cost of computing and memory access. Furthermore, due to the complex, sequential, and highly data-dependent characteristics of the essential algorithms in AVC, both pipelining and parallel processing techniques are difficult to be directly employed. The hardware utilization and throughput are also decreased because of the inherent block/macroblock/frame-level reconstruction loops.

In addition to the complexity issue, another big headache is the power consumption, especially for mobile applications. Allocating powerful hardware resources for real-time computing is the basic requirement. At the same time, we also have to consider the power optimization. For battery-powered appliances, the energy is limited, and no one can accept a product that needs to be charged frequently. An integrated circuit (IC) consuming too much power will also cause the heat dissipation and stability problem.

In this chapter, three hardwired H.264/AVC encoder/decoder design cases are presented. The first design is an HDTV720p 30 frames/s AVC encoder.[1,2] From the discussions of this design, we can see how hardware resources are efficiently allocated to fulfill the computing requirement. The second case is the design of a low-power and power-aware AVC encoder.[3] Here, the target specification of video resolution is smaller, and the design puts more emphasis on power issues. Low-power techniques are discussed, and the power-awareness concept is realized in this power-scalable architecture. In the third reference design, the architecture of a hybrid task-pipelining AVC decoder[4,5] is described. The major design issue is to arrange decoding functions into proper pipelining schedules, and to reduce the external and internal memory bandwidth.

Before detailed discussions of the three design cases, a brief introduction and analysis of the H.264/AVC algorithm is given in section 8.1.

8.1 H.264/AVC Overview

Figures 8.1 and 8.2 show the functional block diagrams and features of the H.264/AVC encoder and decoder. Similar to its predecessors of MPEG and H.26x standards, the basic framework is still based on a motion-compensated transform coding. In AVC, many new features are added to achieve much better compression performance and subjective quality. To remove more spatial redundancy, H.264/AVC intraprediction suggests 13 prediction modes to improve intracoding. To remove more temporal redundancy, interprediction is enhanced by motion estimation (ME) with quarter-pixel accuracy, variable block sizes (VBSs), and multiple reference frames (MRFs). Moreover, the advanced entropy coding tools are content adaptive to reduce more statistic redundancy. The perceptual quality is also improved by the in-loop deblocking filter. Simulation results[6] show that H.264/AVC can save 25–45 percent and 50–70 percent of bit rates compared with MPEG-4 advanced simple profile (ASP) and MPEG-2, respectively. Readers can refer to references 7–9 for a more detailed introduction of the H.264/AVC video coding standard.

High coding performance comes with high computational complexity. According to the instruction profiling, the H.264/AVC decoding of HDTV size videos requires 83 giga-instructions per second (GIPS) of computation and 70 GB/s of memory access. As for the encoding process, up to 3,600 GIPS and 5,570 GB/s are required. General-purpose processors have difficulties handling such high computational complexity. For real-time applications, highly parallel hardwired architecture is usually preferable. The stream-like video data can be more efficiently processed by parallel and pipelined array processing elements. Even for a processor-based implementation of a video codec, dedicated accelerating architecture of some critical operations or modules, such as motion estimation, is a common element to assist and off-load the processor.

Taking a closer look of the design difficulties of H.264/AVC, in addition to ultra-high computational complexity and memory access, the coding path is very long, as it includes intra- and interprediction, block and macroblock (MB), frame-level reconstruction loops, entropy coding, and an in-loop deblocking filter. The reference software[10] adopts sequential processing of many blocks in one MB. The sequential approach restricts the parallel architecture design and is harmful to hardware performance. The block-level reconstruction loop caused by intraprediction induces bubble cycles and decreases the hardware utilization and throughput. Some coding tools have multiple modes. More hardware resource is required if multiple processing elements (PEs) are designed for every mode without resource sharing and data reuse. Some coding tools involve many

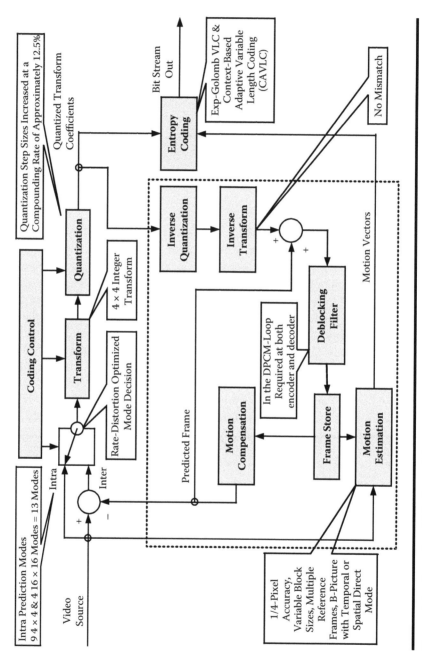

Figure 8.1 Functional blocks and features of H.264/AVC encoding.

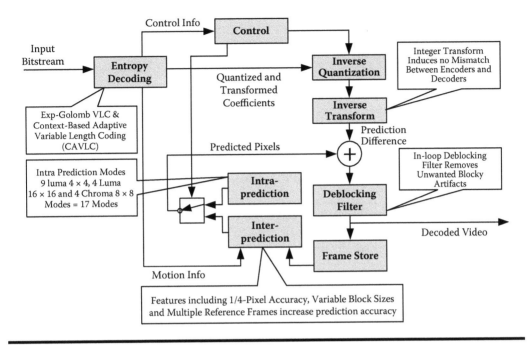

Figure 8.2 Functional blocks and features of H.264/AVC decoding.

data dependencies to enhance the coding performance. Considerable storage space is therefore required to store correlated data during the coding process.

8.1.1 Algorithm Complexity Analysis

Software profiling is usually the first step to get more insights to algorithm complexity and to understand algorithm characteristics of computing and memory load/store behavior. It helps to find the critical parts of the whole design, and then we can spend more effort on its optimization. To focus on the target specification, here a software C model is developed by extracting all baseline profile compression tools from the reference software, and IPROF,[11] an instruction-level profiler, is used for the complexity analysis.

The focused design case is targeted at SDTV (720 × 480, 30 fps)/HDTV720p (1,280 × 720, 30 fps) videos with a maximum reference frame number of 4/1 and a maximum search range of H[−64, +63] and V[−32, +31]. According to the simulation results, the computational complexity and memory access of SDTV/HDTV720p coding are 2,470/3,600 GIPS and 3,800/5,570 GB/s. It is about ten times more complex than that of MPEG-4 Simple Profile.[12] This is mainly due to the introduction of MRF-ME and VBS-ME in interprediction. For the full-search algorithm, the complexity of integer ME (IME) is proportional to the number of reference frames, while that of fractional ME (FME) is proportional to the MB numbers constructed by variable blocks and the number of reference frames. The huge computational loads are far beyond the capability of today's general-purpose processors. The runtime percentages of H.264/AVC P frames encoding are shown in figure 8.3. Interprediction occupies 97.32 percent of computation and is the processing bottleneck of the AVC interframe coding. Mode decision and intraprediction dominate the rest and occupy 77 percent of computation of intraframe coding.

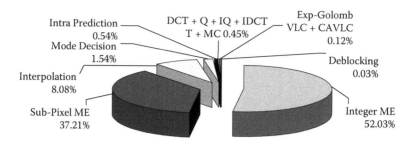

Figure 8.3 Runtime profiling of H.264/AVC (baseline profile) interframe coding.

As for the decoder with HDTV1024p (2,048 × 1,024, 30 fps) specification, 83 GIPS and 70 GB/s of computation and memory access are required. The complexity of AVC baseline profile decoding is about two to three times that of MPEG-4 SP. The runtime percentages of several main tasks are shown in figure 8.4. The interprediction and deblocking filter contribute the most computation time (39 and 38 percent, respectively), while IQ/IDCT, entropy decoding, and intra-prediction occupy the rest. Note that the complexity of the encoding process is much higher than that of the decoding process.

8.1.2 Design Challenges

The first major design challenge of an H.264/AVC hardwired architecture is parallel processing under the constraint of sequential flow. A high degree of parallel processing is required, especially for high-resolution videos, because there are very limited cycles but many operations. However, the H.264/AVC reference software[10] adopts many sequential processes to enhance the compression performance. It is difficult to efficiently map a sequential algorithm to parallel hardware architecture. For system scheduling, the coding path (which includes intra- and interprediction; block-, MB-, and frame-level reconstruction loops; entropy coding; and in-loop deblocking filters) is very long. The sequential encoding process should be partitioned into several tasks and processed MB by MB in pipelined structure to improve the hardware utilization and throughput.

For module-level architecture design, the problem of the sequential algorithm is critical for ME, because it is the most computationally intensive part and requires the greatest degree of parallelism. The FME must be done after the IME. In addition, in FME, the quarter-pixel-precision refinement

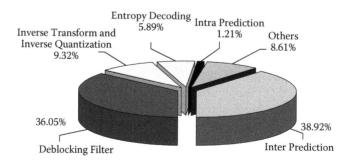

Figure 8.4 Runtime profiling of H.264/AVC (baseline profile) decoding.

must be processed after the half-pixel-precision refinement. Moreover, the inter-Lagrangian mode decision takes motion vector (MV) costs into consideration, which also causes inevitable sequential processing. The modified hardware-oriented (hardware-friendly) algorithms provide benefits to enable parallel processing without noticeable quality drop. The analyses in processing loops and data dependencies are also helpful to map the sequential flow into parallel hardware architecture.

The second design challenge of a hardwired H.264/AVC coder is the problem caused by reconstruction loops. In addition to the frame-level reconstruction loop for ME and motion compensation (MC), the intraprediction induces the MB and block-level reconstruction loops. Because the reconstructed pixels of the left and upper neighboring MBs/blocks are required to predict the current MB/block, the intraprediction of the current MB/block cannot be performed until the neighboring MBs/blocks are reconstructed. The reconstruction latency is harmful to hardware utilization and throughput if the intraprediction and reconstruction engines are not jointly considered and well scheduled.

Data dependencies are the third design challenge. By extracting more data correlations, the new coding tools remove more temporal, spatial, and statistic redundancies. The algorithm, therefore, causes many data dependencies during coding. The frame-level data dependencies contribute to considerable increments of system bandwidth. The dependencies between neighboring MBs constrain the solution space of MB pipelining, and those between neighboring blocks limit the possibility of parallel processing. Because a great deal of data and coding information may be required by the following encoding and decoding procedures, the storage space of both off-chip memories and on-chip buffers is largely increased. To reduce the chip cost, the functional period or lift time of these data must be jointly considered with the system architecture and processing schedule.

The fourth problem is abundant modes. Many coding tools of H.264/AVC have multiple modes. For example, there are 17 different modes for intraprediction and 259 kinds of partitions for interprediction. Six kinds of two-dimensional (2-D) transform functions, $4 \times 4/2 \times 2$ DCT/IDCT/Hadamard transforms, are involved in reconstruction loops. Adaptive filter taps and two filter directions also must be supported for in-loop deblocking filters. Reconfigurable processing engines and reusable prediction cores are important to efficiently support all these functions.

Last but not least, the bandwidth requirement of the H.264/AVC encoding system is much higher than those of previous coding standards. The MRF-ME contributes the heaviest traffic for loading reference pixels. Neighboring reconstructed pixels are required by intraprediction and deblocking filters. Lagrangian-mode decision and context-adaptive entropy coding have data dependencies between neighboring MBs, and transmitting related information contributes considerable bandwidth as well. Hence, an efficient memory hierarchy combined with data sharing and data reuse schemes must be designed to reduce system and local memory bandwidth.

8.2 An HDTV720p 30 fps H.264/AVC Encoder

This section describes the architecture design of an H.264/AVC encoder for HDTV-size videos. At the system level of this encoder, different from the traditional two-stage MB pipelining approach, where prediction (ME only) and block engine are the two stages, this design extracts five major functions and maps them into a four-stage MB pipelining architecture with well-arranged task scheduling. At the module design level, the design consideration and optimization for some key modules are described. With these techniques, an efficient implementation of an H.264/AVC encoding system can be achieved.

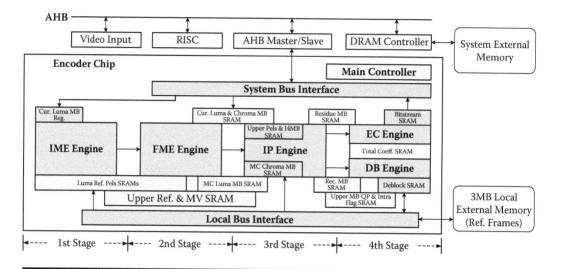

Figure 8.5   Block diagram of the four-stage pipelined H.264/AVC encoder architecture.[2] Five major tasks, including integer motion estimation (IME), fractional ME (FME), intraprediction with reconstruction loop (IP), entropy coding (EC), and deblocking filter (DB), are partitioned from the coding procedure and processed MB by MB in a pipelined structure.

8.2.1   Four-Stage MB Pipelining Architecture

The block diagram of the four-stage pipelined AVC encoder is shown in figure 8.5. Five major tasks, including IME, FME, intraprediction with reconstruction loop (IP), entropy coding (EC), and in-loop deblocking filter (DB), are partitioned from the sequential encoding procedure and processed MB by MB in a pipelining fashion. Several issues of designing this system pipelining are described as follows. The prediction, which is ME only in previous video coding standards, now includes IME, FME, and intraprediction in H.264/AVC. Because of the algorithm diversities and different computational complexity, it is difficult to implement IME, FME, and intraprediction by the same hardware. Putting IME, FME, and intraprediction in the same MB pipelined stage leads to very low utilization. Even if resource sharing is achieved, the operating frequency becomes too high due to sequentially processing. Therefore, FME is first pipelined MB by MB after IME to double the throughout.

As for intraprediction, because of MB-level and block-level reconstruction loops, it cannot be separated from the reconstruction engine. Besides, the reconstruction process should be separated from ME and pipelined MB by MB to achieve the highest hardware utilization. Therefore, engines of intraprediction together with forward, inverse transform, and quantization should be located in the same stage: the IP stage. In this way, the MB-level and block-level reconstruction loops can also be isolated in this pipeline stage. The EC encodes the MB header and residues after mode decision and prediction. The DB generates the standard-compliant reference frames after reconstruction. Because the EC and DB can be processed in parallel, they are placed at the fourth stage. The reference frame will be stored in an external memory for the coding of the next frame, which constructs the frame-level reconstruction loop. Please note that the luma MC is placed in the FME stage for reuse of *Luma Ref. Pels SRAMs* and interpolation circuits. The compensated MB is transmitted to the IP stage for generation of residues after intra- or intermode decision. On the

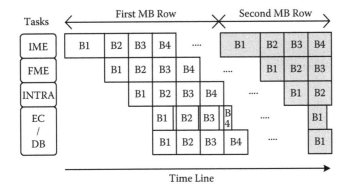

Figure 8.6   The schedule of the four-stage MB pipelining.

other hand, chroma MC is implemented in the IP stage because it can only be executed after intra- or inter-mode decision. In summary, five main functions extracted from the coding process are mapped into the four-stage MB pipelined structure. The processing cycles of the four stages are balanced with different degrees of parallelism to achieve high utilization. MBs within one frame are coded in raster order with the schedule shown in figure 8.6.

As for the reduction of system bandwidth, many on-chip memories are used for three purposes. First, to find the best-matched candidate, a huge amount of reference data is required for both IME and FME. Because pixels in neighboring candidate blocks are considerably overlapped, and so are the search windows of neighboring current MBs, the bandwidth of system bus can be greatly reduced by designing local buffers to store reusable data. Second, rather than transmitted through system bus, the raw data such as luma motion-compensated MBs, transformed and quantized residues, and reconstructed MBs are shifted forward via shared memories. Third, because of data dependency, a MB is processed according to the data of the upper and left MBs. Local memories are used to store the related data during the encoding process. For a software implementation, the external bandwidth requirement is up to 5,570 GByte/s. As for a hardware solution with local search window buffer embedded, the external bus traffic is reduced to 700 MByte/s. After all three techniques are applied, the final external bandwidth requirement is about 280 Mbyte/s.

8.2.2   Low-Bandwidth Parallel IME Architecture

IME requires the most computing power and memory bandwidth in H.264/AVC. A large degree of parallelism is required, or ME will not be able to complete operations within limited cycles, which is constrained by real-time requirements. However, the sequential Lagrangian mode decision flow makes it impossible to design parallel architecture for IME. Therefore, techniques on algorithmic and architectural levels are used to enable parallel processing and to reduce the required hardware resources. Also, efficient memory hierarchy and data reuse schemes are applied jointly to greatly reduce the memory bandwidth requirement.

Motion estimation is the most critical module in a video codec, so there are many research works about parallel ME architecture designs. Interested readers can refer to Huang et al.[13] for a survey on block-matching motion estimation algorithms and architectures.

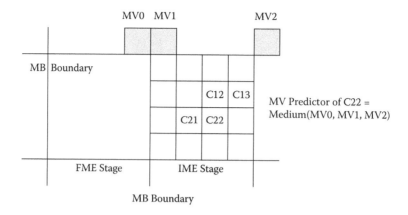

Figure 8.7   Modified motion vector predictor (MVP). To facilitate parallel processing and MB pipelining, MVPs of all 41 blocks are changed to the mediums of MV0, MV1, and MV2.

8.2.3   Hardware-Oriented Algorithm

The MV of each block is generally predicted by the medium values of MVs from the left, top, and top right neighboring blocks. The rate term of the Lagrangian cost function can be computed only after MVs of the neighboring blocks are determined, which causes an inevitable sequential processing. That is, the blocks and subblocks in an MB cannot be processed in parallel. Moreover, when an MB is processed at the IME stage, its previous MB is still in the FME stage. The MB mode and the best MVs of the left blocks cannot be obtained in the four-stage MB pipelined architecture. To solve these problems, the modified motion vector predictor (MVP) is applied for all 41 blocks in MB, as shown in figure 8.7. The exact MVPs of variable blocks, which are the mediums of MVs of the top left, top, and top right blocks, are changed to the mediums of MVs of the top left, top, and top right MBs. For example, the exact MVP of the C22 4 × 4 block should be the mediums of the MVs of C12, C13, and C21. We change the MVPs of all 41 blocks to the mediums of MV0, MV1, and MV2 to facilitate the parallel processing and MB pipelining.

As for the searching algorithm, full-search ME is adopted to guarantee the highest compression performance. The regular searching pattern is suitable for parallel processing. In addition, full search can effectively support VBS-ME by reusing sums of absolute differences (SADs) of 4 × 4 blocks for larger blocks. Pixel truncation[14] of 5-bit precision and subsampling[15] of the half-pixel rate are applied to reduce the hardware cost. Moreover, adaptive search range adjustment[16] is also applied to save the computations. These modifications combined with full-search pattern will not cause noticeable quality loss for HDTV/SDTV videos.

8.2.4   Architecture Design of IME

Figure 8.8 shows the low-bandwidth parallel IME architecture, which mainly comprises eight *PE-Array SAD Trees*. The current MB (CMB) is stored in *Cur. MB Reg.* The reference pixels are read from external memory and stored in *Luma Ref. Pels SRAMs*. Each PE array and its corresponding 2-D SAD tree compute the 41 SADs of VBS for one searching candidate at each cycle. Therefore, eight horizontally adjacent candidates are processed in parallel. All SAD results of VBS are input to the *Comparator Tree Array*. Each comparator tree finds the smallest SAD among the eight search points and updates the best MV for a certain block size.

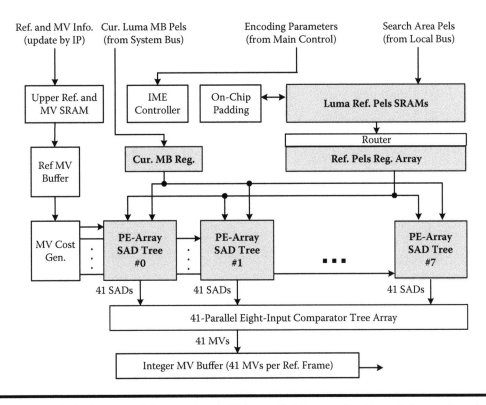

Figure 8.8   Block diagram of the low-bandwidth parallel IME engine.[2] It mainly consists of eight *PE-Array SAD Trees*, and eight horizontally adjacent candidates are processed in parallel.

Because search windows of neighboring current MBs are considerably overlapped, and so are the pixels of neighboring candidate blocks, a three-level memory hierarchy, including external memory, *Luma Ref. Pels SRAMs*, and *Ref. Pels Reg. Array*, is used to reduce bandwidth requirement by data reuse. Three types of data reuse are implemented—MB level, intercandidate, and intracandidate. The *Luma Ref. Pels SRAMs* are first embedded to achieve MB-level data reuse. When the ME process is changed from one CMB to another CMB, there is the overlapped area between neighboring search windows. Therefore, the reference pixels of the overlapped area can be reused, and only a part of the search window must be loaded again from system memory. The system bandwidth can thus be reduced.[17] The *Ref. Pels Reg. Array* acts as the temporal buffer between *PE-Array 2-D SAD Tree* and *Luma Ref. Pels SRAMs*. It is designed to achieve intercandidate data reuse. Figure 8.9 shows the M-parallel PE-Array SAD Tree architecture. A horizontal row of reference pixels, which are read from SRAMs, is stored and shifted downward in *Ref. Pels Reg. Array*. When one candidate is processed, 256 reference pixels are required. When eight horizontally adjacent candidates are processed in parallel, not (256×8) but ($256 + 16 \times 7$) reference pixels are required. Besides, when the ME process is changed to the next eight candidates, most data can be reused in *Ref. Pels Array*. This parallel architecture achieves intercandidate data reuse in both horizontal and vertical directions and reduces the on-chip SRAM bandwidth.

Figure 8.10 shows the architecture of *PE-Array SAD Tree*. The costs of sixteen 4×4 blocks are separately summed up by 16 *2-D Adder Subtrees*, and then reused by one *VBS Tree* for larger blocks. This is so-called intracandidate data reuse. All 41 SADs for one candidate are simultaneously

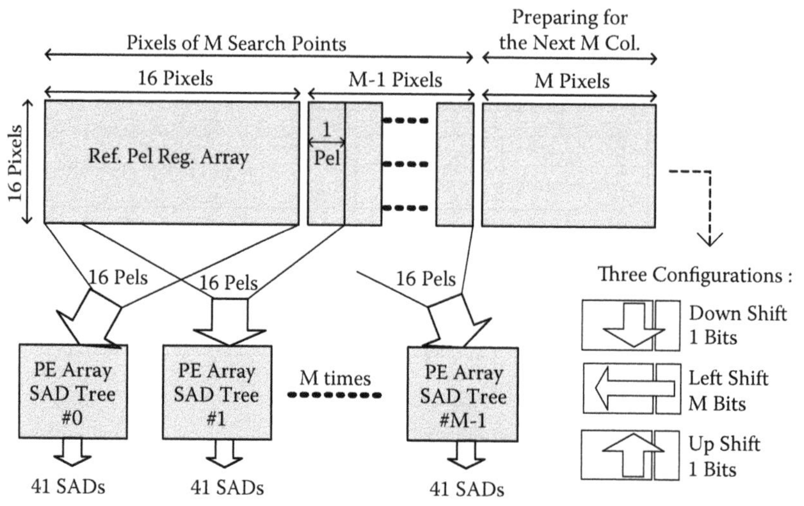

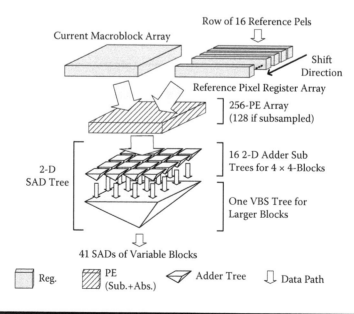

Figure 8.9   M-parallel *PE-Array SAD Tree* architecture.[2] **The intercandidate data reuse can be achieved in both horizontal and vertical directions with *Ref. Pels Reg. Array*, and the on-chip SRAM bandwidth is reduced.**

generated and compared with the 41 best costs. No intermediate data is buffered. Therefore, this architecture can support VBS without any partial SAD registers.

Figure 8.11 summarizes the bandwidth reduction techniques and performance of the IME design with SDTV specification and four reference frames. Five cases are discussed. The first case is only one Reduced Instruction Set Computer (RISC) in the hardware without any local memory.

Figure 8.10   *PE-Array SAD Tree* architecture.[2] **The costs of sixteen 4 × 4 blocks are separately summed up by 16 2-D Adder Subtrees and then reduced by one VBS Tree for larger blocks.**

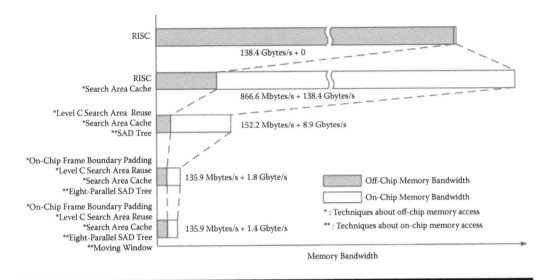

Figure 8.11 **Bandwidth reduction techniques in the IME engine:[2] 99.9 percent system band-width is reduced with the search area cache memories and the MB-level data reuse; 98.99 percent on-chip memory bandwidth is saved with the proposed parallel hardware.**

The reference pixels are accessed directly from external memories and no on-chip memory band-width is required. The second case is one RISC with *Luma Ref. Pels SRAMs*. By MB-level data reuse of the on-chip memories, 866.6 Mbyte/s of the system bandwidth is required, but the on-chip memory bandwidth is increased to 138.4 GByte/s. This trade-off is worthwhile because the system bandwidth is limited. The third case is one *PE-Array SAD Tree* with MB-level data reuse. The system bandwidth is reduced to 152.2 MByte/s. Because of the intracandidate data reuse, the on-chip memory bandwidth is only 8.9 GByte/s. In the fourth case, the system bandwidth is reduced by on-chip frame boundary padding and intercandidate data reuse. The on-chip memory bandwidth is reduced to 1.8 GByte/s. Finally, the modified algorithm of the moving window is applied in the fifth case, and the on-chip memory bandwidth is reduced to 1.4 GByte/s; 99.90 percent of system bandwidth is reduced compared to the first case. Furthermore, 98.99 percent on-chip memory bandwidth is saved with the proposed parallel hardware, compared to the second case.

8.2.5   Parallel FME with Lagrangian Mode Decision

The main design challenge of FME is to achieve parallel processing under the constraints of the sequential FME procedure. In this section, a new VLSI architecture for FME is discussed. Seven processing loops are used to represent the FME procedure, and two decomposing techniques are adopted to parallelize the algorithm. With these techniques, the hardware architecture is designed with a regular schedule, fully pipelined structure, and high utilization.[18]

8.2.6   Analysis of FME Loops

For simplification, we decompose the entire FME procedure into seven iteration loops, as shown in figure 8.12(a). The first two loops are for the multiple reference frames and the 41 variable blocks, respectively. The third loop is the refinement procedure of half-pixel precision and then

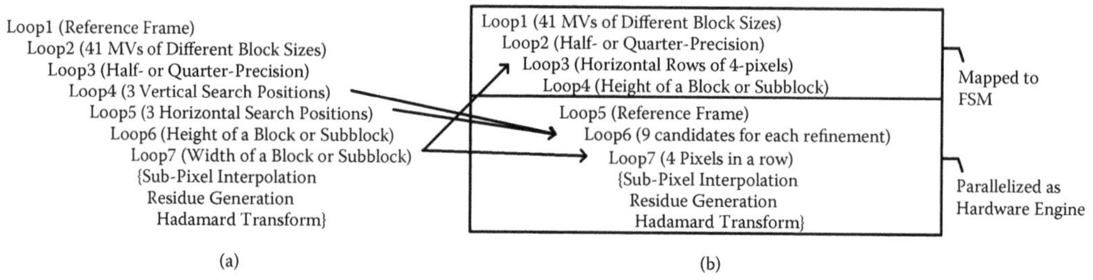

Loop1 (Reference Frame)
Loop2 (41 MVs of Different Block Sizes)
Loop3 (Half- or Quarter-Precision)
Loop4 (3 Vertical Search Positions)
Loop5 (3 Horizontal Search Positions)
Loop6 (Height of a Block or Subblock)
Loop7 (Width of a Block or Subblock)
{Sub-Pixel Interpolation
Residue Generation
Hadamard Transform}

(a)

Loop1 (41 MVs of Different Block Sizes)
Loop2 (Half- or Quarter-Precision)
Loop3 (Horizontal Rows of 4-pixels)
Loop4 (Height of a Block or Subblock)
Loop5 (Reference Frame)
Loop6 (9 candidates for each refinement)
Loop7 (4 Pixels in a row)
{Sub-Pixel Interpolation
Residue Generation
Hadamard Transform}

Mapped to FSM

Parallelized as Hardware Engine

(b)

Figure 8.12 (a) Original FME procedure. (b) Rearranged FME procedure. FME is mapped to parallel hardware with 4 × 4 block decomposition and efficient scheduling for vertical data reuse.

quarter-pixel precision. The next two loops are for the 3×3 candidates of each refinement process. The last two loops are iterations of pixels in one candidate, and the number of iteration ranges from 16×16 to 4×4. Main tasks inside the most inner loop are the fractional pixel interpolation, residue generation, and Hadamard transform. The residue generation performs the subtraction operation between current and reference pixels. The interpolation requires a 6-tap FIR filter in both the horizontal and vertical directions. The Hadamard transform is a 2-D 4×4 matrix operation. These three main tasks have different I/O throughput rates, which makes it quite challenging to achieve parallel processing and high utilization at the same time.

To meet the real-time constraint, some loops must be unrolled and mapped into hardware for parallel processing. The costs of a certain block in different reference frames are processed independently. Therefore, the first loop has no sequential issues and can be easily unrolled for parallel processing. The second loop is not suitable to be unrolled because 41 MVs of VBS-ME may point to different positions. The memory bitwidth of the search window will be too large if reference pixels of VBS-ME are read in parallel. Besides, there is an inevitable sequential processing order among VBS-ME for taking MV costs into consideration. The third loop cannot be parallelized because the searching center of quarter-resolution refinement depends on the result of half-resolution refinement. Similar to the first loop, the costs of 3×3 candidates are processed independently and are feasible for parallel processing. Most interpolated pixels can be reused by the neighboring candidates to save redundant computations and processing cycles. The iteration count of the last two loops depends on the block size, which ranges from 4 to 16. The parallelization strategy will affect the hardware utilization for supporting both 16×16 and 4×4 blocks. In addition, the 6-tap 2-D FIR filter and 4×4 Hadamard transform must also be taken into consideration.

Two techniques here are proposed with hardware considerations. The first technique is 4×4 block decomposition. The 4×4 block is the smallest common element of blocks with different sizes, and the sum of absolute transformed differences (SATD) is also based on 4×4 blocks. That is, every block and subblock in an MB can be decomposed into several 4×4 elements with the same MV. Therefore, we can concentrate on the design of a 4×4 element processing unit (PU) and then apply the folding technique to reuse the 4×4 element PU for different block sizes. Figure 8.13(a) takes the 16×8 block as an example. One 16×8 block is decomposed into eight 4×4 element blocks. These 4×4 element blocks are processed in sequential order, and the corresponding SATDs are then accumulated for the final costs.

The second technique is efficient scheduling for vertical data reuse. After the 4×4 block decomposition, redundant interpolating operations appear in the overlapped area of adjacent

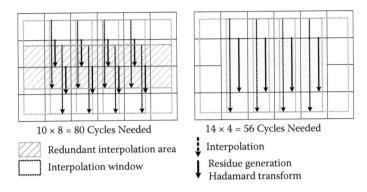

Figure 8.13   Main concept of FME design. (a) 4 × 4 block decomposition. (b) Efficient scheduling for vertical data reuse.

interpolation windows, which is shown in figure 8.13(a). As figure 8.13(b) shows, the interpolation windows of vertically adjacent 4 × 4 elements are integrated. Both the hardware utilization and the throughput are increased. Please note that each 4 × 4 element PU is arranged with four degrees of parallelism to process four horizontally adjacent pixels for residue generation and Hadamard transform in parallel. Most horizontally adjacent integer pixels can be reused for the horizontal FIR filters, and the on-chip memory bandwidth can be further reduced.

8.2.7 *Architecture Design of FME*

Figure 8.14 shows the parallel FME architecture. The IME generates 41 × 4 of IMVs. The FME then refines these MVs to quarter pixel resolution, and MC is performed after intermode decision. The outputs include the best prediction mode, the corresponding MVs, and the MC results. The Luma Ref. Pels SRAMs storing the reference pixels are shared with the IME pipeline stage to reduce the system bandwidth. There are nine 4 × 4 block PUs to process nine candidates around the refinement center. Each 4 × 4 *block PU* is responsible for the residue generation and Hadamard transform of each candidate. The 2-D *interpolation engine*–generating reference pixels in half-pixel or quarter-pixel resolution is shared by all 4 × 4 *block PUs* to achieve data reuse and local bandwidth reduction. The *rate-distortion optimized mode decision* is responsible for the sequential procedures of the first through the fourth loops in figure 8.12(b).

The architecture of each 4 × 4 *block PU* is shown in figure 8.15. Four subtractors generate four residues in parallel and transmit them to the *2-D Hadamard transform unit*. The *2-D Hadamard transform unit*[19] contains two *1-D Hadamard* units and transpose shift registers. The first 1-D Hadamard unit filters the residues row by row in each 4 × 4 block, while the second 1-D Hadamard unit processes column by column. The data flow of the transpose registers can be configured as a rightward shift or downward shift. Two configurations interchange with each other every four cycles. First, the rows of 1-D transformed residues of the first 4 × 4 block are written into the transpose registers horizontally. After four cycles, the columns of the 1-D transformed residues are read vertically for the second 1-D Hadamard transform. Meanwhile, the rows of 1-D transformed residues of the second 4 × 4 block are written into transpose registers vertically. In this way, the *2-D Hadamard transform unit* is fully pipelined with residue generators. The latency of the *2-D Hadamard transform unit* is four cycles, and there is no bubble cycle.

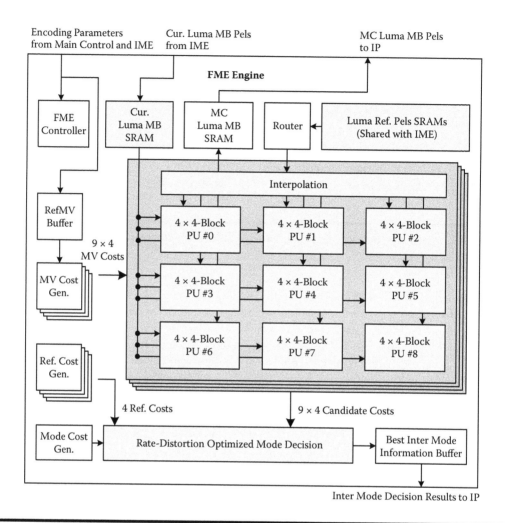

Figure 8.14   Block diagram of the FME engine.[2] There are nine 4 × 4 block PUs to process nine candidates around the refinement center. One 2-D interpolation engine is shared by all 4 × 4 block PUs to achieve data reuse and local bandwidth reduction.

Figure 8.16(b) shows the 2-D interpolation engine. The operations of the 2-D FIR filter are also decomposed to two 1-D FIR filters with the shift buffers located in *V-IP units*. A row of ten horizontally adjacent integer pixels is fed into the five 6-tap horizontal FIR filters (HFIRs) for interpolating five horizontal pixels in half-pixel resolution. These five half pixels and six neighboring integer pixels are then shifted downward in the *V-IP unit*, as shown in figure 8.16(a). After the six-cycle latency, the 11 vertical FIR filters (VFIRs) in the *V-IP units* generate vertical pixels in half-pixel resolution by filtering the corresponding columns of pixels in *V-IP units*. The dotted rectangle in the bottom of figure 8.16(b) stands for all predictors needed each cycle for residue generation in half-pixel refinement. As for quarter-pixel refinement, another bilinear filtering engine getting data from the dotted rectangle is responsible for quarter-pixel generation. The efficient vertical scheduling can reuse interpolated pixels in the *V-IP units*, and 26 percent of cycles can be saved.

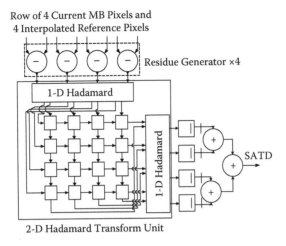

Figure 8.15   Block diagram of 4 × 4 block PU.[2] The *2-D Hadamard transform unit* is fully pipelined with residue generators.

8.2.8   *Reconfigurable Intrapredictor Generator*

The intraprediction supports various prediction modes, which includes four I16 MB modes, nine I4 MB modes, and four Chorma intramodes. If a RISC-based solution is adopted, where the prediction values are generated sequentially for each mode, the required operation frequency will

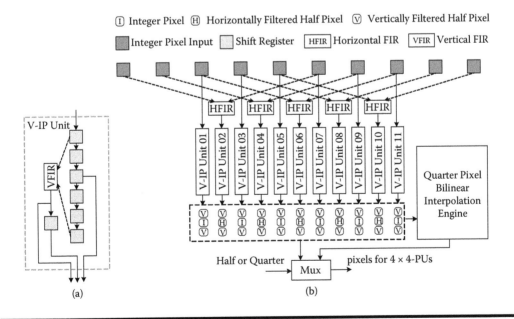

Figure 8.16   (a) Block diagram of *V-IP unit*.[2] (b) Block diagram of 2-D interpolation engine. The operations of 2-D FIR are decomposed to two 1-D FIRs with an interpolation shift buffer. The efficient vertical scheduling can reuse interpolated pixels in the *V-IP units*, and on-chip memory bandwidth of cycles can be saved.

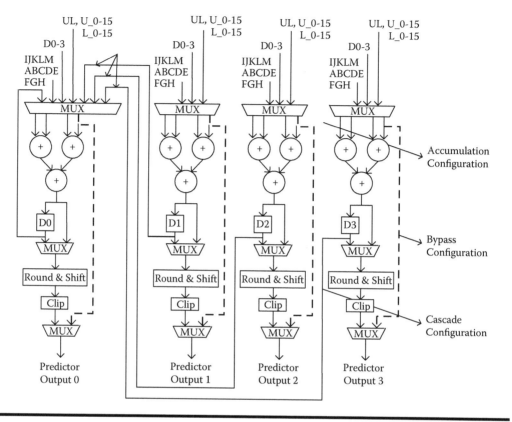

Figure 8.17 Four-parallel reconfigurable intrapredictor generator.[2] Four different configurations are designed to support all intraprediction modes in H.264/AVC.

be too high. On the other hand, if the dedicated hardware is adopted, 17 kinds of PEs for the 17 modes lead to high hardware cost. Therefore, the reconfigurable circuit with the resource sharing for all intraprediction modes is an efficient solution.[20]

The hardware architecture of the four-parallel reconfigurable intrapredictor generator is shown in figure 8.17. Capital letters (A, B, C, …) are the neighboring 4 × 4 block pixels. UL, L0-L15, and U0-U15 denote the bottom right pixel from the upper left MB, the 16 pixels of the right-most column from the left MB, and the 16 pixels of the bottom row from the upper MB, respectively. Four different configurations are designed to support all intraprediction modes. First, the I4 MB/I16 MB horizontal/vertical modes use the bypass data path to select the predictors extended from the block boundaries. Second, multiple PEs are cascaded to sum up the DC value for I4 MB/I16 MB/chroma DC mode. Third, the normal configuration is used for I4 MB directional modes 3–8. The four PEs select the corresponding pixels multiple times according to the weighted factors and generate four predictors independently. Finally, the recursive configuration is designed for I16 MB plane prediction. The predictors are generated by adding the gradient values to the result of the previous cycles.

8.2.9 Dual-Buffer CAVLC Architecture with Block Pipelining

The symbols of the 4 × 4 blocks are coded by context-based adaptive variable-length coding (CAVLC) through two phases: a scanning phase and a coding phase. In the scanning phase, the residues are read from the *residue MB memory* in a backward zigzag order. The run-level symbols and required information are extracted. In the coding phase, the symbols are translated into codewords with the corresponding class of tables. Different from traditional fixed VLC tables, CAVLC utilizes the intersymbol correlation to further reduce the statistical redundancy. This means the selection of VLC tables depends on the related statistics, such as total coefficient number and total run number. Not until the scanning of a 4×4 block is finished can we know the statistic information. The scanning and coding phases of each block must be processed in sequential order, which leads to low hardware utilization.

Figure 8.18(a) shows the proposed dual-buffer architecture and the corresponding block-pipelined scheme.[21] There is a pair of ping-pong mode statistic buffers. After the scanning phase of the first 4×4 block, the run-level symbols and related statistic information are stored in the first buffer, and the coding phase starts. At the same time, the scanning phase of the second 4×4 block is processed in parallel by use of the second buffer. As shown in figure 8.18(b), by switching the ping-pong mode buffers, scanning and coding of the 4×4 blocks within a MB can be processed simultaneously with the interleaved manner. In this way, both the throughput and the hardware utilization are doubled.

To further improve our design, a zero skipping technique is applied. When the residues within an 8×8 block are all zeros, the 4×4 blocks inside do not have to be coded. In this situation, the scanning process can be skipped early. The symbol of code block pattern (CBP) in the MB header is used for the skipping decision.

Figure 8.19 shows the number of processing cycles required by the single-buffer architecture,[20] the dual-buffer block-pipelined architecture, and the one with zero skipping technique. Compared with the single-buffer architecture, the dual-buffer block-pipelined architecture processes two 4×4 blocks in parallel, and thus enhances the hardware utilization and throughput. It can reduce up to half of the processing cycles at high-bit-rate situations. At low-bit-rate situations,

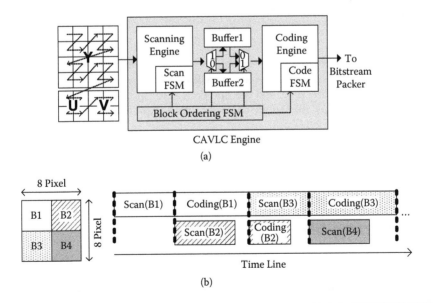

Figure 8.18 **(a) Dual-buffer architecture of CAVLC engine. (b) Block-pipelined schedule by switching the ping-pong mode buffers, scanning and coding of the 4×4 blocks can be processed simultaneously, and the throughput and hardware utilization are doubled.**[2]

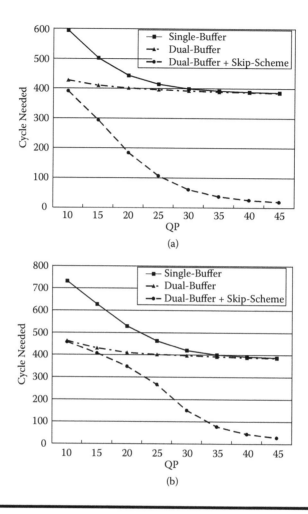

Figure 8.19   Numbers of processing cycles per MB required by the single buffer architecture,[20] the dual-buffer architecture, and the one with zero skipping technique. (a) Foreman video sequence, QCIF 30 fps. (b) Mobile calendar video sequence, QCIF 30 fps.

most residues are zero, and the scanning phase dominates the processing cycles. The zero skipping technique can further improve the design by eliminating the redundant scanning process, and up to 90 percent of the cycles are reduced.

8.2.10   Deblocking Filter

The deblocking filter is employed in-loop at 4×4 block boundaries to reduce block artifacts in H.264/AVC. In the reference software, the processing order of block boundaries is done vertically and then horizontally in the order shown in figure 8.20 (v1, v2, v3, …, v23, v24, and then h1, h2, h3, …, h23, h24). Figure 8.21 shows the proposed hardware architecture. Before deblocking an MB, the MB data and adjacent block data are prepared in the *deblock SRAM*. An *eight-parallel 1-D*

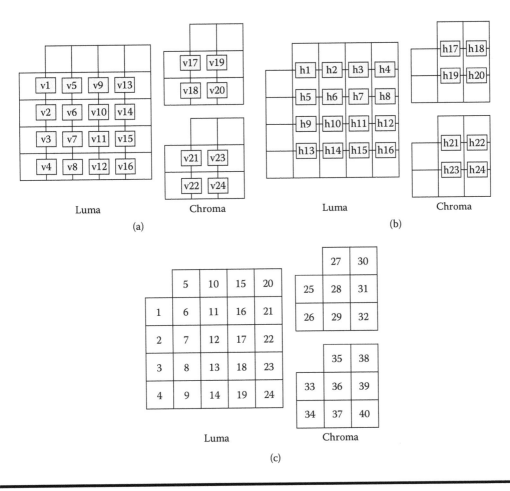

Figure 8.20   Processing order of boundaries: (a) Horizontal filtering across vertical edges, (b) Vertical filtering across horizontal edges, (c) Block index.

filter can process eight pixels for one edge in parallel. The 8 × 4 unfiltered pixels of two adjacent 4 × 4 blocks are stored in the 8 × 4 pixel array with reconfigurable data paths to support both horizontal and vertical filtering with the same 1-D filters.

The main feature is described as follows. The processing order of both vertical and horizontal boundaries is modified to the transpose order (v1, v5, v9, ..., v22, v24, and then h1, h5, h9, ..., h22, h24, as shown figure 8.20).[22] This modification can achieve considerable data reuse without affecting the data dependency defined by the H.264/AVC standard. As shown in figure 8.20, after boundary v1 is horizontally filtered, we only have to write block 1 from the array to the deblock SRAM. As for block 6, we can directly send it back to the filter with block 11 from SRAM to process boundary v5. This data reuse scheme can be applied by both horizontal and vertical filters. About 50 percent bandwidth of deblock SRAM can be reduced, and the hardware utilization and throughput are increased.

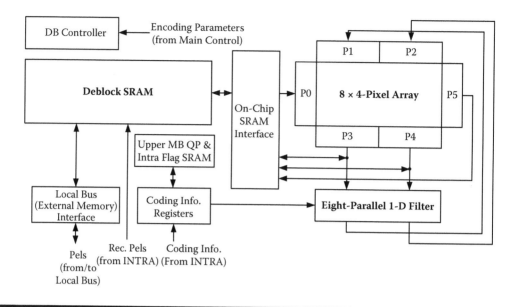

Figure 8.21   Block diagram of the deblocking filter engine.[2]

8.2.11   Implementation

The specification of this H.264/AVC encoder is baseline profile with a level up to 3.1. The maximum computational capability is to real-time encode SDTV 30fps video with four reference frames or HDTV720p 30fps video with one reference frame. Table 8.1 shows the logic gate count profile synthesized at 120 MHz. The total logic gate count is about 922.8K. The prediction engine, including IME, FME, and IP stages, dominates 90 percent of logic area. As for on-chip SRAM, 34.88 KB is required. The chip is fabricated with UMC 0.18 μm 1P6M CMOS process. Figure 8.22 shows the chip micrograph. The core size is 7.68×4.13 mm^2. The power consumption is 581 mW for SDTV videos and 785 mW for HDTV720p videos at 1.8 V supply voltage with 81/108 MHz operating frequency. Table 8.2 shows the detailed chip features.

The encoded video quality of our chip is competitive with that of reference software, in which full search is implemented with Lagrangian mode decision. As shown in figure 8.23, with

Table 8.1   Hardware Cost of the H.264/AVC Encoder

Module Name	Gate Counts	Memory (KByte)
Integer ME (IME)	305,211	13.71
Fractional ME (FME)	401,885	13.82
Intraprediction (IP)	121,012	5.01
Entropy coding (EC)	29,332	1.27
Deblocking filter (DB)	20,125	0.91
Others	45,176	0
Total	922,768	34.72

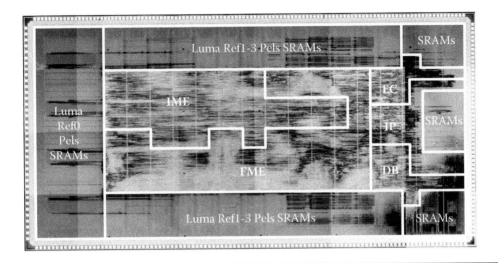

Figure 8.22   Chip micrograph of the H.264/AVC encoder.[23]

improvement of the Lagrangian multipliers, our compression performance is even better at high bit rate.

8.2.12   Summary

In this section, the analysis and architecture design of an H.264/AVC baseline profile encoder is presented. The four-stage MB pipelined architecture can encode HDTV720p 30fps videos in real-time at 108 MHz. The pipelined architecture doubles the throughput of the conventional two-stage MB pipelined architecture with high hardware utilization. Five main functions of IME, FME, intraprediction with reconstruction loops, CAVLC, and the deblocking filter are discussed in detail. In summary, parallel processing and pipelining techniques are used to reduce the frequency and increase the utilization, while folding and reconfigurable techniques are applied to reduce the chip area. With this hardwired encoder, full-search quality can be achieved with 1,200 times of speed-up compared with the reference software executed on a PC with a Pentium IV 3GHz CPU.

Table 8.2   Specification of the H.264/AVC Baseline Profile Encoder Chip

Technology	UMC 0.18μm 1P6M CMOS
Pad/core voltage	3.3/1.8V
Core area	7.68 × 4.13 mm²
Logic gates	992.8K (2-input NAND gate)
SRAM	34.72 KB
Operating frequency	81/108 for D1/HDTV720p
Power consumption	581/785 mW for D1/HDTV720p
Encoding features	All baseline profile compression tools
Maximum number of reference frames	4/1 for D1/HDTV720p
Maximum search range (ref. 0)	H[−64, +63] V[−32, +31]
Maximum search range (ref. 1–3)	H[−32, +31] V[−16, +15]

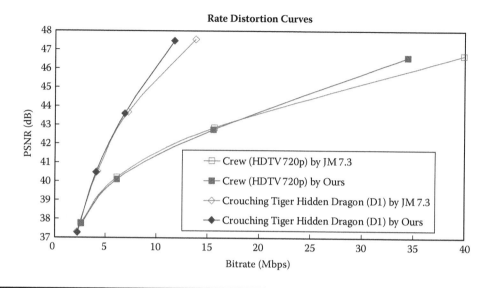

Figure 8.23   Rate-distortion curves. Comparison of coding quality with the H.264/AVC reference software.

8.3   A 2.8 to 67.2 mW Low-Power and Power-Aware H.264/AVC Encoder

Power consumption is a critical design issue of modern ICs, especially for mobile devices. Classical hardware architecture focuses more on area (hardware complexity) and speed (latency and throughput). Now, power-oriented design methodology has emerged as a key factor. Low power and power awareness[24–26] are the most important issues in the power-oriented design. A low-power video encoder can provide the maximum compression performance under the specific power constraint. Beyond low power, the feature of a power-aware video encoder can further extend the battery life by varying the compression performance and power consumption according to the system power status. It is a power-rate-distortion or complexity-rate-distortion optimization concept.[27,28]

Figure 8.24 shows the battery discharging effects of three kinds of video encoders. In practice, a battery has two important nonideal properties: rate capacity effect and recovery effect.[29]

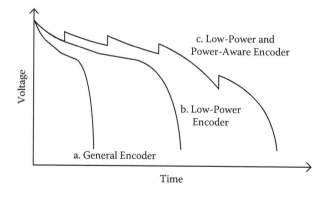

Figure 8.24   Battery discharging effects of three kinds of video encoders.

The capacity of a battery is dependent on the discharging rate, and a battery with an intermittent loading may have a larger capacity than a battery with a continuous loading. In figure 8.24, with low-power encoders, we can reduce the loading to have a longer battery lifetime because halving the current achieves more than double the battery lifetime. With power-aware encoders, we can further extend the battery lifetime by gradually stepping down the power dissipation because the battery capacity can be recovered with a lower loading.

Furthermore, mobile multimedia devices are diverse with different applications. Different systems have their characteristics in terms of transmission bandwidth, storage size, display quality, power capacity, and so on. A power-aware encoder can compromise transmission rate and power consumption with quality distortion through the adjustments of quantization parameter and encoding parameters. The design target is to achieve maximum system performance with the power-rate-distortion optimization.

In this section, a low-power and power-aware AVC encoder design is discussed. This design demonstrates the power scalability through a flexible system hierarchy that supports content-aware algorithms and a module-wire gated clock. Part of the module designs are based on the architecture presented in the previous section, while some modules are with architecture refinement and more power issues are addressed.

8.3.1 Low-Power ME

According to the computational complexity analysis of the H.264/AVC reference software, over 90 percent of the computational loading is contributed by the ME with the new techniques of quarter-pixel resolution, variable block sizes (VBSs), multiple reference frames (MRFs), and Lagrangian mode decision. The development of a low-power ME module is therefore the most critical in a low-power AVC encoder. In AVC, the ME consists of integer ME (IME) and fractional ME (FME). The IME searches for the initial prediction in coarse resolution. Then, FME refines this result to the best match in fine resolution. In the following, we will first describe the low-power techniques at the algorithm and architecture levels and define the problem. Then, the ME engines are developed through algorithm and hardware co-optimization for low-power considerations.

In ME algorithms, the current frame is partitioned into macroblocks (MBs). For each current MB (CMB) in the current frame, one best-matched block, which is the most similar to the current MB, is searched within a search window in a reference frame. The ME calculates the matching costs of the candidates in the search window, and the candidate with the smallest matching cost is the best match. For IME and FME, the criteria of the matching costs are the sum of absolute differences (SAD) and the sum of absolute transformed differences (SATD) between current pixels of CMB and reference pixels of each candidate.

In a typical ME module, reference pixels of the search window are stored in local memories, and matching costs are calculated by parallel processing elements (PEs). The power consumption of the ME module mainly comes from two parts. One is contributed by data access to read reference pixels from memories, and the other is contributed by computations to calculate matching costs.

Two techniques are used to reduce the power consumption. At the architecture level, because the reference pixels of neighboring candidates are considerably overlapped, reference pixels read from local memories can be stored in registers and reused by parallel PEs. Data access power can thus be reduced. At the algorithm level, fast algorithms can be applied to reduce the candidate number, and both the data access power and the computational power can thus be saved.

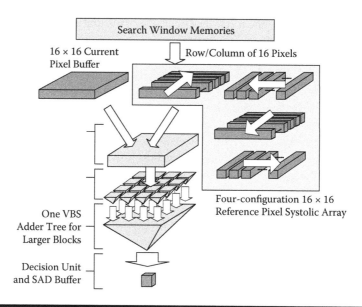

Figure 8.25   Integer motion estimation (IME) architecture.

An optimal low-power ME engine should be the parallel architecture that supports the fast algorithm with efficient data reuse. This is not easy because fast ME algorithms are difficult to be realized on parallel architectures due to their irregular and sequential natures. Generally, the parallel hardware-supporting fast algorithm has poor data reuse efficiency. The power reduction at the algorithm level constrains that at the architecture level.

8.3.2   Fast IME with Efficient Data Reuse

Figure 8.25 illustrates the architecture of the IME engine.[30] To reduce the power consumed by data access, reference pixels read from search window local memories are stored and reused in *systolic register array* (SRA). A four-step search (4SS) fast algorithm, as shown in figure 8.26,

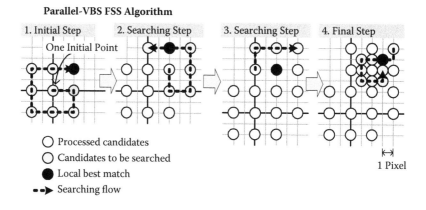

Figure 8.26   Four-step search ME algorithm and the searching flow.

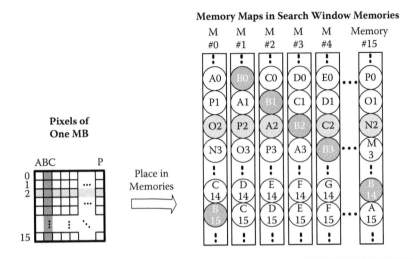

Figure 8.27   Ladder-shaped data arrangement for search window memories.

is adopted. The fast ME algorithm reduces computational complexity but has an irregular searching flow.

The inefficient data reuse problem for fast algorithms mainly comes from the access restriction in search window memories. To efficiently combine these two techniques, the ladder-shaped data arrangement (LSDA) shown in figure 8.27 is adopted for the search window memories. Traditionally, the horizontally adjacent pixels are interleavingly arranged in different search window memories. Under this arrangement, reference pixels can only be read row by row in parallel, but not column by column. In the LSDA way, the *n*th rows are rotated by (*n* − 1) pixels. For example, the second row is rotated by one pixel, while the third row is rotated by two pixels. In this way, the reference pixels of A2–P2 and B0–B15 are both arranged in different memories. Both the horizontally and vertically adjacent reference pixels can be accessed in parallel.

The systolic register array is further designed with four configurations for different searching directions: up-, down-, left-, and right-shift, as shown in figure 8.25. Then, the searching flow is designed to string up all candidates to be searched. For example, in figure 8.26, after step 1, reusable pixels are stored in the systolic array. During step 2, the shift direction of systolic array is successively configured as down-, right-, up-, and left-shift configurations. The corresponding rows and columns of pixels are read from search memories, shifted in systolic register array, and then reused by the 16 × 16 PE array. Without data reuse, 256 pixels have to be read from memories for each candidate. With data reuse, only 16 pixels are required for each candidate. Memory power of IME is thus reduced.

8.3.3   One-Pass FME with Efficient Data Reuse

The previous FME fast algorithm involves sequential processing, which constrains the parallel processing and data reuse. After the IME, the half-pixel MV refinements are performed around the best integer search positions with ±½-pixel search range. The quarter-pixel ME, as well, is then performed around the best half-search position with ±¼-pixel search range. To calculate the

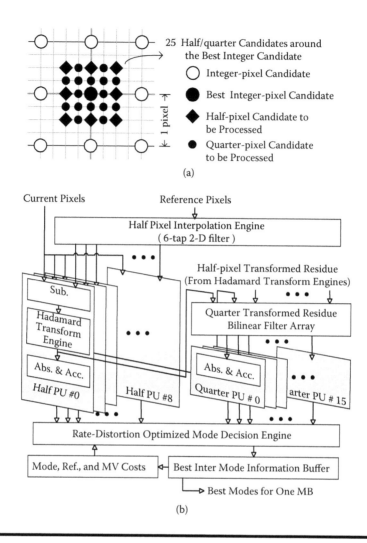

Figure 8.28   Low-power FME architecture.[3] (a) Hardware-oriented one-pass FME algorithm. (b) Block diagram of FME architecture.

matching cost, the fractional pixels are interpolated from reference pixels read from search window memories, and then the SATDs are calculated.

Figure 8.28 shows the low-power FME engine. To facilitate parallel processing and data reuse, a hardware-oriented algorithm is proposed. Different from the sequential half-then-quarter MV-refinement algorithm, the proposed one-pass algorithm searches 25 half-pixel and quarter-pixel candidates around the best integer MV. In this parallel architecture, 25 processing units (PUs) process 25 half-pixel and quarter-pixel candidates for the selected variable-sized block in parallel. The half pixels interpolated by the 6-tap 2-D interpolation engine are shared by the nine half PUs, and 89 percent of memory access and 6-tap filter power are saved for the half-pixel candidates. Then, the residues of half-pixel candidates in the transformed domain are reused by the bilinear filter array to generate the residues on quarter-pixel candidates in the transformed domain for

16 quarter PUs. All memory access, 6-tap filter, and Hadamard transform power are saved for the quarter-pixel candidates.

8.3.4 Power-Aware Algorithm and Flexible Encoder System

Processor-based implementation usually has better flexibility but is less power efficient than customized hardwired architecture. Hardwired design is viewed as a fixed-function one, but it is possible to provide some extent of flexibility for the consideration of power awareness. That is, even though it is a dedicated circuit, part of the design can have reconfigurable architecture and have multiple adjustable parameters and power modes for power adaptation.

Based on the low-power ME engine, this hardwired encoder further demonstrates the capability of power scalability with the flexible encoder system and power-aware algorithm. A good power-aware encoder must be based on a low-power consumption platform and can be judged according to the quantity and quality—whether it has a wide range of power scalability, and whether it can have efficient trade-off between the power consumption and the compression performance. In the following, we will describe a power-aware algorithm based on content awareness (CA). A flexible system based on low-power considerations is developed with power scalability.

8.3.5 Power-Aware Algorithm

Figure 8.29 shows the proposed power-aware coding algorithm. A wide range of power scalability is achieved with this algorithm performed on the flexible system and reconfigurable processing

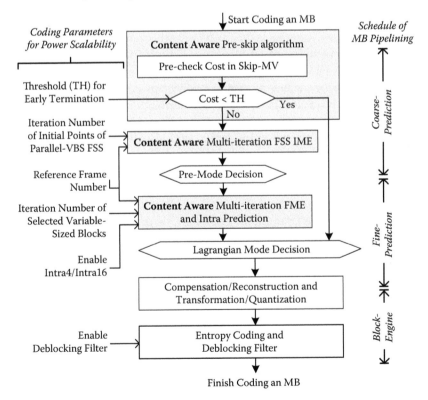

Figure 8.29 Power-aware algorithm based on content awareness.[3]

engines (PEGs). H.264/AVC involves various compression tools, and each tool is useful for different video features. In this encoder, many parameters of all the encoding features can be dynamically adjusted. These encoding parameters include threshold for early termination, the number of reference frames, variable-sized block number, and searching iterations for IME and FME, enable or disable flag in intraprediction and deblocking filter.

An effective power-quality trade-off is achieved with content awareness. A content-aware scheme guides the encoder to adaptively spend the limited power budget on the more power-quality-efficient modes, which is realized by dynamically adjusting parameters according to previous coding information. In the content-aware preskip algorithm, because best interprediction modes of a significant number of MBs are the skip modes, the algorithm monitors this event in the early stage to skip the regular IME and FME operations. At least 20 percent of encoder power is saved with this approach.

Furthermore, the neighboring motion activities are used as content awareness in IME. If the neighboring MVs are quite different, there should be several objects moving toward different directions. In this situation, more initial searching points are generated according to these MVs. As for the content awareness in FME, most of the variable-sized blocks are rejected early according to the searching results in IME. That is because if the matching cost of a block in integer-pixel resolution is too large, such block is less likely to be the final best mode in quarter-pixel resolution.

8.3.6 Flexible System

Figure 8.30 shows the system architecture. The encoder has three MB pipeline stages: coarse prediction, fine prediction, and block engine, with the tasks shown in figure 8.29. Three MBs are pipelined and simultaneously processed. To provide the flexibility in this hardwired encoder, unlike the previous system hierarchy,[23,32] in which each pipeline controller is tightly coupled with its corresponding processing engines in each pipeline stage, this system hierarchy separates processing engines from pipeline controllers. Processing engines can thus be flexibly reused between different pipeline stages.

There are two main advantages. First, it provides flexibility in system scheduling. Compared with Huang et al.,[23] our system has a more compact and balanced schedule, which reduces pipeline stages and subsequently saves power for data pipelining. Moreover, it provides the flexibility for algorithm development. For example, to enable the preskip algorithm as shown in figure 8.29, the FME, which normally operates in the fine-prediction stage, can also calculate the matching cost associated with the motion vector predictor, which is the motion vector of skip mode, at the beginning of the coarse-prediction stage.

The gated clock technique is essential for a power-aware encoder design. Based on the system hierarchy, a fine-grained module-wise gated clock scheme is implemented to precisely turn off the clocks of static register files and inactive PEGs. According to the power profile, the hardware still consumes considerable power even if it does not function. In an encoder, to reduce the off- and on-chip memory bandwidth, many registers are used in the memory hierarchy. In addition, the power-aware feature makes PEGs idle in various time periods. The content of the large register files in system and pipeline controllers is seldom changed. With this system hierarchy and module-wise gated clock, 18.8 to 33.2 percent power can be saved.

8.3.7 Low-Power CAVLC Architecture

In the AVC baseline profile, there are two entropy coding tools: Exp-Golomb coding and CAVLC, which are for the macroblock (MB) headers and transformed prediction residues, respectively.

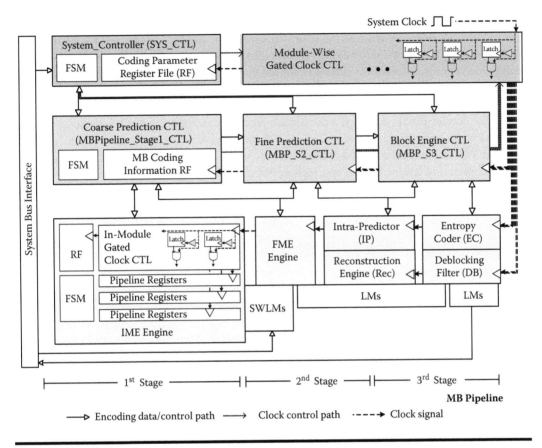

Figure 8.30   Flexible H.264/AVC encoder system with module-wise gated clock scheme.[3]

Because Exp-Golomb coding has far lower power consumption than CAVLC, only the design of CAVLC is introduced here.

The CAVLC defines six encoding symbols, including *TotalCoeff*, *TrailingOnes*, *TrailingOnes-Sign*, *Level*, *TotalZeros*, and *RunBefore*. First, all residues are scanned in the inverse zigzag order for extracting the symbols. The definitions for all symbols are listed below. Note that TotalCoeff and TrailingOnes form a joint symbol for coding.

TotalCoeff: Number of total nonzero residues.
TrailingOnes: Number of consecutive ±1 from the beginning, considering only the non-zero residues.
TrailingOnesSign: Sign of TrailingOnes, one for negative and zero for positive.
Level: Value of nonzero residues.
TotalZeros: Zeros after the first nonzero residue.
RunBefore: Zeros between current and previous nonzero residues, recorded from the second nonzero residue till all zeros have been counted.

The TotalCoeff symbol, as an important factor of context-based adaptivity in CAVLC, is the first one among all symbols to be coded. This is because other CAVLC symbols are coded with adaptive table selection depending on the value of TotalCoeff, which can help to compress the statistical redundancies efficiently. In the previous two-pass CAVLC schemes, the CAVLC engine first needs to sequentially scan all transformed prediction residues in the residue SRAM, and online extracts the symbols into buffers. This is called the scan pass. Note that the buffers cannot be discarded because all symbols must wait for the TotalCoeff symbol for the context-based adaptive coding. Only after the scan pass is finished can the TotalCoeff be obtained and the lookup table (LUT) pass start to code symbols. This forms the single-block two-pass CAVLC architecture.[20] To increase the throughput, a dual-block-pipelined architecture was proposed in Chen et al.,[21] which can perform the scan and LUT simultaneously.

These two schemes are not good enough in the low-power performance due to their large power consumption in the residue SRAM and register-based symbol buffers. The side information–aided (SIA) symbol-look-ahead (SLA) one-pass CAVLC scheme[33] is proposed to reduce the large power consumption in the residue SRAM and symbol buffers efficiently. In this architecture, the side information registers help to minimize the residue SRAM accesses by indicating the locations of nonzero residues, and meanwhile provide more information for one-pass symbol look-ahead and coding. This means the symbols can be obtained and immediately coded in one pass, before a complete scan of the residue SRAM.

The first type of adopted side information is called nonzero flags, which are composed of a 16×16 2-D array of one-bit registers. The value of its entry is assigned to 1 if the corresponding residue is nonzero valued, and 0 is assigned otherwise. The second side information type is abs-one flags. If the residue value equals ±1, its corresponding entry is set to 1. With the nonzero and abs-one flags, the low-power CAVLC scheme is proposed, and listed as follows. Note that the SIA flags are also accessed in the inverse zigzag scan order during executing the scheme.

> *TotalCoeff*: Sum up all nonzero flags.
> *TrailingOnes*: Sum up the abs-one flags located before the first one of XOR of nonzero and abs-one flags.
> *TrailingOnesSign*: Read residue SRAM with address generated by the nonzero flag's position.
> *Level*: Same as TrailingOnesSign.
> *TotalZeros*: Sum up the inverted nonzero flags located after the first nonzero flag.
> *RunBefore*: Subtract previous nonzero flag's position from current nonzero flag's position.

The nonzero flags are greatly helpful for improving the total CAVLC performance. Most symbols can be obtained from it, or from the residue SRAM with reading address, generated by its information. The abs-one flags in cooperation with the nonzero flags can make the TrailingOnes symbol coding also one-pass. Without abs-one flags, a residue SRAM scan is required to calculate the number of trailing ones among all residues, and the SLA scheme will fail. With both flags as the side information, the low-power CAVLC scheme can minimize the residue SRAM accesses and eliminate the symbol buffers, such that the total EC power can be greatly reduced.

Figure 8.31 shows the low-power entropy coding architecture. In the CAVLC engine, an SLA module is implemented with the SIA flags as the input to execute the symbol look-ahead operations. All symbols are extracted and fed to the LUT modules in one pass without buffering. Finally, the bit stream packer packs all coded bits into a bus-width-aligned bit stream for output.

The SLA module is the key component to support the low-power CAVLC scheme. Figure 8.32 shows the proposed reconfigurable architecture. It is designed to exploit the hardware resource sharing of common circuits in the SLA module. A leading detector is employed to generate the value of the first one's position in a 16-bit binary string. The masking unit generates a 16-bit binary mask and performs bitwise logic operations between the mask and its input. With this

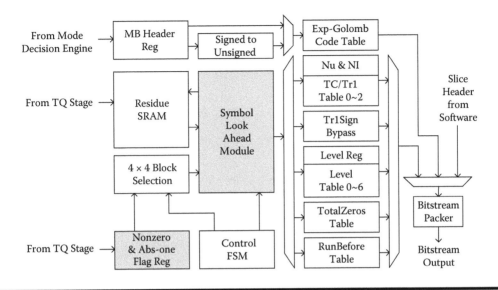

Figure 8.31   Proposed low-power architecture of H.264/AVC baseline profile entropy coding. Shaded modules are the key features.[33]

SLA module, the implementation result shows that the total entropy coding power consumption is reduced by 69 percent.

8.3.8   Implementation Results

Table 8.3 shows the chip features of this low-power and power-aware H.264/AVC encoder. The core size is 12.84 mm² with 0.18 μm 1P6M CMOS technology. The design costs 452.8K logic gates and 16.95 KB SRAMs. Power dissipation is 9.8 to 43.3 mW at 1.3/1.8 V supply voltage and 13.5/27 MHz working frequency for 30fps CIF video. Figure 8.33 shows the chip micrograph. Figure 8.34 shows the power-distortion curves. On average, this encoder consumes 22.1 percent power compared to the full-search H.264/AVC ASIC encoder[23] with similar

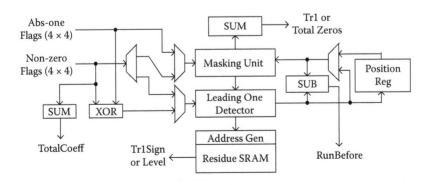

Figure 8.32   Reconfigurable SLA module.[33]

Table 8.3   The Chip Specification of the Power-Aware H.264/AVC Encoder

Technology	TSMC 0.18μm 1P6M CMOS
Pad/core voltage	3.3 V (core)/1.8 V (I/O)
Core area	3.47 × 3.70 mm²
Logic gates	452.8K (2-input NAND gate)
SRAM	16.95 KByte
Maximum number of reference frame	2
Maximum horizontal/vertical search range	[−32, +31]/[−16, +15]
Power consumption	33.5–67.2 mW for SDTV, 1 ref @ 54 MHz, 1.8 V
(Measured results)	40.3 mW for CIF, 2 ref @ 27 MHz, 1.8 V
	9.8–15.9 mW for CIF, 1 ref @ 13.5 MHz, 1.3 V
	8.7 mW for QCIF, 2 ref @ 6.25 MHz, 1.3 V
	2.8–4.3 mW for QCIF, 1 ref @ 3.125 MHz, 1.3 V
Power consumption	9.1–16.3 mW for SDTV, 1 ref @ 54 MHz, 1.3 V
(Simulated results with	12.9 mW for CIF, 2 ref @ 27 MHz, 1.3 V
TSMC 0.13μm process)	5.1–8.2 mW for CIF, 1 ref @ 13.5 MHz, 1.3 V
	4.5 mW for QCIF, 2 ref @ 6.25 MHz, 1.3 V
	1.5–2.2 mW for QCIF, 1 ref @ 3.125 MHz, 1.3 V

compression performance. In addition, the encoder makes a 1.96 dB quality improvement with about 5 mW extra power compared to the low-power MPEG-4 encoder.[32] With the power scalable capability, this encoder has a 2.4 dB quality variation with associated power consumption from 9.8 to 43.3 mW for CIF 30fps video. With low-power consumption, the encoder can extend the

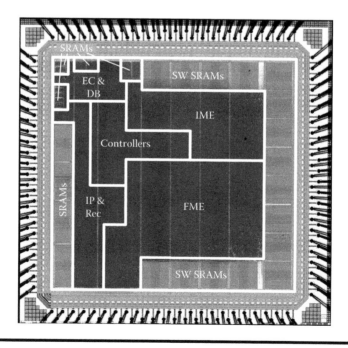

Figure 8.33   The power-aware H.264/AVC encoder chip micrograph.[3]

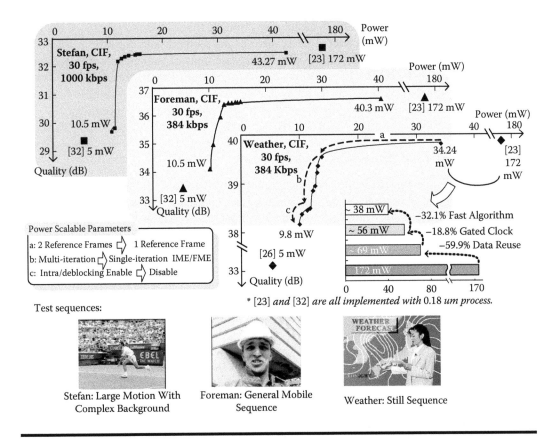

Figure 8.34   Power-distortion curves. The power-scalable performance of the H.264/AVC encoder.

operation time of portable devices. With power-aware functionality, the encoder can trade certain compression performance for less power consumption according to power status and application requirements for mobile applications.

8.3.9   Summary

In this section, we see the design of a low-power and power-aware H.264/AVC encoder. Parallel architecture along with fast algorithms and a data reuse scheme enable about 78 percent power savings compared with the first design case discussed in this chapter. Power scalability is provided through a flexible system hierarchy and reconfigurable architecture that supports multiple power modes, content-aware algorithms, and module-wise clock gating.

8.4   An H.264/AVC Decoder with Hybrid Task Pipelining

Typical video coding algorithms adopt asymmetric-complexity architecture. That is, the implementation complexity of a decoder is much less than that of an encoder. It is based on the idea that for broadcasting or streaming applications, video is compressed once and decoded many times at the consumers' side. We have seen from complexity profiling that motion estimation is the most time-consuming process in an encoder. This resource-eating monster does not exist in a decoder.

In this section, an H.264/AVC decoder with hybrid task pipelining architecture[5] is presented. The specification is up to HDTV1024p 30fps. As for specifications down to QCIF (176 × 144) 15Hz fps, the design delivers real-time decoding at 725 KHz with a low-power feature. In the following, the scheduling as well as the key modules of this decoding system are discussed.

8.4.1 Hybrid Task Pipelining Architecture

Figure 8.35 shows the system architecture of the decoder. The sequence parameter set, picture parameter set, and slice headers are parsed by the system processor. The MB-level information, including MB headers and transformed/quantized residues, is decoded by the *Parser* engine. The predicted pixels are generated by the *Inter Pred.* engine or *Intra Pred.* engine according to the MB mode. The residues are recovered by the *IQ/IT* engine. The MB is reconstructed by the *Sum and Clipping* engine. Finally, the *Deblocking* engine filters MB pixels and outputs them to the external memory. The buffers between the processing engines are required to separate pipelining stages. Different from the MB pipelining scheme, the decoder architecture here is based on a hybrid task pipelining scheme including 4 × 4 block-level pipelining, MB-level pipelining, and frame-level pipelining.

There are several reasons to consider this hybrid scheme. In H.264/AVC, 4 × 4 block is the smallest common element of the prediction block mode. The transforms and entropy coding are

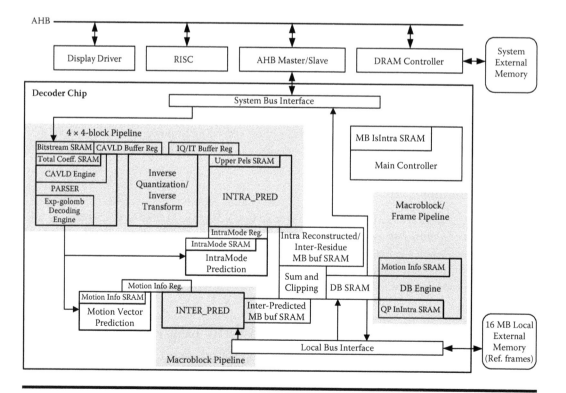

Figure 8.35   Hybrid task pipelining decoder system block diagram.[5]

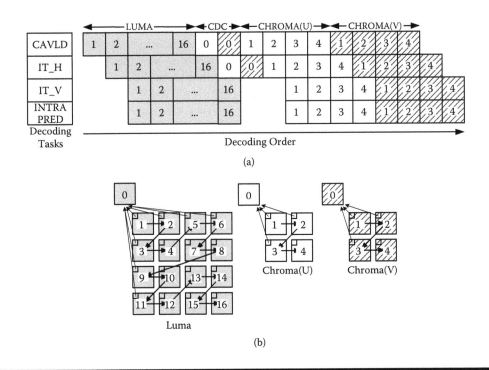

Figure 8.36 (a) 4 × 4 block pipelining scheme. Each box represents a 4 × 4 block, and the number on each 4 × 4 block is the 4 × 4 block index. (b) Double-z-scan order of 4 × 4 blocks.

also based on 4 × 4 blocks. Therefore, a 4 × 4 block pipelining scheme can be designed for context-adaptive variable-length code decoder (CAVLD), inverse quantization/inverse transformation, and intraprediction with the benefit of less coding latency. It requires about 1/24 the buffer size as traditional MB pipelining architecture. As with intraprediction, the basic processing element of interprediction is also a 4 × 4 block.

Figure 8.36(a) explains the proposed 4 × 4 block pipelining scheduling. Each box labeled with a number represents a 4 × 4 block, the horizontal axis represents time, and the boxes of the same time slot denote concurrent pipelining tasks. Following the decoding order shown in Figure 8.36(b), the decoder processes 1 luma DC block (if any), 16 luma blocks, 2 chroma DC blocks, 4 chroma-u blocks, and 4 chroma-v blocks. As for interprediction, it produces the predicted MB pixels from previously decoded reference frames. Due to the 6-tap FIR filter for interpolation, 9 × 9 integer reference pixels are required for a current 4 × 4 block. If the blocks of prediction mode are larger than 4 × 4, overlapped reference frame pixels of these 4 × 4 blocks can be reused to reduce the system bandwidth. The inherent order of 4 × 4 blocks in the bit stream is the double-z-scan order, as illustrated in figure 8.36. Reference frame data reuse will be less efficient if the Inter Pred. engine adopts the 4 × 4 block pipelining scheme and follows the double-z-scan order. For example, blocks 3 and 9 in figure 8.36(b) cannot share reference frame data efficiently. Therefore, the Inter Pred. engine should be scheduled to MB-level pipelining with a customized scan order to exploit the reference frame data reuse. All reference pixels necessary to predict a MB are read from memory at once to reduce memory bandwidth.

Deblocking filter is another special case that does not suit the 4 × 4 block pipelining double-z-scan order. The deblocking engine filters the edges of each 4 × 4 block vertically, then horizontally.

In addition, one 4 × 4 block cannot be completely filtered until its neighboring blocks are reconstructed. This data dependency makes it impractical to fit the deblocking operations into a 4 × 4 block pipelining, because the buffer cannot be efficiently reduced and serious control overhead is required. Therefore, the MB pipelining schedule is usually adopted for deblocking filter. However, if the decoder has to support flexible macroblock ordering (FMO) and arbitrary slice ordering (ASO), where MBs in a frame may not be coded in raster-scan order, the DB engine has to be scheduled to frame-level pipelining because the filtering order of one frame cannot be violated in MB boundaries.

8.4.2 Multisymbol CAVLC Decoder

The processing cycles of the CAVLC decoder vary with the symbol count. In the situation of the highly textured video or high-quality applications, the real-time constraint may be violated. A multisymbol decoding engine with low area cost is adopted. In general, the parallelization in the sequential CAVLD algorithm will cause the exponentially increased area. Therefore, instead of decoding all possible successive symbols in parallel, the proposed multisymbol decoding engine supports the decoding of multiple *level* or *run* symbols in one cycle. Because these two symbols occur consecutively most frequently, the method can greatly improve the hardware throughput with acceptable area overhead. The simulation results are shown in figure 8.37.

8.4.3 Low-Bandwidth Motion Compensation Engine

According to the analysis in system-level design, motion compensation should be scheduled in MB-level pipelining with a customized scan order to exploit the reference frame data reuse. We first adopt the 4 × 4 based MC. All VBSs are decomposed into several 4 × 4 element blocks, and processed sequentially by the 4 × 4 based MC engine with full hardware utilization. The straightforward memory access scheme processes every decomposed 4 × 4 element block independently, and always loads 9 × 9 pixels from the external memory for interpolation, as shown in figure 8.38(a). The system bandwidth requirement of 4 × 4 based MC can be reduced by two bandwidth reduction techniques.[35]

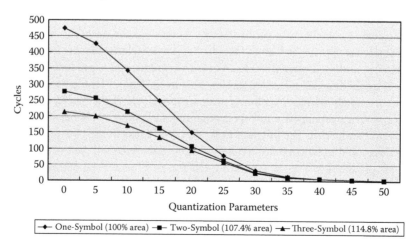

Figure 8.37 Simulation results of one-symbol and multisymbol decoding of CAVLC.

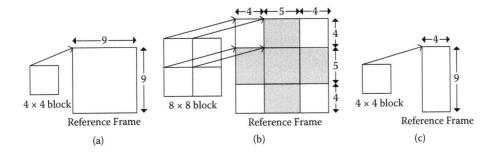

Figure 8.38 **(a) General-case interpolation window. (b) Four interpolation windows for an 8 × 8 block (shaded region means reusable). (c) Interpolation window when MV pointing to horizontal integer pixels.**

The first technique is interpolation window reuse (IWR). As shown in figure 8.38(b), there are overlapped regions between interpolation windows for neighboring 4 × 4 element blocks when the block mode is larger than 4 × 4. The shaded regions represent the reference pixels that can be reused. The second scheme is interpolation window classification (IWC). The interpolation window is not always (X + 5) × (Y + 5) for an X × Y block. As shown in figure 8.38(c), a 4 × 4 block with integer MV in the horizontal direction does not require horizontal filtering. Only a 4 × 9 interpolation window is required to be loaded in this situation. In brief, the IWR and IWC schemes aim to precisely control the MC hardware to load a smaller and exact interpolation window.

Figure 8.39 shows the MC architecture. An efficient scheduling for vertical data reuse is applied with *downshift register array* to support vertical IWR. In addition, a *horizontal reuse memory* is designed for horizontal IWR. The IWC is implemented by *control FSM* and *address generator*.

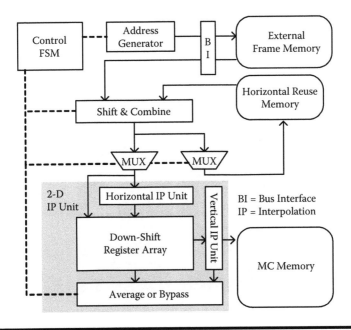

Figure 8.39 **Block diagram of MC hardware.**[5]

Table 8.4   Gate Count Profile of the H.264/AVC Decoder

	Gate Count	%
Entropy decoder	21,121	9.7
Interprediction	69,695	32.1
Deblocking filter	35,437	16.3
Intraprediction	28,707	13.2
IQ/IT	19,792	9.1
Control engine	22,695	10.4
SRAM BIST	8,937	4.1
Miscellaneous	11,043	5.1
Total	217,428	100

The *shift and combine* circuit packs the required integer pixels' input from the external frame memory and horizontal reuse memory. The 2-D IP unit performs the interpolation, and the compensated MB is buffered in the MC memory. The 2-D IP unit is the key part of the MC module. The proposed techniques can provide about 60 to 80 percent bandwidth reduction for the 4 × 4 based MC. After integrated into an H.264/AVC HDTV1024p decoder, the MC engine can reduce 40 to 50 percent of the total system bandwidth.

8.4.4 Deblocking Filter

The in-loop deblocking filter is employed at 4 × 4 block boundaries to reduce block artifacts. In the reference software, the processing order of block boundaries is done vertically and then horizontally in the order shown in figure 8.20 (v1, v2, v3, ..., v23, v24, and then h1, h2, h3, ...,

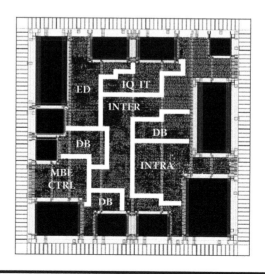

Figure 8.40   The chip layout of H.264/AVC decoder.

Table 8.5   Detailed Specification of the H.264/AVC Decoder

Technology	TSMC 0.18µm 1P6M CMOS
Chip area	3.24×3.24 mm
Core area	2.19×2.19 mm
Gate count	217,428 (SRAM excluded)
On-chip memories	128×32 dual-port SRAM $\times 1$
	160×32 dual-port SRAM $\times 1$
	$1,024 \times 32$ single-port SRAM $\times 1$
	128×12 single-port SRAM $\times 1$
	128×16 single-port SRAM $\times 1$
	256×20 single-port SRAM $\times 1$
	512×30 single-port SRAM $\times 1$
	96×32 single-port SRAM $\times 2$
	96×40 single-port SRAM $\times 2$
	Total : 79,872 bits (9,984 bytes)
Clock rate	120 MHz
Processing capability	$2,048 \times 1,024$ 30fps @ 102 MHz
Power consumption	186.4 mW @ 120 MHz, 1.8 V

h23, h24). In this decoder design, the same deblocking filter architecture (figure 8.21) as that in the encoder is adopted. Before deblocking an MB, the MB data and adjacent block data are prepared in the deblock SRAM. An eight-parallel 1-D filter can process eight pixels for one edge in parallel. The 8×4 unfiltered pixels of two adjacent 4×4 blocks are stored in an 8×4 pixel array with reconfigurable data paths to support both horizontal and vertical filtering with the same 1-D filters.

There have been many papers about deblocking filter architecture design with refinement in different aspects. Interested readers can refer to some of them[37-42] for more details. For example, in Shih et al.[37] the column addressing pixels in MB is used after horizontal filtering to favor the direction of vertical filtering. In Liu et al.,[38] the horizontal and vertical edges are filtered interleaving, and more data reuse can be achieved.

8.4.5   Implementation

Table 8.4 shows the logic gate count synthesized at 120 MHz. The total logic gate count is about 217K. The chip is fabricated with TSMC 0.18µm 1P6M CMOS process. The core size is about 2.19×2.19 mm. It can support AVC decoding at a baseline profile level of 4.1 with five reference frames. In total, about 10 KByte of on-chip SRAMs are required. Figure 8.40 shows the layout of the decoder, and the detailed specifications are listed in table 8.5.

8.4.6   Summary

In this section, architecture design for an H.264/AVC decoder is discussed. The major idea is to arrange decoding functions into proper pipelining schedules. Degrees of parallelism are chosen to meet real-time requirements. The external memory bandwidth is greatly reduced by an efficient data reuse scheme. Simulation results show that the processing capability of this architecture can support up to 2,048 × 1,024 30fps videos at 120 MHz with 1.8V power supply and 186.4 mW power consumption. For QCIF 15 Hz video, the design can support real-time decoding at 725 KHz with 1.8 V power supply and consumes power of about 1.18 mW.

References

1. Y.-W. Huang. 2004. Algorithm and architecture design for motion estimation, H.264/AVC standard, and intelligent video signal processing. PhD dissertation, Graduate Institute of Electronics Engineering, National Taiwan University, Taipei.
2. T.-C. Chen, S.-Y. Chien, Y.-W. Huang, C.-H. Tsai, C.-Y. Chen, T.-W. Chen, and L.-G. Chen. 2006. Analysis and architecture design of an HDTV720p 30 frames/s H.264/AVC encoder. *IEEE Transactions on Circuits and Systems for Video Technology* 16:673–88.
3. T.-C. Chen, Y.-H. Chen, C.-Y. Tsai, S.-F. Tsai, S.-Y. Chien, and L.-G. Chen. 2.8 to 67.2mW low-power and power-aware H.264 encoder for mobile applications. In *Proceedings of 2007 Symposium on VLSI Circuits,* Kyoto, 222–223.
4. T.-W. Chen. 2005. Design and implementation of an H.264/AVC decoder for 2048 × 1024 30fps videos. Master thesis, Graduate Institute of Electronics Engineering, National Taiwan University, Taipei, Taiwan.
5. T.-W. Chen, Y.-W. Huang, T.-C. Chen, Y.-H. Chen, C.-Y. Tsai, and L.-G. Chen. 2005. Architecture design of H.264/AVC decoder with hybrid task pipelining for high definition videos. In *Proceedings of 2005 International Symposium on Circuits and Systems (ISCAS 2005),* Kobe, 3, 2931–34.
6. T. Wiegand, H. Schwarz, A. Joch, F. Kossentini, and G. J. Sullivan. 2003. Rate-constrained coder control and comparison of video coding standards. *IEEE Transactions on Circuits and Systems for Video Technology* 13:688–703.
7. T. Wiegand, G. J. Sullivan, G. Bjontegaard, and A. Luthra. 2003. Overview of the H.264/AVC video coding standard. *IEEE Transactions on Circuits and Systems for Video Technology* 13:560–76.
8. J. Ostermann, J. Bormans, P. List, D. Marpe, M. Narroschke, F. Pereira, T. Stockhammer, and T. Wedi. 2004. Video coding with H.264/AVC: Tools, performance, and complexity.' *IEEE Circuits Syst. Mag.* 4:7–28.
9. A. Puri, X. Chen, and A. Luthra. 2004. Video coding using the H.264/MPEG-4 AVC compression standard. *Signal Processing: Image Communication* 19:793–849.
10. Joint Video Team. 2004. Reference software JM8.5.
11. IPROF. Portable instruction level profiler by Peter M. Kuhn. http://www.lis.ei.tum.de/research/bv/topics/method/e_iprof.html.
12. H.-C. Chang, L.-G. Chen, M.-Y. Hsu, and Y.-C. Chang. 2000. Performance analysis and architecture evaluation of MPEG-4 video codec system. In *Proceedings of IEEE International Symposium on Circuits Systems (ISCAS'00),* Geneva, 2, 449–52.
13. Y.-W. Huang, C.-Y. Chen, C.-H. Tsai, C.-F. Shen, and L.-G. Chen. 2006. Survey on block matching motion estimation algorithms and architectures with new results. *Journal of VLSI Signal Processing Systems* 42:297–320.
14. Z.-L. He, C.-Y. Tsui, K.-K. Chan, and M.-L. Liou. 2000. Low-power VLSI design for motion estimation using adaptive pixel truncation. *IEEE Transactions on Circuits and Systems for Video Technology* 10:669–78.
15. B. Liu and A. Zaccarin. 1993. New fast algorithms for the estimation of block motion vectors. *IEEE Transactions on Circuits and Systems for Video Technology* 3:148–57.

16. S. Saponara and L. Fanucci. 2004. Data-adaptive motion estimation algorithm and VLSI architecture design for low-power video systems. *IEE Proceedings on Computers and Digital Techniques* 151:51–59.
17. J.-C. Tuan, T.-S. Chang, and C.-W. Jen. 2002. On the data reuse and memory bandwidth analysis for full-search block-matching VLSI architecture. *IEEE Transactions on Circuits and Systems for Video Technology* 12:61–72.
18. T.-C. Chen, Y.-W. Huang, and L.-G. Chen. 2004. Fully utilized and reusable architecture for fractional motion estimation of H.264/AVC. In *Proceedings of IEEE International Conference on Acoustics, Speech, and Signal Processing (ICASSP'04)*, 5, 9–12.
19. T.-C. Wang, Y.-W. Huang, H.-C. Fang, and L.-G. Chen. 2003. Parallel 4 × 4 2D transform and inverse transform architecture for MPEG-4 AVC/H.264. In *Proceedings of IEEE International Symposium on Circuits and Systems (ISCAS'03)*, Bangkok, 2, 800–3.
20. Y.-W. Huang, B.-Y. Hsieh, T.-C. Chen, and L.-G. Chen. 2004. Analysis, fast algorithm, and VLSI architecture design for H.264/AVC intra frame coder. *IEEE Transactions on Circuits and Systems for Video Technology* 15:378–401.
21. T.-C. Chen, Y.-W. Huang, C. Y. Tsai, and L.-G. Chen. 2006. Architecture design of context-based adaptive variable length coding for H. 264/AVC, *IEEE Transactions on Circuits and Systems II*, 53, 832–36.
22. Y.-W. Huang, T.-W. Chen, B.-Y. Hsieh, T.-C. Wang, T.-H. Chang, and L.-G. Chen. 2003. Architecture design for deblocking filter in H.264/JVT/AVC. In *Proceedings of IEEE International Conference on Multimedia and Expo (ICME'03)*, Baltimore, 1, 693–96.
23. Y.-W. Huang, T.-C. Chen, C.-H. Tsai, C.-Y. Chen, T.-W. Chen, C.-S. Chen, C.-F. Shen, S.-Y. Ma, T.-C. Wang, B.-Y. Hsieh, H.-C. Fang, and L.-G. Chen. 2005. A 1.3TOPS H.264/AVC single-chip encoder for HDTV applications. In *IEEE International Solid-State Circuits Conference 2005 (ISSCC'05)*, San Francisco, 1, 128–30. Digest of Technical Papers.
24. T. Sakurai. 2003. Perspectives on power-aware electronics. In *IEEE International Solid-State Circuits Conference 2003 (ISSCC'03)*, San Francisco, 1, 26–29. Digest of Technical Papers.
25. L.-G. Chen. 2006. Power-aware multimedia. In *IEEE International Solid-State Circuits Conference 2006 (ISSCC 2006)*, 17. SE2 Power-Aware Signal Processing, Evening Session, Digest of Technical Papers.
26. M. Bhardwaj, R. Min, and A. P. Chandrakasan. 2001. Quantifying and enhancing power awareness of VLSI systems. *IEEE Transactions on Very Large Scale Integration (VLSI) Systems* 9:757–72.
27. Z. He and S. K. Mitra. 2005. From rate-distortion analysis to resource-distortion analysis. *IEEE Circuits and Systems Magazine* 5:6–18.
28. Z. He, Y. Liang, L. Chen, I. Ahmad, and D. Wu. 2005. Power-rate-distortion analysis for wireless video communication under energy constraints. *IEEE Transactions on Circuits and Systems for Video Technology* 15:645–58.
29. D. Linden. 1995. *Handbook of batteries*. 2nd ed. New York: McGraw-Hill.
30. T.-C. Chen, Y.-H. Chen, S.-F. Tsai, S.-Y. Chien, and L.-G. Chen. 2007. Fast algorithm and architecture design of low power integer motion estimation for H.264/AVC. *IEEE Transactions on Circuits and Systems for Video Technology* 17:568–77.
31. T.-C. Chen, Y.-H. Chen, C.-Y. Tsai, and L.-G. Chen. 2006. Low power and power aware fractional motion estimation of H.264/AVC for mobile applications. In *Proceedings of IEEE International Symposium on Circuits and Systems (ISCAS2006)*, Kos, Greece, 4, 21–24.
32. C.-P. Lin, P.-C. Tseng, Y.-T. Chiu, S.-S. Lin, C.-C. Cheng, H.-C. Fang, W.-M. Chao, and L.-G. Chen. A 5mW MPEG4 SP encoder with 2D bandwidth-sharing motion estimation for mobile applications. In *Proceedings of 2006 Internal Solid-State Circuits Conference (ISSCC2006)*, San Francisco, 1, 1626–35.
33. C.-Y. Tsai, T.-C. Chen, and L.-G. Chen. 2006. Low power entropy coding hardware design for H.264/AVC baseline profile encoder. In *Proceedings of IEEE International Symposium on Multimedia & Expo (ICME2006)*, Toronto, 1941–44.

34. H.-Y. Kang, K.-A. Jeong, J.-Y. Bae, Y.-S. Lee, and S.-H. Lee. 2004. MPEG4 AVC/H.264 decoder with scalable bus architecture and dual memory controller. In *Proceedings of International Symposium on Circuits and Systems (ISCAS'04)*.

35. C.-Y. Tsai, T.-C. Chen, T.-W. Chen, and L.-G. Chen. 2005. Bandwidth optimized motion compensation hardware design for H.264/AVC HDTV decoder. In *Proceedings of 2005 International Midwest Symposium on Circuit and Systems (MWSCAS'05)*, Cincinnati, 2, 1199–1202.

36. Y.-W. Huang, T.-C. Wang, B.-Y. Hsieh, T.-C. Wang, T.-H. Chang, and L.-G. Chen. 2003. Architecture design for deblocking filter in H.264/JVT/AVC. In *Proceedings of IEEE International Conference on Multimedia and Expo (ICME2003)*, 1693–96.

37. S.-Y. Shih, C.-R. Chang, and Y.-L. Lin. 2004. An AMBA-compliant deblocking filter IP for H.264/AVC. In *IEEE International Symposium on Circuits and Systems (ISCAS'05)*, Kobe, 5, 4529–32.

38. T.-M. Liu, W.-P. Lee, and C.-Y. Lee. 2005. An area-efficient and high-throughput de-blocking filter for multi-standard video applications. In *Proceedings of IEEE International Conference on Image Processing (ICIP'05)*, Genova, 3, 1044–47.

39. C.-C. Cheng, T.-S. Chang, and K.-B. Lee. 2006. An in-place architecture for the deblocking filter in H.264/AVC. *IEEE Transactions on Circuits and Systems II: Express Briefs* 53:530–34.

40. G. Khurana, A. A. Kassim, T. P. Chua, and M. B. Mi. 2006. A pipelined hardware implementation of in-loop deblocking filter in H.264/AVC. *IEEE Transactions on Consumer Electronics* 52:536–40.

41. Y.-C. Chao, J.-K. Lin, J.-F. Yang, and B.-D. Liu. 2006. A high throughput and data reuse architecture for H.264/AVC deblocking filter. In *IEEE Asia Pacific Conference on Circuits and Systems 2006 (APCCAS'06)*, Singapore, 1260–63.

42. H.-Y. Lin, J.-J. Yang, B.-D. Liu, and J.-F. Yang. 2006. Efficient deblocking filter architecture for H.264 video coders. In *Proceedings of IEEE International Symposium on Circuits and Systems (ISCAS'06)*, Kos, Greece, 2617–20.

Chapter 9

Low-Power Analog-to-Digital Converters (ADCs) for Mobile Broadcasting Applications

J. Li, X. Zeng, and Y. Guo

Contents

Keywords

analog-to-digital converter, ADC, ADC architectures, low power, mobile TV

9.1   Introduction

Digital broadcasting technologies for mobile TV have started a new era in the field of consumer electronics. These technologies include Terrestrial–Digital Multimedia Broadcasting (T-DMB)[1] in Korea, Integrated Service Digital Broadcasting–Terrestrial (ISDB-T)[2] in Japan, Digital Video Broadcasting–Handheld (DVB-H)[3] in Europe, and Media Forward Link Only (FLO)[4] technology introduced by Qualcomm. Table 9.1 shows the typical channel bandwidth of these standards. In all the above technologies, analog-to-digital converters (ADCs) are indispensable and key components in the receiver signal chains to perform the analog-to-digital signal conversions. For example, figure 9.1 shows a simplified receiver architecture for the technologies. In figure 9.1, the radio frequency (RF) signal from the antenna goes to the tuner, which downconverts the RF signal to an intermediate frequency (IF) or a zero-IF signal, depending on the receiver architecture. The signal is filtered by a low-pass or band-pass filter to limit the bandwidth of the signal. An ADC digitizes the filtered band-limited signal. The digitized signal is then processed by the digital demodulator.

After the introduction, section 9.2 describes basic fundamentals and some important specifications of ADC. Section 9.3 reviews the most popular ADC architectures. Section 9.4 discusses the ADCs used for the digital TV applications and gives design examples. The summary is given in section 9.5.

9.2   Introduction and Characterization of ADCs

When an ADC is used in the communication application it is important to know how it works and how it will influence the performance of the entire system. Therefore, the ADC needs to be understood and characterized. This section discusses the fundamentals of the ADC and introduces a number of commonly used dynamic performance measures.

Table 9.1   Typical Channel Bandwidth of Various Digital TV Broadcasting Standards

	T-DMB	ISDB-T	DVB-H	FLO
Typical channel bandwidth (MHz)	1.7	6, 7, 8	5, 6, 7, 8	5, 6, 7, 8

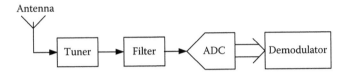

Figure 9.1   A simplified receiver architecture for the technologies.

9.2.1   *Fundamental Processes of ADC*

Figure 9.2 shows a block diagram of a general ADC. The applied analog input signal is first sampled and held for the analog-to-digital conversion by a sample-and-hold circuit. A quantizer then quantizes the sampled signal. The quantization results are then encoded by the digital processor to produce the corresponding digital bits. Generally, there is a filter, which is usually called an antialiasing filter, at the input of the ADC to avoid the aliasing (which will be explained later).

The operation of ADCs can be better understood by looking at the ADC input-output transfer curve. Figure 9.3 shows the input-output characteristic for an ideal 3-bit ADC. The input of an ADC is an analog signal, and the output is a digital code, which corresponds to the relative value of the input analog signal. Popular digital codes used for ADCs include binary, thermometer, Gray, and two's complement. However, the most widely used digital code is the binary code, which is shown in the figure. The resolution of the ADC, which is typically given in the number of bits N, indicates the smallest analog change that can be distinguished by an ADC. The value of the least significant bit (LSB) of an N-bit ADC is $FS/2^N$, where FS is the full scale of the input signal.

9.2.2   *Low-Pass Sampling and Band-Pass Sampling*

It is important to understand the frequency response of the ADC. Assume that the analog input signal, whose spectrum is shown in figure 9.4(a), is band limited with a maximum frequency of fmax. When the analog input signal is sampled at a frequency of fs, its spectrum is repeated at integer multiples of the sampling frequency fs, as shown in figure 9.4(b). When the fmax of the original analog signal is larger than 0.5 fs, the spectra begin to overlap (often called aliasing), as shown in figure 9.4(c). When aliasing happens, it is impossible to recover the original signal. Consequently, an antialiasing filter is usually needed at the input of the ADC to avoid the aliasing.

A sampling rate of two times the highest-frequency component of the analog signal is called the Nyquist sampling rate. Based on the rate at which the input signals sampled relative to its bandwidth, ADCs are partitioned into two groups: Nyquist-rate ADCs and oversampling ADCs. The types of ADCs that can operate at the Nyquist sampling rate are called Nyquist analog-to-digital converters. The types of ADCs that operate at a sampling rate much larger than the Nyquist sampling rate are called oversampling analog-to-digital converters.

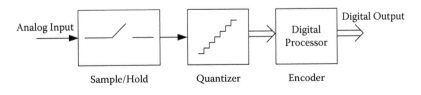

Figure 9.2   General block diagram for an ADC.

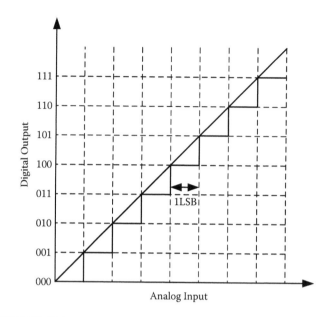

Figure 9.3   Input-output transfer curve of an ideal 3-bit ADC.

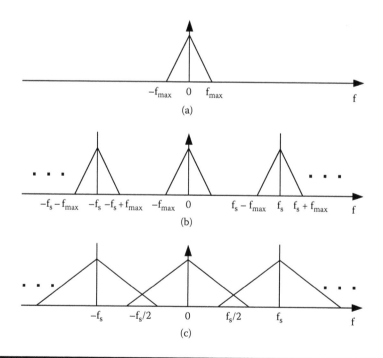

Figure 9.4   (a) Spectrum of the analog input signal. (b) Spectrum of the sampled data. (c) Case where fmax is larger than 0.5 fs.

The sampling process described above is called Nyquist sampling or low-pass sampling, in which the signal to be digitized is a low-pass band-limited signal, which has no frequency components above a certain frequency fmax. In low-pass sampling, the sampling rate needs to be at least two times the highest signal frequency fmax (i.e., $fs \geq 2fmax$) to prevent aliasing.

When the signal to be digitized is a band-pass signal, which has no frequency components below a certain frequency fl and above a certain frequency fh, the sampling process is called band-pass sampling. For band-pass sampling, the sampling frequency, which allows for an exact reconstruction of the information content of the analog signal, does not need to be larger than two times the highest signal frequency fmax. The minimum requirements on the sampling rate for band-pass sampling have to be at least two times the bandwidth of the information to prevent aliasing (i.e., $fs \geq 2[fh - fl]$) and the sampling frequency fs must satisfy[5]

$$\frac{2fh}{k} \leq fs \leq \frac{2fl}{k-1},$$ (9.1)

where k is restricted to integer values that satisfy

$$2 \leq k \leq \frac{fh}{fh - fl}$$ (9.2)

and $fh - fl < fl$.

If the output signal of the tuner (the input of the ADC) in figure 9.1 is a low-pass signal, then the ADC is required to perform low-pass sampling. If the output signal of the tuner (the input of the ADC) in figure 9.1 is a band-pass signal, then the ADC is required to perform band-pass sampling. In particular, band-pass sampling can be used to downconvert a signal from a band-pass signal at an RF or IF to a band-pass signal at a lower IF, which behaves like a mixer. Therefore, the sampling process can eliminate traditional analog down conversion components and so could reduce the system costs.

9.2.3 Important Specifications

The characteristics of ADCs can be divided into static and dynamic properties. However, for ADCs used in communications applications, the dynamic properties are more important. In digital TV broadcasting applications, the most critical criteria are that the ADC shall not significantly degrade the signal-to-noise ratio (SNR) of the system and that the ADC shall not introduce any large spurs or distortions. Here, a number of important dynamic specifications, which are determined by a single-tone sinusoidal input signal, are described. For a complete set of specifications, one is referred to van de Plassche[6] and Gustavsson et al.[7] and manufacturers' data books. Figure 9.5 shows a FFTspectrum of a 10-bit nonideal ADC when an input signal is a single-tone sinusoidal. The input signal appears as the fundamental in the FFT spectrum. The quantization error, which will be described next, generates a white noise floor. The spurs above the noise floor are caused by the nonlinearities of the ADC. Based on the FFT in figure 9.5, a number of dynamic specifications can be defined.

9.2.3.1 Signal-to-Noise Ratio (SNR)

The most important dynamic specification of an ADC is the signal-to-noise ratio (SNR). The input-output characteristic for an ideal ADC in figure 9.3 shows that the analog signal cannot be represented exactly with only a finite number of discrete amplitude levels. Therefore, some error, which is called quantization error, is introduced into the quantized signal. So an ideal ADC

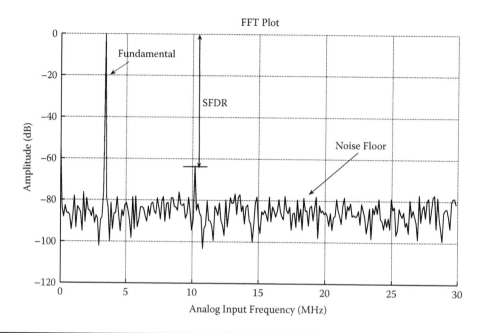

Figure 9.5   FFT spectrum of a nonideal 10-bit ADC.

introduces quantization noise in the converted signal due to the quantization. For simplicity, assume the error signal is uniformly distributed within a quantization level. Then the mean square value of quantization noise power can be expressed as

$$Pn = \frac{LSB^2}{12}.$$
(9.3)

If the analog input is a sinusoid with full scale, that is, $(2^{N-1} \cdot LSB)$, the total signal power Ps is equal to $(2^{N-1} \cdot LSB)^2/2$. Thus, the peak signal-to-noise ratio is

$$SNR = \frac{Ps}{Pn} = \frac{(2^{N-1} \cdot LSB)^2 / 2}{LSB^2 / 12} = 1.5 \cdot 2^{2N},$$
(9.4)

which, when expressed in decibels, becomes

$$SNR(dB) = 10 \cdot \log\left(\frac{Ps}{Pn}\right) = 6.02N + 1.76 dB$$
(9.5)

If oversampling is used, which means that the sampling rate *fs* is much larger than the signal bandwidth fin, the quantization noise floor has dropped. The *SNR* is still the same as before, but the noise energy has been spread over a wider frequency range. By following with a digital filter, the quantization noise power is less, because most of the noise passes through the digital filter. This enables oversampling ADCs to achieve wide dynamic range from a low-resolution ADC. Now the signal-to-noise ratio becomes

$$SNR(dB) = 10 \cdot \log\left(\frac{Ps}{Pn}\right) = 6.02N + 1.76 + 10\log(OSR) dB,$$
(9.6)

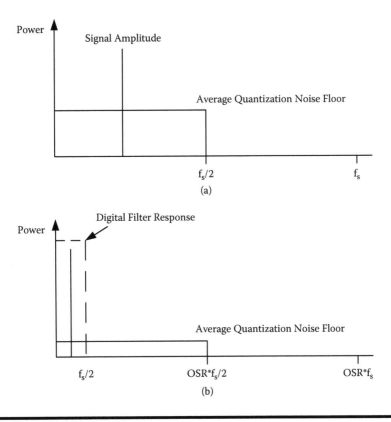

Figure 9.6 **(a) Spectrum diagram of an ADC with a sampling frequency fs. (b) Spectrum diagram of an ADC with a sampling frequency OSR*fs together with a digital filter.**

where the oversampling ratio (*OSR*) is given by *OSR* = *fs*/2/*fin*. The effect of oversampling on quantization noise in the frequency domain is illustrated in figure 9.6.

The *SNR* of an actual ADC includes nonideal effects of the ADC. For single tone measurements the *SNR* is the ratio of the power of the fundamental to the total noise power within half the sampling frequency, that is,

$$SNR = 10 \cdot \log\left(\frac{Signal\ Power}{Total\ Noise\ Power}\right). \qquad (9.7)$$

9.2.3.2 *Spurious Free Dynamic Range (SFDR)*

The spurious free dynamic range (*SFDR*) is defined as the ratio of the signal power to the power of the largest spurious within a certain frequency band. *SFDR* is usually expressed in *dBc* as

$$SFDR(dBc) = 10 \cdot \log\left(\frac{Signal\ Power}{Largest\ Spurious\ Power}\right). \qquad (9.8)$$

For an exact *SFDR* definition, the power level of the fundamental signal relative to the full scale must also be given. Normally the limiting factor of the SFDR in ADCs is harmonic distortion.

9.2.3.3 Total Harmonic Distortion (THD)

The harmonic distortion power of the total harmonic distortion (*THD*) is the ratio of the total harmonic distortion power to the power of the fundamental in a certain frequency band, that is,

$$THD = 10 \cdot \log\left(\frac{Signal\ Power}{Total\ Harmonic\ Distortion\ Power}\right) \tag{9.9}$$

9.2.3.4 Signal-to-Noise and Distortion Ratio (SNDR)

A more realistic figure of merit for an ADC is the signal–to–noise and distortion ratio (*SNDR* or *SINAD*), which is the ratio of the signal power to the total error power, including all spurs and harmonics, that is,

$$SNDR = 10 \cdot \log\left(\frac{Signal\ Power}{Noise\ Plus\ Distortion\ Power}\right). \tag{9.10}$$

9.2.3.5 Dynamic Range (DR)

The dynamic range is the ratio of the power of a full-scale sinusoidal input to the power of a sinusoidal input for which SNDR is 0 dB. DR is the range from full scale to the smallest detectable input. DR can be obtained by measuring the SNR as a function of the input power.

9.2.3.6 Effective Number of Bits (ENOB)

The effective number of bits (*ENOB*) of an ADC with a full-scale sinusoidal input is defined as

$$ENOB = \frac{SNDR - 1.76}{6.02}. \tag{9.11}$$

The *ENOB* provides an easy way to compare ADC with the same number of bits, but due to different circuit designs having different performance. For example, a 10-bit ADC with a 50 dB *SNDR* shows an 8-bit *ENOB*.

9.3 Overview of ADC Architectures

There are various ADC architectures with different properties. Before choosing a proper architecture for a particular application, their limitations must be known. In this section we review some of the most important architectures, which include the Nyquist ADCs (flash, two-step, pipelined, SAR) and the oversampling ADC (delta-sigma ADC). Their limitations and possible applications are also discussed.

9.3.1 Flash ADC

The fastest (with highest sampling rate) ADC is the flash ADC, which is also known as parallel ADC. The basic principle of the flash converter is illustrated in figure 9.7. The architecture can

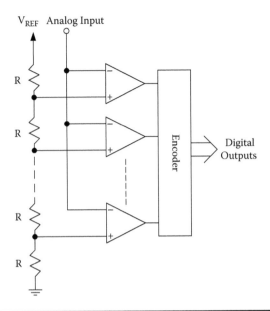

Figure 9.7   The flash ADC.

achieve very high sampling rates because the only analog building block is the comparator. The analog input signal is first sampled and then compared with different reference levels. There is one comparator for each decision level. The reference levels are usually generated by a resistor string, as shown in figure 9.7. The output of the comparators is a thermometer-coded representation of the input signal. An encoder is used to generate a more convenient representation, like binary coded, at the output.

The main problem with this architecture is that for N-bit ADC the number of comparators is $2^N - 1$, which could be very large when N is large. Furthermore, the offsets of all comparators need to be much less than the LSB of the converter, which makes the comparator design difficult and power consumption large when the resolution of the ADC is high. Due to the large number of high-resolution comparators, the number of bits of the flash ADC is limited to eight or less. Another problem caused by the large number of comparators is the large input capacitance, which must be handled by the ADC driving circuit. However, there are at least two advantages of this architecture: its fast conversion rate, and the latency through the converter is only one clock cycle. Latency in this case is defined as the difference between the time an analog sample is acquired by the ADC and the time when the digital data is available at the output. For applications requiring data immediately (for example, if the ADC is within a feedback control loop) or requiring very large bandwidths, a flash converter is generally the choice of architecture. Application examples include data acquisition, satellite communication, radar processing, sampling oscilloscopes, high-density disk drives, and so on.

9.3.2   Two-Step ADC

The two-step ADC, which is also referred to as half flash, could reduce the number of the comparators significantly. The architecture is shown in figure 9.8. The analog input is first sampled by a sample-and-hold (S/H) circuit; during the hold period, the first flash ADC1 performs a coarse

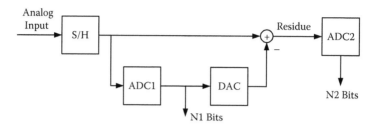

Figure 9.8   The two-step ADC.

quantization on the held signal. The resolved digital data is converted back to an analog value using a digital-to-analog converter (DAC). The held input signal is then subtracted from the analog output of the DAC; the residue of the subtraction is then quantized by ADC2 to perform a fine quantization. The number of the comparators for N-bit two-step ADC is now $2^{N1} + 2^{N2} - 2$, where $N1$ and $N2$ are the number of bits resolved in ADC1 and ADC2, respectively, and the sum of them is N. Although this architecture still requires the low-offset comparator with the full resolution of the converter, the number of low-offset comparators required is reduced significantly.

Actually, an interstage gain amplifier can be used (shown in figure 9.9) to amplify the residue signal to relax the comparator offset in the ADC2. Also, the accuracy requirement of the comparator in the ADC1 can be relaxed by using the digital error correction technique.[8,9] The basic idea of digital error correction is that by reducing interstage gain, the offset and nonlinearities caused by ADC1 can be tolerated and finally corrected in the digital domain by the digital error correction logic, which is composed by digital delays and adders. In practice, the gain is usually reduced by 1/2, which is realized by amplifying the residue signal by 2^{N1-1} instead of 2^{N1}, as shown in the figure. By using concurrent processing, the throughput of this architecture can sustain the same as the flash ADC. The performance and speed of this architecture largely depend on those of the S/H, DAC, and subtraction circuits. The converted outputs now have a latency of two clock cycles due to the extra stage, instead of one clock cycle in the flash one. However, if the system can tolerate latency of converted signal, two-step ADC can achieve higher resolution or smaller die size and power for a given resolution than the flash ADC. Two-step ADC is a cross between a flash ADC

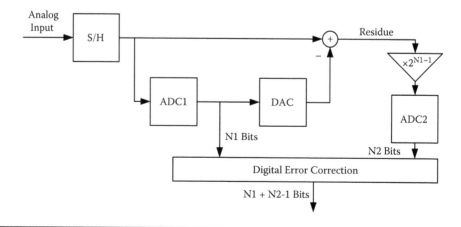

Figure 9.9   The two-step ADC with residue amplifier and digital error correction.

and pipelined ADC, which will be described next. So its applications are also a cross between the flash and pipelined ADC.

9.3.3   Pipelined ADC

In the two-step ADC, the number of comparators still grows exponentially with the overall number of bits. Also, the interstage gain block is not easy to build for high-resolution ADC. To further reduce the hardware cost, thus power consumption and die size at high resolution, one can reduce the per-stage resolution and cascade more stages to get the full resolution. This particular architecture is called pipelined. Figure 9.10 shows the conventional pipelined ADC architecture block diagram. As in previous architecture cases, the input signal is quantized to the resolution of the stage by ADCi, which is usually called sub-ADC, and concurrently is also subtracted from the DAC output of the present-stage digital output. The residue is then amplified and sampled and held for further conversion by the next stage. Identical operation is performed for each stage, and the digital outputs are combined properly by the digital error correction logic to achieve the required ADC resolution. The highest accuracy is required for the front-end S/H and MDAC in the first stage. After the first stage a reduced accuracy can be applied without influencing the overall converter accuracy too much, because of the gain factor in each stage.

By pipelining operation, the pipelined ADC offers a very high throughput and approximately linear hardware cost with resolution. The converted outputs now have a latency of several clock cycles that is linearly dependent on the number of the pipelining stages. But in most applications this is not a severe limitation. In most pipelined ADCs designed with CMOS technology, the S/H, DAC, summation node, and gain amplifier are usually implemented as a single switched-capacitor circuit block called a multiplying DAC (MDAC), as shown in figure 9.10. The major factor limiting MDAC accuracy is the inherent capacitor mismatch. In general, for about 12 bits of accuracy or higher, some form of capacitor/resistor trimming or calibration is required, especially for the first couple of stages. Up to now, the pipelined ADC has become the most popular ADC architecture for high-speed (sampling rates from a few megasamples per second (MS/s) up to 100 MS/s

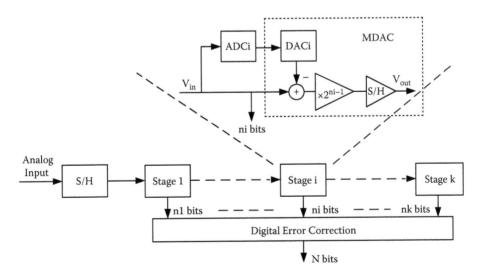

Figure 9.10   The pipelined ADC.

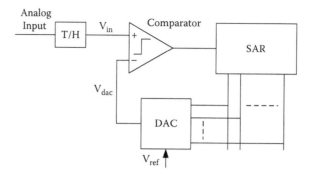

Figure 9.11   The SAR ADC.

and above) and high-resolution (8 to 16 bits) applications. These kinds of resolutions and sampling rates cover a wide range of applications, including charge-coupled device (CCD) imaging, digital receiver, base station, digital video (for example, HDTV), and fast Ethernet.

9.3.4   SAR ADC

The basic architecture of the successive approximation register (SAR) ADC is shown in figure 9.11. It consists of a track/hold (T/H) circuit, a comparator, a DAC, and a SAR. The analog input voltage (Vin) is held on the T/H. The register first forces the DAC output (Vdac) to be Vref/2, where Vref is the reference voltage provided to the ADC. A comparison is then performed to determine if Vin is less than or greater than Vdac. If Vin is greater than Vdac, the comparator output is logic high, or 1, and the most significant bit (MSB) of the register remains at 1. Conversely, if Vin is less than Vdac, the comparator output is logic low and the MSB of the register is cleared to logic 0. The process is repeated for all the bits. Once this is done, the conversion is complete, and the digital word is available in the register.

Many SAR ADCs use a capacitive DAC that provides an inherent track/hold function, so the T/H circuit can be embedded in the DAC and may not be an explicit circuit. The comparator must resolve small differences in Vin and Vdac within the specified time; in other words, the comparator needs to be as accurate as the overall system. The linearity of the overall SAR ADC is largely limited by the linearity of the DAC. Like the pipelined ADC, the major factor limiting DAC accuracy is the inherent capacitor mismatch. So in general, for 12 bits of accuracy or higher, some form of trimming or calibration is required. The SAR ADC could achieve a similar resolution as the pipelined ADC but has a much lower conversion speed (typically less than 5 MS/s) due to its serial operation. However, the SAR ADC has much lower power consumption than a pipelined ADC because it only needs one comparator and no amplifier is needed. Also, a SAR has only one clock cycle of latency. These features make their applications, including portable instruments, pen digitizers, industrial controls, and data/signal acquisition.

9.3.5   Delta-Sigma ADC

Oversampling delta-sigma ADC, which trades signal bandwidth for resolution, is insensitive to circuit imperfections and component mismatch, and thus could take good advantage of the enhanced density and speed of scaled digital VLSI circuits.[10-13]

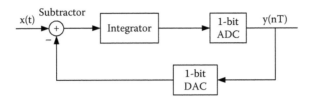

Figure 9.12   First-order delta-sigma modulator.

It was shown in (9.6) that oversampling can improve the SNR of an ADC despite circuit implementation. Thus, oversampling ADCs are more suitable for high-resolution (18 bits and more) implementations. Among all oversampling ADCs, the delta-sigma ADC is the most widely used. This is because it is the most robust ADC against circuit imperfections. Delta-sigma ADC employs the oversampling technique and the noise-shaping technique, which will be described as follows. A delta-sigma ADC consists of an analog delta-sigma modulator and the digital and decimation filter. A first-order delta-sigma modulator is shown in figure 9.12. It contains an integrator, a 1-bit ADC, a 1-bit DAC, and a subtractor. The name first-order stems from the fact that there is only one integrator in the circuit.

The input signal is fed to the ADC (quantizer) via an integrator. The quantized output feeds back to subtract from the input signal. When the integrator output is positive, the ADC output is 1 and a positive reference voltage is fed back and subtracted from the input signal to move the integrator output in the negative direction. Similarly, when the integrator output is negative, the ADC output is 0 and a negative reference voltage is fed back and subtracted from the input signal. As the input signal goes more positive, the number of 1 increases and the number of 0 decreases. Likewise, as the input signal goes more negative, the number of 1 decreases and the number of 0 increases. In fact, the feedback forces the average ADC output to track the average value of the input signal.

A linear model of the modulator is shown in figure 9.13. The z-domain output signal is given by

$$Y(z) = z^{-1} \cdot X(z) + (1 - z^{-1}) \cdot Q(z), \tag{9.12}$$

where $X(z)$, $Y(z)$, and $Q(z)$ are the z-transforms of the modulator input and output, and the quantization error of the ADC. Note that z^{-1} represents a unit delay. On the other hand, $(1 - z^{-1})$ has high-pass characteristics. Hence, the input signal is only delayed while the quantization noise is high-pass filtered. The filtering function $(1 - z^{-1})$ is called noise transfer function (*NTF*). In the frequency domain, z is replaced by, $e^{j2\pi fT}$ where $T = 1/fs$ is the sampling period. The magnitude of

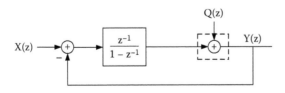

Figure 9.13   Linear model for the first-order delta-sigma modulator.

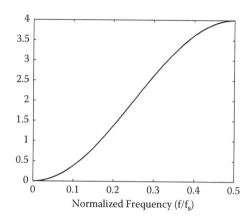

Figure 9.14   Square magnitude of the NTF.

the noise transfer function is found to be

$$| NTF | = | 1 - e^{-j2\pi fT} | = 2\sin(\pi fT).$$ (9.13)

The square magnitude of the *NTF* as a function of frequency is illustrated in figure 9.14.

As figure 9.14 shows, the *NTF* is a high-pass filter function. If the signal bandwidth is small compared to the sampling frequency, when a low-pass digital filter is followed, only a small portion of the quantization noise appears in the signal band and a high resolution results. Any nonlinearity of the ADC is suppressed in-band the same way as that for the quantization noise. However, because the DAC resides in the feedback path, its nonlinearity is an additive signal to the input of the modulator and is not reduced by the negative feedback.

The output of the delta-sigma modulator is a 1-bit data stream at the sampling rate. The delta-sigma modulator is followed by the digital low-pass and decimation filter (figure 9.15), which extracts information from this data stream and reduces the data rate to a more useful value. The digital filter averages the 1-bit data stream, improves the ADC resolution, and removes quantization noise that is outside the band of interest. It determines the signal bandwidth, settling time, and stop-band rejection. It also occupies most of the die area and consumes most of the power.[13] The digital filter also introduces inherent pipeline delay. So the delta-sigma ADC has conversion delay. The signal bandwidth of delta-sigma ADCs is generally less than that of the Nyquist ADCs, because of the oversampling operation. Delta-sigma ADCs are widely used in low-power, high-resolution applications, which include audio, instrumentation, sonar, telecommunications, and wireless systems. The other advantage of the delta-sigma ADCs is that the inherent oversampling greatly relaxes the requirements on the ADC antialiasing filter.

Figure 9.15   Delta-sigma modulator with digital filter.

9.4 ADC Design for Digital TV Broadcasting Applications

Before designing an ADC, the specifications such as the sampling rate, resolution, or dynamic range should be given. Depending on the actual implementation of the digital TV receiver, the ADC requirements may vary dramatically. In this section, we discuss the general considerations of ADC in the digital TV receivers. We also give the ADC design examples and the application considerations.

9.4.1 Considerations of ADC in Digital TV Receivers

The dynamic range, which could be converted to ENOB according to (9.11), and sampling rate are the two most important requirements in the receiver ADC design. The digital TV standards define the maximum allowable transmission error, often in the form of bit rate (BER), for a reliable communication channel. To fulfill the BER speciation, the noise floor in the receiver must be suppressed, and enough samples of the received signal need to be provided to guarantee proper demodulation performance. The dynamic range and sampling rate of the ADC, which is the last analog block in a receiver signal chain, depend strongly on the specifications and architecture of the receiver. The minimum signal-to-noise ratio (SNRm) at the ADC output is essentially set by the desired bit error rate (BER) of the reception. As the ADC itself is not allowed to deteriorate the SNR, its noise contribution must be well below the SNRm for a particular BER level. Also, the upper limit of SNR is set by the power dissipation of the ADC and DSP. According to the sample theorem, the sample rate must be at least twice the signal bandwidth. As pointed out in (9.6), the SNR of ADC is raised when OSR is increased. So a high sampling rate is preferred, while the power dissipation, including also the DSP, sets the upper limit for the maximization. The sampling rate needs to satisfy not only the sampling theorem for signal recovery, but also the sampling requirement in the demodulation. Usually a demodulator model based on the TV standard is developed to help to find the SNRm and sampling frequency fs required to meet the BER requirement in the standard. The ADC dynamic range needs to consider both the SNRm requirement in the demodulator and the received signal dynamic range DRsig at the input of the ADC.[14] DRsig is largely dependent on the gain in the receiver front end. If a gain control system, like a variable gain amplifer (VGA,) exists before the ADC, the DRsig could be reduced and therefore relaxes the dynamic range required in the ADC. However, this will increase the system cost because of the additional block. But with the absence of a gain control circuit, the ADC needs to provide a large dynamic range to accommodate the received signal. According to Schreier et al.,[15] the ADC in a universal TV receiver needs to provide a SNR of 55 dB to fulfill the requirement for "perfect picture" in analog TV. And many ADCs used in digital TV receivers need to provide 10-bit[16–19] or above[15,20,21] resolution.

The power consumption of the ADC should be minimized to extend the battery time, which is very important in mobile applications. The ADC also should occupy a small area and should be compatible with the CMOS process to be easily integrated with digital circuits, and hence reduce the system cost. Table 9.2 shows the ADC requirements for the digital TV broadcasting applications.

The ADC requirements listed in table 9.2 are basic ones; the real specification of the ADC used in a particular system is affected by many other factors and is finally decided by the system engineer. The design of the ADC is directly related to the given ADC specification. Although different system engineers may give different ADC specifications for one particular digital TV broadcasting application, the following shows two design cases.

Table 9.2 ADC Requirements for the Digital TV Applications

Minimum sampling rate	At least twice the signal bandwidth
Minimum resolution	10 bits
Latency requirement	No
Technology	Standard CMOS process
Power consumption	As low as possible
Area	As small as possible

9.4.2 Design Examples

9.4.2.1 A 10-Bit Pipelined ADC in a DVB-H Demodulator

Figure 9.16 shows the DVB-H receiver system. The RF signal is downconverted to 36 MHz IF by the tuner; then the IF signal is filtered with a band-pass surface acoustic wave (SAW) filter to remove out-of-band noise and interference. The output signal of the SAW is amplified by a voltage-controlled variable gain amplifier (VGA) to fully utilize the resolution of the ADC. A 10-bit ADC directly samples the VGA output at a clock frequency of 28.8 MHz. The signal is then downconverted to baseband by a digital mixer and filtered by a pair of low-pass image rejection filters. The filtered signal is further demodulated. Because there is a gain control system before the ADC, the DRsig could be reduced and therefore relaxes the dynamic range required in the ADC. So a high-performance, 10-bit ADC with the ENOB larger than 9 bits for the highest input signal frequency (i.e., $36 + 4 = 40$ MHz) can satisfy the system. So the design target is a low-power, 10-bit ADC that runs at tens of megahertz clock frequency and can deliver the required performance in a direct IF sampling architecture.

As shown in the above section, flash ADC is impractical for 10-bit resolution because too many precision comparators are needed. Also, SAR or delta-sigma ADC is not practical for such a high-speed operation. For such kinds of resolution and sampling rate, the best candidate may be the pipelined ADC, known to be more power and area efficient at the expense of conversion latency. However, this parameter is not a limitation in this application. Also, pipelined ADCs are easily integrated in CMOS by using switched-capacitor design.

Section 9.3.3 described the operation principle of the pipelined ADC. By using digital error correction, the requirement for the sub-ADC, which is composed by comparators, can be greatly relaxed. This means that the power consumption and design complexity of the comparators used in sub-ADC are reduced greatly. The most critical blocks are the front-end S/H, which samples and holds the analog input signal, and the MDAC, which performs the DAC conversion, subtraction,

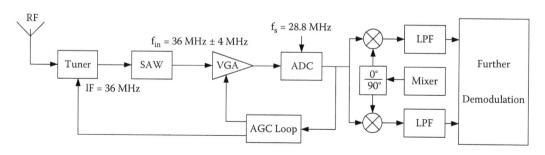

Figure 9.16 A DVB-H receiver system.

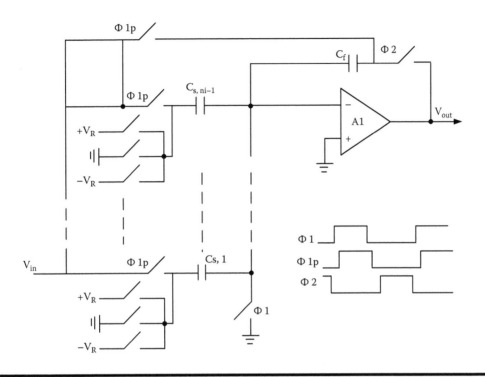

Figure 9.17   A switched-capacitor MDAC.

sample and hold, and amplification of the residue. They consume most of the power and determine the linearity of the whole ADC. Figure 9.17 shows the general MDAC implemented by a switch capacitor (SC) circuit as a single-ended configuration for simplicity.

During one clock phase the input signal is sampled on a set of capacitors. During the other clock phase, the DAC function is performed by charging/discharging a set of capacitors with a reference voltage source; depending on the resolved bits by the sub-ADC of that stage, one of the capacitors is moved as a feedback capacitor of the amplifier, while the others are connected to the $+V_R$, ground, or $-V_R$ source to perform the subtraction and amplification for the residue signal, which is passed to the next stage for more fine conversion.

Assume the operational amplifer (opamp) A1 is a single-pole amplifier; the closed-loop bandwidth (BW) of the circuit is given by

$$BW = (Gm/C_L)*f, \tag{9.14}$$

where Gm is the transconductance of the opamp, C_L is the output load capacitance, and f is the feedback factor. The feedback factor f of the circuit is approximately $1/2^{ni-1}$, where ni is the resolved bits by the stage. Then the BW is

$$BW = (Gm/C_L)/2^{ni-1}. \tag{9.15}$$

Based on the above equation, decreasing the per-stage resolution ni gives a larger feedback factor and thus a larger bandwidth. Thus, the minimum per-stage resolution (i.e., 1.5 bits/stage) architecture maximizes the bandwidth of the SC circuit, which limits the overall conversion rate, and so can reduce the power consumption.[22]

From the operation of the MDAC described above, it can be seen that the opamp, which consumes most of the power and area, is used in only one clock phase. During the sampling phase, the opamp is idle. The opamp-sharing architecture[23,24] is based on this observation and shares one opamp between two consecutive stages during different clock phases. The opamp-sharing architecture only needs half the number of opamps, and thereby could reduce significant power consumption.

There are many other schemes used in the pipelined architecture to reduce the power consumption of the ADC. However, the time-interleaving architecture[25,26] could not show a good dynamic performance due to the mismatches between the time-interleaving channels. Also, the pseudodifferential architecture[27] could not provide a good dynamic performance for the relatively high-frequency inputs due to its nondifferential nature, and hence the higher even-order harmonics compared with that of the fully differential one. Consequently, the ADC in this case is based on a 1.5 bits/stage pipeline with amplifier-sharing architecture.[23,24]

9.4.2.1.1 Opamp in the ADC

The amplifier, which is the core of an MDAC, is the most critical block of a pipelined stage. The resolution, speed, and power consumption of the whole ADC are usually determined by the amplifier. The opamp finite-open-loop DC gain Ao limits the settling accuracy of the amplifier output and causes a gain error in the MDAC. The settling error resulting from the finite-open-loop DC gain Ao is approximately given by

$$\varepsilon_{Ao} = \frac{1}{Ao \cdot f},$$

(9.16)

where f is the feedback factor and can be approximated to equal $1/2^{ni-1}$.

The total equivalent settling error to the input of the whole ADC caused by the finite DC gain of the amplifers in all pipelined stages is

$$\varepsilon_{tot} = \frac{\varepsilon_{A_{o,1}}}{2^{n1-1}} + \frac{\varepsilon_{A_{o,2}}}{2^{n1-1}\, 2^{n2-1}} + \cdots + \frac{\varepsilon_{A_{o,m}}}{2^{n1-1}\, 2^{n2-1}\cdots 2^{nm-1}}.$$

(9.17)

It can be seen that the error caused by the later pipelined stages is reduced by the interstage gain of the preceding stages. The total equivalent settling error must be less than LSB/2, which corresponds to $1/2^N$ for an N-bit ADC. For the 1.5 bits/stage in the case, the resolved bits per stage is $n_1 = n_2 = \cdots = nm = 2$. Assume the DC gain of the amplifers in all pipelined stages is the same and the caused error is ε_{Ao}. Then, when N is large,

$$\varepsilon_{tot} \approx \varepsilon_{Ao}\left(\frac{1}{2} + \frac{1}{4} + \cdots \frac{1}{2^N}\right) = \varepsilon_{Ao}\left(1 - \frac{1}{2^N}\right) < \frac{1}{2^N}.$$

(9.18)

Substituting (9.16) into (9.18), the Ao must hold:

$$Ao > 2^{N+1}.$$

(9.19)

In this case, the resolution N is 10 bits, so $Ao > 2,048$, which is about 67 dB.

The settling time of the MDAC is determined first by the slew rate (SR) and finally by the gain bandwidth (GBW) of the amplifier. From the operation of the MDAC, it can be seen that the settling time should be within half of the clock cycle (T/2). Let us reserve half of the settling time for the SR-limited part and half for the GBW-limited part. The *SR* of a single-stage amplifer is linearly dependent on the maximal current Imax, which equals two times the drain current *Id*, charging and discharging the load capacitance *Cl*, and is given by

$$SR = \frac{2Id}{Cl}. \tag{9.20}$$

Assume that the input signal ranges from $-V_{FS}/2$ to $+V_{FS}/2$. For a worst-case slewing of the full-scale voltage V_{FS}, the *SR* time should be within half of the settling time (*T*/2). That is,

$$\frac{T}{4} \cdot SR = \frac{T}{4} \cdot \frac{2Id}{Cl} > V_{FS}. \tag{9.21}$$

So the drain current should be

$$Id > 2 \cdot V_{FS} \cdot Cl \, / \, T = 2 \cdot V_{FS} \cdot Cl \cdot fs. \tag{9.22}$$

The closed-loop bandwidth (*BW*) of MDAC, assuming a single-pole circuit, is given by (9.14). The error ε_s caused by the incomplete exponential settling of the single-pole circuit during *T*/4 (1/4*fs*) is given by

$$\varepsilon_s = e^{-\frac{BW}{4fs}}. \tag{9.23}$$

Similar to the finite-open-loop gain error, this error also causes a gain error at the output of the MDAC under the assumption that the circuit is linear. So following the similar derivations for the open-loop gain error, assuming all stages are identical (1.5 bits/stage), the closed-loop bandwidth (BW) must hold:

$$BW > 4 \cdot fs \cdot N \cdot \ln 2. \tag{9.24}$$

The *GBW* of the amplifier thus must hold:

$$GBW = BW/f > 4 \cdot fs \cdot N \cdot \ln 2 \cdot 2 = 5.6 \cdot N \cdot fs \tag{9.25}$$

for 10-bit ADC running at 30 MHz clock frequency. The GBW should be larger than 1680 Mrad/s, which is 267 MHz.

Based on the requirement derived above, the operational amplifier could be designed. In our implementation, the amplifiers used in all ADC stages are based on a gain-boosting telescopic architecture[28] to get a high open-loop gain and excellent bandwidth with less power consumption.

9.4.2.1.2 The Passive Sampling Network

From the operation of the MDAC, it can be seen that the signal from the previous stage or the input of the ADC first needs to be sampled by a passive sampling network. Figure 9.18 shows the passive sampling network as a single-ended configuration for simplicity used in the S/H and all the MDAC blocks. There are two switches for each sampling capacitor Cs, where the switch controlled by Φ1p is opened slightly before that of Φ1 to perform the bottom-plate sampling[29] operation. When the switch controlled by ϕ1p is opened, the sampled charge on Cs cannot change. Therefore, the charge injection due to the switch controlled by ϕ1 will not influence the charge.

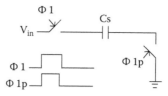

Figure 9.18 The passive sampling network.

The most critical requirement for the passive sampling network happens in the front end of the ADC, where an analog signal is sampled by the network. The switches on-resistance together with the sampling capacitor Cs in the front end of ADC form a RC low-pass network, which determines the maxima analog signal bandwidth. To deal with high-frequency inputs, it is required that the input sampling switch (the switch controlled by $\phi1$) with very low and constant on-resistance. Conventional CMOS switches hardly meet the requirements. Bootstrapping circuits could reduce signal distortion by keeping the gate-source voltage of the sampling switches constant independently of input signal levels. So the bootstrap switch in Dessouky and Kaiser[30] is used as the input sampling switch.

The value of the sampling capacitor Cs is obtained from noise and matching constrains and is different for each stage. In the sampling network, the switch on-resistance causes the so-called KT/C noise. The mean squared noise voltage is[31]

$$Vn^2 = \frac{k \cdot T}{Cs}.$$
(9.26)

where $k = 1.38 \times 10^{-23}$ J/K is the Boltzman constant, and T is the absolute temperature. This value should be less than the quantization noise power, that is,

$$\frac{k \cdot T}{Cs} < \frac{LSB^2}{12}.$$
(9.27)

Then the value sampling capacitor should hold

$$Cs > 12 \cdot k \cdot T \cdot \left(\frac{2^N}{V_{FS}} \right)^2.$$
(9.28)

For $N = 10$-bit ADC with $V_{FS} = 1$ V at $T = 300$ K, the Cs should be larger than 0.052 pF.

On the other hand, the gain of the MDAC is usually decided by a capacitor ratio. Thus, accurate capacitor matching is required to design a high-resolution pipelined ADC. A large area, which means large capacitor value, and a large perimeter-to-area ratio improve the matching.[7] In our design, the capacitor value is limited by matching rather than noise. According to our technology-matching data, the sampling capacitors were chosen at about 1pF for the first stage.

9.4.2.1.3 The Sub-ADC

The sub-ADC used in a pipelined stage is a flash ADC that consists of parallel comparators. By using the digital error correction technique,[9] the accuracy requirement for the sub-ADC is determined by the resolution of the pipelined stage rather than the whole ADC. So the accuracy requirement for the comparator is greatly relaxed. For a 1.5 bits/stage architecture, the largest comparator offset voltage allowed is $\pm V_R/4$. If the reference voltage V_R is 1 V, then the largest allowable comparator offset is ± 250 mV. This allows the use of the low-power dynamic comparators[32] without any preamplifier to cancel the offset voltage in the sub-ADC. In addition to the quantizing comparators, a thermometer-to-binary encoder and a small decoding logic to generate the control signals for the MDAC are needed.

9.4.2.1.4 Other Components

For high-performance ADCs, high-quality reference voltages and currents are usually needed. Bandgap reference circuits can generate a supply voltage that is process and temperature independent. The reference current could be generated by adding an external precision resistor with a low temperature coefficient to the precision reference voltage.[7] Buffer amplifiers should be used to drive the positive and negative reference voltages for the MDACs. The output impedance of the buffers should be low enough for driving the load capacitors in MDACs. The clock generation of the ADC is implemented with five local clock generators driven by a single master clock to coincide between stages, to avoid any loss of clock period due to skews related to the layout.

Although in practice the real circuit may be more complex than that considered above, for example, the frequency response of the amplifer may be more complicated than that produced by just one pole, and there are parasitic capacitors at the input of the amplifiers, and so on, that the analysis and derivation presented above have been useful in the design of at least two pipelined ADCs[16,17] for digital TV applications. More analysis, circuit implementation, and architectural and experimental results can be found in Li et al.[16,17]

9.4.2.2 Delta-Sigma ADC in Digital TV Receiver

In some TV receiver designs, the ADC is required to provide a much higher resolution[15,20,21] than that in table 9.2. This is because the use of high-bandwidth and high-resolution ADCs allows part of the channel filtering and VGA functions to be performed in the digital domain, which is more suitable for process scaling. When the resolution of the ADC is higher than 12 bits, straight pipelined ADC may not be suitable, because the accuracy of the capacitor ratios is restricted for a given technology, and the accuracy of the capacitor ratios influences the overall ADC linearity. Various calibration techniques[33–35] may be necessary to alleviate the limitation. However, these techniques generally require a large power consumption, occupying a huge area, which is not suitable to be integrated with the digital circuits.

Delta-sigma ADC is very suitable to achieve high resolutions in the modern digital-driven process technologies. Section 3.5 described the operation principle of the first-order delta-sigma ADC. For the first-order modulator, the resolution increases by 1.5 bits for every doubling of the sampling frequency. To get a high resolution, a very high OSR is needed, which is not suitable for applications where the signal bandwidth is large. The required OSR can be reduced by using a second-order modulator, which can be constructed by inserting an additional integrator into the feed-forward path of the modulator. The NTF for the second-order modulator is the square of that for the first-order one. So there is an increased attenuation of quantization noise at low frequencies. Now the resolution increases by 2.5 bits for every doubling of the sampling frequency. In general, an Lth-order modulator can be obtained by placing L integrators in the forward path. For an Lth-order modulator we gain L + 0.5 bits for every doubling of the sampling frequency.[12]

However, there are problems with increasing the order of the modulator due to stability. Another architecture, which eases the stability problems associated with high-order modulators, is the cascade modulator, also called the MASH modulator.[12] In this type of converters several low-order modulators are cascaded to achieve a high-order noise-shaping function. In many applications (such as the digital TV receiver), where the signal bandwidth is large, low OSR is desirable. But if the required resolution is high, low OSR would imply a high-order modulator. It is shown in (9.5) that each additional bit of the quantizer improves the SNR by an additional 6 dB. So an alternative way of implementing high-speed, high-resolution delta-sigma ADCs is to employ a multibit quantizer, instead of a single-bit quantizer, in the modulator loop. As mentioned earlier, the nonlinearities of the DAC of the delta-sigma ADC are not suppressed by the negative feedback, and thus result in comparable nonlinearities for the overall system. Many techniques have been suggested in the literature to solve the problem of DAC nonlinearity inherent in the use of multibit quantization. By using high-order and multibit quantizer modulators, high-speed, high-resolution delta-sigma ADCs can be implemented.[11,12]

In the past, the primary application of commercial delta-sigma A/D converters was in narrowband systems such as digital audio. However, as we benefit from the increased speed of submicron devices, the delta-sigma modulators are exploited for wider-band systems. Recently the delta-sigma ADCs were successfully implemented in the modern CMOS process for digital TV applications.[15,21] The detailed implementation and experimental results can be found in Schreier et al.[15] and Fujimoto et al.[21]

9.4.3 ADC Application Considerations

Although an ADC may be designed very well, it cannot show optimum performance if it is not used correctly. Figure 9.19 shows the typical block diagram of the ADC integrated with the DSP part.

The ADC integrated with the DSP usually has the analog input(s) Ai, analog power supply Va, digital power supply Vd, and clock signal input Clock, and may have the reference input Vref. Careful attention must be paid to these signals. The analog input signal Ai, which may be single ended or differential, should be as clean as possible. Any fast-switching digital signal lines that

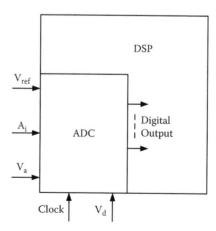

Figure 9.19   The block diagram of the ADC integrated with the DSP.

may couple into the analog input path should be kept away from the analog input, because any noise on the Vref will directly influence the accuracy of the ADC. The reference input Vref should be treated as another analog input and should be kept as clean as possible. The Vref is a fixed DC voltage, so large decoupling capacitors could be added to the reference pin(s) to keep Vref clean. These decoupling capacitors should be placed as close to the ADC as possible and should avoid them to be connected through vias to reduce the inductance of the line and vias, which will reduce the capacitor's effectiveness. Most ADCs have separate power supply inputs, Va for the analog circuits and Vd for the digital. It is recommended to add adequate decoupling capacitors in a way similar to that for Vref. For high-speed and high-resolution applications, the clock signal is very critical and may limit the performance of the ADC. The clock signal jitter, which is the deviation from the ideal timing period, causes uncertainty in the sampling time. Uncertainty in the sampling time creates uncertainty in the signal amplitude. So the jitter of the clock signal can degrade the *SNR* of the ADC. For Nyquist ADCs the degradation of SNR due to jitter when sampling a sine wave input is given by[36]

$$SNR = -20 \log(2 \cdot \pi \cdot fa \cdot \Delta t), \qquad (9.29)$$

where *fa* is the analog input frequency, and Δt is the root mean square (rms) clock jitter. For a 70 MHz analog input, the clock jitter must be less than 2 ps to achieve an *SNR* equivalent to 10 bits. In an oversampling ADC, the jitter error is removed by the decimation filter at frequencies above the baseband. Because the clock jitter could be assumed to be white, the total power of the jitter error is reduced by the oversampling ratio in the decimator process.[10] But the in-band jitter noise can still degrade the *SNR*. So to prevent the clock signal from degrading the system *SNR*, its jitter should be low enough.

9.5 Summary

In this chapter, we have discussed the basic fundamentals and some important specifications of ADC. Many popular ADC architectures, their limitations, and possible applications were briefly reviewed. General requirements of ADC in digital TV receivers and detailed design examples and analyses were given. Depending on the receiver architecture, pipelined ADC and delta-sigma ADC, which are easy to be integrated with CMOS digital circuits, may be the best architectures for the digital TV applications.

References

1. What's DMB? http://www.kbs.co.kr.
2. ARIB. 1998. *Terrestrial integrated services digital broadcasting (ISDB-T) specifications of channel coding, framing structure, and modulation.*
3. DVB-H standard. Draft. http://www.dvb.org.
4. FLO™ Technology Overview. http://www.qualcomm.com/mediaflo.
5. J. A. Wepman. 1995. Analog-to-digital converters and their applications in radio receivers. *IEEE Common Magazine* 33:39–45.
6. R. J. van de Plassche. 1994. *Integrated analog-to-digital and digital-to-analog converters.* Boston: Kluwer Academic Publishers.
7. M. Gustavsson, J. J. Wikner, and N. N. Tan. 2002. *CMOS data converters for communications.* Boston: Kluwer Academic Publishers.

8. B. Ginetti and P. Jespers. 1990. A 1.5 Ms/s 8-bit pipelined RSD A/D converter. In *European Solid-State Circuits Conference Digest of Technical Papers*, 137–40.

9. S. H. Lewis and P. R. Gray. 1987. A pipelined 5-Msamples/s 9-bit analog-to-digital converter. *IEEE Journal of Solid-State Circuits* SC-22:954–61.

10. B. E. Boser and B. A. Wooley. 1988. The design of sigma-delta modulation analog-to-digital converters. *IEEE Journal of Solid-State Circuits* 23:1298–308.

11. S. R. Norsworthy, R. Schreier, and G. C. Temes. 1997. *Delta-sigma data conveners: Theory, design and simulation*. IEEE Press.

12. R. Schreier and G. C. Temes. 2004. *Understanding delta-sigma data conveners*. Wiley-IEEE Press.

13. P. E. Allen and D. R. Holberg. 2002. *CMOS analog circuit design*. 2nd ed. Oxford: Oxford University Press.

14. B. Xia. 2004. Analog-to-digital interface design for wireless receivers. Doctoral thesis, Texas A&M University.

15. R. Schreier, N. Abaskharoun, H. Shibata, I. Mehr, S. Rose, and D. Paterson. 2006. A 375-mW quadrature bandpass $\Delta\Sigma$ADC with 8.5-MHz BW and 90-dB DR at 44 MHz. *IEEE Journal of Solid-State Circuits* 41:2632–40.

16. J. Li, J. Zhang, B. Shen, X. Zeng, Y. Guo, and T. Tang. 2005. A 10BIT 30MSPS CMOS A/D converter for high performance video applications. In *Proceedings of the European Solid-State Circuits Conference (ESSCIRC)*, 523–26.

17. J. Li, X. Zeng, L. Xie, J. Chen, J. Zhang, and Y. Guo. 2006. A 1.8-V 22-mW 10-bit 30-MS/s subsampling pipelined CMOS ADC. In *Proceedings of IEEE Custom Integrated Circuits Conference (CICC)*, 513–16.

18. H. C. Choi, J. H. Kim, S. M. Yoo, K. J. Lee, T. H. Oh, M. J. Seo, and J. W. Kim. 2006. A 15mW 0.2mm² 10b 50MS/s ADC with wide input range. In *2006 IEEE International Solid-State Circuits Conference on Digital Technology Papers*, 842–51.

19. O. A. Adeniran and A. Demosthenous. 2006. A 19.5mW 1.5V 10-bit pipeline ADC for DVB-H systems in 0.35µm CMOS. In *Proceedings of IEEE International Symposium, Circuits and Systems (ISCAS)*, 5351–54.

20. I. Mehr et al. 2005. A dual-conversion tuner for multi-standard terrestrial and cable reception. In *Symposium on VLSI Circuits on Digital Technology Papers*, 340–43.

21. Y. Fujimoto, Y. Kanazawa, and P. Lore. 2006. Masayuki Miyamoto: An 80/100MS/s 76.3/70.1dB SNDR delta–sigma ADC for digital TV receivers. In *2006 IEEE International Solid-State Circuits Conference on Digital Technology Papers*, 201–10.

22. S. H. Lewis. 1992. Optimizing the stage resolution in pipelined, multistage, analog-to-digital converters for video-rate applications. *IEEE Transactions on Circuits and Systems II* 39:516–23.

23. P. C. Yu and H. S. Lee. 1996. 2.5-V, 12-b, 5-Msamples/s pipelined CMOS ADC. *IEEE Journal of Solid-State Circuits* 31:1854–61.

24. K. Nagaraj, H. Fetterman, J. Anidjar, S. Lewis, and R. Renninger. 1997. A 250-mW, 8-b, 52-Msamples/s parallel-pipelined A/D converter with reduced number of amplifiers. *IEEE Journal of Solid State Circuits* 32:312–20.

25. J. Arias, V. Boccuzzi, L. Quintanilla, L. Enríquez, D. Bisbal, M. Banu, and J. Barbolla. 2004. Low-power pipeline ADC for wireless LANs. *IEEE Journal of Solid-State Circuits* 39:1338–40.

26. S. Limotyrakis, S. D. Kulchycki, D. K. Su, and B. A. Wooley. 2005. A 150-MS/s 8-b 71-mW CMOS time-interleaved ADC. *IEEE Journal of Solid-State Circuits* 40:1057–67.

27. D. Miyazaki, S. Kawahito, and M. Furuta. 2003. A 10-b 30-MSs low-power pipelined CMOS AD converter using a pseudodifferential architecture. *IEEE Journal of Solid-State Circuits* 38:369–73.

28. K. Bult and G. Geelen. 1990. A fast-settling CMOS opamp for SC circuits with 90-dB DC gain. *IEEE Journal of Solid-State Circuits* 25:1379–84.

29. K. Y. Kim, N. Kusayanagi, and A. A. Abidi. 1997. A 10-b, 100-MS/s CMOS A/D converter. *IEEE Journal of Solid-State Circuits* 32:302–11.

30. M. Dessouky and A. Kaiser. 1999. Input switch configuration for rail-to-rail operation of switched opamp circuits. *Electronics Letters* 35:8–10.
31. D. A. Johns and K. Martin. 1997. *Analog integrated circuit design.* New York: John Wiley & Sons.
32. L. Sumanen, M. Waltari, and K. Halonen. 2000. A mismatch insensitive CMOS dynamic comparator for pipeline A/D converters. In *Proceedings of the IEEE International Conference on Circuits and Systems (ICECS'00)*, I-32–35.
33. S.-H. Lee and B.-S. Song. 1992. Digital-domain calibration of multistep analog-to-digital converters. *IEEE Journal of Solid-State Circuits* 27:1679–88.
34. Y.-M. Lin, B. Kim, and P. R. Gray. 1991. A 13-b 2.5-MHz self-calibrated pipelined A/D converter in 3-μm CMOS. *IEEE Journal of Solid-State Circuits* 26:628–36.
35. K. Iizuka, H. Matsui, M. Ueda, and M. Daito. 2006. A 14-bit digitally self-calibrated pipelined ADC with adaptive bias optimization for arbitrary speeds up to 40 MS/s. *IEEE Journal of Solid-State Circuits* 41:883–90.
36. M. Shinagawa, Y. Akazawa, and T. Wakimoto. 1990. Jitter analysis of high speed sampling systems. *IEEE Journal of Solid-State Circuits* 25:220–24.

Chapter 10

Application Layer Forward Error Correction for Mobile Multimedia Broadcasting

Thomas Stockhammer, Amin Shokrollahi, Mark Watson,
Michael Luby, and Tiago Gasiba

Contents

Keywords

raptor code, UMTS, forward error correction, FEC, MBMS, cross-layer optimization, reliable multicast transmission, application layer FEC, fountain codes, mobile TV, video streaming

Abstract

Application Layer Forward Error Correction (AL-FEC) is an innovative way to provide reliability in mobile broadcast systems. Conventional data such as multimedia files or multimedia streams are extended with repair information that can be used to recover lost data at the receiver. AL-FEC is integrated into content delivery protocols (CDPs) to support reliable delivery. Several standardization committees such as 3GPP and DVB have recognized the importance of AL-FEC and have standardized Raptor codes as the most powerful AL-FEC codes to be used for such applications. The major characteristics of Raptor codes are channel efficiency, low complexity, and flexibility. An important consideration when using AL-FEC is system integration. With the right system design including AL-FEC, the efficiency or quality of delivery services can be significantly enhanced. This work shows these benefits in selected use cases, specifically focusing on 3G-based multimedia broadcasting within the MBMS standard.

10.1 Introduction

The primary objective of Mobile TV is to bring TV-like services to mobile phones. However, handheld devices in use today differ significantly from traditional TV equipment. For example, mobile phones integrate two-way communication network connections and flexible operating systems as well as powerful hardware platforms which enables the use of smart and powerful software applications and tools. With these valuable additions, mobile TV users can enjoy personalized and interactive TV with content specifically adapted to the mobile medium. In addition to traditional live TV channels, mobile TV delivers a variety of services including video on demand and video downloading services, and the content may be delivered to a mobile user either on-demand or by subscription.

From a delivery perspective there are to date two different approaches to delivering mobile TV services. It is noteworthy that currently more than 90 percent of commercially deployed mobile TV services run over two-way cellular networks such as UMTS, CDMA2000, WiMAX, or extensions of those. However, more recently, unidirectional broadcast technologies such as DVB-H, DMB/DAB and MediaFLO are attracting significant attention. Furthermore, two-way cellular networks are currently being extended with IP multicast transport, e.g., with 3GPP Multimedia Broadcast/Multicast Services (MBMS) or 3GPP2 BroadCast MultiCast Services (BCMCS), which provide the possibility to distribute IP multicast data over point-to-multipoint radio bearers and therefore increase efficiency and allow delivery of more revenue-generating services.

The increasing demand of mobile users for multimedia information in many different application scenarios leads to significant challenges. End users having experience with TV-grade video signals also expect high quality from mobile applications. Furthermore, network and service operators expect high efficiency and low costs in terms of infrastructure and hardware, while still providing the highest customer satisfaction. Whereas personalized services can be handled by improving point-to-point distribution, popular content requires more efficient broadcast distribution. However, the reliable delivery of large files in podcasting or clipcasting-like service or simultaneous live video broadcasting to many users over unreliable and bandwidth-limited networks is an extremely challenging task.

Many of the challenges arise on the physical and medium access control layers. However, there is a general tendency to reuse as much as possible of the existing network infrastructure to avoid huge upfront investments. For example, DVB-H relies heavily on DVB-T infrastructure, and 3GPP MBMS reuses existing signal processing and network infrastructure from point-to-point UMTS. Therefore, the optimization potentials on the layer below IP are limited and IP multicast transmission is in general neither reliable nor completely optimized. Therefore, content delivery protocols (CDPs) play an important role in the successful service delivery over wireless and mobile channels via IP multicast.

A full specification of a CDP usually consists of a collection of different tools. Many commercial standards bodies look first for standardized protocols in the Internet Engineering Task Force (IETF). The IETF has defined protocols which provide delivery of files or streaming content to many users, and these have become important components in commercial standards. The most popular protocol for file delivery over IP multicast is the File Delivery over Unidirectional Transport (FLUTE) protocol [1]. The IETF is in the process of specifying a similar framework for the delivery of streaming media [2]. In both frameworks, the integration of application layer forward error correction (AL-FEC) [3], [4] is the primary technology used to overcome IP packet losses for the provisioning of a reliable service. The guidelines on how to use AL-FEC to provide support for reliable delivery of content within the context of a CDP are provided in [3]. The IETF is in the process of defining several AL-FEC schemes, e.g. [5], whereby especially Raptor codes [6] have been selected in different CDP definitions recently by commercial standards. For a detailed description on Raptor codes we refer to [7], but a summary of the codes can be found along with some implementation details and performance results in Section 10.2.

3GPP MBMS uses IETF standardized protocols, including the Raptor code, for streaming and file delivery. MBMS extends the existing architecture by the introduction of an MBMS Bearer Service and MBMS User Services [8]. The former is provided by the packet-switched domain to deliver IP multicast datagrams to multiple receivers using minimum radio and network resources. The Bearer service re-uses many existing components of the UMTS system such as radio access including physical layer coding based on Turbo codes. The end user is provided with two MBMS user services [9], download delivery for reliable multicasting of files as well as streaming delivery for real-time multicasting of multimedia streams. Those two services make use of different CDPs. The end-point of the CDP on the network side is a new architectural component, the Broadcast Multicast Service Center (BM-SC). It provides the MBMS User Services to the User Equipment (UE). Both delivery methods in MBMS mandate that the UE supports Raptor codes. We discuss the integration of AL-FEC into the two delivery methods and provide an overview over the MBMS system as an exemplary multimedia broadcast system in Section 10.3.

The assessment of mobile multimedia broadcasting services is quite complex. This is especially the case because features which can provide reliability are separated in the overall protocol stack, for example physical layer forward error correction (PHY-FEC), power control, and AL-FEC. Also, mobility aspects are fairly complex to handle as long-term signal variations significantly influence the performance of a system. Furthermore, the criteria for performance evaluation are quite different for different services. Whereas file delivery services usually have relaxed timing constraints, for real-time streaming delivery and live Mobile TV aspects such as delay, latency, or channel switching times are very important. Therefore, in Section 4 we introduce a realistic system level approach that allows assessing the performance of mobile multimedia broadcast applications. Selected simulation results for different services are discussed in Section 10.5. These show the benefits of AL-FEC in mobile broadcasting services. Some discussions and optimization potentials are presented. The final section summarizes our results and provides further conclusions.

10.2 Standardized Raptor Code

Raptor codes were introduced by Shokrollahi in 2001 [10] and and a comprehensive overview is provided in [7]. They are an extension of LT-codes, introduced by Luby [11]. Raptor codes have been standardized to address the needs of compliant implementations in many different environments for efficiently disseminating data over a broadcast network. The major standardization work has been done in 3GPP and the standardized Raptor specification is provided in the MBMS specification [9], which is identical to the Raptor specification in [6] and [12]. Raptor codes provide improved system reliability, while also enabling a large degree of freedom in the choice of transmission parameters. Raptor codes are fountain codes; therefore, as many encoding symbols as desired can be generated by the encoder on the fly from the source symbols of a source block of data. The decoder is able to recover the source block from any set of encoding symbols only slightly more in number than the number of source symbols. As a result, Raptor codes operates very closely to an ideal fountain code, which would require only the exact number of source symbols for recovery.

The following subsections are designed to familiarize the reader with the main concepts behind Raptor codes, their operational use, and efficient encoding and decoding algorithms. To fix notation, we assume that we send a piece of content consisting of k symbols over an unreliable channel in which symbols may get lost. In our context a symbol is a collection of bits; it can be as small as one bit, or as large as a transmission packet over the Internet. We denote the vector of symbols by $x = (x_1, x_2, \ldots, x_k)$, and we assume that all the symbols in this vector have the same size (in bits). We call vector x the source block, the vector of source symbols, or simply the source symbols. The encoding procedures we outline below use the simple procedure of XOR on the symbols; the XOR of two symbols x_i and x_j is a symbol whose ℓth bit is the XOR of the ℓth bit of x_i and the ℓth bit of x_j, respectively. We denote the XOR of x_i and x_j by $x_i \oplus x_j$. If a is in $GF(2)$, then we denote by ax the symbol in which the ℓth bit is the binary AND of a and the ℓth bit of x. Using this notation, if $a_1,\ldots,a_k$ are elements of $GF(2)$, then the expression $\oplus_{i=1}^{k} a_i x_i$ is a well-defined symbol.

10.2.1 Fountain Codes

Fountain codes are a novel and innovative class of codes designed for transmission of data over time varying and unknown erasure channels. They were first mentioned without an explicit construction in [13], and the first efficient construction was invented by Luby [14]. A fountain code designed for k source symbols is specified by a probability distribution $\mathcal{D}$ on the set of binary strings of length k. Operationally, a fountain code can produce from the vector x a potentially limitless stream of symbols $y_1, y_2, y_3, \cdots$, called output symbols, satisfying several fundamental properties:

1. Each output symbol can be generated according to the following probabilistic process: the distribution $\mathcal{D}$ is sampled to yield a vector $(a_1,\ldots,a_k)$, and the value of the output symbol is set to be $\oplus_{i=1}^{k} a_i x_i$. This process is referred to as encoding, and the vector $(a_1,\ldots,a_k)$ is called the mask corresponding to the output symbol.
2. The output symbols can be independently generated.
3. The source symbols can be recovered from any set of n output symbols, with high probability. The recovery process is usually called decoding, and the number $n/k-1$ is called the overhead of the decoder. The probability that the decoder fails is called the error probability of the code.

The third condition shows that fountain codes are robust against erasures, since only the number of received output symbols is important for decoding. Different fountain codes differ in terms of their overhead for a given error probability. But they also differ in terms of the computational efficiency of the encoding and decoding processes. To fix notation, we call the expected number of XORs that is required to produce an output symbol the encoding cost of a fountain code. The expected number of XORs required to decode the source symbols from the received output symbols is called the decoding cost. In terms of the computational complexity the best type of fountain codes one can envision have a constant encoding cost (independent of k), and a decoding cost that grows linearly with k. As a caveat, we would like to mention that considering the computational complexity in isolation does not make much sense; generally one has to look at all the parameters of a fountain code, i.e., overhead, computational complexity, and the error probability of the decoder. We briefly elaborate on this issue later when comparing LT-codes and Raptor codes.

In operation the output symbols need to contain indications that allow the receiver to recover the mask of each of these symbols. This is accomplished by equipping output symbols with Encoding Symbol ID's (ESI's). In the standardized Raptor code, an ESI is a 16-bit integer which facilitates the creation of the mask associated to an output symbol. Details are described in [9].

The conceptually simplest form of decoding a fountain code is the following: the receiver recovers for every received symbol y_i its corresponding mask $(a_{i1},\ldots,a_{ik})$, and sets up the following system of linear equations:

$$\begin{pmatrix} a_{11} & a_{12} & \cdots & a_{1k} \\ a_{21} & a_{22} & \cdots & a_{2k} \\ \vdots & \vdots & \ddots & \vdots \\ a_{n1} & a_{n2} & \cdots & a_{nk} \end{pmatrix} \begin{pmatrix} x_1 \\ x_2 \\ \vdots \\ x_k \end{pmatrix} = \begin{pmatrix} y_1 \\ y_2 \\ \vdots \\ y_n \end{pmatrix}. \tag{10.1}$$

In effect, all decoding methods for fountain codes try to solve this system of equations, either implicitly or explicitly. The task of the code designer is to design the fountain code in such a way that a particular (low-complexity) decoding algorithm performs very well. The following subsections give examples of such codes.

10.2.2 LT-Codes

LT-codes, invented by Luby [11], are the first realization of fountain codes. LT-code exhibit excellent overhead and error probability properties. For LT-codes the probability distribution $\mathcal{D}$ has a particular form which we describe by outlining its sampling procedure. At the heart of LT-codes is a probability distribution Ω on the integers $1,\ldots,k$. This distribution is often called the *weight* or *degree* distribution of the LT-code. To create an output symbol, the following procedure is applied:

1. Sample from Ω to obtain an integer $w \in \{1,\ldots,k\}$. The number w is called the *weight* or *degree* of the output symbol.
2. Choose a binary vector $(a_1,\ldots,a_k)$ of Hamming weight w uniformly at random.
3. Set the value of the output symbol to $\oplus_{i=1}^{k} a_i x_i$.

An LT-code as described above is determined by its parameters (k, Ω). As outlined above, the output symbol is given an ESI which enables the recreation of its mask.

As with other fountain codes, LT-codes can be decoded by solving the system (1). However, in many applications, straightforward solution of this system using, e.g., a naive Gaussian elimination, is prohibitively expensive. It is therefore imperative to employ faster elimination algorithms, and design the distribution Ω such that these decoding algorithms have low overhead while maintaining stringent bounds on their error probabilities. One of the simplest elimination algorithm one can envision is the greedy one. We describe it using a graph terminology.

Upon reception of output symbols $y_1,\ldots,y_n$, we arrange them in a bipartite graph with the output symbols forming one side, and the source symbols $x_1,\ldots,x_k$ the other side. We connect an output symbol y to all the input symbols of which y is the XOR. So, if for example $y = x_1 \oplus x_5 \oplus x_9$, we connect y to the source symbols x_1, x_5, and x_9.

The decoding algorithm is a modification of the one presented in [15] and proceeds in rounds. At each round, we search for an output symbol of degree one, and copy its value into the value of its unique neighbor among the source symbols. We then XOR the value of the newly found source symbol into all the neighbors of the source symbol among the output symbols, and delete all edges emanating from the source symbol. We continue the procedure until we cannot find an output symbol of degree one. If at this point not all the source symbols are recovered, then we declare a decoding error.

In applications it is often advantageous to not perform the XOR operations in this algorithm immediately. Instead, one would use the decoding algorithm outlined to create a "schedule" (as proposed in [9]) which stores the order in which the XORs are performed. Such a schedule has a number of advantages. For example, when interleaving is used to create multiple symbols with the same mask to be packed into a transmission packet, scheduling needs to be done only once, amortizing the cost of scheduling over the interleaving depth.

Figure 10.1 provides a toy example of an LT-code giving its associated graph, and a schedule which provides an algorithm for recovering the source symbols from the received output symbols.

It is by no means certain that the greedy decoding algorithm succeeds. In fact, in a well-defined sense, almost all choices for the distribution Ω would lead to algorithms with very large error probabilities even with large overheads. It can be very easily seen that if the decoding

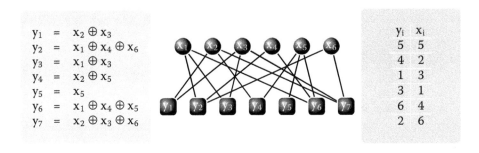

Figure 10.1 Toy example of an LT-code. The collected output symbols are shown on the left. In the middle part, these symbols are transformed into a graph. The array on the right side gives a "mini-schedule" for recovering the source symbols: output symbol y_5 recovers x_5; thereafter, y_4 recovers x_2, y_1 recovers x_3, etc.

algorithm is to have an error probability that decays inversely proportional to k, then the encoding cost associated with the distribution Ω has to be of the order $O(\log(k))$ [7,11], and the average decoding cost of a successful algorithm is of the order $O(k \log(k))$. It is remarkable that this bound can be matched with a specific design, called the "robust soliton distribution", which asymptotically guarantees small error probabilities with an overhead of the order $O(\log^2(k)/\sqrt{k})$ [11].

10.2.3 Nonsystematic Raptor Codes

Despite the excellent performance of LT-codes, it is not possible to give a construction with constant encoding and linear decoding cost without sacrificing the error probability. In fact, a simple analysis shows that to obtain constant encoding cost with reasonable overheads, the error probability has to be constant as well.

An extension of LT-codes, Raptor codes are a class of fountain codes with constant encoding and linear decoding cost. Compared to LT-codes, they achieve their computational superiority at the expense of an asymptotically higher overhead, although in most practical settings Raptor codes outperform LT-codes in every aspect. In fact, for constant overhead ε one can construct families of Raptor codes with encoding cost $O(\log(1/\varepsilon))$, decoding cost $O(k \log(1/\varepsilon))$, and a decoding error probability that asymptotically decays inversely polynomial in k [7].

Raptor codes achieve their performance using a simple idea: the source x is precoded using a linear code C of dimension k and block-length m. The encoding of x with C produces a vector $z = (z_1, \dots, z_m)$ of symbols called *input symbols*. Often a systematic encoding is used for C, in which case $z = (x_1, \dots, x_k, z_1, \dots, z_{m-k})$, where $z_1, \dots, z_{m-k}$ are redundant symbols. A suitably chosen LT-code of type (m, Ω) is then applied to z to create output symbols $y_1, y_2, \dots$ The characterization of a Raptor code can be determined by its parameters (C, k, Ω).

A toy example of a Raptor code is provided in Figure 10.2. In this example the check matrix of the precode C is equal to

$$H = \begin{pmatrix} 1 & 0 & 1 & 1 & 0 & 1 & 1 & 0 \\ 1 & 1 & 0 & 1 & 1 & 1 & 0 & 1 \end{pmatrix}.$$

Note that LT-codes form a special subclass of Raptor codes: for these codes the precode C is trivial. At the other extreme there are the *precode-only* (PCO) codes [7] for which the degree

$$
\begin{aligned}
y_1 &= x_1 \\
y_2 &= x_1 \oplus x_5 \\
y_3 &= x_1 \oplus x_3 \oplus x_4 \\
y_4 &= x_1 \oplus x_3 \oplus x_4 \oplus x_6 \\
y_5 &= x_4 \oplus z_2 \\
y_6 &= x_5 \oplus z_2 \\
y_7 &= x_4 \oplus x_5 \oplus x_6 \oplus z_1
\end{aligned}
$$

$$
\begin{aligned}
0 &= x_1 \oplus x_3 \oplus x_4 \oplus x_6 \oplus z_1 \\
0 &= x_1 \oplus x_2 \oplus x_4 \oplus x_5 \oplus x_6 \oplus z_2
\end{aligned}
$$

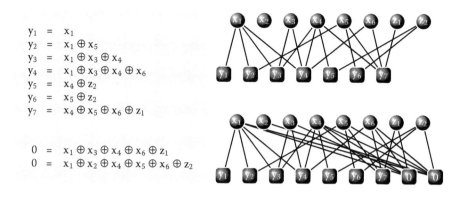

Figure 10.2   Toy example of a Raptor code. The received output symbols are shown on the left, together with the relations among the input symbols dictated by the precode. The top graph is the one between the dynamic output symbols and the input symbols. The input symbols are divided into the source symbols $x_1,\dots x_6$ and the redundant symbols z_1, z_2. As can be seen, node x_2 is not covered and cannot be recovered. In the lower graph the static output symbols are added to the graph. The node x_2 is covered now.

distribution Ω is trivial (it assigns a probability of 1 to weight 1, and zero probability to all other weights). All Raptor codes in use are somewhere between these two extremes: they have a non-trivial (high-rate) precode, and they have an intricate (though low-weight) degree distribution.

Raptor codes can be decoded in a variety of ways. The conceptually simplest decoder sets up a system of linear equations and solves the system using Gaussian elimination. The system to set up has the following shape: suppose that the code C has a check matrix H with m columns and $m - k$ rows. Moreover, suppose that each collected output symbol y_i has mask $(a_{i1},\dots,a_{mi})$, recovered using the ESI of the output symbol. In addition, let $(z_1,\dots,z_m)$ denote the input symbols of the LT-code. Recovering these input symbols is tantamount to the recovery of the source symbols. (This is obvious if C is systematic, and is very easy to see in general as well.) The input symbols can be recovered by solving the system of linear equations

$$
\underbrace{\begin{pmatrix}
a_{11} & a_{12} & \cdots & a_{1m} \\
a_{21} & a_{22} & \cdots & a_{2m} \\
\vdots & \vdots & \ddots & \vdots \\
a_{n1} & a_{n2} & \cdots & a_{nm}
\end{pmatrix}}_{H}
\begin{pmatrix}
z_1 \\ z_2 \\ \vdots \\ z_m
\end{pmatrix}
=
\begin{pmatrix}
y_1 \\ y_2 \\ \vdots \\ y_n \\ 0 \\ \vdots \\ 0
\end{pmatrix}
\tag{10.2}
$$

One can employ the Gaussian elimination algorithm to decode. This decoder is optimal as far as the success of the recovery procedure is concerned: decoding (by means of *any* algorithm) fails if and only if the Gaussian elimination decoder fails. However, the running time of this decoder is prohibitively large.

A different decoder with much lower complexity operates in the same manner as the greedy algorithm for LT-codes: The matrix in (2) is interpreted as the connection matrix between the m

input symbols, and $n+m-k$ output symbols. There are n dynamic output symbols corresponding to the collected output symbols. The last $m-k$ static output symbols correspond to the precode, and the values of these symbols are set to zero. The greedy algorithm of Section 2.2 can be applied to this graph to recover the values of the input symbols. A modification of this algorithm has been completely analyzed in [7] and designs have been presented which show that the failure probability of the algorithm is very small even for small overheads, if k is in the range of tens of thousands.

The superior computational performance of the greedy decoding algorithm comes at the expense of large overheads for small values of k. This can be explained by the fact that for small k the variance of the decoding process is too large compared to k, and hence decoding fails more often than for large k. It seems hard to be able to control the variance for small values of k. To remedy this situation, a different decoding algorithm has been devised [16]. Called *inactivation decoder*, this decoder combines the optimality of Gaussian elimination with the efficiency of the greedy algorithm.

Inactivation decoding is useful in conjunction with the scheduling process alluded to in Section 2.2 and outlined in [9]. The basic idea of inactivation decoding is to declare an input symbol as *inactivated* whenever the greedy algorithm fails to find an output symbol (dynamic or static) of weight 1. As far as the algorithm is concerned, the inactivated symbol is treated as decoded, and the decoding process continues. The values of the inactivated input symbols are recovered at the end using Gaussian elimination on a matrix in which the number of rows and columns are roughly equal to the number of inactivations. One can view Gaussian elimination as a special case of inactivation decoding in which inactivation is done at every step. Successful decoding via the greedy algorithm is also a special case: here the number of inactivations is zero.

If the number of inactivations is small, then the performance of the algorithm does not differ too much from that of the greedy algorithm; at the same time, it is easy to show that the algorithm is optimal in the same sense as Gaussian elimination.

The design problem for Raptor codes of small length which do not exhibit a large number of inactivations is tough, but solvable to a large degree. An application of the theoretical tools used for such a design is the standardized Raptor code which is discussed in the next section, along with a description of the systematic version of these codes.

10.2.4 The Systematic Standardized Raptor Code

In a variety of applications it is imperative to have the source symbols as part of the transmission. A systematic fountain code is a fountain code which, in addition to the three conditions given in Section 2.1 satisfies the following properties:

1. The original source symbols are within the stream of transmitted output symbols. The output symbols not belonging to the set of source symbols are called *repair symbols*.
2. For all $0 \leq \ell \leq m$ all the source symbols can be recovered from any set of ℓ of the source symbols and any set of $m - \ell$ repair symbols, with high probability.

The straightforward idea of sending the source symbols alongside the normal output symbols of a nonsystematic Raptor code fails miserably. This is because there is large discrepancy between the statistics of the source symbols and that of the repair symbols. Instead, what is needed is a method which makes the source symbols indistinguishable from the other output symbols. With such a method, the distinction between the two disappears, and it does not matter which portion of the received symbols is source.

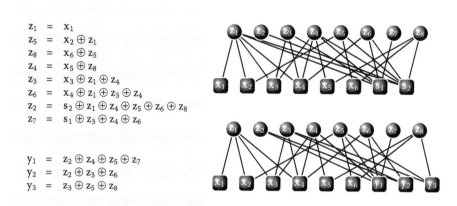

$$z_1 = x_1$$
$$z_5 = x_2 \oplus z_1$$
$$z_8 = x_6 \oplus z_5$$
$$z_4 = x_5 \oplus z_8$$
$$z_3 = x_3 \oplus z_1 \oplus z_4$$
$$z_6 = x_4 \oplus z_1 \oplus z_3 \oplus z_4$$
$$z_2 = s_2 \oplus z_1 \oplus z_4 \oplus z_5 \oplus z_6 \oplus z_8$$
$$z_7 = s_1 \oplus z_3 \oplus z_4 \oplus z_6$$

$$y_1 = z_2 \oplus z_4 \oplus z_5 \oplus z_7$$
$$y_2 = z_2 \oplus z_3 \oplus z_6$$
$$y_3 = z_3 \oplus z_5 \oplus z_8$$

Figure 10.3   Toy example of a systematic Raptor code. The source symbols are $x_1,\ldots, x_6$. The nodes with labels s_1, s_2 are obtained from the relations dictated by the precode, and their values are 0. In a first step, the intermediate symbols $z_1,\ldots, z_8$ are obtained from the source symbols by applying a decoder. The sequence of operations leading to the z_i is given on the left. Then the output symbols are generated from these intermediate symbols. Examples for three output symbols y_1, y_2, y_3 are provided. Note that by construction the x_i are also XORs of those z_i to which they are connected.

Such a method has been outlined in [7] and in [17]. The main idea behind the method is the following: we start with a nonsystematic Raptor code, and generate k output symbols. We then run the scheduling algorithm to see whether it is possible to decode the input symbols using these output symbols. If so, then we identify these output symbols with the source symbols, and decode to obtain a set of m *intermediate symbols*. The repair symbols are then created from the intermediate symbols using the normal encoding process for Raptor codes.

An example of a systematic Raptor code together with its encoding procedure is provided in Figure 10.3.

The crux of this method is the first step in which k output symbols need be found which are "decodable." This corresponds to decoding with zero overhead. A variety of methods can be employed to do this. The output symbols generated by these methods differ in terms of the error probability and complexity of the decoder. The computations corresponding to these symbols can be done offline, and the best set of output symbols can be kept for repeated use. What is then needed is an efficient method to re-produce these output symbols from a short advice, for example a 16-bit integer. The standardized Raptor code [9] does exactly this, and provides for any length k between 1 and 8192 a 16-bit integer, and a procedure to produce the k output symbols from this integer.

Figure 10.4 gives a brief description of the standardized Raptor code in terms of the precode and the probability distribution Ω.

10.2.5   Performance of Standardized Raptor Codes

Several aspects need to be considered in the assessment of the power of Raptor codes. These include coding performance, complexity (especially at the decoder), and the flexibility to different use cases. A reasonable insight is obtained by comparing the standardized Raptor code to an ideal

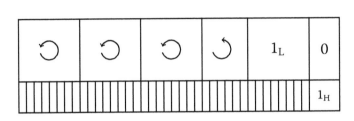

i	Ω_i
1	0.00971
2	0.4580
3	0.2100
4	0.1130
10	0.1110
11	0.0797
40	0.0156

Figure 10.4 The check matrix of the precode and the LT-degree distribution Ω for the standardized Raptor code. The check matrix consists $L + H$ rows, where L is the smallest prime greater than or equal to $X + \lfloor 0.01k \rfloor$ where X is the smallest integer such that $X(X - 1) \geq 2k$. H is the smallest integer such that $\binom{H}{\lceil H/2 \rceil} \geq L + k$. The check matrix is composed of an $L \times (k + L + H)$ matrix consisting of block-circulant matrices of row-weight 3, and block size L, an $L \times L$-identity matrix 1_L, and an $L \times H$-matrix consisting of zeros. The last circulant matrix appearing before the identity matrix may need to be truncated. The lower $H \times (k + L + H)$-matrix consists of binary vectors of length H and weight $\lceil H/2 \rceil$ written in the ordering given by a binary reflected Gray code, followed by an $H \times H$-identity matrix 1_H. The distribution Ω for the LT-code is given on the right. Ω_i is the probability of picking the integer i.

fountain code. An ideal fountain code has the property that for any number k of source symbols it can create any number m of repair symbols with the property that any combination of k of the $k + m$ source and repair symbols is sufficient for the recovery of the k source symbols. Thus, an ideal fountain code has zero reception overhead: the number of received symbols needed to decode the source symbols is exactly the number of source symbols independent of which symbols are received. Simulation results for Raptor codes provided for example in [18,19] show that Raptor codes have reception overhead very close to ideal fountain codes.

For a file delivery session using AL-FEC, the *transmission overhead* is defined as $100 * (N / K - 1)$, where N is the number of encoding packets transmitted in the file delivery session and K is the number of source packets in the original file (all packets are equal size). Thus, the transmission overhead is the amount of repair data sent for the file delivery measured as a percent of the file size. During the standardization phase of MBMS, 3GPP extensively tested different alternatives to provide reliability for download delivery in 3GPP systems and measured transmission overheads. An exemplary result is provided in Figure 10.5: which shows the decoding failure probability versus transmission overhead when transmitting a 3MByte file encoded with the Raptor code over a MBMS UMTS bearer at different link layer loss rates p compared to an ideal fountain code. The recommended parameter settings according to [9] have been used. It is clear from the results that for these conditions the Raptor codes perform basically as good as ideal fountain codes for all loss rates. It is worth noting that the encoding symbol loss rates in general are higher than the link layer loss rates as the mapping of IP-packets to link layer packets is in general not aligned.

The performance of Raptor codes compared to an ideal fountain code has been investigated further. The reception overhead performance of an AL-FEC code can be expressed by the decoding failure probability $P_f(n, k)$ as a function of the source block size k and the number of received

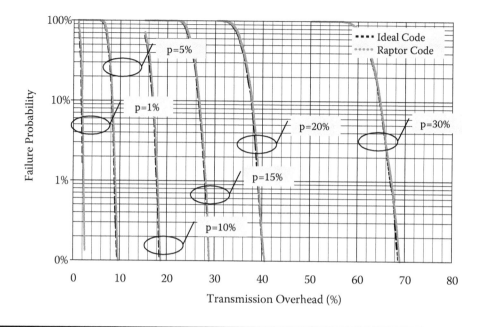

Figure 10.5 **Failure probability versus transmission overhead when transmitting a 3MByte file encoded with the Raptor code over a MBMS UMTS bearer at different link layer unit loss rates *p* compared to an ideal fountain code.**

symbols n. An interesting and quite powerful estimation for the reception overhead of standardized Raptor codes has been determined as

$$P_f(n, k) = \begin{cases} 1 & \text{if } n < k, \\ 0.85 \times 0.567^{n-k} & \text{if } n \geq k, \end{cases}$$

Figure 10.6 plots the failure probability $P_f(n, k)$ versus $n - k$, where $n - k$ is the reception overhead. The figure also contains selected simulation results, which verify that the above reception overhead performance estimate is quite accurate. An experiment has been carried out where 40 percent of the source data was dropped randomly and then a random $0.4k + (n - k)$ repair symbols were chosen and the decoding failure probability was evaluated. It is observed that for different values of k, the equation almost perfectly emulates the actual reception overhead performance. For the Raptor code the failure probability for $n \geq k$ decreases exponentially with an increasing number of received symbols. The increase is so fast, that for only about 12 additional symbols the failure probability is 0.1% and for 24 additional symbols the failure probability is 0.0001 percent. For a typical source block sizes of $k \geq 1000$ symbols then the overhead for a 0.1 percent failure probability is below 1.2 percent and the overhead for a 0.0001 percent failure probability is below 2.4 percent.

The average reception overhead of an erasure code, ε, is the average amount of additional data necessary to recover a source block. In a practical scenario, this would correspond to a receiver that requests encoding symbols as long as decoding is not successful. Based on the function $P_f(n, k)$,

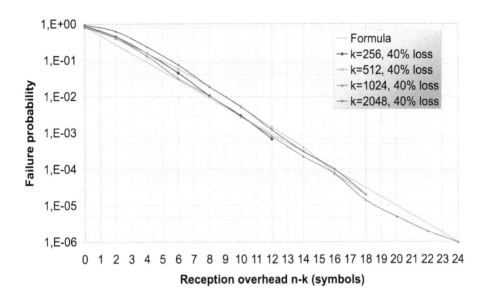

Figure 10.6 **Raptor source block loss rate for different source block size *k*, reception overhead *n* − *k*, and symbol loss rate 40%.**

the average reception overhead, ε, depending on k, results in

$$\varepsilon(k) = \frac{1}{k}\sum_{i=0}^{\infty} i(P_f(k+i-1,\,k) - P_f(k+i,\,k)) = \frac{0.85}{k}\sum_{i=0}^{\infty} i(0.567^{i-1} - 0.567^{i}) = \frac{0.85}{(1-0.567)k}.$$

The final expression is approximately $2/k$. Therefore, the number of additional symbols is on average 2, independent of k. The average reception overhead decreases with increasing k, and for example, for typical values of $k \geq 1000$ it is at most 0.2 percent.

In terms of complexity, the standardized Raptor codes are quite attractive. The complexity has been evaluated in the 3GPP MBMS standardization effort. For example, on a 206 MHz ARM platform, decoding speeds of more than 25 Mbps can be supported. Compared to Reed–Solomon codes, which operate on nonbinary symbols, the computational complexity of Raptor codes is orders of magnitude less. Further advantages of the Raptor codes are that the complexity is linear in the size of source data and the complexity is independent of the observed packet loss rate. The memory requirements for Raptor codes are also very attractive, as for both encoding and decoding only slightly more memory is needed than the source block size.

10.3 Multimedia Delivery Services in MBMS

10.3.1 MBMS Architecture

MBMS is a point-to-multipoint service in which data is transmitted from a single-source entity to multiple recipients. Transmitting the same data to multiple recipients allows network resources to be shared. The MBMS bearer service offers a broadcast mode and a multicast mode. The MBMS architecture enables the efficient usage of radio network and core network resources, with an

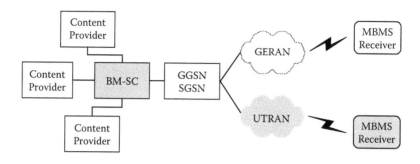

Figure 10.7 Simplified MBMS system architecture.

emphasis on radio interface efficiency. In the bearer plane, this service provides delivery of IP multicast datagrams to User Equipments (UEs). A new functional entity, the Broadcast Multicast Service Center (BM-SC) provides a set of functions for MBMS user services. The system architecture is shown in Figure. 10.7.

MBMS user service architecture is based on an MBMS receiver on the UE side and a BM-SC on the network side. Reception of an MBMS multicast service is enabled by different phases such as subscription, joining, data transfer, and leaving. In this work we focus on the data transfer phase exclusively. Furthermore, we concentrate on MBMS delivery over UTRAN, and specifically focus on the mobile radio efficiency.

10.3.2 MBMS Protocol Stack

MBMS defines two delivery methods—download and streaming delivery. MBMS delivery methods make use of MBMS bearers for content delivery but may also use the associated delivery procedures for quality reporting and file repair. A simplified MBMS protocol stack focusing on data delivery aspects for streaming and download delivery is shown in Figure 10.8.

Streaming data such as video streams, audio programs, or timed text are encapsulated in RTP and then transported over the streaming delivery network. In this case, AL-FEC is applied on UDP flows, either individually or on bundles of flows. The streaming framework provides significant flexibility in terms of code rates, protection periods, and so on. [9]. Discrete objects such as still images, multimedia streams encapsulated in file formats, or other binary data are transported using the FLUTE protocol (RFC 3926[1]) when delivering content over MBMS bearers. In both delivery services the resulting UDP flows are mapped to MBMS IP multicast bearers.

The MBMS bearer services reuse most of the legacy UMTS protocol stack in the packet-switched domain. Only minor modifications are introduced to support MBMS. The IP packets are processed in the Packet Data Convergence Protocol (PDCP) layer, where, for example, header compression might be applied. In the Radio Link Control (RLC) the resulting PDCP- Protocol Data Units (PDUs), generally of arbitrary length, are mapped to fixed length RLC-PDUs. The RLC layer operates in unacknowledged mode as feedback links on the the radio access network are not available for point-to-multipoint bearers. Functions provided at the RLC layer are for example segmentation and reassembly, concatenation, padding, sequence numbering, reordering and out-of-sequence and duplication detection. The Medium Access Control (MAC) layer maps and multiplexes the RLC-PDUs to the transport channel and selects the transport format

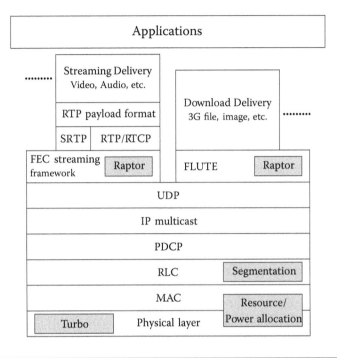

Figure 10.8 MBMS protocol stack.

depending on the instantaneous source rate. The MAC layer and physical layer can appropriately adapt the RLC-PDU parameters to the expected transmission conditions by applying, among others, channel coding, power and resource assignment, and modulation.

10.3.3 MBMS Bearer Service over UMTS

The UMTS bearer provides services with different QoS that are fundamental to support the MBMS broadcast mode. Radio bearers are specified, among others, by the data throughput, data transport format, PHY-FEC, rate matching, and power allocation. MBMS uses the *Multimedia Traffic Channel* (MTCH), which enables point-to-multipoint distribution. This channel is mapped to the *Forward Access Channel* (FACH), which is finally mapped to the *Secondary-Common Control Physical Channel* (S-CCPCH) [20]. Among others, an MBMS radio bearer is defined by the transport format size and number of transport blocks that are to be protected by PHY-FEC at every transmission time interval (TTI). The TTI is transport channel specific and can be selected from the set {10 ms, 20 ms, 40 ms, 80 ms} for MBMS. Thereby, higher values are accomplished by longer interleaving or longer codeword sizes of channel code, but at the expense of higher latencies.

Refer again to Figure 10.8. After RLC layer processing the resulting RLC/MAC blocks are mapped into the transport blocks according to the specified radio bearer settings and a 16-bit CRC is appended. The resulting blocks might be concatenated and then further segmented into code blocks such that the maximum length of a code block does not exceed 5,114 bits. [21,22]. This limitation comes from the restrictions on complexity, memory, and power consumptions of turbo decoders in handheld devices. After turbo coding is applied, the resulting blocks are

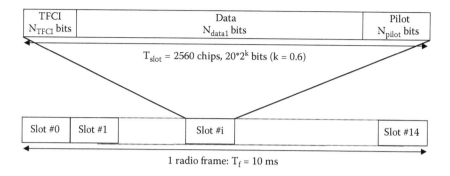

Figure 10.9   Structure of the UMTS radio frame.

concatenated, interleaved, and eventually rate matching is performed such that the Turbo code rate, r_{inner}, can be set in the range $[\frac{1}{3};1]$. By repetition of Turbo FEC parity bits, even lower code rates than $r_{inner} = \frac{1}{3}$ can be supported. The resulting codeword is mapped to one or more transmission slots that are finally mapped onto radio frames, as shown in Figure 10.9. A radio frame consists of 15 slots, whereby the number of bits per slot depends on the applied spreading code. Finally, radio frames are transmitted every 10 ms using a chip-rate of 3.84 Mcps and QPSK modulation.

For interoperability and testing purposes a number of MBMS reference radio bearers have been defined as a preferred configuration [23] for a specific data rate. The configurations define a single default radio parameter set including Turbo code rates, transport format, transmission time interval, symbol rate, and so on. However, for our purposes to leave some flexibility in investigating the trade-off between the Turbo code and the Raptor code we only fix a subset of parameters of selected MBMS bearers in Table 10.1. In contrast to specified bearers, we allow the turbo code rate, r_{inner}, to be adjusted so that trade-offs between the Turbo code and the Raptor code can be analyzed. Note that MBMS allows use of configurations other than those specified in Reference [23], and in particular all configurations we apply in the following fully comply with the specification.

To appropriately compare different settings, we specify an MBMS bearer not by its data rate at the RLC layer, but instead by the symbol rate at the physical layer. The considered bearers with their respective settings are provided in Table 10.1. The variation of the Turbo code rate results in varying RLC bit rates: higher Turbo code rates offer higher data rates but also result in higher RLC-PDU loss rates, whereas lower Turbo code rates result in lower loss rates, but also offer higher RLC data rates. A primary purpose is to investigate combinations of code rate settings for Raptor codes used at the application layer and Turbo codes used at the physical layer that optimize the overall use of the network.

Table 10.1   Bearer Physical Layer Parameters

Channel Bit Rate	SF	Bits Per Slot	Data Bits Per Slot
120 kbps	64	80	72
240 kbps	32	160	152
480 kbps	16	320	312
960 kbps	8	640	632

At the receiver side, the inverse operations are applied. Specifically, if CRC for a transport block fails after turbo decoding, the block is considered erased. All correct data is delivered to the upper layers. At the RLC layer, only correctly received RLC-SDUs (usually containing an entire IP packet) are delivered to the higher layer. Therefore, any IP packet that is partly contained in an erroneous transport block is lost and is not available at the receiver.

MBMS is generally applied in a multicellular environment. If the same MBMS services are offered synchronously in not just a single cell, but in an entire area, then the receiver performance, especially at cell edges, can be improved. A mobile terminal can perform combining of different signals, either on the physical layer in the form of *soft combining* or on the RLC layer, referred to as *selective combining*. In these cases an MBMS UE listens to more than one Node-B signals simultaneously, and for selective combining, it individually decodes the streams in the hope that for at least one signal passes the CRC such that the correct RLC-PDU can be forwarded to the upper layer. However, any combining, scheme increases the complexity and costs of a receiver and therefore might not be used in initial deployments of MBMS.

Mobile radio bearers can be configured in a quite flexible manner. For the IP multicast bearers in MBMS, the parameters in Table 10.1 provide insight into the configuration. On top of this, as already mentioned, the turbo code rate r_{inner} can also be modified. The quality of such bearer configurations can basically be evaluated by their supported bitrate on RLC layer and the observed RLC-PDU loss rate. Whereas the bit rate is a transmitter configuration, for example a combination of the bearer parameter in Table 10.1 and the turbo code rate, the observed loss rate significantly depends on the position and mobility of the user under investigation. Furthermore, the observation window for the loss rate measurement is quite important in the interpretation of the loss rate. The long-term loss rates might be quite different than those observed over a shorter period.

With the use of Raptor codes, higher loss rates can be compensated by the transmission of additional repair symbols, ensuring that all transmitted data that is useful for recovery of the original source data. In this case a good measure for the overall network performance is the so-called *goodput*, defined as the supported bit rate multiplied by 1 minus the packet loss rate of the user measured over some window of time. A goodput measurement represents the average received amount of data over a window of time, and a sequence of goodput measurements can be continually varying depending on the changing average loss rate within different windows of time. Variations of goodput measurements depend not only on the transmitter configuration and user mobility, but also on the observation window for measuring the goodput: in general, the smaller the observation window for measuring the goodput the higher the variance of the measured goodputs. Some selected measurements of goodput distributions for different MBMS bearer settings are provided in Section 10.5.

10.3.4 MBMS Download Delivery Service

To deliver a file[*] in a broadcast session, FLUTE provides mechanisms to signal and map the properties of a file to the Asynchronous Layered Coding (ALC) [24] protocol such that receivers can assign these parameters to the received files. The file is partitioned into one or several *source blocks*. Each source block consists of k *source symbols*, each of length T except for the last source symbol, which can be smaller. Both parameters T and k are signaled in the session setup and are fixed for

---

[*] For simplicity we continue with the notion of a file in the following though the ALC/LCT concept uses the more general terminology *transport object*.

one session. For each source block, additional repair symbols can be generated by applying Raptor encoding as explained in detail in Section 10.2. Each encoding symbol, that is a source symbol or a repair symbol, is assigned a unique encoding symbol ID (ESI), to identify the symbol and its type. With respect to the symbol type, if the ESI is less than k then it is a source symbol; otherwise, it is a repair symbol. Let us denote the total number of encoding symbols to be transmitted as n and define the resulting Raptor code rate as $r_{outer} = k/n$. One or more encoding symbols of the same type with consecutive ESIs are placed in each FLUTE packet payload. The source block number as well as ESI of the first encoding symbol in the packet are signaled in the FLUTE header. FLUTE packets are encapsulated in UDP and distributed over the IP multicast MBMS bearer.

Receivers collect received FLUTE packets containing encoding symbols, and with the information available in the packet headers and the file session setup, the structure of the source block can be recovered. If no more encoding symbols for this source block are expected to be received, the Raptor decoder attempts to recover the source block from all received encoding symbols. Due to heterogeneous receiving conditions in a broadcast session, the amount as well as the set of received encoding symbols differs among the receivers. If all source blocks belonging to the file are recovered at a receiver, the entire file is recovered. If file recovery fails, a post-delivery repair phase might be invoked. With the download delivery protocols in place, different services can now be realized.

10.3.4.1 Scheduled Distribution without Time-Limits

In a scheduled broadcast service, files are distributed once within a session and all users join the session at the very beginning. The costs of such a service in mobile cellular systems is appropriately measured by the consumed resources on the physical layer, which comprise of the bandwidth share, the transmit power, and duration of the session. For simplicity, we consider the case of distributing a single file. Assuming that we fix the bandwidth share, a suitable single measure for the costs relevant for transmitter is the product of the assigned transmit power for such services and the "on air time" for the distribution of the file. The product results in the necessary energy, E, to distribute the file. Secondary aspects such as the experienced "download time", that is how long it takes to receive the file, are generally not essential in this use case as it is assumed that the distribution is not time-critical.

In terms of user perception, file download delivery is, to a large extent, binary, that is for each user the file is either fully recovered and the user is satisfied or the file is not fully recovered and the user is unsatisfied. Clearly, not all users can always be satisfied, and file distribution services are usually operated such that a certain percentage of users are satisfied. Unsatisfied users are not necessarily excluded in the MBMS download service, and may rely on post-delivery methods to complete the file recovery. We evaluate the necessary system resources in terms of the required energy to satisfy at least a certain percentage of the user population in the MBMS service area. As a reasonable number, the support of 95 percent of the user population is the objective.

10.3.4.2 Time-Constrained Distribution

In a second service scenario we consider scheduled distribution with the additional constraint that files of a certain aggregate size need to be distributed within a certain amount of time. For simplicity, we consider the case of distributing a single file. In this scenario, it is of interest is to evaluate the radio resources required to transport a file of a certain size, and/or the maximum size file that can be transmitted in a certain amount of time. In the latter case, the ratio of the maximum supported file size and the allowable delivery duration also expresses the maximum supported bit rate within the time and resource constraints.

An example for such as service is the following: Assume that a provider offers the possibility to purchase a song that is played on a regular analog or digital radio program right after the song is played. To enable this, the song is distributed via a download delivery bearer, for example within an MBMS system, and is available to all users. The user can then select to purchase the song, that is unlock it, or not. In this case the song must be delivered within the on-air time. For the case that radio resources are restricted to a certain maximum, the efficiency of the system determines the maximum bit rate of the compressed song, which relates to the quality of the media stream.

10.3.4.3 Carousel Services

File delivery using a carousel is a possibly time-unbounded file delivery session in which a fixed set of files are delivered. Two types of carousel services are distinguished, static and dynamic. Whereas the former delivers only fixed content within the file, in dynamic file carousels individual files may change dynamically. When using FLUTE the file delivery carousel is realized as a content delivery session whereby file data tables and files are sent continuously during a possibly time unbounded session. In case no AL-FEC is available, the data must be repeated. In the case that the Raptor code is used, the data transmitted for a given file generally includes repair symbols generated by Raptor encoding in addition to the original source symbols. In particular, file reception time is minimized if symbols are never repeated until all 65,536 possible symbols (source and repair) have been sent. With this, the fountain property of the Raptor codes can be optimally exploited.

In terms of system configuration, the transmitter has only limited options; basically only the transmit rate can be selected. However, of interest for the receiver is the amount of time it takes to acquire the file. The objective is to minimize the time that it takes for a receiver to acquire all the files, with a given probability, when joining the stream at some random time. Typically, the acquisition times for 95 percent of the users are a reasonable measure, but in this case, the average reception time also provides an interesting service quality measure. The performance of carousel services and also more advanced carousel services which allow Video-on-Demand-like services over broadcast channels have for example been assessed in [25].

10.3.5 MBMS Streaming Delivery Service

Real-time MBMS streaming services mainly target classical mobile TV services. For these services the MBMS FEC streaming framework, including the Raptor codes, [9,2] plays an important role (see Figure 10.8). The FEC streaming framework operates on RTP packets or, more precisely, on UDP flows, incoming on the same or different UDP ports. In video streaming applications, these RTP packets generally include H.264 Network Abstraction Layer (NAL) [26] units or audio packets. It has been proven beneficial to apply NAL unit fragmentation such that RTP packets do not exceed the size of the underlying RLC frame size. As shown in Figure 10.8, the FEC streaming framework is based on top of the UDP layer. The legacy RTP packets and the UDP port information are used to generate Raptor repair symbols. Original UDP payloads become source packets by appending a 3 byte FEC source payload ID field at the end of each UDP payload. These packets are then UDP encapsulated and transported on the IP multicast bearer.

As shown in Figure 10.10, a copy of these packets is forwarded to the Raptor encoder and arranged in a source block with row width T bytes with each consecutive packet starting at the first empty row. The source symbol starts at the beginning of a new row, but it is preceded by a

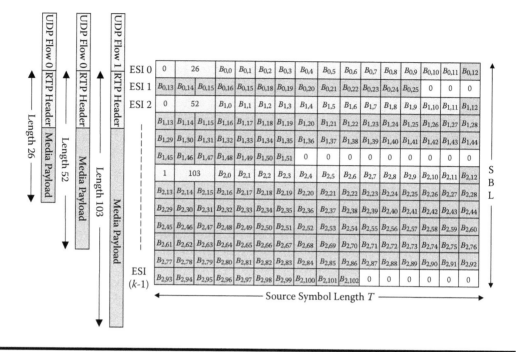

Figure 10.10 MBMS FEC streaming framework.

3-byte field containing the UDP flow ID (1 byte) and the length field (2 bytes), both of which are part of the source symbol. If the length of the packet plus the 3-byte field is not an integral multiple of the symbol length then the remaining bytes in the last row are padded out with zero bytes. The source block is filled up to k rows, where the value of k is flexible and can be changed dynamically for each consecutive source block. The selection of k depends on the desired delay, the available terminal memory and other service constraints.

After collecting all packets to be protected as a single source block, the Raptor encoder generates $n - k$ repair symbols of size T as described in Section 10.2, where the selection of n depends on how much loss is to be protected against. The generated Raptor repair symbols can be transmitted individually or as blocks of symbols in the payloads of UDP packets, called repair packets. Each source and repair packet contains sufficient information such that a receiver can use Raptor decoding to recover a source block if enough encoding symbols are received for that source block.

A large number of system parameters can be adjusted to fulfill certain utility functions. There are a significant number of options for a resource and quality-of-service optimized system configuration for an operator that runs Mobile TV services using MBMS. In terms of system *resources*, an operator can choose radio bearer configurations as discussed previously.

In addition to the radio parameters for the IP multicast bearer, within the streaming delivery service, one can basically select the following parameters:

■ The settings of the Raptor code parameters, mainly (1) the Raptor code rate, $r_{outer} = k/n$, which determines, together with the physical layer settings, the available bit rate for the application, and (2) the protection period T_{pp}, which influences the efficiency of the code, and also impacts the end-to-end and tune-in delay.

- ■ The video coding parameters, mainly determined by the bit rate and encoding quality Q_{enc}, as well as the error resilience and tune-in properties determined by the random access point frequency, and in case of H.264/AVC determined by the instantaneous decoder refresh (IDR) frame distance, T_{IDR}.

The selection of the parameters should be such that user satisfaction is maximized, whereas the usage of system resources is minimized. The target for an operator is user satisfaction for as many users as possible in the serving area, whereby the environment, as well as user behavior such as mobility, also influences the reception quality.

In contrast to the download delivery service, the definition of user satisfaction is more complex for mobile TV services. The service quality from the user perspective is mainly determined by the video quality, whereby it is essential to understand that both the *error-free video quality*, Q_{enc}, and the degradation due to errors matter. A reasonable service quality is only achieved if the encoded video has at least a certain encoded video quality $Q_{enc,min}$ and if errors only occur infrequently, that is if the *video quality degradation*, $\overline{D}_{dec}$, due to errors does not exceed a certain value $\overline{D}_{dec,max}$.

If only a single service is offered, the user perception might be slightly influenced by the tune-in time, but this aspect is usually of less relevance. However, in case MBMS is used for mobile TV services with multiple channels, then an important service parameter is also how long it takes to tune into a program or how long it takes to switch between different channels of the mobile TV service. For our system configurations, tune-in and zapping times are identical, and therefore we focus on the notion of tune-in time, $T_{tune-in}$.

10.4 MBMS System-Level Simulation

10.4.1 *Motivation*

As discussed in Section 10.3, mobile broadcast services, just as any mobile multimedia service, allow for a significant amount of system parameter settings. However, in p-t-p transmission systems the concept of quality of service (QoS) provision of the lower layers for the higher layers is quite established and also reasonable as the quality on the lower layers can be controlled by frequent feedback messages, adaptation to changing channel conditions, retransmissions, acquisition control, or other means. Therefore, it is quite reasonable to optimize each layer individually, or at least it is not necessary to do a full end-to-end and across-layer evaluation and optimization. In contrast, mobile broadcast systems do not provide any of these fast QoS control mechanisms. The overall performance depends significantly on the settings on different layers, and efficient and optimized parameter configurations in different layers can only be obtained by understanding the service from the end-to-end perspective and across all layers, from the physical layer up to the media coding layer. Mobile broadcast systems require cross-layer evaluation and optimization to fully exploit their potential.

In addition to the comprehensive end-to-end approach, user behavior and mobility primarily resulting in varying channel conditions in various time scales need to be taken into account. As in general many users consume a multimedia broadcast service in parallel, the heterogeneity of the reception conditions of different users influences the service quality. To meet the requirements and expectations of this rather complex system design, comprehensive end-to-end system level simulations are necessary. We have taken this approach to motivate the benefits of Raptor codes in mobile multimedia broadcast systems, specifically in MBMS. The basic concepts, the applied simulation framework, as well as individual components of the system-level simulation are introduced in more detail in the following.

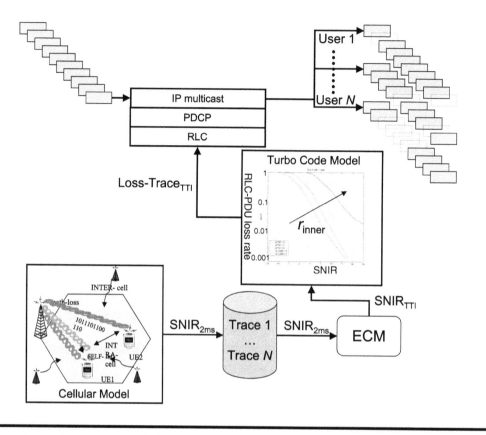

Figure 10.11 Concept of simulation setup for MBMS IP multicast delivery over UMTS.

10.4.2 Modeling and Simulation of MBMS IP Multicast Bearer

For the modeling of the IP multicast bearer a comprehensive approach on propagation, interference, multiusers, physical layer, and protocol stack modeling is proposed. Figure 10.11 provides insight into this approach. The simulator is composed of different modules that simulate and model different components of the entire system. It is divided into two blocks, the mobile cellular channel model and the radio protocol stack including the Turbo code. The cellular channel model generates traces for the carrier, interference, and noise observed at the mobile terminal, and also the orthogonality factor (required to compute self-interference), with a resolution of 2 ms and for as many as N different users with different random initial positions, whereby each trace corresponds to values captured over 10 minutes. The number of users in our case is $N = 500$.

These traces are generated for normalized transmit power and no spreading gain and are subsequently modified to obtain an effective SINR for each TTI by applying some appropriate combining, referred to *as equivalent SNR method based on convex metric* (ECM), power assignment and spreading code assignment. A resulting SINR is obtained for each TTI, which is converted to a sequence of RLC-PDU loss traces by applying a suitable table lookup for the Turbo code. The resulting RLC-PDU loss traces for each individual user are then applied to an IP multicast stream.

10.4.2.1 Mobile Radio Channel in Multicellular Environment

The mobile radio channel places fundamental limits on the performance of a wireless communication system. Unlike wired channels which are more stationary and predictable, radio channels, show extremely varying behavior. In fact, the radio transmission path between the transmitter and the receiver can vary from a simple line-of-sight to one that is severely obstructed by buildings, mountains, foliage, and so on. Also, the speed of the mobile terminal has a great impact on the received radio signal. In order to evaluate the impact of the mobile radio in a multicellular environment, a channel model that uses standard models and techniques has been defined and developed in 3GPP, [27], partly based on real measurements. In particular, effects such as *path loss, Doppler spreads, shadowing, antenna radiation pattern* and *interference* are taken into account in this simulation setup.

This channel model allows simulation of *pedestrian* and *vehicular* mobile users within a cell, whereby the main difference between these users is the speed at which they move within the cell, but also their power-delay spectra. Figure 10.12 shows examples of movements of users in a cell for different speeds and different starting points. The applied movement model is based on random walk with high directional correlation. The users do not leave the area, but bounce at the cell edges. However, possible handover effects are simulated as the signal is not necessarily received from the base station assigned to the hexagonal cell, but from the strongest one. The figure shows the position of four different users for a time of 10 min. Notice that the vehicular user undergoes a much larger distance due to his speed of 30 km/h, while the pedestrian users at 3 km/h cover less distance in the same amount of time. The right-hand side shows the signal to interference and noise ratio (SINR) for each of the users. The SINR is varying due to large scale effects such as attenuation and shadowing on both useful signals and interferers, as well as due to short-term effects such as fading and Doppler. Note also that for the vehicular user, the SINR shows faster variations than for the pedestrians.

10.4.2.2 ECM Method

The generic mobile radio simulator computes the signal-to-interference and noise ratio every 2 ms, which is, for example, the TTI in HSDPA. However, the investigated MBMS bearers use transmission time intervals of up to 80 ms. Therefore, an appropriate conversion of the effective SINR for every TTI is required. This is achieved using a *link error prediction method* called *equivalent SNR method based on convex metric* (ECM), as defined in reference [27]. This technique allows the combination of several SINR values into a single effective SINR that is equivalent to the channel decoder in case interleaving over these multiple channel access slots is provided. The method is based on Shannon's channel capacity formula and is processed in the following steps:

1. Compute the channel capacity C_i for every TTI ($i = 1, 2, \ldots, n$)
2. $\bar{C} = \frac{\alpha}{n} \sum_{i=1}^{n} C_i$
3. Compute SNR_{eff} such that $C(\text{SNR}_{eff}) = \bar{C}$

whereby the factor α is a correction factor that depends on the mobile speed, the interleaver, and so on. For low mobile speeds and the almost ideal UMTS interleavers it has been found [27] that $\alpha = 1$ is an appropriate value.

10.4.2.3 Power and Spreading Code Assignment

As already mentioned, the traces are generated for a normalized transmit power. However, the MBMS bearer might get assigned different power, resulting in different SINR values. Appropriate

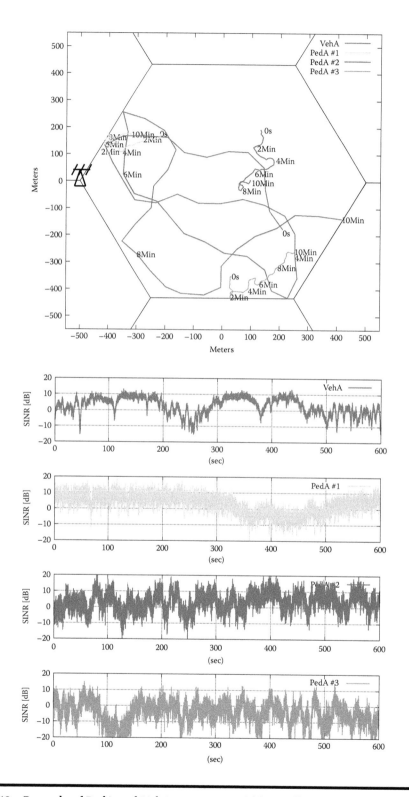

Figure 10.12 Example of PedA and VehA movement within a cell.

transmit power adjustments results in increase or decrease of the effective SINR. Furthermore, changing the MBMS radio bearer parameters, in particular the spreading factor, also leads to different SINR values. Because the individual values for the carrier, the interference, and the noise are stored in the trace files, it becomes quite easy to update the carrier power and recompute the effective SINR by

$$\mathrm{SINR}_{\mathrm{eff}}(C,\ I,\ N,\ OF,\ SF) = 10 \log_{10} \frac{C \times SF}{C \times (1 - OF) + I + N},$$

whereby C represents the carrier power, I the interference power, N the noise power, OF the orthogonality factor and SF the spreading factor.

10.4.2.4 Physical Layer FEC Modeling

In system-level simulation, the loss probability of the Turbo code is determined by table lookups that map the effective SINR to the loss probability. Based on this loss probability, a random generator decides whether the included RLC/MAC block is decodable. However, for codes with different code rates, this still requires a significant amount of link-level simulations, as each possible code rate needs to be simulated.

Luckily, turbo codes as applied in UMTS have the property that for a given SINR and a given code rate r_{inner}, the decoder is either almost always able to decode or it almost always fails. The so-called waterfall region of long turbo codes is rather narrow. The waterfall region for practical turbo codes coincides quite well with the computational cutoff rate $R_0(\mathrm{SINR}) = 1 - \log_2(1 + e^{-\mathrm{SINR}})$ in a sense that if the rate of the code is below the cutoff rate for this specific SINR, decoding is successful, otherwise it fails. Note the above equation is valid for BPSK transmission as well as for each component in case of QPSK transmission. Therefore, after each TTI, the mobile channel simulator computes the effective SINR. Based on this value and the applied code rate, the RLC-PDU are either assumed error-free or are lost.

10.4.3 System-Level Simulation of MBMS Download Delivery

For the simulation of MBMS Download Delivery, the MBMS Download Delivery CDP including the Raptor code is simulated over different MBMS IP multicast bearer as shown in Figure 10.13. The service quality, the transport of a file using the FLUTE protocol and Raptor is simulated. Thereby, for each of the N users, and for different Raptor code rates, it is evaluated if the file can be recovered. More precisely, for each of the N users, it is evaluated, how many repair symbols are necessary to send for each of the N users to recover the file. As soon as sufficient user satisfaction is achieved, for example as in our case 95 percent of the users have recovered the file, we assume that the distribution of the file is stopped. For each of the different IP multicast bearer configurations, this value is evaluated and the necessary energy, i.e., the download time of the 95 percent user multiplied by the power, is used as a criterion for the goodness of the configuration. An important aspect in the assessment of the service is also the file size that may vary from for example 32 kByte up to several tens of MBytes. A representative but still rather small value has been selected, namely a 512 KByte file in our simulations. This might correspond to a short multimedia clip, a still image or a reasonably sized ring tone.

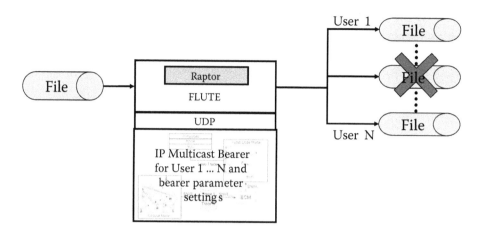

Figure 10.13 Simulation setup for MBMS download delivery.

10.4.4 System-Level Simulation of MBMS Streaming Delivery

In a manner similar to that for download delivery, streaming delivery over MBMS is evaluated. The concept of the simulation approach is shown in Figure 10.14. The MBMS IP multicast bearer simulation is composed in the same way as MBMS download delivery. The only difference is the CDP simulation.

For the evaluation of the MBMS streaming user service, several 3GPP tools available in reference [28] have been used. The following procedure is applied: Initially, for a certain setting of IDR frame distance and a target quality, an encoded video stream encapsulated in RTP is generated and stored in RTP dump format. The different IDR frame distances result in different bit rates, but the streams have the same quality. The applied encoding follows a rather strict constant bit rate control, but the bit rate still might fluctuate in ranges of several percent within each group

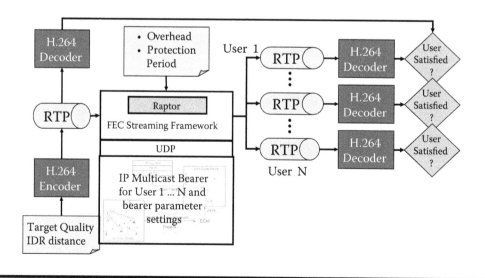

Figure 10.14 Concept of simulation setup for MBMS streaming delivery over UMTS.

of picture (GOP). Raptor encoding is applied to the generated RTP streams to generate a certain number of repair symbols for each source block consisting of a certain number of source symbols. Thereby, the Raptor code rate r_{outer} is selected such that together with the setting of the Turbo code rate, r_{inner}, the bearer resources are optimally used. A protection period, T_{pp}, is selected such that RTP packets within a protection period are collected in a single source block, and the source block size k may vary slightly depending on the video statistics. After applying the IP packet loss pattern resulting from the MBMS bearer configuration, the resulting video stream is decoded and is compared to the reconstructed stream without any errors to obtain the percentage of Degraded Video Duration (pDVD) $\overline{D}_{dec}$ for this stream. The pDVD $\overline{D}_{dec}$ is used to check if the user is satisfied: the pDVD $\overline{D}_{dec}$ shall not exceed 5 percent. This experiment is repeated for all N users to obtain the percentage of satisfied users for the specific parameters applied. In addition, the received stream is evaluated in terms of average tune-in delay by assessing tune-in at each RLC-PDU and measuring the resulting necessary delay to display the first correct IDR frame and to ensure the display of all remaining frames without any jitter. The resulting average tune-in delay $\overline{T}_{tune-in}$ is obtained by averaging over all RLC-PDU positions and all satisfied users.

10.5 Selected Simulation Results

The simulation concept and details presented in Section 10.4 allow simulation of the performance of download and streaming delivery services in MBMS. Extensive system level simulations are performed in order to evaluate system performance and specifically the trade-off between AL-FEC based on Raptor codes and PHY-FEC based on Turbo codes.

10.5.1 *Performance Evaluation of Radio Bearer Settings*

To get some insight in the performance of different radio bearer settings, we evaluate the distribution of the goodput for different system parameters assuming users randomly placed and randomly moving in the service area. In Figure 10.15 we show the cdf of the goodput for a bearer with 240 ksps, different transmit powers, and turbo code rate $r_{inner} = 0.33$ and $r_{inner} = 0.67$ for an observation window of 10 minutes. The receivers do not use any receiver combining. It can be observed from the figures that the maximum value of the goodput is determined by the Turbo code rate, as expected. Higher turbo code rates result in higher throughputs at the expense of higher error rates. However, the error rates are not that severe and comparing the two diagrams and 95 percent of the users satisfied, with $r_{inner} = 0.67$ instead of $r_{inner} = 0.33$, the same goodput can be achieved with $P = 2W$ instead of $P = 16W$. Therefore, if a CDP can make use of these bearer and physical properties, significant system benefits can be expected.

Figure 10.16 shows the corresponding results when selective combining is used. It is clear that the goodput is significantly improved, but similar to the case without combining, the same goodput can be achieved by using higher turbo code rate of $r_{inner} = 0.67$, and the power can be reduced to $P = 1W$ to have a 95 percent support, which is as good as for a code rate $r_{inner} = 0.33$ and $P = 16W$. These findings are exploited in the cross-layer design for the delivery services in the following.

10.5.2 *Performance of Download Delivery*

To assess the performance of download delivery, the approach as described in subsection 10.3 has been applied to a selected parameter set. The chosen bearer supports 240 ksps at the physical layer. Simulations are run for $N = 500$ users whereby their starting position is randomly and uniformly

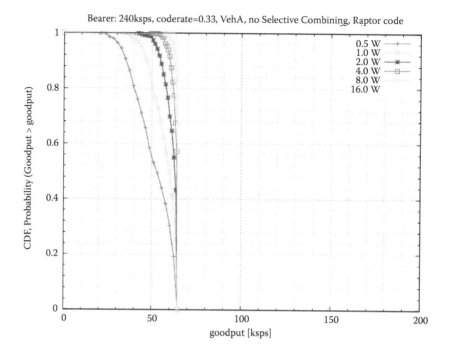

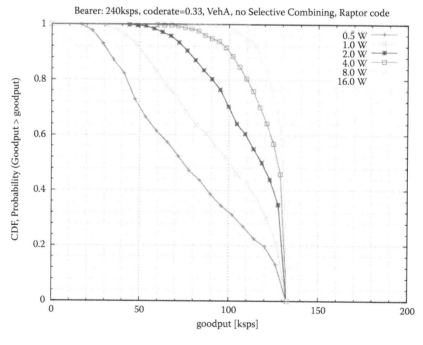

Figure 10.15 **CDF of goodput for a 240ksbs bearer with code rate $r_{inner} = 0.33$ (top) and code rate $r_{inner} = 0.67$ (bottom) for Vehicular A mobility model and for Raptor decoding without combining.**

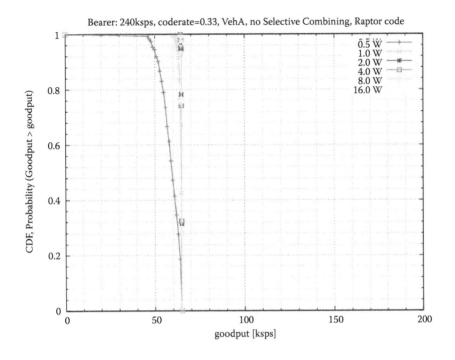

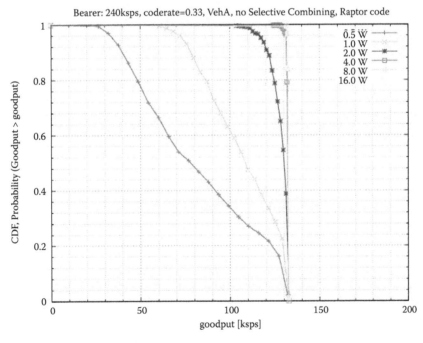

Figure 10.16   CDF of goodput for a 240ksbs bearer with code rate r_{inner} = 0.33 (top) and code rate r_{inner} = 0.67 (bottom) for Vehicular A mobility model and for Raptor decoding with selective combining.

distributed over the cell area. These users are simulated for vehicular and pedestrian mobility and propagation model. We also compare receiver performance with and without selective combining. Noteworthy, many more simulations than shown in this section have been carried out, and the results show a reasonable and representative selection.

In the assessment of different system configurations, basically two aspects are of major interest, user perception of the multimedia delivery and the resources consumed on the physical layer. The latter is most suitably expressed by the necessary energy, E, to distribute the file. We evaluate the necessary system resources in terms of the required energy to satisfy at least 95 percent of the user population for different system parameter configurations. We investigate different settings of Raptor code rates and Turbo code rates and transmit power assignments. For intuitive interpretation of the results, we present the RLC-PDU loss rate of the worst-supported user and the necessary energy to support this user.

Figure 10.17 and Figure 10.18 show the necessary energy over the resulting RLC-PDU loss rate for different transmit power assignments for the MBMS service. Vehicular users only and pedestrian users only, both with and without selective combining at the receiver, are assessed. The curves are generated by applying different inner code rates r_{inner} and as much Raptor encoding as necessary to ensure that 95 percent of the users are satisfied. The curves generally terminate on the left due to the restriction on the Turbo code rate of 0.33; the leftmost point corresponds to the lowest RLC-PDU loss rates, and therefore to a lower Turbo code rate, while the rightmost point of the curve corresponds to a higher RLC-PDU loss rate, and therefore a higher Turbo code rate.

From the simulation results it is apparent that there are some optimum system configurations that minimize transmit energy. Generally, the optimum is at rather high RLC-PDU losses and are not achieved when using the lowest Turbo code rate 0.33. For example, in Figure 10.17, if the system allocates 4W of transmit power for MBMS service, the optimal RLC-PDU loss rate for minimal required delivery energy is about 40%. If stronger Turbo coding is applied, the RLC-PDU loss rate decreases. However, the throughput at the RLC layer also decreases, as already elaborated in the goodput evaluation results. This leads to an increased download delivery time and consequently to more required energy.

If a Turbo code rate of 0.33 is chosen, the required energy for successful delivery is about 60 percent higher than in one case of the optimum configuration. However, if the Turbo code rate is too high then the resulting higher bit rates cannot be compensated by the increasing RLC-PDU loss rate, that is this leads to increased download delivery time, and consequently higher required energy. These results suggest that using the Raptor code with a low code rate at the application layer and working at rather high RLC-PDU loss rates is overall very beneficial for the system resources and reduces the overall required energy for the file distribution. By the use of Raptor coding the goodput maximization can be exploited. Another interesting observation is that transmission with lower transmission power is advantageous. In all, the presented results, transmission with 0.5W, always result in the minimal required energy. Although even lower transmit powers might provide better performance, other effects such as frequent loss of synchronization or very long on-air times would be counterproductive.

Selective combining, if applicable, has impact on the required energy and increases the system capacity significantly. In Figure 10.17, right, the minimum required delivery energy for 0.5W is less than half the energy required for the corresponding case without selective combining. This was also already predicted by the goodput results. The RLC-PDU loss rates for optimal energy delivery with selective combining are lower mainly due to lower download time, not the use of a different Turbo code rate. Therefore, receivers with and without selective combining can quite well coexist and should be operated with similar system parameters. Note, however, that the loss

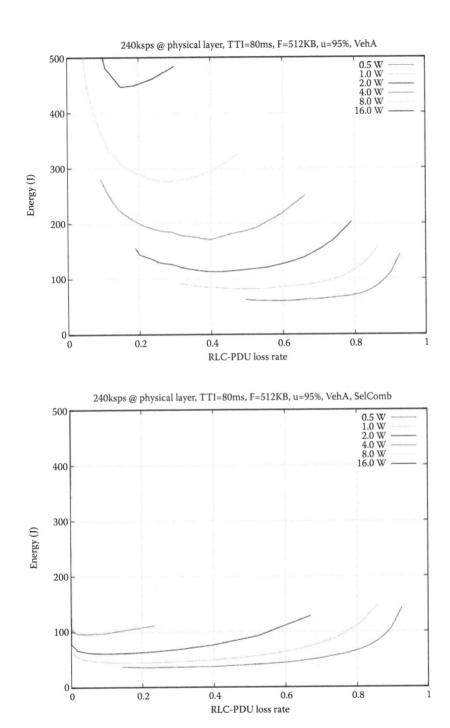

Figure 10.17   Simulation Results for a 240 ksps bearer, vehicular A mobility model, without combining (top) and with selective combining (bottom).

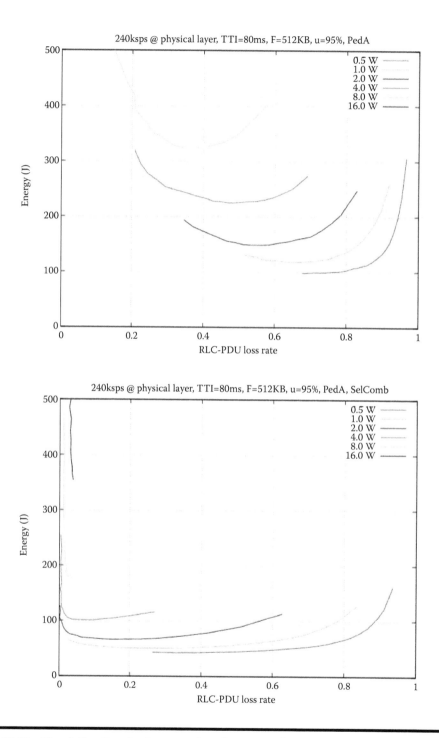

Figure 10.18   Simulation Results for a 240 ksps bearer, pedestrian A mobility model, without combining (top) and with selective combining (bottom).

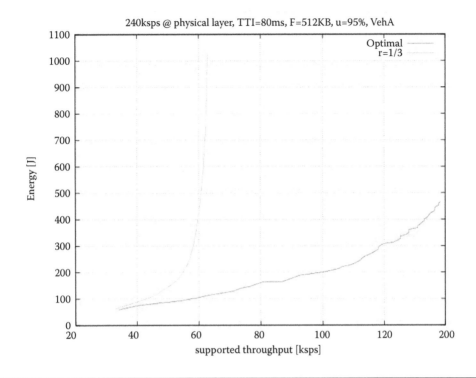

Figure 10.19 **Required transmission energy to deliver a 512 KB file versus the supported bit-rate over a bearer type of 240 ksps, whereby 95 percent of users are satisfied and follow a Vehicular A mobility model.**

rates for optimal system operation points with the use of selective combining are still in the range of 15 percent to 25 percent.

When comparing vehicular and pedestrian mobility scenarios, we conclude that less energy is required to deliver a file if the users are moving at higher speeds, that is, use the vehicular model. This can be explained as higher mobility results in higher diversity gains. When a pedestrian user is in a deep fade, it remains in this situation for longer time than a vehicular user, which moves faster.

10.5.3 Time-Constrained Download Delivery

In this section we consider the scenario where the broadcast of the file takes place over a limited amount of time. A similar setup as considered for the previous simulations is assumed. Figure 10.19 shows simulation results for time-constrained broadcast of a 512 KB file over a 240 ksps bearer, whereby the users follow the vehicular A mobility model. Specifically, the figure shows the required transmission energy to deliver a 512 KB file as a function of the media bit rate for the same channel and mobility models for a conservative setting of the Turbo code rate $r_{inner} = 0.33$ and an optimized setting. To deliver a file with a certain bit rate, for the case of higher Turbo code rate, significantly less energy is necessary. Note also that the conservative setting limits the bit rate of the file to 64 kbit/s, whereas the optimized setting can easily provide at least twice the bit rate.

10.5.4 Streaming Delivery and Mobile TV Services

For streaming delivery, similar simulations as for the file delivery case have been performed. Tradeoffs in resource allocation have been evaluated to obtain suitable system configurations. Still, the variability of the system only allows to study selected use cases and only selected but also representative performance results are reported. In the following we briefly describe the parameters applied for the following results. For the results the bearer parameters for bearer 2 in Table 10.1 was used. The applied video sequence is the sequence *party* from [28] in QCIF resolution and 12 fps. The 30 seconds sequence was looped 15 times such that basically the transmission of a 9 minutes video stream was simulated. The sequence was encoded with IDR frame distances $T_{IDR} = \{2\}$ seconds* and to achieve a target quality of average PSNR of at least 32dB. The resulting bit rate is approximately 100 kbit/s. The applied protection periods for the Raptor code were $T_{pp} = \{4,8,16\}$ seconds. The Raptor code rate was selected to optimally fill the IP bearer for a chosen Turbo code rate of $r_{inner} = \{0.24, 0.245, 0.26, 0.3, 0.5, 0.7, 0.9\}$ which results in Raptor code rates of of $r_{outer} = \{1.0, 0.99, 0.91, 0.79, 0.47, 0.33, 0.26\}$. Transmit powers of $P_{Tx} = \{2,4,8,16\}$W were applied, but in contrast to the download delivery case, it turned out that only full power of 16W provides satisfactory results. In total, each experiment was carried out for $N = 500$ users which all are assumed to move at speed 30km/h in the serving area using the a vehicular model or with speed 3km/h using the pedestrian channel model. For the video quality evaluation, a pDVD of $\bar{D}_{dec,max} = 5$ percent was considered as satisfying quality. In any case we do not use any combining technology in the physical layer.

In a first experiment, the benefits of Raptor codes to the system is investigated along with the influence of the protection period. Figure 10.20 shows the percentage of satisfied users versus the Raptor code rate for constant system resources, IDR frame distance 2 seconds, and different protection periods compared to no AL-FEC. The results are for vehicular users. Along with the different configurations for the protection periods, also the average tune-in delays are reported.

Without AL-FEC and using only PHY-FEC, the performance of the system is pretty low, only 60 percent of the users can be supported despite the application of a quite low Turbo code rate. With the use of Raptor codes, significantly more users can be supported. For a fixed protection period of, for example, four seconds, and using the right combination of Turbo coding and Raptor coding, the number of number of non-satisfied users decreases tremendously. A reasonably good operation point is when the Turbo code and the Raptor code use about the same code rate of 0.5. If the Turbo code rate is set higher then the Raptor code rate must be set lower and the performance decreases again. It is also clear from the results that with longer protection periods, more and more users can be supported. With a 16-second protection period and code rate of 0.5 for each code, almost all users observe satisfying quality. However, the introduction of the Raptor code as any application layer error recovery mechanism increases the tune-in delay, as can see from the values. This trade-off needs to be taken into account in the system design.

In a second set of experiments, also pedestrian users have been included. In addition, a second mode has been introduced, which ensures that the start of an AL-FEC source block is always aligned with an IDR frame. The results for these additional experiments are shown in Figure 10.21. It is observed that the alignment is beneficial in performance, as the size of the source block size is less variable. However, the tune-in delay reductions are not that significant, as the chosen IDR frame frequency of two seconds does not provide significant misalignment. Furthermore, it can

---

* larger values $T_{IDR} = \{4,8,16\}$ seconds have been checked, but the bit rates gains were only in the range of 5 percent, such that sacrificed tune-in delay is not justified and the 2 seconds value was used.

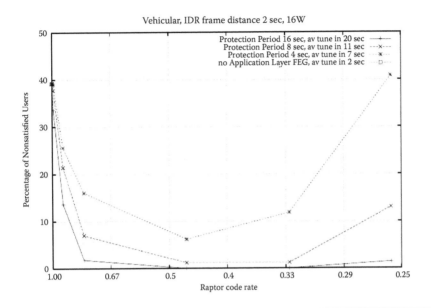

Figure 10.20 Percentage of satisfied users versus Raptor code rate for constant system resources, IDR frame distance 2 seconds, and different protection periods compared to no AL-FEC. Also reported are the average tune-in delays.

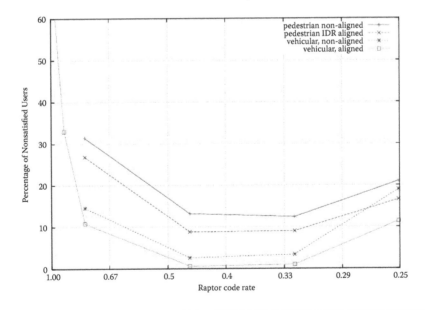

Figure 10.21 Percentage of satisfied users versus Raptor code rate for constant system resources, IDR frame distance 2 seconds, 8 seconds protection period, min-buffer time 14 seconds resulting in 11 seconds tune-in delay for different IDR frame alignment and different mobility models.

be seen that the support of pedestrian users is more difficult, as the channel variations are slower, and therefore less time diversity in the same time frame can be exploited. Still, the same beneficial tendencies of using Raptor codes for faster-moving users still applies to slower moving users.

10.6 Discussions and Optimizations

The usage of long time diversity and AL-FEC is very beneficial and basically essential, as seen from the MBMS performance results. However, the time diversity can only be fully exploited if longer protection periods are applied. If conventional sending arrangements and stringent playout strategies are applied as done for the above simulations, then the protection period also influences the channel switching times. This is of special relevance for the case of linear broadcast video delivery in mobile TV environments. Therefore, work and improvements on zapping times are necessary. Several methods have been proposed and discussed for this purpose, for example combinations of unicast and multicast delivery, provision of low-resolution fast-switching channels, or smart combinations of AL-FEC and media playout, see for example [29].

In conjunction with AL-FEC, several aspects of improving switching times and efficiency have been proposed, for example in [30]. We highlight one variant in the following. The basic idea is shown in Figure 10.22: A continuous data stream (yellow) is partitioned into source blocks of certain size such that an AL-FEC encoding strategy can be applied. The source symbols and the generated repair symbols from a single source block are distributed over multiple transmission slots as for example typical in DVB-H because of time-slicing. Two different sending arrangements are discussed: Sending arrangement 1 distributes the source symbols and the repair symbols sequentially over the transmission slots. This scheme is applied for the results in the previous section. Sending arrangement 2 distributes the source symbols and the repair symbols in such a way that each burst contains a mixture of source and repair symbols. Both arrangements have

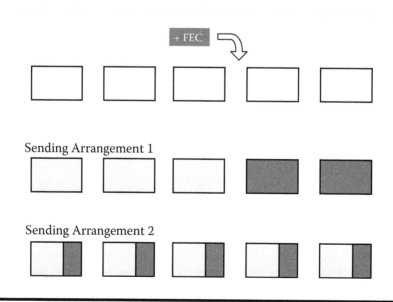

Figure 10.22 Zapping-optimized sending arrangement.

advantages and drawbacks. For sending arrangement 1, in good channel conditions one might be able to ignore bursts containing only repair symbols, thus leading to power savings, see [30]. However, sending arrangement 1 can also result in increased tune in delays. For example, if the user happens to tune in to a burst of repair symbols, if there are not enough repair symbols to decode then since the corresponding source symbols were sent in earlier bursts these repair symbols are discarded and the display of the video can only commence after reception of subsequent bursts for subsequent source blocks.

Sending arrangement 2 sends source symbols interleaved with repair symbols, such that fast tuning is supported, because immediate access to source symbols is possible. For example, as soon as a burst is received without loss containing source symbols and the source symbols correspond to a random access point to the media stream, the data can be immediately decoded and displayed. By these means, the channel switching times can be reduced. However, fast switching relies on no loss in the initial received bursts. This can cause problems as once being tuned to a service and staying with the program, the AL-FEC is quite likely required at some later point in time when there is packet loss. In a simple receiver implementation, the video decoder would then just apply rebuffering, once the AL-FEC is required. However, the video and audio decoders can easily and without perceptual degradation slow down the media playout. This concept is known as adaptive media playout (AMP), see for example, reference [31]. Therefore, it is reasonable that after switching, the media decoder slows down the playout by, for example 25 percent, such that a buffer for AL-FEC decoding can be built up for some time. With a slow down of 25 percent and for a AL-FEC delay of 10 seconds, the AL-FEC can be fully exploited within 40 seconds. If the AL-FEC needs to be used more quickly, more aggressive strategies might be used, which might lead to some small initial degradation, but losses can be compensated.

These sending strategies may also be applied for MBMS as the sending order for MBMS is not prescribed.

10.7 Conclusions

In this chapter we have introduced and investigated MBMS download and streaming delivery services in UMTS systems considering a comprehensive analysis by applying a detailed and complex channel model and simulation setup. A significant part of MBMS is AL-FEC based on Raptor codes, which have been standardized for MBMS for the broadcast delivery of multimedia content and integrated in CDPs. A thorough review of the Raptor codes and some implementation guidelines are provided. Their benefits are manifold, but the use of Raptor codes for applications in mobile broadcast environments is a perfect match, mainly due to their excellent performance, being close to ideal fountain codes, their low computational complexity, and their flexibility. Despite the detailed analysis of Raptor codes in the MBMS standardization efforts, no full system level evaluation of AL-FEC, and especially Raptor codes has been previously done from a comprehensive and realistic system-wide perspective. Therefore, we have provided an accurate and comprehensive simulation model that takes into account the effects of different layers in the protocol stack and also evaluates the services for the two most important metrics, user experience and radio resource consumption.

Of specific interest in the evaluation is the trade-off of code rates and resources being used in the physical layer, compared to the case where the resources are spent on the application layer. The results clearly indicate that a trade-off and thorough balancing of the overhead is necessary. In contrast to some beliefs and conjecture that all problems can be solved on the physical layer, our results clearly show that only a well-designed system that considers combinations of settings of the parameters at the different protocol layers can optimize system resources and user perception. In particular, it was shown that for file delivery a well-designed system should use less physical layer Turbo code protection and much more application layer Raptor code protection than considered in the MBMS standardization process. Raptor codes can spread protection over long intervals of time whereas Turbo codes only provide protection over very short intervals of time. Because channel conditions have less variance when measured over longer periods of time than shorter periods of time, the Raptor codes are more efficient at recovering losses averaged over long intervals of time than the Turbo codes are at preventing losses over short intervals of time. Thus, it turns out to be beneficial to use less Turbo code protection and accept the consequent higher RLC-PDU loss rates that can be more efficiently protected using Raptor codes. This shows that packet loss is not per se a bad thing and, counter-intuitively, high rates of packet loss can be a fundamental property of a well-designed system. The principle findings have been verified for different system parameter settings such as different power assignments, different bit rates, different mobility models, as well as advanced receiver techniques such as selective combining.

Similar results and findings have been provided for streaming delivery. However, in this case, the protection period must be lower to support the real-time delivery of the service with small channel change times. The system design here needs to consider not only the FEC on different layers, but also the video coding parameters. The trade-offs of different settings have been shown, and the reported gains when using AL-FEC make the solution very attractive despite a possible increase in channel switching times. However, with smart sending arrangements and media play-out schemes, these drawbacks can be compensated to a large extent.

Although details are bound to be different, we hypothesize that the system-level benefits of using AL-FEC (and in particular Raptor codes) and the system-wide trade-offs between AL-FEC and PHY-FEC shown for MBMS will also translate to other broadcast and multicast channels and services. As an example, the benefits of using Raptor codes for file delivery within the DVB-H IPDC standard have been demonstrated and the standardized Raptor codes have also been adopted by that standard.

Acknowledgments

The authors thank Waqar Zia from Munich University of Technology for assisting in the streaming delivery simulation. Also the support of the staff of Nomor Research in the generation of this work, specifically Eiko Seidel for providing useful, constructive, and insightful comments on the manuscript.

References

1. T. Paila, M. Luby, R. Lehtonen, V. Roca, and R. Walsh, 2004. FLUTE—File delivery over unidirectional transport, RFC 3926, Tech. Rep., IETF 2007.
2. M. Watson, Forward error correction (FEC) framework, Internet Engineering Task Force (IETF), draft-ietf-fecframe-framework-00.txt 2007.
3. M. Watson, M. Luby, and L. Vicisano, 2007. Forward error correction (FEC) building block, IETF, RFC 5052.
4. M. Luby, L. Vicisano, J. Gemmell, L. Rizzo, M. Handley, and J. Crowcroft, The Use of Forward Error Correction (FEC) in Reliable Multicast, Internet Engineering Task Force (IETF), RFC3453, Dec. 2002.

5. M. Watson, "Basic forward error correction (FEC) schemes, Internet Engineering Task Force (IETF), draft-ietf-rmt-bb-fec-basic-schemes-revised-03.txt.

6. A. Shokrollahi, M. Watson, M. Luby, and T. Stockhammer 2007. Raptor forward error correction scheme for object delivery, Internet Engineering Task Force (IETF), RFC 5053, Oct. 2007.

7. A. Shokrollahi 2006. Raptor codes, *IEEE Transactions on Information Theory*, 52, 2551–2567.

8. *Technical specification group services and system aspects; Multimedia broadcast/multicast service; Architecture and functional description* 3GPP TS 23.246 V6.9.0 2005.

9. *Technical specification group services and system aspects; Multimedia broadcast/multicast service; Protocols and codecs*. 3GPP TS 26.346 V6.1.0 2005.

10. A. Shokrollahi, S. Lassen, and M. Luby, Multi-stage code generator and decoder for communication systems, June 27, 2006, u.S. Patent No. 7,068,729.

11. M. Luby 2002. LT codes, In *Proceedings 43rd Annual IEEE Symposium on Foundations of Computer Science*.

12. ETSI TS 102 472 v1.2.1, *IP Datacast over DVB-H: Content Delivery Protocols*, Mar. 2006, technical Specification, http://www.dvb-h.org.

13. J. Byers, M. Luby, M. Mitzenmacher, and A. Rege, A digital fountain approach to reliable distribution of bulk data, in *proceedings of ACM SIGCOMM '98*, 1998.

14. M. Luby, Information additive code generator and decoder for communication systems, October 23 2001, u.S. Patent No. 6,307,487.

15. M. Luby, M. Mitzenmacher, A. Shokrollahi, and D. Spielman, Efficient erasure correcting codes, *IEEE Transactions on Information Theory*, vol. 47, pp. 569–584, 2001.

16. A. Shokrollahi, S. Lassen, and R. Karp, Systems and processes for decoding chain reaction codes through inactivation, 2005, u.S. Patent number 6,856,263.

17. A., Shokrollahi and M., Luby. Systematic encoding and decoding of chain reaction codes, U.S. Patent 6,909,383.

18. A. Shokrollahi, Raptor codes, Digital Fountain, Tech. Rep. DR2003-06-001, Jun. 2003.

19. M. Luby, T. Gasiba, T. Stockhammer, and M. Watson, Reliable multimedia download delivery in cellular broadcast networks, *IEEE Transactions on Broadcasting*, vol. 53, no. 1, pp. 235–246, Mar. 2007.

20. *Technical specification group radio access network; Introduction of the multimedia broadcast multicast service (MBMS) in the radio access network (RAN)*. 3GPP TS 25.346 V7.0.0 2006.

21. *Technical specification group radio access network; Multiplexing and channel coding (FDD)*. 3GPP TS 25.212 V7.0.0 2006.

22. *Technical specification group radio access network; Multiplexing and channel coding (TDD)*. 3GPP TS 25.222 V7.0.0 2006.

23. *Technical specification group radio access network; Typical examples of radio access bearers (RABs) and radio bearers (RBs) supported by universal terrestrial radio access (UTRA)*. 3GPP TS 25.993 V6.13.0 2006.

24. M. Luby, J. Gemmell, L. Vicisano, L. Rizzo, M. Handley, and J. Crowcroft. 2002. Asynchronuous layered coding (ALC) protocol Instantiation, RFC 3451, Tech. Rep., IETF, 2002.

25. T. Stockhammer, T. Gasiba, W. Samad, T. Schierl, H. Jenkac, T. Wiegand, and W. Xu, Nested harmonic broadcasting for scalable video over mobile datacast channels, *Wiley Journal - Wireless Communications and Mobile Computing, Special Issue on Video Communications for 4G Wireless Systems*, vol. 7, no. 2, pp. 235–256, Feb. 2007.

26. S. Wenger, T. Stockhammer, M. Hannuksela, M. Westerlund, and D. Singer. 2004. RTP payload format for H.264 video, RFC 3984, IETF.

27. *Link error prediction for E-DCH*, PSM SWG. 3GPP TSG-RAN WG1 R1-030984. Seoul, South Korea: Oct. 2003.

28. *TR26.902 Video Codec Performance*, 3GPP, June 2007. [Online]. Available: http://www.3gpp.org

29. *TD00096, Fast channel changing in RTP, Internet Streaming Media Alliance*, ISMA, June 2007. [Online]. Available: http://www.isma.tv/technology/TD00096-fast-rtp.pdf
30. D. Gomez-Barquero and A. Bria, Application Layer FEC for Improved Mobile Reception of DVB-H Streaming Services, in *Proceedings IEEE VTC Fall*, Montreal, CA, Sept. 2006.
31. M. Kalman, E. Steinbach, and B. Girod, Adaptive media playout for low delay video streaming over error-prone channels, *IEEE Trans. on Circuits and Systems for Video Technology*, June 2004.

TECHNOLOGY

Chapter 11

Time and Frequency Synchronization Schemes for OFDM-Based Mobile Broadcasting

Xianbin Wang, Yiyan Wu, and Jean-Yves Chouinard

Contents

Keywords

orthogonal frequency division multiplexing (OFDM), carrier frequency offset, timing offset, intersymbol interference, intercarrier interference, synchronization

Orthogonal frequency division multiplexing (OFDM) is the primary modulation technique for digital broadcasting, including the Digital Video Broadcasting (DVB)[1] and Digital Audio Broadcasting (DAB)[2] systems. OFDM is a type of multichannel modulation that divides a broadband wireless channel into a number of parallel subchannels, or subcarriers, so that multiple symbols are sent in parallel. Earlier overviews of the OFDM system and its applications can be found in Bingham[3] and Zou and Wu.[4]

OFDM has received considerable attention during the last two decades due to its robustness against intersymbol interference (ISI) and multipath distortion, low implementation complexity, and high spectral efficiency. With the introduction of the parallel transmission concept, the symbol duration in OFDM becomes significantly longer, compared to single-carrier transmission with the same channel condition and given data rate. Consequently, the impact of the intersymbol interference in the OFDM system is substantially reduced. This is why OFDM became the primary technology for broadcasting, where multipath distortion is very common. The type of OFDM that we will describe in this chapter for mobile broadcasting uses the discrete Fourier transform (DFT)[5] with a cyclic prefix.[6] The DFT (implemented with a fast Fourier transform [FFT]) and the cyclic prefix have made OFDM both practical and attractive to the broadcasting system designer. A similar multichannel modulation scheme, discrete multitone (DMT) modulation, has been developed for static channels such as the digital subscriber loop.[7] DMT also uses DFTs and cyclic prefixes but has the additional feature of bit loading, which is generally not used in OFDM, although related ideas can be found in Wesel.[8]

One of the principal disadvantages of OFDM is its sensitivity to synchronization errors, characterized mainly by the so-called frequency and timing offsets. Frequency offset causes a reduction of desired signal amplitude in the output decision variable and introduces intercarrier interference (ICI) due to the loss of orthogonality among subcarriers. Timing offset results in the rotation of the OFDM subcarrier signal constellation. As a result, an OFDM system cannot recover the transmitted signal without a near-perfect synchronization, especially when high-order quadrature amplitude modulation (QAM) is used. As such, OFDM-based mobile broadcast systems are very sensitive to synchronization errors. In this chapter, the impact of the synchronization errors, including carrier frequency and timing offsets, will be analyzed. Various techniques for the estimation and tracking of the frequency and timing offsets will be overviewed and discussed.

The organization of this chapter is as follows. A brief introduction to OFDM systems is first presented. Then the generation, equalization, and demodulation of the OFDM signals for mobile broadcasting are overviewed. The impact of the synchronization errors on the performance of OFDM systems is then considered. Synchronization techniques for OFDM-based broadcast systems are analyzed, with special emphasis on the DVB–Terrestrial (DVB-T) and DVB–Handheld (DVB-H) standards' pilot and frame structures.

11.1 Overview of OFDM Principles

Digital broadcasting involves the transmission of information in digital form from an information source to one or several destinations. In an ideal channel, there is no intersymbol interference (ISI) caused by multipath channel distortion, and error-free transmission can be achieved. However, this condition could not be satisfied with channel distortion as in a digital broadcasting environment due to the multipath propagation effects, including reflection and scattering. Equalization and channel control coding methods can be applied to achieve robust transmission. A time domain equalizer could be used to shorten the effective channel impulse response duration, or length, of a dispersive channel,[9] and whose coefficients are updated with an adaptive algorithm like that of Kalman filtering or the gradient algorithm, for instance. However, adaptive equalization could considerably increase the system implementation complexity, and the convergence of such an equalizer is not guaranteed.

This critical problem associated with time domain equalization convinced researchers to investigate other modulation schemes. Because the maximum delay present in the channel is fixed, a solution to overcome the spreading of the channel impulse response and ISI would consist in using several carriers in parallel instead of one, as shown in figure 11.1. The main feature of multicarrier modulation (MCM) techniques is to divide a wideband channel into a number of orthogonal narrowband subchannels. This is accomplished by modulating parallel information data at a much lower rate on a number, N, of subcarriers. Because the symbol duration for each subcarrier is multiplied by this factor N, the ratio of the maximum delay to the modulation period can be reduced significantly with a large number N of subcarriers.[4]

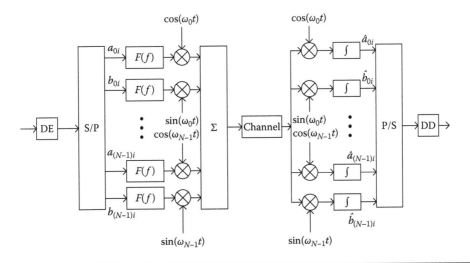

Figure 11.1   Principle of multicarrier modulation.

In early MCM technologies like conventional frequency division multiplexing, there is a guard band between adjacent subcarriers so as to be able to isolate them at the receiver using conventional bandpass filters. However, the bandwidth can be used much more efficiently in multicarrier modulation systems if the spectra of subcarriers are permitted to overlap. By using subcarriers separated by a frequency difference that is the reciprocal of the symbol duration, the orthogonality between the multiplexed tones can be realized. In this context, the multicarrier modulation is called orthogonal frequency division multiplexing (OFDM).

Orthogonal frequency division multiplexing is a form of multicarrier modulation that was first introduced more than three decades ago.[10–12] The first multichannel modulation systems appeared in the 1950s as military radio links, which were systems best characterized as frequency division multiplexed systems. The first OFDM schemes were presented by Chang[11] and Saltzberg.[12] Actual use of OFDM was limited, and the practicability of the concept was questioned. However, OFDM was made more practical through the work of Chang and Gibby,[13] Weinstein and Ebert,[5] Peled and Ruiz,[6] and Hirosaki.[14] OFDM embodies the use of parallel subchannels to transmit information over channels with impairments. There are two main features of this technique. One is that it can increase the bit rate of the channel because of its high spectral efficiency. The other is that it can mitigate intersymbol interference and impulsive noise effectively because the symbol duration in OFDM is much longer than it is in single-carrier modulation with the same data rate. This technique has been known by many names: multicarrier modulation (MCM), orthogonally multiplexed QAM, digital multitone (DMT), parallel data transmission, and so on.

11.1.1 Principles of DFT-Based OFDM System

Implementation of the OFDM system can be achieved through different approaches, including frequency division multiplexing, discrete Fourier transformation, as well as wavelet transformation. However, the majority of the OFDM systems today are based on inverse fast Fourier transform (IFFT) and FFT for modulation and demodulation, respectively. Using this method, both transmitter and receiver can be implemented using efficient FFT techniques, which reduce the number of operations from N^2 to $N\log_2 N$.

Consider the IDFT/DFT-based OFDM system in figure 11.2.[15] First, the serial binary data stream passes the data encoder (DE), which is used to map $\log_2 M$ binary data onto a two-dimensional M-ary digital modulation signal constellation. The resultant symbol (i.e., M-ary signal) stream is grouped into blocks, each block containing N symbols. Thus, an M-ary signal (data) sequence $(d_0, d_1, d_2, \ldots, d_{N-1})$ is produced, where d_k is a complex number $d_k = a_k + jb_k$. Then, the N serial data symbols are converted to parallel and an inverse Fourier transform is performed. The output of the inverse discrete Fourier transform (IDFT) is

$$S_n = \sum_{k=0}^{N-1} d_k e^{-j(2\pi nk/N)} = \sum_{k=0}^{N-1} d_k e^{-j2\pi f_k t_n}, \tag{11.1}$$

where $f_k = k/(NT)$, and $t_n = nT$, where T is an arbitrarily chosen symbol duration of the serial data sequence d_k. The real part of the vector S_n has components

$$S_n' = \sum_{k=0}^{N-1} (a_k \cos 2\pi f_k t_n + b_k \sin 2\pi f_k t_n). \tag{11.2}$$

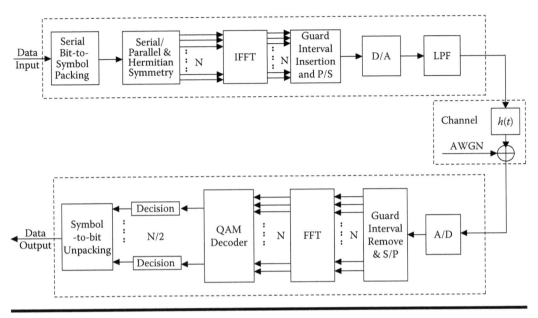

Figure 11.2   Block diagram of an IFFT/FFT-based OFDM system.

If these components are sent to a low-pass filter at fixed time intervals *T*, the desired OFDM signal is obtained as

$$S(t) = \sum_{k=0}^{N-1} (a_k \cos 2\pi f_k t + b_k \sin 2\pi f_k t). \tag{11.3}$$

If we consider an infinite transmission time, the OFDM signal becomes

$$S(t) = \sum_{j=-\infty}^{\infty} \sum_{k=0}^{N-1} (a_{kj} \cos 2\pi f_k t + b_{kj} \sin 2\pi f_k t) \Pi(t - jT), \tag{11.4}$$

where $\Pi(t) = \begin{cases} 1, & 0 \le t \le T \\ 0, & \text{elsewhere} \end{cases}$, is a unit rectangular window function.

Figure 11.3 gives an example of the construction of an OFDM signal in which the emitted symbols are from an alphabet of a quadrature phase shift keying (QPSK) constellation, that is, $\{1 + j, 1 - j, -1 + j, -1 - j\}$. Figures 11.3(a) and (b) show, respectively, the data for the real and imaginary parts of complex data $c_k = a_k + j b_k$. Figure 11.3(c) depicts their corresponding waveforms for each subchannel. From (11.3) and (11.4), the signal amplitude spectrum of each subchannel can be shown in figure 11.4: the subchannels do overlap. When the emitted symbols are independent and have equal probabilities, the corresponding power spectral density of the OFDM signal can be easily calculated. Assume that all the carriers are modulated in an independent way. The power spectral density of the transmitted signal is obtained by the sum of the power spectral density of all the subcarriers.

The Fourier transform of each carrier is the convolution of Fourier transformation of a sine wave with the Fourier transform of rectangular function $\Pi(t)$. An example is given in figure 11.5

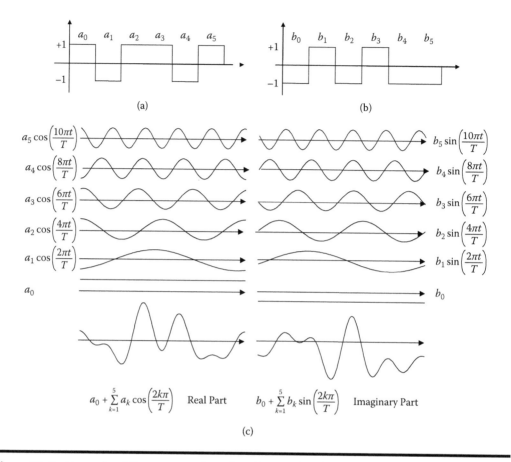

$$a_0 + \sum_{k=1}^{5} a_k \cos\left(\frac{2k\pi}{T}\right) \quad \text{Real Part} \qquad b_0 + \sum_{k=1}^{5} b_k \sin\left(\frac{2k\pi}{T}\right) \quad \text{Imaginary Part}$$

(c)

Figure 11.3   Example of the construction of an OFDM signal.

for the case of $N = 32$. It should be noted that the value of N considered in figure 11.5 is used to make the diagram clear: in practice, the value of N is considerably larger. It should also be noted that even if the secondary sidelobes have a high amplitude, their width is proportional to $1/NT$, and their relative widths therefore decrease rapidly as N increases. The spectrum of an OFDM signal then tends asymptotically toward an ideal rectangular spectrum.

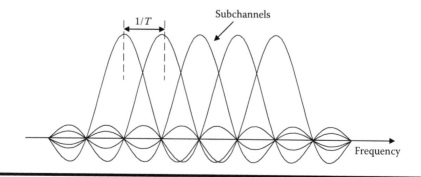

Figure 11.4   Signal amplitude spectrum of the subchannels in an OFDM system.

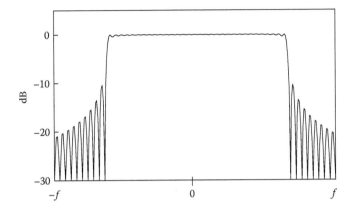

Figure 11.5   Power spectral density of an OFDM signal (number of carriers *N* = 32).

The demodulation process is based on the following orthogonal conditions:

$$\int_0^T a_k \cos(2\pi f_k t) \cos(2\pi f_{k'} t)\,dt = \begin{cases} 0 & \text{if } k' \neq k \\ a_k \dfrac{T}{2} & \text{if } k' = k, \end{cases} \tag{11.5}$$

$$\int_0^T \cos(2\pi f_k t) \sin(2\pi f_{k'} t)\,dt = 0. \tag{11.6}$$

However if a phase shift, ϕ_k, is introduced to the nonideal channel, the above equations will become

$$\int_0^T a_k \cos(2\pi f_k t + \phi_k) \cos(2\pi f_{k'} t)\,dt = \begin{cases} 0 & \text{if } k' \neq k \\ \dfrac{T}{2} a_k \cos(\phi_k) & \text{if } k' = k, \end{cases} \tag{11.7}$$

and

$$\int_0^T a_k \cos(2\pi f_k t + \phi_k) \sin(2\pi f_{k'} t)\,dt = \begin{cases} 0 & \text{if } k \neq k' \\ a_k \dfrac{T}{2} \sin(\phi_k) & \text{if } k = k'. \end{cases} \tag{11.8}$$

Obviously, the loss of orthogonality will cause intrachannel interference (ICI) between the in-phase and quadrature components of each subcarrier. However, ICI can be eliminated through channel estimation and equalization.

The implementation of the DFT-based OFDM can be efficiently realized with the fast Fourier transform (FFT). An example of an eight-point radix-2 FFT is illustrated in figure 11.6, where *W* is the *twiddle coefficient*.[16]

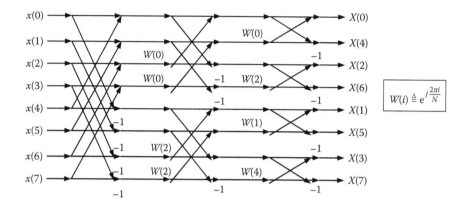

Figure 11.6   Implementation of the DFT: 8-point Radix-2, DIF FFT.[16]

11.1.2   ISI Mitigation Through Cyclic Prefix

In the presence of ISI caused by the transmission channel, the properties of orthogonality between the subcarriers are no longer maintained. In this situation, it is impossible to isolate an interval of T seconds in the received signal containing information from only one symbol. The signal is thus corrupted by intersymbol interference.

One can approach asymptotically toward a solution to the ISI problem by increasing indefinitely the number of subcarriers N. This would give rise to an increased symbol duration for a specific channel with a given data rate. However, this method is limited by technological limitations such as phase noise affecting the oscillators at the receiver and nonpractical implementation complexity. Another solution is to deliberately sacrifice some of the emitted energy by preceding each OFDM symbol with a guard interval, to eliminate the ISI problem, as explained in the following.[17] The duration of each symbol is changed to $(T - \tau_p)$ seconds with a guard interval as indicated in figure 11.7. When the guard interval is longer than the duration of the channel impulse response, or the multipath delay, then ISI can be eliminated.

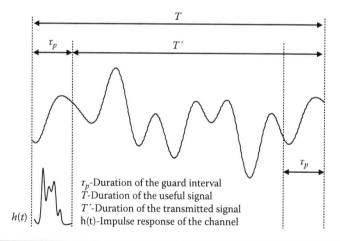

Figure 11.7   Principle of the intersymbol interference mitigation through the use of a cyclic prefix.

At the receiver, only the "useful" signal is demodulated and the guard interval is discarded. The use of a guard interval results in a loss in transmission capacity of $10\log_{10}(T'/T)$, with $T' = T - \tau_p$. This capacity reduction can in practice be kept below 1 dB and can be largely compensated by the system advantages, such as high bandwidth efficiency and ISI-free transmission.

In practice, another widely used method to combat intersymbol interference consists in adding a cyclic prefix to an OFDM symbol.[18] The reason for its popularity is that it is easy to implement in digital form. In this approach, the guard interval is a *cyclic extension* of the IFFT output sequence. If N is the original OFDM block length and the channel's impulse response $h(n)$ has length G, and assuming that the length of the added cyclic extension is also G, then the cyclically extended OFDM block has a new length of $(N + G)$. The original symbol sequence S is cyclically extended to form the new symbol sequence S^g with a cyclic prefix as follows:

$$S = \{S_0, S_1, S_2, \ldots, S_{N-1}\},$$
(11.9)

$$S^g = \{S_{N-G}, \ldots, S_{N-1}, S_0, S_1, S_2, \ldots, S_{N-G}, \ldots, S_{N-1}\}$$
(11.10)

For instance, for $N = 6$ and $G = 3$, we have a new cyclically extended symbol sequence of length $(N + G) = 9$ described by the above equation. Thus, as seen by the finite-length impulse response of the channel $h(n)$ of length G, each extended symbol sequence S^g appears as if we had repeated the original symbol sequence S periodically. The cyclically extended sequence S^g convolved with the impulse response of the channel sequence $h(n)$ appears as if it was convolved with a periodic sequence consisting of repeated Ss. Therefore, using the cyclic extension, the new OFDM symbol of length $(N + G)$ sampling periods suffers no longer from ISI. For the previous example with $N = 6$, $G = 3$, and $N + G = 9$, we can obtain a subset of six samples at the receiver as follows:

$$\underline{R} = \underline{S} \cdot \begin{bmatrix} 0 & h_0 & h_1 & h_2 & 0 & 0 \\ 0 & 0 & h_0 & h_1 & h_2 & 0 \\ 0 & 0 & 0 & h_0 & h_1 & h_2 \\ h_2 & 0 & 0 & 0 & h_0 & h_1 \\ h_1 & h_2 & 0 & 0 & 0 & h_0 \\ h_0 & h_1 & h_2 & 0 & 0 & 0 \end{bmatrix},$$
(11.11)

where $\underline{S}$ and $\underline{R}$ are transmitted and received signal vectors (both of them are length N row vectors), and $h = (h_0, h_1, h_2)$ is the impulse response of the channel. Equivalently, there exists a cyclic convolution between $\underline{S}$ and h, and the following DFT transform pair holds:

$$\underline{S} \otimes h \Leftrightarrow DFT[\underline{S}] \cdot H(k),$$
(11.12)

where $\otimes$ denotes the cyclic convolution operation, and $H(k)$ is the Fourier transform of h. With the knowledge of $H(k)$ at the kth subcarrier, we can mitigate the intrachannel interference coming from $H(k)$ inside each symbol.

11.1.3 Bit Rate of OFDM Systems

If we consider that each carrier conveys a symbol taken in a two-dimensional constellation with 2^a points and is modulated during T seconds, the bit rate can be shown to be:[17]

$$D = \frac{Na}{T} \quad \text{[bit/s]}. \tag{11.13}$$

As indicated in figure 11.4, the frequency spacing between two adjacent subcarriers is $1/T$. Three sidelobes are also counted at each side of the OFDM spectrum border when we determine the bandwidth of the signal. Therefore, the total bandwidth occupied by the N carriers is then given by[4,17]

$$W = \frac{N-1}{T} + 2\frac{3}{T} = \frac{N+5}{T}. \tag{11.14}$$

Thus, the spectral efficiency is

$$\eta = \frac{D}{W} = a\frac{N}{N+5} \quad \text{[bit/s/Hz]}. \tag{11.15}$$

Asymptotically, η tends toward a bits/s/Hz when N increases, and OFDM can be considered an optimum modulation for spectral efficiency. If we take the guard interval τ_p into consideration, W and η will become

$$W' = \frac{N-1}{T'} + \frac{6}{T'} = \frac{N+5}{T-\tau_p}, \tag{11.16}$$

$$\eta' = \frac{D}{W'} = \frac{aN(T-\tau_p)}{(N+5)T} \quad \text{[bit/s/Hz]}. \tag{11.17}$$

11.1.4 In-Band Pilots and OFDM Channel Estimation

In-band pilots, that is, subcarriers modulated with symbols known to the receiver, are normally used for channel estimation purposes in conventional OFDM systems. Channel response at pilot frequency is obtained at the receiver side by demodulating the pilot. Assume we have an OFDM symbol **x** denoted by 1 from the transmitter; the signal at the kth subcarrier after FFT is

$$Y_k = X_k H_k + W_k, \quad 0 \le k \le N-1 \tag{11.18}$$

The estimate of the channel frequency response, at pilot subcarrier p based on least square estimation, is given by

$$\hat{H}_p = \frac{Y_p}{X_p}. \tag{11.19}$$

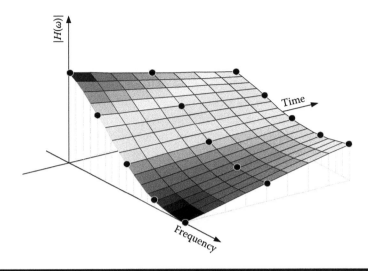

Figure 11.8   Estimation and interpolation of time-frequency-selective multipath channel using in-band pilots.

The overall frequency response of the channel for a given OFDM symbol is obtained by the interpolation of the channel responses at all pilot frequencies, as shown in figure 11.8. Due to the varying nature of the channel in both frequency and time domains, different pilot patterns can be used to improve the performance of the corresponding channel estimator. The design of pilot patterns for a given channel relies mainly on the time and frequency selectivity of the channel. Four different pilot patterns normally used in OFDM systems are illustrated in figure 11.9. The criteria for choosing a specific pilot pattern for an OFDM system rely mainly on the time and frequency selectivity of the channel model. For instance, the top-right pilot pattern shown in

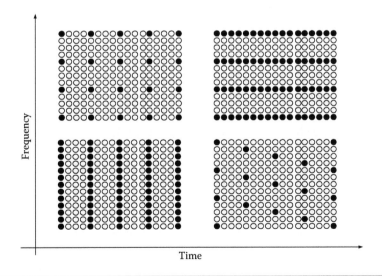

Figure 11.9   Four different pilot patterns commonly used in OFDM systems.

figure 11.9 is suitable for low to medium frequency-selective fast fading channels, whereas the pattern at the bottom left of the figure performs better in severely frequency-selective and slow fading channels. Hybrid pilot patterns can also be used in certain applications; for instance, DVB-T employs pilot arrangements as those shown in the top-left and bottom-right areas of figure 11.9. As each pilot can be regarded as one sample of the channel response in the frequency domain at a given time, the sampling theorem can then be applied for the design of the pilot patterns for OFDM systems. For an accurate channel estimation, the interval between the two adjacent pilots must be limited by the two-dimensional version of the sampling theorem.

11.1.5 Modulation Schemes for Subchannels

OFDM systems can be classified into two categories according to the modulation schemes for each subchannel: (1) coherent OFDM systems for which coherent modulation such as M-ary phase shift keying (MPSK) and M-ary quadrature amplitude modulation (MQAM) are used as subchannel modulation schemes, and (2) noncoherent OFDM systems for which noncoherent modulation such as differential phase shift keying (DPSK) is used as the subchannel modulation scheme.

Several factors influence the choice of a digital modulation scheme for an OFDM system. A desirable modulation scheme provides low-bit-error rates at low received signal-to-noise ratios (SNR), performs well in dispersive channel conditions, occupies a minimum of transmission bandwidth, and is not complex and cost-effective to implement. Existing modulation schemes do not simultaneously satisfy all of these requirements: trade-offs have to be made depending on the requirements of the particular application when selecting a digital modulation scheme. In the following, two digital modulation schemes are briefly introduced, and these schemes will be used as subcarrier modulation schemes in the OFDM systems, whose performance will be studied.

- **M-ary phase shift keying (MPSK) and M-ary differential phase shift keying (MDPSK):** In digital phase modulation, the signal waveforms are represented as[14]

$$s_m(t) = \mathrm{Re}[g(t)\exp(j2\pi f_c t + \theta_m)], \qquad \text{for } m = 1, 2, \ldots, M, \quad \text{and} \quad 0 \le t \le T, \quad (11.20)$$

  where $g(t)$ is the signal pulse shape, and T is the symbol duration. $\theta_m = 2\pi(m-1)/M$ are the M possible phases of the carrier that convey the transmitted information. Digital phase modulation is usually called phase shift keying (PSK). The mapping, or assignment of k information bits to the $M = 2^k$ possible phases, can be done in such a way that the most likely errors caused by noise will result in a single bit error. This mapping scheme is called Gray bit mapping. The signal space diagrams for $M = 2$, 4, and 8 with Gray bit mapping are illustrated in figure 11.10. A differentially encoded phase-modulated scheme is called M-ary differential phase shift keying (MDPSK).

- **M-ary quadrature amplitude modulation (MQAM):** It is to be noted here that MQAM is employed for each subchannel of the OFDM system. The signal space diagram is rectangular, as shown in figure 11.11 for 16QAM. The bandwidth efficiency of MQAM modulation is identical to that of MPSK modulation. In terms of power efficiency, MQAM is superior to MPSK because MQAM efficiently uses the signal constellation space to increase the distance between constellation points (and hence reduce the probability of error detection). MQAM is widely used in mobile wireless and fixed line communication systems.

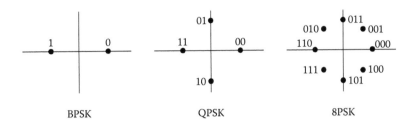

BPSK   QPSK   8PSK

Figure 11.10   Signal space diagrams for PSK-based modulation schemes.

11.1.6   OFDM Error Probability Performances

The error probability performance of OFDM systems is closely related to the Euclidean distances between the points in the signal constellation. Each carrier is modulated with signal points taken from a two-dimensional signal constellation. The constellation can be different from one carrier to another, and the Euclidean distance between the points will establish the robustness of each subcarrier. The more spread the constellation is (i.e., with the maximum Euclidean distance for a given energy and given number of points), the better the system performance is.

To improve the power efficiency of OFDM systems, square constellations are often used. An important point to determine is the relation between the distance d separating two adjacent signal points and the mean energy of the constellations, which is defined as[17]

$$\bar{E}_c^2 = \frac{1}{2} \sum_{k=0}^{2^c-1} \frac{a_k^2 + b_k^2}{2^c} = \bar{a}_k^2 = \bar{b}_k^2, \tag{11.21}$$

where c is the number of the constellation points. An error will occur if the noise component is larger than half the distance between two points in the constellation in each subchannel,[17] as shown in figure 11.12.

Let A be the emitted signal point and B be the detected signal point. n_k is the noise vector and is represented by an in-phase component n_k^c and quadrature component n_k^s. It can be shown that the noise samples n_k^c and n_k^s are always uncorrelated. As a result, the error probability for QAM symbols can be determined from the symbol error probability of the two pulse amplitude modulation (PAM) systems. Detailed analysis can be found in de Couasnon et al.[17] and Proakis.[19]

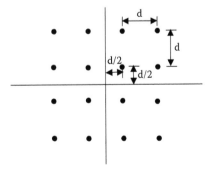

Figure 11.11   Signal space diagram for 16QAM modulation.

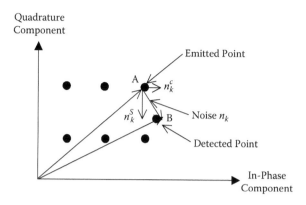

Figure 11.12   Phasor diagram of an error event with Gaussian noise.

11.2   OFDM-Based Mobile Broadcasting Standards

The digitization of traditional broadcast systems has made significant progress in recent years. Recently the broadcast industry has turned its eyes to mobile multimedia broadcasting, including mobile TV. Mobile TV still has many obstacles to overcome, but at the moment it looks very likely to be the *next killer application* in broadcast industry. There are several OFDM-based systems that can provide multimedia broadcasting services, including DVB-H,[20] MediaFLO,[21] and T-DMB.[22] So far, there have only been a few real implementations and many pilot projects. This is mainly due to the fact that there are a couple of competing technologies and the *winning technology* is yet to be determined. DVB-H technology is the leader at the moment, as it has several major industry players backing it and running pilot projects around the world. This section describes the three most promising mobile multimedia broadcasting technologies at the current time: DVB-H, MediaFLO, and T-DMB.

11.2.1   DVB-H Digital Mobile Broadcasting System

The first mobile broadcast technology discussed here is the DVB-H standard. A comprehensive overview of the DVB-H system can be found in ETSI EN 302 304[20] and Kornfeld and May.[23] DVB-H is a new standard that emphasizes mobile features for terrestrial digital video broadcasting (DVB-T). It is the latest development from the European DVB standard family, targeted for handheld devices like mobile phones and personal digital assistants (PDAs). The enhancement of DVB-T with DVB-H introduces the timing-slicing technique to save battery power, improved performance with Multi-Protocol Encapsulation–Forward Error Correction (MPE-FEC), and the hybrid networks for mobile handheld reception. As DVB-H is built upon DVB-T, an overview of DVB-T will be given first.

11.2.1.1   DVB-T System

The DVB-T system was developed by a European consortium of public and private sector organizations—the Digital Video Broadcasting Project. The DVB-T specification is part of a family of specifications also covering satellite (DVB-S) and cable (DVB-C) operations. This family allows for digital video and digital audio distribution as well as transport of forthcoming multimedia services. For terrestrial broadcasting, the system was designed to operate within the existing UHF spectrum allocated to analog PAL and SECAM television standard transmissions. Although the system was developed for 8 MHz channels, it can be scaled to different channel bandwidths, that is, 6, 7, or 8 MHz, with corresponding scaling in the data capacity. The net bit rate available in the 8 MHz channel ranges between 4.98 and 31.67 Mbits/s, depending on the choice of channel coding parameters, modulation types, and guard interval duration.

The system was essentially designed with built-in flexibility, to be able to adapt to different types of channels. It is capable of coping not only with Gaussian channels, but also with Ricean and Rayleigh channels. The system is robust to interference from delayed signals, with echoes resulting from either terrain or building reflections. The system uses OFDM modulation with a large number of carriers per channel modulated in parallel via an FFT process. It has two operational modes: a *2k mode*, which uses a 2k FFT, and an *8k mode*, which requires an 8k FFT. The system makes provisions for selection between different levels of QAM modulation and different inner code rates and also allows two-level hierarchical channel coding and modulation. Moreover, a guard interval with selectable width separates the transmitted symbols, which allows the system to support different network configurations, such as large-area single-frequency networks (SFNs) and single-transmitter operation. The 2k mode is suitable for single-transmitter operation and small-SFN networks with limited distance between transmitters, whereas the 8k mode can be used for both single-transmitter operation and small- and large-SFN networks.

The DVB-T standard was first published in 1997 and was not targeted for mobile receivers. Nevertheless, following positive results, DVB-T mobile services were launched in Singapore and Germany. Despite the success of mobile DVB-T reception, its major downfall has been the battery life. The current and projected power consumption is too high to support mobile devices that are supposed to last a long period with a single battery charge. Another issue for DVB, which is improved in DVB-H, is IP Datacasting.* IP Datacasting will facilitate the interoperability of telecommunications and broadcasting networks, a complex topic involving detailed work on the interface at different service levels. Although DVB-T is the world's most used digital terrestrial television system, the current situation is most obviously going to be changed by DVB-H or MediaFLO.

11.2.1.2 DVB-H System Overview

The objective of DVB-H is to provide an efficient way for carrying multimedia data over digital terrestrial broadcasting networks to handheld terminals. It is the latest development within the set of DVB transmission standards. The DVB-H transmission system is built from the capabilities of DVB-T, but it overcomes the two key limitations of DVB-T technology. It extends the battery life of the handheld device and improves the robustness of the mobile reception in fading environments. DVB-H uses a power-saving technique based on the time-multiplexed transmission of different broadcast services. The technique, called time slicing, results in a large battery power-saving effect. Additionally, time slicing allows soft handover if the receiver moves from one broadcast cell to another with only one receiving front end. For reliable transmission in poor signal reception conditions, an enhanced error-protection scheme on the link layer is introduced. This scheme is called MPE-FEC. MPE-FEC employs powerful channel coding on top of the channel coding included in the DVB-T specification and offers a degree of time interleaving. DVB-H also

---

* IP Datacasting: Internet Protocol Datacasting.

provides an enhanced signaling channel for improving access to the various services. DVB-H, as a transmission standard for mobile broadcasting, also specifies the physical layer as well as the elements of the lowest protocol layers.

Furthermore, the DVB-H standard features an additional transmission mode, the *4k mode*, offering additional flexibility in designing single-frequency networks (SFNs), which are still well suited for mobile reception. DVB-H allows this additional 4k mode to be used, which is created via a 4,096-point inverse discrete Fourier transform (IDFT) in the OFDM modulator. The 4k mode represents a compromise solution between the DVB-T 2k and 8k modes. It allows for a doubling of the transmitter distance in SFNs compared to the 2k mode and, when compared to the 8k mode, is less susceptible to the impairment effects caused by Doppler shifts in the case of mobile reception. The 4k mode also offers a new degree of network planning flexibility. Because DVB-T does not include this mode, it may only be used in dedicated DVB-H networks.

11.2.1.3 Time-Slicing Technique for Power Reduction

A major problem for mobile broadcasting receivers is the limited battery capacity. Therefore, being compatible with DVB-T would place a burden on the DVB-H receiver because demodulating a high-data-rate stream like the DVB-T involves significant power dissipation at the mobile receiver. The major disadvantage of DVB-T is the fact that the whole data stream has to be decoded before any one of the services (i.e., TV programs) of the multiplexed DVB-H data stream can be accessed. Time slicing is a power-saving technique that takes advantage of the fact that the service the user wants to watch or listen to is only transmitted for a fraction of the time since there are multiple services carried in a multiplexed stream. This allows for the RF front end to be turned off when the desired service signal is not being transmitted. This allows a significant amount of power to be saved, since the RF front end's amplifiers are relatively inefficient in terms of power consumption because OFDM reception requires highly linear RF amplifiers, and the higher the required linearity of the amplifier, the lower the power efficiency. With DVB-H, service multiplexing is performed in a pure time division multiplex fashion, as illustrated in figure 11.13. The data of one particular service is therefore not transmitted continuously but in compact periodical bursts with transmission interruptions in between. Multiplexing of several services leads again to a continuous, uninterrupted transmitted stream with a constant data rate.

Consequently, battery power saving is made possible through receiving the broadcast services in short time burst signals. The terminal synchronizes to the bursts for the wanted service but switches to a power-save mode during the time intervals when other services are being transmitted. The power-save time between bursts, relative to the on-time required for the reception of an individual service, is a direct measure of the power saving provided by DVB-H. Bursts entering the receiver have to be buffered and read out at the service data rate. The position of the bursts is signaled in terms of the relative time difference between two consecutive bursts of the same service. A lead time for powering up the front end of the mobile receiver, for resynchronization and channel estimation, and so on, has to be taken into account. Depending on the ratio of on-time/power-save time, the resulting power saving may be more than 90 percent.

In addition, time slicing offers another benefit for the design of mobile terminals. The long power-save periods may be used to search for other channels in neighboring radio cells offering the selected service. Smooth channel handover can be achieved at the border between two cells

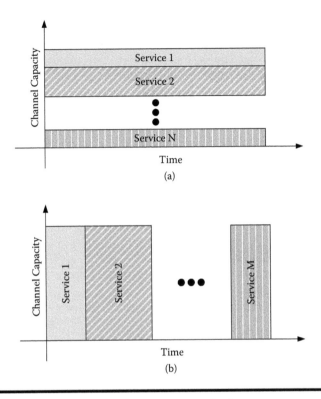

Figure 11.13 Time-slicing principle for the DVB-H standard.

providing the mobile broadcast service. Both the monitoring of the services in adjacent cells and the reception of the selected service data can be realized with the one receiving front end.

11.2.1.4 Multi-Protocol Encapsulation–Forward Error Correction (MPE-FEC)

Different from other DVB transmission systems that are based on the DVB transport stream from the MPEG-2 standard, the DVB-H system is using the Internet Protocol (IP) as the interface to higher protocol layers, which allows the DVB-H system to be combined with other IP-based networks. The introduction of MPE-FEC[23] also improves the system performance under mobile receiving conditions. DVB-H terminals are expected to be used in various situations of reception: indoor/outdoor while the user is static, a pedestrian, or mobile. This time-slicing technique imposes the implementation of a long time interleaver to mitigate the deep-fading impairments experienced in mobile reception. For this purpose, DVB-H defined an additional protection through MPE-FEC. Nevertheless, the MPEG-2 transport stream is still used by the base layer. The IP data is embedded into the transport stream by means of Multi-Protocol Encapsulation (MPE), an adaptation protocol defined in the DVB data broadcast specification. On the level of the MPE, an additional stage of forward error correction is added. MPE-FEC complements the physical layer FEC of the underlying DVB-T standard, with the purpose of reducing the signal-to-noise ratio requirements for reception by a handheld device. Intensive testing of DVB-H, which was carried out by DVB member companies, showed that the use of MPE-FEC results in a coding gain of about 7 dB over DVB-T systems.

MPE-FEC processing is located on the link layer at the level of the IP input streams before they are encapsulated by means of the MPE. MPE-FEC, MPE, and the time-slicing technique were defined jointly and directly aligned with each other. The IP input streams provided by different sources as individual elementary streams are multiplexed according to the time-slicing method.[23] The MPE-FEC error protection is first calculated separately for each individual elementary stream. Then IP packets are encapsulated and embedded into the transport stream. All relevant data processing is carried out before the transport stream interface to guarantee compatibility to a DVB-T transmission network.

MPE-FEC and time slicing are closely related techniques. Both are applied on the elementary stream level, and one time-slicing burst includes the content of exactly one MPE-FEC frame. Separating the IP data and parity check bytes of each burst makes the use of MPE-FEC decoding in the receiver optional because the application data can be recovered while ignoring the parity information when the reception condition is good.

11.2.1.5 Transmission Parameter Signaling (TPS) for DVB-H

Transmission parameter signaling (TPS) in DVB standards creates a reserved information channel that provides tuning parameters to the receiver. In DVB-T, 23 of 68 TPS bits in a frame are currently used to carry information about the transmission mode and a cell identifier. The signaling of parameters of the DVB-H elementary streams in the multiplex uses an extension of the TPS. The new elements of the TPS channel provide the information that time-sliced DVB-H elementary streams are available in the multiplex and indicate whether MPE-FEC protection is used in at least one of the elementary streams.[23] In addition, broadcasting of the cell identifier known as an optional element of DVB-T is made mandatory for DVB-H. The availability of this cell identifier simplifies the detection of neighboring network cells in which the selected same service is available.

11.2.2 MediaFLO Digital Mobile Broadcasting System

MediaFLO is a technology developed by Qualcomm based on FLO (Forward Link Only) technology. It is an orthogonal frequency division multiplexing (OFDM)–based air interface designed specifically for multicasting a significant volume of multimedia content to wireless handsets.[21] The MediaFLO system consists of two components: FLO technology and Media Distribution System (MDS). FLO technology is designed for markets where regulations permit high-power transmission from a single tower or a small number of towers. FLO can also be deployed across wide-area regions using a network of transmitters, spaced 60 km apart.

The MDS enables the efficient delivery of high-quality network-scheduled video content for viewing by masses of mobile subscribers. The MDS seamlessly handles multiple content streams from multiple sources and plays them on client software in the most popular video and audio formats. Because MDS is air interface independent, it can be deployed on any IP packet data network or current point-to-point third-generation (3G) wireless networks, and will scale easily for tomorrow's multicast networks. The MDS provides the tools to assimilate and aggregate content, bundle channels into subscription packages, and ultimately merchandise and deliver this content securely to wireless operator target subscribers. Wireless operators can also leverage additional

MDS features to entice users to engage with other media (video on demand, music on demand, ring tones, games, etc.) over their 3G networks while viewing content delivered over FLO.

11.2.3 T-DMB Digital Mobile Broadcasting System

The DMB system can operate via satellite (S-DMB) or terrestrial (T-DMB) wireless links. DMB is based on the Eureka 147 standard, also known as Digital Audio Broadcasting (DAB) standard, and shares some similarities with DVB-H.[22] It is operated in band III from 174 to 230 MHz and in L band from 1452 to 1492 MHz. It is a narrowband solution for mobile broadcasting. T-DMB services started in South Korea in December 2005, and some pilots began in Europe in 2006, for instance, in Germany, France, and the United Kingdom. DAB technology has a vast amount of users and about 800 services worldwide. Most of them are directed to mobile radio users and will not affect mobile TV users.

11.3 Frequency and Timing Synchronization for OFDM-Based Broadcast Systems

At the front end of the mobile terminal, the received broadcast signals are subject to synchronization errors due to the variation of oscillator frequency and sample clock differences. The demodulation of the received OFDM signal to baseband, possibly via an intermediate frequency, involves oscillators whose frequencies may not be perfectly aligned with the transmitter frequencies. This results in a carrier frequency offset. Also, demodulation of the OFDM signal usually introduces phase noise acting as an unwanted phase modulation of the carrier wave. Both carrier frequency offset and phase noise degrade the performance of an OFDM system. The most important effect of a frequency offset between transmitter and receiver is a loss of orthogonality between the subcarriers, resulting in intercarrier interference (ICI). The characteristics of this ICI are similar to white Gaussian noise and lead to a degradation of the SNR.[24] For both AWGN and fading channels, this degradation increases with the square of the number of subcarriers. Like frequency offsets, phase noise and sample clock offsets cause ICI, and thus a degradation of the SNR. However, for a DVB-like OFDM system, Muschallik[25] concludes that phase noise is not performance limiting in properly designed consumer receivers for OFDM.

When the baseband signal is sampled at the analog-to-digital (A/D) converter, the sample clock frequency at the receiver may not be the same as that at the transmitter side. This sample clock offset not only causes errors, but also may cause the duration of an OFDM symbol at the receiver to be different from that at the transmitter. If the OFDM symbol synchronization is derived from the sample clock, this generates variations in the symbol clock, that is, timing offsets. Because the receiver needs to determine accurately when the OFDM symbol begins for proper demodulation with the FFT, a symbol synchronization algorithm at the receiver is usually necessary. Symbol synchronization also compensates for propagation delay changes in the channel. The effects of synchronization errors have been investigated by, among others, Moose,[26] Wei and Schlegel,[27] and Garcia Armada and Calvo.[28] The degradation due to symbol timing errors is not graceful. If the length of the cyclic prefix exceeds the length of the channel impulse response, a receiver can capture an OFDM symbol anywhere in the region where the symbol appears to be cyclic, without sacrificing orthogonality. A small timing error only appears as pure phase rotations of the data symbols and may be compensated by a channel equalizer, still preserving the system's orthogonality. A large timing error resulting in capturing a symbol outside this allowable interval, on the other hand, causes ISI and ICI and leads to performance degradation. Pollet et al.[29] showed that the degradation due to a sample clock frequency offset differs from subcarrier to subcarrier—the highest subcarrier experiencing the largest SNR loss.

Summarizing, oscillator phase noise and sample clock variations generate ICI but seldom limit the system performance. Frequency offsets and timing offsets (symbol clock offsets), however, generally need to be tracked and compensated at the receiver. We now give a brief review of some recently proposed frequency and timing estimators for OFDM, and then describe one of these methods, based on the cyclic prefix, in more detail.

11.3.1 Timing Offset Estimation Techniques

Timing offset estimators have been addressed in a number of publications (see references 26, 27, and 30–37). We divide these estimators conceptually into two approaches. For the first approach, the authors of references 26 and 30–32 assume that transmitted data symbols are known at the receiver. This can in practice be accomplished by transmitting known pilot symbols according to some protocol. The unknown symbol timing and carrier frequency offset may then be estimated from the received signal. The insertion of pilot symbols usually implies a reduction of the data rate. An example of such a pilot-based algorithm is found in Warner and Leung.[30] Joint time and frequency offset estimators based on this concept are described in Classen and Meyr[31] and Schmidl and Cox,[32] and in Moose[26] the repetition of an OFDM symbol supports the estimation of a frequency offset.

The second approach, considered by the authors of references 33, 34, 36, and 37, uses statistical redundancy in the received signal. The transmitted OFDM signal is modeled as a Gaussian process. The offset values are then estimated by exploiting the intrinsic redundancy provided by the L samples constituting the cyclic prefix. The basic idea behind these methods is that the cyclic prefix of the transmitted OFDM signal yields information about where an OFDM symbol is likely to begin. Moreover, the transmitted signal's redundancy also contains useful information about the carrier frequency offset. Tourtier et al.[33] observe that the statistic

$$\xi(m) = \sum_{k=m}^{m+L-1} |r(k) - r(k+N)| \qquad (11.22)$$

contains information about the time offset. This statistic, implemented with a *sliding sum*, identifies samples of the cyclic prefix by the sum of L consecutive differences. The statistic is likely to become small when index m is close to the beginning of the OFDM symbol. Sandell et al.,[35] van de Beek et al.,[36] and later Lee and Cheon[37] use the statistic

$$\gamma(m) = \sum_{k=m}^{m+L-1} r(k) \cdot r(k+N)^*, \qquad (11.23)$$

where $r(k+N)^*$ is the complex conjugate of $r(k+N)$, to estimate the time offset. This statistic is the sum of L consecutive correlations, and its magnitude is likely to become large when the index m is close to the start of the OFDM symbol.

However, the above time synchronization techniques can only provide a coarse synchronization time at integer signal samples. Therefore, residual timing offset is unavoidable after the synchronization using either (11.22) or (11.23). A residual time offset Δn, which is normalized to the sampling interval, is considered next to evaluate its impact on the system performance of the OFDM-based broadcasting system.

To estimate the impact of the timing offset Δn, consider the following OFDM symbol given by the N-point complex modulation sequence:

$$x_n = \frac{1}{\sqrt{N}} \sum_{k=-K}^{K} X_k e^{j2\pi \frac{nk}{N}}, \qquad n = 0,1,2,\dots,N-1 \tag{11.24}$$

It consists of $2K+1$ complex sinusoids or subcarriers, which have been modulated with $2K+1$ complex data symbols X_k. The subcarriers are mutually orthogonal within the symbol interval, that is,

$$\sum_{n=0}^{N-1} x_{nk} x_{nl}^* = \frac{1}{N} |X_k|^2 \delta_{kl}, \tag{11.25}$$

where δ_{kl} is the Kronecker Delta function. After passing through a bandpass channel, the complex envelope of the received sequence can be expressed as[26]

$$r_n = \frac{1}{\sqrt{N}} \sum_{k=-K}^{K} X_k H_k \cdot e^{j2\pi \frac{(n+\Delta n)k}{N}} + w_n \quad n = 0,1,\dots,N-1; \quad N \geq 2K+1, \tag{11.26}$$

where H_k is the channel transfer function at the kth carrier frequency, Δk is the relative frequency offset (the ratio of the actual frequency offset to the subcarrier spacing), Δn is the relative timing offset (the ratio of the timing offset to the sampling interval), and w_n is the sample of a complex Gaussian random variable with zero mean and variance $\sigma_w^2 = \frac{1}{2} E[|w_n|^2]$. After the DFT demodulation, the kth element of the DFT sequence R_k is[26]

$$R_k = DFT_N\{r_n\}$$
$$= X_k H_k \cdot e^{j2\pi k \Delta n/N} + W_k \tag{11.27}$$

W_k denotes the Gaussian noise component for the kth subcarrier. The impact of the timing offset can be evaluated by the above equation. A phase shift will be introduced to all the subcarriers. With the help of the in-band pilot, the phase shift corresponding to the residual timing offset for each subcarrier can then be easily estimated and compensated.

11.3.2 Frequency Offset and Estimation Techniques

As discussed earlier, frequency offset is caused by a carrier frequency mismatch between the transmitter and receiver oscillators. Frequency offset is especially problematic in OFDM systems compared to single-carrier systems. To achieve a negligible bit error rate (BER) degradation, the tolerable frequency offset should be within the order of 1 percent of the subcarrier spacing, which is unlikely achievable in an OFDM system using low-cost commercial crystals without applying any frequency offset compensation techniques. The frequency offset is divided into an integer part and a fractional part, that is,

$$\Delta f_c T = K_{\Delta f} + \Delta k, \qquad (11.28)$$

where $K_{\Delta f}$ is an integer, and $\Delta k \in (-1/2, 1/2)$. The integer part $K_{\Delta f}$ can be found by a simple frequency domain correlation between the demodulated OFDM symbol and the in-band pilot. The fractional part is estimated by correlating the signal samples at an offset T (OFDM symbol duration) using the cyclic prefix

$$\Delta \hat{k} = \frac{1}{2\pi} \arg \left\{ \sum_{n=0}^{L-1} r(n) r^*(n+N) \right\} \qquad (11.29)$$

because the signal samples of the cyclic prefix and its counterpart at offset T are identical except for a phase factor caused by the frequency offset.

11.3.3 Joint Estimation of the Frequency and Timing Offsets

The estimation techniques discussed in the previous sections deal with the frequency and timing offsets separately. In practice, joint estimation of the two different offsets is often used to improve the estimation accuracy. Let us consider an OFDM symbol with a normalized frequency offset Δk and a timing offset Δn. After passing through a bandpass channel, the complex envelope of the received sequence for the OFDM symbol under consideration can be expressed as[26,38,39]

$$r_n = \frac{1}{\sqrt{N}} \sum_{k=-K}^{K} X_k H_k \cdot e^{j2\pi \frac{(n+\Delta n)(k+\Delta k)}{N}} + w_n \quad n = 0,1, \quad ,N-1; \quad N \geq 2K+1 \qquad (11.30)$$

where w_n is the sample of a complex Gaussian random variable with zero mean and variance $\sigma_w^2 = \frac{1}{2} E[|w_n|^2]$. After the demodulation, the kth element of the DFT output R_k is[26]

$$R_k = DFT_N \{r_n\}$$

$$= X_k H_k \left\{ \frac{\sin(\pi \Delta k)}{N \sin(\pi \Delta k/N)} \right\} \cdot e^{j\pi \Delta k(N-1)/N} \cdot e^{j2\pi k \Delta n/N} \cdot e^{j2\pi k \Delta n/N} \qquad (11.31)$$

$$+ I_k + W_k,$$

where I_k denotes intercarrier interference (ICI) caused by the frequency offset,

$$I_k = \sum_{\substack{l=-K \\ l \neq k}}^{K} X_l H_l \left\{ \frac{\sin[\pi \Delta k]}{N \sin[\pi(l-k+\Delta k)/N]} \right\} \cdot e^{j\pi(N-1)(l-k+\Delta k)/N} \cdot e^{j2\pi k \Delta n/N} \cdot e^{j2\pi \Delta k \Delta n/N}, \qquad (11.32)$$

and W_k denotes the Gaussian noise component in the frequency domain for the kth subcarrier after the demodulation. Because each intercarrier interference (ICI) sample is the summation of $(N-1) \times N$ samples, and N is usually sufficiently large, the central limit theorem can be used to approximate its statistics. Consequently, the ICI can be regarded as Gaussian distributed. The demodulation decision variable can be expressed as

$$R_k = S_k + V_k, \qquad (11.33)$$

where $V_k = W_k + I_k \cdot I_k$ is a two-dimensional Gaussian distributed variable because it is the summation of two independent Gaussian distributed random variables. Because the in-phase and quadrature components of V_k are mutually independent, the joint probability density function of the in-phase and quadrature components of the kth subcarrier can be written as[39]

$$q_{\Re[R_k],\Im[R_k]} = \frac{1}{2\pi\sigma_V^2} \exp\left[-\frac{(\alpha - \Re[S_k])^2 + (\beta - \Im[S_k])^2}{2\pi\sigma_V^2} \right], \qquad (11.34)$$

where $\Re[R_k]$, $\Im[R_k]$ and $\Re[S_k]$, $\Im[S_k]$ denote the real and imaginary parts of R_k and S_k, respectively. Now convert the above equation to a polar coordinate system to obtain the magnitude and phase information, Γ_k and Φ_k; we have

$$q_{\Gamma_k,\Phi_k} = \frac{\gamma_k}{2\pi\sigma_V^2} \exp\left[-\frac{\gamma_k^2 + \varepsilon_k - 2\gamma_k\sqrt{\varepsilon_k}\cos\varphi_k}{2\sigma_V^2} \right], \qquad (11.35)$$

where

$$\varphi_k = \tan^{-1}\left(\frac{\Im[R_k]}{\Re[R_k]} \right) - \tan^{-1}\left(\frac{\Im[S_k]}{\Re[S_k]} \right), \qquad (11.36)$$

$$\gamma_k = |R_k| = \sqrt{(\Re[R_k])^2 + (\Im[R_k])^2}, \qquad (11.37)$$

and

$$\varepsilon_k = |S_k|^2 \qquad (11.38)$$

The probability density function (*pdf*) of the phase φ_k can be obtained by integrating the above equation with respect to γ_k:[38,39]

$$q_{\Phi_k} = \frac{e^{-\lambda_k^2}}{2\pi} + \lambda_k \frac{\cos\varphi_k}{2\sqrt{\pi}} e^{-(\lambda \sin\varphi_k)^2} \{1 + \text{erf}(\lambda_k \cos\varphi_k)\}, \tag{11.39}$$

where

$$\lambda_k = \sqrt{\frac{\varepsilon_k}{2\sigma_V^2}} \tag{11.40}$$

The following *pdf* $p(\varphi_k)$ well approximates q_{Φ_k},[39] which is symmetric and bell shaped over the range of interest of λ_k. That is,

$$p(\varphi_k) = \frac{1}{\sigma_{\varphi_k}\sqrt{2\pi}} \exp\left[-\frac{\varphi_k^2}{2\sigma_{\varphi_k}^2}\right], \tag{11.41}$$

with

$$\sigma_{\varphi_k} = \sqrt{\frac{\sigma_V^2}{\varepsilon_k}} \tag{11.42}$$

The frequency and timing offsets after the acquisition are to be estimated by demodulating the synchronization preamble or the pilot tones. The observed phase of the rotated constellation in the demodulated OFDM synchronization symbol for the *k*th subcarrier with frequency and timing offset is represented as

$$\phi_k = \frac{\pi\Delta k(N-1) + 2\pi k\Delta n + 2\pi\Delta k\Delta n}{N} + \varphi_k, \tag{11.43}$$

where φ_k is the phase shift error due to ICI and Gaussian noise. As the product of the frequency and timing offsets is usually very small in practice, (11.43) can be well approximated as

$$\phi_k \approx \frac{\pi\Delta k(N-1) + 2\pi k\Delta n}{N} + \varphi_k \tag{11.44}$$

Define a linear observation model for frequency and timing offset estimation, in the form of

$$\bar{\phi} = H\bar{\theta} + \bar{\varphi}, \tag{11.45}$$

where

$$\bar{\theta} = \left[\frac{\pi \Delta k (N-1)}{N} \quad \frac{2\pi \Delta n}{N} \right]^{\mathrm{T}}$$

(11.46)

$$H = \left[\begin{matrix} 1 & 1 & 1 \\ -K & -K+1 & K \end{matrix} \right]^{\mathrm{T}},$$

(11.47)

with [ ]^{T} being the transpose of the matrix. Then, the least squares line fitting of θ is given by[40]

$$\bar{\theta} = (H^{\mathrm{T}} H)^{-1} H^{\mathrm{T}} \bar{\phi} = \left[\begin{matrix} \dfrac{\sum_{k=-K}^{K} \phi_k}{2K+1} \\ \dfrac{\sum_{k=-K}^{K} k \phi_k}{\sum_{k=-K}^{K} k^2} \end{matrix} \right]$$

(11.48)

This estimator is unbiased, that is, the mean of the estimated frequency and timing offsets is zero. Therefore, the residual frequency and timing offsets after synchronization can be regarded as Gaussian random variables with the same variances as the estimated frequency and timing offsets $\Delta \hat{k}$ and $\Delta \hat{n}$, which can be found through the above equation. Let

$$\Delta k' = \Delta k - \Delta \hat{k}$$

(11.49)

and

$$\Delta n' = \Delta n - \Delta \hat{n}$$

(11.50)

The variances of the residual frequency and timing offsets can be easily determined as[40]

$$\sigma_{\Delta k'}^2 = \frac{N^2 \sum_{k=-K}^{K} \sigma_{\varphi_k}^2}{\pi^2 (N-1)^2 (2K+1)^2}$$

(11.51)

and

$$\sigma_{\Delta n'}^2 = \frac{N^2 \sum_{k=-K}^{K} k^2 \sigma_{\varphi_k}^2}{4\pi^2 \left(\sum_{k=-K}^{K} k^2 \right)^2}$$

(11.52)

11.3.3.1 OFDM System Performance with Residual Frequency and Timing Offsets

After the estimation and compensation of the frequency and timing offsets, residual offsets still exist in both frequency and time domains due to the presence of the interference during the synchronization process. To evaluate the impact of such residual offsets on the OFDM system performance, denote the residual frequency and timing offsets as $\Delta k' = \Delta k - \Delta \hat{k}$ and $\Delta n' = \Delta n - \Delta \hat{n}$, respectively. Consider the demodulated output of the synchronized OFDM receiver,

$$R_k = X_k H_k \cdot \left\{ \frac{\sin(\pi \Delta k')}{N \sin(\pi \Delta k' / N)} \right\} \cdot e^{j\pi \Delta k'(N-1)/N} \cdot e^{j2\pi k \Delta n'/N} \cdot e^{j2\pi \Delta k' \Delta n'/N} + I_k + W_k \quad (11.53)$$

where

$$I_k = \sum_{\substack{l=-K \\ l \neq k}}^{K} (X_l H_l) \cdot \left\{ \frac{\sin(\pi \Delta k')}{N \sin(\pi (l - k + \Delta k') / N)} \right\} \cdot e^{j\pi (N-1)(l-k+\Delta k')/N} \cdot e^{j2\pi k \Delta n'/N} \cdot e^{j2\pi \Delta k' \Delta n'/N} \quad (11.54)$$

Before a decision is made, the decision variable is usually normalized according to the modulation scheme at the transmitter side as

$$R_k = X_k \cdot e^{j\pi \Delta k'(N-1)/N} \cdot e^{j2\pi k \Delta n'/N} \cdot e^{j2\pi \Delta k' \Delta n'/N} + \frac{I_k + W_k}{H_k \{\sin(\pi \Delta k') / (N \sin(\pi \Delta k' / N))\}}$$

$$= X_k + n_{syn} + I'_k + W'_k \quad (11.55)$$

where n_{syn} is defined as

$$n_{syn} = \sum_{n=1}^{\infty} \frac{[(j\pi \Delta k'(N-1) + j2\pi k \Delta n' + j2\pi \Delta k' \Delta n')/N]^n}{n!} \cdot X_k \quad (11.56)$$

$$I'_k = \sum_{\substack{l=-K \\ l \neq k}}^{K} \frac{X_l H_l}{H_k} \cdot \frac{\sin(\pi \Delta k' / N)}{\sin(\pi (l - k + \Delta k')/N)} \cdot e^{-j\pi (l-k)/N} \cdot e^{j\pi \Delta k'(N-1)/N} \cdot e^{j2\pi k \Delta n'/N} \cdot e^{j2\pi \Delta k' \Delta n'/N} \quad (11.57)$$

and

$$W'_k = \frac{W_k N \sin(\pi \Delta k' / N)}{H_k \sin(\pi \Delta k')} \quad (11.58)$$

Because $\Delta k'$ and $\Delta n'$ are usually very small, the variance of I'_k and V'_k can be approximated as ($\langle \cdot \rangle$ is the time average operator)

$$\sigma^2_{I_{k'}} = \sum_{\substack{l=-K \\ l \neq k}}^{K} \left\langle \left(\frac{X_l H_l}{H_k} \right)^2 \right\rangle \cdot \left\langle \frac{\sin^2(\pi \Delta k' / N)}{\sin^2(\pi(l-k+\Delta k')/N)} \right\rangle$$

$$\approx \sum_{\substack{l=-K \\ l \neq k}}^{K} \left\langle \left(\frac{X_l H_l}{H_k} \right)^2 \right\rangle \cdot \left\langle \frac{\pi^2 \sigma^2_{\Delta k'}}{N^2 \sin^2(\pi(l-k+\Delta k')/N)} \right\rangle \qquad (11.59)$$

$$\sigma^2_{W_{k'}} = \frac{\sigma^2_W}{|H_k|^2} \qquad (11.60)$$

$$\sigma^2_{n_{syn}} = \left| \sum_{n=1}^{\infty} \frac{[(j\pi\Delta k'(N-1) + j2\pi k\Delta n' + j2\pi\Delta k'\Delta n')/N]^n}{n!} \right|^2 E(X_k) \qquad (11.61)$$

It is obvious in (11.55) to (11.61) that the decision variable R_k can be reduced as the summation of the desired signal component, and that the variance of Gaussian noise can be determined from (11.59) to (11.61).

To study the impact of $\Delta k'$ and $\Delta n'$ on the symbol error rate (SER) performance, an OFDM system with QAM modulation scheme is considered. For rectangular signal constellations with $L = 2^{B_k}$, the QAM signal is equivalent to two pulse amplitude modulation (PAM) signals modulated on quadrature carriers, each with $\sqrt{L} = 2^{B_k/2}$ signal levels.[19] Because the signals in the in-phase and quadrature components can be perfectly separated at the demodulator, the error probability for QAM is easily determined from the probability of error for the constituent PAM signals. Specifically, the symbol error rate of the $\sqrt{L}$-ary PAM for the kth subcarrier can be estimated by[39]

$$P'_k(x) = 2\left(1 - \frac{1}{\sqrt{L}}\right) Q\left[\sqrt{\frac{3}{L-1} \frac{P_s(k)}{\left(\sigma^2_{I_k} + \sigma^2_{W'_k} + \sigma^2_{n_{syn}}\right)}}\right] \qquad (11.62)$$

where $Q(\alpha)$ is the error function

$$Q(\alpha) = \int_{\alpha}^{\infty} \frac{1}{\sqrt{2\pi}} e^{-\frac{y^2}{2}} dy \qquad (11.63)$$

and $P_s(k)$ is the signal power in the decision variable of $\sqrt{L}$-ary PAM:

$$P_s(k) = E\left\{a_k^2\right\} = \frac{1}{2}E\left\{a_k^2 + b_k^2\right\}$$

(11.64)

The symbol error rate for the kth subcarrier of the OFDM system is

$$P_k(x) = 1 - (1 - P_k')^2$$

$$\approx 4\left(1 - \frac{1}{\sqrt{L}}\right)Q\left[\sqrt{\frac{3}{L-1}\frac{P_s(k)}{\left(\sigma_{I_k'}^2 + \sigma_{W_k'}^2 + \sigma_{n_{syn}}^2\right)}}\right]$$

(11.65)

The overall SER with the impact of residual synchronization error can then be evaluated from the error probability contribution for each subcarrier, $P_e(k)$, as

$$P_e = \frac{1}{N}\sum_{k=-K}^{K} P_e(k)$$

(11.66)

The joint estimation of frequency and timing offsets as well as the impact of the residual synchronization errors has been studied in Wang et al.[39] The root mean square (RMS) values of the residual estimation errors against the received signal-to-noise ratio (SNR) are shown in figures 11.15 and 11.16, respectively. At high SNRs, it is observed that the estimation error levels off where the intercarrier interference is dominant over Gaussian noise due to the loss of orthogonality among the OFDM subcarriers. The SER of different subcarriers in the presence of the residual frequency and timing offsets is plotted in figures 11.17 and 11.18 for 256 OFDM subcarriers. The initial frequency and timing offsets from the coarse acquisition, Δn and Δk, are set to moderate values.[41] As expected, the impact of the residual offsets on the OFDM system performance is different from one subcarrier to another. The higher the subcarrier index is (with reference to the center of the channel), the poorer the performance that the subcarrier has. Note that the reference

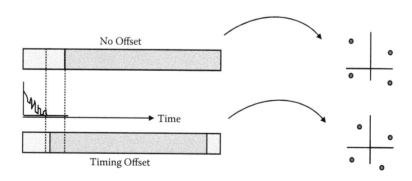

Figure 11.14 Impact of the timing offset on OFDM symbol demodulation.

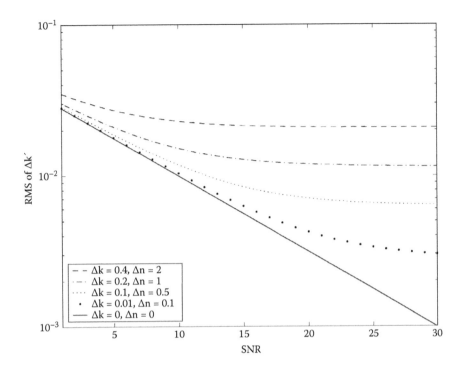

Figure 11.15 Root mean square (RMS) value $\sigma_{\Delta k'}$ of the residual frequency offset. Δk and Δn for curves 1 to 5 are 0.4, 0.2, 0.1, 0.01, and 0, and 2, 1, 0.5, 0.1, and 0, respectively. The number of the subcarriers is 256.

of the subcarrier index in figures 11.17 and 11.18 is at the center of the channel. It can be seen, from figures 11.17 and 11.18, that the SER increment among all the subcarriers increases with the SNR, that is, the sensitivity of the OFDM system to synchronization offsets increases with the SNR. A minimum of the symbol error rate can also be observed in the middle of the subcarriers. This is based on the assumption that the carrier frequency is in the middle of the OFDM signal spectrum. Variances of residual frequency and timing offsets should be determined using (11.51) and (11.52). It is also observed that the overall SER of the OFDM system is dominated by the subcarriers having larger indexes. The system performance is simulated based on the assumption of an ideal frequency domain equalizer to remove $A(k)$. An imperfect frequency domain equalizer will lead to deterioration of the system performance. Because the sensitivity of the OFDM system to synchronization offsets increases with SNR, higher-order QAM modulation constellations, which require higher SNR, need better and more accurate synchronization systems.

11.3.4 Fast Synchronization for DVB-H System

Various synchronization algorithms have been discussed in the previous sections. As an OFDM-based system, DVB-H can achieve synchronization using these techniques. However, an extra synchronization requirement is needed for the optimal performance of burst transmission of DVB-H. In this section, a fast synchronization technique for DVB-H system is discussed.[42]

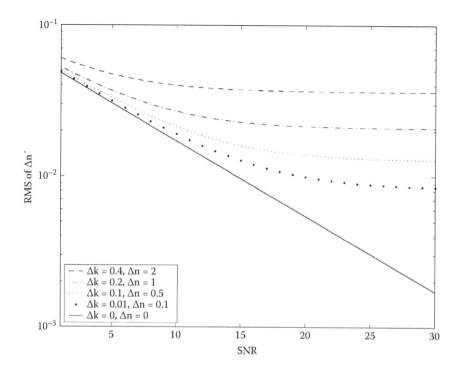

Figure 11.16 Root mean square (RMS) value $\sigma_{\Delta n'}$ of the residual frequency offset. Δk and Δn for curves 1 to 5 are 0.4, 0.2, 0.1, 0.01, and 0, and 2, 1, 0.5, 0.1, and 0, respectively. The number of the subcarriers is 256.

One major problem for the design of DVB-H terminals is the limited battery life. Being compatible with DVB-T would place a burden on the DVB-H terminal because demodulating and decoding a broadband, high-data-rate stream like the DVB-T stream involves constant power dissipation for the tuner and demodulator. A considerable drawback for battery-operated DVB-T terminals is that the whole data stream has to be decoded before any one of the multiplexed data streams can be accessed. In DVB-H, service multiplexing is performed in a pure time division multiplex fashion. The data for a particular service is therefore not transmitted continuously, but instead in short periodic bursts with interruptions in between. This bursty signal can be received *time selectively*: the terminal synchronizes to the bursts of the wanted service but switches to a power-save mode during the intermediate time when other services are transmitted. This technique is called the *time slicing* technique, described in section 11.2.1.3. The power-save time between bursts, relative to the on-time required for the reception of an individual broadcast service, is a direct measure of the power saving provided by the DVB-H receiver. Time slicing offers another benefit for the receiving terminal architecture. The rather long power-save periods may be used to search for channels in neighboring radio cells offering the selected service. This way, a channel handover can be performed at the border between two cells, which remains imperceptible to the user. Both the monitoring of the services in adjacent cells and the reception of the selected service data can then be realized with the same front end.[43]

When the handheld terminal requires a constant lower bit rate, the duration of each burst will be very short. Therefore, the synchronization times of the DVB-H receiver must be rigorously

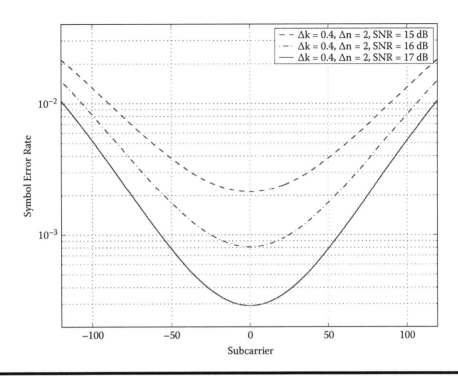

Figure 11.17 **Impact of the residual frequency and timing offsets on the SER performance of 16QAM-OFDM at various SNRs. Δk and Δn are 0.4 and 2, respectively. The number of the subcarriers is 256.**

minimized to fully exploit the benefits of time slicing. For the conventional DVB-T receiver, synchronization is usually achieved in two steps: the coarse time acquisition based on cyclic prefix and the removal of the residual timing offset by comparing the demodulated and transmitted pilots. However, identification of pilots is usually based on transmission parameter signaling (TPS) pilots for the DVB system, which introduces a long delay of up to 68 OFDM symbols. In Schwoerer and Vesma,[43,44] two fast synchronization schemes were proposed for DVB-H based on the scattered pilots in the frequency domain. However, the power-based scattered pilot synchronization may not work effectively in frequency-selective channels. For the correlation-based scattered pilot approach, four OFDM symbols are needed before the synchronization can be achieved.

In Wang et al.,[42] a fast time synchronization technique for DVB-H using only one OFDM symbol is proposed. This new technique is based on the time domain correlation between the received signal and four local reference symbols generated from in-band pilots, including the scattered pilots and continual pilots, as shown in figure 11.19. To reduce the computational complexity, a *coarse time window* can be derived from the cyclic prefix. The proposed time synchronization method can be achieved in two steps: a coarse OFDM symbol synchronization using the cyclic prefix, and a pilot symbol synchronization based on a time sequence generated from scattered pilots. Consider a DVB-H symbol given by the N-point complex modulation sequence:

$$s(n) = \frac{1}{\sqrt{N}} \sum_{k=0}^{N-1} X_k e^{j\frac{2\pi nk}{N}}, \qquad n = 0, 1, 2, \ldots, N-1 \qquad (11.67)$$

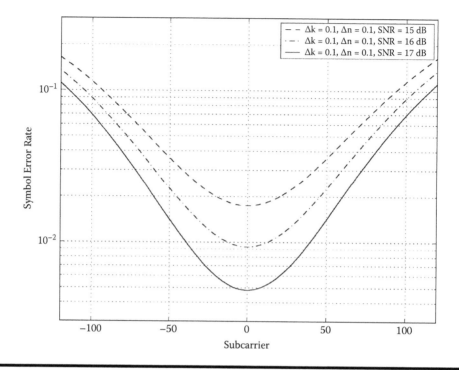

Figure 11.18 **Impact of the residual frequency and timing offsets on the SER performance of 64QAM-OFDM at various SNRs.** Δk **and** Δn **are both 0.1. The number of the subcarriers is 256.**

The OFDM symbol consists of N complex sinusoids or subcarriers modulated with the complex data X_k, which can be divided into two different sets: the data symbol to be transmitted and the pilot symbols for channel synchronization and estimation. A cyclic prefix is inserted to protect the OFDM signal from intersymbol interference. The cyclic nature of the OFDM signal provides a straightforward way to achieve the coarse time synchronization using

$$C = \sum_{m=0}^{L-1} r(m)r^*(m+N) \tag{11.68}$$

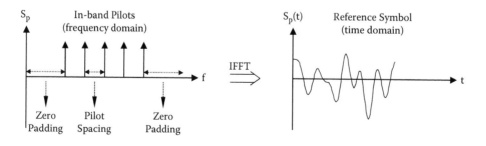

Figure 11.19 **Construction of the reference symbol in time domain.**

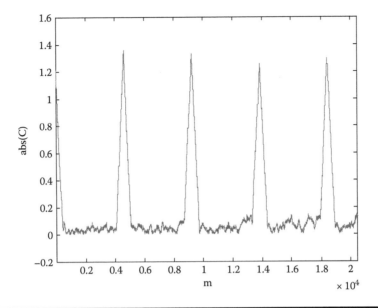

Figure 11.20   Absolute value of the correlation function using the cyclic prefix.[42]

The correlation function in (11.68) has a triangular shape but can be corrupted by other interferences, as illustrated in figure 11.20.

Once the coarse timing is achieved, the position of embedded pilots has to be identified so that the channel can be estimated and the residual time offset can be derived more accurately using the in-band pilots. Here, the timing offset is obtained from the time domain correlation of the received signal and four local references generated from the scattered pilots. Denote the subcarrier sets for these two symbol sets as data carrier $\mathcal{D}$ and pilots $\mathcal{P}$, respectively. Equation (11.67) may be reorganized as[42]

$$s(n) = d(n) + p_i(n), \tag{11.69}$$

where the data symbols are

$$d(n) = \frac{1}{\sqrt{N}} \sum_{k \in D} X_k e^{j\frac{2\pi nk}{N}} \tag{11.70}$$

and the pilots are

$$p_i(n) = \frac{1}{\sqrt{N}} \sum_{k \in P_i} X_k e^{j\frac{2\pi nk}{N}}, \tag{11.71}$$

where i is the index of the pilot p and ranges from 0 to 3 due to the shifting of the scattered pilots. As a result, p will repeat itself every four OFDM symbols. After passing through a multipath channel characterized by its complex impulse response $h(n)$, the received signal $r(n)$ can be written as

$$r(n) = d(n) \otimes h(n) + p_i(n) \otimes h(n) + w(n) \qquad (11.72)$$

Identification of the scattered pilots is based on the correlation of the received signal and the four local pilot sequences p. When the correct local reference is selected, a correlation peak will be observed. Under this condition, the correlation function between the received signal $r(n)$ and the time domain pilot sequence $p_i(n)$ is given by[42]

$$R_{rp} = R_{pp} \otimes h(n) + R_{dp} \otimes h(n) + w(n) \otimes h(n), \qquad (11.73)$$

where R_{xy} denotes the correlation between signals x and y. Using the central limit theorem (assuming a sufficiently large value of N), R_{dp} can be approximated as a Gaussian-distributed variable with mean zero and a variance of $\sigma_d^2 \sigma_p^2 / N$. Similarly, the autocorrelation function of $p(n)$ can be formulated as

$$R_{pp}(m) = \frac{1}{N} \sum_{n=0}^{N-1} p_i(n) p_i^*(n-m),$$

$$R_{pp}(m) = \begin{cases} \dfrac{1}{N} \displaystyle\sum_{n=0}^{N-1} p_i(n) p_i^*(n) \approx \sigma_p^2, & m = 0 \\ w(m), & m \neq 0 \end{cases} \qquad (11.74)$$

where $w(m)$ is a Gaussian noise with a variance of σ_p^4 / N. Note that the convolution of a Gaussian noise with $h(n)$ is still Gaussian distributed. For convenience of the analysis, we assume that R'_{rp} has the same duration as $h(n)$:

$$R'_{rp}(n) \approx w_1(n) + h(n)\sigma_p^2, \qquad (11.75)$$

where $w_1(n)$ is the combined interference from the second term in (11.73) and $w(m)$ in (11.74). An example of the correlation function of the scattered pilot sequence is plotted in figure 11.21.

The scattered pilot position is therefore identified with (11.74) when it achieves its maximum with one of the four local pilot sequences. The channel estimation normalized to the main path can therefore be obtained through the cross-correlation between the received signal and the time domain pilot sequence as

$$\hat{h}(n) = \frac{R'_{rp}(n)}{R'_{rp,\max}} = \frac{R'_{rp}(n)}{h_{\max}\sigma_p^2} + w_2(n), \qquad (11.76)$$

where the subscript *max* denotes the index of the main path, and $w_2(n)$ is the interference term from $w_1(n)$. One example for the proposed fast synchronization technique can be found in figures 11.22 and 11.23. Once the scattered pilots are identified, the residual timing offset can be easily determined by the phase shift of the FFT output for the scattered pilots.

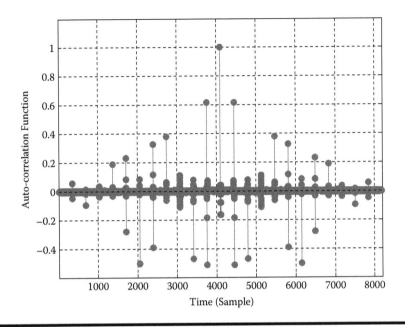

Figure 11.21   Autocorrelation function of the scattered pilot sequence.[42]

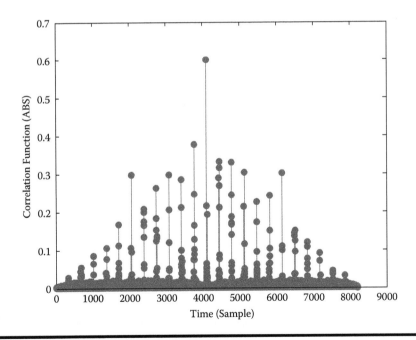

Figure 11.22   Cross-correlation function between the DVB-H signal and the scattered pilot sequence.

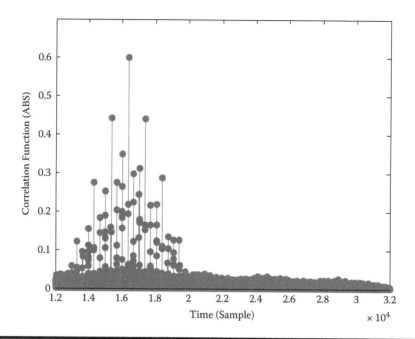

Figure 11.23   Cross-correlation function between the received signal (four OFDM symbols) and the scattered pilot sequence.

11.4   Summary

Orthogonal frequency division multiplexing (OFDM) is the primary modulation technique for digital mobile broadcast systems. The OFDM-based mobile broadcast standard is overviewed in this chapter. The principles of OFDM are presented, including the intersymbol interference mitigation through cyclic prefix and various channel estimation techniques. One of the principal disadvantages of OFDM systems is their inherent sensitivity to synchronization errors, caused mainly by the so-called frequency and timing offsets. Frequency offset causes a reduction of desired signal amplitude in the output decision variable and introduces intercarrier interference (ICI) due to the loss of orthogonality among subcarriers. Timing offset results in the rotation of the OFDM subcarrier constellation. As a result, an OFDM system cannot recover the transmitted signal without a near-perfect synchronization. In this chapter, the impacts of the synchronization errors, including carrier frequency and timing offsets, were analyzed. Various techniques for the estimation and tracking of the frequency and timing offsets were overviewed and discussed. The effects of the synchronization errors on the system performance of OFDM are also considered. Synchronization techniques for OFDM-based DVB-T and DVB-H broadcast systems are analyzed, with special consideration given to the DVB pilot and frame structures.

References

1. European Telecommunications Standards Institute. 1997. *Radio broadcasting systems; Digital audio broadcasting (DAB) to mobile, portable and fixed receivers*. ETS 300 401, 2nd ed. Valbonne, France.
2. European Telecommunications Standards Institute. 1997. *Digital video broadcasting (DVB); Framing structure, channel coding and modulation for digital terrestrial television*. ETS EN 300 744, v.1.1.2.

3. J. A. C. Bingham. 1990. Multicarrier modulation for data transmission: An idea whose time has come. *IEEE Communications Magazine* 28:5–14.

4. W. Y. Zou and Y. Wu. 1995. COFDM: An overview. *IEEE Transactions on Broadcasting* 41:1–8.

5. S. B. Weinstein and P. M. Ebert. 1971. Data transmission by frequency-division multiplexing using the discrete Fourier transform. *IEEE Transactions on Communications* 19:628–34.

6. A. Peled and A. Ruiz. 1980. Frequency domain data transmission using reduced computational complexity algorithms. In *Proceedings of the IEEE International Conference on Acoustics, Speech, and Signal Processing (ICASSP'80)*, 964–67.

7. ANSI. 1995. *Network and customer installation interfaces—Asymmetric Digital Subscriber Line (ADSL) metallic interface.* ANSI standard T1.413.

8. R. Wesel. 1995. Fundamentals of coding for broadcast OFDM. In *Proceedings of the 29th Asilomar Conference on Signals, Systems & Computers*, ACM, Pacific Grove, CA, 2–6.

9. B. R. Saltzberg. 1998. Comparison of single-carrier and multitone digital modulation for ADSL applications. *IEEE Communications Magazine* 36:114 –21.

10. H. F. Harmuth. 1960. On the transmission of information by orthogonal time functions. *AIEE Transactions* I:248–55.

11. R. W. Chang. 1966. Synthesis of band-limited orthogonal signals for multichannel data transmission. *Bell System Technical Journal* 45:1775–96.

12. B. R. Saltzberg. 1967. Performance of an efficient parallel data transmission system. *IEEE Transactions on Communications Technology* COM-15:805–11.

13. R. W. Chang and R. A. Gibby. 1968. Theoretical study of performance of an orthogonal multiplexing data transmission scheme. *IEEE Transactions on Communications* 16:529–40.

14. B. Hirosaki. 1981. An orthogonally multiplexed QAM system using the discrete Fourier transform. *IEEE Transactions on Communications* 29:982–89.

15. Y. Wu and W. Y. Zou. 1995. Orthogonal frequency division multiplexing: A multi-carrier modulation scheme. *IEEE Transactions on Consumer Electronics* 41:392–99.

16. A. V. Oppenheim and R. W. Schafer. 1989. *Discrete signal processing.* Englewood Cliffs, NJ: Prentice Hall.

17. T. de Couasnon, R. Monnier, and J. B. Rault. 1994. OFDM for digital TV broadcasting. *Signal Processing* 39:1–32.

18. A. Ruiz, J. M. Cioffi, and S. Kasturia. 1992. Discrete multiple tone modulation with coset coding for the spectrally shaped channel. *IEEE Transactions on Communication* 40:1012–29.

19. J. G. Proakis. 1995. *Digital communications.* 3rd ed. New York: McGraw-Hill.

20. *Digital video broadcasting (DVB): Transmission system for handheld terminals (DVB-H).* ETSI EN 302 304, v.1.1.1.

21. M. R. Chari, F. Ling, A. Mantravadi, R. Krishnamoorthi, R. Vijayan, G. K. Walker, and R. Chandhok. 2007. FLO physical layer: An overview. *IEEE Transactions on Broadcasting* 53:145–60.

22. S. Cho, G. Lee, B. Bae, K. Yang, C.-H. Ahn, S.-I. Lee, and C. Ahn. 2007. System and services of terrestrial digital multimedia broadcasting (T-DMB). *IEEE Transactions on Broadcasting* 53:171–78.

23. M. Kornfeld and G. May. 2007. DVB-H and IP datacast-broadcast to handheld devices. *IEEE Transactions on Broadcasting* 53:161–70.

24. T. Pollet, M. van Bladel, and M. Moeneclaey. 1995. BER sensitivity of OFDM systems to carrier frequency offset and Wiener phase noise. *IEEE Transactions on Communications* 43:19193.

25. C. Muschallik. 1995. Influence of RF oscillators on an OFDM signal. *IEEE Transactions on Consumer Electronics* 41:592–603.

26. P. H. Moose. 1994. A technique for orthogonal frequency division multiplexing frequency offset correction. *IEEE Transactions on Communications* 42:2908–14.

27. L. Wei and C. Schlegel. 1995. Synchronization requirements for multiuser OFDM on satellite mobile and two-path Rayleigh fading channels. *IEEE Transactions on Communications* 43: 887–95.

28. A. Garcia Armada and M. Calvo. 1998. Phase noise and sub-carrier spacing effects on the performance of an OFDM communication system. *IEEE Communications Letters* 2:11–13.

29. T. Pollet, P. Spruyt, and M. Moeneclaey. 1994. The BER performance of OFDM systems using non-synchronized sampling. In *Proceedings of the IEEE GLOBECOM'94*, San Francisco, 253–57.

30. W. D. Warner and C. Leung. 1993. OFDM/FM frame synchronization for mobile radio data communication. *IEEE Transactions on Vehicular Technology* 42:302–13.

31. F. Classen and H. Meyr. 1994. Frequency synchronization algorithms for OFDM systems suitable for communication over frequency-selective fading channels. In *Proceedings of the IEEE Vehicular Technology Conference (VTC'94)*, Stockholm, 1655–59.

32. T. M. Schmidl and C. Cox. 1997. Robust frequency and timing synchronization for OFDM. *IEEE Transactions on Communications* 45:1613–21.

33. P. J. Tourtier, R. Monnier, and P. Lopez. 1993. Multicarrier modem for digital HDTV terrestrial broadcasting. *Signal Processing: Image Communication* 5:379–403.

34. F. Daffara and O. Adami. 1995. A new frequency detector for orthogonal multicarrier transmission techniques. In *Proceedings of the Vehicular Technology Conference (VTC'95)*, Chicago, 804–9.

35. M. Sandell, J. J. van de Beek, and P. O. Börjesson. 1995. Timing and frequency synchronization in OFDM systems using the cyclic prefix. In *Proceedings of the IEEE International Symposium on Synchronization*, 16–19.

36. J. J. van de Beek, M. Sandell, and P. O. Börjesson. 1997. ML estimation of time and frequency offsets in OFDM systems. *IEEE Transactions on Signal Processing* 45:1800–5.

37. D. Lee and K. Cheon. 1997. A new symbol timing recovery algorithm for OFDM systems. *IEEE Transactions on Consumer Electronics* 43:767–75.

38. K. W. Kang, J. Ann, and H. S. Lee. 1994. Decision-directed maximum-likelihood estimation of OFDM frame synchronization offset. *Electronics Letters* 30:2153–54.

39. X. Wang, T. T. Tjhung, Y. Wu, and B. Caron. 2003. SER performance evaluation and optimization of OFDM system with residual frequency and timing offsets from imperfect synchronization. *IEEE Transactions on Broadcasting* 49:170–77.

40. J. L. Melsa and D. L. Cohn. 1978. *Decision and estimation theory*. New York: McGraw-Hill.

41. H. Minn, M. Zeng, and V. K. Bhargava. 2000. On timing offset estimation for OFDM systems. *IEEE Communications Letters* 4:242–44.

42. X. Wang, Y. Wu, and J.-Y. Chouinard. 2006. A fast synchronization technique for DVB-H System using in-band pilots and cyclic prefix. In *Proceedings of the IEEE ICCE*, 407–8.

43. L. Schwoerer. 2004. Fast pilot synchronization schemes for DVB-H. In *Proceedings of the 4th International Multi-Conference Wireless and Optical Communications*, 420–24.

44. L. Schwoerer and J. Vesma. 2003. Fast scattered pilot synchronization for DVB-T and DVB-H. Paper presented at Proceedings of the 8th International OFDM Workshop, Germany.

Chapter 12

Antenna Diversity Schemes Suitable for Orthogonal Frequency Division Multiplexing (OFDM)- Based Standards

D. A. Zarbouti, D. A. Kateros, D. I. Kaklamani, and G. N. Prezerakos

Contents

Keywords

MIMO, OFDM, space-time codes, space-frequency codes, space-time-frequency codes, flat-fading, selective-fading

12.1 Introduction

Orthogonal frequency division multiplexing (OFDM) has drawn much attention as a promising modulation scheme for the broadband wireless communications, and it is a strong candidate for the next-generation communication system. The application of the OFDM technique in a wideband and, as a result, frequency-selective wireless channel leads to the division of the initial signal bandwidth into several narrowbands and, as a result, flat fading subchannels. Moreover, the basic principle of OFDM is to split a high-rate data stream into a number of lower-rate substreams that are simultaneously transmitted over the OFDM carriers. This way the symbol duration of the substreams increases and the delay spread caused by multipath has a minor effect on the received signal. Finally, the guard interval that is introduced for every OFDM symbol decreases further the intersymbol interference while eliminating the intercarrier interference. A more thorough analysis concerning the OFDM basic principles can be found in Prasad.[1]

The rest of the chapter is dedicated to the presentation of several antenna transmit diversity techniques, which appear in the references. Special interest is given to schemes that concern selective fading channels because they dominate the broadband wireless technology.

12.2 System Model

12.2.1 MIMO-OFDM

The transceiver of a MIMO-OFDM system can be viewed in figure 12.1. The transmitter is equipped with M_t antennas while the receiver is equipped with M_r antennas. Each transmitter chain involves an OFDM modulator that is fed by the output of the appropriate encoder used. The serial data stream leads N_s data symbols into the encoder, which transforms them into a

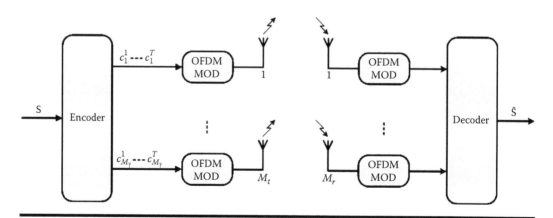

Figure 12.1 Digital entertainment delivery in a wireless home.

$N \cdot T \times M_t$ code matrix. In the last equation N are the available carriers provided by the IDFT module of the transmitter, and T is the duration of each time slot. Let us consider $\mathbf{S} = [s_1, s_2, ..., s_{N_s}]$, the modulated data symbols that enter the encoder, as it is depicted in figure 12.1; the produced codeword is given by

$$\mathbf{C} = \begin{pmatrix} \mathbf{C_1} \\ \vdots \\ \mathbf{C_T} \end{pmatrix}, \tag{12.1}$$

where C_i, $i = 1, ..., T$, is a $N \times M_t$ matrix that corresponds to the transmission codeword at each slot and given by:

$$\mathbf{C_i} = \begin{pmatrix} c_{11} & \cdots & c_{1M_t} \\ \vdots & \ddots & \vdots \\ c_{N1} & \cdots & c_{NM_t} \end{pmatrix}. \tag{12.2}$$

The coded symbol $c_{j,k}$ of the above matrix is transmitted by the kth antenna with the jth carrier during the ith time slot.

In the following paragraphs we will present three basic performance parameters of a system that are used for comparing encoding schemes.

12.2.1.1 Code Rate

A simple parameter of comparison for different MIMO-OFDM encoding techniques is the code rate that they achieve. Generally, the code rate of an encoding scheme is a metric that states the amount of useful information that is transmitted during each transmission burst. In the general case of the MIMO-OFDM system described above, the N_s useful data symbols are sent through $N \cdot T$ channels (transmission units). As a result the code rate in this case is

$$R = \frac{N_S}{N \cdot T}. \tag{12.3}$$

12.2.1.2 Diversity Gain

Another crucial parameter in the performance investigation of a space diversity encoding technique is the achievable diversity gain.[6,7] To avoid the reception of poor-quality signals in the receiver, wireless communication systems exploit the spatial, spectral, or temporal resources to send replicas of the transmitted signal. In this way the receiver can improve its performance, because the probability of simultaneous fading is extremely low. In case of flat fading MIMO channels, the maximum diversity gain that can be reached is $M_t \cdot M_r$, while in cases of selective fading, the diversity gain can be increased to $M_t \cdot M_r \cdot L$, where L are the channel paths of independent fading. We must note at this point that in cases where the antennas elements are too dense or the wireless channel is degenerated, that is, because of a keyhole phenomenon, the achievable diversity gain is obviously lower than the aforementioned upper bounds. A study on the achievable diversity gain of a MIMO channel can be found in Paulraj et al.[6] and mainly in Foschini and Gans.[8]

12.2.1.3 Decoding Complexity

Finally, an encoding scheme is investigated under the decoding complexity that it demands. Obviously, special interests exhibit the schemes that involve low-complexity receivers with fast decoding methods like maximum likelihood (ML) and maximum ratio combining (MRC). All the antenna diversity schemes that will be presented in the following will be commented toward the aforementioned parameters.

12.3 Antenna and Temporal Diversity Techniques for OFDM

In this section we will present encoding techniques that offer spatial and temporal diversity gain in flat and selective fading MIMO channels. In the bibliography, two kinds of codes are investigated, the trellis and block encoding space-time (S-T) schemes. However, because of the high decoding complexity induced by trellis codes, we conduct a more extensive analysis on block codes. Specifically, we mainly focus on orthogonal codes that offer full transmission rate (one), as they have attracted the attention of the research community to a larger extent.

12.3.1 Trellis Codes for OFDM

Space-time trellis codes (STTCs)[19] are a type of space-time code used in multiple-antenna wireless communications. This scheme involves the transmission of multiple, redundant copies of a trellis code distributed over time and space (a number of antennas). These multiple copies of the data are used by the receiver to enhance its capability to reconstruct the actual transmitted data. The number of transmit antennas must necessarily be greater than 1, but only a single receive antenna is required. However, multiple receive antennas are often used, when it is feasible, because this improves the performance of the system. In MIMO-OFDM systems STTCs are mainly considered in an OFDM framework where the incoming information symbols are trellis coded across both the OFDM subchannels and transmit antennas, to obtain the additional multipath diversity. In the following section we present examples of STTCs for MIMO-OFDM systems found in the references.

Tarokh et al.[19] present STTCs for 4-PSK and 8-PSK constellations, specifically, 4-PSK codes for transmission of 2 b/s/Hz using two transmit antennas and 8-PSK codes for transmission of 3 b/s/Hz using two transmit antennas. Assuming one receive antenna, these codes provide a diversity advantage of 2. In figure 12.2 we include the 4-PSK codes assuming 8 and 16 states.

In Lu and Wang[20] the authors make use of the STC encoder depicted in figure 12.3. The encoder makes use of memory registers and the application of complex channel frequency responses for each subcarrier. The response $\mathbf{H}_i$ for subcarrier i is obtained assuming a tapped-delay line model. The encoder produces two copies of the input signal, which are fed to two 4-PSK mappers.

Lastly, Blum et al.[21] present codes that are shown to outperform the 2-antenna, 16-state code given in Tarokh et al.[19] and shown in figure 12.2 and the codes given in Blum et al.[21] and Yan and Blum.[22] The suggested codes are 4-antenna and 16- and 256-state, designed using an ad hoc approach. The generator matrices for the aforementioned codes are shown figure 12.4.

12.3.1.1 Alamouti Technique

The breakthrough in recent space-time coding was the Alamouti OSTBC design.[1] The Alamouti design aimed to flatten fading channels and is a simple transmit diversity technique that considers two transmit antennas. The Alamouti technique achieves full-diversity order without the need of channel state information at the transmitter. The scheme can be applied in

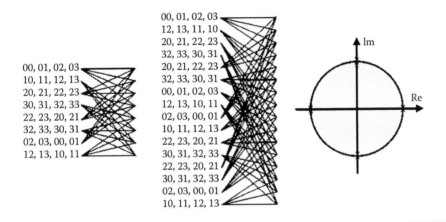

Figure 12.2   4-PSK, 8- and 16-state two space-time codes, 2 b/s/Hz.

cases with a single receiver or with multiple receivers. In figure 12.5 the Alamouti transmitter is depicted in the case of one receive antenna.

The concept of Alamouti is rather simple: two coded symbols, c_0 and c_1, are launched and transmitted simultaneously from the two transmit antennas during the first time slot, while the encoded symbols $-c_1^*$ and c_0^* are transmitted during the second time slot. In table 12.1 the encoding scheme across time is presented. In order for a space-time encoding scheme to be shortly described, we use the code matrix. The code matrix of a general single-carrier orthogonal code is a $T \times M_t$ matrix, with T the number of the time slots and M_t the number of transmit antennas. In case that $T = M_t$, the produced schemes are of rate 1, while in cases that $T > M_t$, the schemes

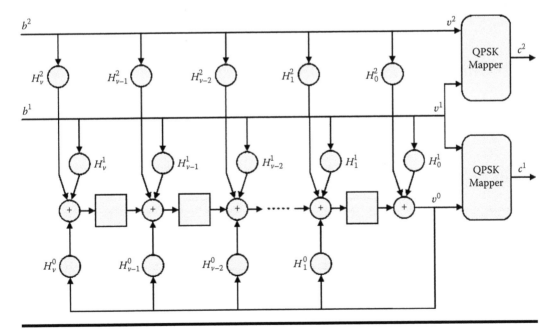

Figure 12.3   STC encoder structure of Lu and Wang.[20]

$$\begin{pmatrix} D + 2D^2 & 1 + 2D^2 \\ 2D & 2 \end{pmatrix}$$

4-PSK, 8 States, 2
Transmit Antennas

$$\begin{pmatrix} D + D^2 & 2 + D \\ 2 + D & 2 + 2D^2 + 2D^2 \end{pmatrix}$$

4-PSK, 8 States, 2
Transmit Antennas

$$\begin{pmatrix} 2D^2 & 2 + D + 2D^2 \\ 2 + D & 2D + 2D^2 \end{pmatrix}$$

4-PSK, 8 States, 2
Transmit Antennas

$$\begin{pmatrix} (1 + a) + D & a + (1 + a)D & a + D & 1 + (1 + a)D \\ a + (1 + a)D & a + D & 1 + (1 + a)D & 1 + (1 + a)D \\ a + D & 1 + (1 + a)D & 1 + (1 + a)D & (1 + a) + aD \\ 1 + (1 + a)D & 1 + (1 + a)D & (1 + a) + aD & 1 + aD \end{pmatrix}$$

4-PSK, 16 States, 4 Transmit Antennas

$$\begin{pmatrix} (1 + a) + (1 + a)D + aD^2 & (1 + a)D + aD^2 & 1 + D^2 & 1 + D^2 \\ (1 + a)D + aD^2 & 1 + D^2 & 1 + (1 + a)D + (1 + a)D^2 & (1 + a) + aD + (1 + a)D^2 \\ a + D^2 & 1 + (1 + a)D + (1 + a)D^2 & (1 + a) + aD + (1 + a)D^2 & (1 + a) + aD \\ 1 + (1 + a)D + (1 + a)D^2 & (1 + a) + aD + (1 + a)D^2 & (1 + a) + aD & D + D^2 \end{pmatrix}$$

4-PSK, 256 States, 4 Transmit Antennas

Figure 12.4   Two space-time trellis codes, 4-PSK, 8 and 16 states, 2b/s/Hz. Space-time orthogonal codes for flat fading channels.

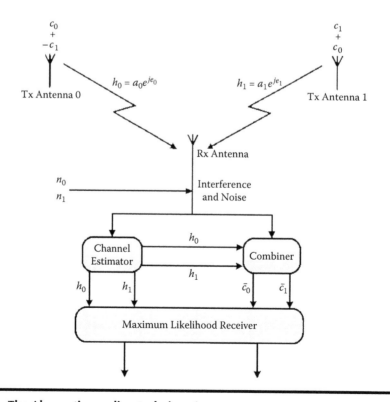

Figure 12.5   The Alamouti encoding technique.[1]

Table 12.1 Alamouti Technique in Case of OFDM S-T Coding

	Antenna 1	Antenna 2
T	c_1	c_2
$T + \tau$	$-c_2^*$	$-c_2^*$

sacrifice their rate for orthogonality reasons, achieving rates lower that 1. Obviously, the Alamouti scheme is an orthogonal scheme of rate 1.

In the following we will give the simple procedures that take place in a receiver to prove the full-diversity possibility of this technique.

The channel gains for time slot t_0 for the two antennas are $h_0(t)$ and $h_1(t)$, while for the next time slot the channel gains are $h_0(t+T)$ and $h_1(t+T)$. According to table 12.1, the received signals during both time slots are

$$y_0 = \sqrt{\frac{E_s}{2}} h_0 c_0 + \sqrt{\frac{E_s}{2}} h_1 c_1 + n_0 \tag{12.4}$$

$$y_1 = -\sqrt{\frac{E_s}{2}} h_0 c_1^* + \sqrt{\frac{E_s}{2}} h_1 c_0^* + n_1. \tag{12.5}$$

We note that channel gains h_0 and h_1 are considered constant over the two time slots, which follow the initial assumption of the slow and flat fading channel.

Equations (12.4) and (12.5) can be written as

$$y = \sqrt{\frac{E_s}{2}} \begin{bmatrix} h_0 & h_1 \\ h_1^* & -h_0^* \end{bmatrix} \begin{bmatrix} c_0 \\ c_1 \end{bmatrix} + \begin{bmatrix} n_0 \\ n_1^* \end{bmatrix}. \tag{12.6}$$

We can observe that in equation (12.6) the $\mathbf{H}_{eff} = \begin{bmatrix} h_1 & h_2 \\ h_2^* & -h_1^* \end{bmatrix}$ matrix, which characterizes the channel, is orthogonal. If instead of vector $\mathbf{y}$ we use the equivalent vector $\mathbf{z}$, given in equation (12.7),

$$\mathbf{z} = \mathbf{H}_{eff}^H \mathbf{y} = \sqrt{\frac{E_s}{2}} \mathbf{H}_{eff}^H \mathbf{H}_{eff} \mathbf{c} + \mathbf{H}_{eff}^H \mathbf{n} \Rightarrow \mathbf{z} = \sqrt{\frac{E_s}{2}} \|h\|_F^2 \, I_2 \mathbf{c} + \mathbf{H}_{eff}^H \mathbf{n}, \tag{12.7}$$

the output of a simple ML decoder is given by

$$\hat{\mathbf{C}} = \arg\min_{\mathbf{C}} \left\| \mathbf{Z} - \|h_{eff}\|_F^2 \, \mathbf{C} \right\|_F^2 = \arg\min_{\mathbf{C}} \left\| \mathbf{Z} - \left(|h_1|^2 + |h_2|^2 \right) \mathbf{C} \right\|_F^2. \tag{12.8}$$

The decoding procedure appears to be a linear process at the receiver, while at the same time the $\mathbf{H}_{eff}$ matrix retains its orthogonality regardless of the $\mathbf{H}$ channel matrix. The last two observations have made the Alamouti concept for achieving diversity the dominant philosophy in STBC techniques.

The diversity order that this scheme can achieve is $2M_r$.

12.3.1.2 Extended Alamouti Schemes

The simple scheme described above has been extended for more than two transmit antennas.[3] In fact, we can produce orthogonal designs for any number of transmit antennas. The designs appearing in literature may be classified into two main categories: the real orthogonal designs, constructed out of real code matrices, and the complex orthogonal designs, constructed out of complex code matrices.

Real constellations (pulse amplitude modulation [PAM]) lead to real orthogonal designs. In a real orthogonal design the code matrix consist of the real entries $\pm c_1, \pm c_2, \ldots, \pm c_N$ and the problem of existence is a Hurwitz-Radon problem.[4]

In case a square code matrix is required, the limitations are many, so only three full-rate schemes can be produced. Specifically, real orthogonal designs of a full-rate and square code matrix can only be implemented for two, four, and eight transmit antennas. However, generalized real orthogonal schemes of full-rate and $T \times M_t$ code matrices can also be produced; they are called delay-optimal designs and offer easy decoding schemes for $M_t \leq 8$.

In the case of complex orthogonal designs, the code matrices consist of the entries $\pm c_1, \pm c_2, \ldots, \pm c_{M_t}$ and their conjugates $\pm c_1^*, \pm c_2^*, \ldots, \pm c_{M_t}^*$. The Alamouti scheme described above belongs to this category. We note that the construction of orthogonal encoding schemes of any number of transmit antennas is possible; however, the rate of such codes is lower than 1.

12.3.2 S-T Codes for Selective Fading Environments

All the discussion up to now has concerned diversity encoding schemes for flat fading channels. Nevertheless, broadcasting technologies such as Digital Media Broadcasting–Terrestrial (DMB-T) must deal with frequency-selective channels, so the construction of STBC for these kinds of channels is of great concern in this chapter. Frequency-selective channels destroy the orthogonality of those techniques. Diversity techniques for selective fading environments follow.

The simplest space-time coding scheme for this kind of channels is presented in Lee and Williams.[5] The OFDM technique is used to transform the frequency-selective channel to multiple flat fading ones; then the implementation of a typical space-time code is straightforward.

The proposed transmitter consists of two antennas with an OFDM chain attached to each. The block diagram is shown in figure 12.6.

Let us assume a DFT module of N carriers. The serial-to-parallel converter produces $2 \cdot N$ data symbols per time slot, so the $N \times 1$ data vectors $\mathbf{X_o}$ and $\mathbf{X_e}$ are constructed and transmitted by antennas 1 and 2. During the next time slot the data vectors $-\mathbf{X_e^*}$ and $\mathbf{X_o^*}$ are transmitted by antennas 1 and 2 correspondingly. The scheme appears to be analogous to the simple Alamouti scheme described above.

The channel can be modeled as a diagonal matrix $N \times N$. The diagonal entries of the channel matrix are the channel gains ($h_1(n)$, $h_2(n)$ for $n = 1, \ldots, N$) of each carrier, which are subject to flat fading. Let $\mathbf{H_1}$ and $\mathbf{H_2}$ be the channel matrices, and let us assume that the channel is subject to slow fading; in that case, $\mathbf{H}_1(T_0) = \mathbf{H}_1(T_0 + \tau)$ and $\mathbf{H}_2(T_0) = \mathbf{H}_2(T_0 + \tau)$. The vectors ($\mathbf{Y_1}$ and $\mathbf{Y_2}$) received during the two time slots are given by the equations:

$$\mathbf{Y}_1 = \mathbf{H}_1 \mathbf{X}_0 + \mathbf{H}_2 \mathbf{X_e} + \mathbf{Z}_1 \qquad (12.9)$$

$$\mathbf{Y}_1 = \mathbf{H}_1 \mathbf{X}_0 + \mathbf{H}_2 \mathbf{X_e} + \mathbf{Z}_1. \qquad (12.10)$$

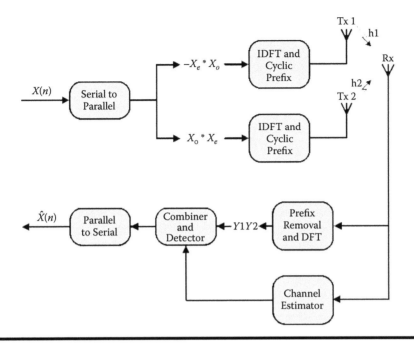

Figure 12.6   Block diagram of the Tx.

Assuming perfect channel knowledge at the receiver, the following estimations of the transmitted vectors can be made (equations [12.11] and [12.12]). This way, the described scheme requires a simple decoding process:

$$\hat{\mathbf{X}}_0 = \mathbf{H}_1^* \mathbf{Y}_1 + \mathbf{H}_2 \mathbf{Y}_2^* \tag{12.11}$$

$$\hat{\mathbf{X}}_e = \mathbf{H}_2^* \mathbf{Y}_1 - \mathbf{H}_1 \mathbf{Y}_2^*. \tag{12.12}$$

Obviously, the diversity order of this scheme is $2 \cdot M_R$, which indicates that this scheme does not exploit the multipath diversity that the frequency-selective channel offers. As it was proved in Paulraj et al.,[6] the maximum diversity gain over a selective fading MIMO channel is $M_r \cdot M_t \cdot L$ where L is the number of delay taps of the channel impulse response.

The next space-time block code that we describe achieves a diversity gain of $M_r \cdot M_t \cdot L$. The basic advantage of this algorithm is the fast decoding process that can be implemented. The code structure of this code is based on general OSTBC designing[3] applied to single-carrier systems. Specifically, if $\mathbf{G}$ is the $T \times M_t$ code matrix of a general OSTBC code scheme, then equation (12.13) stands:

$$\mathbf{G}^H \cdot \mathbf{G} = (|c_1|^2 + |c_2|^2 + \cdots + |c_{N_s}|^2) \cdot \mathbf{I}_{M_t}. \tag{12.13}$$

In (12.13) $\mathbf{G}^H$ is the Hermitian of $\mathbf{G}$. The code structure that is proposed in Zhang et al.[9] has the following code matrix:

$$\mathbf{C} = \sqrt{\gamma} \mathbf{G}' \otimes \mathbf{1}_{\Gamma \times 1}. \tag{12.14}$$

Table 12.2

	Antenna 1	Antenna 2
1st OFDM symbol	c'_1	c'_2
2nd OFDM symbol	c'^*_2	c'^*_1

In equation (12.14) $\otimes$ is the symbol of the Kronecker product, γ is a scalar that ensures the $\|\mathbf{C}\|^2_F = T \cdot N \cdot M_t$ power constraint, 1 is a $\Gamma \times 1$ vector of all 1's, and $\mathbf{G}'$ is the aforementioned $\mathbf{G}$ code matrix, but each entry c_i of the original $\mathbf{G}'$ matrix is replaced be a symbol vector $\mathbf{c}'_i = [c_i(1) \cdots c_i(\frac{N}{\Gamma})]^T$ of dimensions $\frac{N}{\Gamma} \times 1$. The elements $c_i(j)$, $j = 1, \ldots, \frac{N}{\Gamma}$, of the $\mathbf{c}_i$ vector are members of any kind of modulation constellations, and $\Gamma = 2^{\lceil \log_2(L') \rceil}$ in case that $L' < L$. In Zhang et al.[9] the verification of (12.13) for the $\mathbf{C}$ code matrix is presented.

For simplicity and understanding purposes, we cite the code matrix in case of two transmit antennas in equation (12.15) and the transmission scheme in table 12.2. Obviously, the simple S-T encoding scheme presented before is a special case of (12.15) in case of $\Gamma = 1$:

$$\mathbf{C} = \begin{pmatrix} \mathbf{c}'_1 & \mathbf{c}'_2 \\ -\mathbf{c}'^*_2 & \mathbf{c}'^*_1 \end{pmatrix} \otimes \mathbf{1}_{\Gamma \times 1} = \begin{pmatrix} \mathbf{c}_1 & \mathbf{c}_2 \\ -\mathbf{c}^*_2 & \mathbf{c}^*_1 \end{pmatrix}. \tag{12.15}$$

The OSTBC code of equation (12.14) can offer a code rate that is Γ times less than the code rate off the code described with $\mathbf{G}$. However, it achieves a diversity gain of $M_t \cdot M_r \cdot L'$, and in cases where $L' = L$, a full-diversity gain is provided.

12.4 Antenna and Frequency Diversity Techniques for OFDM

The simplest way to produce algorithms of this category is by applying the same space-time coding schemes across space and frequency. However, in cases of broadband MIMO systems the algorithms produced fail to exploit the frequency diversity introduced by those systems. In the following we present one scheme for flat fading channels, while special interest is attributed to space-frequency (S-F) schemes appropriate for selective fading environments. The design criteria of S-F codes can be found in Bolcskei and Paulraj.[10]

12.4.1 Alamouti Technique

The concept of this design is rather simple: the coded symbols are spread over two OFDM carriers instead of two time slots.[7] A clarifying depiction of that is shown in figure 12.7.

Although simple, this technique does not achieve full-diversity gain, which is possible in the case of broadband technologies.

12.4.2 S-F Codes for Selective Fading Environments

In Leeand and Williams[11] a code design is presented that achieves full frequency and spatial diversity gain. The scheme concerns OFDM systems exclusively and is based on the FFT matrix usage.

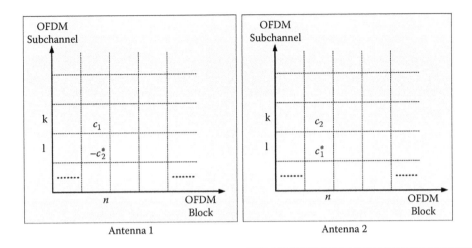

Figure 12.7   Alamouti scheme as a space-frequency encoding technique.

Specifically, the scheme is addressed to a system of M_t transmit antennas and M_r receive antennas, while the channel considered has L independent paths. The elements of the $N \times N$ FFT matrix ($\mathbf{F}$) are given in equation (12.16):

$$[F]_{m,n} = \frac{1}{\sqrt{N}} \exp\left(-j2\pi\frac{mn}{N}\right), m,n = 0,1,\ldots,N-1. \tag{12.16}$$

As a result, $\mathbf{F}$ is a unitary matrix with its columns orthogonal to each other. The proposed $N \times M_t$ code matrix is provided by

$$\mathbf{C}^T = [\mathbf{F}_1 \cdot \mathbf{c} \; \mathbf{F}_2 \cdot \mathbf{c} \cdots \mathbf{F}_{M_t} \cdot \mathbf{c}]. \tag{12.17}$$

In equation (12.17) $\mathbf{c}$ is a $K \times 1$ vector of modulated symbols extracted from the constellation alphabet used and $\mathbf{F}_i$, $i = 1, \ldots, M_t$, are $N \times K$ matrices containing vectors of the $\mathbf{F}$ matrix. The vectors that are used for the construction of the $\mathbf{F}_i$ matrices are selected in a way that the following criteria are satisfied:

■ $\mathbf{F}_i \cdot \mathbf{F}_i^H = \mathbf{I}_K$, $i = 1,\ldots,M_t$, or none of the matrices $\mathbf{F}_i$ and $\mathbf{F}_j$ can have two identical columns.

■ $\mathbf{F}_i \cdot \mathbf{F}_j^H = \mathbf{0}_K$, for $i \neq j$, or the $\mathbf{F}_i$ and $\mathbf{F}_j$ matrices should not share the same column.

■ $\mathbf{F}_i^H \cdot \mathbf{D} \cdot \mathbf{F}_i = \mathbf{0}_K$, $i = 1,\ldots,M_t$, or none of the $\mathbf{F}_i$ and $\mathbf{F}_j$ matrices should contain two neighboring columns of the $\mathbf{F}$ matrix.

■ $\mathbf{F}_i^H \cdot \mathbf{D}^H \cdot \mathbf{F}_j = \mathbf{0}_K$, for $i \neq j$, or the matrices $\mathbf{F}_i$ and $\mathbf{F}_j$ cannot share two neighboring columns of the $\mathbf{F}$ matrix.

The diversity gain of this scheme is $M_t \cdot M_r \cdot L$, but the achievable symbol rate remains under $\frac{1}{M_t \cdot L}$.

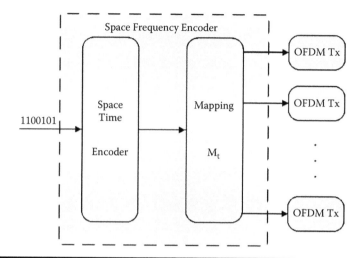

Figure 12.8   Space-frequency encoder proposed in Su et al.[12]

Another approach that achieves full diversity in selective fading MIMO environments can be found in Su et al.[12] The authors in this work prove that the coding schemes that provide full-diversity gain in flat fading environments can be used to construct space-frequency codes that provide full-diversity gain in selective fading environments. In other words, the proposed schemes consist of a typical space-time encoder, producing the $T \times M_t$ codeword, and a mapper that introduces the L channel taps. The proposed S-F encoder concept is shown in figure 12.8. We note that the encoder is suitable for space-time block codes as well as space-time trellis codes with the produced codeword to alter correspondingly.

Let $\mathbf{g} = [g_1 \, g_2 \cdots g_{M_t}]$ be the $1 \times M_t$ vector of the encoded symbols generated by the space-time encoder. The mapper of figure 12.8 performs the following mapping:

$$(g_1 \, g_2 \cdots g_{M_t}) \rightarrow \mathbf{1}_{l \times 1}(g_1 \, g_2 \cdots g_{M_t}),$$ (12.18)

where $\mathbf{1}$ is a $l \times 1$ vector of 1s, and l is a number that $1 \le l \le L$. Obviously, the output of this mapper is an $l \times M_t$ matrix. We choose the integer k so $k = \lceil \frac{N}{l \cdot M_t} \rceil$, and we call $\mathbf{G}$ the space-time encoder matrix of dimensions $k \cdot M_t \times M_t$. The proposed S-F coding scheme constructs the following $N \times M_t$ codematrix:

$$\mathbf{C} = \begin{bmatrix} \mathbf{M}_l(\mathbf{G}) \\ \mathbf{0}_{(N-klM_t) \times M_t} \end{bmatrix},$$ (12.19)

where $\mathbf{M}_l(\mathbf{G}) = [\mathbf{I}_k \, M_t \otimes \mathbf{1}_{l \times 1}] \cdot \mathbf{G}$ and the symbol $\otimes$ stands for the Kronecher product. The output of the encoder is actually the $\mathbf{G}$ matrix repeated l times and adding zeros. The zero padding is necessary for retaining the $N \times M_t$ matrix dimensions of the codeword and can also be used in cases of space-time trellis initial encoding for driving the trellis encoder to the zero state. The diversity gain of this code is $M_r \cdot M_t \cdot l$.

A new work of the same authors[13] presents space-frequency schemes that achieve full-diversity gain as well as full-symbol rate (code rate 1). The authors proceed to the decomposition of the diversity gain as the product of the intrinsic and extrinsic diversity products. The intrinsic diversity product depends on the signal constellations and the code design, while the extrinsic diversity product depends on the applied permutation and the power delay profile of the channel. The application of an optimum permutation strategy leads to the maximization of the extrinsic diversity product.

Another effort for producing full-diversity S-F codes with high code rate, equal to the number of transmit antennas, is presented in Kiran and Sundar Rajan.[14] The authors establish a set of more general design criteria for space-frequency codes, and based on them they define a new class of space-frequency codes, the structure of which is provided in Su et al.[13]

The code structure is based on code matrices of the following form:

$$\mathbf{C} = \begin{bmatrix} \mathbf{G}_1^T & \mathbf{G}_2^T & \cdots & \mathbf{G}_P^T & \mathbf{0}_{N-PK}^T \end{bmatrix} \tag{12.20}$$

In equation (12.20) $K = l \cdot M_t, 1 \le l \le L$ and $P = [\frac{N}{K}]$ the **0** matrix is used for zero padding as before, when the numbers of subcarriers are not a multiple of K. The matrices $\mathbf{G}_P$ are of dimensions $K \times M_i$ and follow the structure of equation (12.21):

$$G(\mathbf{X}_1, \mathbf{X}_2, ..., \mathbf{X}_{M_t}) = \begin{bmatrix} \mathbf{X}_{11} & \phi \mathbf{X}_{21} & \phi^2 \mathbf{X}_{31} & \cdots & \phi^{M_t-1} \mathbf{X}_{M_t,1} \\ \phi^{M_t-1} \mathbf{X}_{M_t,2} & \mathbf{X}_{12} & \phi \mathbf{X}_{22} & \cdots & \phi^{M_t-2} \mathbf{X}_{(M_t-1)2} \\ \phi^{M_t-2} \mathbf{X}_{(M_t-1)3} & \phi^{M_t-2} \mathbf{X}_{M_t,3} & \mathbf{X}_{13} & \cdots & \phi^{M_t-3} \mathbf{X}_{(M_t-2)3} \\ \vdots & \vdots & \vdots & \ddots & \vdots \\ \phi \mathbf{X}_{2M_t} & \phi^2 \mathbf{X}_{3M_t} & \phi^3 \mathbf{X}_{4M_t} & \cdots & \mathbf{X}_{1M_t} \end{bmatrix} \tag{12.21}$$

The $\mathbf{X}_i$ columns of equation (12.21) are constructed according to

$$\mathbf{X}_i = \begin{bmatrix} X_{i1}^T & X_{i2}^T & \cdots X_{iM_t}^T \end{bmatrix}^T \in \mathfrak{R} = \{\Theta s \mid s \in S^{K \times 1}\} \tag{12.22}$$

where Θ is a $K \times K$ matrix, and s is the signal constellation set used. The code rate of the proposed code is $\frac{PKM_t}{N}$, and in the case where N is a multiple of K, the rate is M_t.

The matrix Θ is chosen so that the difference vector $\mathbf{X} - \bar{\mathbf{X}}$ has a Hamming weight equal to K for any $\mathbf{X} \ne \bar{\mathbf{X}} \in \mathfrak{R}$. The interested reader can look in Su et al.[13] for a summary of such matrices. Apart form the Θ matrix, the complex number ϕ is another innovation of this algorithm. ϕ is chosen according to the delay profile and the matrix Θ. Obviously, this code requires channel knowledge at the transmitter. Kiran and Sundar Rajan[14] give several examples of the Θ matrix construction and the ϕ number choice, taking into consideration different delay profiles.

12.5 Antenna, Time, and Frequency Diversity Techniques for OFDM

This group of codes incorporates all kinds of diversity possibilities and is applied especially to selective fading channels. The goal of these codes is the achievable diversity gain whose maximum value is the product of transmit antennas, receive antennas, and channel length. Moreover, the full-diversity gain employing space-frequency encoding techniques is usually of prohibitive complexity. In the following, a number of STF encoding schemes are thoroughly analyzed and the weaknesses of each are pointed out.

The STF code proposed in Liu et al.[15] is based on the concept of grouping the correlated OFDM carriers into groups of carriers. In this way, the system is divided into groups of STF (GTFM) subsystems within which the proposed encoding STF scheme is applied. Liu et al.[15] prove that the proposed concept of grouping retains the maximum diversity gain while involving simplified encoding and decoding architectures.

Because the information exchanged between the Tx and Rx in case of an STF coding scheme can be expressed as a point in a three-dimensional space, we cite a system model suitable for this kind of scenario. The **H** channel matrix elements for each OFDM carrier are given by

$$H_{i,j}(p) = \sum_{l=0}^{L} h_{ij}(l) \cdot e^{-j(2\pi/N) \cdot l \cdot p}$$

(12.23)

Equation (12.23) provides the channel gain between the ith transmit and the jth receive antenna for the pth OFDM carrier, while equation (12.24) shows the link level model for the MIMO OFDM system:

$$y_n^j(p) = \sum_{\mu=1}^{M_t} H_{ij}(p) x_n^i(p) + w_n^j(p), \ j=1,...,M_r, \ p=0,...,N-1$$

(12.24)

In equation (12.24) $x_n^i(p)$ is the transmitted symbol from the ith antenna during the nth time slot on the pth subcarrier, and $w_n^j(p)$ is the additive white noise. The symbol $x_n^i(p)$ is produced by the STF code. The codeword of such a code is expected to be three-dimensional with space, time, and frequency dimensions. Let M_t be the transmit antennas, N the carriers, and N_x the available time slots. The total coded symbols of the produced codeword are $M_t \cdot N N_x$, and the encoding scheme under investigation performs the mapping $\psi : s \rightarrow X$, where s is the $N_s \times 1$ symbol vector that has already been defined in the previous paragraphs. The graphical representation of the **X** codeword in space, time, and frequency is shown in figure 12.9. Obviously, the dimensions of **X** are $M_t \times N \cdot N_x$.

Next, we will describe the carriers' grouping process in the case of STF coding. First, a number of carriers N_c multiple of the channel length is chosen, which represents the number of groups:

$$N_c = N_g \cdot (L+1)$$

(12.25)

The codeword **X** must be split into N_g subgroups and the codewords $\mathbf{X_g}$ of $M_t \times N_x \cdot (L+1)$ dimension must be produced. Specifically, the produced codewords are $\mathbf{X_g} = [X_g(0), X_g(1),..., X_g(L)]$ for $g = 0,...,N_g - 1$, and they must follow the rule

$$X_g(l) := X_g(N_g l + g)$$

(12.26)

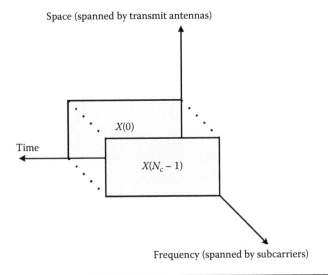

Figure 12.9   The spanned codeword in time and frequency.

The grouping of the carriers leads to the construction of a new simplified code mapping, which is symbolized as $\psi_g : \mathbf{s}_g \to \mathbf{X}_g$. Obviously the symbol vector $\mathbf{s}_g$ is now a $N'_s \times 1$ vector, where $N'_s = \frac{N_t}{N_g}$.

Even though we do not cite the way the coding design criteria apply in this STF technique, we present some interesting remarks that have arisen in Liu et al.[15] The rank criterion is satisfied only in the case where $N_x \geq M_t$. Under this constraint, it is obvious that the minimum dimension of the codeword $\mathbf{X}_g$ is $M_t \times M_t \cdot (L+1)$ while the minimum codeword size in the case of S-F coding is no less than $M_t \times N$ with N the total number of carriers. Because the size of the codewords affects the design complexity, the proposed STF structure involves a simpler design process in comparison with the aforementioned S-F tactics. Exploiting again the application of design criteria into the proposed encoding scheme, it is deduced that the maximum diversity advantage of each GSTF subsystem is $M_t \cdot M_r \cdot (L+1)$, which equals the diversity gain provided by the STF system without channel grouping. Finally, Liu et al.[15] point out that although the proposed GSTF does retain the diversity and code gain of the simple STF codes, the BER might be deteriorated.

The structure of the code can follow two steps: constellation precoding, which provides frequency diversity, and space-time component encoding, which provides spatial diversity. This two-step encoding process is translated in a low-complexity, two-stage optimal decoding, as it is well established in the same work of the authors.

According to Tarokh et al.[3] every generalized complex orthogonal design is characterized by an $M_t \times N_d$ matrix $\mathbf{O}$ with its nonzero entries taken from the set $\{d_i, d_i^*, i = 0, ..., N_t - 1\}$. The $\mathbf{O}$ matrix follows the rule

$$\mathbf{O}_{M_t}^H \mathbf{O}_{M_t} = a \sum_{i=0}^{N-1} |d_i|^2 \, \mathbf{I}_{M_t} \tag{12.27}$$

In equation (12.27) a is a positive constant. Every code design that is produced by this equation is of size (N_s, N_d). A generalized orthogonal complex design of size (N_s, N_t) exists only if N_d and N_s depend on the transmit antenna M_t according to the following rule:

$$(N_s, N_d) = \begin{cases} (2,2), & \text{if } M_t = 2 \\ (3,4), & \text{if } M_t = 3,4 \\ (M_t, 2M_t), & \text{if } M_t > 4 \end{cases} \tag{12.28}$$

After this short reference to the generalized orthogonal designs we continue with the specific STF design. First, the parameter N_s' is chosen according to the equation $N_s' = N_t(L+1)$ and the original $\mathbf{s_g}$ symbol vector is demultiplexed into N_t subgroups in a way that

$$\mathbf{s}_g := \left[\mathbf{s}_{g,0}^T, \dots, \mathbf{s}_{g,N_t-1}^T \right] \tag{12.29}$$

In equation (12.29) each element $\mathbf{s}_{g,i}$ is a $(L+1) \times 1$ complex vector that must be submitted to the precoding process that will distribute the information symbols over multiple subcarriers. The precoding process leads to the precoded blocks $\tilde{S}_{g,i}$ that are $(L+1) \times (L+1)$ complex matrices that are produced by $\tilde{S}_{g,i} := \Theta \mathbf{s}_{g,i}$. In the last equation, Θ is a $(L+1) \times (L+1)$ complex matrix, which denotes the square constellation precoder. At the second stage the $\tilde{s}_{g,i}$ symbol matrix is transformed into the $\mathbf{X}_g$ codeword with the space-time component.

In the precoding constellation stage the Θ matrix is produced in relation to the modulation applied and the L value. The analytical construction of Θ matrix is presented in Xin et al.[16]; however, we present here an example for easier understanding. In case of QPSK modulation and $L = 1$, the Θ matrix is given by

$$\Theta = \frac{1}{\sqrt{2}} \begin{bmatrix} 1 & e^{j(\pi/4)} \\ 1 & e^{j(5\pi/4)} \end{bmatrix} \tag{12.30}$$

After having specified the precoding matrix Θ and the symbol matrices $\tilde{s}_{g,i} := \mathbf{s}_{g,i}$, the codeword construction follows.

The matrix $\mathbf{O}_{M_t}$ that describes a general complex orthogonal design can be represented as

$$\mathbf{O}_{M_t} = \sum_{i=0}^{N_t-1} \mathbf{A}_i d_d + \mathbf{B}_i d_i^* \tag{12.31}$$

In equation (12.31) the real $N_d \times M_t$ matrices satisfy the following equations:

$$A_i^T A_{i'} + B_i^T B_{i'} = a I_{M_t} \delta(i - i') \tag{12.32}$$

$$A_i^T B_{i'} = 0 \tag{12.33}$$

The pairs $\{A_i, B_i\}_{i=0}^{N_s-1}$ are used in the construction of the codeword $\mathbf{X}_g$:

$$X_g^T(l) = \sum_{i=0}^{N_s-1} (A_i \tilde{s}_{g,i,l} + \mathbf{B}_i \tilde{s}_{g,i,l}^*)$$ (12.34)

Equation (12.34) shows that the $\tilde{s}_{g,i,l}$ variables are to $\mathbf{X}_g$ what d_i variables are to $\mathbf{O}_{M_t}$.

The next STF code that we will present is analyzed in Zhang et al.[18] It proposes a systematic design of full-diversity STF codes based on layered algebraic design. Most of the encoding schemes of this category consider a maximum diversity gain of $M_t \cdot M_r \cdot L$. However, the encoding scheme of Zhang et al.[18] provides a maximum diversity gain of $M_r \cdot M_t \cdot L \cdot M_b$, where M_b is the number of the fading blocks of the frequency-selective fading MIMO channel. Contrary to the proposed S-F encoding schemes, which are addressed to quasi-static fading channels, the authors suggest that in block fading channels the coding across multiple fading blocks can offer extra diversity advantage. Moreover, the proposed encoding scheme offers full-rate-M_t under all circumstances, which makes this algorithm different from the already introduced ones in cases where $M_b = 1$.

Before the code presentation a brief reference to the system model used is necessary. The $N \times M_t \cdot M_b$ codeword of the system is produced by mapping N_s data symbols and is written as

$$\mathbf{C} = [\mathbf{C}^1 \ \mathbf{C}^2 \ ... \ \mathbf{C}^{M_b}]$$ (12.35)

In equation (12.35) each element of $\mathbf{C}$ is given by $\mathbf{C}^t = [\mathbf{c}_1^t \ \mathbf{c}_2^t \ ... \ \mathbf{c}_{M_t}^t]$. Each of the column vectors of $\mathbf{C}^t \ \mathbf{c}_j^t$, are sent as an OFDM block to the jth transmit antenna. The received signal after the cyclic prefix removal and the FFT module for the ith receive antenna for the t fading block is given by

$$\mathbf{Y}_i^t = \sum_{j=1}^{M_t} \text{diag}(\mathbf{c}_j^t) \cdot \mathbf{H}_{i,j}^t$$ (12.36)

In equation (12.36) $\mathbf{H}_{i,j}^t$ is the frequency response of the channel and is provided by the following equation:

$$\mathbf{H}_{i,j}^t = \mathbf{F} \cdot \mathbf{h}_{i,j}^t$$ (12.37)

In equation (12.37) $\mathbf{F} = [\mathbf{f}_0 \ \mathbf{f}_1 \ ... \ \mathbf{f}_{L-1}]$. and each column vector of $\mathbf{F}$ is given by $\mathbf{f}_l = [1 \ \omega_l \ ... \ \omega_l^{N-1}]^T$, where $\omega_l = \exp(-j2\pi \frac{\tau_l}{T_s})$, and T_s is the OFDM symbol duration. In the same equation, $\mathbf{h}_{i,j}^t$ is the $L \times 1$ impulse response vector for the channel with each element $h_{i,j,l}^t, l = 0,...,L-1$ being the complex amplitude of the lth channel path.

To give a compact equation for the channel model, let $\mathbf{D}_l = \text{diag}(\mathbf{f}_l), l = 0,...,L$ and $D_l \mathbf{c}_j^t = \text{diag}(\mathbf{c}_j^t) \cdot f_l$. If we use equation (12.37) in (12.36) we find that

$$\mathbf{Y}_i^t = \sum_{j=1}^{M_t} \left[\mathbf{D}_0 \mathbf{c}_j^t \ \mathbf{D}_1 \mathbf{c}_j^t \ \cdots \ \mathbf{D}_{L-1} \mathbf{c}_j^t \right] \cdot \mathbf{h}_{i,j}^t$$ (12.38)

Equation (12.38) after some matrix permutations becomes

$$\mathbf{Y}_i^t = \sum_{l=1}^{L} \left[\mathbf{D}_l \mathbf{c}_1^t \ \ \mathbf{D}_l \mathbf{c}_2^t \ \ \cdots \ \ \mathbf{D}_l \mathbf{c}_{M_t}^t \right] \cdot \mathbf{h}_{i,l}^t = \sum_{l=1}^{L} \mathbf{D}_l \cdot \mathbf{C}^t \cdot \mathbf{h}_{i,l}^t \tag{12.39}$$

We note $\mathbf{X}_t = [\mathbf{D}_0 \mathbf{C}^t \ \ \mathbf{D}_1 \mathbf{C}^t \ \ \cdots \ \ \mathbf{D}_{L-1} \mathbf{C}^t]$ and $\mathbf{h}_i^t = \left[\mathbf{h}_{i,0}^t{}^T \ \ \mathbf{h}_{i,1}^t{}^T \ \ \cdots \ \ \mathbf{h}_{i,L-1}^t{}^T \right]^T$ in order for (12.39) to be written as

$$\mathbf{Y}_i^t = \mathbf{X}_t \cdot \mathbf{h}_i^t \tag{12.40}$$

for $t = 1, 2, \ldots, M_b$ and $i = 1, 2, \ldots, M_r$. Finally, the following notations are used to obtain the system model presented in equation (12.41):

$$\mathbf{Y} = \left[\mathbf{Y}_1^{1^T} \cdots \mathbf{Y}_1^{M_b^T} \cdots \mathbf{Y}_{M_r}^{1^T} \cdots \mathbf{Y}_{M_r}^{M_b^T} \right]^T$$

$$\mathbf{h} = \left[\mathbf{h}_1^{1^T} \cdots \mathbf{h}_1^{M_b^T} \cdots \mathbf{h}_{M_r}^{1^T} \cdots \mathbf{h}_{M_r}^{M_b^T} \right]^T$$

$$\mathbf{X} = \mathbf{I}_{M_r} \otimes \mathrm{diag}(\mathbf{X}_1 \ \ \mathbf{X}_2 \ \cdots \mathbf{X}_{M_b})$$

$$\mathbf{Y} = \sqrt{\frac{\rho}{M_t}} \mathbf{X} \cdot \mathbf{h} + \mathbf{n} \tag{12.41}$$

In equation (12.41) $\mathbf{Y}$ and $\mathbf{n}$ are complex $N \cdot M_b \cdot M_r \times 1$ vectors representing the received vector and the noise vector correspondingly, $\mathbf{X}$ is a $N \cdot M_b \cdot M_r \times M_t \cdot L \cdot M_b \cdot M_r$ complex matrix of the transmitted signal, and $\mathbf{h}$ is the $M_t \cdot L \cdot M_b \cdot M_r \times 1$ vector of the channel impulse response.

In the following we will present the code structure that Zhang et al.[18] suggest. N_s data symbols form the $N \cdot M_t \cdot M_b \times 1$ symbol vector and are parsed into $J = N/K$ subblocks, where $K = 2^{\lceil \log_2(M_t \cdot L) \rceil}$. Each of the parsed data vectors $S_i \in A^{KM_t M_b}$ ($i = 1, 2, \ldots, J$) are encoded into a STF-coded matrix $\mathbf{B}_i \in C^{K \times M_t M_b}$. Let $K = N_p \cdot N_q$ so $N_p = 2^{\lceil \log_2 L \rceil}$ and $N_q = 2^{\lceil \log_2 M_t \rceil}$. In case that $L < 1$ then $M_t > 1$. Additionally, because N is a power of 2, according to the used FFT, J is always an integer.

Because the same STF encoding scheme is applied in every $\mathbf{B}_i$ matrix, the code structure will be presented only for the case of $\mathbf{B}_i$. The concept of the proposed scheme is also depicted in figure 12.10, where the STF design along with its layered algebraic nature is shown.

$$\mathbf{B}_i = \begin{pmatrix} \bar{\mathbf{X}}_1^1 & \bar{\mathbf{X}}_1^2 & \cdots & \bar{\mathbf{X}}_1^{M_b} \\ \bar{\mathbf{X}}_2^1 & \bar{\mathbf{X}}_2^2 & \cdots & \bar{\mathbf{X}}_2^{M_b} \\ \vdots & \vdots & \ddots & \vdots \\ \bar{\mathbf{X}}_{N_p}^1 & \bar{\mathbf{X}}_{N_p}^2 & \cdots & \bar{\mathbf{X}}_{N_p}^{M_b} \end{pmatrix} \tag{12.42}$$

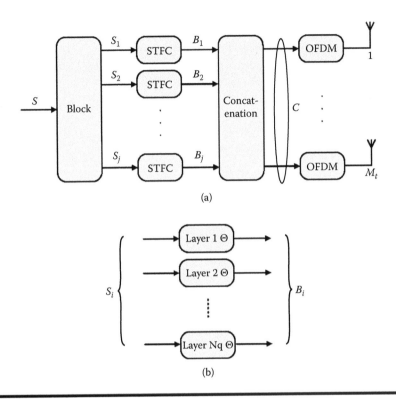

Figure 12.10 The STF code based on layered algebraic design.

Because $\mathbf{B}_i \in C^{K \times M_t M_b}$ each element of (12.42) is a $N_q \times M_t$ complex matrix that is presented in (12.43):

$$\bar{\mathbf{X}}_m^t = \begin{pmatrix} X_{m,1}^t(1) & \phi X_{m,2}^t(1) & \cdots & \phi^{M_t-1} X_{m,M_t}^t(1) \\ \phi^{N_q-1} X_{m,N_q}^t\left(\left\lfloor \dfrac{M_t}{N_q} + 1 \right\rfloor\right) & X_{m,1}^t(2) & \cdots & \phi^{M_t-2} X_{m,M_t-1}^t(2) \\ \vdots & \vdots & \ddots & \vdots \\ \phi X_{m,2}^t(M_t) & \phi^2 X_{m,3}^t(M_t) & \cdots & \phi^{\left(1-\left\lfloor \frac{M_t}{N_q} \right\rfloor\right)M_t} X_{m,\left(1-\left\lfloor \frac{M_t}{N_q} \right\rfloor\right)M_t+1}^t(M_t) \end{pmatrix} \quad (12.43)$$

In equation (12.43) ϕ is a complex that is introduced below. The diagonal layers of the matrix in (12.43), and consequently the matrix as a whole, are constructed by the proposed design. For simplicity, we represent each diagonal layer of (12.43) with the row vector of (12.44):

$$\bar{\mathbf{X}}_{m,n}^t = \begin{bmatrix} X_{m,n}^t(1) & X_{m,n}^t(2) & \cdots & X_{m,n}^t(M_t) \end{bmatrix} \quad (12.44)$$

In equation (12.44) $n = 1, \ldots, N_q$. We note that because $N_q = 2^{\lceil \log_2 M_t \rceil}$, obviously $N_q > M_t$. This way, the N_q diagonal layers represent the whole matrix of (12.43). Because each layer given by (12.44) is independent of m and t, it can be represented by equation (12.45) and can be produced with the usage of the Θ matrix, as is shown in equation (12.46):

$$\bar{X}_n = \left[\bar{X}_{1,n}^1 \quad \cdots \quad \bar{X}_{N_p,n}^1 \quad \bar{X}_{1,n}^2 \quad \cdots \quad \bar{X}_{N_p,n}^2 \quad \cdots \quad \bar{X}_{1,n}^{M_b} \quad \cdots \quad \bar{X}_{N_p,n}^{M_b} \right] \tag{12.45}$$

$$\bar{\mathbf{X}}_n = \Theta \cdot \mathbf{S}_n. \tag{12.46}$$

If $N = N_p \cdot M_t \cdot M_b$, then Θ is a unitary matrix that is given by equation (12.47):

$$\Theta = \mathbf{F}_N^H \operatorname{diag}(1, \theta, \ldots, \theta^{N-1}). \tag{12.47}$$

In equation (12.47), θ is algebraic over K with degree at least $N_p N_q M_b$, and K is the field extension of the original field Q that contains all the entries of Θ, the signal alphabet, and the $e^{-j2\pi \frac{l \cdot l}{s}}$, $(l = 0, 1, \ldots, L-1)$. F_N is the $N \times N$ DFT matrix while $\phi = \theta^{1/N_q}$. The diversity achieved with the STF described above is $M_t M_r M_b L$ while the M_t rate is also achieved.

References

1. R. Prasad. 2004. *OFDM for wireless communications systems*. Boston: Artech House.
2. S. M. Alamouti. 1998. A simple transmit diversity technique for wireless communication. *IEEE J. Select. Areas Commun.*, 16:1458.
3. V. Tarokh, H. Jafarkhani, and A. R. Calderbank. 1999. Space-time block codes from orthogonal designs. *IEEE Trans. Inform. Theory* 45:1456–67.
4. A. V. Germita and J. Seberry. 1979. *Orthogonal designs, quadratic forms and hadamard matrices.* Lecture Notes in Pure and Applied Mathematics, vol. 43. New York: Marcel Dekker.
5. K. F. Lee and D. B. Williams. 2000. A space-time transmitter diversity technique for frequency selective fading channels. In *Proceedings of the IEEE Sensor Array and Multichannel Signal Processing Workshop*, Cambridge, MA, 149–152.
6. A. Paulraj, R. Nabar, and D. Gore. 2003. Introduction to space-time wireless communications. Cambridge: Cambridge University Press.
7. W. Zhang, X. Xia, and K. Letaief. 2007. Space-time-frequency coding for MIMO-OFDM in next generation broadband wireless systems. *IEEE Wireless Commun. Magazine*, 14, 34–42.
8. G. J. Foschini and M. J. Gans. 1998. On limits of wireless communications in a fading environment when using multiple antennas. *Wireless Personal Communications* 6:311–35.
9. W. Zhang, X. Xia, and P. C. Ching. 2007. Full-diversity and fast ML decoding properties of general orthogonal space-time block codes for MIMO-OFDM systems. *IEEE Transactions on Wireless Communications*, 6(3): 311–335.
10. H. Bolcskei and A. J. Paulraj. 2000. Space-frequency coded broadband OFDM systems. In *Proceedings of IEEE WCNC-62000*, Chicago, 1, 1–6.
11. K. F. Leeand and D. B. Williams. 2000. A space frequency transmitter diversity technique for OFDM systems. In *Proceedings of IEEE Global Communications Conference*, San Francisco, 3, 1473–77.
12. W. Su, Z. Safar, M. Olfat, and K. J. R. Liu. 2003. Obtaining full-diversity space-frequency codes from space-time codes via mapping. *IEEE Trans. Signal Processing*, 51:2905–16.

13. W. Su, Z. Safar, M. Olfat, and K. J. R. Liu. 2005. Full-rate full-diversity space–frequency codes with optimum coding advantage. *IEEE Trans. Inform. Theory* 51:229–49.

14. T. Kiran and B. Sundar Rajan. 2005. A systematic design of high-rate full-diversity space-frequency codes for MIMO-OFDM systems. In *Proceedings of the IEEE International Symposium of Information Theory*, Adelaide, 2075–79.

15. Z. Liu, Y. Xin, and G. B. Giannakis. 2002. Space-time-frequency coded OFDM over frequency-selective fading channels. *IEEE Trans. Signal Processing* 50:2465–76.

16. Y. Xin, Z. Wang, and G. B. Giannakis. 2002. Space-time diversity systems based on linear constellation precoding. *IEEE Trans. Wireless Commun.* 2(2):294–309.

17. W. Zhang, X. G. Xia, and P. C. Ching. 2005. High-rate full-diversity space-time-frequency codes for MIMO multipath block-fading channels. In *Proceedings of IEEE Global Communications Conference (GLOBECOM2005)*, St. Louis, 3, 1587–91.

18. W. Zhang, X. G. Xia, and P. C. Ching. 2007. High-rate full-diversity space-time-frequency codes for broadband MIMO block-fading channels. *IEEE Trans. Commun.* 55:25–34.

19. V. Tarokh, N. Seshadri, and A. R. Calderbank. 1998. Space-time codes for high data rate wireless communication: Performance criterion and code construction. *IEEE Trans. Inform. Theory* 44:744–65.

20. B. Lu and X. Wang. 2000. Space-time code design in OFDM systems. In *Proceedings of IEEE Global Communications Conference*, San Francisco, 1000–4.

21. R. S. Blum, Y. Li, J. H. Winters, and Q. Yan. 2001. Improved space-time coding for MIMO-OFDM wireless communications. *IEEE Trans. Commun.* 49:1873–78.

22. Q. Yan and R. S. Blum. 2000. Optimum space–time convolutional codes. In *Wireless Communications and Networking Conference*, Chicago, 3, 1351–55.

Chapter 13

Soft Handover Techniques for Time-Slicing-Based Broadcast Systems

Gunther May

Contents

Keywords

handover, mobility, time slicing, DVB-H, mobile broadcast

Traditionally, broadcast equipment was mainly used in fixed environments, such as TV sets with a rooftop antenna. With the development of new broadcast systems specifically targeted at reception with small, handheld terminals, new challenges are faced. One of these challenges is to ensure service continuity when a user moves with his terminal from one network cell to another. In such situations, a *handover* is needed to switch the reception of the signal from the previous cell to the signal from the new one, ideally without any perception of the user.

In this chapter, techniques for realizing handovers with time-sliced-based broadcast systems such as Digital Video Broadcasting–Handheld (DVB-H)[1] will be presented and evaluated. It will be shown that with such systems, loss-free handovers are possible with only a single receiver front end.

This chapter starts with an overview of handovers in broadcast systems and their critical aspects. Afterwards, strategies to avoid handovers are discussed. Synchronization techniques are evaluated in terms of how well data loss may be avoided using them. An approach to realize soft handovers is presented afterwards. Simulations are used to verify the theoretical models derived before. The chapter closes with a short conclusion and an outlook.

13.1 Handovers in Broadcast Systems

Handovers are well known from mobile communications systems. However, handovers in broadcast systems have to be dealt with differently due to the unidirectional nature of broadcast systems. The network infrastructure, taking care of deciding whether a handover is necessary in mobile communications systems, is not aware of the terminals consuming the transmitted broadcast content or their reception status. Therefore, broadcast terminals have to perform the handovers on their own. Within the DVB Project, such handovers are called *passive*.

With the development of *hybrid networks* incorporating both a broadcast channel and an interactivity channel (e.g., using a mobile communications network), handover of terminals may also be actively supported or even controlled by the network infrastructure. Such handovers are called *active*.

13.2 Data Loss in Context of Handovers

When designing a handover mechanism for mobile broadcast systems, several critical aspects have to be kept in mind. One of the most important aspects is to avoid data loss. Data loss in this context means that the terminal loses parts of the data stream belonging to the currently consumed service while performing a handover. Typically, such data loss implies a negative effect on the quality of service being delivered to the user. This effect may be very severe (e.g., while performing an important file download) or just annoying (e.g., when watching a TV program).

Data loss may basically occur due to two different reasons in relation with handovers. First, the handover process itself may cause data loss. This may happen, for instance, because of the switching time from one network cell to another in which the terminal is not able to receive the signal from either the previous cell or the new cell. Such data loss may be avoided by suitable design of the handover mechanism or by integrating a second receiver front end.

Second, data loss may occur due to unexpected breakdowns in signal strength, which are typical for mobile scenarios in the border regions of cells. Such data loss may be reduced by using *soft handover* techniques, known, for example, from Universal Mobile Telecommunications System (UMTS), where the terminal employs a diversity reception to receive the signals of two or more cells and chooses the better one.

13.3 Handover Avoidance Strategies

Like indicated previously, performing handovers usually implies the risk of having a negative effect on the quality of service, as data loss may occur. Therefore, it should be investigated how handovers may be avoided.

Classic analog broadcast networks are exclusively set up using a *multifrequency network* (MFN) approach. In these networks, adjacent transmitters make use of different transmission frequencies to avoid interferences. However, this implies that every transmitter of the network forms an individual network cell, and at the borders between those cells, handovers would be necessary to ensure service continuity.

With digital broadcast systems such as Digital Audio Broadcasting (DAB), digital video broadcasting–terrestrial (DVB-T), Digital Video Broadcasting–Handheld (DVB-H), and others, *single-frequency networks* (SFNs) may be established. Within these networks, all transmitters transmit exactly identical signals on the same transmission frequency. Therefore, no handovers are required in the coverage area of such a network. However, SFNs are limited in their size due to several constraints.

One of these constraints is the so-called *guard interval*. This interval was introduced to reduce interference caused by different latencies between the individual transmitter signals to the receiver. The duration of the guard interval, and therefore the acceptable latency differences, is limited. Thus, this results in a limitation of the SFN size. However, several approaches allow an increase of the size of an SFN despite this constraint.[2]

The distribution of local content in SFNs is problematic. Within an SFN, all transmitters have to transmit exactly the same content. Therefore, content with only a local scope would have to be transmitted over the whole SFN coverage area, which would result in a very low efficiency in case the scope of the service in question is only small compared to the whole network area. Regarding this problem, a solution for the broadcast system DVB-H has been proposed that allows the transmission of local content under certain constraints.[3] The principle is to introduce intervals for local content and decouple those intervals by so-called adaptation intervals from global content, which is transmitted in the whole SFN coverage area. This way, local content may be transmitted and, for global content, the SFN gain still exists.

A practical constraint regarding the setup of SFNs is the availability of transmission frequencies. Due to regulatory reasons, typically it is not possible to have the same transmission frequency available for the planned coverage area of a network.

Therefore, in practice, most big broadcast networks will make use of a cellular structure composed of different SFNs covering subregions and single transmitters. Between those cells, handovers will be required for continous reception of services. The basic architecture of such a network cell structure is shown in figure 13.1.

13.4 Synchronization Techniques to Avoid Data Loss in Handovers

The terrestrial digital broadcast systems DAB and DVB-T both offer handover mechanisms. However, neither is loss-free in the typical case that the receiver only contains a single radio frequency (RF) front end. The reason for this is that the service data streaming of both systems is approximately continuous. Therefore, all interceptions caused by switching from one frequency

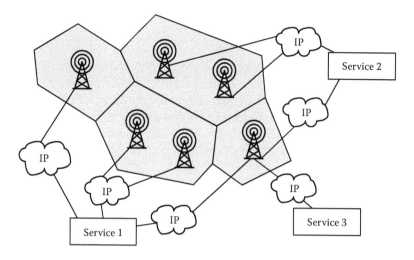

Figure 13.1   Scenario with a network consisting of SFN and single-transmitter cells and services with different scopes.

to another, to perform a handover or monitor the signal on this other frequency, result in a loss of data of the currently consumed service.

Some broadcast systems, such as DVB-H, make use of a time-slice-based transmission. The principle is to transmit data belonging to one service not continuously multiplexed together with the data for other services, but in time slices carrying exclusively data for this single service (figure 13.2).

This scheme was introduced to allow terminals to save power by being able to switch off their front ends between the bursts belonging to the service that is currently consumed. The so-called off-time may also be used to perform handover, and therefore to avoid data loss, as no data relevant for the service in question is transmitted within this period. This principle has been shown in Väre and Puputti.[4]

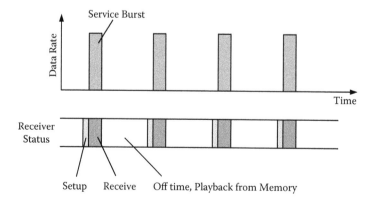

Figure 13.2   Time slicing in DVB-H shown for a single service (not to scale).

For the monitoring of signals from other cells before actually performing a handover, avoiding data loss is always possible using this approach if the off-time is long enough (which is typically the case; otherwise, no power savings would be possible either). However, the data loss inferred by the handover itself may only be avoided if the actual switching is performed in both the off-time of the signal from the current cell and the off-time of the signal from the new cell. Otherwise, parts of a time slice from the new signal may be lost.

To guarantee that the switching is performed in the off-time of both signals, the demand for loss-free handovers with a single front end requires a synchronization of the signals of adjacent network cells. The most obvious approach for this synchronization would be to use an in-phase synchronization, ensuring that time slices from one service are always transmitted at the same points of time in adjacent cells. This way, the off-time in both cells would be identical, and therefore may be used for performing the handover.

However, to ensure loss-free handover in this case, the contents of the time slices sent in parallel in adjacent cells were required to be absolutely identical. Otherwise, seamless continuity when performing handovers could not be guaranteed. An example would be that in one cell the IP packets numbered n to $n + m$ are transmitted within one time slice, while an adjacent cell transmits the IP packets up to number $n + m + 1$ within this time slice. In this case, when performing a handover from the first to the second cell after this time slice, one IP packet would be lost as the new time slice starts with IP packet number $n + m + 2$ and not $n + m + 1$. Therefore, content synchronization of the time slices in adjacent cells would also be needed. This is difficult to achieve if both cells have different service portfolios, and therefore have different IP encapsulators for generating the DVB-H data stream. One reason for this issue is that the IP feeding networks used to carry the services from the service providers to the playout equipment have different delays. Also, different IP encapsulators utilize different algorithms for creating the time slices.

Another technique for synchronization is to use the *phase-shifting* approach.[5] In this approach, a static phase shift between the signals of the two cells is applied (figure 13.3). The phase shift should be large enough so that there is no overlapping between the time slices of adjacent cells and the necessary synchronization time for the terminal to change from one signal to another. If the phase shift is configured accordingly, there is an overlapping of the IP packets contained in the time slices between two consecutive time slices of two different cells. This ensures, even with significant tolerance toward IP feeding stream delays, seamless loss-free handovers. It can be seen in the figure that this principle works not only for changes between cell 1 and cell 2, but also in the opposite direction.

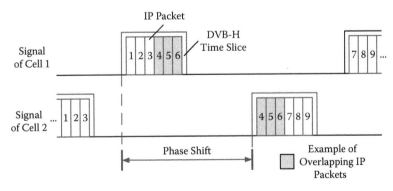

Figure 13.3 **The principle of phase shifting shown for a single service (not to scale).**

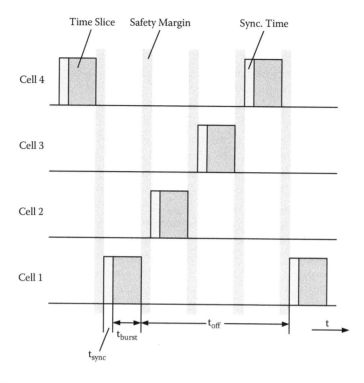

Figure 13.4   Phase shifting as a four-color problem.

In real networks, more than two cells have common borders, so more than two different phase shifts are needed to ensure seamless service continuity between any two cells. The various phase shifts that are necessary for *n* adjacent cells can be interpreted mathematically as colors in a graph coloring problem, never allowing the same color twice in adjacent nodes. This problem is very common, for example, in frequency planning of cellular communications networks. In theory, it leads to a challenge similar to the four-color problem.[6] With four different phase shifts, loss-free handover between any two cells will be possible, no matter what the shape of the cells might be. Depending on the cell shape, for example, with an idealized hexagonal one, it might be possible to use less phase shifts.

Figure 13.4 illustrates how the service can be phase shifted in four adjacent cells to allow seamless handovers between any two of those cells. The synchronization time of the receiver has been taken into account. Additionally, safety margins are required as DVB-H encoders usually introduce a time-slice jitter.

Phase shifting is entirely backwards compatible with the DVB-H specification. The approach may be implemented in the IP encapsulators used in the transmission chain. For time synchronization, approaches similar to those for setting up SFNs may be applied.

13.5   Soft Handovers

In real-life networks, the borders between the coverage areas of the individual cells are not sharp. Typically, overlapping areas between the cells exist. Thus, reception of two or even more signals is possible in these areas. However, usually the signal strengths are relatively low and the reception of each individual signal tends to be error-prone. On the other hand, due to different propagation channels and transmission frequencies, received noise and interferences affecting the signals from the different network cells often have a low correlation only.

So, if a user wants to receive a service in the border area of two cells while the service is available in both of them, the individual signals might be too weak for an error-free reception. In this case, a diversity reception of both signals for improving the quality of the service looks promising. This approach is known from the *Universal Mobile Telecommunications System* (UMTS), where the term *soft handover*[7] is used. In UMTS, signals from adjacent cells from one network are transmitted on the same frequency. Distinguishing between the signals is possible using the code division multiple access (CDMA) technique. For broadcast systems such as DVB-H, adjacent cells use different frequencies. Therefore, the proposed approach for realizing soft handovers with such broadcast systems[10] is rather different.

In theory, the different signals of the two cells could be received simultaneously and afterwards combined at the RF level to obtain an improved signal-to-noise ratio of the combined signal. However, this would introduce significant additional complexity to the system due to the fact that two front ends and a suitable signal combiner would be necessary. Additionally, it would cause severe requirements regarding the transmitted signals. The signals would have to be exactly identical, at least for the periods in which the data of the service in question is transmitted. Unless all services would be identical in all cells (which is not realistic due to the different signaling and different local scope of services; this would be the SFN approach), this is very difficult due to the multiplexing of services, as well as interleavers, forward error corrections (FECs), and other mechanisms applied in the transmission chain spreading the data of the different services across the stream.

However, the packet streams of the services themselves, that is, in the case of DVB-H-IP packets, usually are identical in different network cells, as they are provided by a single content provider. Therefore, a diversity functionality may be implemented on the packet level. If different signals contain bit errors and therefore packet errors that do not have a strong correlation, the merged packet stream would contain significantly less packet errors if erroneous packets from one stream are replaced by error-free packets from the other.

To implement a terminal capable of diversity reception according to the described approach, usually two receiver modules would be needed to allow independent reception of the two signals. With DVB-H, a single receiver front end may be sufficient if the phase-shifting technique is applied. As discussed in the previous section, phase shifting ensures that the transmitters of adjacent cells never transmit data belonging to one service at the same point of time. Therefore, the terminal would be able to switch between the signals of the different cells and receive the whole packet stream for the service from each of them in parallel (figure 13.5).

To achieve gain from this type of diversity reception, the lost packets from one packet stream need to be replaced by packets from another stream. Whether a certain packet is corrupt may be derived from the CRC checksum of the MPE datagram sections. The merging of the packet streams to replace erroneous packets from one stream with error-free ones from another may be performed at several reference points in the DVB-H receiver protocol stack. Figure 13.6 shows different options.

For the merging of the packet streams, it is necessary to know which packet from one stream is congruent to which of another, to replace erroneous packets from one stream with error-free ones from another. The easiest way for implementation would be to have sequence numbers in the packets. However, the header of TS packets only provides a continuity counter of 4 bits, which allows using packet numbers between 0 and 15. For the present purpose, this counter is not sufficient, as DVB-H time slices may contain more than 1,000 TS packets. MPE datagram sections do not contain a sequence number field in their headers at all.

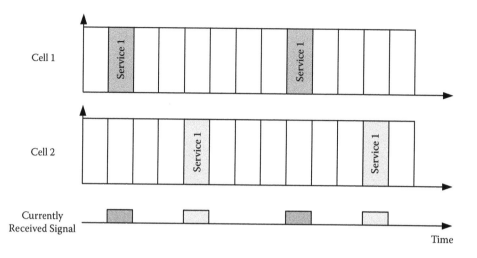

Figure 13.5   Principle of packet-level diversity reception for soft handover support.

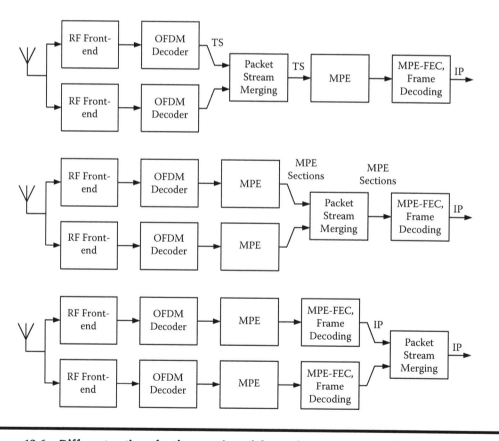

Figure 13.6   Different options for the merging of the packet streams according to May.[10] Please note that the entities symbolizing the diversity reception (e.g., two RF front ends) are only logical. Physically, only a single entity is existent that may be used for several signals in parallel due to the time-slicing and phase-shifting properties.

If the merging is done on the IP level, sequence numbers used in the higher-layer protocols, like the Real-Time Protocol (RTP),[8] may be employed that are used for transmission according to the IP Datacast standard.[9] Therefore, merging would be relatively easy. For merging on the MPE and TS levels, different options have been discussed in May.[10] However, this is more difficult than implementation on the IP level.

Time slicing has been introduced to reduce the power consumption of DVB-H terminals by switching off the RF front end in periods between two time slices of a service. The power consumption of the RF front end depends on the time the terminal has to switch on the RF front end in relation to the whole operating time. As the RF front end would have to be switched on n times in each cycle time if n signals were received in parallel, the power consumption of the RF front end would ascend linearly with the number of signals received.

However, handover situations in DVB-H are relatively rare due to the typically bigger cell size compared to mobile communications systems. Additionally, current DVB-H receiver modules have a power consumption of about 40 mW in total, including the RF front end. Because usually other components in a terminal such as the display have a much higher power consumption, an increase to about 80 mW caused by the diversity reception during periods within the border areas of cells would not be too severe. For especially strong power-efficiency demands, this diversity reception could be limited, for example, to important file downloads.

13.6 Simulation and Implementation Results

To verify and evaluate the proposed approaches, simulations have been performed.

13.6.1 Synchronization Techniques for Loss-Free Handover Support

For simulation of the synchronization techniques, a simulation using the OPNET Modeler simulation platform has been conducted. The parameters shown in table 13.1 have been assumed.

Table 13.1 Simulation Parameters

Parameter	Value
DVB-H mode	8K 16-QAM Viterbi code rate: 2/3 MPE-FEC-CR: 3/4
Channel model	TU-6, 10 Hz Doppler
Time-slice cycle time	2 s
Burst size	0.5 Mbit
IP packet size	1,000 bytes
Synchronization time	100 ms

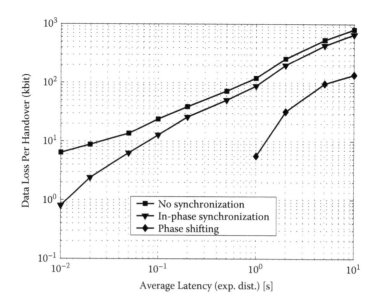

Figure 13.7   Simulation results for data loss for different synchronization schemes.

As the feeding network delay to the different network cells has been identified as a critical factor for realizing loss-free handovers, the packet loss per handover for different synchronization schemes as a function of the feeding network delay was analyzed. As a model for the feeding networks, a delay with an exponential distribution function, independent for the feeding path to each network cell, has been assumed.

The simulation results from figure 13.7 confirm the theoretical considerations discussed earlier in this chapter. When no synchronization is performed (not shown here due to the logarithmic scaling), handovers already show an average loss of about 4 kbit for the assumed parameters, even without any feeding network latencies. For higher feeding network latencies, this data loss increases quickly.

In-phase synchronization (without content synchronization) results in a loss-free handover for ideal feeding networks. However, already for an average feeding network delay of 10 ms, a remarkable risk for data loss exists.

The phase-shifting approach only shows data loss when the feeding network latency is in the order of one second or higher.

13.6.2 Soft Handover

For simulation of soft handovers, transport stream (TS) packet error patterns were generated using a COCENTRIC System Studio environment, simulating the physical layer capabilities of DVB-H bit accurately. These TS packet error patterns were fed into the OPNET Modeler simulation, modeling the soft handovers.

It was assumed that the error patterns of the signals of adjacent cells were independent, as the transmission frequencies and the transmitter locations are different. The IP packet error rate has

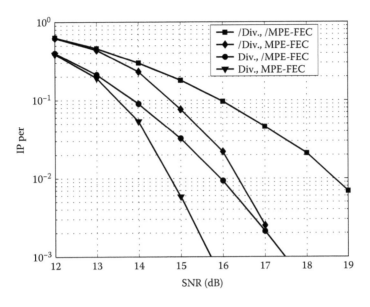

Figure 13.8 Simulation results for soft handovers.[10]

been monitored for single-signal and packet-level diversity reception as a function of the signal-to-noise ratio. For both of the signals received in the diversity case, an equal signal-to-noise ratio was assumed. This is the average case; for a weak main signal, the gain from soft handover is higher, otherwise lower. It was assumed that the packet stream combining was performed on the IP packet level.

Figure 13.8 shows that the gain is approximately 2.5 to 3 dB in the case that no MPE-FEC is used, and in the order of 1 to 1.5 dB with the MPE-FEC. In May[10] it was shown that the gain for the case with MPE-FEC is higher if the stream combining is performed before applying the MPE-FEC decoding, which is, however, significantly more difficult to implement.

13.6.3 Real-Life Implementation

To evaluate the handover algorithms in a real-life environment, an implementation of terminal handover support functionality has been conducted on a Dell PDA equipped with a DVB-H receiver plug-in card. The card, manufactured by DIBCOM, uses an secure digital (SD) slot interface to connect to the PDA. In figure 13.9, the basic software architecture used for the implementation is shown.

The main functionality is encapsulated in a *dynamic link library* (DLL). The DLL connects to the lower DVB-H protocol layers via a device driver provided by DIBCOM with the receiver card. To the higher layers, the DLL provides an *application programming interface* (API), which is based on the *Media-Independent Handover* (MIH) specification, IEEE 802.21.[11] The MIH specification, originally designed for supporting especially mobile communications systems, WiFi (IEEE 802.11), and WiMAX (IEEE 802.16) systems, provides an abstraction layer independent from the underlying physical layer system, and therefore enables the reusage of handover functionality

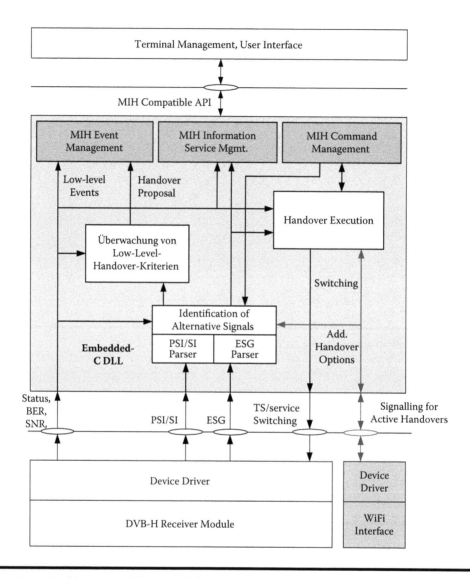

Figure 13.9 Architecture of the sample handover support implementation.

entities and easy intersystem handovers. However, as the MIH specification requires a bidirectional link between the terminal and the network infrastructure, only a subset of the API has been implemented. With the availability of an interactivity channel integrated in the terminal, this functionality may be enhanced to support active handovers also.

Nevertheless, the implementation already provides without an interactivity channel a good support of handover functionality to the higher protocol layers implemented on the terminal. We were able to demonstrate loss-free handovers between different DVB-H network cells using this terminal.

13.7 Summary and Outlook

Handovers are a crucial part of mobile broadcast systems. Although handovers are well known from mobile communications systems, in broadcast systems they are rather different and new challenges have to be faced. In this chapter, it was shown how data loss may be avoided with time-slice-based broadcast systems such as DVB-H. Unlike other approaches, only a single receiver front end is needed even to allow soft handovers.

In the future, mobile broadcast systems and mobile communications systems will converge more and more. It may be assumed that this convergence also benefits the mobility support of broadcast systems due to the availability of an interactivity channel, enabling new possibilities to perform handovers.

Links

1. DVB Project: http://www.dvb.org.
2. DVB-H: http://www.dvb-h.org.

References

1. ETSI. 2004. *Digital Video Broadcasting (DVB); Transmission system for handheld terminals (DVB-H)*. ETSI EN 302 304, V1.1.1.
2. A. Mattsson. 2005. Single frequency networks in DTV. *IEEE Transactions on Broadcasting Band* 51:413–22.
3. G. May, P. Unger. 2006. A new approach for transmitting localized content within digital single frequency broadcast networks. In *IEEE International Symposium on Broadband Multimedia Systems and Broadcasting*.
4. J. Väre, M. Puputti. 2004. Soft handover in terrestrial broadcast networks. In *Proceedings of IEEE International Conference on Mobile Data Management*, 236–42.
5. G. May. 2005. Loss-free handover for IP datacast over DVB-H networks. In *Proceedings of IEEE International Symposium on Consumer Electronics (ISCE)*, 203–8.
6. A. B. Kempe. 1879. On the geographical problem of the four colors. *Am. J. Math.* 193–200.
7. ETSI. 2006. *Universal mobile telecommunications system (UMTS); Radio resource control (RRC); Protocol specification*. 3GPP TS 25.331, version 7.3.0, release 7; ETSI TS 125 331, V7.3.0.
8. H. Schulzrinne et al. 2003. *RTP: A transport protocol for real-time applications*. IETF RFC 3550.
9. ETSI. 2006. *Digital Video Broadcasting (DVB); IP datacast over DVB-H: Architecture*. ETSI TR 102 469, V1.1.1.
10. G. May. 2007. Packet-level diversity reception in cell border regions with DVB-H. In *IEEE Consumer Communications and Networking Conference, "Seamless Consumer Connectivity" (CCNC)*.
11. IEEE. 2006. *Draft IEEE standard for local and metropolitan area networks: Media independent handover services*. IEEE 802.21/D01.00.

Chapter 14

Transmission Aspects and Service Discovery Strategy for Seamless Handover in DVB-H

Jani Väre and Tommi Auranen

Contents

Keywords

DVB-H, service discovery, signaling, PSI/SI, seamless handover

Digital Video Broadcasting–Handheld (DVB-H)[1] is one of the first standards developed to provide broadcast digital multimedia content to handheld devices. From the very beginning, the mainstream content within DVB-H has been the Internet Protocol (IP)–based services such as MobileTV, composed of digital video and audio. The DVB-H standard specifies only Open Systems Interconnection (OSI) layers 1 and 2, and hence DVB developed another standard, IP Datacast (IPDC) over DVB-H,[2] which enhanced the DVB-H standard by defining a complete end-to-end DVB-H system for IP-based MobileTV services. This chapter focuses on the issues related mostly to the DVB-H-specific part, that is, OSI layers 1 and 2, of such a system.

DVB-H had already been a popular research topic before the first publication of the standard. For example, a generic algorithm for seamless handover in DVB-H was introduced for the first time in Väre and Puputti.[3] Since then, topics relating to mobility issues in DVB-H have been studied within several publications. Also, at the time of this writing, the DVB ad hoc group Convergence of Broadcast and Mobile Services (CBMS) was finalizing the first European Telecommunications Standards Institute (ETSI) draft of the mobility implementation guidelines of the IPDC over DVB-H standard.

The intention of this chapter is not just to repeat the topics and views already discussed in the standards, standard guidelines, and academic publications; instead, we aim to give our view of signaling and receiver implementation-related issues, based on many years of experience within different areas of DVB-H implementation and standardization work.

In this chapter we will clarify the fundamentals needed to accomplish seamless handover. The covered topics include signaling, service discovery, and definition of an algorithm for seamless handover in DVB-H. The signaling and service discovery–specific part is mainly focused on transmission and usage of program specific information (PSI)/service information (SI),[4] which has a major role in the service discovery within DVB-H. The handover algorithm–specific part is focused on defining the technical background for the definition of a seamless handover algorithm as well as providing a step-by-step approach to the example implementation.

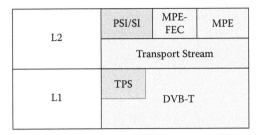

Figure 14.1   The DVB-H protocol stack.

14.1   The Service Discovery Signaling Principle within DVB-H

Service discovery signaling within DVB-H is considered to consist of two parts: transmission parameter signaling (TPS) and PSI/SI. The real-time signaling carried within MPE and Multi-Protocol Encapsulation–Forward Error Correction (MPE-FEC) headers is considered to be part of the service access rather than service discovery, and hence it is not elaborated further within this work. Figure 14.1 illustrates the DVB-H protocol stack, where TPS is carried as part of the physical layer and the PSI/SI is carried within the data link layer.

14.1.1   TPS

TPS consists of L1 service discovery parameters and is carried within the orthogonal frequency division multiplexing (OFDM) frames. It is defined over 68 consecutive OFDM symbols per one OFDM frame (see figure 14.2). One OFDM superframe is composed of four sequential OFDM frames, and one TPS bit is conveyed within each OFDM symbol. TPS bits are categorized as follows: 1 initialization bit, 16 synchronization bits, 37 information bits, and 14 redundancy bits for error protection.

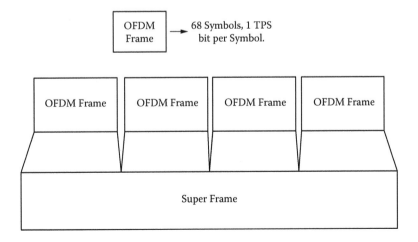

Figure 14.2   Relation of OFDM frames, superframes, symbols, and TPS bits.

In addition to the DVB-H modulation parameters, which consume 27 bits, 10 bits are defined as follows:

- **One bit for DVB-H indicator:** Indicates whether DVB-H is supported within the associated signal.
- **One bit for MPE-FEC indicator:** Indicates whether MPE-FEC is supported within the associated signal.
- **Eight bits for cell identification (cell_id):** The total size of cell_id is 16 bits, but the signaling is optimized by signaling one half of the cell_id, that is, 8 bits, in every first and third frame, while the other half of the cell_id is signaled within every second and fourth frame of the superframe.

TPS is an important part of the service discovery within DVB-H. The DVB-H indicator bit is used for an early elimination of signals that are not supporting DVB-H. The MPE-FEC indicator bit enables the receiver to prepare different reception strategies, depending on whether the MPE-FEC is supported. Cell identification, in turn, is used by the receiver to discard signals that have the same frequency as the targeted signal but are actually part of a different cell or network. The frequencies are associated with geographical location and frequencies within PSI/SI. Based on the information acquired from the PSI/SI and the TPS bits, the receiver can validate the network and cell of each candidate signal that may be found during the handover process.

The access time for the TPS bits depends on symbol speed, which in turn is affected by the modulation used. TPS bits are received slightly prior to synchronization to the signal, and hence it is possible to achieve TPS lock before the actual signal lock, which makes the TPS inspection faster.

14.1.2 PSI/SI

PSI/SI signaling consists of *tables* that are carried over transport streams (TSs) of the IPDC over the DVB-H network. The PSI/SI signaling needed for the service discovery in IPDC over DVB-H consists of the following tables: Network Information Table (NIT), Program Association Table (PAT), Program Map Table (PMT), IP/MAC Notification Table (INT), and Time and Date Table (TDT). Each table, in turn, excluding PAT and TDT, carries a number of different *descriptors* that contain most of the actual information that is carried within the tables. The following applies to all PSI tables:

- The section number of the first section of each subtable is 0x00.
- Excluding PMT, the section number is incremented by 1 with each additional section of a subtable.
- Any addition, removal, or change in content of any section within a subtable affects the version number change. Two sequential transmissions of a subtable using the same version number have the same number of sections, and the content and order of sections are identical.
- Elementary streams that contain PSI/SI sections do not contain any MPE or MPE-FEC sections.

14.1.2.1 The Concept of Tables, Subtables, and Sections

A table is a high-level entity that consists of a collection of subtables identified with the same table_id. The standard definition of tables allows tables to be global, network specific, and transport stream specific. Each subtable, in turn, may be composed of one or more sections

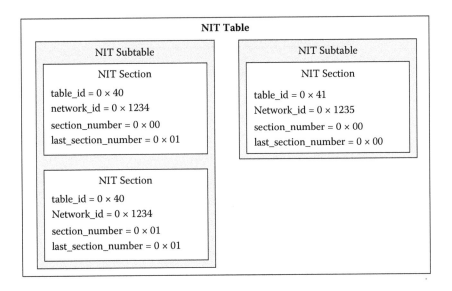

Figure 14.3 An example of table-subtable-section relation of NIT.

with the same table_id_extension and version_number. Figure 14.3 illustrates an example of a table-subtable-section relation, where one NIT table has been split into two NIT subtables. The figure depicts the NIT_actual subtable on the left and the NIT_other subtable on the right. The NIT_actual subtable is identified with table_id = 0x40, and the NIT_other subtable is identified with table_id = 0x41. Furthermore, the NIT_actual subtable is composed of two sections, while the NIT_other subtable has only one section.

The Program Map Table (PMT) and Time and Data Table (TDT) are exceptions to the table concept discussed above. The PMT subtable may always contain only one PMT section. Hence, the section_number and last_section_number within PMT sections must always be set to 0. The TDT, in turn, is unorthodox as regards the generic stable syntax and is only 8 bytes long.

The section structure of PSI/SI subtables, excluding TDT, is in conformance with the structure defined in table 14.1. The latter means that each section of a subtable always has a similar header, even though the carried signaling information is different between sections.

The semantics of the data fields in table 14.1 are as follows:

table_id: Identifies the table that the corresponding section belongs to.
section_syntax_indicator: Indicates whether the syntax in the current section is according to DVB specifications.
section_length: Specifies the section length in bytes. The length is calculated for the fields followed by this field. The maximum section length is 4,096 bytes for the INT and 1,024 bytes for the other DVB-H-specific section types.
table_id_extension: Carries a table-specific identifier, which can be used for unique detection of a particular section. The table_id_extension types for the different tables are as follows:

> PAT: transport_stream_id
> PMT: program_number

Table 14.1   Generic Structure of PSI/SI Section

Syntax	No. of Bits
table_id	8
section_syntax_indicator	1
reserved_future_use	1
reserved	2
section_length	12
table_id_extension	16
reserved	2
version_number	5
current_next_indicator	1
section_number	8
last_section_number	8
[table specific]	
CRC_32	32

INT: platform_id_hash + action_type

NIT: network_id

version_number: Provides the version number of the associated subtable. Each time a subtable is updated, a value within this field is incremented by 1. After the version number exceeds the maximum value of 31, the counting is restarted from zero.

current_next_indicator: Indicates whether the received subtable is valid immediately or in the future. The subtable is valid immediately if this field is set to 1. If this field is set to 0, the subtable is not yet valid.

CRC_32: Indicates the cyclic redundancy check (CRC) value of the section.

Generally, most of the data in the PSI/SI tables is contained in descriptors, which in turn are allocated within descriptor loops. In addition to descriptor loops, tables may also contain other loops that carry data, such as the transport stream loop in the NIT. Most such loops have a loop length field, which indicates the length of each loop. The latter allows the receiver to determine when each loop ends and the next field begins. The length of these loop length fields is 12 bits. If specific loop length fields do not exist (e.g., in the case of the last loop of PMT), the receiver may use the previous length field.

Descriptors (see table 14.2) have two common fields followed with type-specific content. These fields are descriptor_tag and descriptor_length.

descriptor_tag: An 8-bit field that identifies each descriptor. The purpose of the descriptor_tag is to distinguish descriptors from each other.

descriptor_length: An 8-bit field specifying the total number of bytes of the data portion of the descriptor following the byte defining the value of this field.

Each subtable and section are encapsulated into transport stream packets and can be identified by means of a packet identifier (PID), which uniquely refers to the elementary stream. The PID value associated with the elementary stream of the given PSI/SI table can be either static or dynamic, depending on the table_id, which uniquely determines the table type. The PMT and INT tables

Table 14.2 Generic Structure of IPDC Over DVB-H-Specific Descriptors

Syntax	No. of Bits
descriptor_tag	8
descriptor_length	8
[descriptor specific]	

have dynamically allocated PIDs, while the other subtables have static PID values. The transmission interval of different subtables varies accordingly to different table type. The minimum transmission interval for a section is equal to the minimum transmission interval of the associated subtable. Table 14.3 lists table_ids, table_id_extensions, maximum section sizes, transmission intervals, and PID allocations for each IPDC over DVB-H-specific PSI/SI subtable.

The NIT is used within DVB-H for two main purposes. First, it provides information of the available cells that may be located within actual or other networks. Each cell may carry up to two different transport streams, which may be transmitted through one or more transmitters throughout the cell coverage area. The second purpose of the NIT is to provide linkage for locating IP platforms, each representing a single INT subtable. Finally, the NIT available within each cell may be composed of one NIT_actual subtable and several NIT_other subtables. An NIT_actual subtable describes the current network, while one NIT_other subtable can be used for describing other existing networks. NIT_other and NIT_actual can be mutually distinguished based on the different table_id. A separation between two NIT_other subtables can be done based on the network_id, which is a table_id_extension for an NIT subtable. The DVB-H-specific descriptors within the NIT are as follows:

linkage_descriptor: Used to provide linkage to the parameters that enable the discovery of the elementary stream carrying the INT subtable through the PAT and PMT. The linkage descriptor may contain linkages to one or more INT subtables available within one TS.

Table 14.3 Table_ids, Table_id_Extensions, Max. Section Sizes, Transmission Intervals, and PID Allocations of IPDC Over DVB-H-Specific PSI/SI Tables

Table	Table_id	Table_id_extension	Max. Section	Transmission Interval	PID
PAT	0x00	Transport_stream_id	1,024	25–100 ms	0x0000
PMT	0x02	program_number	1,024	25–100 ms	Allocated by PAT[a]
NIT	0x40, 0x41	network_id	1,024	25 ms–10 s	0x0010
INT	0x4C	Platform_id_hash & action_type	4,096	25 ms–30 s	Allocated by PAT[a]
TDT	0x70	—	1,024	25 ms–30 s	0x0014

[a] Any value between 0x0020 and 0x1FFE.

network_name_descriptor: Used to announce the network name of the actual network. The network name may be up to 256 characters long (i.e., 256 bytes). The latter equals 17 bytes for the name that the receiver can discover from this descriptor. The network name has no significance regarding the service discovery itself. It is more of an informative parameter that can, for example, be displayed to the end user.

cell_list_descriptor: Used to map the cell identifiers with the geographical areas covered by each cell. The cell coverage area is defined by four parameters: cell_longitude, cell_latitude, cell_extent_of_latitude, and cell_extent_of_longitude. The latter parameters refine the rectangle that encloses approximately the area covered by the transmitted signals of the particular cell. Furthermore, the identification and coverage area for a subcell are also provided within this descriptor. A subcell consists of the coverage area of the transposer, which transmits the same content to that of the cell, but on a different frequency. Similarly as the cell, the subcell is defined with four parameters: subcell_longitude, subcell_latitude, subcell_extent_of_latitude, and subcell_extent_of_longitude. A subcell is always included inside the coverage area of the cell. Figure 14.4 illustrates an example of cell coverage area where the large circle represents the actual cell coverage area and the rectangle represents the signaled cell coverage area. The small circle, in turn, represents the coverage area of a subcell provided by a transposer.

The receiver may use information carried within this descriptor to limit handover candidates announced in the NIT. For example, the receiver could discard all such cells that do not have overlapping rectangles or are adjacent to the rectangle of the current cell.

time_slice_fec_identifier_descriptor: Used for mapping transport streams with time-slicing and MPE-FEC parameters. When located in the first loop of the NIT, the descriptor applies to all elementary streams within all transport streams announced within the subtable. Any time_slice_fec_identifier_descriptor occurring after this descriptor within the second loop of the NIT or within the INT overwrites the information signaled by this descriptor. Figure 14.5 illustrates the principle of the use of the time_slice_fec_identifier_descriptor, whereas:

A. The time_slice_fec_identifier_descriptor is located in the first loop of the NIT subtable. All elementary streams that are carried within transport streams A, B, and C and which contain IPDC services are mapped with parameters announced within the descriptor.

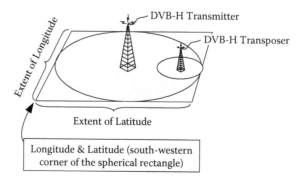

Figure 14.4 The actual and signaled coverage area of a cell that includes one subcell.

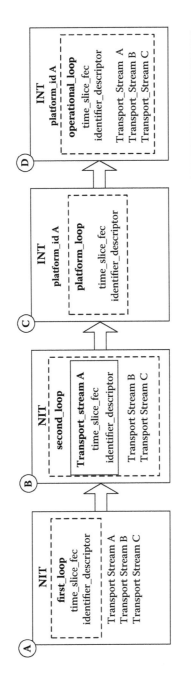

Figure 14.5 **The principle of use of the time_slice-fec_identifier descriptor.**

B. The time_slice_fec_identifier_descriptor is included again in the second loop of the NIT subtable, in iteration where transport stream A is announced. All parameters mapped with the time_slice_fec_identifier_descriptor within the first loop of the NIT subtable are overwritten with the parameters announced within the time_slice_fec_identifier_descriptor included within this loop.

C. The next occurrence of the time_slice_fec_identifier_descriptor is in the platform_loop of the INT subtable. The time_slice_fec_identifier_descriptor now applies to all elementary streams referenced within the NIT subtable. Hence, even the parameters mapped for transport streams B and C are now overwritten.

D. The last occurrence of the time_slice_fec_identifier_descriptor is in the operational_loop of the INT subtable. Again, all previous information announced with the time_slice_fec_descriptor is overwritten.

terrestrial_delivery_system_descriptor: Associates the tuning parameters with each transport stream announced within the NIT. One such descriptor is provided by the transport stream. In addition to announcing the tuning parameters, this descriptor also indicates whether the associated transport stream carries DVB-H services, and further, whether the carried DVB-H services support MPE-FEC.

cell_frequency_link_descriptor: Associates each transport stream with all cells and frequencies where it is available.

The PAT and PMT tables provide the connection for the parameters carried within different tables, and ultimately the link between IP streams and elementary streams. The PAT carries no descriptors; the DVB-H-specific descriptors within the PMT are as follows:

stream_identifier_descriptor: Used to associate the component_tag with the elementary stream.

data_broadcast_id_descriptor: Used to associate the INT subtable, that is, platform_id, with the elementary stream. Furthermore, through this descriptor the receiver is able to check the version information of the associated INT subtable.

The PAT and PMT are used as a chain, where the PAT maps each listed service_id with the PMT subtable. The PMT subtables map each service_id with the PID value of the elementary stream. Finally, each elementary stream may comprise one or more DVB-H service or INT subtables. Figure 14.6 illustrates the mapping of INT through the NIT, PAT, and PMT.

Each transport stream contains one INT subtable for each platform. Moreover, each INT provides the mapping for the location of IP streams of the corresponding platform, within transport streams available in the current and adjacent cells (see figure 14.7).

As seen in figure 14.7, the mapping between IP streams and PIDs of the associated elementary streams is almost similar to that of mapping in the case of the INT. The only difference between the mappings of the INT and DVB-H service is the use of the component_tag. In the case when PID values of multiple elementary streams are announced within one PMT, the component_tag is used for distinguishing these from each other. The DVB-H-specific descriptors used within the INT are as follows:

target_IPv4_descriptor: In general, used to announce IP addresses of the IP streams within the INT. This descriptor is the simplest form of target descriptors within the INT, which announces only the destination IP address of the associated IP streams. A separate descriptor is used in the case of IPv4 and IPv6.

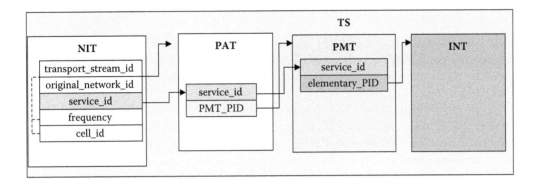

Figure 14.6 The mapping of INT through NIT, PAT, and PMT.

target_IPvx_slash_descriptor: Announces the destination IP addresses of the associated IP streams. In addition, it announces slash masks for each announced IP address. A slash mask is an 8-bit field that indicates the IP address mask in short form notation. A separate descriptor is used in the case of IPv4 and IPv6.

target_IPvx_source_slash_descriptor: Similar to that of the target_IPvx_slash_descriptor, but in addition to destination addresses and corresponding slash masks, it also announces source addresses and their slash masks. A separate descriptor is used in the case of IPv4 and IPv6.

IP/MAC_stream_location_descriptor: Associates the IP addresses announced with one or more target descriptors described above. As a result of the association between this descriptor and the target descriptors, each IP stream will be mapped to the one or more transport streams available within one or more networks. Moreover, for each IP stream, a mapping is provided with parameters enabling access to the elementary streams through the PAT and PMT.

time_slice_fec_identifier_descriptor: See previous section describing NIT.

Finally, the TDT is used to signal the UTC time of the associated network.

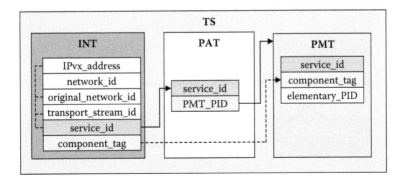

Figure 14.7 The mapping of IP streams.

14.1.2.2 Robustness of PSI/SI

Studies[6,7] have shown that the DVB-H networks are able to provide robust PSI/SI transmission when network configuration and receiver implementation have been carefully considered. The studies have shown that the selected configuration may significantly influence the robustness level of the transmission.

The impact of robustness within PSI/SI transmission is different than that of the transmission of services. PSI/SI does not have a FEC method, such as MPE-FEC, for improving robustness in the transmission of services. Instead, PSI/SI has a cyclic redundancy check (CRC), which can be used for detecting errors but not for the error correction, that is, improving robustness. In the case of PSI/SI, the impact of the robustness level is not on the consumption of the service. Instead, the robustness level has a direct impact on receiver latency. Hence, in the both cases, robustness level has an impact on the end-user experience, but in a slightly different way.

The main reason for the increase in receiver latency, due to poor robustness within PSI/SI transmission, is in the structure of PSI/SI signaling. Signaling information within PSI/SI is usually scattered into multiple subtables, each of which may be divided further into multiple sections. By default, the reception of all needed tables may take a long time, especially if the maximum repetition intervals are used. In cases where some of the sections are corrupted, the receiver needs to wait until the next transmission to receive the missing sections, because it has no means to repair the corrupted section (see figure 14.8).

In [6] and [7] it was discovered that the robustness of PSI/SI transmission can be improved, and hence receiver latency reduced, by using the correct combination of section sizes and repetition intervals. It was also discovered that receiver latency can be further decreased by *intelligent* receiver implementation.

The optimal size for the section depends on the amount of needed signaling, and hence on the resulting subtable size. There is no single section size that is optimal for all configurations. However, based on the calculations presented within the studies, it was clear that the standard maximum section sizes recommended, for example, for INT, 4,096 bytes, should not be used. Generally, small sections are less prone to corruption than larger ones. Each configuration should be inspected case sensitively.

The studies also revealed that the recommended maximum repetition intervals are not reasonable for all network configurations. For example, the maximum repetition interval for INT, 30 s,

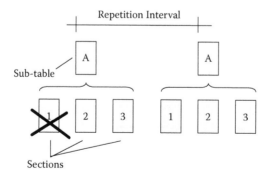

Figure 14.8   The reception of a PSI/SI subtable in a situation where one of the sections is corrupted within the first reception.

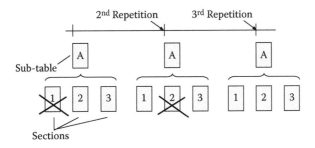

Figure 14.9   An example of the transmission of a PSI/SI subtable where different sections of the subtable are corrupted within the first and second repetitions, and finally uncorrupted within the third repetition.

could be easily decreased without significant influence on the total network capacity. The network capacity still remained under 1 percent within all inspected configurations, even when the total subtable size was increased due to smaller sections.

Finally, it was discovered that receiver implementation also influences latency time, in cases where PSI/SI sections are received in randomized order. The latter situation is possible, especially when one or more sections are corrupted within each repetition of a subtable and the receiver needs to collect different sections from different repetitions of a subtable. Such intelligent receiver implementation should be encouraged rather than settling for an implementation in which the receiver is able to receive the entire subtable only when all sections are uncorrupted within a single transmission. Figure 14.9 illustrates an example of the PSI/SI transmission where an intelligent receiver is able to receive the entire subtable already within the second repetition of the subtable. A less intelligent receiver would need to wait until the third repetition of the subtable, where all sections of the subtable are uncorrupted.

14.2   Time Slicing

The main purpose of time slicing is to enable the receiver to switch the power off when services are not received. Hence, power consumption of the receiver can be decreased up to 90 percent. Furthermore, another advantage of time slicing comes forward in the implementation of the handover. In Väre and Puputti,[3] an algorithm for utilizing off-periods for seamless handover due to time slicing was introduced. Figure 14.10 illustrates the time-slicing principle, in which the receiver may perform handover activities, for example, investigate signals available in the adjacent cells during off-periods.

14.3   Service Discovery

Service discovery in DVB-H means resolving the mapping of IP addresses to the logical channels carried within signals available in different cells and networks. Service discovery is needed to enable the consumption of the selected service within the current location and to maintain the service consumption when the receiver switches to another signal, that is, performs a handover.

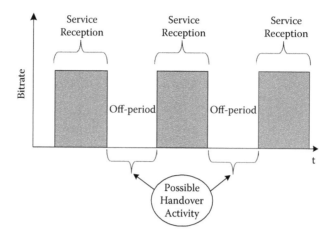

Figure 14.10 Time-slicing and handover principle.

Such a seamless handover is possible only if signaling is provided for the services available within the neighboring cells. Moreover, the receiver needs to be capable of collecting the needed signaling to be able to make a seamless handover.

In the following sections, the basic principles of service discovery within DVB-H are explained. First, a high-level description of the signal scan procedure is given in section 14.3.1. A more detailed explanation of the PSI/SI parameter discovery procedure is given in section 14.3.2.

14.3.1 Signal Scan

Signal scan is mostly a physical layer procedure that needs to be performed only rarely. At a minimum, it should be performed once, to obtain information of all the signals available within the current location. However, within DVB-H, it is not mandated to transmit NIT_ other subtables describing the contents of the other networks within the current network; that is, when the receiver is moving into a new cell, there is always the possibility that a new, previously unfound network is available. Hence, signal scan is the only method to fully guarantee that all possible signals are available after the receiver enters a new cell. The fundamentals for the signal scan within DVB-H were first introduced within Väre and Puputti.[3] The purpose of signal scan is to collect information of the frequencies available within the current location and within the found networks. The signal scan procedure utilizes receiver RF functionality, physical layer signaling, and PSI/SI signaling. The following steps of the signal scan dataflow are described in figure 14.11.

> **Step 1:** In this step, the receiver attempts to tune to the given frequency. The used frequency range may be implementation specific. However, the exhaustive frequency range covers frequencies from 474 to 858 MHz, with the offset determined by the used bandwidth within the area. The possible bandwidth used within DVB-H may be 5, 6, 7, or 8 MHz.
>
> **Step 2:** TPS lock is achieved prior to the complete signal lock. Through the inspection of DVB-H bit value within the TPS, the receiver is able to quickly drop such signals

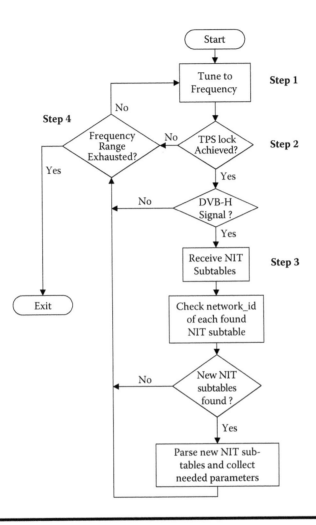

Figure 14.11   The generic dataflow for signal scan.

that do not carry DVB-H services. That is, if the DVB-H signaling bit is set to 0, the originating signal carries at least one DVB-H service.

Step 3: In case the DVB-H signal is found, the receiver tunes to it and starts to receive the NIT_actual and NIT_other subtables to collect information on the signals available within the current location and elsewhere within the network. The amount of stored information at the time may be considered, for example, on the basis of available memory capacity. To ensure the minimum functionality for a seamless handover, the receiver should collect information on the current signal and on those adjacent and neighboring signals that carry currently consumed services. The receiver may skip such NIT subtables that it has already accessed and which are unchanged. All found NIT subtables with a previously unknown network_id or version_number are considered new and hence processed.

Step 4: Once the defined frequency range has been exhausted, the signal scan procedure is completed.

14.3.2 PSI/SI Parameter Discovery

Through the PSI/SI, the receiver is able to maintain up-to-date service discovery information of the desired services available within current and neighboring cells. In most cases, more than one PSI/SI table needs to be sought before the requested information can be discovered. Usually most of the information carried within the PSI/SI is a combination of two or more PSI/SI parameters. In what follows, the relation of parameters carried within different PSI/SI tables is described by means of the three procedures that form the core part of the DVB-H service discovery. These three procedures are INT access discovery, network discovery, and INT discovery. To highlight the mutual connection of each parameter and to be able to make an unambiguous interpretation of the procedures, exemplary parameter values are used within each procedure. It is assumed that the receiver is tuned to the frequency of 498 MHz, which is covering the area of the cell identified with cell_id = 0x0010.

14.3.2.1 INT Access Discovery

INT access discovery is needed to discover the access parameters for the INT of the requested platform. The INT discovery procedure starts from the first loop of NIT, where one linkage descriptor is provided for each transport stream of the network carrying INT subtables. More than one INT subtable, that is, IP platform, may be announced within a single linkage_descriptor. Figure 14.12 illustrates an example of parameter mapping in the INT discovery procedure where an IP platform with platform_id = 0x000001 is associated with the INT subtable carried within the elementary stream with a PID value of 0x0025.

The INT access discovery illustrated in figure 14.12 is described in three steps:

1. The program_number with the same value as that of the service_id announced within the linkage_descriptor and associated with the platform_id 0x000001, that is, service_id = 0x0015, is sought from the PAT.
2. The PAT associates program_number = 0x0015 with program_map_PID = 0x0022. Hence, the PMT subtable identified with program_number = 0x0015 is sought from the elementary stream with the PID value of 0x0022.
3. The PMT associates elementary_PID = 0x0025 with platform_id = 0x000001. Hence, the INT subtable identified with platform_id = 0x000001 is sought from the elementary stream with the PID value = 0x0025.

14.3.2.2 Network Discovery

Network discovery consists of the inspection of NIT subtables that enable the discovery of the parameters needed to access the physical frequencies, that is, signals. The NIT associates these signals with geographical coverage areas, that is, cells, and with transport streams. Other information that may be provided by the NIT includes network name and time-slicing and MPE-FEC-specific parameters. By default, each signal carrying DVB-H services always carries NIT_actual. However, depending on the network operator, NIT_other may also be available. The receiver may access one or more of the available NIT subtables. Figure 14.13 illustrates an example of the parameter mapping in the network discovery procedure, where two cells are associated with two different transport streams.

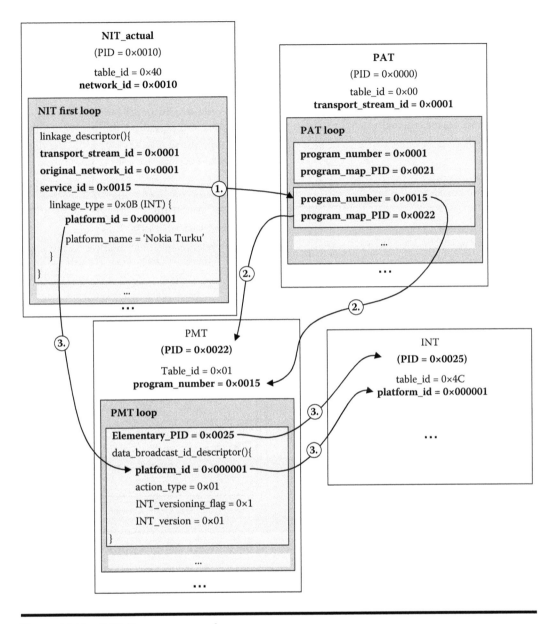

Figure 14.12   INT discovery procedure.

The network discovery procedure according to figure 14.13 is as follows:

1. All information announced within this subtable is associated with network_id = 0x0010, which is associated with the network name NOKIA.
2. The cell_list_descriptors announce cells with cell_id = 0x0010 and 0x0011. Furthermore, the cell with cell_id 0x0010 is associated with the subcell identified by cell_id_extension = 0x01. The cell coverage area of the latter two cells and one subcell is as illustrated in figure 14.14.

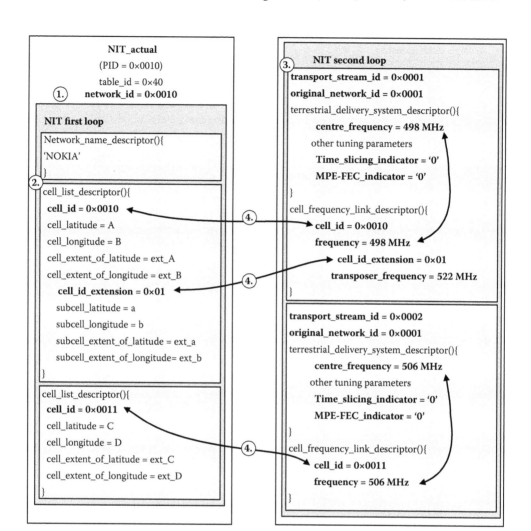

Figure 14.13 Parameter mapping and four steps of the network discovery procedure.

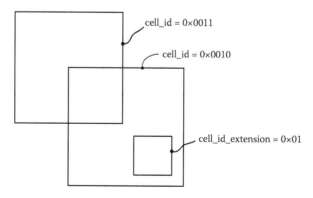

Figure 14.14 Cell coverage areas of the cells and one subcell announced within the NIT according to the network discovery procedure.

3. The second loop of the NIT associates transport streams with cells, subcells, frequencies, and other tuning parameters. First, the transport stream identified with transport_stream_id = 0x0001 and original_network_id = 0x0001 is associated with the frequency of 498 MHz and other tuning parameters carried within terrestrial_delivery_system_descriptor. Next, the same transport stream is further associated with cell_id = 0x0010 by the cell_frequency_link_descriptor. Also, the subcell with cell_id_extension = 0x01 and transposer_frequency = 522 MHz is associated with the given transport stream and cell.

   Another transport stream identified with transport_stream_id = 0x0002 and original_network_id = 0x0001 is associated, by following a principle similar to the first-mentioned transport stream, with the frequency of 506 MHz, other tuning parameters, and cell_id = 0x0011.

   Also, for both transport streams, the time_slicing_indicator and MPE-FEC_indicator are set to 0, which indicate that the associated transport streams carry at least one DVB-H service that has MPE-FEC support.

4. As the cell_list_descriptor defines the coverage areas according to figure 14.14, the receiver is able to deduce that the cell with cell_id = 0x0011, the cell with cell_id = 0x0010, and the subcell with cell_id_extension = 0x01 are all neighboring, and hence also potential handover candidate, regardless of the current location of the receiver. Based on the mapping with the information acquired from the NIT second loop, the receiver has complete tuning information for these potential candidates.

14.3.2.3 IP Service Discovery

IP service discovery is a procedure that maps IP addresses to the elementary streams. The mapping procedure is partially similar to that of INT access discovery, where the IP platform is associated with the particular elementary stream carrying the corresponding INT subtable. Due to handover support, the mapping is needed for the services available in the transport stream of the current cell and also for the services that are available in the transport streams located within neighboring cells. Figure 14.15 illustrates seven steps of the IP service discovery in the case where four IP streams are associated with two transport streams.

The network discovery procedure is as follows:

1. The table_id and platform_id within the INT subtable header identify the INT subtable.
2. The first loop of the INT carries the platform_name descriptor, which indicates the name of the platform associated with platform_id = 0x000001. In addition, the first loop also carries the time_slice_fec_identifier_descriptor, which associates all IP streams announced within this table with the time-slicing and MPE-FEC-specific parameters. First, time slicing and MPE-FEC indicators are set to the value 0x1, which indicates that all IP streams announced within this table are DVB-H services that also support MPE-FEC. The other parameters associated with the descriptor are frame_size, max_burst_duration, max_average_rate, and time_slice_fec_id. The latter four parameters provide information on time slicing and MPE_FEC and may be used by the receiver.
3. The second loop of the INT associates the addresses of the target IP streams with the PSI/SI parameters, which further associate the IP streams with the elementary streams carried within one or more transport steams. It is mandatory for each INT subtable to associate IP streams with the current transport stream and with the transport streams available in the neighboring cells. Moreover, the time_slice_fec_identifier_descriptor is also located within the beginning of the

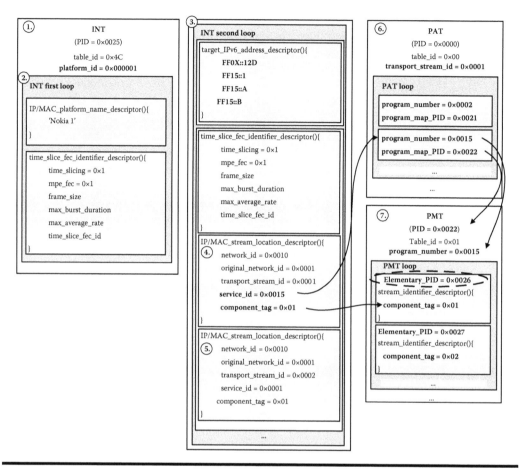

Figure 14.15 The parameter mapping and steps of the IP service discovery procedure where four IP streams are associated with two transport streams.

loop, overriding the information associated with the same descriptor previously. The parameters are valid until the next occurrence of the same descriptor.

4. The first IP/MAC_stream_location_descriptor announces the parameters of the transport stream identified with original_network_id = 0x0001 and transport_stream_id = 0x0001. The network discovery procedure in figure 14.13 shows that the associated transport stream is the transport stream carried with the signal that the receiver has currently tuned. As well, the same transport stream is available within a subcell of the current cell, that is, on the frequency 522 MHz.

5. The second IP/MAC_stream_location_descriptor announces parameters of the transport stream identified with original_network_id = 0x0001 and transport_stream_id = 0x0002. According to the parameter mapping in the network discovery procedure (see figure 14.13), the transport stream associated within this descriptor is available within a neighboring cell. With the service_id and component_tag, the receiver may discover a corresponding elementary stream through the PAT and PMT available within the neighboring cell, similarly as was done in the current transport stream in step 4.

6. The program_number corresponding with service_id = 0x0015 is sought from the PAT, which associates it with program_map_PID = 0x0022.
7. The PMT subtable identified with program_number = 0x0015 is sought from the elementary stream with a PID value of 0x0022. Finally, the elementary stream carrying the associated four IP streams is found based on the component_tag associated with the service_id in step 4. The receiver can now access the four IP streams announced within the INT by filtering the elementary stream with a PID value of 0x0026.

14.4 Network Topology Scenarios

The DVB-H system allows several different network topologies, which need to be taken into account in the receiver implementation, and particularly in the design of the handover algorithm. The network topology may be a composition of single transmitter cells forming a multifrequency network (MFN) and may be composed of one or more single-frequency network (SFN) areas. Finally, the network topology may consist of only a single SFN area, where handover is not needed. The first two cases are, from the handover algorithm point of view, identical in the sense that the network consists of one or more cells with different frequencies. However, there are some physical characteristics in the SFN topology that may also need to be considered within the receiver implementation. These characteristics relate to the selection of guard interval length and the sizing of SFN areas. The echoes limit the maximum size of the SFN area. The echoes must arrive to the receiver within the guard interval period. Otherwise, the echo power is destructive, not constructive. If the echo arrives after the guard interval, the echo level should be low, in theory less than the required contrast-to-noise (C/N) level for the DVB-H mode (assuming a simple receiver). The maximum possible SFN area sizes are presented in table 14.4.

A longer guard interval leads naturally to a bigger SFN area, but the negative effect is a lower bit rate. If a larger SFN area compared to the guard interval length is needed, the first transmitter (B) outside the SFN site area must be far enough away compared to the serving transmitter (A). The signal from (B) must attenuate (free space loss) below the required C/N for the mode. With very low power transmitters this is possible, but it leads to very small SFN cells and a very high number of transmitters/gap fillers. In practice, this kind of network is not used.

Table 14.4 Maximum SFN Sizes

FFT	Guard Interval μs		SFN Site Area Size km
8K	1/4	219	65.7
	1/8	107	32.1
	1/16	51	15.3
	1/32	23	6.9
2K	1/4	51	15.3
	1/8	23	6.9
	1/16	9	2.7
	1/32	2	0.6
4K	1/4	107	32.1
	1/8	51	15.3
	1/16	23	6.9
	1/32	9	2.7

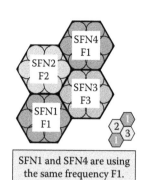

Figure 14.16 SFN area reuse where one allotment per SFN is used.

In practice normal cell reuse patterns are used for SFN areas. One SFN area consists of several transmitters using the same frequency. Frequency planners call this allotment planning, which means that there is one SFN per allotment. Figure 14.16 illustrates an example of SFN area reuse, where SFN1 and SFN2 are using the same frequency.

The following sections describe examples of network topologies that can be considered within the design of the handover algorithm.

14.4.1 Networks in Major Cities

In cases where DVB-H networks are limited to the major cities and passage routes between them, such as motorways and railways, the DVB-H networks will be quite "spotty." This means that the resulting topology does not have uniform coverage. Figure 14.17 describes an example of such a nonuniform SFN network, where two main cities and one town are covered only by a DVB-H network, but the passage route between them does not have coverage.

The networks inside cities are based on SFN areas with some reuse pattern. In practice, inside a city the alternative frequencies adjacent to the current cell are limited to a maximum of our to six, assuming one service operator. If one operator has several multiplexes, the number of frequencies might be double, assuming that the service is in both multiplexes. However, this is improbable because covering even a large city usually takes no more than three frequencies.

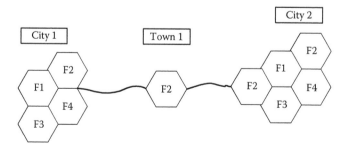

Figure 14.17 DVB-H networks in major cities.

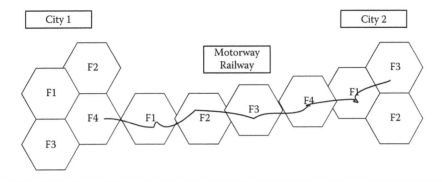

Figure 14.18   Network covering motorway.

The practical SFN area size with 4K modulation is roughly 25 km, while with 8K modulation the SFN area could be up to 50 km. The important point to note is that many areas between the cities do not have DVB-H coverage at all. This would happen, for example, in a journey from city 1 to city 2. Reception during this time is naturally not possible. The handover algorithm must be able to cope with the nonreception areas.

14.4.2 Network Covering Main Highway and Railway

In some cases a motorway between two major cities may be covered by several transmission sites. This kind of scenario allows uninterrupted service consumption even when moving between two cities. Figure 14.18 illustrates an example of such a scenario where uninterrupted reception is possible during a journey from city 1 to city 2.

On high-speed motorways, 2K or 4K mode could be used to achieve greater speeds and the practical SFN area sizes are roughly 10 km at the minimum. Assuming that the car is moving at 120 km/h, it passes through a cell with a 5 km radius in 2.5 min. The latter could be considered the shortest interval for handover. Considering that delta-t cycle time would be 3 s, a handover would occur after a reception of roughly 50 time-slicing frames.

In motorway network planning, the cell sizes are extended as much as possible, for example, by using directional antennas. Because of this, in practice the SFN area can be made much bigger than one having a 5 km radius.

14.4.3 SFN Gap Fillers and Transposers

DVB-H networks have two types of amplifiers, which are used to amplify the signals transmitted by the main transmitters. These are called gap fillers and transposers. Gap fillers amplify the transmission of the main transmitters using the same frequency, and hence they do not affect the handover procedure. Transposers, in turn, amplify the signal transmitted by the main transmitters at a different frequency. Transposers have an impact on the handover, because when the receiver moves to the area covered by the transposer signal from the area of the main transmitter signal, it needs to perform a handover. This is also considered within the signaling, where each area covered by the transposer signal is identified as a subcell. A network using RF distribution is an example of the usage of transposers. The idea in a network using RF distribution is to have one high-power central transmitter. This transmitter feeds medium-power transposers, which operate within

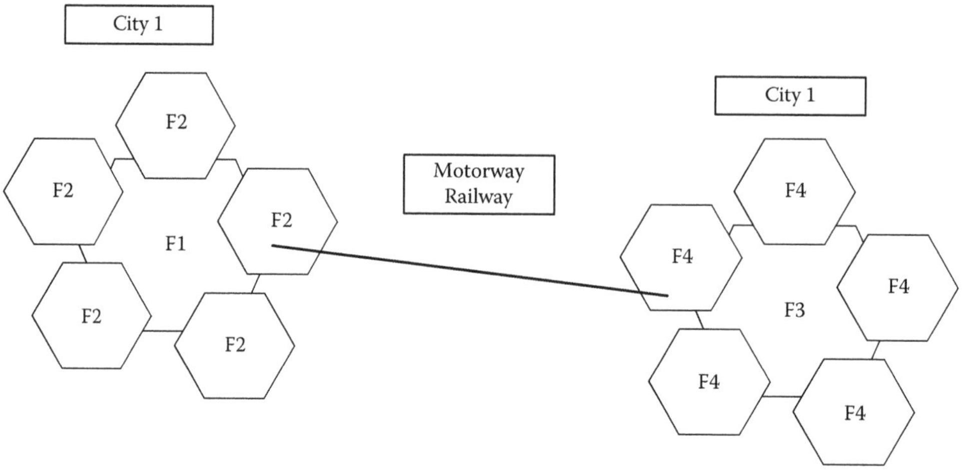

Figure 14.19 Network using RF distribution.

different frequencies (see figure 14.19). The transposers receive the signal using high-gain antennas from the high-power transmitter. The SFN area therefore consists of two frequencies, which are then reused normally in the network planning.

In the handover, this network topology has a special effect. The handover also needs to be done inside one SFN area between the two frequencies (main transmitter and gap fillers). The cell sizes are similar to those of the normal SFN network.

14.4.4 SFN Area with Holes

The practical SFN areas do not have guaranteed coverage for all the areas. The SFN area might have a coverage hole somewhere within the cell. Figure 14.20 illustrates an example of an SFN area where a hole is located in the middle. Because there is no signal coverage within the hole, it needs to be taken into account in the handover design.

14.5 Seamless Handover Algorithm

Implementation of a receiver that is capable of a seamless handover in DVB-H is relatively straightforward from the signaling point of view, and especially if the reception conditions are optimal. However, the physical layer characteristics and variety of different network topologies also need

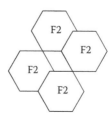

Figure 14.20   SFN area with hole in the middle of coverage area.

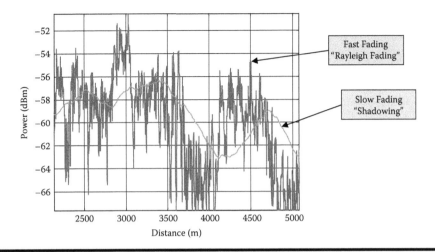

Figure 14.21 The difference between Rayleigh fading and shadowing.

be considered in the design of a handover algorithm. The following sections describe the design principles and finally the step-by-step implementation of a handover algorithm within DVB-H, which enables a seamless handover.

14.5.1 Design Principles and Preconditions

In what follows, the design principles and preconditions assumed for the receiver and network implementation are defined. The following principles and preconditions are taken into account in the design of the handover algorithm.

Fast fading and averaging: The signal strength change over the distance traveled is called fading. It consists of fast (Rayleigh) and slow (shadowing) fading. Rayleigh fading changes very rapidly, but the shadowing change speed is much slower. Fast fading is caused by the varying additive and subtractive multipath signal components as the mobile terminal moves. The distance between two successive fast fading maxima or minima is typically in the order of a half-wavelength. Slow fading is caused by the dynamic changes of propagation paths: new paths arise and old paths disappear as the mobile terminal moves.

Rayleigh fading should be averaged out from the handover signal strength measurements, because it would produce totally random signal strength values for the decisions. The algorithm should follow long-term signal strength changes, not instantaneous changes. This can easily be done, for example, with moving-average calculation or with digital filtering for the signal strength value. Figure 14.21 shows the difference between Rayleigh fading and shadowing. The shadowing curve is constructed by calculating the moving average over several samples.

Adaptive interval of measurements: This means that intervals of the signal measurements are adapted accordingly to the used signal margin and candidate order. The signal margin is determined based on the signal strength of the received signal. The candidate order, in turn, is determined based on the probability and quality.

Fallback mode. The algorithm has an exception handling when the signal is lost for some reason.

Hysteresis. This needs to be defined to avoid a ping-pong effect, which results in unnecessary handovers.

Service discovery. The receiver needs to have up-to-date information on the services within current and neighboring cells to be able to support seamless handover.

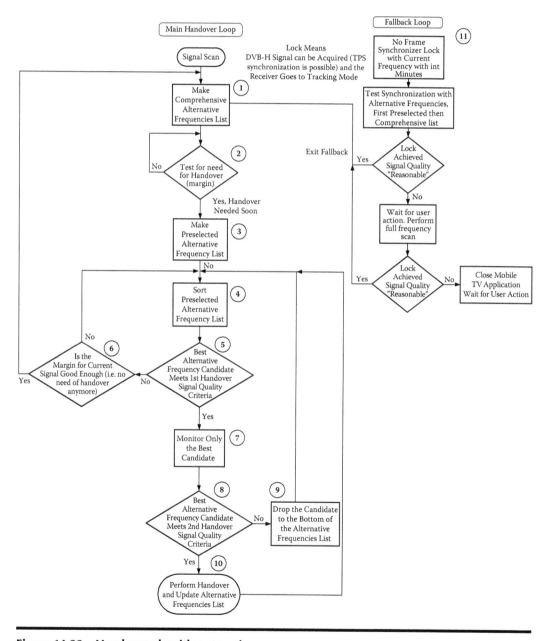

Figure 14.22 Handover algorithm, top view.

14.5.2 Top View

A top view of the handover algorithm is presented in figure 14.22. The handover algorithm consists of two main parts: main handover loop and fallback loop. The main handover loop should be able to cope with most of the practical network situations. However, making a foolproof handover algorithm is very difficult; therefore, a fallback loop has to be implemented. The fallback loop is able to recover the reception in case something goes wrong in the main handover loop. It is also able to cope with various special cases, like entering a tunnel.

The flowchart presents the main states in the algorithm. All the decision points and branching points are presented. The following sections will include a detailed explanation of the various states. The algorithm states are numbered in the top view, and the same numbers will be used in the detailed explanations. In the detailed descriptions, the top-level algorithm state "box" is copied for clarity. After that, the more detailed contents of the state box are presented.

14.5.3 Algorithm State 1

Algorithm state 1 is depicted in figure 14.23. The idea is to make a comprehensive alternative frequencies list. The list can be based on a combination of several parameters, like: contains the wanted service, geographical place, contains similar service, and so on. The most common method is to select only the cells that are adjacent and contain the currently received service. In this case, the handover has priority for one service, which is followed.

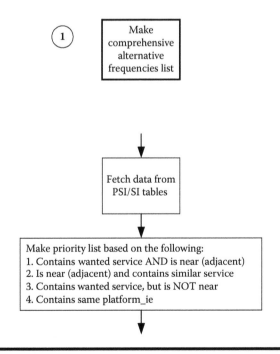

Figure 14.23   Algorithm state 1.

First, that the alternative frequency cell is geographically nearby is ascertained. This check is needed to limit the number of possible alternative cells. The cell_frequency_link_descriptor announces *all* the available alternative frequencies (in all networks if NIT_other is used).

The alternative frequencies are signaled in the cell_frequency_link_descriptor. The geographical position can be calculated based on the information presented in the cell_list_descriptor. The currently received cell can be identified by the cell_id, which is transmitted in the DVB-H TPS bits.

The cell_latitude and cell_longitude parameters in the cell_list_descriptor describe the southwest corner of the cell. Southern latitudes and western longitudes are negative numbers. The numbers are coded as two's complement. For example, the cell_longitude and cell_latitude parameters are calculated with the following equations from the real longitude and latitude values (in degrees):

$$cell\_longitude = \frac{longitude}{180} \cdot 2^{15} \qquad (14.1)$$

$$cell\_latitude = \frac{latitude}{90} \cdot 2^{15}. \qquad (14.2)$$

The cell size is defined by cell_extent_of_latitude and cell_extent_of_longitude.

At the equator, the actual "length of degree" can be calculated from the following equation:

$$length = \frac{deg}{360} \cdot 2\pi \cdot 6371000. \qquad (14.3)$$

The cell_extent_of_latitude and cell_extent_of_longitude parameters can then be calculated from

$$extent\_of\_latitude = \frac{deg}{90} \cdot 2^{15} \qquad (14.4)$$

$$extent\_of\_longitude = \frac{deg}{180} \cdot 2^{15}, \qquad (14.5)$$

where deg means the length of the cell in degrees. Usually the extent can be observed directly from a network coverage map. The operator must enter this data correctly into the SI database.

The distance between the center of two cells can be approximated in the following way:

■ The coordinate parameters cell_longitude, cell_latitude, extend_of_latitude, and extent_of_longitude are converted to degrees with equations 1, 2, 3, and 4 (solve longitude and latitude). Remember that two's complement notation is used for the parameters.
■ The center of the first cell is calculated by adding half of the extent to the original southwest (bottom left) coordinate in degrees:

$$X_{1,center} = X_1 + \frac{X_{1,extent\_of\_longitude}}{2} \qquad Y_{1,center} = Y_1 + \frac{Y_{1,extent\_of\_latitude}}{2}. \qquad (14.6)$$

- The center of the second cell can be calculated the same way.
- The formula for the distance between the two center points of the cells is ($X_{1,center} = x_1$, $Y_{1,center} = y_1$, $r = 6{,}371{,}000$ m):

$$z = \arccos(\cos(x_1)\cos(y_1)\cos(x_2)\cos(y_2) + \cos(x_1)\sin(y_1)\cos(x_2)\sin(y_2)$$

$$+ \sin(x_1)\sin(x_2)) \cdot \frac{1}{360} \cdot 2 \cdot \pi \cdot r.$$

(14.7)

where x_1, x_2, y_1, and y_2, in degrees, and r = 6,371,000 m, west longitude and south latitude, are negative values.

- In case the above formula cannot be used, the following (brutal) approximation can be used for calculating the distance (in degrees):

$$Z = \sqrt{(X_{1,center} - X_{2,center})^2 + (Y_{1,center} - Y_{2,center})^2}$$

(14.8)

- The degrees can be converted to meters using the following approximation:

$$length = \frac{deg}{360} \cdot 2\pi \cdot 6371000$$

(14.9)

The distances from the current cell to the alternative cells (center points) should be calculated. Cells where the distance is bigger than, for example, 150 km can be dropped from the candidate list.

The list should be sorted so that the most interesting signals are in the top priority. The obvious way for the sorting is to put signals that contain the currently received service to top priority. The sorting based on signal quality will be made later on.

14.5.4 Algorithm State 2

Algorithm state 2 is depicted in figure 14.24. The RSSI is averaged over several measurements. The averaging is needed to filter out Rayleigh fading. Only the longer-term variation is interesting for the handover. The averaging can be done with a simple first-order digital low-pass filter. The filtering formula is given below:

$$F_{average} = \frac{F_{new} + F_{old} \cdot (n-1)}{n}$$

(14.10)

$$F_{old} = F_{average}$$

Figure 14.25 depicts the practical field/signal strength measurement. In the figure, the signal strength value without filtering and with filtering is presented. A good initial value for the filtering depth is n = 10. The depth should be a programmable parameter, and it should finally be selected based on field trials.

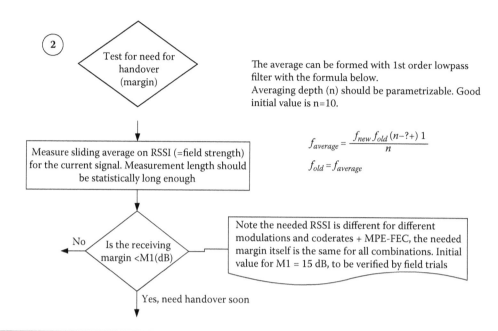

The average can be formed with 1st order lowpass filter with the formula below.
Averaging depth (n) should be parametrizable. Good initial value is n=10.

$$f_{average} = \frac{f_{new} \, f_{old} \, (n-?+) \, 1}{n}$$

$$f_{old} = f_{average}$$

Note the needed RSSI is different for different modulations and coderates + MPE-FEC, the needed margin itself is the same for all combinations. Initial value for M1 = 15 dB, to be verified by field trials

Figure 14.24 Algorithm state 2.

The measurement interval should be a programmable parameter. The interval should be set by a time-slicing period or internal timer. A good initial value for the maximum measurement interval is 2 s. If the time-slicing period is longer than 2 s, the timer should set the interval; otherwise, the delta-t cycle time can be used. It must be ensured in the implementation that the measurement is not done during the normal reception time. The optimum timing for the measurements is directly after (or before) the received signal. This method saves the time overhead for powering the RF up for handover measurements.

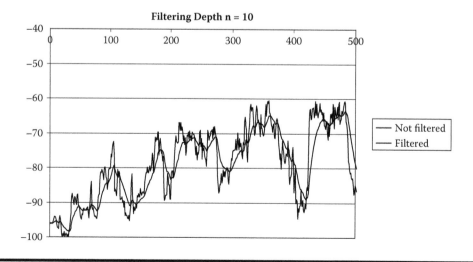

Figure 14.25 Signal strength filtering.

Table 14.5   Sensitivity for Different DVB-H Modes

Modulation	Code Rate	C/N for QEF	Sensitivity
QPSK	1/2	5.6	−93.6
QPSK	2/3	7.4	−91.8
QPSK	3/4	8.4	−90.8
16QAM	1/2	11.3	−87.9
16QAM	2/3	13.7	−85.5
16QAM	3/4	15.1	−84.1
64QAM	1/2	17	−82.2
64QAM	2/3	19.2	−80.0
64QAM	3/4	20.8	−78.4

The need for handover is tested. The receiving margin is calculated with the following formula:

$$M1[dB] = RSS1[dBm] - Sensitivity[dBm] \qquad (14.11)$$

In the formula, RSSI is the received signal strength in dBm, and sensitivity is the required signal strength for QEF reception. The sensitivity is different for different DVB-H modes. Sensitivity is tabulated in Table 14.5.

A good initial value for the margin threshold M1 is 15 dB. Margin M1 should be a programmable parameter.

14.5.5   Algorithm State 3

Algorithm state 3 is depicted in figure 14.26. The signals are sorted by signal/field strength. For the signal strength measurement, the same averaging method as in state 2 is used (digital filtering with n = 10). Averaging over ten measurements should be sufficient to obtain reliable signal strength values. With a measurement interval of 2 s, this takes 20 s. The study showed that in the worst case

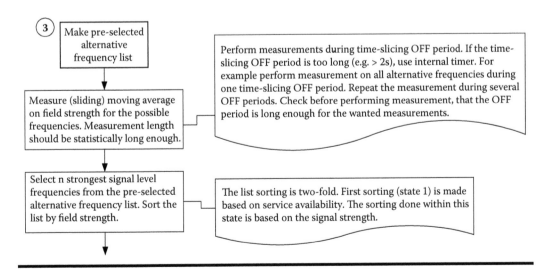

Figure 14.26   Algorithm state 3.

the time available for the handover is 150 s (motorway with 120 km/h, half of the cell radius assumed for the handover area). A 20 s measurement time should therefore be on the safe side.

N strongest signals are left to the preselected alternative frequency list. A good initial value for the N is 6.

14.5.6 Algorithm State 4

Algorithm state 4 is depicted in figure 14.27. This state is entered after the preselected and once-sorted alternative frequency list is made in the previous state.

In this state the monitoring and list sorting is continued until a good-enough alternative frequency is found. The monitoring period can be adaptive. The monitoring frequency is proportional to both the receiving margin and alternative frequency signal strength. The idea is that while we still have quite a lot of margin over the sensitivity limit, the monitoring is less frequent to save power. When the margin drops more and the need for handover becomes evident, the measurement frequency is increased. The second parameter affecting the monitoring frequency is signal strength. The strongest alternative frequency is the most probable, and therefore it is monitored most frequently. The second strongest is monitored less frequently, and so on.

Good initial values for the monitoring interval parameters are N = 1, M = 2, and L = 3. The shortest monitoring period initial value could be 2 s. Therefore, the most frequent measurement frequency is ½*1 = 0.5 Hz. The second is ½*1/2 = 0.25 Hz, and the third is ½*1/3 = 0.167 Hz. The initial value for K could be K = 1 if the margin is <10 dB, K = 2 if the margin is <15 dB, and K = 3 if the margin is >15 dB.

The averaging is done similarly as earlier. Five to ten measurements should be collected before sorting the list, which means 10 to 20 s measurement time (when K = 1 and the period is 2 s). Accordingly to state 3, the measurement time is 20 + 10 = 30 s. This should still be fast enough

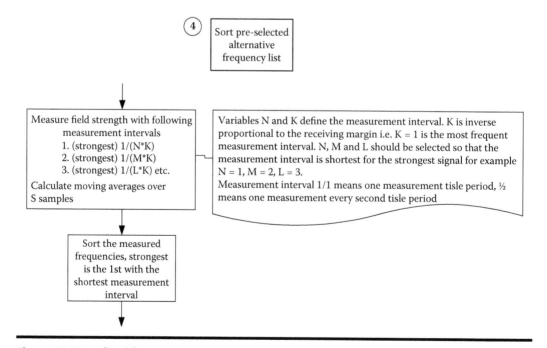

Figure 14.27 Algorithm state 6.

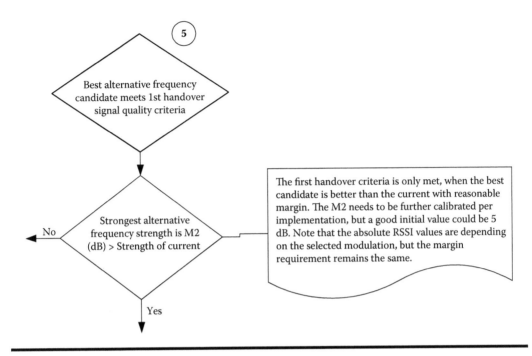

Figure 14.28 Algorithm state 5.

for the worst-case handover speed. In the motorway case there should be only two alternative frequencies.

The strength of the current signal is monitored in the same way as the alternative frequencies. The strength of the current can be naturally monitored during every time slice, because it can be done at the same time as the normal reception.

After stable signal strength values have been calculated, the list can be sorted. The strongest alternative frequency is made the first priority, and so on.

14.5.7 Algorithm State 5

Algorithm state 5 is depicted in figure 14.28. The signal strength of the strongest alternative frequency is evaluated. The actual frequency change procedure (i.e., handover) is only started when the strength of the alternative frequency is M2 dB stronger than the current. The handover is therefore only made when there is a better signal available. This way, unnecessary handovers are avoided and hysteresis is built into the system. A good initial value for the M2 threshold is 5 dB. This gives 10 dB hysteresis, which should be enough, together with the Rayleigh fading filtering.

14.5.8 Algorithm State 6

Algorithm state 6 is depicted in figure 14.29. The handover algorithm exit point is in this state. The exit can happen if the strength of the current goes back to good enough even before changing the frequency. Some additional margin M4 is needed to avoid reentering the handover process

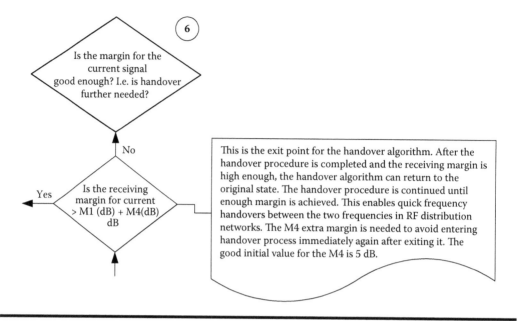

Figure 14.29   Algorithm state 6.

immediately after completing it. The M4 initial value is 5 dB. The "normal" exit happens when the old current and best alternative frequency have been changed and the margin of the new current is big enough.

14.5.9 Algorithm State 7

Algorithm state 7 is depicted in figure 14.30. Now the final alternative frequency candidate is selected. Only this frequency (+ current frequency) is monitored. So far, the only parameter that has been monitored has been receiving the margin (RSSI). Now the TS–packet error rate (PER) is also monitored. The idea is to ensure that the final candidate quality is good before switching the frequency. A minimum of three measurements (initial value) should be done (using the

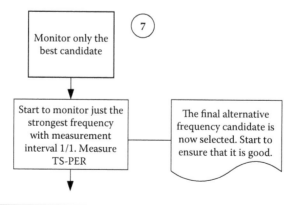

Figure 14.30   Algorithm state 7.

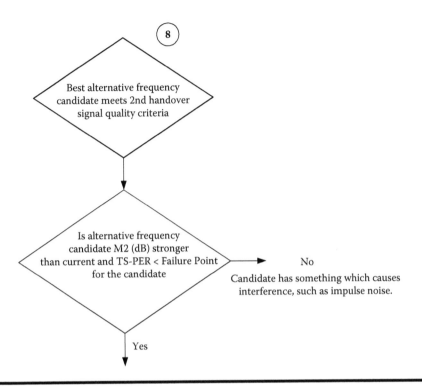

Figure 14.31   Algorithm state 8.

averaging method presented earlier) before going to the next state. The TS-PER measurement can be a simple average over a minimum of three samples.

14.5.10   *Algorithm State 8*

Algorithm state 8 is depicted in figure 14.31. In this state, the quality of the best-alternative-frequency candidate is finally evaluated. It is ensured that the best candidate is M2 (initial value of 5 dB) stronger than the current, and that the TS-PER is smaller than the failure point. A good initial value for the failure point is zero, that is, no TS packet errors.

14.5.11   *Algorithm State 9*

Algorithm state 9 is depicted in figure 14.32. In rare cases, the TS packet error rate can be greater than the failure point even if signal strength is good. This could happen if the channel is corrupted by impulse noise. This signal cannot be accepted for the new current. The candidate should be dropped to the bottom of the preselected alternative frequency list. It is not wise to drop the candidate completely because the impulse noise corruption might stop.

One practical possibility for implementing the functionality presented above is to set the averaged signal strength of the impulse noise corrupted signal to –100 dBm. Then the candidate automatically drops to the bottom of the list in the next sorting round. The handover algorithm selects the second alternative frequency if its quality is good enough. If the signal strength of the impulse noise-corrupted candidate remains high, the rank in the list will rise slowly.

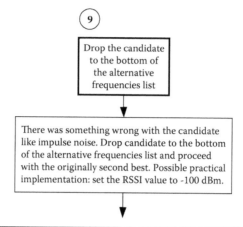

Figure 14.32 Algorithm state 9.

14.5.12 Algorithm State 10

Algorithm state 10 is depicted in figure 14.33. In this state we know the current and the best candidate where the handover will be made. Now the correct time for the handover needs to be decided. Support for the phase-shifted network should be implemented. The easiest way to do this

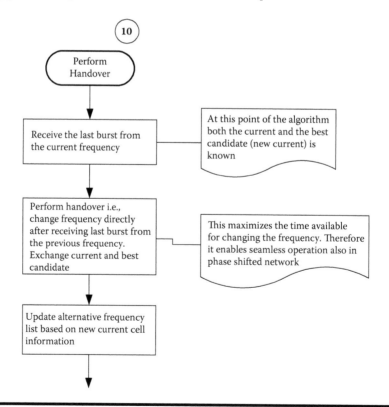

Figure 14.33 Algorithm state 10.

is to use the method presented in figure 14.33, where the last burst is received from the current signal, and then the receiver switches to another signal and continues the reception from the next burst. The alternative frequency list should be updated after handover.

The algorithm returns to state 4. The final exit from the handover procedure is done via state 6. The hysteresis principle, where the candidate must be M2 dB stronger than the current, avoids multiple handovers. However, until a big-enough margin for current is achieved, that is, M1 + M4 dB, it is wise to stay in the handover monitoring state to enable fast handovers in cell overlapping areas.

14.5.13 Fallback Loop

The fallback loop is depicted in figure 14.34. The fallback loop is a process parallel to the main handover loop. The fallback loop helps in case something goes wrong in the main handover loop or some special network case, such as traveling through a tunnel without indoor coverage.

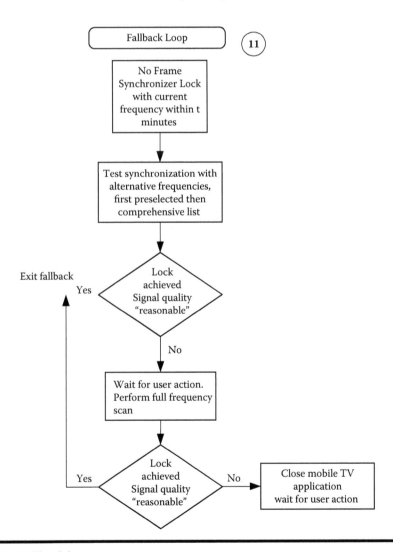

Figure 14.34 Fallback loop.

First, in the fallback loop the receiver tries to reacquire the current signal. If the current is not able to achieve lock within t minutes, other frequencies should be tried. First, the preselected alternative list should be tried, then the comprehensive alternative frequencies. If that does not help, a full-frequency scan should be performed. For the frequency scan, user interaction would most probably be required. Finally, if nothing helps, user interaction is requested. If any of the resynchronization trials succeeds, the algorithm returns to the start of the main loop.

14.6 Summary

The DVB-H standard is quite young in the commercial markets. However, the development of DVB-H dates back to the beginning of this century. A lot of effort has been put into the technical features, but also for the exploration of the end-user views in the usage of the services offered by the system. Seamless handover is one of the features that has been of interest in academic research, and also one of the challenges, and has been proven to work and bring added value to the standard, and hence ultimately to the end users.

This chapter provided some thorough details of the end-to-end service discovery process in the DVB-H system, and explained one realization of handover implementation by which also seamless handover can be accomplished.

Links

1. DVB: http://www.dvb.org/.
2. Global DVB MobileTV: http://www.dvb-h.org/.

References

1. ETSI. 2004. *Digital video broadcasting (DVB); Transmission system for handheld terminals (DVB-H)*. EN 302 304 v1.1.1.
2. DVB. 2005. *IP datacast over DVB-H: Set of specifications for phase 1*. A096.
3. J. Väre and M. Puputti. 2004. Soft handover in terrestrial broadcast networks. In *Proceedings of the 2004 IEEE International Conference on Mobile Data Management*, 236–42.
4. ETSI. 2006. *IP datacast over DVB-H: PSI/SI*. TS 102 470 v1.1.1.
5. ETSI. 2005. *DVB-H validation task force report*. TR 102 401 v1.1.1.
6. J. Väre, J. Alamaunu, H. Pekonen, and T. Auranen. 2006. Optimization of PSI/SI transmission in IPDC over DVB-H networks. In *Proceedings of the 56th Annual IEEE Broadcast Symposium*.
7. T. Jokela and J. Väre. 2007. Simulations of PSI/SI transmission in DVB-H systems. In *Proceedings of the 2007 IEEE International Symposium on Broadband Multimedia Systems and Broadcasting*.

Chapter 15

Radio Resource Management Schemes for Mobile Terrestrial and Satellite Broadcasting

Nikos Dimitriou

Contents

Keywords

resource management, scheduling, handover, QoS, coding

15.1 Introduction

The role of radio resource management (RRM) is to define the way radio resources are allocated among different users, services, and subsystems with the objective of achieving the optimum use of the available bandwidth and power without compromising the specific quality-of-service requirements. The problem of resource allocation concerns practically all layers of a protocol architecture, and the schemes employed on each layer are driven by the quality-of-service (QoS) requirements of the specific services that are provided to the end user.

Therefore, radio resource management is supposed to include various techniques such as call admission control, scheduling, power control, channel allocation, handoff, adaptive coding and modulation, and so on. These techniques have an impact on the individual user connections (call establishment, link parameters) and on the total system capacity (a system can be viewed as a single cell/spot beam or a multicellular coverage area, with unidirectional or bidirectional links). They effectively reduce the effect of multiple-access interference and link impairments, allowing for a greater spare capacity to be exploited for accommodating additional users or for new services to be provided within the network.

The objective of this chapter will be to present an overview of the RRM schemes used in typical broadcasting systems (terrestrial and satellite) and to point out the criteria based on which these schemes operate and the limitations that they have.

15.2 Definition of a Radio Resource

Future and emerging wireless communication systems will introduce the wide use of multimedia (speech, video, WWW) services with different characteristics (activity cycle, average bit rate, call duration) and requirements (e.g., packet/bit error rate, delay), providing a variety of coverage areas, such as indoor, small range (e.g., WLAN), cellular range (e.g., GPRS, UMTS, WiMAX, DVB-T, DVB-H), and satellite (e.g., DVB-RCS, S-DMB).

Each of the provided services will require a portion of the available radio frequency carrier capacity of each system, which has soft bounds and in each case is determined in a different way and is affected by various factors (multiple-access scheme, statistical multiplexing, interference, propagation channel conditions, and so on).

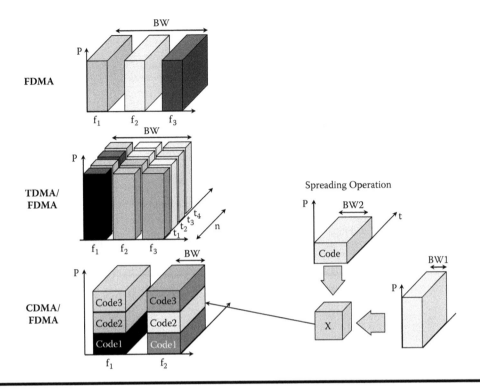

Figure 15.1 Radio resources in T/F/CDMA.

The concept of a radio resource can include a number of different characteristics that reflect different aspects of a radio frequency (RF) carrier or a communication node (terminal or base station).

First, according to the specific multiple-access scheme that is used, a radio resource can be defined as the means by which a communication link can be established without affecting (to a certain extent) other active communication links. Therefore, the radio resource in case of frequency division multiple access (FDMA) is a frequency band (belonging to the pool of the available frequency regions), in the case of time division multiple access (TDMA) is a time slot (during which all other links are considered to be idle), and in the case of code division multiple access (CDMA) is the specific spreading/scrambling code that isolates each user signal from all other transmissions (figure 15.1).

Naturally, the bandwidth requirements of the specific service determine the actual required allocated band in FDMA, the required slot duration in TDMA, and the required spreading code length in CDMA. There also can be cases where a combination of the above access schemes leads to the appropriate definition of a resource combining the characteristics of the merged systems (i.e., in Global System for Mobile [GSM] [TD-FD-MA] the resource is a time slot on a specific frequency carrier, and in the forward link of Universal Mobile Telecommunications System (UMTS) [FD-CD-MA] the resource is a specific spreading code on a specific RF carrier).

Apart from the scheme that isolates the user signal of interest from the interference at the receiver, another factor that is related to the radio resource definition is linked to the transceiver capabilities and especially the power requirements. The available power at the terminal is of

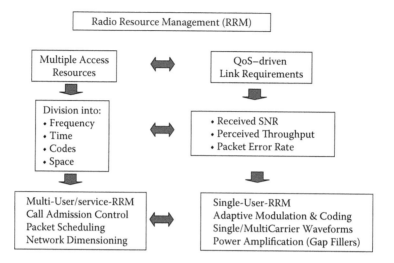

Figure 15.2   Broadcasting systems.

critical importance and can be considered as a resource, because it is limited by the following factors:

■ Limited battery standby time (terminals)
■ Emerging multimedia applications that require more processing power than the usual voice service
■ Requirement for terminals to be robust under severe channel conditions
■ Minimization of generated interference

The concept of RRM also includes ways and techniques that guarantee minimum power dissipation and emission levels in the terminal. Those techniques can be applied to various Open Systems Interconnection (OSI) layers (figure 15.2).

In the physical layer, many algorithms can be found in literature that allow the terminal to adapt to the emerging channel conditions by modifying the parameters of specific baseband blocks, such as adaptive modulation and coding. In that sense, the terminal can switch between different modes to achieve the required QoS with the minimum possible processing and transmitted power. Furthermore, the terminal may be able to reconfigure some of its baseband blocks to switch to another standard that again will ensure the seamless provision of the required QoS with the minimum power requirements.

In the upper layers (2, 3) one can find numerous techniques and algorithms employed in the radio resource controller for allocating radio resources (e.g., power) to each user. In the case of interference-limited systems, it is important to regulate the transmitted power of each terminal in a multiuser environment to avoid the generation of excessive interference to the other users, which will result in outage effects.

It is obvious that optimum solutions can be developed only when the physical layer exchanges information with the radio resource controller. In that sense, terminal power and energy minimization can be achieved as a result of the interaction and the cross-layer optimization of the first three (at least) OSI layers.

15.3 Characteristics of Broadcasting Systems

Broadcasting systems (figure 15.3) consist of a broadcasting station (a base station or a satellite) that transmits the same signal to a large number of terminals within the station coverage area (which can have the size of a standard UMTS cell up to a satellite spot beam). It is a unidirectional, point-to-multipoint communication that has the following characteristics:

- Because the same signal is destined for all users, it has to be designed to be received with the desired QoS by the users with worse channel conditions (usually the users close to the edge of the coverage area).
- Because there is no return channel from each user to the transmitting station, there is no way for a user to send feedback information and ask for either higher power (power control), stronger coding (adaptive modulation and coding), or packet retransmission (ARQ schemes). Each user terminal has to exploit the signal as it is received.
- Because the same signal is transmitted to all users, it can be carried by a single RF carrier within the coverage area. When having contiguous coverage areas of the same system, there is the option to assign different frequencies to each area to avoid interference to users close to the area borders. However, there are also solutions that can be used to allow the use of a single frequency among also neighboring coverage areas, allowing for a more efficient use of the allocated spectrum.
- The services provided by broadcasting systems are mainly based on video (video streaming, video on demand), with very high requirements in terms of QoS metrics (video quality, frame error rate, etc.). Additionally, the user terminals in the emerging broadcasting systems (e.g., DVB-H) will have to support mobility (speeds up to 50 kmph) and portability (small dimensions similar to those of a 3G mobile phone, large battery life). These requirements dictate the use of very advanced technological solutions in terms of all the receiver parts (RF, baseband processing, etc.).

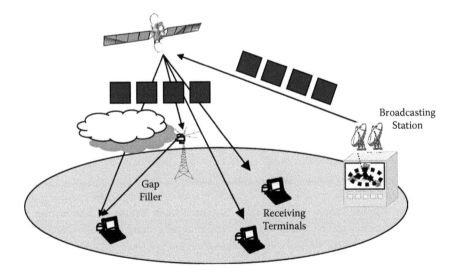

Figure 15.3   Radio resource management functionalities.

15.4 Broadcasting Systems to be Examined

15.4.1 DVB-T

The terrestrial digital video broadcasting (DVB-T) has been designed with the objective to be as much as possible similar with the DVB-S standard. DVB-T is a technical standard developed by the DVB Project that specifies the framing structure, channel coding, and modulation for digital terrestrial television (DTT) broadcasting. It is a flexible system that allows networks to be designed for the delivery of a wide range of services, from HDTV to multichannel SDTV, fixed, portable, mobile, and even handheld reception (especially when used in conjunction with DVB-H). The DVB Project has now begun work on developing a next-generation terrestrial specification that will take advantage of advances in modulation technology and is destined to meet the requirements of the broadcasting industry following the analog switch-off.[1]

15.4.2 DVB-S2

The satellite component of the Digital Video Broadcasting (DVB-S) standard was introduced in 1994. The DVB-S standard specifies QPSK modulation and concatenated convolutional and Reed–Solomon channel coding, and is now used by most satellite operators worldwide for television and broadcasting services. In 1997 the Digital Satellite News Gathering component of DVB (DVB-DSNG) introduced, additionally, the use of 8PSK and 16QAM modulation for satellite news gathering and distribution services.[2,3]

15.4.3 DVB-H

To broadcast DVB-H services over the terrestrial channels, DVB-H reuses the well-known DVB-T standard. This allows DVB-H services to share the MPEG-TS multiplex with traditional MPEG-2-based broadcast services. In case of dual transmission over the DVB-T network, both DVB-T and DVB-H receivers will be supplied with the same signal, allowing DVB-T receivers to decode the whole services, while permitting the DVB-H receivers to benefit on the DVB-H transmission features.[4–6]

15.4.4 SATIN

The system investigated by the FP5-IST project SATIN (Satellite UMTS IP-based Network) is similar to the frequency division duplex (FDD) wideband code division multiple access (WCDMA)–based air interface standardized within the Third Generation Partnership Project (3GPP) initiative (UMTS terrestrial radio access [UTRA], FDD).

The satellite system under consideration is effectively unidirectional. The space segment consists of a geostationary satellite that covers the EU area with several beams corresponding to different linguistic groups. The satellite features a transparent digital processing payload with multiple beams. This choice provides the desired flexibility in updating/enhancing the system throughout its life and is accompanied by reduced technology and investment risk. The satellite system component is closely integrated into the packet-switched domain of UMTS. A return link is provided via the terrestrial mobile networks (T-UMTS). Central to the system concept is the use of terrestrial gap fillers, also called intermediate module repeaters (IMRs).[7]

15.4.5 MBSAT

MBSAT is a commercial communications satellite system. It delivers digital multimedia information services such as CD-quality audio, MPEG-4 video, and data to mobile users throughout Japan and Korea. These satellite digital broadcasting services for mobile and personal users will be conveyed by the 2.6 GHz radio frequency band. This service consists of a large number of various multimedia broadcasting programs such as high-quality digital audio programs, video programs, and data. The available receivers include small portable devices that can be used in pedestrian or vehicular environments. To realize this service, a geostationary satellite with high-power transponders and a high-gain large deployable S-band antenna 12 m in diameter is used. Terrestrial repeaters, which transmit broadcasting signals to the areas where the satellite signal is blocked by obstacles such as tall buildings, inside tunnels, and subways, are also used.[8]

15.5 Classification of Schemes

One of the metrics related to the QoS provided in a wireless link between a broadcasting transmitter and a receiver can be expressed in terms of the received signal-to-interference-plus-noise ratio (SNIR). According to the link budget expressions, the SNIR is equal to

$$SNIR_{RX} = P_{TX} + G_{TX} - L_{FSL} - L_S - L_F + G_{RX} - (I + N)$$

P_{TX}, G_{TX} are the transmitted power and the antenna gain of the transmitter, respectively. L_{FSL}, L_S, L_F are the free space (path) loss, the shadowing, and the fast fading margins, respectively. G_{RX} is the receiving antenna gain and $(1 + N)$ is the sum of the interference and noise power that are also received.

Additionally, the received SNIR should be greater than or equal to the minimum acceptable value that is determined by the requirements (bit error rate [BER] or packet error rate [PER]) of the specific service that is provided by the system and the physical layer characteristics (signal bandwidth, bit rate, code rate, modulation index) of the transmission link:

$$SNIR_{RX} \geq \left(\frac{E_s}{I_0 + N_0} \right) \cdot \frac{R_s}{B_T}$$

where $\left(\frac{E_s}{I_0 + N_0} \right)$ is the ratio of the transmitted symbol energy over the interference and noise density, R_S is the symbol rate, and B_T is the transmission bandwidth.

The symbol rate can be further expressed as $R_S = \frac{R_b}{r \cdot m}$, where R_b is the useful data rate, r is the FEC code rate (ratio of the number of data bits over the number of data + parity bits in each frame), and m is the modulation index (number of bits per modulation symbol).

All the RRM techniques that will be mentioned in the following sections affect some of the abovementioned parameters.

In the physical layer, the variable coding and modulation schemes provide flexibility in choosing a code rate and modulation size, therefore the ratio of bandwidth allocated to parity overhead, to meet the BER or PER requirements of the received signal. Additionally, the layered and hierarchical coding and modulation schemes effectively split the allocated bandwidth in two parts, which could contain either the same signal with different bit rates (a high- and a low-rate version) or two different signals.

In the upper (2, 3) layers, Multi-Protocol Encapsulation (MPE) is a method of multiplexing in time packet streams, belonging to the same or different programs. Furthermore, with the use of MPE-FEC, a flexible amount of the transmission capacity is allocated to parity overhead, to improve the SNR and Doppler performance in highly mobile channels and to improve the tolerance to impulse interference.

Another technique that aims at reserving terminal power and increasing the terminal standby time is discontinuous reception, with which it is also possible for the terminal to monitor transmissions from stations in adjacent coverage areas and to perform handover when their SNR is higher than the SNR of the serving station.

Additionally, the user group scheduling provides a way of regulating the admission of services, program broadcasts via the same broadcast station, without exceeding the related power limits and minimizing the total interference.

Finally, in the case of satellite systems that are characterized by large path losses, the received SNR of the broadcast signal can be further improved with the use of terrestrial stations (gap fillers) that relay the broadcast signal to the mobile terminals.

15.5.1 Coding and Modulation

15.5.1.1 DVB-S2: Variable and Adaptive Modulation and Coding

DVB-S2 includes a number of advances on the transmission techniques of DVB-S: new coding schemes were introduced and combined with higher-order modulation constellations for increased efficiency. Additionally, the concept of variable coding and modulation (VCM) was introduced, to provide different levels of error protection to different service components. In the case of interactive and bidirectional applications, the VCM can be combined with the option of employing a feedback channel, to achieve adaptive coding and modulation (ACM), that is regulated by the channel quality information that is carried by the feedback link to the transmitting station from each user. This may be used to change coding and modulation to optimize the experienced throughput by each user. DVB-S2 uses a powerful FEC system based on concatenation of Bose–Chaudhuri–Hocquenghem (BCH) with low-density parity check (LDPC) inner coding. The result is performance that is at times only 0.7 dB from the Shannon limit. The choice of FEC parameters (available rates 1/4, 1/3, 2/5, 1/2, 3/5, 2/3, 3/4, 4/5, 5/6, 8/9, 9/10) depends on the system requirements. With VCM and ACM, the code rates can be changed dynamically, on a frame-by-frame basis. There are four available modulation modes: QPSK and 8PSK for broadcast applications through nonlinear satellite transponders driven near to saturation, and 16APSK and 32APSK for professional applications requiring semilinear transponders. The latter schemes trade off power efficiency for much greater throughput.[3]

15.5.1.2 DVB-T: Hierarchical Coding and Modulation

DVB-T includes a variety of punctured convolutional codes, based on a baseline convolutional code of rate 1/2 with 64 states. This allows the proper selection of the level of error correction for a given service or data rate. The supported code rates are 2/3, 3/4, 5/6, and 7/8. Additionally, all data carriers in one OFDM frame are modulated using either QPSK, 16QAM, 64QAM, nonuniform 16QAM, or nonuniform 64QAM constellations. This variety of code rates and modulation sizes is used to trade off throughput versus robustness.

Hierarchical channel coding and modulation may also be implemented. Two independent MPEG transport streams, one considered the high-priority stream and the other the low-priority stream, can be mapped separately onto the chosen constellation. This allows for a program to be simulcast in the form of two parallel streams, one a low-bit-rate, robust version, and another of higher bit rate and less robustness. This added flexibility/layering in the modulation and coding design results in addressing the needs of both users with good channel conditions that will exploit the high-bit-rate version, and of users close to the edge of coverage that will still be able to receive the low-bit-rate stream. The only additional requirement from the receiver is the ability for the demodulator/de-mapper to produce one stream selected from those mapped at the sending end. This method can also be used to multiplex two different programs on these parallel streams.[1]

15.5.1.3 DVB-H: Multi-Protocol Encapsulated Data–Forward Error Correction (MPE-FEC)

DVB-T and DVB-H are based on multiplexing parallel streams (audio, video, data) by transforming each stream into transport packets, each having an attached specific program ID (PID). The user then filters this stream of packets and selects the packets with a specific PID and then organizes, synchronizes, and uses them to produce the desired content (e.g., a movie) as shown in figure 15.4.

MPEG-2 was developed by the Motion Pictures Expert Group to define the format of each component of a multimedia program and the way these components are combined to form a single transmission bit stream. The most basic component is termed an *elementary stream*. Each program (for example, a TV show or a DVD track) comprises a number of elementary streams (typically a stream for video, audio, control data, subtitles, etc.).

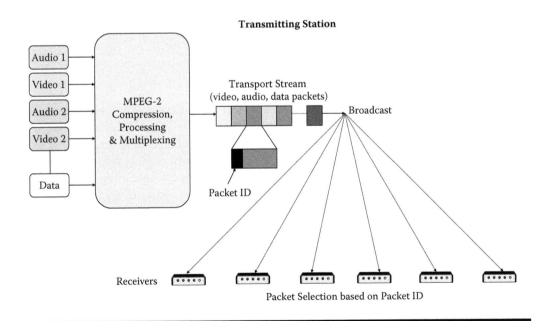

Figure 15.4 MPEG-2 multiplexing of elementary streams.

In case of wireless applications with a probability of packet loss, the MPEG-2 elementary streams are organized to a transport stream, a series of fixed-sized transport packets.

The transport stream of digital video broadcasting (DVB) in particular consists of a sequence of transport packets of 188 bytes in length (184 bytes of payload and 4 bytes of header). The header includes the *packet identifier*, which is used to assemble the contents of a specific program at the receiver. When the various parts of the elementary streams of a program are retrieved and decoded, tight synchronization is required to guarantee their parallel reproduction (e.g., audio and video streams).[15]

An additional feature introduced in DVB-H is MPE-FEC. The objective of the MPE-FEC is to improve the SNR and Doppler performances in highly mobile channels and to improve the tolerance to impulse interference. (Impulse interference is generated from electrical arcing, which can occur on the opening of switch contacts, especially those responsible for switching inductive components, such as motors or transformers. An impulse interference source could be a radiating electrical device or the ignition system of a car.)

With this scheme, a Reed–Solomon (RS) block code is used to protect the IP datagrams. The parity data (RS data) calculated from the IP datagrams of the burst is added in separate MPE-FEC sections. In that way, the IP datagrams can be recovered without any errors at the receiver, even under bad reception conditions.

With MPE-FEC, a flexible amount of the transmission capacity is allocated to parity over-head. A parity overhead of up to 25 percent of the frame size results in an achieved performance at the receiver with MPE-FEC similar to that of a receiver without MPE-FEC, but with the use of additional antenna diversity. The MPE-FEC overhead can be traded off with the proper choice of a lower inner convolutional code rate.

DVB-H is designed to be backward compatible with DVB-T; therefore, receivers not employing MPE-FEC (but who are MPE capable) can receive the data stream in a fully backward-compatible way, provided they do not reject the stream_type information. MPE-FEC, together with the virtual time interleaving, can result in a considerable reduction in the required SNR on mobile channels, and the resulting performance is similar to the performance measured in the case of using antenna diversity.[9]

15.5.1.4 SATIN: Layered Coding

In a GEO-based broadcast-multicast system such as SATIN, either open- or closed-loop power control is not generally possible. Hence, the communication system may have to rely solely on forward error correction (FEC) coding techniques. On the one hand, coding may be designed to address the worst-case fading scenario, which leads to unnecessary receiver processing complexity for the majority of users. Alternatively, coding may address an average fading scenario, which cannot provide a hard guarantee for the quality of service of every addressed user. Ideally, coding should allow a user with a good channel to recover the information with low complexity, while a user with a bad channel should still be able to achieve an acceptable bit error rate (BER) at the cost of increased complexity or of some extra decoding delay and power consumption. This is the type of scenario where the concept of layered coding comes in handy.

Layered coding (LC) is related to the structure of the FEC scheme, which consists of an inner and an outer code that can be separated and can operate in two different ways (or three, if we also consider the uncoded case). According to the required maximum BER and the received SNR, the user terminal can choose to operate with the received symbols that have not been encoded, with

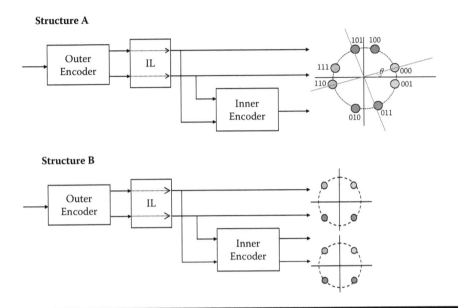

Figure 15.5   Layered coding structures.

the symbols encoded using only the outer encoder (convolutional encoder), or with the symbols encoded using the concatenation of the inner and outer encoders (SCCC). The penalty associated with the last choice is decoding complexity and delay.

Two different structures for LC have been considered in SATIN. Their main difference is that structure A uses a code of overall rate 1/3 with 8PSK while structure B uses a code of rate ¼ and QPSK.[10,11]

Figure 15.5 illustrates structures A and B. Structure A is based on mapping the coded bits either onto an 8PSK symbol (outer and inner codes, low-SNR option) or onto a QPSK symbol (outer code, high-SNR option).

Structure B is based on mapping the coded bits either onto two consecutive QPSK symbols with different error protection (outer and inner codes, low-SNR option) or onto one QPSK symbol (outer code, high-SNR option).

In the receiver side, two soft-input/soft-output (SISO) decoders, one interleaver and one de-interleaver, are used for iterative decoding. First, the SISO decoder of the inner code calculates a set of probabilities that are fed to the de-interleaver.

The SISO outer decoder uses the output of the de-interleaver to extract the first set of the overall decoder outputs. These outputs are interleaved and then fed back to the input of the inner SISO decoder for the next iteration.[12,13]

The use of the layered-coding scheme improved the system performance compared to the convolutional code, offering at the same time the flexibility of using either the option of the convolutional code for higher bit error rate and low complexity, or the option of SCCC scheme for lower bit error rate and higher complexity.

Additionally, the option of the SCCC scheme allows more users to be served as the system load increases, while the convolutional code's performance would result in outage conditions for many users of the satellite multicast/broadcast system. Thus, the option of layered coding keeps the coverage ratio at high values even for high system loads.[14] This is analyzed in more detail at the end of the chapter.

15.5.2 User Group/Service Scheduling

15.5.2.1 SATIN: Service Scheduling

In the system envisaged by SATIN, which involved broadcasting one-to-many links without any feedback return channel, there was no chance for the satellite to get any information regarding the forward link quality. Therefore, there was no ability to perform power control or channel quality-based scheduling. Therefore, in terms of RRM, apart from the concept of layered coding described earlier (to compensate in a sense the absence of power control), the idea of service multiplexing and scheduling was further developed.

User group/service scheduling was done via the following RRM functions: admission and load control, packet scheduling, and radio bearer allocation and mapping.

The admission control regulated the admission of new user group/service requests. A new request was accepted, provided there were enough network resources to accommodate the request and at the same time no other existing broadcast or multicast scheduled transmission would be affected (by the added interference of the new service and the resource [bandwidth] reservation it would require).

The objective of load control was to constantly monitor the resource usage and proactively avoid cases of overload that would degrade the quality of service. In cases of unavoidable overloads, load control would reactively take the necessary measures that would bring the system back to stability and QoS to the required level.

The packet scheduler aimed at multiplexing in time flows of different QoS requirements into physical channels of fixed spreading factor, in a way that can satisfy these requirements. It adjusted the transmitted powers of the flows, not on the basis of channel feedback information, but based on the packet/transport block size to be served or the knowledge of the expected audience distribution within the beam. The scheduler treated independently at the TTI level each physical channel. The exact number of physical channels at a specific time instance and the corresponding mapping of transport channels onto the code channels were defined by the radio bearer allocation mapping and admission control, depending on the RRM operational mode.

The objective of radio bearer allocation was to dimension the system and to define acceptable aggregations of different traffic flows, assuming that related information for each traffic flow (mean arrival rate, mean service duration, and requested rate for each type of service) was available.[7]

15.5.3 Discontinuous Operation and Handover

15.5.3.1 DVB-H: Time Slicing

The objective of time slicing is to reduce the average power consumption of the terminal and enable smooth and seamless service handover. Time slicing consists of sending data in bursts using a significantly higher instantaneous bit rate than the bit rate required if the data was transmitted using traditional streaming mechanisms (figure 15.6). To indicate to the receiver when to expect the next burst, the time reference to the beginning of the next burst is included within the burst currently being received.

Between the bursts, data of other programs can be sent; therefore, the DVB-H programs are time multiplexed and the DVB-H receivers just have to synchronize themselves to the appropriate bursts to receive the desired elementary stream/program. Therefore, with time slicing the DVB-H receiver can operate in a discontinuous mode, reserving energy and thus achieving increased battery autonomy, which enables the terminal portability.

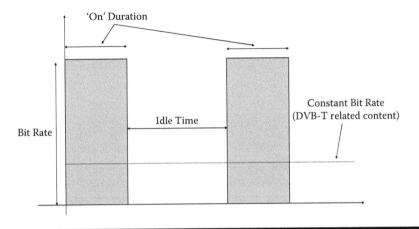

Figure 15.6 Time-slicing concept in DVB-H.

Additionally, the receiver can utilize the periods of inactivity to monitor the transmissions from neighboring stations, and can thus perform a handover between adjacent cells broadcasting the same content by just resynchronizing to the time pattern of the bursts transmitted by the new base station.[5,9]

There are various types of handover in DVB-H:

- A terminal may hand over from one cell to an adjacent cell that uses the same frequency. Therefore, the terminal just continues to receive the same transport stream by a different base station (SFN handover).
- The terminal may move to a cell that uses a different frequency (MFN handover).
- The terminal may also move to the coverage area of a different DVB-H network.

To be able to operate in such a situation, the receiver needs to regularly monitor adjacent cells that broadcast transport streams. Each of these IP transport streams is associated with the cell ID and the RF carrier frequency.

The topology of a DVB-H network will highly depend on the services it will be designed to provide. In case there will be a provision of localized services, the network will have to consist of a large number of small-sized cells operating on a different frequency and broadcasting different programs destined to their respective geographical locations. If no localized services are scheduled to be provided, the DVB-H network can consist of a number of SFN cells, all broadcasting the same content. The DVB-H cell size will be smaller than DVB-T cells, due to the fact that DVB-H will have to support mobile receivers with low antenna gains, located inside buildings that will have to be able to receive a fast fading signal.

Because the nature of the communication in DVB-H is unidirectional (one to many), the handover procedure will be different, compared to the handover procedure in cellular bidirectional systems. The handover procedure in DVB-H has to be initiated by the user terminal, based on measurements conducted during the periods of inactivity of the home cell broadcasting (due to the time-slicing principle). Because the terminal can monitor both the transmission of the home cell and the transmission of the neighboring cells (the bursts are separated in TDM fashion and are identified by their respective cell_id), it can perform soft handover. The absence of a feedback channel and a dedicated forward channel implies that the terminal has to take the handover decision based on its local measurements and without any interaction with the DVB-H network; this handover is termed passive handover.[16]

15.5.4 Coverage Extension

15.5.4.1 DVB-T/H: OFDM Size Related to Coverage/Mobility

The transmission modes that are provided in DVB-T (2K, 8K) and the one introduced in DVB-H (4K) offer the same range of transmission capacity, which only depends on the combination of the modulation constellation per subcarrier, the chosen FEC coding rate, and the guard time interval.

The difference among these three modes lies on the actual coverage size and the terminal mobility characteristics they can support.

The OFDM signal that is broadcast to all users consists of orthogonal subcarriers that are properly spaced in frequency to avoid intercarrier interferences. The transmission channel has a fixed bandwidth (either 6, 7, or 8 MHz); therefore, the number of subcarriers governs the intercarrier spacing (i.e., ~1/~2/~4 KHz for 8K/4K/2K modes, respectively).[9] The performance of the transmission for mobile terminals (i.e., tolerance to Doppler impairments) increases with the intercarrier spacing.

As a consequence of the orthogonality law (i.e., $\Delta f = 1/\Delta T$) the durations of both the OFDM symbol and guard interval are inversely proportional to the intercarrier spacing (i.e., ~1/~0.5/~0.25 ms for 8K/4K/2K modes, respectively). Accordingly, the resilience of the transmissions against the echoes (i.e., the intersymbol interferences) decreases with the intercarrier spacing.[4,5,9]

Clearly, the 2K mode allows four times the speed of the 8K mode; on the other hand, the 2K mode offers a maximum transmission cell size four times smaller than the 8K mode. It is clear that the 4K mode breaks the gap between the 2K and 8K modes.

Therefore, the additional 4K transmission mode is an interpolation of the parameters defined for the 2K and 8K transmission modes. It aims to offer an additional trade-off between transmission cell size and mobile reception capabilities, providing an additional degree of flexibility for network planning.

15.5.4.2 SATIN and MBSAT: Coverage Extension via Gap Fillers

An important design aspect considered in modern satellite communication system concepts is the use of terrestrial intermediate module repeaters (IMRs) that receive the same downlink signal from the satellite and retransmit it to their respective coverage areas (figure 15.7). With this approach, the user terminal receives a number of delayed replicas of the same broadcast signal. This multipath signal can be exploited by an appropriate Rake receiver that will equalize the various paths by using maximal ratio combining to offset the channel effects on each of the paths.

The use of the IMR stations was proposed for four different environments (urban, vehicular, UMTS coverage areas, and ships or airplanes). Additionally, three different functionalities were suggested, (simple IMR-like booster, IMR with node B functionalities, and IMR with node B and RNC functionalities). To reduce the deployment costs, IMRs were proposed to be collocated with UMTS base stations; therefore, the IMR network topology would match the topology of a coverage area consisting of UMTS cells.[7]

The MBSAT system topology consists of a broadcasting center, a broadcasting satellite with its control ground stations, a network of terrestrial gap fillers, and various types of user receivers.

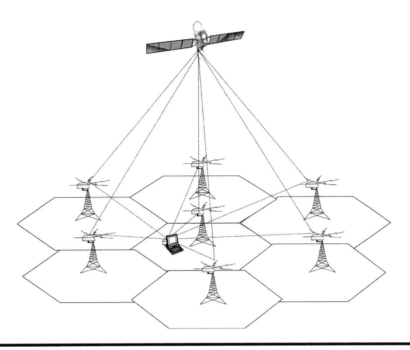

Figure 15.7   Gap filler concept.

The signal received by the satellite consists of a TDM and a CDM signal. The CDM signal is converted into an S-band downlink signal by the satellite transponder and transmitted to Japan for direct reception by mobile terminals. The TDM signal is converted into a 12 GHz downlink signal by the satellite transponder and transmitted via the gap fillers to the end users The gap fillers (due to their location, which has a line of sight [LoS] with the satellite) can mitigate the blocking effects caused by large buildings and vegetation. They receive the TDM signal from the satellite, convert it to CDM, which has the same contents as the CDM satellite signal, and transmit it to cover the area where the satellite signal is blocked by the obstacles.

Therefore, with the combination of the direct and indirect (via the gap fillers) signals there is a ubiquitous coverage all over Japan in all sorts of terrains with various population densities.

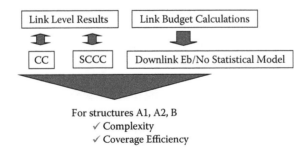

Figure 15.8   Complexity and coverage efficiency calculation methodology.[14]

The following procedure was followed to design the gap filler network:

■ *Estimation of blind spots*: Information is gathered regarding the terrain and urban/suburban/rural morphology that causes signal obstructions and shadowing, such as buildings (height, shape), hilly terrain, vegetation, and so on. This enables the prediction of spots where there is no LoS link between the ground terminals and the satellite and which could benefit from the installation of a gap filler station.

■ *Preliminary positioning*: Using the data collected in the aforementioned step, a set of wide-range omnidirectional gap fillers is installed. These repeaters have a coverage range of up to 3 km, and a series of field measurements is conducted.

■ *Adjustment of positioning*: After launching the satellite, additional field measurements are conducted, and in cases where there are some coverage gaps, additional (of narrower beam width and range) repeaters are installed.[8]

15.6 Summary

The aim of this chapter was to point out the various cases of radio resource management across multiple layers for broadcasting systems and the associated criteria that are considered in each case. Because the nature of the point-to-multipoint communication does not allow for feedback (reverse) links, the main functionalities of RRM are limited to the service/user group scheduling (at the broadcasting station), the receiver-enhanced functionalities (with variable coding and modulation, layered/hierarchical coding, discontinuous reception, Multi-Protocol Encapsulation–Forward Error Correction [MPE-FEC]), and the coverage enhancement techniques, via the use of gap fillers (signal amplifiers) or various OFDM sizes for a trade-off between coverage and mobility.

All the above techniques aim at improving the QoS of the received signal at the user terminal and effectively impact the coverage efficiency and capacity of the broadcasting systems.

Appendix: Layered Coding Coverage and Complexity Analysis

The objective of this section[14] is to perform a system-level study concerning the coverage efficiency and complexity of the proposed layered coding schemes. The basic principle of layered coding is that the receiver complexity can be traded off with the received signal strength (bit error rate). Therefore, a user receiving the multicast signal with high E_b/N_0 can just use the outer convolutional decoder, whereas a user who experiences worse channel and interference conditions will have to resort to the more complex serially concatenated code (which involves using both inner and outer convolutional decoders for a number of iterations).

The complexity is related to the number of decoder memory elements, the type of decoding (Viterbi, MAP, forward-backward), and the number of iterations. The complexity measure characterizes the efficiency of the receiver decoder in terms of processing delay and power consumption. Therefore, it is a crucial criterion in designing the FEC scheme and should also be taken into account along with the FEC performance results (bit error rate versus E_b/N_0).

The preceding sections showed the BER performance of the layered coding structures under the effect of the correlated multipath fading channel. These link-level simulations were done taking into account the short-term fast fading processes. For a system-level study, the long-term slow fading (shadowing) process also has to be taken into account. According to Lutz et al.[17] and

Taaghol and Tafazolli,[18] the shadowing loss in both terrestrial and satellite cases follows a lognormal distribution with exponentially correlated samples in time.

The following downlink budget expression shows the received power at the terminal:

$$P_{received} = EIRP_{tx} - L_p - L_s + G_{rx} \qquad (15.1)$$

where $EIRP_{tx}$ is the effective isotropic radiated power of the transmitting satellite or IMR, L_p is the free space loss (satellite or terrestrial), L_s is the loss due to shadowing (lognormal random variable), and G_{rx} is the antenna gain of the receiver terminal. It is evident that, assuming a specific position of the user terminal within the coverage area, the received power from each IMR and the satellite is also a lognormal random variable.

Moreover, the received E_b/N_0 at the terminal is equal to

$$\left(\frac{E_b}{N_0 + I_0} \right)_{received} = \frac{\sum_{i=1}^{7} P_{received,IMR_i} + P_{received,SL}}{N_0 + I_0} \qquad (15.2)$$

where the nominator includes the sum of the signals transmitted by the seven IMRs and the satellite, and the denominator includes the noise and interference density. The received $E_b/(N_0 + I_0)$ is the sum of multiple lognormal random variables; therefore, it will be also a lognormal random variable. By examining the distribution of the received $E_b/(N_0 + I_0)$, one can calculate the percentage of users that can use the FEC with low complexity and those who have to resort to the solution with high complexity.

Figures 15.9 and 15.10 illustrate the distribution among the FEC schemes (CC and SCCC) within the coverage area for varying cell loads. The user terminal was assumed to be on the edge of the IMR coverage.

Table A.1 shows the values used for the system-level calculations. The system load is given by the expression[19]

$$X = \frac{I}{N + I} \qquad (15.3)$$

and is a way to refer to the amount of potential capacity used and the corresponding interference generated.

From figures 15.9 and 15.10 it is apparent that for low values of the system load the mean of the received E_b/N_0 is quite high, resulting in a large percentage of user terminals using the low-complexity convolutional decoder. As the system load increases and the interference levels rise, the mean received E_b/N_0 decreases, leading to the need for a larger percentage of users to use the SCCC decoder. Structure B has the best performance in both convolutional and SCCC schemes; therefore, more user terminals can rely on the low-complexity receivers. Additionally, with the same FEC structure and as the sector load increases, a higher percentage of users can still receive the multicast signal with the SCCC, leading to an increased percentage of SCCC receivers compared to structure A1. In the case of structure A2, the convolutional encoder performance is worse than that of structure A1, leading more users to resort to the option of the SCCC scheme.

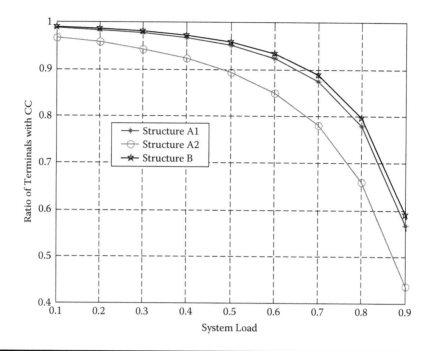

Figure 15.9 **Ratio of CC users within the coverage.**[14]

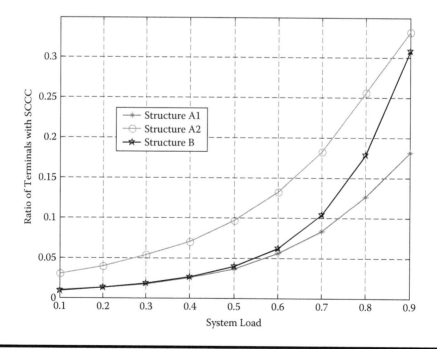

Figure 15.10 **Ratio of SCCC users within the coverage area.**[14]

Table A.1   Link Budget Parameters[14]

Forward link—Nomadic/Vehicular Class Reception	
Service Downlink	
Frequency of operation (GHz)	2.5
Satellite EIRP/traffic code (dBW)	57
IMR EIRP/traffic code	−19
Polarization + pointing losses (dB)	1
Cell radius (m)	400
Terminal antenna gain (dBi)	2
Terminal system temperature (°K)	300
Thermal noise density (dBW/Hz)	−204
Interference density (dBW/Hz)	−209
Spreading factor	8
Standard deviation of shadowing (terrestrial) (dB)	10
Standard deviation of shadowing (satellite) (dB)	8

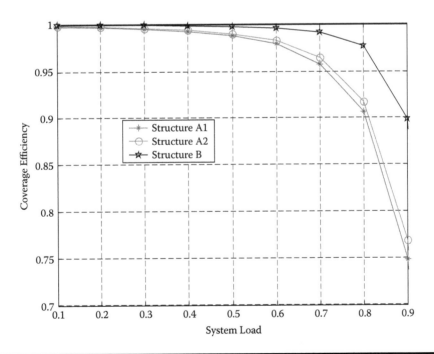

Figure 15.11   Coverage efficiency for varying system loads.[14]

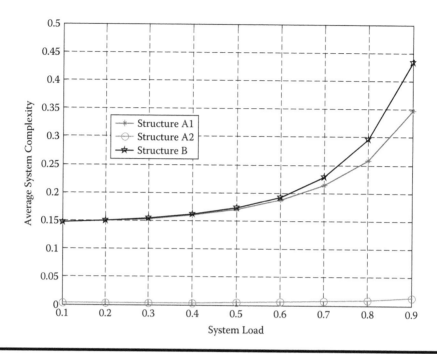

Figure 15.12   Average system complexity for varying system loads.[14]

Furthermore, as figure 15.11 illustrates, for sector loads higher than 50 percent, the coverage efficiency (this includes all the terminals that can, with either CC or SCCC, receive the multicast signal with the required BER) starts dropping from 100 percent and goes down to around 75 percent for 90 percent load and structures A1 and A2, meaning that there is an increasing percentage of user terminals unable to receive the broadcast signal with the required BER using either CC or SCCC. In the case of structure B, because the BER performance is better for SCCC, a larger percentage of the user terminals (90 percent) can still receive the multicast signal with acceptable QoS, leading to a higher coverage efficiency.

Figure 15.12 illustrates the average system complexity for varying cell loads. The average system complexity is calculated as the sum of the complexities of CC and SCCC receivers, weighted appropriately by their ratio and normalized to the complexity of the SCCC.

It is evident that structures A1 and B have a complexity that increases with the sector load, and in a fully loaded system it reaches the value of 44 percent. On the other hand, structure A2, due to having fewer memory elements per encoder, achieves better performance than structure A1 (as shown in the preceding section) with much less complexity. It can be seen that even in a fully loaded system the average system complexity (normalized to the SCCC of structure A1) is just 2.2 percent.

References

1. ETSI. 2004. *Digital video broadcasting (DVB); Framing structure, channel coding and modulation for digital terrestrial television.* EN 300 744 V1.5.1, European standard (Telecommunications series).
2. ETSI. 1999. *Digital video broadcasting (DVB); Framing structure, channel coding and modulation for digital satellite news gathering (DSNG) and other contribution applications by satellite.* EN 301 210 V1.1.1, European standard.

3. ETSI. 2006. *Digital video broadcasting (DVB); Second generation framing structure, channel coding and modulation systems for broadcasting, interactive services, news gathering and other broadband satellite applications.* EN 302 307 V1.1.2, European standard.

4. ETSI. 2005. *Digital video broadcasting (DVB); DVB-H implementation guidelines.* TR 102 377 V1.2.1, Technical report.

5. ETSI. 2004. *Digital video broadcasting (DVB); Transmission system for handheld terminals (DVB-H).* EN 302 304 V1.1.1, European standard (Telecommunications series).

6. M. Kornfeld and U. Reimers. 2005. *DVB-H—The emerging standard for mobile data communication.* EBU technical review.

7. M. Karaliopoulos et al. 2004. Satellite radio interface and radio resource management strategy for the delivery of multicast/broadcast services via an integrated satellite-terrestrial system. *IEEE Communications Magazine*, 108–117.

8. S. Fujita and Y. Yamaguchi. 2003. *Satellite digital broadcasting system and services for mobile and personal users in Japan.* Technical report of IEICE. SAT Vol. 103, No. 385, pp. 65–69.

9. G. Faria, J. A. Henriksson, E. Stare, and P. Talmola. 2006. DVB-H: Digital broadcast services to handheld devices. *IEEE Proceedings*, 94(1):195–209.

10. C. Fragouli and A. Polydoros. 2002. Serially concatenated coding for broadcasting S-UMTS applications. Paper presented at IEEE 7th ISSSTA'02, Prague, Czech Republic.

11. C. Fragouli and A. Polydoros. 2002. Symbol-interleaved serially concatenated trellis codes. Paper presented at 36th Annual Conference on Information Sciences and Systems (CISS 2002), Princeton, NJ.

12. S. Benedetto, D. Divsalar, G. Montorsi, and F. Pollara. 1998. Serial concatenation of interleaved codes: Performance analysis, design, and iterative decoding. *IEEE Transactions on Information Theory* 44:909–26.

13. S. Benedetto, D. Divsalar, G. Montorsi, and F. Pollara. 1997. A soft-input soft-output APP module for iterative decoding of concatenated codes. *IEEE Communications Letters*, 1(1):22–24.

14. A. Levissianos, G. Metaxas, N. Dimitriou, and A. Polydoros. 2004. Layered coding for satellite-plus-terrestrial multipath correlated fading channels. *International Journal of Satellite Communications and Networking* 22:489–502.

15. Digital TV Group. http://www.dtg.org.uk/reference/tutorial_mpeg.html.

16. X. Yang, J. Väre, and T. J. Owens. 2006. A survey of handover algorithms in DVB-H. *IEEE Communications Surveys* 8.

17. E. Lutz, D. Cygan, M. Dippold, F. Dolainsky, and W. Papke. 1991. The land mobile satellite communication channel—Recording, statistics and channel model. *IEEE Transactions on Vehicular Technology*, 40(2):375–386.

18. P. Taaghol and R. Tafazolli. 1997. Correlation model for shadow fading in land-mobile satellite systems. *IEEE Electronics Letters*, 33(15):1287–1289.

19. J. S. Lee and L. E. Miller. 1998. *CDMA systems engineering handbook.* Artech House Publishers.

20. Digital Video Broadcasting Project. http://www.dvb.org/.

21. P. Barsocchi et al. 2005. Radio resource management across multiple protocol layers in satellite networks: A tutorial overview. *International Journal of Satellite Communications and Networking* 23:265–305.

22. L. F. Fenton. 1960. The sum of log-normal probability distributions in scatter transmission systems. *IRE Transactions on Communication Systems* CS-8:57–67.

23. B. Vucetic and J. Yuan. 2000. *Turbo codes: Principles and applications.* Kluwer Academic Publishers.

24. J. Yuan, W. Feng, and B. Vucetic. 2002. Performance of parallel and serial concatenated codes on fading channels. *IEEE Transactions on Communications* 50:1600–8.

25. C. Fragouli and R. D. Wesel. 1999. Semi-random interleaver design criteria. In *Communication Theory Symposium at Globecom'99*, 2352–56. Vol. 5.

26. K. Narenthiran et al. 2004. S-UMTS access network for broadcast and multicast service delivery: The SATIN approach. *International Journal of Satellite Communications and Networking* 22:87–111.

Chapter 16

Dynamic Forward Error Control Schemes

Qingchun Chen and Kam-Yiu Lam

Contents

Keywords

forward error control coding, ARQ, puncturing, extending, DVB-H, MPE-FEC

Abstract

The dynamic error control scheme is highlighted in this chapter. It is shown that a flexible dynamic error control coding becomes more and more desirable in wireless mobile communications to accommodate different error protection requirements, or time-varying channels. The FEC and ARQ/FEC schemes are two important underlying techniques for the realization of the dynamic error control scheme, wherein the CSI and CSI-like information are the prerequisite conditions. It is shown that *puncturing* and *extending* are two basic techniques to realize dynamic code rate. Generally, the puncturing techniques outperform in the higher-code-rate region, while the extending outperforms in the lower-code-rate region. Combining both techniques leads to the best of both worlds. The forward error control coding schemes of both DVB-T and DVB-H are presented to indicate the most important problems when realizing the dynamic error control scheme in mobile broadcasting scenarios. Finally, it is stated that although further investigations are needed, the technique of the dynamic error control scheme has great potential in the practical applications of mobile multimedia broadcasting.

16.1 Error Control Coding for Digital Data Transmission

Reliable and efficient data communication over a noisy channel has been a very challenging topic for over 50 years. The fundamental approach to the problems of efficiency and reliability in communication systems is contained in the noisy channel coding theorem developed by Shannon in 1948. Shannon's theorem states that over a noisy channel if the code rate R is less than the channel capacity C, there exists a coding scheme of code rate R with arbitrarily small error probability. The proof of the theorem is essentially nonconstructive. It shows that, for long block length, almost all codes of rate R ($<C$) would be reliable. However, it does not give an explicit construction of the capacity-approaching codes, nor does it lay out practical decoding algorithms.

Since Shannon determined the capacity of noisy channels, the construction of capacity-approaching coding schemes has been the main goal in coding research. However, it was not until the early 1990s that we saw the first class of the so-called turbo codes, whose performance practically approaches Shannon's theoretical limit. In brief, the novelty of turbo codes lies in the pseudorandom interleaver and its iterative decoding. The pseudorandom interleaver introduces enough randomness to achieve reliable communication at data rates near capacity, yet it has enough structure to allow practical encoding and decoding algorithms. The invention of turbo codes has revolutionized the field of error-correcting codes. It led to the rediscovery of low-density parity check codes, and the discovery of the connection between iterative decoding and belief propagation, together with the inference problems. Based on these insights, different powerful error control coding schemes have been proposed in recent years. Today, it is realized that capacity-approaching coding and decoding techniques are not limited to turbo or turbo-like codes, but encompass significantly larger codes. The advances have had tremendous impacts not only in the field of coding but also in other areas, such as channel equalization, interference cancellation, and multiuser detection, where the turbo principle can also be applied. As a result, people nowadays have a number of powerful error control coding choices to guarantee the reliability of data transmission.

The conventional design of an error control system usually consists of selecting a fixed code rate with a certain error correction capability matched to the protection requirement of all the

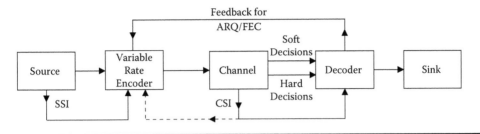

Figure 16.1   Digital transmission with source significance information (SSI) and channel state information (CSI).

data to be transmitted and adapted to the average or worst channel conditions to be expected. This strategy can maintain error probability to be below a predefined value. However, in many cases, one would like to be more flexible because the data to be transmitted may have different error protection needs, and the channel is time varying or has insufficiently known parameters.

As illustrated in figure 16.1,[5] the information to be transmitted might carry source significance information (SSI) indicating different protection requirements (examples include speech or image compression and video codec). On the other hand, the channel characteristics or channel state might vary considerably, as encountered in mobile or multipath radio transmission in a jamming environment. This is indicated by the channel state information (CSI) in cases that the instantaneous CSI is available at the encoder where code adaptation could take place accordingly. Mostly, the receiver can only use a CSI-like fading depth, noise-level variation, short-term signal loss, or jammer activity. The CSI can significantly improve decoder performance together with soft decisions at the receiver. Whenever a feedback communication channel is available, the CSI can be indirectly relayed to the transmitter. Based on the CSI information, both the dynamic forward error correction scheme and the automatic repeat request protocols could be implemented.

In general, error control techniques can be classified as either forward error correction (FEC) or automatic repeat request (ARQ). In the conventional applications of the FEC scheme, redundancy is added to transmitted data blocks to enable the receiver to correct the error patterns caused by noise and interference on the channel. The conventional FEC schemes usually consist of the selection of fixed-rate coding schemes that are well suited to the channel characteristics and the acceptable error rates for the data to be transmitted.[1] This strategy performs well for systems in which a fairly constant level of noise and interference is anticipated on the channel. In this case, enough error correction overhead can be designed into the system to correct the vast majority of error incurred. However, if the channel is not stationary but time varying, the throughput performance of the fixed coding scheme becomes smaller than the achievable throughput by using the optimum code parameters. That is, the fixed powerful low-rate error correction coding will waste capacity when the channel is good. To improve the transmission efficiency of the FEC schemes, an adaptive error control coding should be utilized to dynamically respond to the actual channel error condition by selecting the optimum code rate.

The dynamic forward error control scenarios shown in figure 16.1 require variable codes adapted to the source and channel needs. That is, the code rate (the number of parity check bits), and hence the error correction power of the code, may be changed during transmission of an information frame according to source and channel needs. For practical purposes, instead of switching between a set of encoders and decoders, the dynamic FEC coding is preferred to be implemented by using only one encoder and one decoder without changing their basic structure. Recently, there have been two basic approaches to realize this kind of variable rate encoder/decoder: the *puncturing* and *extending* technologies. In the puncturing implementations, variable code rate can be achieved by not transmitting certain code bits, namely, by puncturing the code.

Mandelbaum was the first to propose punctured codes for transmitting redundancy in incremental steps for Reed–Solomon codes.[13] Later, punctured convolutional codes were introduced to accommodate soft decisions and CSI at the receiver.[6] Hagenauer extended the concept of punctured convolutional codes by adding a rate-compatibility restriction to the puncturing rule.[5] The rate-compatibility restriction implies that all the code bits of a higher-rate punctured code are used by the lower-rate codes, that is, the higher-rate codes are embedded into the lower-rate codes. Rate-compatible codes are constructed from a single rate $1/M$ code, wherein a family of higher-rate codes is formulated by puncturing a successively greater number of code symbols. These codes have practical utility in that the system requires a single $1/M$ encoder. Both the transmitter and receiver need only share a puncturing table to determine which code symbols to transmit at a given time. The receiver may simply insert erasures for all code symbols that have not yet been received, and then performs the decoding algorithm as in the nonpunctured case. For illustration purposes, a rate-compatible punctured $(7, 5)_8$ convolutional code with a punctured period of 4 is illustrated in figure 16.2, where two puncturing tables are employed to dynamically adjust the coder rate and to obtain two different code rates: 4/5 and 4/6. A zero in the puncturing table means that the code symbol is not to be transmitted. In the example, among every four symbols, the fourth bit of the upper branch, the second and third bits, or the third bit, of the lower branch are not transmitted.

Today, successful attempts in generating rate-compatible (RC) codes have used BCH codes,[3] convolutional codes,[5] turbo codes,[7,8] and low-density parity check (LDPC) codes.[14–18] RC codes have many applications in packet data communications, where adaptive coding or unequal error protection is required. As discussed before, one of the main advantages of using RC schemes is that all the codes in the sequence can be encoded and decoded using a single encoder/decoder pair. In addition to having a low complexity, these schemes provide an efficient framework for the transmission of information using the ARQ/FEC protocol. In such a framework, referred to as type-II hybrid ARQ, incremental parity bits of the next-lower-rate code are transmitted in response to a negative acknowledgment (NAK) from the receiver, which indicates the failure of the last code in correcting the errors in the transmitted packet. However, it is shown that the efficiency of

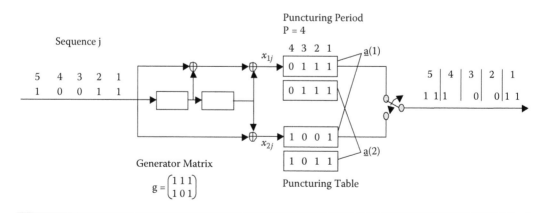

Figure 16.2 A punctured convolutional code with two rate-compatible puncturing tables.

puncturing is limited only to the high-rate range, where the amount of puncturing is small. To extend the dynamic range to lower code rates, the technique of extending was proposed for the LDPC codes to create a set of RC codes from a regular LDPC code.[14–16] Recently, the repetition-based RC codes have been successfully constructed for the convolutional codes to obtain a set of RC codes with a lower code rate than its mother code.[9,10] In fact, the repetition construction may be deemed as the simplest type of extending to construct RC codes.

In ARQ schemes redundancy is added solely for the purpose of error detection. When one or more errors are detected in a received data block, a retransmission request is sent back to the originating transmitter. ARQ schemes provide via retransmission a higher level of reliability than FEC schemes because a fixed amount of redundancy can detect approximately twice as many errors as it can correct. Of course, the paid cost is the reduced throughput. In practical applications, the maximum delay of ARQ schemes can be reduced by limiting the maximum number of retransmissions, yielding the truncated ARQ techniques.[2] Compared with the pure FEC scheme, ARQ protocols have been successfully employed to adapt to changes in the transmission medium. In communication systems with feedback channel, hybrid ARQ/FEC protocols have been established that exploit both the predictable performance of FEC codes and the rate flexibility of ARQ protocols. It is unveiled that truncated hybrid ARQ/FEC could be viewed as FEC that adapts to the instantaneous channel conditions.[24] In the ARQ/FEC scheme, both the puncturing and extending are utilized as well to adjust the coding rate.

With the fast advancement in wireless communications over the past and current decades, the demand for wireless services has been changing from the regular voice telephony services to mixed voice, data, and broadband multimedia services over wireless media. As for the cellular mobile systems, it is expected that multimedia multicast/broadcast services would be a "killer application." Although the allocated spectrum is quite limited, the scarce spectrum resources urge us that, besides the reliability requirement, the transmission efficiency should be the center concern for data transmission design in a wireless system. In a word, high-speed wireless data transmission requires robust and spectrally efficient communication techniques. However, the wireless channel is in nature a hostile transmission medium in the form of rapid time variation, extreme fading, and multipath. For instance, the transmission of signals over the wireless channels is affected by time-varying channel attenuation, called *fading*. The received signal strength can fluctuate over a wide range of 80 dB in the order of milliseconds. On one occasion, the transmission may experience good fading and the transmission error probability will be low. On the other hand, the transmission may experience serious fading at some other occasions and the error probability will be high. Hence, in general, the fading effects of wireless channels impose additional challenges for signal transmissions besides the regular channel noise. Among all promising techniques, the dynamic error control scheme is the key technique to achieve reliable and spectrally multimedia multicast/broadcast communications.

This chapter is organized as follows: after the brief introduction of the development of the dynamic error control scheme in this section, the general dynamic forward error control scheme is presented in section 16.2. The dynamic forward error control schemes based on puncturing and extending techniques are presented in section 16.3. In section 16.4, the error control scheme in mobile multimedia broadcasting (more specifically DVB-H) is addressed. It is shown that although further investigations are needed before the dynamic forward error control scheme could be enabled in mobile broadcasting, the technique of the dynamic error control scheme has great potential in practical applications. Finally, a summary is presented in section 16.5 to conclude this chapter.

16.2 The General Dynamic Error Control Schemes

Figure 16.3 illustrates the general structure of the dynamic error control scheme with channel state information feedback U_1^n. Information message $\omega \in \Omega$ (where Ω is the message set given by $\Omega = \{1, 2, \ldots, M = 2^{NR}\}$) is mapped into a frame of transmitted symbol $X_1^N = \{X_1, \ldots, X_N\}$ using an adaptive encoding function $f_n : \Omega \times U_1^n \to \chi$, where N is the number of symbols in an encoding frame, and R is the encoding rate in terms of bits per symbol. The adaptive encoding function, which is a function of the message index (ω) and causal channel state information at transmitter (CSIT) sequence U_1^n, is given by[12]

$$X_n = f_n\left(\omega, U_1^n\right) \quad \forall n \in [1, N]$$

where $U_1^n = \{U_1, \ldots, U_n\}$ is the causal CSIT sequence available to the transmitter at the nth symbol and $X_n \in \chi$ is the transmitted symbol. The transmitter adapts to channel variations by selecting the proper error protection capability according to the feedback causal CSIT information of $U_1^n = \{U_1, \ldots, U_n\}$. The principle of dynamic error control is to change the coding format in accordance with variations in the channel conditions (or with the signal SSI information), subject to system restrictions. In a system with dynamic error control scheme, users in favorable conditions are typically assigned with higher code rates, while users in unfavorable conditions will be assigned with lower code rates. As discussed in section 16.1, the benefits of the dynamic error control schemes are the increases in the average throughput of the system because a code is chosen adaptively based on the channel estimate.

Figure 16.4 illustrates the general structure of the channel decoder with the channel state information at receiver (CSIR) of $V_1^N = \{V_1, \ldots, V_N\}$. The receiver decodes the message $\hat{\omega}$ based on the entire frame of received symbols, $Y_1^N = \{Y_1, \ldots, Y_N\}$, and the entire sequence of CSIR of $V_1^N = \{V_1, \ldots, V_N\}$. The channel decoder is given by the function

$$\hat{\omega} = g\left(Y_1^N, V_1^N\right)$$

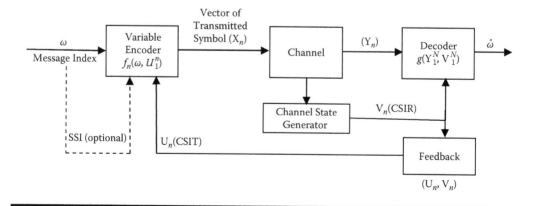

Figure 16.3 General dynamic error control scheme.

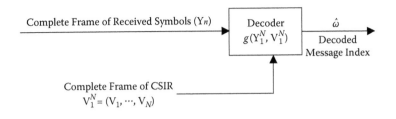

Figure 16.4   A general structure of adaptive channel decoder.

During the decoding procedure, the receiver may derive the utilized error control coding format based on both the CSIR and the agreed adaptive coding rule, or it would be informed by the transmitter via a signaling channel or control header. Decoding error occurs when $\hat{\omega} \neq \omega$. A code rate R is achievable if the error probability $P_e = \lim_{N \to \infty} \Pr[g(Y_1^N, V_1^N \mid \omega) \neq \omega] = 0$. The supreme of the achievable code rate is defined as the *channel capacity*.

Obviously, the CSI at the transmitter (CSIT) is the a priori to realize the dynamic error control scheme. As illustrated in figure 16.3, the channel conditions can be estimated based on feedback from the receiver. However, it is a demanding job to get the knowledge of CSI (at the transmitter or the receiver) in practical applications. The impairment (e.g., the delay or errors) of the channel estimate will give rise to a significant degradation in the achieved performance for the adaptive communications.[11] In view of the information theoretical analysis for the adaptive system, readers are encouraged to read the effect of imperfect CSI knowledge on the channel capacity in Lau and Kwok.[12]

Meanwhile, it must be addressed that in the broadcast/multicast communication scenarios, it is still an open problem for the transmitter to get the CSI knowledge due to its broadcast nature, wherein different receivers may have different processing capabilities and receive completely different signal levels due to their different locations and surroundings. In such case, the hybrid ARQ/FEC scheme is highly recommended for its versatility to enable flexible dynamic error control. A rate-adaptive error control technique for multimedia multicast services in the hybrid satellite and terrestrial network was proposed in Cho,[25] wherein the acknowledgment response from the receiver is used as the CSI. In the scheme, a packet is transmitted via satellite for the user fairness, while the retransmission of the packet is carried out over either terrestrial links or the satellite link based on the number of negative acknowledgments (NAKs) of the packet. If the majority of the receivers get erroneous packets, retransmission is sent over the satellite link; otherwise, it is sent over the terrestrial link. The advantage of the proposed protocol provides a reasonable trade-off between user fairness and the achieved throughput performance.

Basically, the hybrid ARQ/FEC scheme is an implicit link adaptation technique. Explicit CSI measurements or similar measurements are needed in adaptive forward error control to set the coding format, whereas in the hybrid ARQ/FEC scheme, link layer acknowledgments are used for retransmission decisions. Adaptive FEC by itself provides some flexibility to choose an appropriate coding for the channel conditions based on CSI measurements. However, either the inaccurate CSI measurement or the estimate delay will deteriorate the achieved benefit. Therefore, an ARQ protocol in the link layer is still required. ARQ autonomously adapts to the instantaneous channel condition and is insensitive to the measurement error and delay. Combining adaptive FEC with ARQ leads to better performance—adaptive FEC provides the coarse data rate selection, while ARQ provides for fine data rate adjustment based on channel conditions. It is expected that not only the FEC but also ARQ scheme should be carefully devised to make them suited to the dynamic error control requirement for the mobile broadcasting services.

16.3 Dynamic Error Control Coding Schemes

As stated before, puncturing and extending are two basic techniques to adjust error control coding format in the error control coding scheme. In the following, a survey of both techniques will be presented.

16.3.1 Puncturing-Based Dynamic Error Control Coding

The puncturing technique was first proposed by Mandelbaum in the 1970s to achieve variable code rate.[13] Later, puncturing gradually became a primary technique to realize the dynamic error control coding. The primary benefit is its simplicity because all the resultant codeword sets can be encoded and decoded using a single encoder/decoder pair. To enable the incremental redundancy encoding/decoding capability, Hagenauer proposed to include the rate-compatibility rule in the puncturing implementations.[5] Rate-compatible (RC) codes are a family of nested codes where the codeword bits from the higher-rate codes are accurately embedded in the lower-rate codes. In the following, the rate-compatible punctured-based dynamic error control coding is presented for the convolutional code, turbo code, and LDPC code, respectively.

16.3.1.1 Rate-Compatible Punctured Convolutional (RCPC) Codes

A family of RCPC codes is described by the mother code of rate $R = 1/M$ and memory m having the generator tap matrix $g = (g_{i,k})_{M \times (m+1)}$ with the tap connection coefficient $g_{i,k} \in (0,1)$, where a 1 represents a connection from the kth shift register stage to the ith output. Together with M, the puncturing period P determines the range of code rate

$$R = \frac{P}{P+l} \quad l = 1,\dots,(M-1)P$$

between $P/(P+1)$ and $1/M$. The RCPC codes are punctured codes of the mother code with puncturing matrices $a(l) = (a_{i,j}(l))_{M \times P}$ with $a_{i,j}(l) \in (0,1)$, where 0 implies puncturing. The rate-compatibility restriction implies the following rule:

$$if\ a_{i,j}(l_0) = 1,\ then\ a_{i,j}(l) = 1\ for\ all\ l \geq l_0 \geq 1$$

or equivalently,

$$if\ a_{i,j}(l_0) = 0,\ then\ a_{i,j}(l) = 0\ for\ all\ l \leq l_0 \leq (M-1)P-1$$

On the receiving side, the decoder using the Viterbi algorithm (VA) has to know the current puncturing rule $a(l)$. Let $y_{i,j}$ denote the received symbols of $x_{i,j}$. The VA finds the path m with the maximum likelihood path metric by calculating $\max_{m} \sum_{j=1}^{N} \lambda_j^{(m)}$, where the metric increment $\lambda_j^{(m)}$ is given by

$$\lambda_j^{(m)} = \sum_{i=1}^{N} a_{i,j}(l) x_{i,j}^{(m)} y_{i,j} \quad 1 \leq l \leq (M-1)P$$

and $a_{i,j+P}(l) = a_{i,j}(l)$ due to the periodic puncturing. With Viterbi decoding the usual optimality criterion is a large free distance d_{free}, a small number of paths a_d, and a small information error weight c_d on all paths with $d \geq d_{free}$, where a_d is the number of incorrect paths of Hamming weight d for $d \geq d_{free}$ that diverge from the correct path and reemerge with it at some later stage, while c_d indicates the total number of error bits produced by the incorrect paths. Both $\{a_d\}$ and $\{c_d\}$ formulate the so-called distance spectra. More specifically, one has the following upper bounds for the error event probability,

$$P_E \leq \frac{1}{P} \sum_{d=d_{free}}^{\infty} a_d P_d$$

and for the bit error probability,

$$P_b \leq \frac{1}{P} \sum_{d=d_{free}}^{\infty} c_d P_d$$

where P_d is the probability that a wrong path at distance d is selected in the Viterbi decoding and depends on the modulation type and channel characteristics. For the additive white Gaussian noise (AWGN) channel, P_d is given by

$$P_d = Q(\sqrt{2dE_s/N_0})$$

where E_s/N_0 is the signal-to-noise ratio per transmitted channel symbol, and the function $Q(x)$ is defined as

$$Q(x) = \frac{1}{\sqrt{2\pi}} \int_x^{\infty} e^{-y^2/2} dy$$

It could be readily observed that to improve the reliability, the distance spectra $\{a_d\}$ and $\{c_d\}$ should be as small as possible. When puncturing is utilized, the distance spectra depend only on the code parameters of M, m, P, g, and $a(l)$. RCPC codes constitute a specific class of time-varying codes with a fixed generator but periodically time-varying puncturing. Due to the time-varying nature of the RCPC codes, no constructive method is available yet. Recently, effective computer search routines have been developed to get the good puncturing rules in terms of the achieved-best-distance spectra of $\{a_d\}$ and $\{c_d\}$.[5,26] The criterion of goodness for a good RCPC code is the maximum free distance d_{free} and the minimum total number of bit errors c_d produced by all incorrect paths with distance $d \geq d_{free}$ that diverge from the correct path. Based on this optimization criterion, a useful search procedure to find out good puncturing tables with puncturing period P is outlined as follows:[26]

Step 1. Select the best-known rate $1/M$ mother code of memory m and the puncturing period P. Set all elements of puncturing matrix $a(\delta = P)$ to 1, and for $\delta = P-1$, to 1.

Step 2. Select an element $a_{i,j}$, in turn, for $1 \leq i \leq M$ and $1 \leq j \leq P$ and set it to 0.

Step 3. Determine d_{free} and c_d for $d \geq d_{free}$. Select code that gives the maximum d_{free} and the minimum c_d for $d = d_{free}, d_{free} + 1, ...$ Discard catastrophic codes and codes of memory less than m.

Step 4. Repeat steps 2 and 3 with the rate-compatibility restriction, to find other puncturing matrices for the family of the entire RCPC codes.

Figure 16.5 shows the simulated performance of RCPC codes for the Gaussian channel. The simulation results validate that different puncturing gives rise to different error correction capability. In practical applications, a degree of freedom can be gained by increasing the period P. Most of the puncturing tables with favorable distance spectra and the associated RCPC codes can be found in the literature.[5,26]

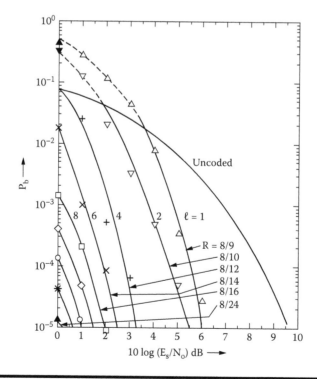

Figure 16.5 **Bit error reliability performance of RCPC codes on Gaussian channels, $M = 3$, $P = 8$, $I = 1, 2, ..., 16$, rate $= P/(P + I)$; simulation with soft decision.**

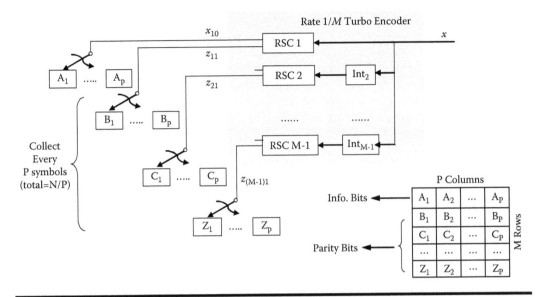

Figure 16.6   General RCPT encoder, 1/M mother code rate, puncturing period P.

16.3.1.2   *Rate-Compatible Punctured Turbo (RCPT) Codes*

Turbo code has gained considerable attention since its invention in 1993 due to its near-capacity performance. The conventional turbo codes are a parallel concatenation of two identical recursive systematic convolutional (RSC) encoders separated by a pseudorandom interleaver. Rate-compatible punctured turbo (RCPT) codes were originally proposed by Barbulescu and S. S. Piertrobon.[7] Given a rate $1/M$ turbo encoder of information block size k and a selected puncturing period P (it is assumed k is an integer multiple of P), figure 16.6. shows the general RCPT encoder's structure, where the puncturing table a is given by

$$
a = \begin{bmatrix}
A_1 & A_2 & \cdots & A_P \\
B_1 & B_2 & \cdots & B_P \\
C_1 & C_2 & \cdots & C_P \\
\vdots & \vdots & \ddots & \vdots \\
Z_1 & Z_2 & \cdots & Z_P
\end{bmatrix}_{M \times P}
$$

with $A_i, B_i, C_i, \ldots, Z_i \in GF(2)$, $i = 1, 2, \ldots, P$, and 0 implies puncturing. Generally speaking, the formulation of the RCPT codes follows that of the RCPC codes almost directly; however, there are several notable distinctions. The most obvious distinction is that the rate $1/M$ convolutional encoder is replaced by a rate $1/M$ turbo encoder, and, of course, the rate $1/M$ soft decision Viterbi decoder is replaced with a bank of soft-in soft-output (SISO) decoders and the associated iterative decoding structure. The more subtle distinction is with the process for selecting the puncturing tables corresponding to different code rates. For RCPC codes, the puncturing tables are selected such that each punctured code with the new lower-rate code yields the greatest possible increment in the free distance. For RCPT codes, however, there are several different criteria for selecting the puncturing patterns owing to the different design objectives, as will be illustrated soon.

In particular, unlike the RCPC codes, the subsets of code symbols that have been transmitted at a given code rate implicitly determine the number of SISO decoders that can be employed in the receiver. Moreover, if nonsystematic parity symbols have not been received for two or more recursive systematic convolutional (RSC) encoders, iterative decoding is not possible. It is unveiled that, in the ARQ/FEC scheme, it is profitable at high coding rates to puncture some of the systematic code symbols so that more nonsystematic symbols may be sent, and thus utilized by more or all of the SISO decoders resident in the receiver.[8]

Traditional techniques for bounding the performance of convolutional codes quickly become intractable, when applied to turbo codes, for sufficiently large block sizes, due to the innumerable state mappings introduced by the random interleaver. However, for the case of additive noise channel, binary antipodal modulation, and the maximum likelihood soft decoding, the achieved bit error rate (BER) performance of turbo codes could be approximated within moderate- and high-SNR regions by using the following union bound:

$$P_b \leq \sum_{w=1}^{k} \frac{w}{k} \sum_{d=d_{free}}^{N} A_{w,d} Q\left(\sqrt{\frac{2RE_b}{N_0} \cdot d}\right) = \sum_{d=d_{free}}^{N} \frac{a_d w_d}{k} Q\left(\sqrt{\frac{2RE_b}{N_0} \cdot d}\right)$$

where k is the size of the input information word, N is the codeword size, and $R = k/N$ is the code rate. The input-output coefficient of $A_{w,d}$ represents the number of codewords with Hamming weight d generated by information words of weight w. Obviously, $\tilde{w}_d$ could be calculated by

$$w_d = \frac{1}{a_d} \sum_{w=1}^{k} w A_{w,d} = \frac{c_d}{a_d}$$

where $\{a_d\}$ and $\{c_d\}$ are the distance spectra of turbo codes. Although the union bound will become weak below the cutoff rate, it still explicitly tells us that the performance of turbo code (especially within the moderate- and high-SNR regions) depends on its distance spectra.

Recently, several algorithms have been developed to calculate the first few terms of the weight distribution of both parallel and serial turbo codes. Both exact algorithms (e.g., references 27 and 28) and approximate algorithms (e.g., references 29–31) have been presented. However, the accurate calculation of the distance spectra for turbo codes is a hard and tedious job for even a sufficiently small information block size k, due to the innumerable permutations introduced by the random interleaver. Besides the accurate or approximate distance spectra calculation, a uniform interleaver approach has been proposed to decouple the RCPT codes designed from the interleaver,[18] where the actual interleaver is replaced with the average interleaver. In fact, the so-called uniform interleaver is nothing but an artificial construction that maps a given input sequence into all possible permutations of the sequence with equal probability. Different design criteria for the puncturing patterns based on the input-output weight enumerating function (IOWEF) employing a uniform interleaver or the accurate distance spectra of the RCPT codes are outlined as follows:

1. **Free distance criterion:** In this criterion, given any code rate, the candidate puncturing patterns yielding the largest free distance will be selected. The free distance criterion is investigated to find out the good rate-compatible puncturing patterns for turbo codes with specific interleaver in the error floor region.[33] The key ingredient is an exact turbo code weight distribution algorithm (like the constrained subcode algorithm[27] and its

improved algorithm[28]) producing a list of all codewords in a turbo code of weight less than a given threshold. Later, a two-step procedure is utilized to calculate the exact free distance of the RCPT code given a candidate puncturing pattern, and then to choose at each iteration step the bit position to be punctured in a sequentially greedy manner. The application of this criteria may be calculation demanding, especially when the information block size is large enough. The free distance criterion based on the uniform interleaver and IOWEF knowledge is an alternative to save the complexity.

2. **Minimum slope criterion:**[8] If one considers an arbitrary turbo decoder error event, the input weight refers to the number of bit errors and the output weight refers to the overall Hamming weight of the error event. If uniform interleaver is assumed, the IOWEF of a turbo code indicates the average number of error events of a given input and output weight. From the IOWEF, one can derive the average weight enumerators (AWEs) of the turbo code. The AWE indicates the average number of error events of a given output weight. Then, the average distance spectrum (ADS) is defined as the set of nonzero AWEs. Although the ADS is most often used in the evaluation of performance bounds, it is also found to be very useful in the search for the puncturing patterns for a given encoder block size, and puncturing period. If a regression line is used to fit to the first 30 or so terms of the ADS, the slope of this fitted line represents a measure of the rate of growth of the AWEs as the output distance increases. The minimum ADS slope criterion implies that the candidate puncturing pattern yielding the minimum slope will be selected. The ADS slope criterion will yield a family of subcodes that perform better at lower SNR where higher distance error events do contribute to error performance. The *minimum ADS slope criterion* yields subcodes that have higher (poorer) error floors, but have better performance at lower SNR. It turns out that the ADS slope criterion is superior for ARQ applications, wherein the RCPT subcodes tend to be employed at their respective low SNR values. Although not directly stated, the accurate distance spectra-based minimum slop criterion should also be applied in the ARQ applications.

3. **Optimization of the sequence (d_w, N_w):**[32] Let d_w denote the minimum weight of codewords generated by input words with weight w, and N_w the number of codewords (multiplicities) with weight d_w. If the pairs of (d_w, N_w) for $w = 2, \ldots, w_{max}$ can be calculated for the uniform interleaver, the candidate puncturing table yielding the optimum values for (d_w, N_w) will be selected in this criterion. More specifically, the puncturing pattern that sequentially optimizes the pairs (d_w, N_w) (first d_w is maximized, and then N_w is minimized) will be selected to create RCPT codes. It is shown that when a uniform interleaver is assumed, this criterion will give rise to the best performance compared with the other two criteria. Illustrated in figure 16.7 is the simulated performance of systematic RCPT codes in terms of E_b/N_0 versus code rate R at BER $= 10^{-5}$ with $k = 100$ (random interleaver with size 100 and 10 decoding iterations). The reason that the optimization criterion of the distance spectra sequences of (d_w, N_w) performs well in creating the RCPT codes could be intuitively understood: it is readily noted that both the maximum free distance criterion and the minimum slope criterion are included in the optimization criterion of the sequence (d_w, N_w). Therefore, it could be expected that the achieved RCPT codes behave well in all punctured code rate ranges. The problem for applying the optimization criterion of the sequence (d_w, N_w) is the computation complexity, if the accurate sequence of (d_w, N_w) is required. However, fortunately, this optimization criterion could be applied separately to the IOWEF of the constituent encoders, and therefore to the search for the "good" rate-compatible puncturing patterns, given the interleaver size. This feature leads to a dramatic reduction in the computational complexity needed.

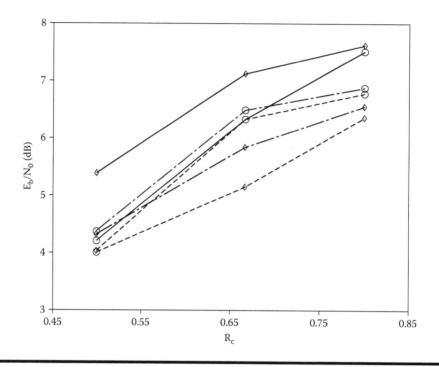

Figure 16.7   The comparison of different design criteria homogeneous (◊) and nonhomogeneous (o) puncturing. Free-distance criterion, dash-dotted curves; minimum slope criterion, solid curves; optimization of the sequence (d_w, N_w), dashed curves.[32]

Besides the application in generating RCPT codes, the puncturing technique could be utilized to realize unequal error protection (UEP), and therefore improve the reliability of turbo codes. It is proposed to identify weak and strong trellis sections in the constituent trellises[34] based on a list of low-weight codewords computed by the constrained subcode algorithm.[27] Each parity bit position in the codewords is associated with a unique trellis section from either the upper or the lower constituent trellis. All trellis sections associated with parity bit positions in the support set of some exhaustive list of low-weight codewords are said to be weak, and a trellis section that is not weak is said to be strong. It is proposed to strengthen weak trellis sections by using repetition coding within some of those sections.[34] To maintain the same overall code rate (and therefore enable the rate allocation among different trellis sections), puncturing is done within strong trellis sections.

16.3.1.3 Rate-Compatible Punctured Low-Density Parity Check (LDPC) Codes

Besides turbo codes, another class of codes that turned out to approach Shannon's bound, namely, *low-density parity check codes* (LDPC codes), were discovered by Robert Gallager in the 1960s. Like turbo codes, LDPC codes have an iterative decoder. The true power of these codes, however, was overlooked, due in part to the lack of sufficient computing power for their implementation at the time. Today, the codes coming closest to the Shannon bound are LDPC codes, and much work has recently been focused on their construction and analysis.

There are two kinds of regular and irregular LDPC codes. A regular LDPC code has parameters (n, k, t, s), which denote the codeword length, data (information) block size, column weight, and row weight of the parity check matrix, respectively. Unlike regular LDPC codes with uniform column/row weights, there are nonuniform column/row weights in irregular LDPC codes. Generally, irregular LDPC codes outperform regular ones in bit error rate, but the difference is marginal for short to moderate code lengths (a few hundred to a few thousand bits).

In general, rate-compatible puncturing could be applied for LDPC code to enable the flexible LDPC code rate.[14] Two methods, the *random puncturing* and *intentional puncturing*, were proposed for puncturing LDPC codes.[16] In the random puncturing method, the punctured bits are chosen randomly. If the puncturing fraction is p, a subset of the bits with cardinality $n \cdot p$ in the LDPC codeword is chosen at random and punctured. In the intentional puncturing method, the puncturing is selected to optimize the selection of puncturing symbols for a code rate (also called a *puncturing distribution*). Based on the degree distributions of irregular LDPC codes, a puncturing method was proposed for irregular LDPC codes.[16] More specifically, variable nodes of the bipartite graph will be grouped in accordance with their degrees. Then a fraction of the nodes in the set of variable nodes of specific degree will be punctured sequentially. It was shown that there exist good (in view of capacity approaching) puncturing distributions over a broad range of code rates. However, these results presume infinite block lengths.

A systematic way was proposed to generate rate-compatible punctured LDPC codes by minimizing the performance loss due to puncturing for either regular or irregular LDPC codes at small block lengths (a few thousand bits).[17] The basic idea is based on the fact that a punctured node (either variable node or check node) will be recovered with reliable messages when: (1) it has more neighboring (check) nodes, and (2) each check node has more reliable neighbors (variable nodes) except for the punctured one. For example, a punctured variable node that has check nodes whose remaining neighboring variable nodes are unpunctured will have nonzero messages from the check nodes in the first iteration. The punctured node that will be recovered in the first iteration is called a one-step recoverable (1-SR). The 1-SR nodes and unpunctured nodes will help recover some of the remaining punctured nodes in the second iteration, and so on. In general, the punctured nodes recovered in the kth iteration are called k-SR nodes. Generally, the greater the number of iterations a punctured node needs for its recovery, the less statistically reliable the recovery message is. Thus, it is better to puncture nodes that require a smaller number of iterations, which results not only in less iterations to decode codewords but also in better performance at a given code rate. Based on this idea, the procedure to generate the rate-compatible punctured LDPC codes may be outlined as follows: (1) Maximize the size of a group containing 1-SR nodes G_1. After that, maximize the size of a group with 2-SR nodes G_2, and so on, till the k-SR nodes' group of G_k. (2) The groups $G_1, G_2, ..., G_k$ will be used to determine the puncturing patterns for increasing the rates in a rate-compatible fashion. Namely, first puncture variable nodes from G_1, then nodes from G_2, and so on, to achieve higher and higher code rates.

Recently, it has been revealed through theoretical analysis[23] that, for any ensemble of LDPC codes, there exists a puncturing threshold p^*. The puncturing threshold means that if the puncturing fraction p is smaller than a threshold value of p^*, then the punctured LDPC code is asymptotically good. In other words, a code from the ensemble can be used to achieve arbitrarily small error probability over a noisy channel while the code rate is bounded away from zero. On the other hand, if $p > p^*$, error probability is bounded away from zero, independent of the communication channel. The puncturing threshold implies that the largest code rate achieved by employing the puncturing technique is not unlimited. Besides the puncturing threshold phenomenon, it is unveiled that, for any rates R_1 and R_2 satisfying $0 < R_1 < R_2 < 1$, the ensemble can be punctured from rate R_1 to R_2, resulting in asymptotically good codes for all rates $R_1 \leq R \leq R_2$. More specifically, theoretical analysis suggests that rates arbitrarily close to 1 are achievable via puncturing, and the punctured LDPC codes behave as good as ordinary LDPC codes, if all the augmented bits from the mother LDPC code ensemble can be located and punctured. For example, let us consider the following parity check equation:

$$c_1 : x_1 \oplus x_2 \oplus x_3 \oplus x_4 \oplus x_5 = 0$$

By splitting the above parity check equation, one gets

$$c_2 : x_1 \oplus x_2 \oplus x_3 \oplus y = 0$$

$$c_3 : x_4 \oplus x_5 \oplus y = 0$$

where y is referred to as an augmented variable node. Figure 16.8 shows the effect of the parity check splitting on the Tanner graph of the node. The variable node y in figure 16.8 could be selected as the punctured variable node. Now consider an LDPC code in which some of the parity check nodes have been split. This code can be considered a punctured code. By splitting, we can make a graph corresponding to a lower-rate LDPC code. When the augmented variables are punctured, we can get a code with the same rate as the original code. Note that the splitting can be performed repeatedly, and a check node that is obtained by splitting can itself be split into more check nodes. However, any check node is split only a finite number of times. It is worth noting that the original LDPC code and its punctured one have the same asymptotic thresholds. However, the two codes can have different finite-length performance. In fact, by a suitable choice of the punctured variable nodes, we may be able to alleviate the destructive effect of short cycles in the Tanner graph. When the code length is short, the short cycles of the graph deteriorate the performance. The cycle lengths can be increased, thus improving the performance of the finite-length codes by using the splitting method.[23]

In this section, a brief survey about the puncturing-based dynamic error control coding is reviewed. It is shown that the puncturing technique is really an effective method to realize dynamic error control coding. Nonetheless, the optimized rate-compatible puncturing is a demanding work, especially for turbo codes and LDPC codes. It must be noted that different optimization criteria have been developed for different purposes. In practical applications, people may select one that can fulfill their requirements.

Rate-compatible punctured codes are of particular interest in packet data systems that allow for retransmission requests such as automatic repeat request with forward error correction systems to achieve desired throughput efficiency with a high degree of flexibility. As illustrated in

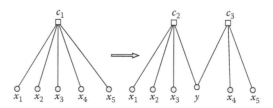

Figure 16.8   Splitting a parity check equation.[23]

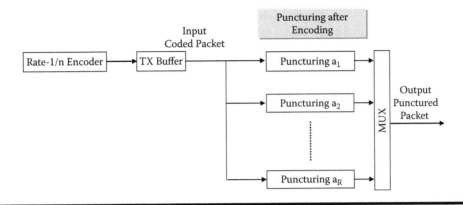

Figure 16.9   The puncturing-based dynamic error control encoder scheme in ARQ/FEC.

figures 16.9 and 16.10, for the hybrid ARQ/FEC scheme, the dynamic forward error control is enabled by selecting one of the rate-compatible puncturing tables of a_1, a_2, ..., a_R with the decreasing code rate (increasing error protection capability) when the receiver fails in recovering the information block. Due to the rate-compatible restriction, incremental redundancy will be sent only when negative acknowledgment (NAK) is received. At the receiver, depuncturing is necessary to insert erasure at those punctured positions before decoding. Code combining could be employed to improve the performance as well.[24] Actually, ARQ/RCPT codes have been adopted as a standard enhanced high-speed packet access (HSPA) technique in the third-generation mobile communications.[11] The dynamic forward error control coding of RC-LDPC codes was utilized in the transmission of progressively coded images over fading channels.[35] Unlike the former application, here the transmitter needs a channel estimator at the receiver to feed back the CSI to adjust

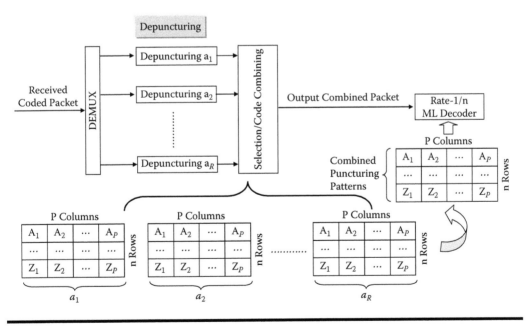

Figure 16.10   The puncturing-based dynamic error control decoder scheme in ARQ/FEC.

the code rate via puncturing. The drawback is that channel estimation error will contribute to degradation in the achieved performance, as discussed in section 16.1.

16.3.2 Extending-Based Dynamic Error Control Coding

Although the rate puncturing technique is effective for generating a set of subcodes with variable code rate, the lowest code rate is the mother code rate of $1/M$. Recently, the extending-based technique was proposed to extend the compatible code rate below the mother code rate limit. Rate-compatible repetition convolutional (RCRC) codes and extending RC-LDPC codes are two examples of the extending-based dynamic error control coding scheme.

16.3.2.1 Rate-Compatible Repetition Convolutional (RCRC) Codes

Samir Kallel and David Haccoun[9] first proposed the RCRC code. The construction method for repetition codes is similar to the one for punctured codes. The main difference is that instead of puncturing the coded bits, the repetition codes are generated by duplicating some code symbols of a parent convolutional code. Kallel and Haccoun proposed to generate the optimum RCRC code from known high-rate punctured convolutional codes.[9] The operation of duplicating the coded bits could also be represented by an $M \times P$ repetition matrix (where P is the repetition period). For example, for a rate 2/5 repetition convolutional code with memory $m = 6$ and repetition period $P = 2$, the optimum repetition matrix is given by

$$\begin{bmatrix} 2 & 1 \\ 1 & 1 \end{bmatrix}$$

The elements of this repetition matrix are greater than or equal to 1, indicating the number of duplications of the corresponding coded bit. Two repetition codes obtained from the same parent code are considered to be rate compatible if the lower-rate code uses all the coded bits of the higher-rate code, plus one or more duplicated bits. This means that all the elements in the repetition matrix of the lower-rate code must be greater than or equal to the corresponding elements in the repetition matrix of the higher-rate code.

Clearly, the repetition code obtained from a noncatastrophic parent code cannot be catastrophic. However, one can choose from among all possible combinations of encoded bits to be duplicated that yield the best repetition code. Lin and Svensson proposed a new algorithm for finding the optimum RCRC codes based on the optimum distance spectrum (ODS) criterion.[10] The searching algorithm is slightly different from the one proposed in Kallel and Haccoun[9] in that it is only constrained by the rate compatibility and the ODS criterion. The construction of RCRC codes starts with the rate $1/M$ parent code. Let $a_{P/MP}$ denote the repetition matrix of the parent code with rate of $1/M$. The construction of good (in terms of the ODS criterion) RCRC codes could be outlined as follows:

> Step 1. **Initialization:** Select a parent code with rate $1/M$ and code memory of m, and select the repetition period P. Set all the elements of the repetition matrix to 1.
>
> Step 2. **Repetition matrix construction:** Let $i = 1$ to i_{max}; for each value of i, $a_{P/(MP+i)} = a_{P/(MP+i-1)}$; let $j = 0$ to $M - 1$, and $k = 0$ to $P - 1$, $a_{P/(MP+i)}(j,k) = a_{P/(MP+i-1)}(j,k) + 1$.

Determine the achieved distance spectra of $\{a_d\}$ and $\{c_d\}$; then select the best repetition matrix as $a_{P/(MP^+i)}$; with the optimum ODS.

In the above algorithm, $a_{P/(MP+i)}$ is the repetition matrix for the rate $P/(MP+i)$ code, and $a_{P/(MP+i)}(j,k)$ refers to the element on the jth row and the kth column of the repetition matrix of $a_{P/(MP+i)}$. i_{max} is the upper limit of i, which can be set to be any integer value larger than or equal to 1. This means that the number of lower rates for RCRC codes has no limit. In summary, the main advantage of the above construction method is its completeness; it does not have any restrictions on the repetition matrix, except that it must be rate compatible. A large family of RCRC codes could be generated from the best-rate $1/M$ codes. The paid cost of the construction is its complexity during the exhaustive searching procedure, wherein the key ingredient is the calculation of the achieved distance spectra given some candidate repetition choice of $a_{P/(MP+i)}(j,k) = a_{P/(MP+i-1)}(j,k)+1$.

It is verified that the hybrid ARQ/FEC scheme based on the dynamic RCRC codes outperforms the conventional hybrid ARQ/FEC scheme in view of its better flexibility and adaptability to channel conditions, even under wide noise variations and severe signal degradations.

16.3.2.2 *Extending-Based Rate-Compatible LDPC Codes*

As discussed in section 16.3.1, the conventional technique of puncturing could be utilized to generate the efficient rate-compatible LDPC codes. Although puncturing provides a viable solution to produce RC-LDPC codes, the efficiency is limited at high-rate ranges where the amount of puncturing is not large. Li and Narayanan[14,15] proposed the extending construction for the RC-LDPC codes. Just opposite to puncturing, the method of extending builds RC codes from high rates to low rates through the addition of more parity bits. A strong motivation for extending comes from the observation that the quality of the initial transmission is the most important to achieve high throughput in ARQ systems. If the initial forward error correction coding transmission corresponds to a nonpunctured LDPC code, and additional parity bits are added to reduce the rate in such a way that the extended code provides sufficiently good performance at the lower rate. In fact, the repetition in RCRC codes is the simplest extending form to construct RC convolutional codes.

The extending-based construction of RC-LDPC codes is illustrated in figure 16.11. The parity check matrix of the highest-rate code has $M_0 = N_0 - K$ rows and N_0 columns with column weight $t \geq 3$ (upper-left part in figure 16.11). The parity check matrix of each lower-rate code is constructed by padding M_i rows and M_i columns, until finally reaching a $(N_0 + \sum_{i=1}^{L} M_i, K)$ code after L levels of padding. A family of RC-LDPC codes of rates $R_0 > R_1 > \cdots > R_L$ thus results, where

$$R_i = \frac{K}{N_0 + \sum_{j=1}^{i} M_j}, 1 \leq i \leq L$$

To embed higher-rate codewords (i.e., rate-compatible restriction) in lower-rate codewords, the upper-right part of each padding must be 0s, as shown. The squares in the bottom-right part have column weight 3 to ensure the resulting parity check matrix also has a column weight of at least 3. The bottom-left area is made reasonably sparse to ease the construction and to save the decoding

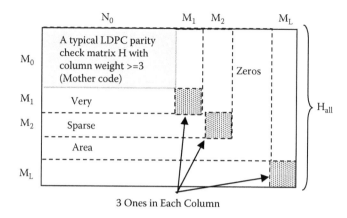

Figure 16.11   Illustration of RC-LDPC codes by extending.[15]

complexity, but at least one 1 is needed for each row to build sufficient dependencies between the code bits of the mother code and the newly added parity bits.

The decoder of extending-based RC-LDPC codes implements the structure of the H_{all} matrix. When decoding the subcode with rate $R_i, i = 0,1,...,L$, only the relevant bits and checks will be used for extrinsic information exchange. As rate decreases, more parity checks and bits join the message exchange process and, through the increased information and enhanced dependencies, offer a better error correction capability. Compared with the puncturing-based RC-LDPC codes, one advantage of the extending-based RC-LDPC code is its low complexity. The puncturing method requires a fixed complexity regardless of channel conditions, whereas the complexity of extending decreases as channel conditions improve. Most of the time, extending involves less decoding complexity than puncturing.[15]

It is shown through computer simulations that extending-based RC-LDPC codes outperform the puncturing-based RC-LDPC codes at low rates, while the situation is just the opposite at high code rates.[15] It follows naturally that efficient RC-LDPC codes should take advantage of both approaches. More specifically, extending is preferred to construct good RC-LDPC codes at low rates, while puncturing is preferred to construct good RC-LDPC codes at high rates.

16.4 Dynamic Forward Error Correction Control for Mobile Broadcasting

In recent years, the wireless industry has seen explosive growth in device capability, especially in relation to mobile cellular phones. Ever-increasing computing power, memory, and high-end graphic functionalities have accelerated the development of new and exciting wireless services. Among all services, the appeal of video and multimedia is enormous. Since its conception in 1993, digital video broadcasting (DVB) offers a groundbreaking infrastructure for future digital TV delivery.[37] The DVB-T (terrestrial) standard provides an efficient way of carrying multimedia broadcasting services over digital terrestrial broadcasting networks, while DVB-H (handheld) is particularly designed for use with handheld terminals.[38] It is fully compatible with DVB-T and

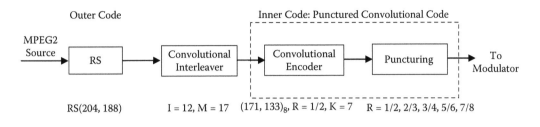

Figure 16.12   Forward error control coding scheme used on DVB-T.

features more power savingd, noise tolerance, and seamless handover. Other mobile broadcasting protocols include DMB (Digital Multimedia Broadcasting), ISDB-T (Integrated Services Digital Broadcasting–Terrestrial), and MediaFLO. The detailed comparative study of these mobile broadcasting protocols is presented elsewhere in this book. This section will focus on the dynamic forward error control scheme for DVB-H. Figure 16.12 shows the block diagram of the error control function with parameters for DVB standards.[37]

First, the MPEG-2 frames shall be organized in fixed-length packets, following the MPEG-2 transport multiplex. The total packet length of the MPEG-2 transport multiplex (MUX) packet is 188 bytes, wherein 1 sync-word byte is included. Then an outer Reed–Solomon RS(204, 188, $t = 8$) shortened code shall be applied to each randomized MPEG-2 transport packet to generate an error-protected packet, where the redundancy of length 16 (= 204 – 188) is computed and appended to the actual 188 bytes so that a block with a length of 204 is formed. The RS(204, 188) shortened code, which allows correction of up to 8 random erroneous bytes in a received word of 204 bytes, is derived from the original systematic RS(255, 239, $t = 8$) code, whose code generator polynomial and field generator primitive polynomial are given by:

$$\text{Code generator polynomial: } g(x) = (x + \beta^0)(x + \beta^1)(x + \beta^2)\cdots(x + \beta^{15}),$$

where $\beta = (00000010)_2 = (02)_{hex}$

$$\text{Extension field generator polynomial: } p(x) = x^8 + x^4 + x^3 + x^2 + 1$$

The RS(204, 188) shortened code may be implemented by adding 51 (= 239 – 188) bytes, all set to zero, before the MPEG-2 information bytes at the input of an RS(255, 239) encoder. After the RS coding procedure these null bytes shall be discarded, leading to an RS codeword of 204 bytes. Then the 204 bytes stream is fed to a convolutional interleaver with depth I = 12 (= 204/17) to increase the robustness to long burst errors, converted from bytes to bits, and processed by a punctured convolutional encoder that further enhances its error-handling capability. The punctured tables employed are given in table 16.1.

It could be readily observed that instead of the fixed puncturing period, the variable puncturing period is employed and the resultant puncturing patterns are not rate compatible compliant.

Because DVB-T was developed originally to transmit the broadcasting signal to the TV set, low power consumption has not been considered. In the link layer of the DVB-H protocol, two techniques—time slicing and Multi-Protocol Encapsulation-Forward Error Correction (MPE-FEC)—are employed. In DVB-H, the time-slicing technique is proposed to achieve less power consumption by time-sliced transmissions (i.e., switching off part of the terminal while maintaining a virtual presence of the service to the user). In the MPE-FEC, additional RS codes with erasure decoding are included to cope with a higher channel error rate in a mobile environment

Table 16.1   Puncturing Patterns for DVB

Code Rates	Puncturing Patterns
1/2	$\begin{bmatrix} 1 \\ 1 \end{bmatrix}$
2/3	$\begin{bmatrix} 1 & 0 \\ 1 & 1 \end{bmatrix}$
3/4	$\begin{bmatrix} 1 & 0 & 1 \\ 1 & 1 & 0 \end{bmatrix}$
5/6	$\begin{bmatrix} 1 & 0 & 1 & 0 & 1 \\ 1 & 1 & 0 & 1 & 0 \end{bmatrix}$
7/8	$\begin{bmatrix} 1 & 0 & 0 & 0 & 1 & 0 & 1 \\ 1 & 1 & 0 & 1 & 0 & 1 & 0 \end{bmatrix}$

in DVB-H. Figure 16.13 shows the forward error control coding scheme used on DVB-H. Besides the additional MPE-FEC RS code using the erasure decoding, the virtual interleaving techniques are employed to provide additional error protection. Here, virtual interleaving means that data is written column-wise and encoded row-wise in the MPE-FEC frame. Figure 16.14 shows the MPE-FEC frame structure specified in the DVB-H protocol.

The MPE-FEC frame is a matrix with 255 columns and a flexible number of rows. The maximum number of rows is 1,024, which makes the total MPE-FEC frame size almost 2 Mbits. The MPE-FEC frame consists of the application data table and RS data table, as shown in figure 16.14. The leftmost 191 columns of the MPE-FEC frame are called the application data table and filled with IP (Internet Protocol) datagrams and padded symbols. The rightmost 64 columns of the MPE-FEC frame are called the RS data table and filled with the parity symbols of RS(255, 191) code.

The application data table is filled with IP datagrams from the empty leftmost column. A new datagram fills after the end of the previous datagram. A datagram continues at the top of

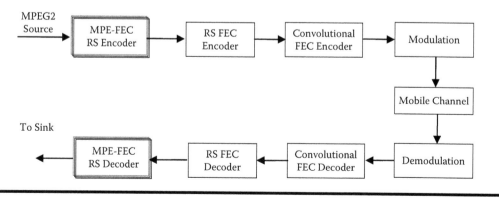

Figure 16.13   Forward error control coding scheme used on DVB-H.

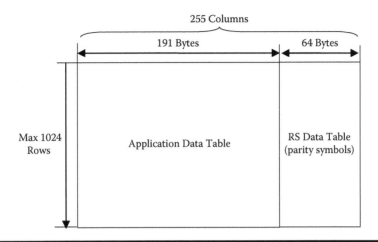

Figure 16.14   The frame structure of MPE-FEC.

the following column if it does not end precisely at the end of a column. Any remaining unfilled columns are padded with zero when all datagrams have entered the application data table, which makes the application data table completely filled. Sixty-four parity symbols can be obtained for each row from 191 data and possibly padded symbols after all of the leftmost 191 columns are filled. The 64 parity symbols fill the RS data table. Therefore, each row is an RS(255, 191) codeword.

One IP datagram corresponds to one MPE section while it is transmitted. On the other hand, one column of the RS data table corresponds to one FEC section. Each section has a start address for the payload in the section header. This address indicates the symbol position in the MPE-FEC frame. Thus, the receiver is able to put the received datagram in the right position in the MPE-FEC frame. All MPE and FEC sections have CRC-32 codes. The received section is marked as reliable if there is no error. Otherwise, it is marked as unreliable. All the symbols in the unreliable section are considered to be erasures. The receiver can correct twice the number of erroneous symbols with erasure decoding becaues the positions of erasure can be found by CRC code check. Thus, RS(255, 191) code can correct errors up to 64 symbols in a 255-symbol codeword by erasure decoding. For illustration purposes, figures 16.15 and 16.16 show the input symbol error rate (SER = number of errored symbols/total number of symbols) versus the output SER for the Reed–Solomon (255, 191, 64) code used for the MPE-FEC and that of the RS(204, 188, 8) code at high SER by DVB-T. It could be readily observed that the MPE-FEC RS code using erasure decoding will improve the reliability of the RS(204, 188, 8) shortened code with no erasure significantly.

The problem of erasure decoding by the MPE-FEC code is the possible waste of error correction capability. In theory, the error-correcting capability of erasure decoding can be doubled, compared to the conventional RS decoding, because the error positions can be found by CRC code. But erasure decoding shows worse error performance than conventional RS decoding in practice (except the very high SER region) because even the correct symbols in an erroneous frame may be considered errors. Therefore, there may be a waste of error-correcting capability because of false erasures. Moreover, the maximum size of one MPE section is 4,096 bytes, which corresponds to four columns. Therefore, there may be much more waste. Lim et al. proposed to replace the MPE-FEC RS code with a LDPC code.[39] Compared with the erasure decoding, there is no waste of error-correcting capability in the LDPC code. And soft decision and iterative decoding are possible in LDPC code. It is shown that the LDPC code has coding gains of about 6 and 31 dB, respectively, compared to the conventional RS and erasure decoding at all mobile speeds

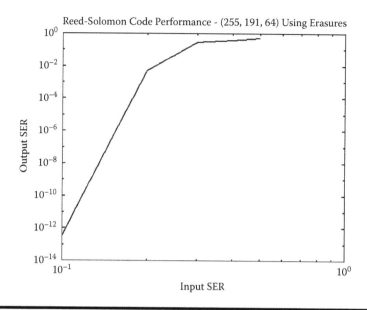

Figure 16.15 **Output versus input SER for MPE-FEC (255, 191, 64) RS code at high SER.**

(20, 80, and 120 km/h).[39] It seems that LDPC code is a promising alternative in the MPE-FEC coding scheme.

There is another problem. Although MPE-FEC gives extra error protection, the standard of DVB-H is designed with fixed-value link layer parameters, that is, fixed code rates in the system.

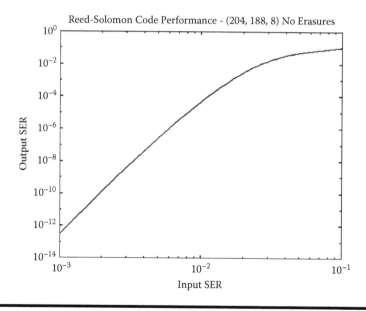

Figure 16.16 **Output versus input SER for (204, 188, 8) RS code at high SER.**

However, as we discussed in section 16.1 in great detail, a mobile network environment tends to change from time to time, so a static error control scheme that behaves well in one case may become unsuitable in another. Recently, a preliminary dynamic error control scheme for the DVB-H was proposed to make the forward error control scheme adapt to the channel conditions.[36] It is assumed that there exists a return channel from all receivers to the broadcast transmitter, and a dynamic MPE-FEC RS code is available. So all end users may periodically report their error rates to the transmitter. At the transmitter, an error observer is employed to aggregate the channel condition reports from receivers. Based on the knowledge of the channel conditions, the transmitter may adjust the Reed–Solomon redundancy accordingly. As expected, simulation results show that dynamic DFEC outperforms nonadaptive systems with the fixed-level FEC scheme.

It must be stated that more research efforts are still needed to make the flexible dynamic error control for mobile broadcasting applications practical and appealing. Some of the problems are listed here for reference:

1. **How do we get the CSI information?** The prospect of interactivity is one great attraction of digital video. To achieve interactivity, a return channel is necessary. Therefore, it is expected that the feedback channels are becoming more and more realistic in mobile applications. So mobile end users may acknowledge their receiving qualities to the broadcast station. However, how do we get the CSI information from dozens, or maybe hundreds and thousands, of feedbacks? Problems arise when multiple end users receive the broadcast signal and report different error rates, as they most likely will experience in reality. Their responses may not be synchronized, and the service provider must try to reconcile the diverse feedbacks to best serve them. The rate-adaptive error control mechanism of the hybrid ARQ/FEC scheme proposed by Cho[25] for the hybrid satellite and terrestrial network offers a preliminary but promising idea. The drawback is the system complexity involved, because both broadcasting communications and traditional point-to-point (PTP) transmission are needed. Meanwhile, it is worthwhile to investigate effective observation fusion schemes for the broadcasting station to know the average receiving signal variations of all mobile users (or at least the variation in its serving coverage area).

2. **How do we respond to the varying conditions via the dynamic error control scheme?** This question is related to not only the theoretical concern, but also the practical interest. In practical applications, there may be a number of mobile users who have subscribed to some kind of broadcasting services. If each mobile user may interact with the broadcasting station, and therefore feed back his or her request to the transmitter, how should the broadcasting station respond to the variations in a timely way? It is shown that the dynamic error control scheme does make sense for mobile broadcasting services if all receivers are cooperative when feeding back CSI information to the transmitter.[36] But only one user will receive a response from the broadcasting station in the dynamic forward error control scheme therein. If multiple CSI feedbacks are assumed to be available, in view of the adaptive coding, how and when should we respond to so many different channel state information feedbacks by mobile end users? Is there any optimal adaptation strategy? This is still an open and unsolved problem for the dynamic forward error control scheme in broadcasting scenarios. Intuitively, this problem is also related to the trade-off between individual quality of service and average service quality (or service fairness). So it seems that the solution to this problem may be service scheduling strategy dependent.

3. **How to incorporate the dynamic forward error control coding scheme with the application layer forward error correction for mobile broadcasting?** Automatic Repeat

Request (ARQ) scheme is a promising technique to solve the reliable broadcast communication problem. Receivers use a feedback channel to send retransmission requests for lost or erroneous packets. However, too much feedback will increase network overhead, cause the feedback implosion as well as wasteful use of bandwidth. In recent years, a new class of sparse-graph codes, i.e., Fountain codes, is developed for protection of data against unknown erasure errors. For any given information symbols, Fountain codes can produce potentially limitless stream of encoding symbols, and each symbol is a linear function of all the information symbols generated independently and randomly according to a degree distribution. And the required number of received symbols for the successful decoding is dependent on the channel conditions. Compared with the ARQ scheme, there is no need for the retransmission via feedback channel. In general, Fountain code behaves well in erasure channels (and the readers are encouraged to read chapter 10 of this book for the details). However, the contaminated signals by the noise and the fluctuations by the fading deteriorate the performance of Fountain codes in wireless scenarios. Besides the improvement by employing the soft decision decoding (e.g., the Belief-Propagation (BP) algorithm), concatenation with some powerful error correction codes like the LDPC codes (e.g., Raptor codes) gives strongly recommended. However, we need more efforts to unveil how to incorporate the dynamic error control coding in the broadcasting coding applications.

16.5 Summary

In this chapter, the dynamic forward error control scheme is highlighted. It is shown that flexible dynamic error control coding becomes more and more desirable in wireless mobile communications to accommodate different error protection requirements or time-varying channels. The FEC and ARQ/FEC schemes are two important underlying techniques for the realization of the dynamic error control scheme, wherein the CSI and CSI-like information are prerequisite conditions. It is shown that puncturing and extending are two basic techniques to realize code rate. Generally, the puncturing technique outperforms in higher-code-rate regions, while extending outperforms in lower-code-rate region. Combining both techniques leads to the best of both worlds. The forward error control coding schemes of both DVB-T and DVB-H are presented to indicate the most important problems when realizing the dynamic error control scheme in mobile broadcasting scenarios. Finally, it is stated that, although further investigations are needed, the technique of the dynamic error control scheme has great potential in practical applications of mobile multimedia broadcasting.

References

1. S. Lin and D. J. Costello Jr. 1984. *Error control coding: Fundamentals and applications*. Englewood Cliffs, NJ: Prentice-Hall.
2. L. Lugand, D. J. Costello Jr., and R. H. Deng. 1989. Parity retransmission hybrid ARQ using rate 1/2 convolutional codes on a non-stationary channel. *IEEE Trans. Commun.* 37:755–65.
3. S. Lin and P. S. Yu. 1982. Hybrid ARQ scheme with parity retransmission for error control of satellite channels. *IEEE Trans. Commun.* 30:1701–19.
4. Y.-M. Wang and S. Lin. 1983. A modified selective-repeat type-II hybrid ARQ system and its performance analysis. *IEEE Trans. Commun.* 31:593–608.

5. J. Hagenauer. 1988. Rate-compatible punctured convolutional codes (RCPC Codes) and their applications. *IEEE Trans. Commun.* 36:389–400.
6. J. B. Cain, G. C. Clark Jr., and J. M. Geist. 1996. Punctured convolutional codes of rate (n—1)/n and simplified maximum likelihood decoding. *IEEE Trans. Inform. Theory* 25:97–100.
7. A. S. Barbulescu and S. S. Piertrobon. 1995. Rate compatible turbo codes. *Electronic Lett.* 31:535–36.
8. D. N. Rowitch and L. B. Milstein. 2000. On the performance of hybrid FEC/ARQ systems using rate compatible punctured turbo (RCPT) codes. *IEEE Trans. Commun.* 48:948–59.
9. S. Kallel and D. Haccoun. Generalized type II hybrid ARQ scheme using punctured convolutional coding. *IEEE Trans. Commun.* 38:1938–46.
10. Z. Lin and A. Svensson. 2000. New rate-compatible repetition convolutional codes. *IEEE Inform. Theory* 46:1651–57.
11. 3GPP. *Physical layer aspects of UTRA high speed downlink packet access.* Release 4. 3GPP TSG-RAN.
12. V. K. N. Lau and Y. K. Kwok. 2005. *Channel adaptation technologies and cross layer design for wireless systems with multiple antennas—Theory and applications.* New York: John Wiley & Sons.
13. D. M. Mandelbaum. 1974. An adaptive-feedback coding scheme using incremental redundancy. *IEEE Trans. Inform. Theory* 20:383–89.
14. J. Li and K. R. Narayanan. 2002. Rate-compatible low density parity check codes for capacity-approaching ARQ scheme in packet data communications. In *Proceedings of the International Conference on Communications, Internet, and Information Technology (CIIT)*, St. Thomas, U.S.V.I., 201–06.
15. J. Li. 2002. Low-compleixty, capacity-approaching coding schemes: Design, analysis and applications. Ph.D. dissertation, Texas A&M University.
16. J. Ha, J. Kim, and S. W. McLauhlin. 2004. Rate-compatible puncturing of low-density parity-check codes. IEEE Trans. Inform. Theory 50:2824–36.
17. J. Ha, J. Kim, D. Klinc, and S. W. McLaughlin. 2006. Rate-compatible punctured low-density parity-check codes with short block lengths. *IEEE Trans. Inform. Theory* 52:728–38.
18. S. Benedetto and G. Montorsi. 1996. Design of parallel concatenated convolutional codes. *IEEE Trans. Commun.* 44:591–600.
19. D. Divsalar, S. Dolinar, R. J. McEliece, and F. Pollara. 1995. *Transfer function bounds on the performance of turbo codes.* JPL TDA Prog. Rep. 42-122. Jet Propulsion Lab., Pasadena, CA.
20. S. Benedetto and G. Montorsi. 1996. Unveiling turbo codes: Some results on parallel concatenated coding schemes. *IEEE Trans. Inform. Theory* 42:409–28.
21. L. C. Perez, J. Seghers, and D. J. Costello Jr. 1996. A distance spectrum interpretation of turbo codes. *IEEE Trans. Inform. Theory* 42:1698–1709.
22. M. R. Yazdani and A. H. Banihashemi. 2004. On construction of rate-compatible low-density parity-check codes. *IEEE Commun. Lett.* 8:159–61.
23. H. Pishro-Nik and F. Fekri. 2007. Results on punctured low-density parity-check codes and improved iterative decoding techniques. *IEEE Trans. Inform. Theory* 53:599–614.
24. E. Malkamaki and H. Leib. 2000. Performance of truncated type-II hybrid ARQ schemes with noisy feedback over block fading channels. *IEEE Trans. Commun.* 48:1477–87.
25. S. Cho. 2000. Rate-adaptive error control for multimedia multicast services in sattelite-terrestrial hybrid networks. In *Proceedings of WCNC*, New Orleans, 1, 446–50.
26. L. H. C. Lee. 1994. New rate-compatible punctured convolutional codes for Viterbi decoding. *IEEE Trans. Commun.* 42:3073–79.
27. R. Garello, P. Pierleoni, and S. Benedetto. 2001. Computing the free distance of turbo codes and serially concatenated codes with interleavers: Algorithms and applications. *IEEE J. Select. Areas Commun.* 19:800–12.
28. E. Rosnes and O. Ytrehus. 2005. Improved algorithm for the determination of turbo-code weight distribution. *IEEE Trans. Commun.* 53:20–26.

29. S. Crozier, P. Guinand, and A. Hunt. 2004. Computing the minimum distance of turbo-codes using iterative decoding techniques. In *Proceedings of the 22th Biennial Symposium on Communications*, Kingston, Ontario, 306–8.

30. R. Garello and A. Vila-Casado 2004. The all-zero iterative decoding algorithm for turbo code minimum distance computation. In *Proceedings of the IEEE International Conference on Communications (ICC)*, Paris, 1, 361–64.

31. C. Berrou and S. Vaton. 2002. Computing the minimum distances of linear codes by the error impulse method. In *Proceedings of the IEEE International Symposium on Information Theory (ISIT)*, Lausanne, 5.

32. F. Babich, G. Montorsi, and F. Vatta. 2004. Some notes on rate-compatible punctured turbo codes (RCPTC) design. *IEEE Trans. Commun.* 52:681–84.

33. E. Rosnes and O. Ytrehus. 2005. On the construction of good families of rate-compatible punctured turbo codes. In *Proceedings of ISIT2005*, Adelaide, 602–6.

34. F. Daneshgaran and P. Mulassano. 2004. The rate-allocation problem for turbo codes. *IEEE Trans. Commun.* 52:861–65.

35. X. Pan, A. H. Banihashemi, and A. Cuhadar. 2006. Progressive transmission of image over fading channels using rate-compatible LDPC codes. *IEEE Trans. Image Processing* 15:3627–35.

36. J. Yao, W.-F. Huang, and M.-S. Chen. 2006. DFEC: Dynamic forward error control for DVB-H. In *Proceedings of IEEE Conference on Sensor Networks, Ubiquitous, and Trustworthy Computing*, Taichung, 172–77.

37. ETSI. 2004. *Digital video broadcasting (DVB): Framing structure, channel coding and modulation for digital terrestrial television.* ETSI EN 300 744.

38. ETSI. 2004. *Transmission system for handheld terminals (DVB-H).* ETSI DVB Document A081.

39. H. T. Lim et al. 2006. Performance improvement of DVB-H system with LDPC code in MPE-FEC frame. In *Proceedings of the 11th IEEE Symposium on Computers and Communications*, Sardinia, 472–6.

Chapter 17

Air Interface Enhancements for Multimedia Broadcast/ Multicast Service

Américo Correia, Nuno Souto, João Carlos Silva, and Armando Soares

Contents

Keywords

link-level simulation, system-level simulator, enhanced UMTS, block error rate, multicasting, broadcasting, video streaming, packet scheduling, multiresolution, macrodiversity combining

Abstract

This chapter first addresses enhancements based on multiresolution schemes due to the use of multicodes, and second, it considers the high-order coded modulations and single-code wideband code division multiple access as a means of increasing the bit rate per user, and as a result increasing the capacity of the digital cellular radio network due to increased spectral efficiency. At last, the spatial dimension by means of MIMO systems, based on spatial multiplexing, promising impressive increases in terms of capacity, will be presented. The combination of the enhancements is accomplished by adaptive transmission techniques. The final section of this chapter presents the use of several adaptive transmission techniques, namely, link adaptation and macrodiversity-combining techniques, reducing the required transmitted power of the digital cellular radio network.

17.1 Introduction

Multimedia Broadcast and Multicast Service (MBMS), introduced by Third Generation Partnership Project (3GPP) in Release 6, is intended to efficiently use network/radio resources (by transmitting data over a common radio channel), both in the core network and, most importantly, in the air interface of the Universal Mobile Telecommunications Systems Terrestrial Radio Access Network (UTRAN), where the bottleneck is placed to a large group of users. MBMS includes point-to-point (PtP) and point-to-multipoint (PtM) modes. The former allows individual retransmissions but the latter does not.

MBMS is targeting high- (variable-) bit-rate services over a common channel. One of the most important properties of MBMS is resource sharing among many user equipments (UEs), meaning that many users should be able to listen to the same MBMS channel at the same time. So, power should be allocated to this MBMS channel for arbitrary UEs in the cell to receive the MBMS service.

There is still a lot of investigation in ways to improve the delivery of multimedia information. The multimedia paradigm has put pressure in resources optimization, and sharing channels is one of the most important aspects in network optimization. Efficient network resources usage should be the leverage for forthcoming multimedia applications. Besides that, to guarantee scalability, multiresolution schemes have to be considered in Universal Mobile Telecommunications System (UMTS) environments.

For broadcast and multicast transmissions in a mobile cellular network, depending on the communication link conditions, some receivers will have better signal-to-noise ratios (SNRs) than others, and thus the capacity of the communication link for these users is higher. Cover[1] showed that in broadcast transmissions it is possible to exchange some of the capacity of the good communication links to the poor ones, and the trade-off can be worthwhile. Possible methods to improve the efficiency of the wideband code division multiple access (WCDMA) network are the use of multiple codes or multiple antennas at both the transmitter and receiver (multiple-input multiple-output [MIMO]) or hierarchical signal constellations. Each one is able to provide unequal bit error protection. In any case, there are two or more classes of bits with different error protection, to which different streams of information can be mapped. Depending on the channel conditions, a given user can attempt to demodulate only the more protected bits, or also the other bits that carry the additional information. Several papers have studied the use of nonuniform constellations for this purpose.[1,3] Hierarchical 16-QAM and 64-QAM constellations are already incorporated in the Digital Video Broadcasting–Terrestrial (DVB-T) standard.[4]

To accomplish high data rates over wireless links, the use of multiple transmit and receive antennas (MIMO) is an alternative that does not require any extra bandwidth. Also in this chapter we will analyze MIMO associated or not with High Speed Downlink Packet Access (HSDPA), as another radio resource management technique to provide multiresolution.

The HSDPA mode[5] has been standardized for UMTS providing bit rates up to 10 Mbps on a 5 MHz carrier for the best-effort packet data services in the downlink. HSDPA supports new features that rely on, and are tightly coupled to, the rapid adaptation of transmission parameters to instantaneous radio conditions. The specific feature for multiresolution is fast link adaptation. Instead of compensating the variations of downlink radio conditions by means of power control, the transmitted power is kept constant and the modulation and coding of the transport block are chosen every transmission time interval (TTI = 2 ms) for each user. This is called adaptive modulation and coding (AMC). For users in good conditions, 16-QAM can be allocated to maximize throughput, while users in bad conditions are penalized on throughput, reaching a point to which the service can be denied. Up to 30 different channel quality indicators (CQIs) are employed associated with corresponding different bit rates.

The HSDPA as a means to deliver MBMS multiresolution video streaming will be studied with suitable packet scheduler algorithms that try to guarantee the same bit rate for all users, to offer fairness and a good capacity.

In this chapter multiresolution broadcast is the first to be presented, including the multicode packet scheduler. Next, we describe hierarchical (or nonuniform) quadrature amplitude modulation (QAM) constellations. MIMO systems are then introduced as another multiresolution technique. The HSDPA mode is presented and related to a multicast multiresolution system. Soft combining is described as a technique capable of reducing the required transmitting power. Simulation results related to the previous topics without and with soft combining are next included. Finally, the main conclusions are discussed.

17.2 Multiresolution Broadcast

The introduction of multiresolution in a broadcast cellular system deals with source coding and the transmission of the output data streams.

In a broadcast cellular system there is a heterogeneous network with different terminal capabilities and connection speeds. For the particular case of video, there are several strategies presented in the literature to adapt its content within a heterogeneous communications environment.[6–9] In this chapter we chose scalable media.

Scalable media (see figure 17.1)[8] provides a base layer for minimum requirements, and one or more enhancement layers to offer improved qualities at increasing bit/frame rates and resolutions. This method therefore significantly decreases the storage costs of the content provider. Common scalability options are temporal scalability, spatial scalability, and SNR scalability.

Spatial scalability and SNR scalability are closely related, the only difference being the increased spatial resolution provided by spatial scalability. SNR scalability implies the creation of multirate bit streams. It allows for the recovery of the difference between an original picture and its reconstruction. This is achieved by using a finer quantizer to encode the difference picture in an enhancement layer. This additional information increases the SNR of the overall reproduced picture, hence the name SNR scalability. Spatial scalability allows for the creation of multiresolution bit streams to meet varying display requirements/constraints for a wide range of clients. It is essentially the same as in SNR scalability, except that a spatial enhancement layer here attempts to recover the coding loss between an upsampled version of the reconstructed reference layer picture and a higher resolution version of the original picture.

Besides being a potential solution for content adaptation, scalable video schemes may also allow an efficient usage of power resources in MBMS, as suggested in Cover.[1] This is depicted in figure 17.1, where two separate physical channels are provided for one MBMS service (e.g., 256 kbps)—one for the base layer, at half the bit rate of the total bit rate (128 kbps), and with a power allocation that can cover whole cell range; and one for the enhanced layer, also at half the bit rate of the total bit rate (128 kbps), but with less power allocation than that of the base layer.

17.2.1 Multicode

A flexible common channel, suitable for point-to-multipoint transmissions, is already available in UMTS networks, namely, the Forward Access Channel (FACH), which is mapped onto the Secondary Common Control Physical Channel (S-CCPCH).

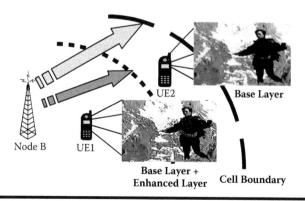

Figure 17.1   Scalable video transmission.

In 3GPP[11] it was shown that without macrodiversity combining, about 40 percent of the sector total power has to be allocated to a single 64 kbps MBMS if full cell coverage is required. This makes MBMS too expensive because the overall system capacity is limited by the power resource. To make MBMS affordable for the UMTS, its power consumption has to be reduced. If MBMS is carried on S-CCPCH, there is no inner-loop power control. Assuming that macrodiversity combining is not used, extra power budget has to be allocated to compensate for the receiving power fluctuations.

Our approach is to consider MBMS video streaming as scalable, with one basic layer to encode the basic quality and consecutive enhancement layers for higher quality. Only the most important stream (basic layer) is sent to all the users in the cell to provide the basic service. The less important streams (enhancement layers) are sent with less amount of power or coding protection, and only the users who have better channel conditions are able to receive that additional information to enhance the video quality. This way, transmission power for the most important MBMS stream can be reduced because the data rate is reduced, and the transmission power for the less important streams can also be reduced because the coverage requirement is relaxed.

The first studied transmission scheme uses a single rate stream (single spreading code with spreading factor SF = 8), which is carried on a single 256 kbps FACH and sent to the whole area in the cell. The second scheme uses a double streaming transmission, that is, two data streams (two orthogonal spreading codes with SF= 16), each FACH with 128 kbps where basic information for basic quality of service (QoS) is transmitted with the power level needed to cover the whole cell, and the second stream conveys additional information to users near node B (base station).

17.2.2 System Model

The system model (illustrated in figure 17.1) consists of two QoS regions, where the first region receives all information while the second receives the most important data. The QoS regions are associated with the geometry factor that reflects the distance of the UE from the base station antenna. The geometry factor G is defined as the ratio of I_{own} interference generated in the one cell to the I_{others} interference generated in the other cells plus P_N thermal noise, that is,

$$G = \frac{I_{own}}{I_{others} + P_N} \tag{17.1}$$

In table 17.1 the chosen G values are shown. For the first region the geometry factor is $G = 0$ dB, and for the second region $G = -3$ dB. Based on the chosen urban macrocellular topology used in our radio subsystem-level simulations, these two geometries values correspond to cell coverage of at least 60 and 80 percent, respectively.

Table 17.1   QoS Region Parameters

QoS Region	UE Capability	Maximum Bit Rate	G (dB)
1	UE1	256 kbps	0
2	UE2	128 kbps	−3

UE2 will receive the most important data (transmitted at 128 kbps) to get a basic video quality service, whereas UE1 will receive all the data to provide a higher-quality reproduction of the input video. Actually, UE2 will also receive both data stream layers, but it is not sure that the block error rate of the received enhancement layer will be below 0.01, and as a result, there will be a reduction in the quality of the video received by this user.

17.2.3 Hierarchical (Nonuniform) QAM Constellations

17.2.3.1 Signal Constellations

Another transmission method that provides a multiresolution system is the use of hierarchical constellations. In hierarchical constellations there are two or more classes of bits with different error protection and to which different streams of information can be mapped. By using nonuniformly spaced signal points (where the distances along the I or Q axis between adjacent symbols are different), it is possible to modify the different error protection levels. In this study we consider the use of 16-QAM nonuniform modulations for the transmission of broadcast and multicast services in WCDMA systems. For 16-QAM, two classes of bits are used. Some modifications to the physical layer of the UMTS to incorporate these modulations were already proposed in Souto et al.[12,13]

16-QAM hierarchical constellations are constructed using a main QPSK constellation where each symbol is in fact another QPSK constellation, as shown in figure 17.2.

The bits used for selecting the symbols inside the small inner constellations are called weak bits, and the bits corresponding to the selection of the small QPSK constellation are called stronger bits. The idea is that the constellation can be viewed as a 16-QAM constellation if the channel conditions are good, or as a QPSK constellation otherwise. In the latter situation the received bit rate is reduced to half. The main parameter for defining one of these constellations is the ratio between d_1 and d_2, as shown in figure 17.2:

$$\frac{d_1}{d_2} = k \tag{17.2}$$

where $0 < k \le 0.5$.

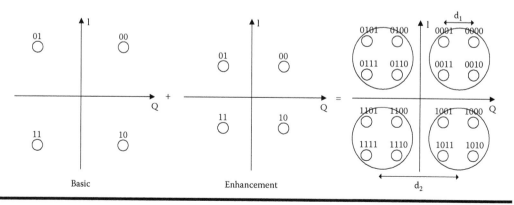

Figure 17.2 Signal constellation for 16-QAM nonuniform modulation.

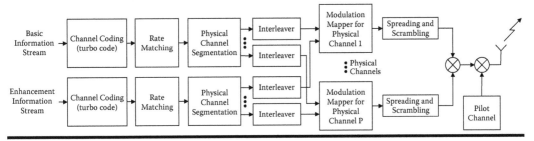

Figure 17.3   Proposed transmitter chain.

Each symbol s of the constellation can be written as

$$s = \left(\pm \frac{d_2}{2} \pm \frac{d_1}{2} \right) + \left(\pm \frac{d_2}{2} \pm \frac{d_1}{2} \right) j \qquad (17.3)$$

If $k = 0.5$, the resulting constellation is a uniform 16-QAM. When k is lower than 0.5, the bit error rate (BER) of the stronger bits decreases while the BER of the weaker symbols increases.

17.2.3.2   Proposed Transmitter Structure

Figure 17.3 shows a simplified transmission chain incorporating 16-QAM hierarchical constellations. In this scheme there are two parallel processing chains, one for the basic information stream and the other for the enhancement information. Each stream is turbo encoded (using the 3GPP rate 1/3 turbo code) and rate matching is performed for achieving the desired bit rate. Then each stream is segmented into P physical channels (each physical channel will be spread by a different orthogonal variable spreading code [OVSF]) that are individually interleaved. The physical channels of the two processing chains are mapped onto the constellation symbols in the modulation mappers according to the importance attributed to the chain. The modulated symbols are spread and scrambled, and the resulting physical channels are summed.

Before the transmission, a pilot channel (named Common Pilot Channel [CPICH] in UMTS) composed of known pilot symbols and spread by a reserved OVSF code with spreading factor 256 is added to the data signal. This pilot channel is orthogonal to all data channels and can be used in the receiver for channel estimation purposes.

17.2.3.3   Structure and Operation of the Iterative Receiver

To support the use of M-QAM hierarchical constellations in typical WCDMA environments, it is necessary to remove the multipath interference (MPI) (in the same path all the signals are orthogonal) composed not only by main users' physical channels but also by other interferers. An iterative receiver that uses feedback information from the turbo decoder for estimating and removing the interpath interference can be employed. This approach, which we refer to as turbo MPIC (multipath interference cancellation), is based on the concept of turbo equalization, where an equalizer and a channel decoder exchange information iteratively for suppressing the intersymbolic interference (ISI) caused by the channel. To accomplish this, it is necessary that the component decoders, in addition to the estimates of the information bits, also output information about the coded bits. This way, the channel decoder can help in the estimation of the transmitted signal, and thus of the MPI, which can be removed from the data channels and from the pilot channel before a new iteration is performed. The application of the MPI cancellation to the pilot channel allows the receiver to reestimate the channel coefficients in each iteration, and thus improve their reliability. To avoid an excessive complexity at the receiver, the turbo MPIC approach is applied only to the main user, while a simple MPIC philosophy is applied for estimating the interferers' signals (i.e., no decoding is performed in each iteration for these signals). The overall estimated MPI is used inside the RAKE and also for improving the channel estimates. This is the idea employed for the design of the proposed iterative receiver, whose structure is shown in figure 17.4.

For the operation of the receiver it is assumed that the receiver has some knowledge about all physical channels transmitted. In each iteration the RAKE performs a maximal ratio combining (MRC) of all despreaded signals processed by the fingers. The result then goes into the sequence of processing blocks that perform all the inverse operations of the transmitter, in the case of the main user's information, or only the despreading operations, in the case of interferers' signals. The demodulator computes the likelihood probabilities of the received coded bits to be used by the turbo decoders. Each turbo decoder has two outputs. One is the estimated information sequence, and the other is the sequence of log-likelihood ratio (LLR) estimates of the code symbols. These can be computed using some appropriate algorithm like MAP (maximum a posteriori) or SOVA (soft-output Viterbi algorithm) These LLRs are passed through the decision device, which outputs either soft-decision or hard-decision estimates of the code symbols. These estimates enter the transmitted signal rebuilder, which performs the same operations of the transmitter. The reconstructed signal then goes into a channel emulator that generates the estimated discrete multipath replicas multiplied by the respective fading coefficients. The estimated multipaths are then fed into the interference canceller, which subtracts the interference from the signals fed to each RAKE finger and to the channel estimator for the next iteration. This interference is composed by the sum of all paths except the one that is going to be extracted by the finger (RAKE) or whose channel coefficients are going to be estimated (channel estimator). Because in the first iterations the reliability of the data estimates is usually low, the interference signals can be weighted before the subtraction.

For the general case of a DS-CDMA system, the tth received signal sample can be expressed as

$$r_t = \sum_{l=1}^{L} \alpha_{l,t-\tau_l} \left(\sum_{p=1}^{P} s_{p,\left\lfloor \frac{t-\tau_l}{SF} \right\rfloor +1} \cdot c_{p,t-\tau_l} + s_{pilot,\left\lfloor \frac{t-\tau_l}{SF_{pilot}} \right\rfloor +1} \cdot c_{pilot,t-\tau_l} \right) + n_t, \tag{17.4}$$

where $\alpha_{l,t}$ and τ_l are the complex-valued channel gain and time delay (in samples) of the lth path, L is the number of resolvable paths, P represents the number of physical channels, SF is the spreading factor, SF_{pilot} is the spreading factor of the pilot channel (256 in the case of the CPICH), $s_{p,t}$ and $c_{p,t}$ represent the modulated symbol and spreading signal waveform of the pth physical channel, and $s_{pilot,t}$ and $c_{pilot,t}$ correspond to the modulated symbols and spreading signal of the pilot channel. The term n_t is the AWGN noise component.

The kth despreaded symbol associated with the lth finger of the pth physical channel is represented as

$$y_{p,l,k} = \frac{1}{SF} \sum_{t=(k-1)\cdot SF+1}^{k\cdot SF} r_{t+\tau_l} \cdot c^{*}_{p,t}. \tag{17.5}$$

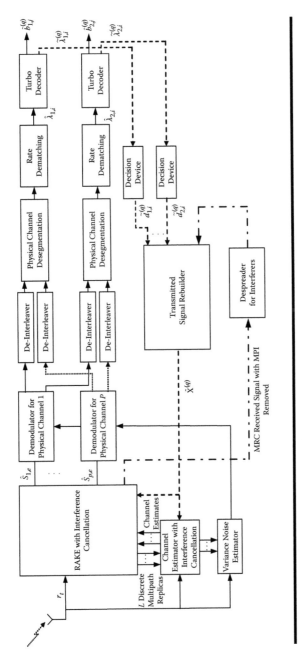

Figure 17.4 Proposed iterative receiver structure.

Assuming perfect channel estimation, the RAKE MRC combined data sequence of the kth symbols of the pth code channel is expressed as

$$\hat{s}_{p,k} = \sum_{l=1}^{L} \hat{\alpha}_{l,k}^{*} \cdot y_{p,l,k} \tag{17.6}$$

where $\hat{\alpha}_{l,k}$ is the estimated channel coefficient for path l in the kth symbol period. The estimated modulated symbols associated with each physical channel are demodulated into LLRs and split into two differently protected streams. After the physical channel desegmentation and rate de-matching, the LLRs of the coded bits, $\hat{\lambda}_{m,i}$ ($m = 1$ or $m = 2$), are fed into the turbo decoder. The turbo decoder performs one decoding iteration and outputs an estimate for each jth information bit, $\hat{b}_{m,j}^{(q)}$, and also the LLR estimates of the code symbols $\tilde{\lambda}_{m,i}^{(q)}$ (q is the iteration number and i is the coded bit number). The decision device then uses these LLRs to estimate the coded bits' values. It can perform either soft decision or hard decision according to

$$\tilde{d}_{m,i}^{(q)} = \tanh\left(\frac{\tilde{\lambda}_{m,i}^{(q)}}{2}\right) \quad \text{for soft decision}$$

and

$$\tilde{d}_{m,i}^{(q)} = \begin{cases} 1, & \tilde{\lambda}_{m,i}^{(q)} \geq 0 \\ -1, & \tilde{\lambda}_{m,i}^{(q)} < 0 \end{cases} \quad \text{for hard decision.}$$

These coded bit values are then modulated into the symbols $\tilde{s}_p(t)$ (p denotes the pth physical channel). The mapping of the bits into the constellation symbols, for a 16-QAM constellation, can be performed independently to the I and Q branch according to

$$\tilde{s}_p = \sum_{l=1}^{2}(-1)^l \frac{D_{3-l}}{2} \prod_{m=1}^{l} \tilde{d}_{m,i}^{(q)} + \sum_{l=1}^{2}(-1)^l \frac{D_{3-l}}{2} \prod_{m=1}^{l} \tilde{d}_{m,i+1}^{(q)} \cdot j \tag{17.7}$$

These symbols are then used to reconstruct the estimate of the transmitted signal

$$\tilde{x}_t^{(q)} = \sum_{p=1}^{P} \tilde{s}_{p,t} \cdot c_{p,t}. \tag{17.8}$$

After this the MPI replica associated with the lth path is estimated as

$$\hat{I}_{l,t}^{(q)} = \alpha_{l,t} \cdot \tilde{x}_t^{(q)}. \tag{17.9}$$

The interference subtracted from the signals fed to each RAKE finger corresponds to the sum of all paths (with their corresponding relative delays), except the one that is going to be extracted by that finger. Thus, the input to the lth finger in the qth MPIC iteration can be represented as

$$r_{t,l}^{(q)} = r_t - w_q \sum_{\substack{j=1 \\ j \neq l}}^{L} \hat{I}_{j,t-\tau_j}^{(q-1)} \tag{17.10}$$

where w_q is a real-valued weight that takes values from the interval [0, 1] and usually increases with the iteration number. This weight factor is used to reduce the impact of possible data decision errors present in the estimated MPI replicas, which are usually higher in the first iterations. After Q iterations the $\hat{b}_{m,j}^{(Q)}$ values are used as final estimates for the information bit streams.

17.2.3.4 Channel Estimation

The transmission of a pilot channel, orthogonal to the data channels, allows simple channel estimation processing at the receiver. To obtain the channel estimates for each path l, the receiver performs the following tasks in the first iteration:

1. Despread the received signal using

$$y_{pilot,l,k'} = \frac{1}{SF_{pilot}} \sum_{t=(k'-1)\cdot SF_{pilot}+1}^{k'\cdot SF_{pilot}} r_{t+\tau_l} \cdot c_{pilot,t}^* \tag{17.11}$$

2. Obtain noisy channel estimates, $\hat{\alpha}_{l,k'}^{noisy}$, in each pilot symbol position, k', by multiplying the despreaded pilot symbols, $s_{pilot,k'}^*$ by their conjugates, with

$$\hat{\alpha}_{l,k'}^{noisy} = \frac{s_{pilot,k'}^*}{\left| s_{pilot,k'} \right|^2} \cdot y_{pilot,l,k'} \tag{17.12}$$

3. The noisy channel estimates are then passed by a moving-average filter with length W, leading to

$$\hat{\alpha}_{l,k'} = \frac{1}{W} \sum_{i=k'-\lfloor W/2 \rfloor}^{k'+\lceil W/2 \rceil -1} \hat{\alpha}_{l,i}^{noisy} . \tag{17.13}$$

This filtering is employed because it has low complexity, does not require knowledge of the fade rate or autocorrelation of the channel, and, for slowly varying channels, can achieve good performances.

4. Because the data channels can have different data rates than the pilot symbol rate, interpolation can be performed over the channel estimates for matching the rates. A simple repeater can be used because it is assumed that the channel is approximately constant during a pilot symbol duration.

After the first decoding iteration the channel estimation can be improved by applying the interference canceller to the pilot channel. Accordingly, using the estimated transmitted signals (17.8) and the received signal with interference cancellation employed (17.10) results in the noisy channel estimates

$$\hat{\alpha}_{noisy,l,t} = \frac{((\tilde{x}_t)^{(q)})^*}{|(\tilde{x}_t)^{(q)}|^2} \cdot (r_{t,l})^{(q)} \tag{17.14}$$

These noisy channel estimates can then be passed by a moving-average filter with length W for obtaining final channel estimates, as in (17.13) (step 3).

Note that the transmitted symbol estimates could also be used as additional pilots for improving the channel estimates, as is usually done for pilot symbol-assisted modulation (PSAM) transmissions. However, because the scheme studied here considers the transmission of a pilot channel in parallel to the data, no significant improvement is achieved with that approach.

17.2.4 MIMO as a Multiresolution System

Multiple-input multiple-output (MIMO) schemes are used to push the capacity and throughput limits as high as possible without an increase in spectrum bandwidth, although there is an obvious increase in complexity. The capacity limit of any CDMA system is taken to be the resulting throughput obtained via the use of the maximum number of codes. As the codes used are orthogonal to each other, there is a strict limit in their maximum number. However, if multiple transmit and receive antennas are employed, the capacity may be raised due to code reusage across transmit antennas. If there are a sufficient number of receive antennas, it is possible to resolve all data streams, as long as the channel correlation between antennas is not too high. For M transmit and N receive antennas, we have the capacity equation[14,15]

$$C_{EP} = \log_2\left(\det\left(I_N + \frac{\beta}{M}\boldsymbol{HH}'\right)\right)_{b/s/Hz} \tag{17.15}$$

where I_N is the identity matrix of dimension $N \times N$, $\boldsymbol{H}$ is the channel matrix, $\boldsymbol{H}'$ is the transpose conjugate of H, and β is the SNR at any receive antenna. Foschini and Gans[14] and Telatar[16] both demonstrated that the capacity grows linearly with $m = min(M, N)$ for uncorrelated channels.

Therefore, it is possible to employ MIMO as a multiresolution distribution system where the concurrent data streams are transmitted by M and received by N ($M \leq N$) different antennas. The downsides to this system are the receiver complexity, sensitivity to interference and correlation between antennas, which is more significant as the antennas are closer together. For the UMTS, it is inadequate to consider more than two or four antennas at the UE/mobile receiver.

Note that, unlike in CDMA where user's codes are orthogonal by design, the resolvability of the MIMO channel relies on the presence of rich multipath, which is needed to make the channel spatially selective. Therefore, MIMO can be said to effectively exploit multipath.

The receiver for such a scheme is obviously complex due to the number of antennas, users, and multipath components. Different receivers were analyzed in Silva et al.,[17,18] to establish the trade-off between performance and complexity for such systems. Two strategies were discussed: minimum mean square error (MMSE) equalization and maximum ratio combining (MRC)–based receivers. The main difference between both strategies is the fact that the equalization receivers operate on the whole block at once, whereas MRC receivers work on the tap/finger level, combining the taps later to form an estimate for the symbols. Thus, best results are obtained with the equalizers, because all sources of interference are considered and removed in a single step (with the help of good channel estimation). Therefore, the MMSE-based receiver will be considered for the MIMO environment.

17.3 Minimum Mean Square Error Algorithm: Design of the System Matrices

Although the RAKE receiver coupled with a MPIC is suitable for SISO due to its low complexity, the MIMO environment has need for a more sophisticated receiver that will solve all interference related with the different transmit antennas.

The design of the system matrices is explained next. A MIMO arrangement is assumed, in which the data streams for each user are either split into different streams for each transmit antenna to increase the bit rate/capacity, or replicated (with interleaving) for each antenna, to increase the transmit diversity, reducing the necessary transmit power for nominal operations.

The transmitted signal associated with the kth spreading code and the tx^{th} transmit antenna is given by

$$x_{tx}^{(k)}(t) = A_{tx}^{(k)} \sum_{n=1}^{N \cdot SF} b_{\lfloor n/SF \rfloor,tx}^{(k)} c_n^{(k)} f_T(t - nT_C) = A_{tx}^{(k)} \sum_n \beta_{n,tx}^{(k)} f_T(t - nT_C), \qquad (17.16)$$

where N is the number of data symbols to be transmitted by each antenna, $A_{tx}^{(k)} = \sqrt{E_k}$ (admitting $ft(0) = 1$, $E[|b^{(k)}|^2]$ and $E[|c^{(k)}|^2] = 1$ with E_k denoting the symbol energy, SF the spreading factor, K the total number of spreading codes per antenna/physical channels (or users, if each user uses only one physical channel), and $RC = 1/TC$ the chip rate. The nth chip associated with the kth user, and the tx^{th} antenna is $\beta_{n,tx}^{(k)} = b_{\lfloor n/SF \rfloor,tx}^{(k)} c_n^{(k)}$. The data symbols to be transmitted at the tx^{th} transmit antenna are $b_{n,tx}^{(k)}$, $n = 1, 2, ..., N$, and the combined spreading and scrambling signature is $c_n^{(k)}$, $n = 1, 2, ..., N \times SF$. $f_T(t)$ is the adopted pulse shape filter (a square-root raised cosine filtering is assumed).

At the receiver we have N_{RX} antennas. The signal for any given receive antenna rx, prior to the reception filter, is given by

$$y_{rx}(t) = \sum_{k=1}^{K} \sum_{n=1}^{N \cdot SF} \sum_{tx=1}^{N_{TX}} \left(x_{n,tx}^{(k)}(t) * h_{n,rx,tx}^{(k)}(t) \right) + n(t), \qquad (17.17)$$

with $n(t)$ denoting the channel noise, assumed Gaussian, with zero mean, and the variance of the real and imaginary components denoted by σ^2. The channel impulse response between the transmit antenna tx and the receive antenna rx is

$$h_{n,rx,tx}^{(k)}(t) = \sum_{l=1}^{L} \alpha_{n,rx,tx,l}^{(k)} \delta\left(t - \tau_l^{(k)}\right), \qquad (17.18)$$

with L denoting the total number of the channel's multipath components, $\alpha_{n,tx,rx,l}^{(k)}$ denoting the complex attenuation (fading) factor for the lth path, and $\tau_l^{(k)}$ representing the propagation delay associated with the lth path (for the sake of simplicity, it is assumed that this delay is constant; the generalization to other cases is straightforward). The received signal associated with each antenna is submitted to a reception filter, with impulse response $f_R(t)$, which is assumed to be matched to $f_T(t)$ (i.e., $f_R(t) = f_T^*(-t)$) leading to the signal

$$r_{rx}(t) = \sum_{tx=1}^{N_{Tx}} \sum_{k=1}^{K} A_{tx}^{(k)} \sum_{l=1}^{L} \sum_{n=1}^{N \cdot SF} b_{\lfloor n/K \rfloor, tx}^{(k)} c_n^{(k)} h_{n,rx,tx,l}^{(k)} p\left(t - nT_C - \tau_l^{(k)}\right) \qquad (17.19)$$

where $p(t) = f_T(t) * f_R(t)$ (for the Nyquist pulses considered in this work $p(t) = f_T(t) * f_R(t) = f_T(t) * f_T^*(-t)$ is such that $p(nT_C) = 0$ for integer $n \neq 0$).

This signal is sampled at the chip rate and the corresponding samples can be written as

$$r_{rx,n} \Delta r_{rx}(nT_C + \tau_0) = \sum_{tx=1}^{N_{Tx}} \sum_{k=1}^{K} A_{tx}^{(k)} \sum_{l=1}^{L} \sum_{n'} \beta_{n',tx}^{(k)} h_{n',rx,tx,l}^{(k)} p\left((n - n')T_C - \left(\tau_l^{(k)} - \tau_0^{(k)}\right)\right)$$

$$= \sum_{tx=1}^{N_{Tx}} \sum_{k=1}^{K} A_{tx}^{(k)} \sum_{l=1}^{L} \sum_{n''} \beta_{n-n'',tx}^{(k)} h_{n-n'',rx,tx,l}^{(k)} p\left(n'' T_C - \left(\tau_l^{(k)} - \tau_0^{(k)}\right)\right) \qquad (17.20)$$

with

$$\beta_{n,tx}^{(k)} = b_{\lfloor n/K \rfloor, tx}^{(k)} c_n^{(k)}. \qquad (17.21)$$

It should be noted that $n'' \in \mathbb{Z}$, but in practical terms it is enough to consider just a few components in the vicinity of $\tau_l^{(k)} - \tau_0^{(k)}$. The above formulas are valid for both the downlink and uplink transmissions. In the case of the uplink transmission, special care must be taken, namely, in the introduction of different delays per user (that can be accounted for in the symbols from different transmit antennas).

17.4 Main System Matrices

Using matrix algebra, the received vector is as follows:

$$r_v = SCAb + n, \qquad (17.22)$$

where S, C, and A are the spreading, channel, and amplitude matrices respectively, built in such a way that the expression in (17.20) is reproduced. The receive vector r_v encompasses the messages for all receive antennas, such that

$$r_v = \begin{bmatrix} r_{rx=1}(t) \\ r_{rx=2}(t) \\ \vdots \\ r_{rx=N_{RX}}(t) \end{bmatrix}$$

Note that the channel matrix encompasses not only the channel coefficients, but the filter's coefficients as well. The spreading matrix accounts for the spreading and scrambling codes, as well as the delays between users and channel replicas. The channel matrix accounts for the fading coefficients for all links between each transmit and receive antenna. For simplicity, the spreading matrix for the downlink will be described, and the assumption that the spreading and scrambling codes are the same for all the transmit antennas will be made, having a direct effect on both the spreading and channel matrices. The other structures remain the same, for both the downlink and uplink transmissions.

17.4.1 Downlink with Equal Scrambling for All Transmit Antennas: S and C Matrices

The downlink spreading matrix $\mathbf{S}$ has dimensions $(SF \cdot N \cdot N_{RX} + \psi_{MAX} \cdot N_{RX}) \times (K \cdot L \cdot N \cdot N_{RX})$ (ψ_{MAX} is the maximum delay of the channel's impulse response, normalized to the number of chips, $\psi_{MAX} = \left\lceil \frac{\tau_{max}}{T_c} \right\rceil$ where T_c is the chip period), and it is composed of submatrices $\mathbf{S}_{RX}$ in its diagonal for each receive antenna $\mathbf{S} = \text{diag}(\mathbf{S}_{RX=1}, \ldots, \mathbf{S}_{RX=N_{RX}})$. Each of these submatrices has dimensions $(SF \cdot N + \psi_{MAX}) \times (K \cdot L \cdot N)$, and they are further composed by smaller matrices $\mathbf{S}_n^L$, one for each bit position, with size $(SF + \psi_{MAX}) \times (K \cdot L)$. The $\mathbf{S}_{RX}$ matrix structure is made of $\mathbf{S}_{RX} = [\mathbf{S}_{\varepsilon,1}, \ldots, \mathbf{S}_{\varepsilon,N}]$, with

$$
\mathbf{S}_{\varepsilon,n} = \begin{bmatrix} 0_{(SF \cdot (n-1)) \times (K \cdot L)} \\ \mathbf{S}_n^L \\ 0_{(SF \cdot (N-n)) \times (K \cdot L)} \end{bmatrix}
$$

The $\mathbf{S}_n^L$ matrices are made of $K \cdot L$ columns

$$
\mathbf{S}_n^L = [\mathbf{S}_{\text{col}(k=1,l=1),n}, \ldots, \mathbf{S}_{\text{col}(k=1,l=L),n}, \ldots, \mathbf{S}_{\text{col}(k=K,l=L),n}]
$$

Each of these columns is composed of

$$
\mathbf{s}_{\text{col}(kl),n} = [0_{(1 \times \text{delay}(l))}, \mathbf{C}_n(k)_{1 \times SF}, 0_{(1 \times (\psi_{MAX} - \text{delay}(l)))}]^T :
$$

where $c_n(k)$ is the combined spreading and scrambling for the bit n of user k.

These $\mathbf{S}^L$ matrices are either all alike, if no long scrambling code is used, or different, if the scrambling sequence is longer than the SF. The $\mathbf{S}^L$ matrices represent the combined spreading and scrambling sequences, conjugated with the channel delays. The shifted spreading vectors for the multipath components are all equal to the original sequence of the specific user

$$
\mathbf{S}_n^L = \begin{bmatrix} \mathbf{S}_{1,1,1,n} & & \cdots & \cdots & \mathbf{S}_{K,1,1,n} & & \\ \vdots & \ddots & \mathbf{S}_{1,1,L,n} & \cdots & \vdots & \ddots & \mathbf{S}_{K,1,L,n} \\ \mathbf{S}_{1,SF,1,n} & & \vdots & \cdots & \mathbf{S}_{K,SF,1,n} & & \vdots \\ & \ddots & \mathbf{S}_{1,SF,L,n} & \cdots & \cdots & \ddots & \mathbf{S}_{K,SF,L,n} \end{bmatrix}.
$$

Note that to correctly model the multipath interference between symbols, there is an overlap between the $\mathbf{S}^L$ matrices of ψ_{MAX}. As opposed to the SISO multipath case presented in Latva-aho and Juntti,[28] the matrix is not trimmed for the last multipath components.

The channel matrix C is a $(K \cdot L \cdot N \cdot N_{RX}) \times (K \cdot N_{TX} \cdot N)$ matrix, and it is composed of N_{RX} submatrices, each one for a receive antenna

$$C = \begin{bmatrix} C_{RX=1} \\ \vdots \\ C_{RX=N_{RX}} \end{bmatrix}.$$

Each C_{RX} matrix is composed of N C^{KT} matrices alongside its diagonals:

$$C_{RX} = \begin{bmatrix} C_{1,RX}^{KT} & & \\ & \ddots & \\ & & C_{N,RX}^{KT} \end{bmatrix}.$$

Each $C_{n,RX}^{KT}$ matrix is $(K \cdot L) \times (K \cdot N_{TX})$ and represents the fading coefficients for the current symbol of each path, user, transmit antenna, and receive antenna. The matrix structure is made up of further smaller matrices alongside the diagonal of $C_{n,RX}^{KT}$, $C_{n,RX}^{KT} = \mathrm{diag}(C_{n,RX,K=1}^T, \ldots, C_{n,RX,K=K}^T)$ with C^T of dimensions $L \times N_{TX}$, representing the combination of fading coefficients and filters' coefficients for the user's multipath and tx^{th} antenna component.

By defining

$$c_{n,rx,tx,l}^{(k)} \triangleq \sum_{n'} \alpha_{n-n',rx,tx,l}^{(k)} p\left(n'T_C - \left(\tau_l^{(k)} - \tau_0^{(k)}\right)\right), \tag{17.23}$$

we have

$$C_{n,RX}^{KT} = \begin{bmatrix} c_{1,RX,1,1} & \cdots & c_{1,RX,N_{TX},1} & & & & \\ \vdots & & \vdots & & & & \\ c_{1,RX,1,L} & \cdots & c_{1,RX,N_{TX},L} & & & & \\ & & & \ddots & & & \\ & & & & c_{K,RX,1,1} & \cdots & c_{K,RX,N_{TX},1} \\ & & & & \vdots & & \vdots \\ & & & & c_{K,RX,1,L} & \cdots & c_{K,RX,N_{TX},L} \end{bmatrix}.$$

The A matrix is diagonal of dimension $(K \cdot N_{TX} \cdot N)$ and represents the amplitude of each user per transmission antenna and symbol,

$$A = \mathrm{diag}(A_{1,1,1}, \ldots, A_{N_{TX},1,1}, \ldots, A_{N_{TX},K,1}, \ldots, A_{N_{TX},K,N},$$

17.4.2 Different Scrambling for All Transmit Antennas and Uplink Modifications: S and C Matrices

The previous S and C matrices assume that all antennas use the same spreading and scrambling code. However, when not operating under full-loading conditions, best results are obtained when different scrambling sequences are used at each transmit antenna to increase diversity.

Therefore, the main changes to the matrices would be the S_n^L submatrix for the S matrix,

$$S_n^L = \begin{bmatrix} S_{1,1,1,1,n} & \cdots & S_{1,1,1,N_{TX},n} & & & \cdots & S_{K,1,1,1,n} & \cdots & S_{K,1,1,N_{TX},n} & & & \\ \vdots & & \vdots & \ddots & S_{1,1,L,1,n} & \cdots & S_{1,1,L,N_{TX},n} & \cdots & \vdots & & \vdots & \ddots & S_{K,1,L,1,n} & \cdots & S_{K,1,L,N_{TX},n} \\ S_{1,SF,1,1,n} & \cdots & S_{1,SF,1,N_{TX},n} & & \vdots & & \vdots & \cdots & S_{K,SF,1,1,n} & \cdots & S_{K,SF,1,N_{TX},n} & & \vdots & & \vdots \\ & \ddots & & S_{1,SF,L,1,n} & \cdots & S_{1,SF,L,N_{TX},n} & \cdots & & & & \ddots & S_{K,SF,L,1,n} & \cdots & S_{K,SF,L,N_{TX},n} \end{bmatrix}$$

and the C^{KT} submatrix for the C matrix,

$$C^{KT} = \begin{bmatrix} c_{1,1,1} & & & & & & & & & & & \\ & \ddots & & & & & & & & & & \\ & & c_{N_{TX},1,1} & & & & & & & & & \\ \vdots & \vdots & \vdots & & & & & & & & & \\ c_{1,L,1} & & & & & & & & & & & \\ & \ddots & & & & & & & & & & \\ & & c_{N_{TX},L,1} & & & & & & & & & \\ & & & \ddots & & & & & & & & \\ & & & & c_{1,1,K} & & & & & & & \\ & & & & & \ddots & & & & & & \\ & & & & & & c_{N_{TX},1,K} & & & & & \\ & & & & \vdots & \vdots & \vdots & & & & & \\ & & & & c_{1,L,K} & & & & & & & \\ & & & & & \ddots & & & & & & \\ & & & & & & c_{N_{TX},L,K} \end{bmatrix}$$

For the uplink transmission, the S matrix should also portray the delays between different users (in this case, different transmit antennas). The S_n^L should thus be adjusted, with the columns shifted either upwards or downwards, depending on the offsets between users. Note that the size (number of lines) of the S_n^L matrix can be increased, to account for the delays between users. Although the final S matrix can have a bigger number of lines than for the downlink case, its overall structure remains the same.

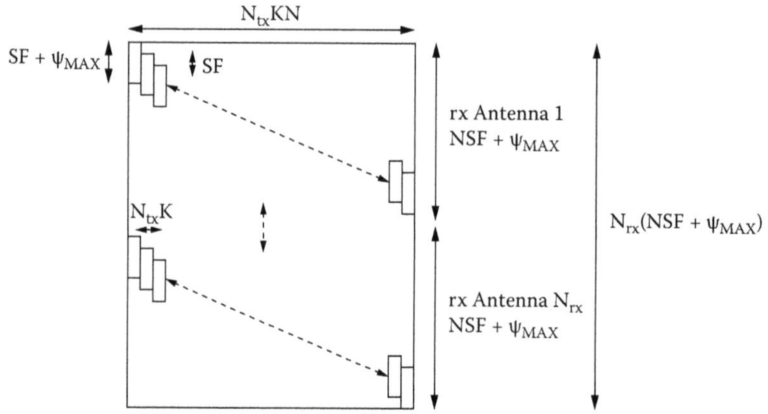

Figure 17.5   Layout of the *SCA* matrix.

17.4.3 Design of the Remaining Structures

The resulting matrix from the SCA operation (henceforth known as *SCA* matrix) is depicted in figure 17.5. It is a $N_{RX} \cdot (N \cdot SF + \psi_{MAX}) \times N_{TX} \cdot K \cdot N$ matrix, and it is the reference matrix for the decoding algorithms. Notice that the *SCA* matrix is sparse in nature.

The resulting *SCA* matrix will have the same size as before; only the number of operations increases while constructing the *SCA* matrix (the SC multiplication has an increase in complexity equal to the number of transmit antennas), because values from different antennas must be treated differently.

Vector b represents the information symbols. It has length $(K \cdot N_{TX} \cdot N)$, and it has the following structure:

$$b = [b_{1,1,1}, \ldots, b_{N_{TX},1,1}, \ldots, b_{1,K,1}, \ldots, b_{N_{TX},K,1}, \ldots, b_{N_{TX},K,N}]^T \qquad (17.24)$$

Note that the bits of each transmit antenna are grouped together in the first level, and the bits of other interferers in the second level. This is to guarantee that the resulting matrix to be inverted has all its nonzero values as close to the diagonal as possible. Also note that there is usually a higher correlation between bits from different antennas using the same spreading code than between bits with different spreading codes.

Finally, the *n* vector is a $(N \cdot SF \cdot N_{RX} + N_{RX} \cdot \psi_{MAX})$ vector with noise components to be added to the received vector r_v, which is partitioned by N_{RX} antennas,

$$r_v = [r_{1,1,1}, \ldots, r_{1,SF,1}, \ldots, r_{N,1,1}, \ldots, r_{N,SF+\psi_{MAX},1}, \ldots, r_{N,1,N_{RX}}, \ldots, r_{N,SF+\psi_{MAX},N_{RX}}]^T, \qquad (17.25)$$

(the delay ψ_{MAX} is only used for the final bit, though its effects are present throughout r_v). Figure 17.6 illustrates the main blocks from which the receiver is compiled.

If transmit diversity is applied, the *b* vector should be arranged accordingly, with replicas (after multiplexing) for each antenna.

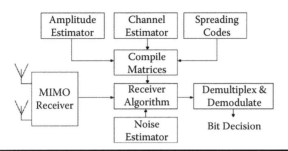

Figure 17.6 MIMO receiver for WCDMA schemes.

17.4.4 Receiver Schemes Using the System Matrices

Equalization-based receivers such as the MMSE and ZF take into account all effects that the symbols are subject to in the transmission chain, namely, the MAI, ISI, and channel effect. Using as its basis the (unnormalized) matched-filter output (obtained by applying the Hermitian to the **SCA** matrix and multiplying by the received vector),

$$y_{MF} = (SCA)^H r_v,$$ (17.26)

and defining **R** as

$$R = A\ C^H\ S^H\ S\ C\ A,$$ (17.27)

the equalization matrix (EM) for MF and ZF can be written as

$$E_{M.MFZF} = R,$$ (17.28)

where the normalized MF estimate is given by

$$y_{NMF} = diag\left(E_{M,MFZF}^{-1}\right)y_{MF},$$ (17.29)

and the ZF estimate as

$$y_{ZF} = E_{M,MFZF}^{-1}\,y_{MF},$$ (17.30)

which is simply applying the inverse of all effects the message was subject to. Once again, to prevent an ill-conditioned matrix for inversion (the EM might become ill-conditioned when the system is fully loaded,[29] depending on the cross-correlations between the users' signature sequences), a small value (e.g., $1E-6$) should be added to all elements in the main diagonal of EM. To avoid round-off problems, the EM should be rounded at a value above the minimum machine precision.

The MMSE estimate aims to minimize $E(|b - \hat{b}|^2)$ From Kay,[30] the EM includes the estimated noise power σ^2, and is represented by

$$E_{M,MMSE} = R + \sigma^2 I. \tag{17.31}$$

The MMSE estimate is thus

$$y_{MMSE} = \hat{b} = E_{M,MMSE}^{-1} y_{MF}, \tag{17.32}$$

Both the ZF- and MMSE-based receivers are seldom used due to their perceived complexity, especially for wideband MIMO systems (with frequency-selective fading channels). Due to the multipath causing ISI, the whole information block is usually decoded at once (although there are some decoding variants in which the block is divided into smaller blocks,[31,32] requiring some overlapping between symbols, to provide best results), requiring the use of a significant amount of memory and computing power for the algebraic operations.

However, if the sparseness of the matrices is taken into account, only a fraction of the memory and computing power is required. As can be inferred from the previously described matrices, all system matrices are sparse in nature and consist of submatrices that are sparse themselves.

The most troublesome matrix to deal with is the EM, due to its inversion (more precisely, the resolution of the equation system leading to the final estimate). Fortunately, the EM is also sparse itself, making it possible to handle with simplicity. For instance, considering a maximum-loading simulation case using 16-QAM modulation, SF = 16 (16 physical channels), L = 2 multipaths (the second multipath with a one-chip delay, resembling the indoor A or pedestrian A channel), a MIMO of a two-transmit/two-receive antenna system, and a block size of 1,024 bits ($N = 256$ symbols, with 4 bits per symbol) per each physical channel of each transmit antenna, the matrix's diagonal width (MDW), which in this case is $(K \cdot L \cdot N_{TX}) \cdot \frac{3}{2} = 96$, is roughly 1.2 percent of the matrix's width (MW) of $(K = 16) \cdot (N_{TX} = 2) \cdot (N = 256) = 8192$ MDW = 1.2 percent MW.

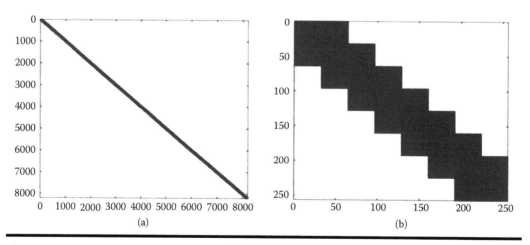

Figure 17.7 **Equalization matrix for a 2-tap channel: (a) K = 16, N_{TX} = 2 and (b) diagonal close-up for the same case.**

Another aspect of the EM is that it is Hermitian positive definite, and thus it can be decomposed using the Cholesky decomposition. Because it is a banded matrix (with all elements concentrated on its diagonal), there is no Cholesky fill-in because the band is dense (cases with small chip delays [figure 17.7]), and thus it presents itself as if the sparse reverse Cuthill–McKee ordering algorithm[27] had been applied to it (a good preordering for LU or Cholesky factorization of matrices that come from long, skinny problems).

17.5 HSDPA Transmission as a Multiresolution Multicast Technique

A flexible shared channel, suitable for MBMS PtP transmissions, is currently available, namely, the High-Speed Downlink Shared Channel (HS-DSCH), which is mapped onto the High-Speed Physical Downlink Shared Channel (HS-PDSCH) (see 3GPP).[23]

HSDPA supports new features that rely on, and are tightly coupled to, the rapid adaptation of transmission parameters to instantaneous radio conditions. These features are:

1. **Fast link adaptation:** Instead of compensating the variations of downlink radio conditions by means of power control, the transmitted power is kept constant and the modulation and coding of the transport block are chosen every TTI = 2 ms for each user; this is called adaptive modulation and coding (AMC). To users in good conditions, 16-QAM can be allocated to maximize throughput, while users in bad conditions are penalized on throughput, reaching a point to which the service can be denied. There is obviously a return channel that indicates the quality of each link.

2. **Fast channel-dependent scheduling:** The scheduler determines to which terminal the shared channel transmission should be directed at any given moment. The term *channel-dependent scheduling* signifies that the scheduler considers instantaneous radio channel conditions. This greatly increases capacity and makes better use of resources. The basic idea is to exploit short-term variations in radio conditions by transmitting to terminals with favorable instantaneous channel conditions.

3. **Fast hybrid ARQ with soft combining:** The terminal (user equipment) can rapidly request retransmission of erroneously received data, substantially reducing delay and increasing capacity (compared to 3GPP Release 99). Prior to decoding, the terminal combines information from the original transmission with that of later retransmissions. This practice, called soft combining, increases capacity and robustness.

All three features are useful for providing multiresolution; in this chapter we will concentrate only on the first one.

Because we have already considered the use of hierarchical 16-QAM modulation and multicode as multiresolution techniques in this section, we will only consider adaptive coding as another multiresolution technique. Here, the first six channel quality indicator (CQI) values of the CQI mapping table[23] will be considered due to use of a single HS-PDSCH (one code and QPSK modulation) with similar bit rates of the MBMS PtM transmission channel (from 64 up to 256 kbps).

Streaming video is the most expected MBMS service. The QoS constraint for such service is often defined by the maximum tolerable delay, which directly translates into the playout buffer size at the mobile receiver. In our study, a packet that includes a basic layer + enhancement layers is fragmented into frames (data streams) of varying sizes due to adaptive coding. For instance, CQI 1 (transport block size of 137 bits) only carries the enhancement layer and plenty of redundant bits (introduced later by the channel encoder). CQI 6 (transport block size of 461 bits) carries the base layer plus five enhancement layers with much less redundant bits. As a result, there is a different energy distribution used for each of the six layers.[21]

In Leitão and Correia[19] and Soares et al.[24] is presented how the scheduling is performed with performance results of throughput, jitter, and probability of receiver buffer underflow of different schedulers. In this chapter we only consider that based on the users' reported CQI in the uplink, they will receive packets with the corresponding transport block size.

17.6 Soft Combining

Soft combining is proposed in 3GPP[11] as an enhancement for MBMS PtM radio transmissions. In a PtM MBMS service the transmitted content is expected to be network specific rather than cell specific, that is, the same content is expected to be multicast/broadcast through the entire network or through most of it. Therefore, a natural way of improving the physical layer performance is to take advantage of macrodiversity. Basically, the diversity-combining concept consists of receiving redundantly the same information bearing signal over two or more fading channels, and combining these multiple replicas at the receiver to increase the overall received signal-to-noise ratio (SNR).

On the network side, this means ensuring sufficient time synchronization of identical MBMS transmissions in different cells; on the UE side, this means the capability to receive and decode the same content from multiple transmitters simultaneously.

This gain of macrodiversity combining through soft combining could be significant, especially when there are a large number of MBMS services offered, and can be obtained with minimum impact on network synchronization requirements.

Logical channels can be multiplexed on the same transport channel for air efficiency. However, because a single turbo encoder is used for one transport channel, the encoded bits are combinations of all logical channels in the transport channel. When the encoded bits are different, they can not be soft combined. Therefore, to support soft combining in the receiver, the information bits encoded by the turbo encoder must be the same. This constraint is illustrated in figure 17.8.

The simplest way is to permit only a one-to-one mapping of MTCH/MCCH to FACH. Here, each MBMS service is mapped to a separate FACH. In addition to the gains from soft combining, another advantage of such a restriction is that it allows a wider class of UE capabilities to receive a given service, because a lower-capability UE may need only to decode the data it has subscribed for. Although this method loses logical channel multiplexing flexibility, transport channel multiplexing can still be efficient.

17.7 Numerical Results

Typically, radio network simulations can be classified as either link level (radio link between the base station and the user terminal) or radio network subsystem system level. A single approach would be preferable, but the complexity of such a simulator—including everything from transmitted waveforms to multicell network—is far too high for the required simulation resolutions and simulation time. Therefore, separate link- and system-level approaches are needed.

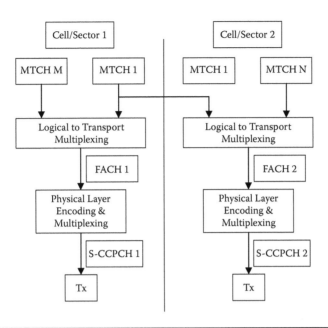

Figure 17.8   Illustration of soft combining for two MBMS channels.

The link-level simulator is needed for the system simulator to build a receiver model that can predict the receiver BLER/BER performance, taking into account channel estimation, interleaving, modulation, receiver structure, and decoding. The system-level simulator is needed to model a system with a large number of mobiles and base stations, and algorithms operating in such a system.

The channel model used in the system-level simulator considers three types of losses: distance loss, shadowing loss, and multipath fading loss. The model parameters depend on the environment. For the distance loss, the COST–Walfisch–Ikegami model, LOS and NLOS, from the COST 231 project was used.

Shadowing is due to the existence of large obstacles like buildings and the movement of UEs in and out of the shadows. This is modeled through a process with a lognormal distribution and a correlation distance.

The multipath fading in the system-level simulator corresponds to the 3GPP channel model, where the pedestrian B and vehicular A (3 km/h) environments were chosen as reference. The latter models are also used in the link-level simulator. In the radio network subsystem (RNS) system-level simulator, only the resulting fading loss of the channel model, expressed in decibels (dB), is taken into account. The fading model is provided by the link-level simulator through a trace of fading values (in dB), one per TTI. For each environment where the mobile speed is the same, several series of fading values are provided for each pair of antennas.

The considered interference is the sum of intracell and intercell interference. Both have a noise-like character. This is mainly due to the large number of sources adding to the signal, which are similar in signal strength.

Link performance results are used as input by the system-level simulator, where several estimates for coverage and throughput purposes can be made by populating the scenario topology uniformly and giving users a random mobility. The estimates are made for every TTI being the packets that

Table 17.2 Urban Macrocellular Parameters

Parameters	Value
Cellular layout	19 cells
Sectorization	Yes, 3 sector/cell
Site-to-site distance	1,000 m
Base station antenna gain + cable losses	14.5 dBi
Antenna beamwidth, –3 dB	70 degrees
Propagation models	Cost 231
Thermal noise DL	–100.3 dBm
Orthogonality factor	0.4
Standard of shadow fading	10 dB
Correlation between sites for slow fading	0.5
Multipath fading	3GPP VehA/PedB 3 km/h
Node B total transmit power (sector)	43 dBm

are received with a BLER over 1 percent considered to be well received. The estimate for coverage purposes is made of an average of five consecutive received packets; if the average received BLER of these packets is above 1 percent, the mobile user is counted as being with coverage. For the throughput calculation, the estimation is made based on each individual packet received with a BLER higher than 1 percent.

17.8 Broadcast Multiresolution Results

As told previously for MBMS PtM transmissions, the available channel is the Forward Access Channel (FACH), which is mapped onto the Secondary Common Control Physical Channel (S-CCPCH).

Results for different multiresolution systems will be analyzed next, associated with this specific transport channel of UMTS.

Tables 17.2 and 17.3 present parameters that are commonly used in the subsequent sections.

17.9 Multicode Results

Figure 17.9 presents the link-level S-CCPCH performance in terms of E_c/I_{or} (dB) representing the fraction of cell transmit power necessary to achieve the corresponding BLER graduated on the vertical axis. For the reference BLER = 10^{-2}, the use of a single spreading code with spreading factor SF = 8 (bit rate of 256 kbps) imposes a geometry factor of 0 dB to achieve E_c/I_{or} less than 80 percent (–1 dB) considering the vehicular A propagation channel. This means that we can only offer such a high bit rate for users located in the middle of the cell, not near the border. The use of two spreading codes, each with SF = 16 (bit rate of 128 kbps), transmitted with different power allows an increase of coverage ($G = -3$ dB) and throughput, as we will next confirm with system-level results. Remember that for multiresolution using multicode, the base layer is transmitted with higher power ($G = -3$ dB) and the enhancement layer with lower power ($G = 0$ dB). Here, we will assume that $E_c/I_{or} = 80$ percent is available for the transmission of MBMS PtM mode, without any macrodiversity combining.

Table 17.3   Link-Level Simulation Parameters

Parameters	Value
S-CCPCH slot format	12, 14 (128 kbps, 256 kbps)
Transport block size and number of transport blocks per TTI	Varied according to information bit rate (128 or 256 kbps) and TTI value
CRC	16 bits
Transmission time interval (TTI)	20, 40, 80 ms
CPICH E_c/I_{or}	−10 dB (10%)
P-SCH E_c/I_{or}	−15 dB (3%)
S-SCH E_c/I_{or}	−15 dB (3%)
Tx E_c/I_{or}	Varied
Orthogonal channel noise simulator (OCNS)	Used to sum the total Tx E_c/I_{or} to 0 dB (100%)
Channel estimation	Enabled
Power control	Disabled
Channels	Pedestrian B, 3 km/h Vehicular A, 3 km/h

In figure 17.10 the average coverage versus E_c/I_{or} is presented. The introduction of multicode allows multiresolution and an increase of average coverage for the same total transmitted power. With multicode, the aggregate bit rate of 256 kbps is achievable with two data streams of 128 kbps for the pedestrian B channel model. One of them is transmitted with $E_c/I_{or1} = 30$ percent, ensuring 62 percent coverage, and the other is transmitted with $E_c/I_{or2} = 50$ percent, offering 85 percent

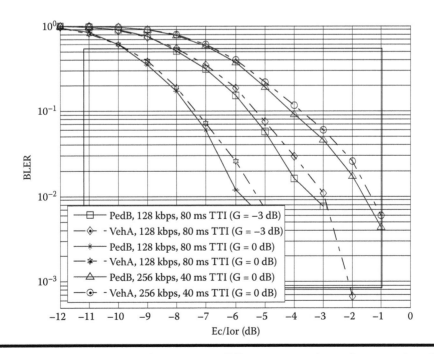

Figure 17.9   BLER versus *Tx* power for S-CCPCH different geometries and propagation channels.

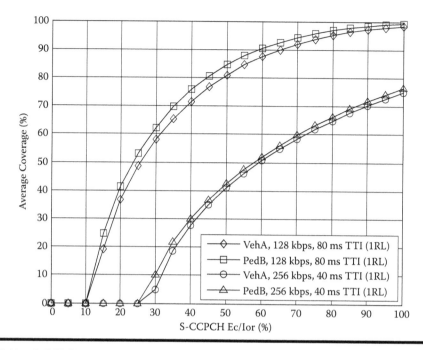

Figure 17.10   S-CCPCH average coverage versus *Tx* **power for one radio link (1RL).**

coverage. For the same total transmitted power, E_c/I_{or} = 80 percent, a single stream of 256 kbps allows a coverage of 66 percent. Similar coverage values (58 and 81 percent for multiresolution and 64 percent for single resolution) are obtained for the vehicular A propagation channel considering the same E_c/I_{or} values.

In figure 17.11 the average throughput versus E_c/I_{or} is presented. Remember that blocks with errors are not retransmitted, in MBMS PtM mode. Taking as reference the same E_c/I_{or} values used for comparison of coverage, namely, 30 + 50 = 80 percent, and the pedestrian B channel, we will check that multicode allows an increase of throughput. For the enhancement data stream E_c/I_{or1} = 30 percent ensures average throughput 80 kbps, and with E_c/I_{or2} = 50 percent the throughput of 110 kbps is achieved for the base data stream. The total throughput is 190 kbps for multiresolution. For the single-resolution stream with E_c/I_{or} = 80 percent the achieved throughput is 172 kbps. The throughput gain is around 10 percent.

Figure 17.12 presents an alternative way of offering multiresolution where the bit rate is 256 kbps using hierarchical 16-QAM modulation and a single spreading code with SF = 16 for *G* = 0 dB and *k* = 0.5. This case is more spectral efficient than the previous one without multiresolution, presented in figures 17.10 and 17.11, because it uses a higher SF. An iterative receiver based on the one described in Souto et al.[12] is employed for decoding both blocks of bits. For the reference value of BLER = 10^{-2}, the difference of total transmitted power between the strong and weak blocks is about 5.5 dB for either vehicular A or pedestrian B. This means that there is a substantial difference in coverage between the two data streams to ensure the reference BLER.

Figure 17.13 presents average throughput and corresponds to figure 17.11. It is obvious that multiresolution using hierarchical modulation provides higher throughput than with single resolution. Taking the same reference, E_c/I_{or} = 80 percent, the achieved throughput for the pedestrian B channel is 240 kbps. There is an obvious increase in throughput obtained with hierarchical

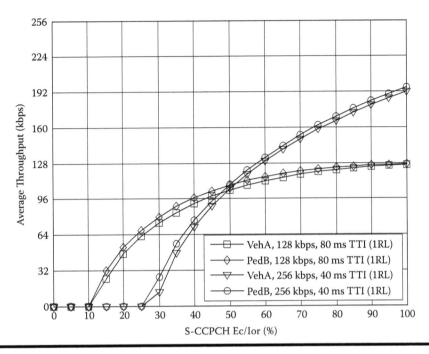

Figure 17.11 **S-CCPCH average throughput versus** *Tx* **power (1RL). Hierarchical modulation results.**

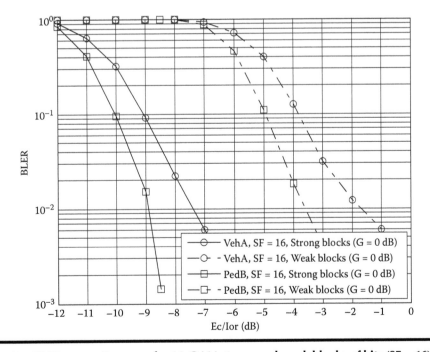

Figure 17.12 **BLER versus** *Tx* **power for 16-QAM strong and weak blocks of bits (SF = 16),** *k* **=0.5.**

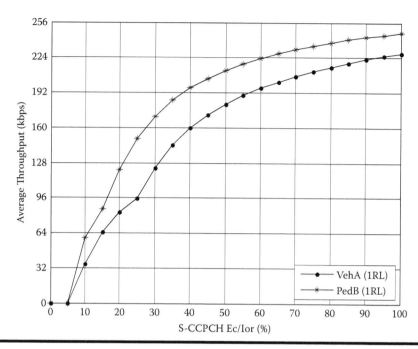

Figure 17.13 **16-QAM average throughput versus *Tx* power (1RL).**

modulation compared to multiresolution by multicode. The throughput gain is 28.3 percent, but there is the disadvantage of requiring a more complex receiver.

17.10 MIMO Results

Figure 17.14 presents an alternative way of offering multiresolution where the bit rate is 256 kbps using multiple transmitting and receiving antennas (MIMO) and a single spreading code with SF = 16 for $G = -3$ dB. The complex correlation coefficients between antennas were taken from 3GPP.[20] This case has the same spectral efficiency as the previous one with multiresolution, presented in figures 17.12 and 17.13, because it uses the same SF. However, it provides higher coverage because the geometry is lower. The base and the enhancement layers are transmitted by different antennas. However, the results presented here only consider the case where the transmitted power per antenna is the same and equal to half of the single transmitting antenna case. If we had considered different transmitted powers per antenna, then we would have a better BLER performance for the base layer than for the enhancement layer stream.

The receiver for both single-resolution SISO (single-input single-output) and multiresolution MIMO is the same as described in Vetro et al.[9] and Samsung.[10] For the reference value of BLER = 10^{-2}, the difference of total transmitted power between the single resolution SISO (1 × 1) and the multiresolution MIMO (2 × 2) schemes is less than 1 dB (vehicular A) or equal to 1 dB (pedestrian B). Considering that we are transmitting at 128 kbps with single resolution and 256 kbps with multiresolution, we would expect double the transmitted power for the latter. However, we can conclude that the multiresolution scheme is much more power efficient than the single-resolution one for BLER < 10^{-2}.

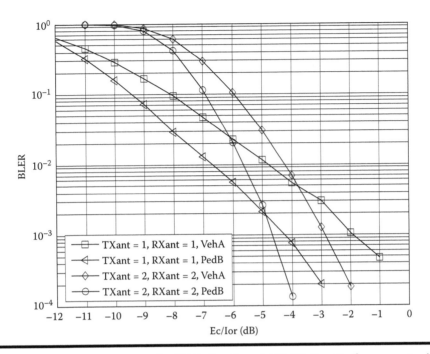

Figure 17.14   BLER versus *Tx* power for SISO (1 × 1) and MIMO (2 × 2); the geometry is *G* = –3 dB.

Figure 17.15 presents average throughput and corresponds to figures 17.11 and 17.13. It is obvious that multiresolution using MIMO provides higher throughput than the previous multiresolution techniques. However, the observation of the curves indicates that there is a decrease in the throughput of the enhanced layer compared to the base layer. For instance, take pedestrian B channel and E_c/I_{or} = 40 percent. The base layer throughput is 128 kbps, but the total throughput (base + enhancement) is around 240 kbps. Considering again the previous reference E_c/I_{or} = 80 percent, the achieved throughput for the pedestrian B channel is 256 kbps. The throughput gain is 32.82 percent relative to 256 kbps single resolution (SF = 8).

17.11   Soft-Combining Results

Figures 17.16 to 17.19 present the macrodiversity gains when soft combining two or three simultaneous S-CCPCH channels.

Comparing coverage results of figure 17.10 for one radio link with the results presented in figure 17.16 for soft combining with two radio links, we can observe a coverage reference of 90 and 70 percent of users, a decrease on the required E_c/I_{or} in the order of 50 to 60 percent for 128 and 256 kbps, respectively. For the case of soft combining, three simultaneous radio links (figure 17.17) for the same reference coverages, we can notice that these are accomplished with less than 10 percent of the total available cell power (E_c/I_{or}) for a bit rate of 128 kbps, and less than 20 percent for the 256 kbps bit rate.

As shown in figure 17.11, bit rates up to 256 kbps could not be achieved using one S-CCPCH radio link, and almost all the available power is necessary for a maximum throughput of 128 kbps.

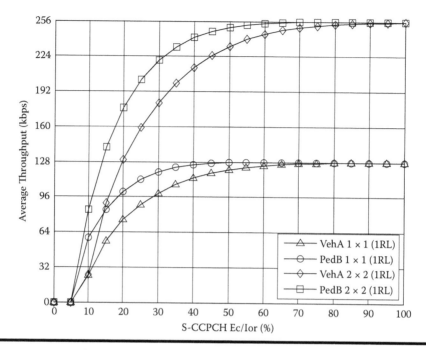

Figure 17.15 Average throughput versus *Tx* power (1RL).

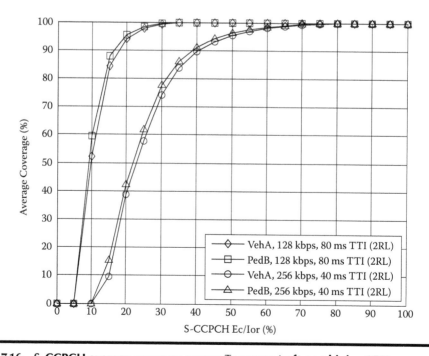

Figure 17.16 S-CCPCH average coverage versus *Tx* power (soft-combining 2RL).

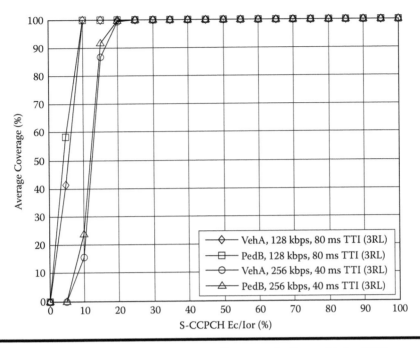

Figure 17.17 S-CCPCH average coverage versus *Tx* power (soft-combining 3RL).

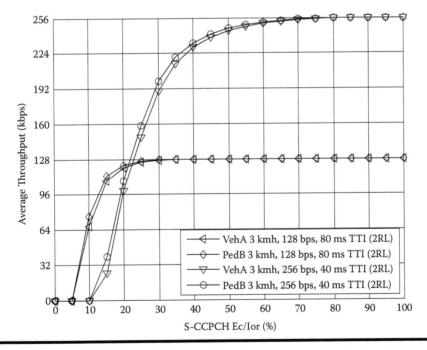

Figure 17.18 S-CCPCH average throughput versus *Tx* power (soft-combining 2RL).

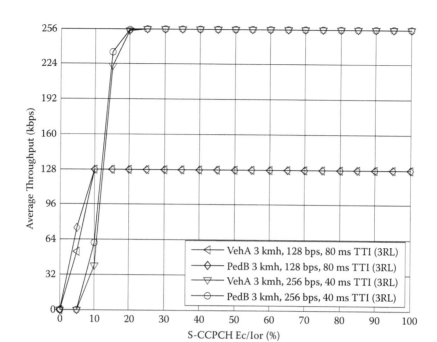

Figure 17.19   S-CCPCH average throughput versus *Tx* power (soft-combining 3RL).

Figures 17.18 and 17.19 show the maximum average throughput expected when soft combining two and three S-CCPCH radio links. Its is clear that better throughput rates are directly related to the better service coverage of applying simulcast transmission with soft combining.

17.12   Multicast Multiresolution Results

17.12.1   HSDPA Results

Figures 17.20 and 17.21 consider the HS-DSCH channel, where BLER performance curves versus E_c/I_{or} are presented for vehicular A (3 km/h) and pedestrian B (3 km/h), respectively. Only the first six CQIs with bit rates from 68.5 up to 230.5 kb/s (QPSK modulated), without and with MIMO, are considered. Multicast multiresolution is achieved by adaptive coding (without MIMO) and by the association of adaptive coding with multiple transmitting and receiving antennas (with MIMO). The geometry chosen was $G = -3$ dB, which corresponds to average coverage greater or equal to 80 percent. It is obvious that depending on the position of the terminal, more or less throughput is achievable for the same transmitted power. E_b/N_0 values can be derived to find the signal-to-interference ratio (SIR) targets for each transmitted bit rate; in our case the E_b/N_0 value is obtained for a BLER of 1 percent, for MBMS services. Table 17.4 presents the HSDPA parameters used in the simulation, and table 17.5 summarizes the results of figure 17.20.

We conclude from the link-level results of table 17.5 that the required fraction of total transmitted power to ensure the reference BLER = 10^{-2} for the geometry $G = -3$ dB increases a lot for increasing bit rates. However, there is an expected substantial reduction of E_c/I_{or} when using MIMO for the same bit rates, for example, compare $R_b = 230.5$ kb/s (CQI 6) of single

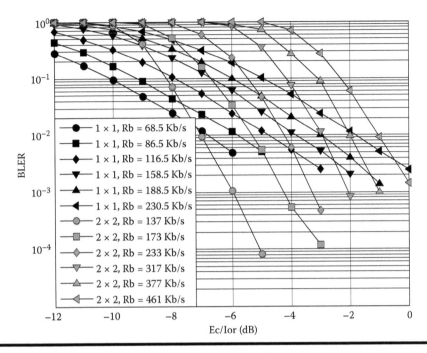

Figure 17.20   HSDPA BLER versus E_c/I_{or} **for vehicular A channel.**

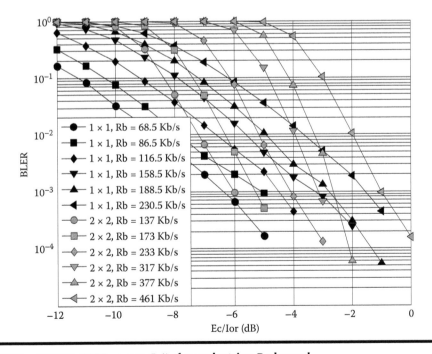

Figure 17.21   HSDPA BLER versus E_c/I_{or} **for pedestrian B channel.**

Table 17.4   HSDPA Parameters

HSDPA used power	38 dBm
Maximum used OVSF codes	1
CQIs	CQI 1→ 6
MBMS sessions	1
HARQ type	Chase combining
Number of retransmissions	1
Link adaptation algorithm	Normal adaptive coding

antenna with R_b = 233 kb/s (CQI 3) of MIMO. In our assumptions, we keep constant E_c/I_{or} = –4.8dB (E_c/I_{or} = 33 percent), which means that only the first CQIs (low resolution) are supposed to offer high coverage. According to table 17.5, it is the multiresolution by the association of adaptive coding and MIMO (2 × 2) that imposes more restrictions of coverage.

The coverage and throughput performance are simulated by system-level simulations for video-streaming multicast using HSDPA; the vehicular A environment (3 km/h) is presented in figures 17.22 and 17.23. As expected, for E_c/I_{or} = 33 percent there are different average coverages for different bit rates. For the same bit rate (same multiresolution), additional coverage is offered with MIMO (2 × 2). Not all the users will get the highest resolution, but the lowest resolution will be offered everywhere. For E_c/I_{or} = 33 percent with MIMO (2 × 2) or without it, there is an increased average throughput only for the first five bit rates. This means that due to the coverage restrictions of the highest bit rate (R_b = 461 kb/s), its expected throughput averaged by all users in the cell is not above 210 kb/s, the average throughput of the previous resolution (R_b = 377 kb/s). However, for users with good channel conditions, the highest resolution offered by R_b = 461 kb/s (with MIMO) is an interesting option.

17.13   Conclusions

In this chapter we have analyzed several air interface enhancements based on multiresolution broadcast systems for wideband code division multiple access (WCDMA) cellular mobile networks, namely, multicode, hierarchical 16-QAM constellations, and multiantenna (MIMO)

Table 17.5   HSDPA with and without MIMO, BLER Target = 1%, E_c/I_{or}, for Different CQI, G = –3dB

CQI	Bit Rate, Rb (kb/s)	E_c/I_{or} (Single Antenna)(dB)	Bit Rate, Rb" (kb/s)	E_c/I_{or} (MIMO) (dB)
1	68.5	–6.9	137.0	–7.0
2	86.5	–5.9	173.0	–5.4
3	116.5	–4.9	233.0	–4.3
4	158.5	–3.9	317.0	–3.0
5	188.5	–3.0	377.0	–2.0
6	230.5	–1.9	461.0	–1.1

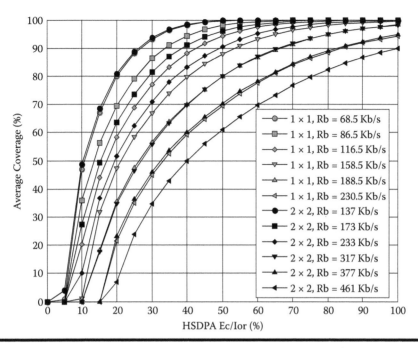

Figure 17.22 HSDPA coverage versus E_c/I_{or} for vehicular A channel.

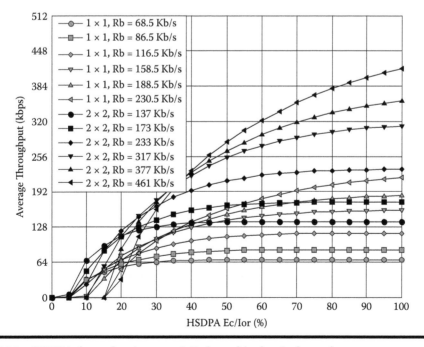

Figure 17.23 HSDPA throughput versus E_c/I_{or} for vehicular A channel.

systems. In addition to the obvious advantage of multiresolution compared to single resolution in terms of graceful degradation of quality, each technique presented here offered coverage and throughput performance gains over conventional single-resolution broadcast systems. We have applied each multiresolution broadcast system to the point-to-multipoint mode of MBMS.

More complex iterative receivers required for hierarchical and MIMO systems were detailed in this chapter.

A comparison between the three specific broadcast multiresolution systems indicates that multicode is the one with less performance gain and has no spectral efficiency gain compared to single resolution (single code). Hierarchical 16-QAM modulation has double spectral efficiency and higher performance gains than multicode. MIMO with two transmitting and two receiving antennas also has double spectral efficiency and offers the highest gains. However, the expected capacity gains that MIMO and hierarchical 16-QAM schemes provide require more complex receivers than multicode or single resolution.

The result of macrodiversity combining was also presented as an effective way to reduce the transmitted power requirements and to guarantee service continuity while the mobile terminals move from one cell to another.

For MBMS transmission using HSDPA channels, the advantages of this transmission were considered for one specific MBMS mode. Namely, HSDPA was analyzed as a multicast multiresolution system (point-to-point MBMS transmission), without or with MIMO. HSDPA by itself (without MIMO) offers much more enhancement layers for multiresolution than the presented broadcast multiresolution systems. Additional multiresolution capability is offered by the association with spatial multiplexing schemes (MIMO).

References

1. T. Cover. 1972. Broadcast channels. *IEEE Trans. Inform. Theory* IT-18:2–14.
2. K. Ramchandran, A. Ortega, K. M. Uz, and M. Vetterli. 1993. Multiresolution broadcast for digital HDTV using joint source/channel coding. *IEEE J. Select. Areas Commun.* 11, 6–23.
3. M. B. Pursley and M. Shea. 1999. Non-uniform phase-shift-key modulation for multimedia multicast transmission in mobile wireless networks. *IEEE J. Select. Areas Commun.* 17, 774–83.
4. ETSI. 1997. *Digital video broadcasting (DVB) framing structure, channel coding and modulation for digital terrestrial television (DVB-T)*. ETSI ETS 300 744.
5. 3GPP. 2004. *High speed downlink packet access (HSDPA) stage 2*. Release 6. 3GPP TS 25.308 V5.4.0.
6. S. Dogan et al. 2004. Video content adaptation using transcoding for enabling UMA over UMTS. Paper presented at Proceedings of Wiamis, Lisbon, Portugal.
7. J. Liu, B. Li, and Y.-Q. Zhang. 2003. Adaptive video multicast over the Internet. *IEEE Multimedia* 10:22–33.
8. W. Li. 2001. Overview of fine granularity scalability in MPEG-4 video standard. *IEEE Trans. CSVT* 11:301–17.
9. A. Vetro, C. Christopoulos, and H. Sun. 2003. Video transcoding architectures and techniques: An overview. *IEEE Signal Processing Mag.* 20:18–29.
10. Samsung Electronics. 2002. Scalable multimedia broadcast and multicast service (MBMS). Paper presented at MBMS Workshop, London.
11. 3GPP. 2005. *S-CCPCH performance for MBMS*. 3GPP 25.803-600 v6.0.0.
12. N. Souto et al. 2005. Iterative turbo multipath interference cancellation for WCDMA systems with non-uniform modulations. Paper presented at Proceedings of the IEEE Vehicular Technology Conference (VTC2005), Stockholm, Sweden.

13. N. Souto et al. 2005. Non-uniform constellations for broadcasting and multicasting services in WCDMA systems. Paper presented at Proceedings of the IEEE IST Mobile & Wireless Communications Summit, Dresden, Germany.
14. G. J. Foschini and M. J. Gans. 1998. On limits of wireless communications in a fading environment when using multiple antennas. *Wireless Pers. Commun.* 6:311–35.
15. I. E. Telatar. 1995. Capacity of multiantenna Gaussian channels. Tech. memo, AT&T Bell Laboratories.
16. I. E. Telatar. 1999. Capacity of multiantenna Gaussian channels. *Eur. Trans. Commun.* 10: 585–95.
17. J. C. Silva et al. 2005. Enhanced MMSE detection for MIMO systems. Paper presented at Proceedings of ConfTele'2005, Tomar, Portugal.
18. J. C. Silva et al. 2005. Equalization based receivers for wideband MIMO/BAST systems. Paper presented at Proceedings of WPMC'2005, Aalgorg, Denmark.
19. F. Leitão and A. Correia. 2006. HSDPA delivering MBMS video streaming using deficit round robin scheduler. Paper presented at Proceedings of ICT'2006, Funchal, Portugal.
20. 3GPP. *Spatial channel model for multiple input multiple output (MIMO) simulations.* 3GPP 25: 996-v6.1.0.
21. M. Sehlstedt and J. P. LeBlanc. 2006. A computability strategy for optimization of multi-resolution broadcast systems: A layered energy distribution approach. *IEEE Trans. Broadcasting* 52:11–20.
22. IST. 2005. *Advanced MBMS for the future mobile world*, C-MOBILE. IST-2005-27423. http://c-mobile.ptinovacao.pt.
23. 3GPP. *Physical layer procedures (FDD).* Release 5. 3GPP 25.214-5b0.
24. A. Soares, J. C. Silva, N. Souto, F. Leitão, and A. Correia. 2006. MIMO based radio resource for UMTS multicast broadcast multimedia services. *Springer Wireless Personal Commun. J.* DOI 10.1007/s 11277-006-9175-x.
25. L. Shoumin and T. Zhi. 2004. Near-optimum soft decision equalization for frequency selective MIMO channels. *IEEE Trans. Signal Processing* 52:721–33.
26. J. C. Silva, N. Souto, A. Rodrigues, A. Correia, F. Cercas, and R. Dinis. 2005. A L-MMSE DS-CDMA detector for MIMO/BLAST systems with frequency selective fading. Paper presented at Proceedings of IEEE IST Mobile & Wireless Communications Summit, Dresden, Germany.
27. A. George and J. Liu. 1981. *Computer solution of large sparse positive definite systems.* Englewood Cliffs, NJ: Prentice-Hall.
28. M. Latva-aho and M. Juntti. 2000. LMMSE detection for DS-CDMA systems in fading channels. *IEEE Trans. Commun.* 48, 194–99.
29. D. Divsalar, M. Simon, and D. Raphaeli. 1998. Improved parallel interference cancellation in CDMA. *IEEE Trans. Commun.* 46:258–68.
30. S. Kay. 1993. *Fundamentals of statistical signal processing: Estimation theory.* Englewood Cliffs, NJ: Prentice-Hall.
31. J. C. Silva, N. Souto, A. Rodrigues, A. Correia, F. Cercas, and R. Dinis. 2005. A L-MMSE DS-CDMA detector for MIMO/BLAST systems with frequency selective fading. Paper presented at Proceedings of IEEE IST Mobile & Wireless Communications Summit, Dresden, Germany.
32. L. Shoumin and T. Zhi. 2004. Near-optimum soft decision equalization for frequency selective MIMO channels. *IEEE Trans. Signal Processing* 52:721–33.

Chapter 18

Optimization of Packet Scheduling Schemes

Hongfei Du, Linghang Fan, and Barry G. Evans

Contents

Keywords

packet scheduling, radio resource management, SDMB, MBMS, cross-layer design, quality of service

Abstract

This chapter presents the optimization techniques on packet scheduling algorithms for supporting multimedia content delivery in mobile Digital Multimedia Broadcasting (DMB). Such services include, for example, e-mail, Web browsing, streaming video/audio, and so on. On one hand, a great deal of research effort has been put on the optimization of packet scheduling schemes for both wired and wireless systems, aimed at maximizing the resource utilization and spectrum efficiency. On the other hand, there has also been work relating to the optimized packet scheduling performance via cross-layer design, obtaining performance gains by actively exploiting the dependence between protocol layers. In this chapter, we take a step in that direction by presenting a literature survey of the packet scheduling optimization techniques in both wired and wireless systems, and by addressing the quality-of-service (QoS) guarantees and performance trade-offs among these schemes. In particular, we consider the development and evaluation of a cross-layer packet scheduling scheme in a uni directional geostationary (GEO) satellite broadcast system, namely, the Satellite Digital Multimedia Broadcasting (SDMB) system, which plays a key role in the convergence of a closer interworking between terrestrial and satellite communications for a more efficient multimedia provisioning. The chapter is organized as follows. In section 18.1, a survey of packet scheduling algorithms in general wired networks is first discussed, with examples drawn from the literature. The adaptation and optimization techniques of packet scheduling algorithms in wireless systems are analyzed in section 18.2. In section 18.3, packet scheduling in SDMB is discussed and existing packet scheduling schemes in such a system are reviewed. In section 18.4, a cross-layer-based packet scheduling scheme, which comprises combined delay and rate differentiation (CDRD) service prioritization and dynamic resource allocation (DRA) algorithms, is presented for QoS provisioning in the SDMB system, followed by the performance evaluation and discussion on the optimized packet scheduling scheme. Finally, we summarize the chapter by raising some opportunities and challenges related to this research, and researchers developing packet scheduling schemes in digital multimedia broadcasting may want to address these as they move forward.

18.1 Introduction to Packet Scheduling

With the tremendous growth of digital multimedia technology, from voice to data to video, and the recent, but growing, demand of supporting diverse QoS guarantees, new demands will be placed on future wireless networks to utilize the available resources in a more efficient and effective way. The key to these demand is the involvement of an efficient resource management scheme, especially the packet scheduling scheme, to provide various QoS supports for multimedia service delivery. In the wake of increasing use of high-speed and high-quality multimedia applications in the past few years, to efficiently distribute the multimedia services over a large number of users, broadcast/multicast transmission is advertised as an attractive approach for achieving less unnecessary transmission and better resource utilization. Due to the unique broadcast nature and ubiquitous coverage of satellite communication systems, the synergy between satellite networks and terrestrial networks provides immense brand new opportunities for delivering point-to-multipoint (or one-to-many) multimedia content to a large number of audiences spreading over extensive geographical coverage. It is expected that the satellite component will play a complementary, but essential, role in delivering multimedia data to those areas where the terrestrial high-bandwidth communication infrastructures are, either economically or technically, unreachable.

Generally speaking, the main functionality of packet scheduling operates at the Medium Access Control (MAC) sublayer of the data link layer, aimed at coordinating the access among competing flows arriving at the queuing buffer at the radio link control (RLC) sublayer of the data link layer. The decision of the packet scheduling is made in coordination with some specific criteria, for example, fairness and QoS requirements, which vary from one scheduling algorithm to another, effectively impacting the overall QoS guarantees and network performance. Therefore, the packet scheduling shall be designed in compliance with the following objectives:

- To coordinate the serving order of contending flows, aimed toward the highest possible degree of resource utilization and spectrum efficiency
- To minimize the transmission power consumption so as to meet the system power constraints
- To achieve the best possible QoS satisfaction in terms of different performance criteria, for example, delay, data rate, on the basis of the service prescribed QoS requirements
- To track the instantaneous traffic dynamics of contending flows and thereby maintain a certain level of performance requirements

Existing literatures have witnessed the diverse effectiveness achieved by different packet scheduling algorithms in various systems, creating crucial hurdles for the design of a feasible and efficient packet scheduling scheme. The following context summarizes representative basic types of packet scheduling schemes in general wired networks, emphasizing the performance trade-offs and QoS guarantees achievable for respective schemes:

- **First-in first-out:** The simplest and most widely deployed packet scheduling discipline in communication systems today is first-in first-out (FIFO) (or first-come first-served), where packets in the queue are served in the order they arrive. It must be noted that this policy describes the packet serving behavior in a single queue dimension. This scheduling discipline is unable to provide any guarantee on delay or bandwidth, and no immunity can be ensured for contending flows.
- **Round-robin:** Another one of the simplest scheduling algorithms is round-robin (RR), where queues are served recursively in their order in a nonpreemptive manner. This scheduling discipline is nonpriority based, thereby offering no differentiation between differentiated service classes. Moreover, the round-robin scheduling is insensitive to packet size; queues with large packet size would be favored over other queues. This algorithm is capable of providing both long-term and short-term fairness and is easy to be implemented. However, it does not consider any differentiation among users; the overall system throughput is fairly low. To ensure a minimum bandwidth allocation and distribution, being regarded as a well-known variation of RR, the weighted round-robin (WRR)[1] assigns a weight to each class. In proportion to the prescribed weights, the available bandwidth is allocated to each class in a round-robin manner. The weight assigned to each class can be regarded as a tunable parameter that can effectively determine the overall performance of each class, for example, delay and throughput.
- **Generalized processor sharing:** Generalized processor sharing (GPS)[2] assumes an idealized scheduling discipline to share the resources in an efficient, flexible, and fair manner based on the fluid model (i.e., infinitesimal packet sizes), which is often not practically

realizable. End-to-end guarantees (i.e., bounded delay) among classes can be provided given that the traffic characteristics of the classes are known ahead of time. Moreover, GPS can ensure fair allocation of bandwidth among all backlogged sessions, which is essential for supporting best-effort and link-sharing services. Due to its desired properties, GPS becomes a foundation of QoS network architecture and a benchmark against which the performance of other packet-based service disciplines can be effectively measured and predicted.

■ **Priority queuing:** To offer service differentiation between two or more classes, priority queue (PRIQ) is introduced, aiming at providing the best performance for the highest-priority class. Packets are maintained in separate queues according to their priorities, queues are served from the highest priority class to the lowest priority class, while packets in each queue are served in a FIFO manner. Packets in lower-priority queues will not be served until all the higher-priority queues are empty. Extra complexity is involved for maintaining prioritized queues. The PRIQ scheduling is capable of offering high-priority queues with the highest possible throughput/bandwidth and lowest possible queuing delay. However, lower-priority queues are served at the mercy of the higher-priority queues; care should be taken so as not to starve lower-priority classes when higher-priority classes saturate the scheduler. Therefore, this type of scheduling discipline cannot provide steadily satisfied performance on real-time scenarios.

■ **Virtual clock:** Virtual clock (VC)[3] time stamp packets with a virtual clock time obtained from the packet size and data rate of the queue. Packets are scheduled based on their time stamps on the head packet of each queue. VC is capable of controlling the average transmission rate of data flows and enforcing user's resource usage according to its reservations. VC achieves worst-case delay for a leaky-bucket controlled session; however, VC cannot bound short-term unfairness.

■ **Fair queuing:** To allow contending flows to fairly share the available link capacity, fair queuing (FQ)[4] estimates the finishing time of all the packets at the head of all nonempty queues based on the arrival time of the packet and the packet size, and selects the queue with the minimum finishing time. The data rate achieved by FQ will not be affected by the packet size. Regarded as a packet approximation of GPS, FQ is capable of providing both bounded delay and fairness in a wired network. To introduce proportional weights to FQ, weighted fair queue (WFQ) was proposed, where the priority is given to the competing flows inversely proportional to the required bandwidth. Unlike FQ, WFQ allows sessions to have different shares on the resource; it can be regarded as a simple approximation of GPS, where the scheduling allows different sessions to occupy proportional resources. By regulating the weight associated with each session, it is possible to guarantee QoS performance, for example, end-to-end delay bound and guaranteed data rate.

■ **Class-based queueing/hierarchical link sharing:** The class-based queueing/hierarchical link-sharing (CBQ/HLS)[5] scheduling algorithm aims to provide flexible link sharing and support multiple queues or classes with bandwidth guarantees. Each class of traffic queue is associated with a link-sharing bandwidth that is aimed to be guaranteed during the scheduling decision. CBQ queues are classified in a hierarchical style; child queues are assigned with some portion of the root queues' bandwidth, that is, resources in root queues are split among their child queues. In this case, priority may be assigned to each queue based on its required bandwidth.

■ **Earliest deadline first:** Earliest deadline first (EDF)[6] associates each packet with a deadline, which is obtained by the sum of the packet arrival time and its associated delay bound for the class that the packet belongs to. It serves queues in ascending order of the deadlines, and therefore provides QoS guarantees for competing sessions. The objective of EDF is to minimize the maximum lateness of packets. As one of the optimized variations of EDF, the service curve-based earliest deadline (SCED)[7] first policy employs service curves to provide a wide spectrum of service characterization by specifying the service using a function. It can be regarded as a generalized policy in that by setting appropriate specification of the service curves, other well-known policies such as VC and EDF can be mapped as special cases. With the ability of allocating and guaranteeing service curves with arbitrary shapes, the SCED is proven to have greater capability to support end-to-end delay-bound requirements than other known scheduling policies.

Of late, cross-layer design has been addressed as a popular subject in research and development activity in the area of communication networks. Although layered architectures have performed well for wired networks, when they are applied to wireless networks, the performance is far from optimum. The reason is that the wireless medium allows richer modalities of communication than wired networks.[8] Performance gains are proved to be achievable by actively exploiting the dependence between protocol layers. Toward this end, a survey of cross-layer design for packet scheduling in wireless systems is presented in the following section, and an optimized packet scheduling scheme utilizing the cross-layer concept is suggested for the satellite broadcast system.

18.2 Packet Scheduling Schemes in Wireless Systems

The aforementioned packet scheduling algorithms can be regarded as universal methodologies that can be applied to any scheduling decision problem. However, when applied directly to wireless networks, their performance is far from optimum, in that wireless transmission features burst-like channel errors and location-dependent link states conditioned by time-varying interference, fading, and shadowing impairments. Therefore, in a wireless environment, mobile stations may experience degraded channel conditions, and therefore will be unable to transmit data effectively. Considering the wireless error-prone characteristics, to design packet scheduling techniques suited for wireless systems, a considerable body of work has been devoted to this research subject in the literature; some of the most popular trends are outlined as follows:

- **Channel-state-dependent scheduling:** To apply packet scheduling to wireless networks, compensation is used for offering differentiated treatments for different channel conditions.[9] Priority is given to users who experience bad channel conditions during the scheduling decision period. This type of scheduling classifies the wireless channel into two states, bad and good, representing the error and error-free channel conditions, respectively. One of the most popular models for emulating the channel state transition procedure is the finite-state Markov channel (FSMC) model[10] with specified error probability associated with the wireless channel.
- **Fair queuing–related scheduling:** Most existing literatures on wireless fair queuing algorithms suggested using the well-known wireline fair queuing algorithms for their error-free service model.[9] Representatives of the channel-state-dependent scheduling are idealized wireless fair queuing (IWFQ),[11] the channel-condition-independent fair queuing (CIF-Q),[12] the service-based fairness approach,[13] and the wireless fair service scheduler.[14] They apply a compensation model on top of classical wireline queuing algorithms; for example, IWFQ uses WFQ or its variants to compute its error-free service, while the

CIF-Q simulates the error-free service by applying a compensation model on top of start-time fair queuing (STFQ),[15] which can be regarded as an enhanced variation of WRR.

■ **Location-dependent scheduling:** One of the key difficulties experienced in wireless networks is that a multimedia session can experience location-dependent channel errors, which may have a significant impact on the amount of data the session can effectively transmit. A representative contribution to this subject is the channel-condition-independent fair (CIF) algorithm proposed in Ng et al.,[12] where delay and throughput are guaranteed for error-free sessions and both long-term and short-term fairness are considered for error sessions.

■ **Max C/I scheduling:** In this scheme, the wireless channel quality in terms of the carrier-to-interference ratios (i.e., C/I values) is estimated by the receiver and reported back to the transmitter through a feedback channel. A most proper modulation and coding scheme is derived for each user based on the reported C/I and system capacity specifications. The max C/I scheduling technique[16] ranks the mobile users in terms of their respective channel quality; users with the best C/I value have the highest rank, and resources are allocated to users according to some predefined criteria. This approach is easy to implement and capable of providing an upper bound of system capacity. However, the performance of the max C/I scheme depends on the distance between mobile users and the base station; the "starvation problem" is more severe for those users near the edge of a cell. Therefore, it can be regarded as one of the most unfair schemes for wireless cellular networks.

■ **Proportional fair scheduling:** Proportional fair (PF)[17,18] packet scheduling is applied to wireless communication systems by scheduling the radio resource according to the preassigned priority associated with each user. This scheme provides better fairness than max C/I and better throughput than round-robin. However, the PF does not necessarily provide a good overall system throughput, for example, it provides a poor delay profile compared to max C/I.[19] It is also proven that the PF could provide a fair output for the wireless end users as time elapses.[18]

■ **Cross-layer scheduling:** The methodology introduced for this type of scheduling uses cross-layer information for the scheduling decision. As proposed in Liu et al.,[20] to achieve more efficient scheduling for diverse QoS guarantees, the interactive queuing behaviors induced by heterogeneous traffic and the dynamic variation of wireless channels are considered in the scheduler design. One of the most popular schemes in this area is to design the adaptive modulation and coding (AMC) scheme at the physical layer in conjunction with the packet scheduling procedure at the data link layer to guarantee the prescribed QoS and achieve efficient bandwidth utilization simultaneously.[21] For example, Liu et al.[22] utilize the channel state information (CSI) estimated at the receiver and select the most appropriate modulation-coding pair, which is sent back to the transmitter through a feedback channel for updating the AMC mode.

■ **Opportunistic scheduling:** The basic idea of opportunistic scheduling is to allocate resource to links according to their experienced channel conditions, which is also referred to as channel-aware scheduling. Opportunistic scheduling introduces an important trade-off between system performance and fairness among users. Numerous research activities on this topic can be found in Liu et al.[24] Although the opportunistic scheduling may be desirable for best-effort traffic in the physical layer, it provides no QoS guarantees for end users.

■ **Utility-based scheduling:** The utility concept was introduced to packet scheduling more than a decade ago.[24] A utility function was proposed to map the service delivered into the performance of the application, aiming at maximizing the performance of the applications.

In wireless networks, the utility can be represented as specific functions, for example, delay,[3] to reflect various performance criteria.

18.3 Packet Scheduling Algorithms for Satellite Broadcasting Systems

18.3.1 System Description

Supporting quality of service in modern telecommunication systems has gained importance over the past decade; the focus of service provisioning is on the final delivered service quality level. The integration of a satellite component to the terrestrial network has become a promising approach for the delivery of Multicast Broadcast Multimedia Service (MBMS).[25] The level of integration between the terrestrial and satellite systems is still under discussion, ranging from the high-level integration where the satellite component is considered an independent network, to the lower-level one where the satellite interface is embedded in the terrestrial network, making the maximum possible reuse of available resources and infrastructure. Among these alternatives, the SDMB system is regarded as one of the most attractive options for this integration, providing wide coverage, low cost, and high QoS guarantees.[26]

As illustrated in Figure 18.1, the SDMB system defines a hybrid satellite-terrestrial communication system, featuring a unidirectional geostationary satellite component that is responsible for the delivery of the point-to-multipoint MBMS services, and provides European coverage by multiple umbrella cells. The SDMB radio interface employs an adaptation of wideband code division multiple access (WCDMA), with the satellite hub (Sat-Hub) hosting both the radio network controller (RNC) and the node B functional entities of the UMTS Radio Access Network. The user equipment (UE) applies the standard 3G terminal enriched with SDMB-enabling functions, which, given the unidirectional nature, are very limited. The terrestrial gap fillers, identified as intermediate module repeaters (IMRs), are co-installed physically at the terrestrial base stations to

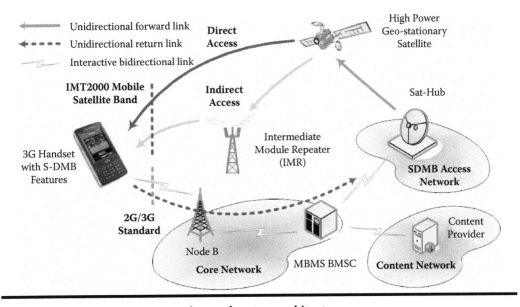

Figure 18.1   SDMB concept overview and system architecture.

enhance the signal reception quality and provide adequate coverage in urban, built-up areas. The SDMB-enabled Broadcast/Multicast Service Center (BMSC), is enhanced with SDMB-specific functions from the standard 3GPP MBMS BMSC.[25] It is noteworthy that no direct satellite return link is envisaged under the baseline SDMB infrastructure; rather, the return path is provided via the terrestrial link if needed. It is assumed that in SDMB, MBMS services are intended for transmission to UEs in either broadcast or multicast mode. In the latter case, service is only delivered to the UEs within a specific multicast group. Packets from the BMSC are first buffered at the Sat-Hub (or node B) in a FIFO manner before being scheduled for transmission over the satellite link. Closely integrated into the baseline architecture of 2.5G/3G mobile cellular networks, the system enjoys maximum reuse of technology and infrastructure and minimum system development cost. The hybrid system takes advantage of the satellite inherent broadcast capability to provide efficient delivery of MBMS contents to the extensive mass mobile market.

Defined by the European Telecommunications Standards Institute (ETSI), the SDMB system provides datacast capacity for various mobile operators. Based on a broadcast nature, the SDMB system offers extensive coverage, low transmission cost for large numbers of terminals, as well as high QoS guarantees for real-time multimedia applications. By employing the WCDMA with frequency division duplexing (FDD), the system can be closely integrated with existing mobile cellular networks, and minimize potential cost impacts on both 3G cellular terminals and network operators. The successful validation and demonstration of the innovative SDMB concept carried out within the European Information Society Technologies (IST) project MoDiS* has pushed the system toward an operational stage. The whole range of issues pertinent to the SDMB system, from system definition to standardization, are addressed in the European IST project MAESTRO.*

In SDMB, the radio resource management (RRM) functionalities comprise three main parts: packet scheduling, radio resource allocation (RRA), and admission control. These functionalities cooperate interactively during the resource allocation procedures.

The RRA entity is responsible for the radio bearer configuration at the beginning of each session, which includes the estimation of the required number of logical/transport/physical channels along with their mappings for each physical channel through the scheme layers/sublayers. The admittance decision of each incoming MBMS session is handled by the admission control function during the phase of service establish/renegotiation, aimed at preserving the required QoS while making efficient utilization of resources.

In SDMB, the nonavailability of a return link via satellite penalizes the system effectiveness and efficiency on short-term resource allocation. No fast power control mechanism is applicable in such a system; therefore, the packet scheduling algorithm becomes the focus of efficient resource allocation.

More specifically, in SDMB, the packet scheduler operates periodically in each transmission time interval (TTI) of the radio bearers, being responsible for two important tasks:

■ Time-multiplexing of service flows with different QoS requirements into physical channels with fixed spreading factor (SF), such as to satisfy these requirements
■ Adjusting the transmit power of the physical channel carrying the data flows on the basis of the required reception QoS in terms of the target block error rate (BLER), and under the constraint that the total available power for all the physical channels within a satellite beam is fixed.

---

* The integrated European IST project MAESTRO (Mobile Application and Services Based on Satellite and Terrestrial Interworking) and MoDiS (Mobile Digital Broadcast Satellite) are partly funded by European Commission under the 6th research framework program.[27,28]

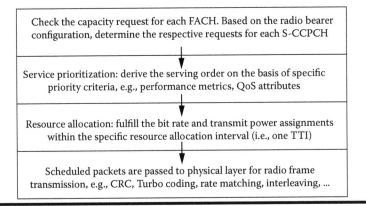

Figure 18.2   Packet scheduling strategy procedures.

As shown in Figure 18.2, the packet scheduling strategy can be conceptualized into the following two main steps:

- **Service prioritization:** The incoming service requests are ordered according to the priority criteria. In selecting the respective criteria, the service attributes are considered to provide dynamic intelligent scheduling tasks while considering multiple essential QoS factors that have a crucial impact on system performance.
- **Resource allocation:** Once all the multiplexed sessions are prioritized, resources are allocated accordingly to these sessions, which consists of bit rate and transmit power assignments within the specific resource allocation interval (i.e., one TTI).

With the unique nature of the SDMB system and growing demand of QoS requirements, the design of a packet scheduler in SDMB becomes a particularly challenging task for the following reasons:

- Given the lack of a return satellite link, the scheduler has to perform the resource allocation without the knowledge of the state of individual channels, that is, channel-state-dependent scheduling is not possible. Even if such information were available, it would have to be exploited in an unconventional manner considering the point-to-multipoint nature of the supported services.
- Wireless multimedia applications feature stringent and diverse QoS requirements; hence, the design of a packet scheduler has to take into account both the differentiation and fulfillment of these requirements.
- The total available power for all the physical channels within a single satellite beam is limited. The packet scheduler has to be designed such as to satisfy the power constraint and minimize unnecessary power consumption.

18.3.2   *Review of Existing Packet Scheduling Schemes*

Given the unidirectional nature of the SDMB system and the point-to-multipoint services it provides, aimed at maximizing spectrum efficiency and satisfying diverse QoS requirements while preserving the radio resources, the design of RRM functionalities implemented at the SDMB access layer proves to be an especially challenging task, and the packet scheduling algorithm, which is the single function performing short-term resource allocation, is the focus of efficient resource allocation.

Although numerous studies on packet scheduling schemes have been proposed in the literature for both wired and wireless networks,[29–33] they cannot be easily applied to the SDMB system because of its unique nature apart from other systems. One interesting research subject in this context is delay differentiated scheduling, where waiting time and queuing delay are considered in packet scheduling, as waiting time priority (WTP) and proportional delay differentiation (PDD) schemes proposed in Dovrolis et al.[29] for terrestrial differentiation networks. Another popular research topic is to exploit the channel quality of fast-varying wireless links for more efficient packet scheduling, for example, adaptive proportional fairness (APF) scheduling was proposed in the High-Speed Downlink Packet Access (HSDPA) system,[31] considering QoS demands for multimedia applications, where the return channel is present and CSI information for individual user is available. However, given the unidirectional nature and long propagation delay, the SDMB system is incapable of tracking real-time channel state information from the mobile terminal side, which makes the channel-state-dependent scheduling not feasible. Even if such information were available, it still has to be exploited in an unconventional manner considering the point-to-multipoint nature of the supported services, that is, increased heterogeneity of users interested in the same content. Furthermore, future multimedia applications feature an increasingly diverse range of capabilities and QoS requirements; hence, the design of the packet scheduling scheme has to take into account both the differentiation and fulfillment of these requirements. Finally, the total available power for all the physical channels within a single satellite beam is limited; the packet scheduling scheme has to be designed so as to optimize the overall transmit power.

Previous studies[31] have systematically addressed the packet scheduling problems in the SDMB system via classical packet scheduling schemes, namely, multilevel priority queuing (MLPQ) and weighted fair queuing (WFQ). However, both of them feature major weaknesses in QoS-differentiated multimedia services provisioning with respect to efficiency and fairness.

WFQ-based scheduling was motivated and developed in the SDMB system based on the well-known WFQ scheme, being capable of guaranteeing a minimum bandwidth per bearer/flux or per set of bearers grouped together for traffic-handling purposes.[31] The WFQ-based scheduler is more specifically based on the *virtual spacing* policy that uses the notion of *virtual time*.[34] The weights are primarily set according to the data rates of the multiplexed service flows rather than their priority. The weight distribution among flows can be adapted in response to new acceptance of a service or variation of channel mapping. The serving orders of the queues are computed depending on the time stamp of the head packet of each queue: queues with the lowest time stamp on their head packet will be served first. The nonpriority nature of this scheduling policy leads to unacceptably long queuing delay in higher-priority queues.

MLPQ-based scheduling[31] is effectively the adaptation of the multilevel, nonpreemptive priority discipline[35] to the SDMB. MLPQ always processes packets starting from those nonempty queues having the highest priority first, with queues having the same priority served in a round-robin fashion. First, MLPQ processes packets in a strict priority-based way; as a result, packets in the lower-priority queues may suffer from a considerably longer queuing delay. This scheme favors the high-priority classes, ensuring a delay bound for their packets, while it provides no guarantees for lower-priority classes. Furthermore, it is generally agreed that background application has no stringent delay constraint, and the only requirement for application in this category is that information should be delivered to the user essentially error-free. In fact, background application still needs a delay constraint (at least an upper bound), because data can be effectively useless if it is received too late for any practical purpose. Finally, MLPQ deals with queues having the same priority in a round-robin fashion. Consequently, there is no differentiation made between queues with the same QoS rank. Therefore, this is not an efficient mechanism. Rather than prioritizing queues in a strict manner, other essential QoS metrics (e.g., delay tolerance and guaranteed data rate) should also be considered in the scheduling discipline design.

- **Delay differentiation queuing:** To achieve better packet scheduling performance in terms of both efficiency and fairness, inherited from the aforementioned PDD scheme in the context of differentiated service networks, a delay differentiation queuing (DDQ) was proposed in Fan et al.,[36] offering improved performance on delay, jitter, and channel utilization. DDQ was proposed for the delay differentiation services in the satellite environment, assuming there are QoS ratios between different traffic priority classes. For each resource allocation interval (e.g., TTI), the serving indexes are obtained based on the average waiting delay for all the packets currently in the queue, the average queuing delay for all the packets that have left the queue, the packet arrival rate, and the QoS ratio. In this scheme, the instantaneous queuing delay is effectively considered for queues with the same QoS rank. Compared with WFQ and MLPQ, DDQ offers improved performance on delay, jitter, and channel utilization. However, DDQ experiences unbalanced performance among multiple QoS attributes, namely, the gain achieved in one performance attribute leads to the performance degradation on other attributes. Furthermore, multimedia services feature differentiated delay constraints; applying the delay constraints for differentiated services in an equal way may lead to poor QoS guarantee for high-priority queues. Therefore, the delay profile has to be considered against the respective delay constraints (i.e., maximum acceptable delay) specified by the class of service. Finally, rather than scheduling competing flows in a static manner, to provide more flexible QoS provisioning and maintain optimal resource utilization, it is highly desired that the scheduler is capable of choosing the best scheduling policy according to diverse QoS preferences of the services and instantaneous performance dynamics.

18.4 Optimization of Packet Scheduling Schemes in the SDMB System

18.4.1 Overview

In this section, we suggest a cross-layer packet scheduling framework, which comprises combined delay and rate differentiation (CDRD) service prioritization and dynamic resource allocation (DRA) algorithms, which are presented for QoS provisioning in the SDMB system. A priority-oriented packet scheduling is considered at the MAC layer for multisession broadcasting with diverse QoS requirements, where each session employs the dynamic rate-matching (DRM) technique at the physical layer. The proposed scheme considers several cross-layer criteria to achieve both efficiency and fairness. First, the service prioritization procedure takes into account the session's traffic priorities and QoS requirements in terms of maximum tolerable queuing delay and required data rate. Second, queuing dynamics are envisaged effectively into service prioritization to track the instantaneous variations at the RLC layer. Furthermore, the instantaneous data rate at the transport channel (TrCH) is proposed as another important criterion to minimize the unnecessary discontinuous transmission (DTX), and thereby optimize the resource utilization.

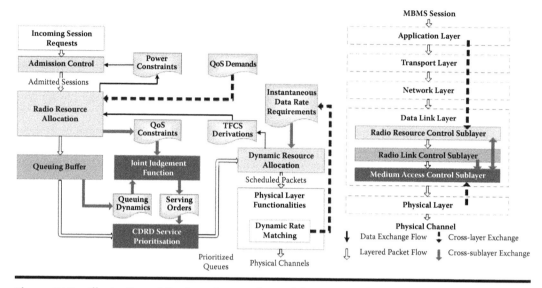

Figure 18.3   Illustration of the layer interactions of the proposed cross-layer scheduling scheme in the SDMB hub.

The proposed cross-layer packet scheduling algorithm is distinct from most existing schemes[29–33] in the following aspects:

1. It guarantees the prescribed QoS requirements.
2. It simultaneously considers multiple cross-layer criteria and balances the priority and fairness.
3. A DRM technique is applied into resource allocation to reduce unnecessary puncturing/repetition and optimize the transmit power.

To the best of the authors' knowledge, this is the first work to consider the cross-layer optimization combining QoS requirements, queuing dynamics, and rate-matching information together into MAC layer scheduling decision. The proposed methodology is envisaged for the SDMB system; however, it can also be applied adaptively to any WCDMA-based broadcast/multicast network.

Figure 18.3 illustrates the interactive framework and the layer/sublayer interactions of the proposed cross-layer packet scheduling scheme. The RRM is mainly handled at the data link layer, which can be further divided into radio resource control (RRC), RLC, and MAC sublayers. As seen from the layer interactions, it is noteworthy that the cross-layer/sublayer correspondence is set up in both upward and downward directions to the packet scheduler at the MAC sublayer of the data link layer, which, in our opinion, is not exploited much in the literature today.

On one hand, the CDRD packet scheduling scheme is employed as the service prioritization algorithm in this cross-layer framework. First, the MBMS service–prescribed QoS demands are passed to the RRA module at the RRC sublayer of the data link layer at the beginning of each admitted session. During the radio bearer configuration, the RRA abstracts the QoS demands of admitted sessions and passes them to the packet scheduler as one set of priority criteria. The queuing dynamics in the RLC buffer are instantaneously monitored and passed to the packet scheduler as another set of priority criteria. A joint judgment function (JJF) is employed in the CDRD packet scheduler to handle the above priority criteria, and therefore derive the serving orders of competing flows in TTI scale.

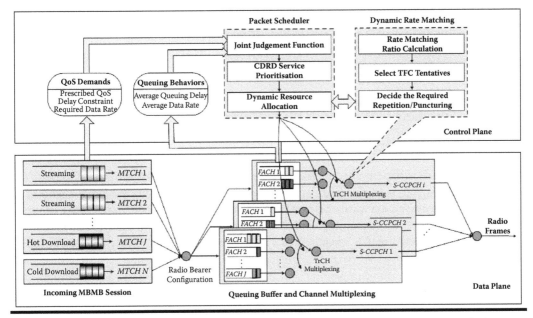

Figure 18.4   The proposed packet scheduling procedure implemented at the SDMB hub.

On the other hand, instantaneous data rate requirements derived from the DRM function at the physical layer are considered another important criterion for the resource allocation procedure implemented in the RLC sublayer.

The parameters considered in the JJF function can be grouped into two categories: (1) QoS parameters and (2) queuing dynamics parameters. The parameters in the first category depend on the application and are derived on a per-session scale, and remain constant during each session transmission. The parameters in the second category track the instantaneous queuing status dynamically at the FACH queue in the RLC buffer in per-TTI scale.

The framework of the proposed packet scheduling scheme in both the data and control plane is illustrated in Figure 18.4. In the SDMB system, the service types considered can be divided into two main classes: streaming and download classes, which correspond to the UMTS streaming and background service classes.* The latter can be further subcategorized into two subclasses according to their sensitivity to delay: hot download and cold download.[37]

For multimedia data transmission in SDMB, there is a one-to-one correspondence between the MBMS session and MBMS point-to-multipoint traffic channel (MTCH) logical channel. The logical channels are then mapped by the RRA, in a one-to-one manner, to the Forward Access Channel (FACH) transport channel. The scheduled packets are delivered to the Secondary Common Control Physical Channel (S-CCPCH) in the form of transport block (TB).[38] One or more FACHs are carried by a single S-CCPCH via TrCH multiplexing at the physical layer.

---

* In UMTS, streaming implies a information transmission technique that allows information entities (e.g., packets) to be transmitted in streams continuously. Through buffering, the streaming technique smooths out the incoming traffic and allows contents to be displayed even before the completion of transmission. Background class, one of the traditional information transmission techniques, has no explicit constraint on transmission delay; contents belonging to this class cannot be displayed until the information transmission is fully completed.

For each active physical channel, the exact format of transport format combination (TFC), which consists of multiple transport block sets (TBSs), is selected (i.e., the amount of data from each transport channel mapped to the physical channel) from the transport format combination set (TFCS).

18.4.2 Combined Delay and Rate Differentiation Service Prioritization

Advances in multimedia applications entail the packet scheduling scheme to support diverse QoS among heterogeneous traffics. To achieve more efficient QoS provisioning among different multimedia services, a novel packet scheduling algorithm, namely, the CDRD algorithm,[39] is described in this section. The proposed algorithm takes into account several key criteria simultaneously for ensuring comprehensive QoS satisfactions. On one hand, rather than only differentiate the competing sessions with respect to their traffic priorities (i.e., service types), the CDRD scheme considers the application-prescribed QoS requirements as a combination of multiple attributes, including the traffic priority, the required data rate, and the queuing delay constraint. On the other hand, the queuing dynamics of the competing flows at the RLC layer are monitored and considered to further optimize the overall system performance.

The CDRD service prioritization scheme is illustrated as Figure 18.5. The accepted ongoing sessions comprise multiple sessions with diverse QoS demands. Each session is assumed to retain an individual FACH queue in the RLC buffer. Packets in the FACH queues are prioritized by CDRD service prioritization according to their respective QoS metrics. Consequently, the packet scheduler handles all the FACH queues in the RLC buffer according to their instant joint priorities rather than the service types of the carried traffics. The prioritized queues are passed to resource allocation and are allocated with the required resources. Because packets belonging to the download class have no explicit delay constraint, the scheduler restricts download services via preserving certain QoS requirements for delay-sensitive streaming service. Nonetheless, as long as there is spare capacity remaining for the streaming services, the packet scheduler will adaptively enable the download services to fill this gap.

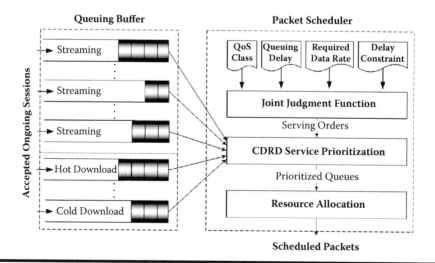

Figure 18.5   CDRD service prioritization scheme.

In CDRD, the JJF function is employed for prioritizing the competing flows based on the criterion factors in (18.1). To make the mathematical expressions of considered performance factors simpler, the index representing current TTI does not appear in the mathematical expressions in this chapter. However, it must be noticed that the evaluation and calculation of joint priority and packet scheduling procedures are excuted recursively in each TTI. For each FACH queue i, the JJF function is defined as

$$Priority^{joint}(i) = Factor^{QoS}(i) \cdot Factor^{delay}(i) \cdot Factor^{data\_rate}(i) \cdot Factor^{delay\_constraint}(i) \quad (18.1)$$

where $Priority^{joint}(i)$ is the priority index for queue i.

$Factor^{QoS}(i)$ is a QoS class factor, which is essentially a time-independent parameter designated for queue i, representing the traffic priority of the service carried by each queue, that is, streaming, hot-download, and cold-download service. Because CDRD assumes that there are QoS ratios between different QoS priority classes, the value of α_i represents the relative traffic priority of the service carried by queue i; that is, streaming service has higher priority than hot-download service, which has higher priority than cold-download service.

$Factor^{QoS}(i)$ defined in (18.2), specifies the delay factor for queue i. This measure describes the delay states of all packets passing through the respective queue, including both the packets that are currently in queue i and the packets that have already been served by queue i till current TTI.

$$Factor^{delay}(i) = \frac{\sum_{j=1}^{N_d} Waiting\_Delay(i,j) + \sum_{k=1}^{N_d} Queuing\_Delay(i,j)}{N_q + N_d} \quad (18.2)$$

where $Waiting\_Delay(i,j)$ is the waiting delay for the jth packet currently in queue i; N_q is the number of packets that are currently in queue i; $Queuing\_Delay(i,k)$ is the queuing delay for the kth packet that has been served by queue i before current TTI; and N_d is the number of packets that have been served by the queue before current TTI.

$Factor^{data\_rate}(i)$ represents the data rate factor for queue i at current time slot. This factor is calculated as the ratio of the service required/guaranteed data rate against the average transmitted data rate until current TTI. The instantaneous joint priority of each queue is affected proportionally by the difference between the average transmitted data rate and the required data rate of each queue. The average transmitted data rate $Average\_Data\_Rate(i)$ or queue i at the current time slot can be expressed as

$$Average\_Data\_Rate(i) = \frac{\sum_{k=1}^{N_d} Packet\_Size(i,k)}{(n-1) \times T_{tti}} \quad (18.3)$$

where $Packet\_Size(i,k)$ is the packet size for the kth packet in queue i, and T_{tti} is the value of TTI, for example, 80 ms in our simulation.

Therefore, the data rate factor $factor^{data\_rate}(i)$ is defined as follows:

$$Factor^{data\_rate}(i) = \frac{Required\_Data\_Rate(i)}{Average\_Data\_Rate(i)} \quad (18.4)$$

where *Required_Data_Rate(i)* is the required/guaranteed data rate specified by the service QoS level. If the average offered data rate is smaller than the required data rate, $Factor^{data\_rate}(i)$ is larger than 1. Thus, the priority index for this queue is increased for this underutilized queue; otherwise, the priority index will be decreased for this overutilized queue. This factor fine-tunes the priority and leads the offered data rate to approach the guaranteed data rate.

$Factor^{delay\_constraint}(i)$ is the delay constraint factor, which depends on the maximum queuing delay tolerated by the corresponding service, which proportionally adjusts itself according to the difference between the average queuing delay for each queue and its delay constraint for queue *i*, reflecting the current queuing delay status. This factor is defined as

$$Factor^{delay\_constraint}(i) = \begin{cases} 2, & \frac{\sum_{j=1}^{N_q} Waiting\_Delay(i,j)}{N_q} \geq Delay\_Threshold(i) \\ 1, & \frac{\sum_{j=1}^{N_q} Waiting\_Delay(i,j)}{N_q} < Delay\_Threshold(i) \end{cases} \quad (18.5)$$

where *Delay_Threshold(i)* is the delay threshold for the service queue *i*.

If the average queuing delay for queue *i* is larger than its delay threshold, the delay constraint factor $Factor^{delay\_constraint}(i)$ doubles the priority of this queue for a better chance to be processed; otherwise, it remains the same. The weight can be chosen as various ratios reflecting the effectiveness of the delay constraint factor in the overall joint judgment function. It is noted that the delay threshold can be chosen as a tunable parameter, which depends on the maximum tolerable delay of the corresponding service. $Factor^{delay\_constraint}(i)$ is only in effect when the average queuing delay is beyond the designated delay threshold, which provides more efficient action to be taken for better QoS provisioning amont differentiated traffic flows.

In each TTI, the scheduler sorts the FACH queues in descending order, according to their instantaneous priorities calculated from the JJF function. Subsequently, the FACH queues with higher priorities will be served ahead of their lower-priority counterparts.

From the viewpoint of implementation, the proposed CDRD service prioritization algorithm poses extra computation complexity.[41] With the input size of *n* (i.e., total number of MTCH/FACH queues*), the computational complexity of the proposed algorithm is derived as $O(n)$, following typical linear statistics. Given the rather small number of input queues, the computational complexity does not involve much constraint.

In this section, a novel service prioritization algorithm is presented for the SDMB system. This algorithm takes into account multiple important performance factors reflecting service QoS demands and queuing behaviors to optimize the overall system performance. Simulation results comparing the proposed scheme with the existing schemes will be given in section 18.4.5. Once the services have been prioritized, the next step is the resource allocation. A novel DRA scheme is presented in the following section, where a DRM technique, is first presented as the fundamental basis.

---

* It is noted that the channel mapping/multiplexing from MTCH logical channels to FACH transport channels can be performed in either a one-to-one or one-to-many way, depending on the channel multiplexing option employed, that is, single-level channel multiplexing or two-level multiplexing.[42] Herein the single-level channel multiplexing scheme is applied, where transport channel multiplexing is not present and there is a simple one-to-one correspondence between the MTCH logical channel and FACH transport channel.

18.4.3 Dynamic Rate Matching

To minimize the DTX insertion in downlink static rate matching (SRM), a novel rate-matching technology, namely, dynamic rate matching, has been proposed for delivering highly rate-variable MBMS service in the SDMB system.[36] The objective of downlink DRM is to minimize the number of DTX bits required for the chosen TFC at a given TTI according to the available physical layer resources. Rather than per-session- and maximum-data-rate-based rate-matching calculation in SRM, in the DRM case, the rate-matching (RM) ratio is calculated in each TTI based on the instantaneous data rate of each TrCH. Therefore, the DRM employs a variable rate-matching ratio to prevent unnecessary DTX insertion. The scenario undertaken is applicable for the TrCHs featuring identical TTI scale. In each TTI, the following phases are performed in DRM:

- **Phase 1 (TFC reordering):** All the TFCs within the TFCS list are reorganized according to their corresponding total data rate (i.e., based on the TFC size).
- **Phase 2 (RM ratio calculation):** The rate-matching ratio is calculated based on the instantaneous data rate (i.e., the TBS size) of each TrCH for each different TFC allowed for a given physical channel.
- **Phase 3 (Bit matching):** According to the selected TFC, a tentative value of repetition/puncturing bits is calculated for each TrCH.

Finally, the rate-matching module performs the tentative value corrections and a rate-matching pattern is generated.

To apply the DRM technique, modifications are identified and applied at both the sending and receiving side in the SDMB system:

- New rate-matching interval in TTI scale, during which rate-matching ratio remains constant.
- Before the transmission starts, rate-matching ratios need to be calculated for all possible TFCs.

By reducing the DTX insertion, the total transmit power can be minimized while the power consumption is optimized. Furthermore, in some cases, DRM can facilitate higher bit repetition, which has an essential impact on improving the bit error rate performance. Finally, the saved transmit power can be utilized by other channels for better transmission capacity; alternatively, it can be used for improving the packet loss rate of allocated channels. The main limitation of DRM lies in that more processing and memory are required for the rate-matching calculation and storage.

18.4.4 Dynamic Resource Allocation

One of the main tasks for packet scheduling besides service prioritization is the resource allocation, which is responsible for selecting the required bit rate and transmit power for all physical channels within the specific resource allocation interval (i.e., one TTI) subject to the QoS guarantees and physical channel constraints. The existing SRM technique at the physical layer has been traditionally designed separately from higher layers. Based on the novel DRM technique at the physical layer, a cross-layer packet scheduling framework is developed. This scheme optimizes the

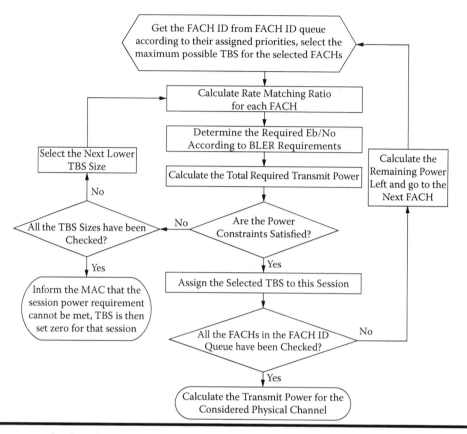

Figure 18.6 The flowchart for the DRA algorithm.

resource utilization at the data link layer, to improve the overall system efficiency, when combined with resource allocation at the data link layer.

Unlike SRM, where the rate-matching ratio is fixed during the session transmission, in DRM, until all the sessions are scheduled, the precise rate-matching ratio for each TrCH will not be known, and the DRM only knows the possible TFC candidates that form a TFC subset. The power requirements are evaluated based on every possible TFC within the TFC subset under the maximum power constraints. In each TTI, a maximum possible TFC is selected for the given physical channel; the DRM will then calculate the number of bits required to be repeated/punctured.

The proposed DRA algorithm, as illustrated in Figure 18.6, relies on the DRM technique to evaluate the required transmit power for the respective TrCH according to its instantaneous data rate requirements. In this case, the physical layer instantaneous supportable data rate effectively corresponds to the maximum TBS size that can be granted subject to the total transmission power constraints. Along with the priority decided by the service prioritization function, the instantaneous data rate information is fed into the resource allocation module for the power estimation; the derived TFC is thereby priority based, DRM associated, and data rate confined, and is then allocated to the corresponding physical channel. In this way, physical layer resources are effectively considered in packet scheduling and overall physical channel utilization is greatly improved.

The procedure for the DRA algorithm based on the DRM technique is described in the following. For each FACH, the packet scheduler scans the TFCS of the physical channel to find all the different

TBS sizes that could be used, namely, serve the whole or part of the queued data in the FACH channel at the queuing buffer. A sorted list of all candidate TBS sizes is created, in decreasing order. The scheduler first seeks to allocate the maximum TBS size to the first FACH in the prioritized queues.

As shown in Figure 18.6, the scheduler first calculates the rate-matching ratio based on (18.6) for all subsets of TFCs from the full TFC set, and these values are stored against each TFC set. Then the scheduler checks which subset the chosen TFC belongs to, and based on this information, the rate-matching ratio is obtained from the stored data and a tentative value determined according to the selected TFC. Following this, the scheduler performs tentative value correction, and then rate-matching patterns are generated.

$$Ratio^{rate\_matching}(p,q) = \frac{Data^{allowed}}{\displaystyle\sum_{k=1}^{\omega}(((length_k + length_{CRC}) \times Coding\_Rate + length_{Tailbits}) \times N_k)}$$

$$for\ 1 \le q \le \omega \qquad (18.6)$$

where p is the SCCPCH ID, q is the TFC ID, and ω is the TFC subset which includes all candidate TBS sizes that have been calculated. $Data^{allowed}$ is the allowed data, $length_k$ is the TB length of FACH k in this TFC, $length_{CRC}$ and $length_{Tailbits}$ are the number of bits of CRC and tail bits, N_k is the number of TB with allocated TBS, and $Coding\_Rate$ is the coding rate.

According to the calculated RM ratio values, the scheduler determines the required E_b/N_0 value according to each session BLER requirement. The scheduler then checks whether the selected TBS size for the new session satisfies the total transmit power criteria.

If this is not satisfied, it will determine whether all the possible TBS sizes have been checked for total transmit power criteria. For the next TBS size (less than the previous one), the above procedure is performed. If none of the TBS sizes satisfy the power criteria, the scheduler assigns TBS to zero for this FACH.

- If the power criteria are met, based on each session and RM combinations, the transmit power for each session is calculated separately according to (18.7) and the highest power requirement assigned as the physical channel transmit power.[36]

$$Trans\_Power^{Required} = \frac{P_N \times Path\_Loss \times \rho \times SF}{R_s \times Coding\_Rate \times Ratio^{rate\_matching}} \qquad (18.7)$$

where $Trans\_Power^{Required}$ is the required transmit power, P_N is the thermal noise, ρ is the E_b/N_0 requirement, $Path\_Loss$ is path loss, $Ratio^{rate\_matching}$ is the rate-matching ratio, R_s is the modulation scheme, and SF is the spread factor.

These procedures are repeated recursively until all the FACHs mapped to each S-CCPCH are assigned.

In conclusion, the DRA algorithm offers two main advantages over the existing SRM-based scheme: (1) it allows better DTX minimization, and (2) it requires less transmit power when the instantaneous data rates are less than the maximum data rate.

In this section, a cross-layer packet scheduling scheme is presented for multimedia delivery in the SDMB system. This scheme not only takes into account the impact of key performance factors reflecting service QoS demands and queuing dynamics, but also utilizes a DRM technique at the physical layer to maximize spectrum efficiency and resource utilization. Simulation has been conducted over extensive scenarios for various performance metrics; discussion on the performance enhancement in terms of queuing delay/jitter, channel utilization, and flexibility is presented in the next section.

18.4.5 Performance Evaluation

18.4.5.1 Simulation Methodology

To demonstrate the performance enhancement of the proposed cross-layer packet scheduling scheme, a system-level simulator implementing the SDMB system has been developed with the aid of the software package ns-2, and a wide range of simulation scenarios have been examined. The packet scheduling scheme is physically implemented in node B (i.e., Sat-Hub) employing the SDMB function, supporting three types of SDMB QoS classes: (1) streaming, (2) hot download, and (3) cold download.[37] The streaming traffic model applies publicly available trace files[42] for video-streaming traffic. Traffic characteristics associated with hot- and cold-download services— or push-and-store services—follow the classical *Pareto* distribution, with different traffic priority assigned. In addition, different guaranteed data rates are selected to examine the performance between users with different rate requirements.

The performance of the proposed scheme and that of previous studies are evaluated via simulations over a wide variety of traffic mixes. The simulation considers individual MBMS sessions with diverse QoS profiles in terms of service type, data rate, and delay constraints for broadcast transmission, each of which is carried by a single FACH queue. Multiple S-CCPCHs are used for carrying heterogeneous multimedia services. A representative traffic mix scenario is identified to demonstrate our research findings. The considered radio bearer mapping scenario is given in Table 18.1.

The considered traffic mix scenario consists of different mixtures of service types, that is, streaming, hot download, and cold download, reflecting diverse QoS demands of heterogeneous multimedia services. Different data rates are also applied for each of these service types, as follows:

- S-CCPCH 1 carries streaming FACHs 2 and 3 with data rate of 256 and 64 kbps, respectively, and hot-download FACH 1 with a data rate of 64 kbps.
- S-CCPCH 2 carries streaming FACHs 4 and 5 with data rates of 256 and 128 kbps respectively.
- S-CCPCH 3 carries cold-download FACH 6 with a data rate of 384 kbps.

Our link-level simulation results provide the E_b/N_0 versus BLER requirements of each TrCH. The radio propagation channel model features either classical Gaussian characteristics for

Table 18.1 Radio Bearer Mapping Configuration (kbit/s)

S-CCPCH id	1			2		3
S-CCPCH bit rate	384			384		384
FACH id	1	2	3	4	5	6
Streaming	—	256	64	256	128	—
Hot download	64	—	—	—	—	—
Cold download	—	—	—	—	—	384

Table 18.2   System Simulation Parameter

Simulation Parameter	Value
Frequency of operation (GHz)	2.5
Chip rate (Mchip/s)	3.84
Spreading factor	8
TTI (ms)	0.08
Modulation	QPSK
Coding	Turbo code
Code rate	1/3
Maximum bit rate (kbit/s)	384
Packet size (bytes)	1,280
Terrestrial channel model	Rayleigh
Satellite channel model	Ricean

satellite-associated path or Rayleigh multipath fading channel for UE-associated path with the consideration of both Doppler effect and propagation impairments. The maximum service data unit (SDU) size is 1,500 bytes. TTI equals 0.08 s and turbo code is applied. The simulation period is set as 1,000 s or 12,500 TTIs.

In the SDMB system, the queuing delay tolerance thresholds are assumed to be 20–100 ms for video streaming, and 200–2000 ms for push-and-store service. Accordingly, various queuing delay threshold values are applied and examined for the specific scenario, showing the performance variation against tuning the delay threshold parameter. A summary of traffic and simulation parameters for our test bed is given in Table 18.2.

18.4.5.2 Performance Evaluation

First, the performance of the CDRD service prioritization is evaluated by investigating the mean queuing delay experienced by packets in each FACH queue at the RLC buffer.

As illustrated in Figure 18.7, download multimedia services (i.e., FACHs 1 and 6) experience much less mean queuing delay in CDRD than MLPQ, while the mean delays experienced by streaming services (i.e., FACHs 2 to 5) feature similar performance between the algorithms. Numerically, hot download and cold download enjoy an average reduction of 32.6 and 23.7 percent on their mean queuing delay respectively, while the maximum increase in the mean queuing delay of streaming service is 7.6 percent. It can therefore be inferred that the significant reduction on the queuing delay of the lower QoS class does not pose dramatic performance degradation on its higher QoS class counterparts, that is, the QoS of streaming application can be safely guaranteed. In the meantime, it also implies that the CDRD enables the download service to efficiently utilize the spare resources of the streaming service without causing significant degradation on the QoS satisfaction of the streaming services.

Figure 18.8 compares the mean queuing jitter, namely, queuing delay variation, experienced by each FACH between MLPQ and CDRD. Obviously, the latter achieves much lower jitter for both streaming and download services. Typically, the average jitter reductions for download and streaming services are 45.5 and 29.1 percent, respectively. It is also worth noticing that the unidirectional streaming service in SDMB is quite sensitive to delay variation; thereby the delay variation of the flows should be limited to preserve the time variation between packets in the stream.[37]

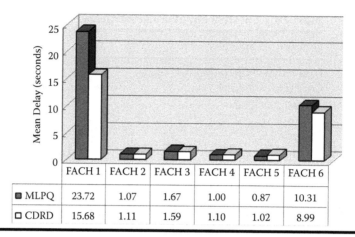

	FACH 1	FACH 2	FACH 3	FACH 4	FACH 5	FACH 6
▣ MLPQ	23.72	1.07	1.67	1.00	0.87	10.31
▢ CDRD	15.68	1.11	1.59	1.10	1.02	8.99

Figure 18.7   Mean queuing delay for MLPQ and CDRD scheduling.

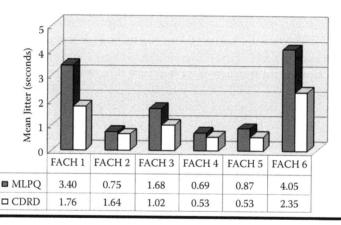

	FACH 1	FACH 2	FACH 3	FACH 4	FACH 5	FACH 6
▣ MLPQ	3.40	0.75	1.68	0.69	0.87	4.05
▢ CDRD	1.76	1.64	1.02	0.53	0.53	2.35

Figure 18.8   The mean queuing jitter for MLPQ and CDRD scheduling.

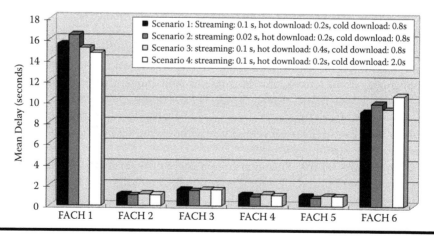

Figure 18.9   The mean queuing delay for FACH queues under different delay thresholds.

The results in figure 18.8 prove that the proposed CDRD scheme provides better delay variation in RLC queues irrespective of their traffic priorities.

As explained earlier, the queuing delay threshold can be tuned as an adjustable parameter indicating the queuing delay tolerance for multimedia traffics. By looking at figure 18.9, tuning the delay thresholds effectively influences the queuing delay performance of the FACH queues. The average queuing delay of a particular FACH queue is affected directly by its delay threshold; that is, more stringent delay thresholds lead to better performance of the corresponding queue while causing longer delays for the other service classes.

The QoS demand expressed by the required data rate is considered another key factor effectively driving the session's instantaneous achieved data rate to the required data rate. This can significantly influence the overall system performance for multimedia content transmissions. Compare figure 18.10(a)

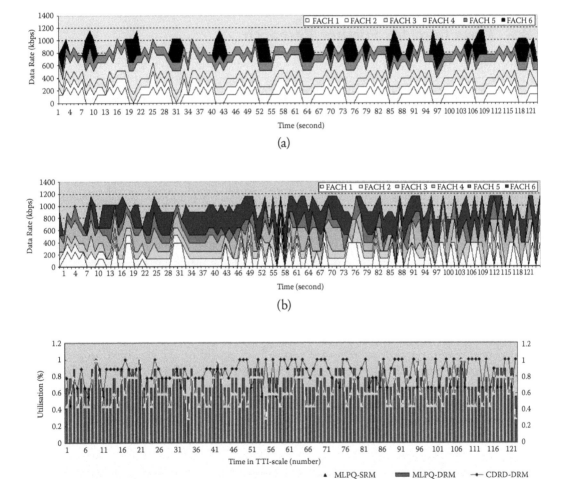

Figure 18.10   Performance evaluation on channel throughput/utilization achieved at the physical layer: (a) instantaneous data rates of FACHs with MLPQ scheduling; (b) instantaneous data rates of FACHs with CDRD scheduling; (c) instantaneous physical channel utilization for MLPQ and CDRD scheduling using SRM and DRM techniques.

and (b), where the instantaneous data rates of individual FACHs are presented cumulatively. By providing better QoS guarantees and considering the physical channel capabilities, the proposed scheme can increase the total mean transmission rate at the physical channels by 12.8 percent. Moreover, the data rate distribution among all the competing flows becomes more even, which means that better fairness with respect to data rate is provided with the CDRD service prioritization than with the MLPQ-based scheduling scheme. Figure 18.10(c) depicts the instantaneous physical channel utilization achieved for different scheduling and rate-matching schemes during a sample simulation period (i.e., around 125 TTIs). The total physical channel utilization achieved by CDRD enjoys noticeable performance gain during the session transmission. Numerically, the mean channel utilization reaches around 83.8 percent in CDRD-based scheduling, compared with 78.2 percent for the MLPQ-based scheduling. On the other hand, with the SRM applied, MLPQ only achieved an average of 62.4 percent channel utilization because of its frequent unnecessary puncturing and repetition.

18.5 Directions for Further Research

In this chapter, we present a survey of literature in the optimization techniques on packet scheduling design, and suggest a cross-layer-based packet scheduling scheme for a satellite broadcast network. The scheme considers the delay and data rate attributes in the scheduling decision. Once the service prioritization process is enriched with other performance metrics, such as throughput and packet loss rate, the performance behavior deserves further investigation. On the other hand, if a finite buffer length is assumed, the impact of buffer length–associated scheduling may lead to further performance improvements.

The optimization technique on packet scheduling is envisaged at the satellite hub via an example of a satellite broadcasting system, namely, SDMB, where unidirectional forward link is assumed and return link via satellite is not applicable. In the presence of a return link, it is interesting to investigate the performance of packet scheduling via utilizing the channel-state-dependent approach. However, the long latency experienced in a GEO satellite link appears to be a key limitation in such a scenario; therein the main concern should be the effective exploitation of usable channel state informantion.

18.6 Summary

The packet scheduling technique plays a crucial role in providing the ever-increasing demands of diverse and stringent QoS for heterogenous multimedia service provisioning over broadcast networks, utilizing the precious communication resources available in a most efficient manner. This chapter has taken stock of the current developement activities in the area of packet scheduling optimization techniques. We have summarized the current state of knowledge in this area via investigating representative packet scheduling schemes for both wired and wireless systems in existing literatures, emphasizing the elaboration and evaluation of an optimized packet scheduling scheme based on the cross-layer concept for a representative unidirectional satellite broadcasting system, namely, SDMB.

First, a proportional differentiation concept is employed into packet scheduling via the introduction of the novel combined delay and rate differentiation (CDRD) service prioritization algorithm. Performance gains are discussed against previous scheduling schemes. It is proved that the CDRD algorithm is capable of offering desired features on delay and data rate performance, and thereby improves the overall achieved QoS.

Furthermore, a cross-layer packet scheduling scheme is presented for multimedia content delivery in the SDMB system. This scheme not only takes into account the impact of multiple key performance factors reflecting service QoS demands and queuing dynamics, but also utilizes the dynamic rate-matching (DRM) technique at the physical layer to maximize spectrum efficiency and resource utilization. Simulation has been conducted over extensive traffic mix scenarios for various performance metrics; results show that the proposed packet scheduling scheme achieves better performance than the existing schemes on delay, jitter, and channel utilization. Discussions on the impact of a tunable threshold parameter further demonstrate the flexibility and scalability achievable in this scheme.

References

1. M. Katevenis, S. Sidiropoulos, and C. Courcoubetis. 1991. Weighted round-robin cell multiplexing in a general purpose ATM switch chip. *IEEE Journal on Selected Areas in Communication* SAC-9:1265–79.
2. A. Parekh and R. Gallager. 1993. A generalized processor sharing approach to flow control—The single node case. *ACM/IEEE Transactions on Networking* 1:344–457.
3. L. Zhang. 1991. VirtualClock: A new traffic control algorithm for packet switching networks. *IEEE/ACM Transactions on Computer Systems* 9:101–24.
4. A. Demers et al. 1990. Analysis and simulation of a fair queuing algorithm. *Journal of Internetworking Research and Experience*, 1, 3–26.
5. S. Floyd and V. Jacobson. 1995. Link-sharing and resource management models for packet networks. *IEEE/ACM Transactions on Networking* 3:365–86.
6. D. Ferrari and D. C. Verma. 1990. A scheme for real-time channel establishment in wide-area networks. *IEEE Journal on Selected Areas in Communications* 8:368–79.
7. H. Sariowan, R. L. Cruz, and G. C. Polyzos. 1999. SCED: A generalized scheduling policy for guaranteeing quality-of-service. *IEEE/ACM Transactions on Networking* 7:669–84.
8. V. Srivastava and M. Motani. 2005. Cross-layer design: A survey and the road ahead. *IEEE Communications Magzine* 43:112–19.
9. V. Bharghavan et al. Fair queuing in wireless networks: Issues and approaches. *IEEE Personal Communications* 6:44–53.
10. H. S. Wang and N. Moayeri. 1995. Finite-state Markov channel: A useful model for radio communication channels. *IEEE Transactions on Vehicular Technology* 44:163–71.
11. S. Lu, V. Bharghavan, and R. Srikant. 1999. Fair scheduling in wireless packet networks. *IEEE/ACM Transactions on Networking*, 7(4):473–89.
12. T. S. E. Ng et al. 1998. Packet fair queueing algorithms for wireless networks with location-dependent errors. *IEEE INFOCOM* 3:1103–11.
13. P. Ramanathan and P. Agrawal. 1998. Adapting packet fair queuing algorithms to wireless networks. *ACM MOBICOM*, 1–9.
14. S. Lu, T. Nandagopal, and V. Bharghavan. 1997. Fair scheduling in wireless packet networks. *ACM SIGCOM*. '97, 63–74.
15. P. Goyal, H. M. Vin, and H. Chen. 1996. Start-time fair queuing: A scheduling algorithm for integrated service access. *ACM SIGCOMM*, 117–130.
16. R. Knopp and P. A. Humblet. 1995. Information capacity and power control in single cell multiuser communications. In *Proceedings of IEEE International Conference on Communications (ICC 2006)*, Istanbul, 331–35.
17. A. Jalalim, R. Padovani, and R. Pankai. 2000. Data throughput for CDMA HDR: A high efficiency-high data rate personal communication wireless system. In *Proceedings of IEEE International Vehicular Technology Conference (VTC 2000)*, Tokyo, 1854–58.

18. A. Pandey et al. Application of MIMO and proportional fair scheduling to CDMA downlink packet data channels. In *Proceedings of IEEE International Vehicular Technology Conference (VTC 2002)*, Birmingham, AL, 2, 1046–50.

19. S. Abedi. 2005. Efficient radio resource management for wireless multimedia communications: A multidimensional QoS-based packet scheduler. *IEEE Transactions on Wireless Communications* 4:2811–22.

20. Q. Liu et al. 2006. A cross-layer scheduling algorithm with QoS support in wireless networks. *IEEE Transactions on Vehicular Technology* 55(3):839–47.

21. Q. Liu et al. 2005. Cross-layer scheduling with prescribed QoS guarantees in adaptive wireless networks. *IEEE Journal on Selected Areas in Communications* 23:1056–66.

22. Q. Liu et al. 2005. Queuing with adaptive modulation and coding over wireless links: Cross-layer analysis and design. *IEEE Transactions on Wireless Communications* 4, 1142–53.

23. X. Liu et al. 2001. Opportunistic transmission scheduling with resource-sharing constraints in wireless networks. *IEEE Journal on Selected Areas in Communications* 19:2053–64.

24. S. Shenker. 1995. Fundamental design issues for the future Internet. *IEEE Journal on Selected Areas in Communications* 13:1176–88.

25. 3GPP. 2005. *Multimedia broadcast/multicast service; Architecture and functional description*. Release 6. 3GPP TS 23.246 v6.8.0.

26. N. Chuberre et al. 2005. Relative positioning of the European satellite digital multimedia broadcast (SDMB) among candidate mobile broadcast solutions. Paper presented at IST Mobile & Wireless Communications Summit, Dresden, Germany.

27. EU FP5 IST MoDiS. http://www.ist-modis.org.

28. EU FP6 IST MEASTRO. http://ist-maestro.dyndns.org.

29. C. Dovrolis et al. 2002. Proportional differentiated services: Delay differentiation and packet scheduling. *IEEE Transactions on Networking* 10:12–26.

30. G. Aniba and S. Aissa. 2004. Adaptive proportional fairness for packet scheduling in HSDPA. In *IEEE Globecom 2004*, Dallas, 4033–37.

31. M. Karaliopoulos et al. 2004. Packet scheduling for the delivery of multicast and broadcast services over S-UMTS. *International Journal on Satellite Communication and Networking* 22:503–32.

32. Y.-J. Choi and S. Bahk. 2006. Delay-sensitive packet scheduling for a wireless access link. *IEEE Transactions on Mobile Computing* 5:1374–83.

33. V. Huang and W. Zhuang, 2004. QoS-oriented packet scheduling for wireless CDMA network. *IEEE Transactions on Mobile Computing* 3:73–85.

34. J. W. Roberts. 1994. Virtual spacing for flexible traffic control. *International Journal of Communication Systems* 7, 307–18.

35. D. Gross and R. Harris. 1985. *Fundamentals of queuing theory*. 2nd ed. New York: Wiley.

36. L. Fan, H. Du, U. Mudugamuwa, and B. G. Evans. 2006. Novel radio resource management strategy for multimedia content delivery in SDMB system. In *Proceedings of AIAA 24th International Communications Sat. Sys. Conference*, San Diego, 5476–84.

37. 3GPP. 2005. *Quality of service (QoS) concept and architecture*. 3GPP TS 23.107 V6.3.0.

38. 3GPP. 2005. *Radio interface protocol architecture*. 3GPP TS 25.301 v6.2.0.

39. H. Du, L. Fan, and B. G. Evans. 2007. Combined delay and rate differentiation packet scheduling for multimedia content delivery in satellite broadcast/multicast systems. Paper presented at Proceedings of IEEE International Conference on Communications (ICC 2007), Glasgow, Scotland.

40. S. Homer and A. Selman. 2001. *Computability and complexity theory*. Springer Verlag, NY.

41. H. Du, L. Fan, and B. G. Evans. 2006. Two-level channel multiplexing: A novel radio resource allocation strategy for SDMB system. In *Proceedings of IEEE International Conference on Communications (ICC 2006)*, Istanbul, 10, 4445–50.

42. Movie trace files. http://www.tkn.tu-berlin.de/research/trace/ltvt.html.

Chapter 19

Finite-State Models for Simulating the Packet Error Behavior of a Mobile Broadcasting System Operating in a Multipath Channel Environment

Jussi Poikonen, Jarkko Paavola, and Valery Ipatov

Contents

Keywords

Broadcasting, communication channels, Markov processes, mobile communication, simulation

19.1 Introduction

Simulations are an essential tool in analysis of the performance of communication systems. With simulations different systems can be compared and the parameters of transmission systems or receiver algorithms can be tuned to maximize performance. Ideally, transmitter and receiver equipment and necessary resources to perform measurements are available, and performance analyses can be accomplished in field conditions or in a laboratory using channel simulator equipment. Still, the resource expenditure related to measurement campaigns is often cost prohibitive, which makes simulations a useful additional analysis tool. Furthermore, simulations can be used to post process measured data to provide additional information for system optimization.

The most typical simulation analyses produce, for example, bit error rate (BER) information as a function of the signal-to-noise ratio (SNR) with a given system configuration. However, increasingly large amounts of data bits should be processed through the entire protocol stack to provide a useful amount of error information to analyze system performance at upper protocol layers such as the transport, network, or application layer. In this kind of analysis it is useful to have models for the packet error process at the protocol layer of interest. Such models implicitly incorporate lower-layer functionalities, such as modulation and channel coding, in addition to physical layer radio wave propagation.

We use the term *packet channel* for a block error process, where a given received data block is labeled as erroneous or correct according to whether it contains bit errors; in our case, the data blocks correspond to specific packets in the system protocol used in simulations. Packet channel models can be implemented using state machines, where the state structure, state transition probabilities, and output symbol emission probability distributions are chosen to provide required accuracy. Furthermore, for practical implementations, packet channel models should be parameterizable. That is, it should be feasible to determine the model parameters from relevant physical layer variables to retain the relation between upper protocol layer performance and physical layer conditions.

The main contribution of this chapter is to provide practical and efficient error models for packet channels. The models considered in this work are targeted for simple and fast simulations while retaining required accuracy. To facilitate the implementation of these models, straightforward procedures are given for calculating the model parameters. The usefulness of the discussed models is considered in simulating mobile reception with Digital Video Broadcasting–Handheld (DVB-H). In this case, the packet channel models describe the behavior of transport stream packets entering the DVB-H link layer or IP packets in the application layer. More specifically, we will consider binary error models realized using aggregated Markov processes with the imposed restriction that the probability distributions of lengths of sequences of consecutive output symbols are independent and time homogenous. This is achieved by setting specific constraints on the model transition probability matrix. The gain of this artificial restriction on the error model is to simplify the estimation of the model parameters, which are then approximated as functions of the time-variant received signal strength and speed of a mobile vehicular DVB-H receiver. It is shown that useful results may be achieved with the described packet error models, especially in simulating mobile reception in field conditions, and several related application approaches are suggested.

The chapter is organized as follows. First, in section 19.2.2, we define the class of finite-state models considered in this study and derive expressions for specific properties of models in this class. These properties are then used in section 19.3 to determine the parameters for three simple models. In section 19.4, measuring of DVB-H error traces (files containing sequences of error indicators) is described; such traces are subsequently utilized in simulations in section 19.5 to evaluate the accuracy of the considered models. In section 19.6 an approach to parameterizing the models is suggested and implemented in simulating mobile DVB-H reception in an urban area. Finally, conclusions are stated in section 19.7.

19.1.1 Related Work

In this chapter, we apply hidden Markov models (HMMs) with the objective to find efficient error models suitable for adaptation to simulating a wide range of realistic communication scenarios. The field of work related to hidden Markov models, even when limited to the application of said models in telecommunication simulations, is rather diverse. We will not attempt to provide a comprehensive overview of the subject, but rather refer to selected classical and contemporary approaches to the problem of finite-state channel modeling. For detailed descriptions of the theory and conventions of hidden Markov modeling and related issues, we refer the reader to Cappé et al.[2] and Ephraim and Merhav,[3] for example.

Several basic structures of generative and descriptive channel models are reviewed in Kanal and Sastry.[4] These include, for example, the classical Gilbert,[5] Gilbert–Elliott,[6] and Fritchman[7] models, which have since been widely used to model various communication channels; in section 19.2, relationships between these models are considered in more detail. Less frequently used structures described in Kanal and Sastry[4] include for example, the infinite-state slowly spreading chain,[8] and a finite-state variant that was recently applied in Tralli and Zorzi[9] to model the block error process in a wideband CDMA system. Other recent generative models for packet channels are presented, for example, in Babich and Lombardi,[10] where the authors extend the Markov model for Rayleigh fading of Wang and Moayeri[11] to modeling block failure processes in slow Ricean fading, and in Konrad et al.[12] where an algorithm for analyzing and reproducing statistical properties of measured packet error traces is presented. Furthermore, in Zhu and Garcia-Frias,[13] models based on stochastic context-free grammars are applied in describing communication channels with bursty error behavior; this model would also seem suitable for modeling packet errors. It is notable that the latter two models require large amounts of empirical data for solving the model parameters, while the approaches used in references[9–11] are similar to the underlying principle of this chapter, where the goal is to parameterize the described packet error models using characteristics of the physical channel. Furthermore, the models described in references 10 and 11 relate the Markov model parameters analytically to physical channel variables for a rather limited fading scenario, while in Tralli and Zorzi[9] and in the following, parameterization of more general models is based on empirical results. We will consider a parameterized error model, which lends itself quite naturally to simulations taking into account the large-scale fading, or shadowing, present in real communication channels. We verify the accuracy of the considered model used in such simulations by utilizing field measurement data on received signal strength as a function of time for specific reception conditions. A statistical approach to simulating large-scale fading combined with a finite-state channel was suggested in Lin and Tseng[14] for bit errors; it would also seem feasible to handle large-scale fading in a manner similar to that applied to the packet error model considered below.

19.2 Model Description

19.2.1 Background

Formally in the following work we consider hidden Markov processes (HMP),[3] or joint discrete-time stochastic processes $\{X_k, Y_k\}_{k=0,1,\ldots}$, where $\{Y_k\}$ is a probabilistic function of the Markov process $\{X_k\}$. Furthermore, we assume both processes to have finite-state spaces. For hidden Markov processes, it is assumed that $\{Y_k\}$ can be observed while generally $\{X_k\}$ cannot; conceptually, a hidden Markov process can be considered a hidden stochastic process $\{X_k\}$ observed through noise in $\{Y_k\}$. One application of hidden Markov processes is to use them to model communication channels with memory. In this context, as with many other applications of HMPs, it is reasonable to speak of hidden Markov models. A classical example is the Gilbert–Elliott model,[6] which consists of two hidden states (good and bad, conceptually) with different probabilities for the output symbols, which typically represent error indicators for data transmitted through the channel.

In the following, our focus will be on aggregated Markov processes (AMPs), whose output $\{Y_k\}$ is a deterministic function of the underlying Markov process $\{X_k\}$. Such models are a subclass of hidden Markov models. Fritchman[7] considered properties of general partitioned-state models that classify as aggregated Markov processes with binary outputs; that is, each state in such models belongs to one of two state groups, where one group represents correct reception of information, and the other represents erroneous reception. As a special case, a simplified model with only one error state, several error-free states, and a specific transition probability structure between the states was suggested and has been applied in many studies since, for example, recently in references 13 and 15. It is relevant to note that any deterministic function of a Markov chain, or AMP, can be described as a trivial finite-alphabet HMP, and any finite-alphabet HMP can be described as a deterministic function of a Markov chain with an augmented state space.[3] We may thus speak of equivalent models, where equivalence means that the probability of any observed output sequence, conditioned on the model structure, is the same for each model.[16]

To examine relationships between the classic examples given above, consider the well-known Gilbert model for binary errors, which can be described as a Gilbert–Elliott model with zero error probability in the good state. This can be represented as an AMP with one error state and two error-free states. Figure 19.1 shows state diagrams of the Gilbert model, and a statistically equivalent

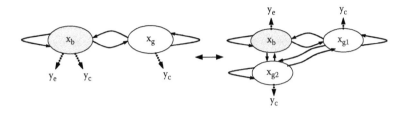

Figure 19.1   Conceptual state diagram of the Gilbert model and an equivalent AMP.

aggregated Markov process; the gray-colored states in the figure indicate bad states, which in this example are states that may emit error symbols, y_e, and the white-colored states indicate good states that emit symbols for correct reception, y_c. Transition and emission probabilities have been omitted from the figure; in fact, with the AMP it is more appropriate to speak of emission functions, because the output is unambiguously defined by the state of the underlying Markov process. It can be observed in figure 19.1 that the AMP is obtained from the Gilbert model by splitting the bad state into two substates, where one outputs error symbols and the other outputs symbols for correct reception. Note that the AMP shown in figure 19.1 is not an example of a typical simplified Fritchman model, where transitions between the two good states would be forbidden. However, as will be discussed in the following, equivalence with such a simplified model is true given certain conditions. Both the simplified Fritchman model and the Gilbert model have the property that their outputs are renewal processes in the sense that the distributions of lengths of sequences, or runs, of a given output symbol are independent and identically distributed (i.i.d.) and time homogenous. With this added distinction we speak of renewal Markov processes.

In the general case the partitioned-state models considered by Fritchman are nonrenewal processes. Again, to present comparison to another classical model, the Gilbert–Elliott model is equivalent to a general partitioned-state model with two error states and two error-free states. Figure 19.2 shows state diagrams of the Gilbert–Elliott model and an equivalent aggregated Markov process. In this example, both the bad state and the good state of the Gilbert–Elliott model are split into two substates to obtain the equivalent AMP. The given simple examples also demonstrate the increase in number of model parameters to be determined to match observed error sequences; although increasing the number of states in the underlying Markov process may provide potential to better fit the model to observed error statistics, it also emphasizes the problem of determining the model parameters.

Generally, a maximum likelihood solution (in terms of finding the model parameters to maximize the conditional probability of a given output sequence) for the transition probabilities of a partitioned-state model can be obtained using the Baum–Welch algorithm[17] or modifications thereof.[16,18] Said modifications take advantage of the fact that, as stated above, any discrete HMM can be

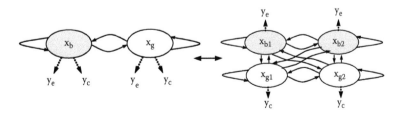

Figure 19.2   Conceptual state diagram of the Gilbert–Elliott model and an equivalent AMP.

modified into an equivalent AMP by increasing the number of states; furthermore, Sivaprakasam and Shanmugan[16] state that, assuming certain conditions (briefly reviewed in section 19.2.2) on the transition probability matrix of an aggregated Markov process, a unique equivalent model can be obtained, where transition probabilities within a group of states corresponding to a given output symbol are represented by a diagonal matrix. For the AMPs shown in figures 19.1 and 19.2, this means removing the transitions between the two good states, and also the two bad states in figure 19.2. This makes the AMP of figure 19.1 a simplified Fritchman model. These properties may be utilized to implement the Baum–Welch algorithm using efficient matrix calculations.[16]

In the following we consider binary aggregated Markov processes, where transition probabilities within a group of states corresponding to a given output symbol are represented by a diagonal matrix, and note that according to reference 16 this partial diagonality of the state transition matrix is not a major limitation on the generality of a model. However, as defined in section 19.2.2, to further simplify the model parameter estimation, we also assume rather strict conditions on the transition probabilities between state groups, which does indeed restrict the generality of the model by artificially forcing the renewal property. In section 19.5, we consider the effects of these limitations on the performance of the defined models. Again, we emphasize that the purpose for these simplifications is to facilitate the construction of high-protocol-layer error models based on physical layer variables such as the carrier-to-noise ratio (CNR) and Doppler frequency.

19.2.2 Definitions

Let $X = (X_0\, X_1, \ldots)$ be a time homogenous Markov chain with a finite-state space S: $\{1,\ldots, N+M\}$, partitioned into groups C: $\{1,\ldots, N\}$ and E: $\{N+1,\ldots, N+M\}$. Also, let the emission function $\phi: S \mapsto \{c,e\}$ be defined as $\phi(i) = c \forall i \in C, \phi(i) = e \forall i \in E$. If we denote the output process of X as $Z = (Z_0, Z_1, \ldots)$ with state space $\{c,e\}$, then X and Z form a hidden Markov model. More specifically, in this case Z is an aggregated Markov process.

In the following we assume the transition matrix $P = [p_{i,j}]_{i,j \in S}$ of X is of the form

$$P = \begin{pmatrix} P_{CC} & P_{CE} \\ P_{EC} & P_{EE} \end{pmatrix},$$
(19.1)

where *(i)* $P_{CC} = diag(\alpha_1,\ldots,\alpha_N)$,

(ii) $P_{EE} = diag(\alpha_{N+1},\ldots,\alpha_{N+M})$,

(iii) $P_{CE} = [p_{ij}]_{i \in C, j \in E} = [(1-\alpha_i)w_j]_{i \in C, j \in E}$,

(iv) $P_{EC} = [p_{ij}]_{i \in E, j \in C} = [(1-\alpha_i)w_j]_{i \in E, j \in C}$,

$$\alpha_i \in [0,1] \forall i$$

and

$$\sum_{i \in C} w_i = \sum_{i=1}^{N} w_i = 1, \quad \sum_{i \in E} w_i = \sum_{i=N+1}^{N+M} w_i = 1.$$
(19.1a)

Conditions (*i*) and (*ii*) above imply that transitions between different correct states (states in group *C*) are forbidden, as well as transitions between error states (states in group *E*). Furthermore, conditions (*iii*) and (*iv*) mean that the conditional probabilities of transitions between correct and error states, or vice versa, do not depend on the current correct (or error) state. As a result of the previous, the probability distributions of the lengths of sequences of consecutive symbols *c* and *e* are independent geometric mixture distributions. As noted in the previous section, according to reference 16, conditions (*i*) and (*ii*) are not a major limitation on the generality of the model. For sake of completeness, we briefly review sufficient conditions on a given binary aggregated Markov process for the existence of an equivalent model fulfilling (*i*) and (*ii*):

1. $\lambda_{CC}, \lambda_{EE} \in \mathbb{R} > 0$,

2. $\pi_i > 0 \forall i$,

3. All elements of $W_{CC}^{-1} P_{CE} W_{EE}$ and $W_{EE}^{-1} P_{EC} W_{CC}$ are positive,

where λ denotes matrix eigenvalue. Condition 1 ensures that the states in the model are sufficiently distinct and reduce the number of transitions between states that correspond to the same observation symbol. Condition 2 means that the stationary probability of each state *i*, π_i, is greater than zero, and the W_{ii} matrices in condition 3 are used in reference 16 to define the equivalent model. More specifically, the equivalent transition probability matrix Λ is defined by $P = W^{-1} \Lambda W$, where $W = diag\ (W_{CC}, W_{EE})$, adapted to the binary model considered here. Condition 3 ensures that the transition probabilities for this equivalent model are nonnegative.

Assuming the previous to hold true, it should be noted that conditions (*iii*) and (*iv*) in (19.1) are a more severe limitation to the generality of the model; the effect of these restrictions will be considered in sections 19.3 and 19.5. In the following, we use (19.1) to first determine the limiting state probabilities for the model, and subsequently derive the probability distributions of the lengths of sequences of consecutive symbols. Finally, to assist in determining the model parameter values in section 19.3, the *k*th derivatives of the probability generating functions (pgf's) of these probability distributions are calculated.

Given (19.1), the limiting probabilities $\vec{\pi} = (\pi_1, ..., \pi_{N+M})$ of *X* are obtained by definition ($\vec{\pi} P = \vec{\pi}$):

$$
\pi_i = \begin{cases}
\alpha_i \pi_i + \displaystyle\sum_{k=1, k \neq i}^{N} 0 \cdot \pi_k + \sum_{l=N+1}^{N+M} (1 - \alpha_l) w_i \pi_l, & i \in C \\[4mm]
\alpha_i \pi_i + \displaystyle\sum_{k=N+1, k \neq i}^{N+M} 0 \cdot \pi_k + \sum_{l=1}^{N} (1 - \alpha_l) w_i \pi_l, & i \in E
\end{cases}
$$

$$
\Leftrightarrow \pi_i = \begin{cases}
\dfrac{w_i}{1 - \alpha_i} A, & i \in C \\[4mm]
\dfrac{w_i}{1 - \alpha_i} B, & i \in E
\end{cases},
$$

where $A = \sum_{i \in E}(1-\alpha_i)\pi_i$, $B = \sum_{i \in C}(1-\alpha_i)\pi_i$. Multiplying the previous by $(1-\alpha_i)$ and summing over $i \in C$, we get on the basis of (10.1a)

$$\sum_{i \in C}(1-\alpha_i)\pi_i = \sum_{i \in C} w_i A \Rightarrow B = A,$$

so that

$$\pi_i = \frac{w_i A}{1-\alpha_i} \; \forall i,$$

and since

$$\sum_i \pi_i = A \sum_i \frac{w_i}{1-\alpha_i} = 1 \Rightarrow A = \frac{1}{\sum_i \frac{w_i}{1-\alpha_i}},$$

we get

$$\pi_i = \frac{w_i}{(1-\alpha_i)\left(\sum_{k=1}^{N+M} \frac{w_k}{1-\alpha_k}\right)} \; \forall i \in \{1,\dots,N+M\}. \tag{19.2}$$

Now let T_C and T_E be discrete random variables corresponding to the dwell times in state groups C and E, respectively. The probability distributions for T_C and T_E are denoted $f_C(n) = P(T_C = n)$ and $f_E(n) = P(T_E = n)$, and defined as

$$f_C(n) = P(X_{k+2} \in C,\dots,X_{k+n} \in C, X_{k+n+1} \notin C \mid X_k \in E, X_{k+1} \in C)$$

$$= \frac{P(X_k \in E, X_{k+1} \notin E, X_{k+2} \in C,\dots,X_{k+n} \in C, X_{k+n+1} \notin C)}{P(X_k \in E, X_{k+1} \in C)}$$

$$= \frac{\sum_{i=N+1}^{N+M} \pi_i \sum_{j=1}^{N} p_{ij} p_{jj}^{n-1}(1-p_{jj})}{\sum_{i=N+1}^{N+M} \pi_i(1-p_{ii})} \tag{19.3}$$

$$= \frac{\sum_{i=N+1}^{N+M} \pi_i(1-\alpha_i) \sum_{j=1}^{N} w_j \alpha_j^{n-1}(1-\alpha_j)}{\sum_{i=N+1}^{N+M} \pi_i(1-\alpha_i)}$$

$$= \sum_{j=1}^{N} w_j \alpha_j^{n-1}(1-\alpha_j) \forall n \in \{1,2,\dots\},$$

and exchanging E and C in the definition of $f_C(n)$,

$$f_E(n) = \sum_{j=N+1}^{N+M} w_j \alpha_j^{n-1}(1-\alpha_j) \forall n \in \{1,2,\dots\}. \tag{19.4}$$

T_C and T_E are clearly independent of each other and of time k, because X was determined to be a time-homogenous stochastic process.

Finally, we find the probability generating functions of (19.3) and (19.4) to be of the form

$$G(z) = \sum_{k=1}^{\infty} z^k \sum_{j} w_j (1-\alpha_j) \alpha_j^{k-1}$$

$$= \sum_{j} \frac{w_j(1-\alpha_j)}{\alpha_j} \sum_{k=1}^{\infty} (z\alpha_j)^k \qquad (19.5)$$

$$= \sum_{j} \frac{w_j(1-\alpha_j)z}{1-z\alpha_j}.$$

Depending on the range of summation in (19.5), we come to either pgf $G_C(z)$ for $f_C(n)$ ($j \in \{1, 2,..., N\}$) or pgf $G_E(z)$ for $f_E(n)$ ($j \in \{N+1,..., N+M\}$).

$G(z)$ is of the form $\sum_j C_j f_j(z)$, where $C_j = \frac{w_j(1-\alpha_j)}{\alpha_j}$ and $f_j(z) = \frac{\alpha_j z}{1-\alpha_j z}$. Now the kth derivative of $f_j(z)$ is obtained as

$$f_j^{(k)}(z) = \frac{d^{k-1}}{dz^{k-1}}[f_j(z)]$$

$$= \frac{d^{k-1}}{dz^{k-1}}\left[\frac{\alpha_j(1-\alpha_j z) - \alpha_j z(-\alpha_j)}{(1-\alpha_j z)^2}\right]$$

$$= \alpha_j \frac{d^{k-1}}{dz^{k-1}}\left[(1-\alpha_j z)^{-2}\right] = \alpha_j^k \cdot k!(1-\alpha_j z)^{-(k+1)}.$$

Thus,

$$G^{(k)}(z) = \sum_{j} \frac{k!\alpha_j^{k-1}w_j(1-\alpha_j)}{(1-z\alpha_j)^{k+1}} \,\forall k \in \{1,2,...\}. \qquad (19.6)$$

Again, different ranges of j produce either $G_C^{(k)}(z)$ or $G_E^{(k)}(z)$. In the following section, we consider methods of determining the transition matrix given in (19.1) from observed error processes. For this we use the general results obtained in this section, namely, equations (19.2) and (19.6).

19.3 Parameter Estimation

We use the AMP described in section 19.2.2 to model observed error processes in given communication systems as follows: let $Q = (Q_1,...,Q_k)$ be the quality of signal reception, defined by a criterion natural to the system under inspection, during K measured sampling intervals. In the following we assume that the signal quality can be quantized so that $Q_n \in \{q_C, q_E\} \forall n \in \{1, ..., K\}$, where q_C and q_E correspond to correct and erroneous reception, respectively. Then one can try to fit the parameters of X to provide best possible consistence between modeled sequences of states $\{c,e\}$ and experimental sequences of $\{q_C, q_E\}$ according to some relevant criteria.

If we define the probability of error in the system under inspection as

$$P(q_E) = \frac{\sum_{i=1}^{K} \delta(Q_i, q_E)}{K},$$

where $\delta(a,b)$ is the Kronecker symbol,

$$\delta(a,b) = \begin{cases} 1, & a = b, \\ 0, & a \neq b, \end{cases}$$

then it is reasonable to require X to be determined so that

$$P(Z = e) = \sum_{i \in E} \pi_i = P(q_E). \tag{19.7}$$

Furthermore, if we denote the length of the ith sequence of consecutive symbols q_E (referred to as error runs) in Q as L_E^i and the length of the jth sequence of consecutive symbols q_C (correct runs) in Q as L_C^j, then it seems natural to approximate L_E^i and L_C^j by T_E and T_C. However, as stated before, T_E and T_C are independent stationary random variables. While we can in many cases assume Q to be stationary in the sense that the distributions of L_E^i and L_C^j are not time dependent (see section 19.4), given the rather strict conditions of (19.1) it is not generally possible to model the correlation properties of L_E^i and L_C^j. Another inherent limitation in the model is the probability distributions of T_E and T_C, which were shown in section 19.2.2 to be geometric mixture distributions. Again, we cannot generally assume that this applies also to L_E^i and L_C^j. However, increasing the number of states in the model, and thus the number of free parameters, potentially improves the accuracy of modeling the observed distributions.

Effects of the limitations described above are considered in more detail in section 19.5; given these limitations, in the following we investigate functional methods of determining the model parameters from observed error processes. We begin by considering a model with only two states ($N = 1$, $M = 1$ as defined in section 19.2.2), and subsequently investigate the advantage of increasing the number of states.

19.3.1 N = 1, M = 1

In the simplest case, we select $C = \{1\}$, $E = \{2\}$, which gives $w_1 = w_2 = 1$ and

$$P = \begin{pmatrix} \alpha_1 & 1 - \alpha_1 \\ 1 - \alpha_2 & \alpha_2 \end{pmatrix}.$$

From (19.2) and (19.7) we immediately obtain the condition

$$P = (Z = e) = \pi_2 = \frac{1}{1 + \frac{1 - \alpha_2}{1 - \alpha_1}} = P(q_E).$$

From (19.6) it is seen that $\mu_C \stackrel{\triangle}{=} \Sigma_n n f_C(n) = G_C^1(1) = \frac{1}{1-\alpha_1}$ and $\mu_E \stackrel{\triangle}{=} \Sigma_n n f_E(n) = G_E^1(1) = \frac{1}{1-\alpha_2}$, so that the previous condition on the AMP is equivalent to

$$\frac{1}{1+\frac{\mu_C}{\mu_E}} = P(q_E). \tag{19.8}$$

Note that regardless of the actual distributions of L_E^i and L_C^j, the following holds:

$$P(q_E) = \frac{\sum_{i=1}^{m} L_E^i}{\sum_{i=1}^{m} L_E^i + \sum_{j=1}^{m} L_C^j},$$

assuming that Q contains m correct and m error runs, and thus $K = \sum_{i=1}^{m} L_E^i + \sum_{j=1}^{m} L_C^j$. If we assume that distributions of L_E^i and L_C^j are not time dependent, as m grows we get

$$P(q_E) \approx \frac{m\overline{L}_E}{m\overline{L}_E + m\overline{L}_C} = \frac{1}{1+\frac{\overline{L}_C}{\overline{L}_E}},$$

where $\overline{L}_C$ and $\overline{L}_E$ denote the sample means of L_C and L_E. Comparing this with (19.8) we see that with the two-state model, reproducing a measured probability of error may be realized by matching the mean run lengths of Z and Q. In practice, this means using the method of moments to fit the run length distributions so that both the probability of error and mean run lengths are accurately modeled. We obtain a very simple form for the parameter estimation:

$$\overline{L}_C \stackrel{\triangle}{=} \mu_C = \frac{1}{1-\alpha_1} \Leftrightarrow \alpha_1 = \frac{\overline{L}_C - 1}{\overline{L}_C}, \tag{19.9}$$

$$\overline{L}_E \stackrel{\triangle}{=} \mu_E = \frac{1}{1-\alpha_2} \Leftrightarrow \alpha_2 = \frac{\overline{L}_E - 1}{\overline{L}_E}. \tag{19.10}$$

Note that this is certainly not the only way of choosing the parameters to match the observed error rate; trivially we could choose $\alpha_2 = P(q_E)$ and $\alpha_2 = 1 - \alpha_2$, which results in Z being equivalent to the error process of a binary symmetric channel. This accurately estimates the observed probability of error, but completely disregards the effect of channel memory.

Furthermore, it can be experimentally shown that the parameter estimation given by (19.9) and (19.10) is not always a good solution in terms of simulation performance. In section 19.5 it is shown through simulations that the following parameter estimation gives a more useful model: let $S_{L_C}^2$ and $S_{L_E}^2$ be the sample variances of the observed run lengths. The model parameters are now selected according to

$$\sigma_C^2 = G_C^{(2)}(1) + G_C^{(1)}(1) - \left[G_C^{(1)}(1)\right]^2 = \frac{\alpha_1}{(1-\alpha_1)^2} \stackrel{\triangle}{=} S_{L_C}^2,$$

$$\sigma_E^2 = G_E^{(2)}(1) + G_E^{(1)}(1) - \left[G_E^{(1)}(1)\right]^2 = \frac{\alpha_2}{(1-\alpha_2)^2} \stackrel{\triangle}{=} S_{L_E}^2,$$

which yield

$$\alpha_1 = \frac{2 + \frac{1}{s_{LC}^2} - \frac{1}{s_{LC}}\sqrt{4 + \frac{1}{s_{LC}^2}}}{2},$$ (19.11)

$$\alpha_2 = \frac{2 + \frac{1}{s_{LE}^2} - \frac{1}{s_{LE}}\sqrt{4 + \frac{1}{s_{LE}^2}}}{2},$$ (19.12)

where use is made of the fact that $\alpha_1, \alpha_2 \in [0,1]$.

However, with only two states in the model, fitting one parameter accurately (in this case the variance) generally results in a poor fit for other parameters (mean run lengths and error probability), due to the lack of degrees of freedom in determining the model. In the following, we increase the number of states to obtain more flexible models for Q.

19.3.2 $N = 2$, $M = 2$

To retain the accuracy of the run length variance estimation—which will be shown to be important in terms of simulation performance—while reproducing also the measured error rate and mean run lengths, additional degrees of freedom in determining the model parameters are required. Therefore, we consider a model with $C = \{1,2\}$ and $E = \{3, 4\}$. Thus,

$$P = \begin{pmatrix} \alpha_1 & 0 & (1-\alpha_1)w_3 & (1-\alpha_1)w_4 \\ 0 & \alpha_2 & (1-\alpha_2)w_3 & (1-\alpha_2)w_4 \\ (1-\alpha_3)w_1 & (1-\alpha_3)w_2 & \alpha_3 & 0 \\ (1-\alpha_4)w_1 & (1-\alpha_4)w_2 & 0 & \alpha_4 \end{pmatrix},$$

and from (19.1a), (19.2) and (19.7),

$$P(Z = e) = \pi_3 + \pi_4 = \frac{w_3}{(1-\alpha_3)(\sum_{k=1}^{4}\frac{w_k}{1-\alpha_k})} + \frac{1-w_3}{(1-\alpha_4)(\sum_{k=1}^{4}\frac{w_k}{1-\alpha_k})}.$$

Keeping for expectations and variations of T_C and T_E the same designations as in the previous subsection, we obtain from (19.6)

$$\mu_C = G_C^{(1)}(1) = \frac{w_1}{1-\alpha_1} + \frac{(1-w_1)}{1-\alpha_2},$$

$$\mu_E = G_E^{(1)}(1) = \frac{w_3}{1-\alpha_3} + \frac{(1-w_3)}{1-\alpha_4},$$

$$\sigma_C^2 = G_C^{(2)}(1) + G_C^{(1)}(1) - \left[G_C^{(1)}(1)\right]^2$$

$$= \frac{w_1\alpha_1(1-\alpha_2)^2 + w_2\alpha_2(1-\alpha_1)^2 + w_1w_2(\alpha_2-\alpha_1)^2}{(1-\alpha_1)^2(1-\alpha_2)^2}$$

and

$$\sigma_E^2 = G_E^{(2)}(1) + G_E^{(1)}(1) - \left[G_E^{(1)}(1) \right]^2$$

$$= \frac{w_3 \alpha_3 (1 - \alpha_4)^2 + w_4 \alpha_4 (1 - \alpha_3)^2 + w_3 w_4 (\alpha_4 - \alpha_3)^2}{(1 - \alpha_3)^2 (1 - \alpha_4)^2}.$$

Using the previous, we can now select the model parameters by solving $\{\alpha_1, \alpha_2, \alpha_3, \alpha_4, w_1, w_3\}$ in the following system of equations:

$$\begin{cases} P(Z = e) = P(q_E) \\ \mu_C = \overline{L}_C \\ \mu_E = \overline{L}_E \\ \sigma_C^2 = S_{L_C}^2 \\ \sigma_E^2 = S_{L_E}^2. \end{cases} \tag{19.13}$$

Although (19.13) is an underdetermined (there are six unknowns, $w_1, w_3, \alpha_1, \alpha_2, \alpha_3, \alpha_4$, but only five equations) system of nonlinear equations, solutions can be found using numerical methods.

19.3.3 Other Numbers of States

It should be noted that if equations (19.9) and (19.11) both result in approximately the same value for α_1, or correspondingly, equations (19.10) and (19.12) both give approximately the same value for α_2, then in the first case only one state with output c and in the second case only one state with output e is needed in the model, assuming the parameter estimation described above. In these cases, the obtained model corresponds to the typical simplified Fritchman model with either a single error-free state or a single error state. Still, the model parameter estimation approach presented in this section can be applied with suitable modifications.

While the moment-based evaluation of the model parameters is efficient and feasible with a four-state implementation of the hidden Markov model structure considered, it is clear that further increasing the number of states complicates the parameter evaluation. Therefore, it is relevant to determine whether there is any advantage gained from such increase in complexity. In section 19.5 it is shown through simulations that with the DVB-H system considered in this chapter, a four-state model produces simulation results that cannot be significantly improved by increasing the number of states, given the restrictions in (19.1). This follows from the renewal property of the model, which means that increasing the number of states only potentially improves the fit of the model output run length distributions to the measured. Simulations imply that the differences between results obtained using a renewal model that closely approximates the measured sample run length distributions, and those obtained using the four-state model are negligible when considering simulation performance. It should be noted that this is a significant difference between the aggregated renewal processes considered in this chapter and general aggregated Markov processes, where increasing the number of states in the model potentially improves the maximum likelihood estimation of the model transition probabilities from the given measurement. In the context of this chapter, we will not consider models with more than four states.

19.4 Description of the Observed Error Process

To evaluate the performance of the error models described in sections 19.2.2 and 19.3, it is necessary to consider realizations of the error process Q. In this chapter, we define Q as packet error indicators for transport stream (TS) packets at the link layer of a DVB-H system, or as packet error indicators for Internet Protocol (IP) packets at the application layer. Details of the DVB-H forward error correction (FEC) for Multiprotocol Encapsulated (MPE) data, denoted MPE-FEC in short, are considered in Faria et al.[1] and Paavola et al.[19] and will not be discussed here. In the following we briefly describe the TS packet error traces required in simulating the DVB-H link layer performance, and determine the required laboratory measurement lengths to obtain sufficient information on the statistics of the channel.

19.4.1 Obtaining Packet Error Traces

In a laboratory setting, DVB-H TS packet error traces were obtained as shown in figure 19.3. Source data was input into a DVB-T/H modulator operating with various combinations of system parameters. The modulated signal was passed through a hardware channel simulator with noise added to the signal to obtain various carrier-to-noise ratio values. The noisy signal was input into a DVB-T/H receiver and subsequent logic analyzer to produce TS packet error traces. These traces correspond to Q as defined in section 19.3, with the observation period corresponding to one TS packet and the quantization of the signal quality performed according to the number of byte errors before physical layer outer decoding in each received TS packet. That is, $Q_n = q_E$ if the nth TS packet contains more than eight byte errors, which is the correction capability of the physical layer outer decoder in DVB-H. Otherwise, the TS packet will be decoded correctly, and by definition, $Q_n = q_C$.

IP packet error traces were generated as in Känkänen[20] by simulating the IP packet error behavior of the DVB-H system based on the above described TS error traces, assuming constant-length (512-byte) IP datagrams. In these link layer simulations, IP datagrams were labeled erroneous if they contained any errors. We refer also to the IP packet error traces obtained by concatenating these packet error indicators as measured error traces, although this is not entirely accurate due to the simulation step involved.

Furthermore, TS packet error traces similar to the ones obtained in the laboratory measurements were recorded also in field conditions. A DVB-H receiver setup and a Global Positioning System (GPS) receiver were used to record synchronized reception information. In this case the

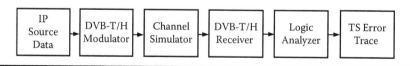

Figure 19.3 **Laboratory measurement setup for obtaining TS packet error traces.**

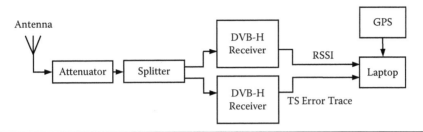

Figure 19.4   Field measurement setup for obtaining synchronized TS packet error and reception condition information.

measurements consisted of synchronized received signal strength indicator (RSSI), receiver location and speed information, and DVB-H TS packet error data. The receiver antenna was placed inside a vehicle moving in an urban environment. The measurement setup is illustrated in figure 19.4; figure 19.5 shows an example of the synchronized data recorded in such conditions.

19.4.2   Required Measurement Lengths

In estimating the error model parameters in section 19.3, we assumed the run length distributions L_C and L_E to be stationary. In the laboratory measurements, the error traces were obtained using an analytical channel model with a fixed carrier-to-noise ratio and Doppler frequency per

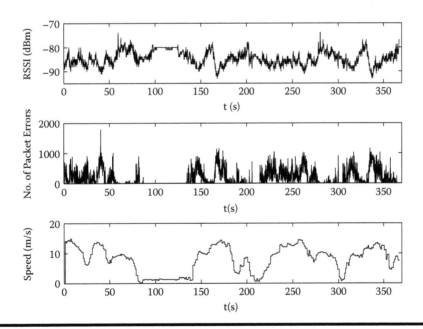

Figure 19.5   Example data from one vehicular urban field measurement.

measurement. Thus, it would seem reasonable to assume that in this case the run length distributions are stationary and, given a sufficiently large observation window, or length K of Q, the measured sample run length distributions are consistent estimators of these stationary distributions. Still, it is not evident how long the measured error traces should be to provide a reliable estimate of the statistics of the channel. In other words, we must investigate the observation length necessary to obtain consistent estimates of the error statistics used to solve the model parameters.

To determine the minimum measurement length required, the variance of sample mean run lengths of DVB-H TS packet error traces was evaluated using the runs test summarized in Konrad et al.[12] As defined in section 19.3, the term *run* refers to a series of consecutive error-free or erroneous packets. The principle of the runs test is to divide the measured error trace into segments of equal lengths, compute the lengths of runs in each segment, count the number of runs of length above and below the median value for run lengths in the trace, and finally compute a histogram for the number of runs counted. Now it was defined that a trace has sufficiently constant mean run lengths with the given segment length, when the number of runs between the 0.05 and 0.95 cutoffs is close to 90 percent.

To approximate the minimum error trace lengths required to obtain sufficiently constant estimates for the relevant error statistics, long measurements (of order >10^6 TS packets) were analyzed by applying the runs test with increasingly long segment lengths, until the number of runs between the specified cutoffs was 90 percent. Figure 19.6 shows the results of this approximation for several values of CNR and Doppler frequency. It is physically obvious and supported by figure 19.6 that the required minimum trace length grows fast with increasing CNR and decreasing Doppler frequency, but with the carrier-to-noise ratios 13–20 dB and Doppler frequencies 5–80 Hz used

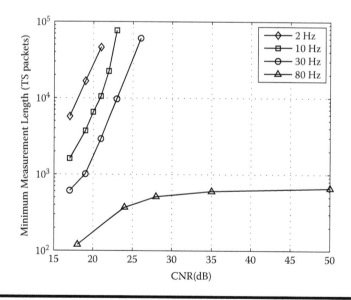

Figure 19.6   Estimations of the required measurement lengths to obtain small variance in sample mean.

in the analyses of this chapter, the required length is clearly below 10^6 TS packets, which was the order of magnitude for the lengths of the measured error traces.

19.5 Evaluation of the Model Performance

19.5.1 DVB-H Link Layer Simulations

In the following we consider the effect of the inherent limitations of the error model as described in section 19.3 on the accuracy of the model in DVB-H link layer simulations. In this case accuracy is measured as the difference in error rates obtained in simulations using measured error traces compared to simulations using error traces generated with stochastic models. More specifically, we consider the deterioration of model accuracy due to the use of independent geometric mixture distributions to approximate measured run lengths.

To evaluate the effect of assuming certain distributions for the run lengths in modeling measured error traces, we would ideally compare the model in question to a model where the generated run lengths are drawn from the underlying run length distributions of the measurements. Of course, it is impossible to sample directly these underlying distributions, but we can find approximations using the sample distributions of L_C and L_E defined in section 19.3. It should be stressed that the objective here is to evaluate the best performance within the class of renewal models, where the run length distributions are independent and time homogenous. For this, we use a modification of the discrete time analog of the inverse transformation technique[21] and draw from a given sample run length distribution for random variable L as follows: denote $P(L = l_j) \triangleq P_j$, and let U be uniformly distributed over $(0, 1)$. Assuming k different lengths of runs are present in the observation, set

$$L = \begin{cases} l_1, & \text{if } U < P_1 \\ l_2, & \text{if } P_1 < U < P_1 + P_2 \\ \vdots \\ l_k, & \text{if } \sum_{j=1}^{k-1} P_i < U < 1. \end{cases} \quad (19.14)$$

Note that P_j, where $j \in \{1,\ldots,k\}$, are easily calculated from measurements as the sample frequencies of lengths of runs. By applying the previous for L_C and L_E until a sufficiently long approximation of Q is obtained, we get error traces with independent time-homogenous run length distributions that closely match the given sample distributions. This can also be seen in figure 19.9, where the sample distribution of error burst lengths of a measurement is compared to that of an error trace generated as described above. The measured error traces used in the following were obtained as laboratory measurements with a 16-QAM modulation mode, 1/4 OFDM guard interval length (8K mode), 1/2 physical layer convolutional code rate, and 10 Hz Doppler frequency for the TU6 (six-tap typical urban) channel model[22,23] applied in the physical layer. The MPE-FEC code rate used was 3/4.

Figures 19.7 and 19.8 show the MPE-FEC frame error rates and IP packet error rates, respectively, obtained with finite-state models described in section 19.3 and the discrete inverse

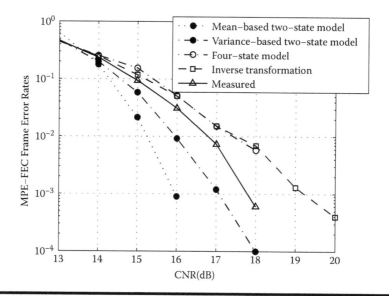

Figure 19.7   Simulated DVB-H MPE-FEC frame error rates.

transformation method described above. The simulation results indicate that with the four-state HMM, the adverse effect of the assumed run length distribution on the simulation accuracy of the model is negligible. Naturally the difference between the two-state models and the inverse transformation is much larger; still, it can be seen that using the variance-based parameter estimation suggested in section 19.3 produces more accurate results than conventional mean-based estimation.

Comparing the performance of the four-state and inverse transformation models, it can be concluded that in the case of DVB-H simulations, very little advantage can be gained by increasing the number of states in an aggregated renewal Markov process beyond four. This is further

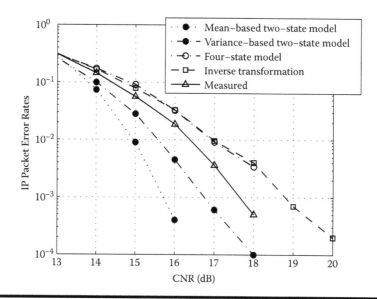

Figure 19.8   Simulated DVB-H IP packet error rates.

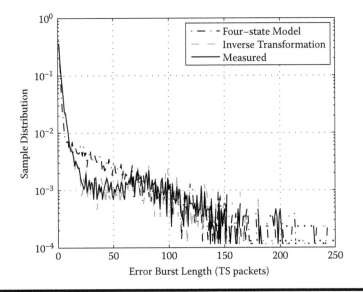

Figure 19.9 **Comparison of sample distributions of TS packet error burst lengths at CNR of 17 dB.**

illustrated in figure 19.9, where it is shown that the four-state AMP produces a good fit to the measured error burst length distribution. Certainly there is margin for improvement in simulation performance, especially with very low error rates, as can be seen in figures 19.7 and 19.8, but this is achieved at the cost of increased model evaluation complexity by using maximum likelihood evaluation of more general (nonrenewal) hidden Markov models.

19.5.2 DVB-H Application Layer Simulations

The hidden Markov models considered above can also be applied in application layer simulations of video streaming with DVB-H; in Känkänen[20] the models were compared with the MTA packet error model[12] and the block error model suggested in Tralli and Zorzi[9] for WCDMA simulations (we will refer to the latter as the S-state model). The considered models are evaluated by applying them in video-streaming simulations, where it is assumed that the bit rate of the video stream is 350 kbps, and the video is transmitted in individual frames, updated 25 times per second. An individual image is then considered faulty if more than two IP packets, out of four, are erroneous. Furthermore, the reception quality is observed in intervals of one second, and a second is labeled erroneous if at least one of the 25 received images is faulty. The term *erroneous seconds ratio* (ESR) is used for the fraction of erroneous seconds in the simulated transmission time. This simplified approach for analyzing video quality was also used in Paavola et al.[19]

Figure 19.10 shows the ESR simulated using measured IP packet error traces, the mean-based two-state HMM (denoted 2-SMM [mean] in the figure), the variance-based two-state HMM (2-SMM [var]), the four-state HMM (4-SMM), the MTA model, and the S-state model (S-SMM). The total simulated video duration is 10 minutes. The results indicate that the four-state AMP produces accurate results in this application. Although the MTA model gives the best results compared to the measurement, it should be noted that compared to the four-state AMP, determining

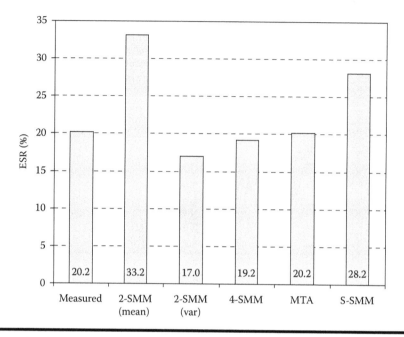

Figure 19.10 Simulated erroneous seconds ratio in a 10-minute video stream.

the MTA model parameters requires considerably more complex analysis of the error traces to be modeled; furthermore, the MTA model yields poorly to model parameterization, as suggested in the following.

19.6 Model Parameterization and Modeling Realistic Field Conditions

The model parameter estimation described in section 19.3 was based on the assumption that the observed run length distributions L_C and L_E are stationary. In section 19.4, this assumption was validated in the case of laboratory measurements with a set carrier-to-noise ratio and Doppler frequency per measurement. However, there is generally no basis for such an assumption in the case of field measurements, because naturally the CNR and Doppler frequency are dependent on the potentially arbitrary movement of the receiver. Therefore, the direct parameter estimation suggested in section 19.3 is not generally justified. However, by considering the field measurement conditions to be piecewise stationary when divided into short time intervals, useful results can be obtained. In the following, we apply a parameterized error model, where the finite-state model parameters are determined as functions of the CNR and Doppler frequency by utilizing an approximation of relevant packet error statistics.

19.6.1 Approximating Relevant Statistics

In section 19.3, the HMM parameters were determined using sample statistics of the observed error process Q. For simulation purposes, it would be more useful to construct models using underlying physical channel variables such as the CNR and Doppler frequency. Because in this case the error model is implemented at a rather high protocol layer, analytical estimation of the effects of the aforementioned variables on the statistics of Q is a nontrivial matter. In the following we use function approximation to find simple relationships between the physical channel variables and the statistics of the error process.

Let $M = \{P(q_E), \bar{L}_C, \bar{L}_E, S^2_{L_C}, S^2_{L_E}\}$ be the set of sample statistics used in evaluating the HMM parameters as described in section 19.3. It was found that good results are obtained by approximating these statistics in the following manner: We obtain laboratory measurements for a suitable range of carrier-to-noise ratios and Doppler frequencies, and calculate the natural logarithm of each of the parameters m_{ρ, f_D} in M_{ρ, f_D} —here subscripts ρ and f_D are used to denote the dependence of the statistics on the CNR and Doppler frequency, respectively. Least square error (LSE) planar fitting is then applied to find the coefficient vector $\bar{C}^*_m = (c_{1m}, c_{2m}, c_{3m})^T$ to minimize $E[|(C_{1m}\rho + C_{2m}f_D + C_{3m}) - \ln(m_{\rho, f_D})|^2]$. For convenience, we summarize a solution: let $\bar{\rho} = (\rho_1, ..., \rho_n)^T$, $\bar{f}_D = (f_D, ..., f_{Dn})^T$, and $\bar{m}_l = (\ln m_1, ..., \ln m_n)^T$, respectively, be vectors composed of the carrier-to-noise ratios, Doppler frequencies, and natural logarithms of error statistics of n different laboratory measurements. Now the objective is to find $\bar{C}^*_m$ from the condition

$$\| (\bar{1} \mid \bar{f}_D \mid \bar{\rho})\bar{C}_m = \bar{m}_l \|^2 = \min_{\bar{C}_m},$$

where $\bar{1} = (1...1)^T$ is an n by 1 vector of ones. Let

$$A \triangleq (\bar{1} \mid \bar{f}_D \mid \bar{\rho}).$$

The LSE coefficients are obtained as

$$\bar{C}^*_m = (A^T A)^{-1} A^T \bar{m}_l.$$

This yields an approximation of the statistics as follows: let $\tilde{M} = \{\tilde{P}(q_E), \tilde{\bar{L}}_C, \tilde{\bar{L}}_E, \tilde{S}^2_{LC}, \tilde{S}^2_{LE}\}$ be the set of approximations given ρ and f_D; we obtain $\tilde{m} \in \tilde{M}$ from

$$\tilde{m} = \exp(C_{1m}\rho + C_{2m}f_D + C_{3m}). \tag{19.15}$$

Table 19.1 contains the LSE approximation coefficients obtained as described above for the statistics of the TS packet error process for DVB-H with 16-QAM modulation mode, 1/4 OFDM guard interval length (8K mode), and physical layer convolutional code rate 1/2. The measurements were performed as specified in section 19.4 using a TU-6 channel model; the measured CNR range was 14–18 dB, and the Doppler frequency range was 5–80 Hz at 5 Hz intervals. Figure 19.11 shows an example of this approximation for $P(q_E)$ as a function of ρ and f_D.

Table 19.1 LSE Coefficients

	$P(q_E)$	$\bar{L}_E$	$\bar{L}_C$	S^2_{LE}	S^2_{LC}
C_{1m}	−0.6068	−0.1288	0.5295	−0.3480	1.0460
C_{2m}	0.0213	−0.0020	−0.0252	−0.0312	−0.0742
C_{3m}	6.0409	4.7484	−2.1428	12.7709	−1.9695

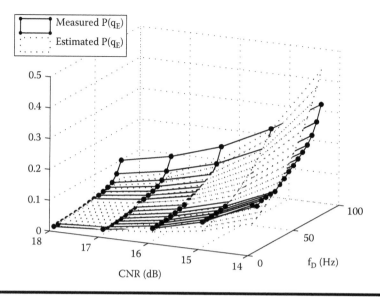

Figure 19.11 **LSE approximation of** $P(q_E)$ **as a function of the CNR and Doppler frequency.**

It is notable that once this parameterization is performed for a given channel model and system parameters, no additional packet error measurements are required in the following application of the model.

19.6.2 Conversion between Measured Instantaneous Reception Conditions and Analytical Model Parameters

One way of utilizing the above described approximation is with field measurements such as those presented in section 19.4, where we have sampled the RSSI and vehicle speed at 100 ms intervals. The measured RSSI values must be converted to corresponding values of ρ in the channel model used in the laboratory measurements (in this case the TU6 model); in our implementation this was done by matching the average TS packet error rates as functions of the RSSI and ρ for the field and laboratory measurements, respectively. It should be noted that this conversion can also be done without explicit information on the TS packet error rates; using noise figures given by the receiver manufacturer and an approximation for the background noise in the measurement environment produced a conversion factor between the RSSI and ρ that differed from the above-mentioned matching by approximately 1.5 dB. Furthermore, the values for f_D were obtained as the maximum Doppler frequency given the measured vehicle speed and carrier frequency of the DVB-H broadcast.

19.6.3 Error Trace Generation

Let N_S be the total number of samples obtained of the RSSI and vehicle speed, and N_{TS} the data-rate-dependent number of received TS packets per sampling interval. We proceed to generate an error trace using a selected model structure by repeating the following steps for each of the N_S sampling intervals:

1. Calculate the values of ρ and f_D corresponding to the measured RSSI and vehicle speed as described above.
2. Using (19.15), determine the values of the packet error statistics needed to solve the finite-state model parameters.
3. Calculate the model parameters, for example, for a four-state model using $\{\tilde{P}(q_E),\ \bar{L}_C, \bar{L}_E, \tilde{S}^2_{L_C}, \tilde{S}^2_{L_E}\}$ and equation (19.13). Use the probabilities given in (19.2) as the initial state probabilities and generate N_{TS} output symbols. Concatenate thus obtained subtrace with previously generated subtraces.

Note that since the individual subtraces generated in step 3 of the above are relatively short, depending of course on the sampling interval of the parameters, it is important to initialize the state probabilities properly. Selecting the initial state arbitrarily may have an adverse effect on the model performance, because the process may not have time to converge to the limiting state distributions within the sampling interval.

19.6.4 Results

Figure 19.12 shows the TS packet error rates as functions of time (averaged over 1 s intervals), obtained from a field measurement and from an error trace generated using the above-described parameterized four-state hidden Markov model. It can be seen that the generated error trace follows the fluctuations in instantaneous packet error rate of the measurement very well. The total average TS packet error rate in the measurement was 23.4 percent, and 23.7 percent in the generated error trace. Furthermore, the simulated average IP packet error rates obtained with the field measurement and with the parameterized four-state model were 21.0 and 20.5 percent, respectively, and the corresponding simulated average MPE-FEC frame error rates were 28.1 and 29.6 percent.

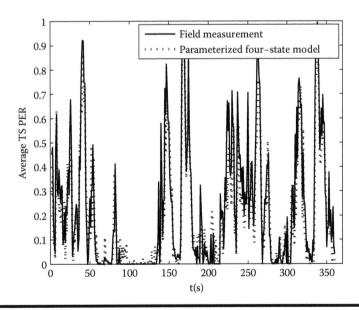

Figure 19.12 Comparison of measured and modeled DVB-H TS packet error rates averaged over 1 s intervals.

The accuracy of the simulation results with the parameterized four-state model is good considering the successive approximations made in the parameterization: the assumption that the TU6 model adequately describes the measured environment, the LSE approximation of the TS packet error statistics given this channel model, and finally the assumptions made in determining the HMM parameters as described in section 19.3. The obtained results suggest that none of these potential sources of inaccuracy severely affect the performance of the parameterized error model.

19.6.5 Potential Applications

In the following, we briefly identify three possible application scenarios, in order of decreasing resource requirements, for the above described parameterized error model. First, the model can be used with received signal strength and vehicle speed measurements as done above to verify the accuracy of the model. This might be considered a trivial application, because field measurements will still have to be performed; however, it can be argued that signal strength and vehicle speed measurements require less resources than constructing a complete DVB-H transmission/reception setup to obtain, for example, TS packet error traces.

Second, the model can be used to simulate arbitrary reception scenarios without performing field measurements. This would require sufficiently detailed and reliable coverage maps of the transmission areas of interest. The modeling principle is illustrated in the simple artificial example of figures 19.13 and 19.14: the former shows a conceptual signal strength map of an area, where the receiver is selected to move along a straight line from the origin to the top right corner. From the coverage map, the normalized signal strength is obtained as a function of the distance traveled, and plotted in the top of figure 19.14. The speed of the receiver is defined as a piecewise-linear function of the distance traveled, and is shown in the bottom of figure 19.14. Based on these estimations, packet error traces could be constructed as specified above. This approach allows for flexible simulation of a wide range of reception scenarios, assuming the coverage maps used are

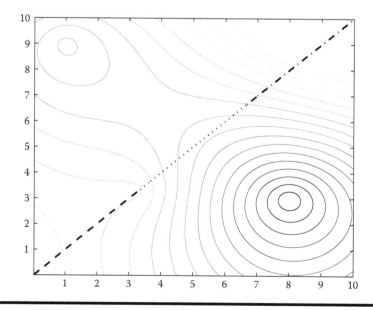

Figure 19.13 Artificial coverage map and selected receiver route.

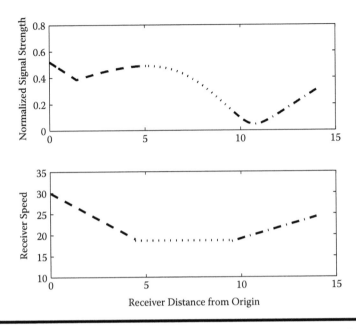

Figure 19.14   Signal strength and receiver speed as functions of distance traveled.

accurate enough. With modern network planning and analysis tools, this is not an unreasonable requirement.

Third, a statistical approach could be used in determining the effect of large-scale fading on the model parameters. A related model is considered in Lin and Tseng,[14] where a two-layer Markov model is used to account for the effects of both large- and small-scale fading on the bit error process of a wireless communication system. The outer layer determines the large-scale fading, which is assumed to follow a lognormal distribution. Using a similar approximation for the received signal strength combined with the parameterized error model considered above would provide a packet error model accounting for statistical large-scale fading.

19.7   Summary

In this chapter we have considered binary error models realized using aggregated Markov processes with the imposed restriction that the probability distributions of lengths of sequences of consecutive output symbols are independent and time homogenous. This is achieved by setting specific constraints on the model transition probability matrix. The gain of this artificial restriction on the error model is to simplify the estimation of the model parameters, which was considered in section 19.3. We found that the method of moments provides a well-justified and simple solution to the parameter estimation for this limited class of hidden Markov models.

In section 19.4, the packet error traces used in subsequent modeling and simulation were presented. Laboratory measurements were used in section 19.5 to evaluate the performance of the considered models in DVB-H link layer simulations. Also, the effects of the imposed model restrictions on the simulation performance were studied by comparing the results obtained with the simplified finite-state model to simulation results obtained using a discrete inverse transformation method.

The results show that the inaccuracy caused by the mismatch in run length distributions is negligible already with a four-state model. Simulation results obtained with direct approximation of the four-state model parameters from measured error traces are accurate enough to be useful for both link layer and application layer simulations.

A more interesting application of the described models was considered in section 19.6, where the finite-state model transition probabilities were determined as functions of the RSSI and vehicle speed measured in field conditions. The packet error traces thus obtained correspond well to the original field measurements. Furthermore, the modeling approach described in section 19.6 offers a tool for simulating various mobile use cases without performing a large number of field measurements. Given an RSSI or CNR map of a specific coverage area, the parameterized error model can be utilized to generate error traces corresponding to a receiver moving arbitrarily within the area in question. Of course, it is reasonable to perform initial measurements to verify the accuracy of the model for the given environment, but still the potential cost reduction is considerable.

References

1. G. Faria, J. A. Henriksson, E. Stare, P. Talmola. 2006. DVB-H: Digital broadcast services to handheld devices. *Proc. IEEE* 94: 194–209.
2. O. Cappé, T. Moulines, T. Rydén. 2005. *Inference in hidden markov models.* New York: Springer Science+Business Media.
3. Y. Ephraim, N. Merhav. 2002. Hidden Markov processes. *IEEE Trans. Inform. Theory,* 48.
4. L. N. Kanal, A. R. K. Sastry. 1978. Models for channels with memory and their applications to error control. *Proc. IEEE* 66: 724–44.
5. E. N. Gilbert. 1960. Capacity of a burst-noise channel. *Bell Systems Tech. J.* 39: 1253–66.
6. E. O. Elliott. 1963. Estimates of error rates for codes on burst-error channels. *Bell Systems Tech. J.* 42: 1977–97.
7. B. D. Fritchman. 1967. A binary channel characterization using partitioned Markov chains. *IEEE Trans. Inform. Theory* IT-13: 221–27.
8. J. G. Kemeny. 1966. Slowly spreading chains of the first kind. *J. Math Anal. Applications* 15: 295–310.
9. V. Tralli, M. Zorzi. 2005 Markov models for the physical layer block error process in a WCDMA cellular system. *IEEE Trans. Veh. Technol* 54: 2102–13.
10. F. Babich, G. Lombardi. 2000. A Markov model for the mobile propagation channel. *IEEE Trans. Veh. Technol.* 49: 63–73.
11. H. S. Wang, N. Moayeri. 1995. Finite-state Markov channel—A useful model for radio communication channels. *IEEE Trans. Veh. Technol.* 44: 163–171.
12. A. Konrad, B. Y. Zhao, A. D. Joseph, R. Ludwig. 2003. A Markov-based channel model algorithm for wireless networks. *Wireless Networks,* New York: Kluwer Academic. 189–99.
13. W. Zhu, J. Garcia-Frias. 2004. Stochastic context-free grammars and hidden markov models for modeling of bursty channels. *IEEE Trans. Veh. Technol.* 53: 666–76.
14. H.-P. Lin, M.-J. Tseng. 2005. Two-layer multistate markov model for modeling a 1.8 GHz narrowband wireless propagation channel in urban Taipei city. *IEEE Trans. Veh. Technol.* 54: 435–46.
15. W. Chang, T. Tan, D. Wang. 2001. Robust vector quantization for wireless channels. *IEEE J. Sel. Areas Common.* 19: 1365–73.
16. S. Sivaprakasam, K. S. Shanmugan. 1995. An equivalent markov model for burst errors in digital channels. *IEEE Trans. Comm.* 43: 1347–55.
17. L. R. Rabiner. 1989. A tutorial on hidden Markov models and selected applications in speech recognition. *Proc. IEEE,* 77: 257–85.

18. W. Turin, M. M. Sondhi. 1993. Modeling error sources in digital channels. *IEEE J. on Sel. Areas in Common.*, 11: 340–47.

19. J. Paavola, H. Himmanen, T. Jokela, J. Poikonen, V. Ipatov. 2007. The performance analysis of MPE-FEC decoding methods at the DVB-H link layer for efficient IP packet retrieval. *IEEE Trans. Broadcasting*, special issue on mobile multimedia broadcasting 53: 263–75.

20. T. Känkänen. 2007. IP packet channel modeling applied in DVB-H datacast system. Master's thesis, University of Turku.

21. S. Ghahramani. 2004. *Fundamentals of probability, with stochastic processes*, 3rd ed., Englewood Cliff, NJ: Prentice Hall.

22. *Digital land mobile radio communications.* COST 207 Official Pub. Eur. Communities, final rep. 1989 Luxembourg.

23. *Digital cellular telecommunications system (phase 2+); Radio transmission and reception.* 3GPP TS 05.05, v.8.20.0. 2005.

Chapter 20

Performance Analysis of the DVB-H Link Layer Forward Error Correction

Heidi Himmanen, Tero Jokela, Jarkko Paavola, and Valery Ipatov

Contents

Keywords

DVB-H, forward error correction, error performance, simulations, link layer

20.1 Introduction

One of the strongest trends in modern telecommunication is the development of mobile wireless multimedia broadcast systems. Digital video broadcasting–handheld (DVB-H) is an example of such a system. It is a data broadcasting standard that enables delivery of various Internet Protocol (IP)–based services to mobile receivers. The standard was ratified by the European Telecommunications Standards Institute (ETSI) in November 2004. By nature, it encompasses various contemporary telecommunication challenges, such as achieving high data rates in wireless networks, implementing power-limited mobile receivers, and the design of bandwidth-efficient single-frequency networks (SFNs). A common factor in all these tasks is the requirement of efficient operation in difficult channel conditions.

The DVB-H standard, which is based on and is compatible with the digital video broadcasting–terrestrial (DVB-T) standard, introduces solutions to problems caused by the mobility of the handheld terminals receiving digital broadcasts. These solutions are required to achieve low power consumption, flexibility in network planning, and good performance in mobile channels. Enhancements to conventional DVB-T systems include the addition of time-slicing and an optional stage of error correction called Multi-Protocol Encapsulation–Forward Error Correction (MPE-FEC) at the link layer. Time-slicing means that the transmission is time division multiplexed, that is, each service is transmitted in bursts separated in time. Power saving is achieved because the receiver can switch off radio components between the bursts. The MPE-FEC utilizes a Reed–Solomon (RS) code combined with time interleaving to combat channel fading. Changes at the DVB-T physical layer consist of a new 4K orthogonal frequency division multiplexing (OFDM) mode, an in-depth interleaver, and utilization of previously unused transmission parameter signaling (TPS) bits informing the receiver on the use of time-slicing and MPE-FEC.

The *DVB-H Implementation Guidelines*[1] define the Reed–Solomon code used in the MPE-FEC and how to puncture or shorten it. This chapter investigates the suitability of MPE-FEC to combat against difficult channel conditions caused by mobility of the receiver. The most serious impairment source is the Doppler phenomenon caused by movement of the receiver. It destroys the orthogonality between the subcarriers in an OFDM system. It will be shown in this chapter that MPE-FEC alleviates problems caused by the Doppler spread. The MPE-FEC decoding method is left open by the standard for each receiver manufacturer to decide. By this reason, the effect of choosing the decoding strategy for MPE-FEC deserves special consideration. In particular, MPE-FEC decoding procedures can differ from each other by the source and utilization of erasure information, with basic Reed–Solomon decoding algorithms remaining traditional. Two different options for obtaining erasure information are considered.

The DVB-H standard defines a very large range of transmission parameters. The physical layer contains the possibility to select from three different modulation modes, three OFDM modes, five convolutional code rates, and four possible values for guard interval. The DVB-H specification adds optional Multi-Protocol Encapsulation–Forward Error Correction (MPE-FEC), which has five commonly considered code rates and four possible MPE-FEC frame sizes. To optimize the operating parameters for a communication system in such circumstances, one must be able to efficiently evaluate the performance of the system with different parameters in different channel conditions. Testing the DVB-H system with actual transmitters and receivers is often time and cost prohibitive. Dedicated system testing apparatus are also expensive and may submit poorly to user modification and research outside the bounds set by the nominal system parameters. For these reasons, efficient software-based simulators were developed and used to explore the system performance under various parameter combinations. They have permitted the singling out some preferable transmission modes based on achievable data rate and error performance.

The chapter is organized as follows. A brief description of the DVB-H system is given in section 20.2. Then, the performance of the DVB-H link layer FEC is analyzed in section 20.3. Based on the performance analysis, some guidelines on the selection of transmission mode are discussed in section 20.4. Section 20.5 is dedicated to reviewing and developing some improvements to link layer FEC coding/decoding presented earlier in Paavola et al.[2]

20.2 DVB-H

A conceptual diagram of the DVB-H system is shown in figure 20.1. The physical layer consists of the DVB-T modulator (shown in figure 20.2) and demodulator, and the link layer consists of the IP encapsulator and decapsulator. Relevant physical layer functions in relation to the analyses performed here are the modulation mode, which can be QPSK, 16-QAM, or 64-QAM, and the convolutional code, which has possible code rates 1/2, 2/3, 3/4, 5/6, and 7/8. Also, the physical layer RS(204, 188) code plays a significant role.

Operations performed by the DVB-H link layer are illustrated in figure 20.3. The IP datagrams are encapsulated column-wise into the MPE-FEC frame. The number of rows of the MPE-FEC frame can be 256, 512, 768, or 1,024, depending on the wanted time-slicing burst size. The data in the frame is encoded row-wise using an RS(255, 191) code. Thus, the encoding also introduces interleaving, referred to in reference 1 as *virtual time interleaving*. The number of data columns is 1 − 191, and the number of redundancy columns is 0 − 64. Different MPE-FEC code rates are achieved with code shortening and puncturing. The code rate is 3/4 (approximately) if all 191 data columns and 64 redundancy columns are used. Other options for the code rate considered in this chapter are 1/2, 2/3, 5/6, 7/8, and 1, the latter representing the uncoded case.

The frame is divided into sections so that an IP datagram forms the payload of an MPE section and an RS redundancy column forms the payload of an MPE-FEC section. When the section header is attached, the CRC-32 redundancy bytes are calculated for each section. The MPE sections are transmitted first, followed by the MPE-FEC sections. Both are transmitted in an MPEG-2 transport stream (TS) format,[3] where a TS packet consists of a 4- to 5-byte TS header and 183–184 bytes of payload. The MPEG-2 format for transport packets is inherited from the DVB-T standard to ensure the compatibility of DVB-H with the existing DVB-T networks.

At the receiver side, link layer parsing, that is, finding the beginning and end of sections from TS packets, and decapsulation of the sections obtained from the received transport stream to restore the MPE-FEC frame are performed. Sections containing erroneous data are detected with the help of the CRC-32 check. In this case, the whole section is marked as erased. Reed–Solomon decoding is performed on each row of the MPE-FEC frame utilizing available erasure information.

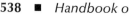

Figure 20.1   A conceptual diagram of DVB-H. (Adapted from Reference 1.)

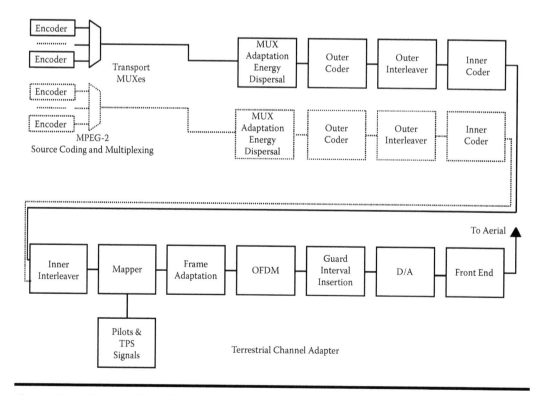

Figure 20.2   DVB-H physical layer. (Adapted from Reference 1.)

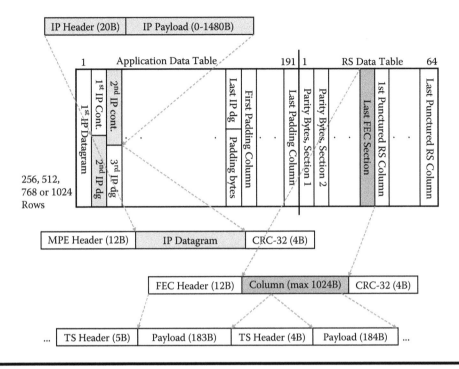

Figure 20.3   The link layer packets of DVB-H.

Decoding based on section level erasure information is in this chapter called section erasure (SE) decoding. It is the suggested method in the *DVB-H Implementation Guidelines*.[1] Another error detection approach and different utilization methods of available erasure information are presented in section 20.5, concentrating on MPE-FEC decoding alternatives.

After MPE-FEC decoding, the IP packets from the frame will be passed to the application layer for further processing. At the link layer, the performance of the receiver can be assessed with comparison of the frame error ratio (FER) without MPE-FEC and the frame error ratio after MPE-FEC decoding (MFER). Frames containing at least one row that cannot be decoded are considered erroneous. The commonly used error criterion for DVB-H, MFER 5 percent, is a useful and easily measured error criterion in laboratory or field tests.[4] However, it is not unambiguous from the application point of view. Therefore, we also apply the IP packet error ratio (IP PER) and byte (symbol) error ratio (SER) as error measures in the simulation analyses.

20.2.1 Description of the Simulator

The efficiency of a software simulator can be enhanced by utilizing appropriate abstraction levels to estimate system performance with suitable accuracy at required protocol levels. For example, to simulate system performance at the link layer, it is generally not necessary to know the waveform or bit-level error behavior of the system; rather, a block error model is quite sufficient for most simulation purposes. Bit-level simulations of thousands of IP packets through the link and physical layers require a huge amount of resources. Therefore, error traces from the physical layer are established here to allow fast simulations at transport stream packet or byte levels. Error traces are series of binary indicators expressing whether a data block contains errors, in this case after the

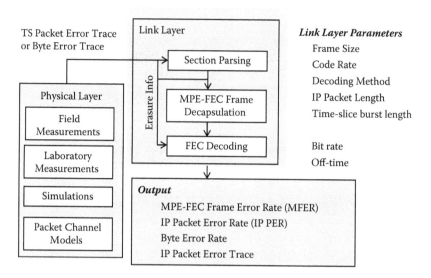

Figure 20.4   The link layer simulator.

physical layer error correction decoding. In principle, error traces can be generated with bit-true simulations at the physical layer, or with actual receiver measurements in real field conditions or in a laboratory using a channel simulator. Aforementioned traces can also be generated with packet channel models, introduced in chapter 19 of this book. The benefit of using error traces is dramatically reduced duration of the link layer simulations. Due to a huge amount of parameter combinations, it would be impossible to analyze the link layer exhaustively without utilizing error traces. It also enables utilization of measurement data in simulations, which provides a way of verifying the simulation results. Two types of error traces were available from physical layer simulations:[5] the byte error indicator stream, which indicates the location of erroneous bytes, and the packet error indicator stream, which indicates the locations of erroneous TS packets. A conceptual description of the simulator is given in figure 20.4.

In the simulation results in the next section, a TS packet error trace was used as link layer input and all bytes in an erroneous TS packet were assumed erroneous. This assumption is justified when using the SE decoding method, which assumes that if an MPE or MPE-FEC section is erroneous, all bytes carried by the section are erased. For the more advanced decoding methods of section 20.5, byte-level error information was used, because performance evaluation is practically impossible without error statistics at this level.

20.2.2 Simulation Parameters

Previous analyses in references 5–7 have shown that good candidates for modulation, providing acceptable error performance in mobile fading channels, are QPSK and 16-QAM used with convolutional code rates 1/2 or 2/3. Therefore, these parameters were also chosen to be investigated further in this chapter. Due to the difficult channel conditions for mobile handheld reception, 64-QAM and convolutional code rates 3/4, 5/6, and 7/8 do not provide acceptable robustness toward errors with a reasonable value of carrier-to-noise ratio (C/N). The OFDM mode used was 8K, with guard interval 1/4. The MPE-FEC frame with 512 rows was used and the IP packet size was assumed to be a constant 512 bytes. The amounts of application data columns and RS data columns used here to

Table 20.1   Obtaining Different MPE-FEC Code Rates

MPE-FEC Code Rate	Data Columns	RS Columns	Total Amount of Columns
1/2	64	64	128
2/3	128	64	192
3/4	191	64	255
5/6	190	38	228
7/8	189	27	216
Uncoded	255	0	255

achieve the analyzed MPE-FEC code rates are presented in table 20.1. The error rates were measured over all services, that is, over the whole transport stream. The services were multiplexed so that one service always uses the whole bandwidth for transmitting the time-slicing bursts.

20.2.3 DVB-H Channel Models

Three different channel models, developed in the Wing TV project,[8] are utilized in this chapter. They are chosen to represent different receiver speeds for the analysis of the Doppler effect. The channel models used in the simulations are named pedestrian outdoor (PO), vehicular urban (VU), and motorway rural (MR), corresponding to the receiver velocities 3, 30, and 100 km/h, respectively. The channel models are based on measurements performed on the corresponding usage scenarios. The channel model taps for the different models are presented in table 20.2.

Two main types of Doppler spectra used in the channel models are Gaussian and classical. The Gaussian Doppler spectrum is described by

$$G(f;\sigma) = \exp\left(-\frac{f^2}{2\sigma^2}\right),$$

(20.1)

Table 20.2   Channel Taps for the Channel Models Presented in Wing TV Project[8]

PO Delay (μs)	Power (dB)	VU Delay (μs)	Power (dB)	MR Delay (μs)	Power (dB)
0	0	0	0	0	0
0.2	−1.5	0.3	−0.5	0.5	−1.3
0.6	−3.8	0.8	−1	1	−3.4
1	−7.3	1.6	−4.1	1.8	−6.8
1.4	−9.8	2.6	−8.8	2.5	−10.2
1.8	−13.3	3.3	−12.6	3.1	−12.9
2.3	−15.9	4.8	−18.6	3.9	−16.3
3.4	−20.6	5.8	−21.6	4.8	−19.5
4.5	−19	7.2	−24.6	5.5	−21.7
5	−17.7	10.8	−20.7	6.4	−23.3
5.3	−18	11.8	−18.2	7	−24.2
5.7	−19.3	12.6	−19.4	9	−25.8

Table 20.3 Doppler Spectra used in the Channel Models

	1st Tap	*Other Taps*
PO	$0.1G(f; 0.08f_D) + \delta(f - 0.5\ f_D)$	$G(f; 0.08f_D)$
VU	$G(f; 0.1f_D)$	$K(f; f_D)$
MR	$G(f; 0.1f_D)$	$K(f; f_D)$

where σ is the standard deviation parameter of the spectrum. The classical Doppler spectrum is given by

$$K(f; f_D) = \frac{1}{\sqrt{1 - (f/f_D)^2}},\qquad(20.2)$$

where f_D is the maximum Doppler frequency. The Doppler spectra proposed in Wing TV project[8] to be used with 12-tap multipath models of table 20.2 are presented in table 20.3. Here $\delta(f)$ is the Dirac delta function.

20.3 Link Layer FEC Performance

20.3.1 MPE-FEC Frame Error Ratios and IP Packet Error Ratios

First, the effect of MPE-FEC is compared to the uncoded case in the three different channels described in the previous section. Physical layer parameters are 16-QAM and convolutional code rate 1/2. The MPE-FEC code rates are 3/4 and 1 (uncoded). The MPE-FEC frame error ratios and IP packet error ratios are presented in figures 20.5 to 20.7. The gain of using MPE-FEC is different depending on the used channel model.

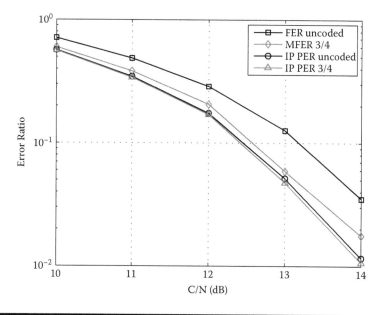

Figure 20.5 MFER and IP PER in the PO channel.

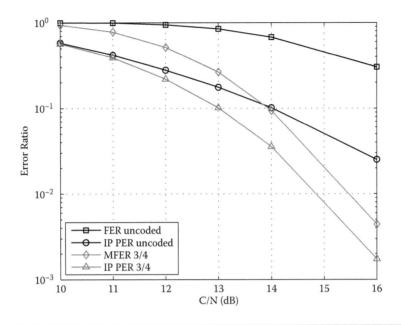

Figure 20.6 MFER and IP PER in the VU channel.

- In the PO channel (figure 20.5), a small gain is observed in MFER. However, when comparing IP PER, there is hardly any advantage of using MPE-FEC.
- The advantage of using link layer coding is greater at higher velocities. In the VU channel (figure 20.6), the gain of using MPE-FEC is clear for both MFER and IP PER for the operational range of C/N values, that is, when the obtained error ratios are below 5 percent.

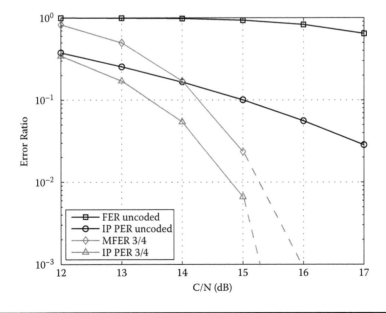

Figure 20.7 MFER and IP PER in the MR channel.

- As figure 20.7 shows, performance of link layer without MPE-FEC appears to be quite poor in the MR channel (the uncoded FER stays close to 100 percent) even for high C/N values, and the positive effect is clearly demonstrated for both MFER and IP PER. (Dashed curve segments in figure 20.7 correspond to the regions of extrapolation.)

Next, simulations were performed to find the C/N operating point to achieve IP PER 1 percent using QPSK and 16-QAM modulation with convolutional code rate 1/2 or 2/3, in combination with all MPE-FEC code rates and the uncoded case (code rate 1). It can be seen from figures 20.5 to 20.7 that the commonly used criterion for operating point MFER 5 percent does not correspond to an unambiguous IP PER value. The relation between MFER and IP PER is dependent on the modulation, code rates, and Doppler frequency. Therefore, we have chosen to compare the different modes at IP packet error ratio 1 percent. For QPSK modulation this criterion corresponds to MFER values close to 5 percent for most code rates in mobile channels. The results are presented in figure 20.8 for QPSK and figure 20.9 for 16-QAM. For some modes, extrapolation was needed to find the C/N at IP PER 1 percent, for example, for 16-QAM 1/2 and MPE-FEC code rate 1 (uncoded), shown in figures 20.6 and 20.7.

- The negligible effect of MPE-FEC in the PO channel, having a very low Doppler shift value, is clearly demonstrated.
- As previously demonstrated, the advantage of using MPE-FEC is significant for transmission experiencing higher receiver mobility. In the VU channel the gain of using MPE-FEC code rate 1/2 compared to MPE-FEC code rate 1 is more than 3 dB for QPSK 1/2 and QPSK 2/3. For 16-QAM 1/2 the maximum gain is about 2.5 dB, and for 16-QAM 2/3 just above 1 dB.
- In the MR channel the gain of using MPE-FEC code rate 1/2 compared to MPE-FEC code rate 1 with QPSK 1/2 is about 4.5 dB. The corresponding gain for QPSK 2/3 and 16-QAM 1/2 is about 5 dB, and for 16-QAM 2/3 about 6 dB.

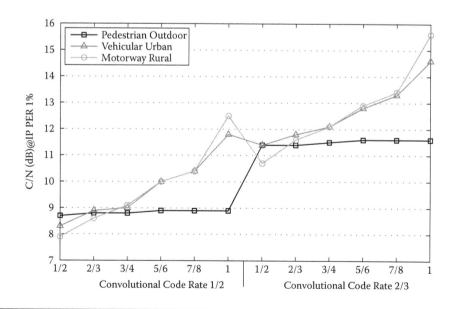

Figure 20.8 Carrier-to-noise ratio required to achieve IP PER 1 percent using QPSK modulation.

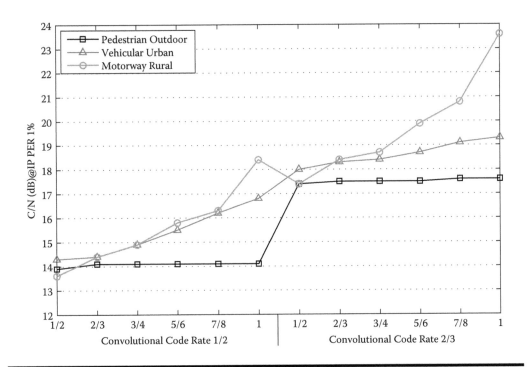

Figure 20.9 Carrier-to-noise ratio required to achieve IP PER 1 percent using 16-QAM modulation.

Using a low MPE-FEC code rate gives a significant gain in C/N, but will also decrease the throughput bit rate. The network operator will likely select the transmission mode based on the desired amount of services, that is, the required throughput bit rate. However, many of the simulated transmission modes give quite similar bit rates, so the choice is not obvious. The optimization of the trade-off between C/N and throughput bit rate is discussed in section 20.4.

20.3.2 Interpretation of Simulation Results

To obtain deeper insight to the functionality of MPE-FEC, the amounts of erasures occurring in the MPE-FEC frames are illustrated for different channel models in figure 20.10, where the distribution of instantaneous IP PER values for each frame is given. The curves represent the situation where average IP PER is 10 percent when MPE-FEC is not utilized (uncoded). The figure shows significant differences in error distributions between the different channel models. The curve of the pedestrian model is very steep, whereas for vehicular speeds there is a large amount of frames with less than 25 percent of the IP packets erased. The different distributions lead to different MPE-FEC performances even though the average IP PER over all frames is equal.

The MPE-FEC decoder can correct as many erasures, that is, unreliable bytes, on each row of the MPE-FEC frame as the number of transmitted redundancy columns. Any code of distance d corrects t_e erasures and t_u errors whenever[9]

$$t_e + 2t_u < d. \tag{20.3}$$

For an RS code, d equals the number of redundancy bytes plus 1. Thus, using code rate (cr) 3/4, the decoding is successful if a row of the MPE-FEC frame contains no more than 64 erasures.

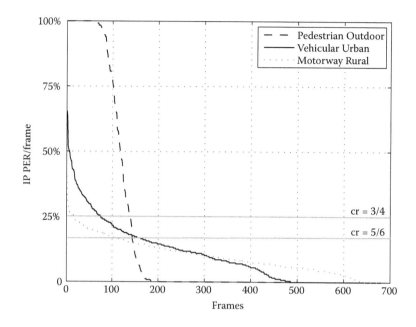

Figure 20.10 IP packet error ratio for each MPE-FEC frame in different channel conditions with 16-QAM modulation and convolutional code rate ½.

The two horizontal lines in figure 20.10 demonstrate the working areas of MPE-FEC code rates 3/4 and 5/6. The decoder can correct any frames with less than 1/4 (25 percent) or less than 1/6 (16.7 percent) of erasures, respectively. In figure 20.10 the number of all frames is 680 for all curves. The number of frames containing errors is 636 in the MR channel, 485 in the VU channel, and 196 in the PO channel. Thus:

■ For the PO channel the amount of corrected frames is 30 percent (57 erroneous frames under the horizontal line for cr = 3/4) of the frames containing erasures for MPE-FEC code rate 3/4 and 25 percent (49 erroneous frames under the horizontal line for cr = 5/6) for code rate 5/6. Slow movement of the receiver causes error bursts to be very long. Thus, most of the frames do not contain errors at all, but when they occur, almost a whole frame is erased. The number of frames containing erasures is approximately 30 percent of all frames.

■ For the VU channel the amount of corrected frames is about 85 percent (412 erroneous frames under the horizontal line for cr = 3/4) of the frames containing erasures for MPE-FEC code rate 3/4 and about 67 percent (323 erroneous frames under the horizontal line for cr = 5/6) for code rate 5/6. The number of frames containing erasures is approximately 70 percent of all frames.

■ In the MR channel the MPE-FEC with code rate 3/4 will correct 97 percent (620 erroneous frames under the horizontal line for cr = 3/4) of the frames containing erasures and 83 percent (527 erroneous frames under the horizontal line for cr = 5/6) for code rate 5/6. The number of frames containing erasures is approximately 90 percent of all frames. For high receiver speed the error bursts are shorter, which makes MPE-FEC more effective.

Thus, MPE-FEC enhances clearly the mobile reception of DVB-H. The reason for this is the error distribution at vehicular speeds, where the error bursts are shorter than in the pedestrian channels, although they occur more often. This will lead to a large amount of MPE-FEC frames with a correctable amount of erasures.

20.4 On the Selection of Transmission Mode

Let us consider a transmission mode to be a combination of modulation, convolutional code rate, and link layer code rate. A good transmission mode is providing the wanted service rate, requiring the lowest C/N to achieve the selected error criterion. One way to compare the modes is presented in the following.

The candidate list consists of all simulated modes from the previous section. The modes are compared based on the C/N needed to achieve IP PER 1 percent (figures 20.8 and 20.9). First, the modes are sorted based on the achieved IP-level bit rate. The comparison is done based on the C/N; modes that require higher C/N to achieve the same or lower bit rate than the other modes are excluded from the list of candidates. To further reduce the number of modes, they are compared based on the transmission efficiency (clarified below).

By only studying the C/N gain at a certain error ratio, it is difficult to make a fair comparison between different transmission modes providing approximately similar error performances. It is necessary to find a way to optimize the trade-off between required C/N and service bit rate. The efficiency of different modulations and code rates can be compared by calculating transmission efficiency, denoted with η, which is defined here as follows:

$$\eta \cong \frac{C/N}{R}, \tag{20.4}$$

where R is the IP-level bit rate.

The mode selection for the vehicular urban channel is given next as an example on how to perform the comparison. The modes are given in table 20.4. They are sorted in ascending order of the IP bit rate. The modes excluded are bold in the C/N column.

- Mode 7 is excluded, as mode 3 gives a higher bit rate with a lower C/N.
- Mode 6 has a higher bit rate than mode 8 with the same required C/N. It also performs better than modes 9 and 13 with the equal bit rate. Thus, modes 8, 9, and 13 can be excluded.
- Mode 14 outperforms modes 12 and 19 in C/N with the equal bit rate. Modes 12 and 19 are excluded.
- Mode 18 outperforms mode 20 in both C/N and bit rate and mode 21 in C/N comparison. Therefore, modes 20 and 21 are excluded.

When the reduction of modes is continued based on the transmission efficiency, five more modes can be excluded. The excluded modes are highlighted in the rightmost column in table 20.5.

- Mode 3 requires a lower value of η than modes 1 and 2.
- Mode 15 outperforms mode 14.
- Mode 24 outperforms modes 22 and 23.

The final result and the comparison for other simulated channel models are presented in table 20.6, where every good mode is marked with X in the different channels. The modes that are not marked with X are not necessary bad modes, but there are other modes with better efficiency.

Table 20.4 C/N-based Mode Comparison for the Vehicular Urban Channel

Mode	Modulation	Convolutional Code Rate	MPE-FEC Code Rate	IP Bit Rate (Mb/s)	C/N (dB) @ IP PER 1%
1	QPSK	1/2	1/2	2.49	8.3
2	QPSK	1/2	2/3	3.32	8.9
7	QPSK	2/3	1/2	3.32	**11.4**
3	QPSK	1/2	3/4	3.73	9.0
4	QPSK	1/2	5/6	4.15	10.0
5	QPSK	1/2	7/8	4.35	10.4
8	QPSK	2/3	2/3	4.42	**11.8**
6	QPSK	1/2	1	4.98	11.8
9	QPSK	2/3	3/4	4.98	**12.1**
13	16-QAM	1/2	1/2	4.98	**14.3**
10	QPSK	2/3	5/6	5.53	12.8
11	QPSK	2/3	7/8	5.80	13.3
12	QPSK	2/3	1	6.63	**14.6**
14	16-QAM	1/2	2/3	6.63	14.4
19	16-QAM	2/3	1/2	6.63	**18.0**
15	16-QAM	1/2	3/4	7.46	14.9
16	16-QAM	1/2	5/6	8.29	15.5
17	16-QAM	1/2	7/8	8.71	16.2
20	16-QAM	2/3	2/3	8.84	**18.3**
18	16-QAM	1/2	1	9.95	16.8
21	16-QAM	2/3	3/4	9.95	**18.4**
22	16-QAM	2/3	5/6	11.06	18.7
23	16-QAM	2/3	7/8	11.61	19.1
24	16-QAM	2/3	1	13.27	19.3

For pedestrian use cases, MPE-FEC coding is not needed. For vehicular urban and motorway rural channels, using MPE-FEC seems reasonable. Thus, if building networks covering all use cases, a high-code-rate MPE-FEC should be considered.

20.5 Methods for Improved MPE-FEC Performance

In this section methods that improve the performance of the MPE-FEC decoding are presented, and their performance is analyzed. These methods, discussed in Paavola et al.,[2] are not in conflict with the present DVB-H standards and differ mainly in the way the erasure information for the MPE-FEC decoder is obtained and utilized. Here, different means for detecting errors (i.e., obtaining the erasure information) for the MPE-FEC are first presented and their reliability analyzed. Then, the methods based on different uses of erasure information are presented and their performance analyzed in both additive white Gaussian noise (AWGN) and mobile multipath channels.

20.5.1 Different Error Detection Methods

There are two possible sources for the erasure information for the link layer RS decoder: CRC-32 of MPE(-FEC) sections and the transport error indicator (TEI) bit in the TS header. CRC-32 was

Table 20.5 Transmission Efficiency-Based Mode Comparison for the Vehicular Urban Channel

Mode	Modulation	Convolutional Code Rate	MPE-FEC Code Rate	IP Bit Rate (Mb/s)	C/N (dB) @ IP PER 1%	η (1/Mbit/s)
1	QPSK	1/2	1/2	2.49	8.3	**2.72**
2	QPSK	1/2	2/3	3.32	8.9	**2.34**
3	QPSK	1/2	3/4	3.73	9.0	2.13
4	QPSK	1/2	5/6	4.15	10.0	2.41
5	QPSK	1/2	7/8	4.35	10.4	2.52
6	QPSK	1/2	1	4.98	11.8	3.04
10	QPSK	2/3	5/6	5.53	12.8	3.45
11	QPSK	2/3	7/8	5.80	13.3	3.68
14	16-QAM	1/2	2/3	6.63	14.4	**4.15**
15	16-QAM	1/2	3/4	7.46	14.9	4.14
16	16-QAM	1/2	5/6	8.29	15.5	4.28
17	16-QAM	1/2	7/8	8.71	16.2	4.79
18	16-QAM	1/2	1	9.95	16.8	4.81
22	16-QAM	2/3	5/6	11.06	18.7	**6.71**
23	16-QAM	2/3	7/8	11.61	19.1	**7.00**
24	16-QAM	2/3	1	13.27	19.3	6.42

discussed above. The TEI bit in each TS packet header indicates whether the physical layer RS(204, 188) decoder was able to correct errors caused by the channel, that is, whether a received RS code word contains less than or equal to eight byte errors. The TEI bit is set to 0 if the physical layer decoder is able to decode the code word. In this case, the TS packet was received correctly. Otherwise, it is set to 1. Other important TS header components for this section are the packet identifier (PID) and the continuity counter. If the PID is incorrect, the TS packet will be lost, because it cannot be recognized as part of the stream. The continuity counter is an incrementing 4-bit number that helps in discovering if a TS packet has been lost. Before proceeding with the different decoding methods, a few words on the reliability of the erasure information obtained by these two different methods are in order.

CRC-32, like any other binary linear code used for error detection, may miss only a fraction 2^{-r} of all possible error patterns, r being the number of redundant bits.[9] For the CRC-32, $r = 32$, and the share of undetectable corrupted section patterns does not go beyond $2^{-32} < 3 \cdot 10^{-10}$. Besides, the probability of an undetected corrupted symbol in a MPE-FEC RS codeword appears to be much smaller than the probability of CRC fault, because in a missed corrupted section not all bytes are necessarily wrong. Therefore, erasure obtained from the CRC-32 can be considered trustworthy.

The reliability of the erasure information generated from the TEI bit in the TS header depends on how efficiently the physical layer RS code can detect errors. One way to estimate the probability of undetected error pattern in maximum distance separable (MDS) codes is studied in Cheung and McEliece,[10] where the results support an intuitive idea that the probability in question (if small enough) may be well approximated by the share of undetectable error patterns:

$$P_{decError} \cong \frac{\text{number of decodable patterns}}{\text{number of patterns}} < q^{-t} 2^{nh\left(\frac{t}{n}\right)}, \tag{20.5}$$

where t is the q-ary code correction capability, n is the code length, and $h(x)$ is the binary entropy.[9] The calculation is presented more thoroughly in Paavola et al.[2] This result can be used for any

Table 20.6 The Simulated Modes

Mode	Modulation	Convolutional Code Rate	MPE-FEC Code Rate	IP Bit Rate (Mb/s)	Pedestrian Outdoor	Vehicular Urban	Motorway Rural
1	QPSK	1/2	1/2	2.49			
2	QPSK	1/2	2/3	3.32			X
3	QPSK	1/2	3/4	3.73		X	X
4	QPSK	1/2	5/6	4.15		X	X
5	QPSK	1/2	7/8	4.35		X	X
6	QPSK	1/2	1	4.98	X	X	
7	QPSK	2/3	1/2	3.32			
8	QPSK	2/3	2/3	4.42			
9	QPSK	2/3	3/4	4.98			X
10	QPSK	2/3	5/6	5.53		X	X
11	QPSK	2/3	7/8	5.80		X	X
12	QPSK	2/3	1	6.63	X		
13	16-QAM	1/2	1/2	4.98			
14	16-QAM	1/2	2/3	6.63			
15	16-QAM	1/2	3/4	7.46		X	X
16	16-QAM	1/2	5/6	8.29		X	X
17	16-QAM	1/2	7/8	8.71		X	X
18	16-QAM	1/2	1	9.95	X	X	X
19	16-QAM	2/3	1/2	6.63			
20	16-QAM	2/3	2/3	8.84			
21	16-QAM	2/3	3/4	9.95			
22	16-QAM	2/3	5/6	11.06			X
23	16-QAM	2/3	7/8	11.61			X
24	16-QAM	2/3	1	13.27	X	X	X

MDS code, including shortened RS codes (such as the physical layer RS(204,188) code). For $t = 8$, $n = 204$, and $q = 256$: $nh(t/n) - t\log_2 q \approx 204 \cdot 0.24 - 64 \approx -15$ so that $P_{\text{decError}} \approx 2^{-15} \approx 3 \cdot 10^{-5}$. This shows that any error pattern of weight greater than t will almost certainly (i.e., with the probability greater than $1 - 3 \cdot 10^{-5}$) be detected in the course of physical layer decoding. Therefore, also erasure information based on the TS header TEI can be considered trustworthy.

20.5.2 Decoding Methods

For Reed–Solomon codes, decoding with errors or erasures is possible. The advantage of the erasure decoder is that it can correct more incorrectly received code symbols than a decoder operating without erasure information. In Joki and Paavola[11] two decoding methods based on correcting both errors and erasures were proposed. The proposed decoding is combined with hierarchical decapsulation, which means that possibly erroneous data is also inserted into the MPE-FEC frame, in contradiction to the method suggested in the implementation guidelines,[1] where all unreliable data is discarded. This way the amount of successfully retrieved data from the MPE-FEC frame grows significantly. It is also possible to ignore available erasure information and use pure error RS decoding. Regardless of the source of erasure information (CRC-32 or TS packet header), the RS erasure decoding procedure itself may be performed based on the well-known algebraic algorithm described, for example, in Blahut.[9] In erasure decoding, an erasure info table (EIT), which is a matrix of the same size as the MPE-FEC frame, is used to keep track of the reliability of each byte in the frame. In the following description it is assumed that 1 in the EIT denotes an erased, or unreliable, byte in the MPE-FEC frame. A reliable byte is denoted with 0.

20.5.2.1 Section Erasure Decoding

In SE decoding, which was the utilized decoding method in the previous sections, the bytes of a section are marked as reliable or unreliable depending on the CRC-32 decoding. If the CRC detects errors, the bytes are marked with 1 in the EIT. Otherwise, they are marked with 0. SE decoding is the suggested method in the *DVB-H Implementation Guidelines*.[1] In practice, SE decoding is not optimal, because all the bytes of a section in which CRC detects errors are marked as unreliable even though many of them may be correct.

20.5.2.2 Transport Stream Erasure Decoding

A more efficient way than section erasure decoding is to use transport stream erasure (TSE) decoding, that is, to extract erasure information from the TS packet header. The physical layer RS(204, 188) decoder sets the TEI bit to 1 if it is unable to decode the 204-byte code word, that is, it contains more than eight byte errors. In this case 1 is written to the EIT for the bytes carried in the TS packet. Otherwise, bytes are marked with 0.

20.5.2.3 Section Error and Erasure Decoding

The decoding method proposed in Joki and Paavola[11] also utilizes erasure information. The main difference compared to the methods presented is that possibly erroneously received packets are also decapsulated into the MPE-FEC frame, unlike in the SE and TSE methods, where all unreliable data is discarded. The proposed decapsulation method is called hierarchical decapsulation, which is presented next.

20.5.2.3.1 Hierarchical Decapsulation

Hierarchical decapsulation aims at losing as small an amount of data as possible. In hierarchical decapsulation the reliability of a byte in the EIT can have three different values corresponding to packet states reliable (0), unreliable (X), or lost (1). The decoding strategy utilizing section erasure information this way is referred to as hierarchical section erasure (HSE). The EIT is formed as follows:

■ Reliable sections are those whose CRC does not indicate errors. Hence, the bytes carried in reliable sections are denoted with 0 in the EIT.
■ Unreliable sections are those in which CRC detects errors. When the CRC indicates that sections contain errors, they are not dropped, as is recommended in the standard. Instead, they are decapsulated into the MPE-FEC frame with low priority and marked with X in the EIT. In this context, low priority means that a later decapsulated reliable section can overwrite a low-priority section. A low-priority section is dropped if a reliable section is already decapsulated into the MPE-FEC frame at the same location as specified in the header of a low-priority section. In this case, it is very probable that an error has occurred in the section header.

- A section is lost if decapsulation fails. The location of a lost section can be found using the section number field in the section header. If the counter indicates that a section is lost, the byte positions of that section are marked with 1 in the EIT. To facilitate the assignment of the following sections into the MPE-FEC frame, the location of a lost section is filled with dummy data symbols.

The decoding method using three levels of error information is called hierarchical decoding.

20.5.2.3.2 Hierarchical Decoding

Hierarchical decoding is based on pure erasure decoding or error and erasure decoding, depending on which is more efficient for the erasure pattern in the codeword. Decoding is performed row-wise to the received MPE-FEC frame, so that each row is treated as its own codeword. Decoding of each row is carried out based on the EIT information from the hierarchical decapsulation as follows:

- If the total number of byte positions in the codeword (the current row of the MPE-FEC frame) marked by 1 or X is less than or equal to the number of RS check symbols, all these marked bytes are treated as erased and pure erasure decoding is carried out.
- If the codeword contains more symbols marked with 1 or X than RS check symbols, pure erasure decoding cannot be fulfilled. Then all symbols marked with 1 are treated as erasures and all other symbols in the codeword are treated as possibly correct or erroneous. Then, the conventional RS decoding with errors and erasures is performed according to well-known algebraic algorithms (see, for example, Blahut[9]).
- If the number of byte positions marked by 1 exceeds the number of RS check symbols, erasure/error correction is impossible, which is signaled by the decoder.

Example

The minimum distance of the RS(255, 191) code with code rate 3/4 is $d = 65$. Let a row of the EIT contain 32 bytes marked with 1, 33 bytes marked with X, and 190 bytes marked with 0. Because there are 65 bytes marked with 1 or X, pure erasure decoding is not possible. On the other hand, the number of bytes marked by 1 is less than 65; thus, the conventional error/erasure decoding is activated assuming $t_e = 32$. If among all nonerased bytes the number of erroneous ones $t_u \leq 16$, this decoding will produce a true codeword. Otherwise, RS decoding either outputs a false word, and thereby incorrect frame, or reports about a detected error.

20.5.2.4 Transport Stream Error and Erasure Decoding

The other decoding method proposed in Joki and Paavola[11] utilizes transport stream erasure information. The erasure information provided by CRC is ignored. Otherwise, all operations are equal to those presented previously. The hierarchical decapsulation is performed as follows:

- Reliable TS packets have the value of TEI = 0 in the packet header. These bytes are marked with 0 in the EIT.
- Unreliable TS packets have the value of TEI = 1, which indicates that the packets contain errors. Again, these packets are not dropped, but decapsulated into the MPE-FEC frame with low priority and marked with X in the EIT. The low priority is interpreted the same

way as with section erasure information. The decapsulated data cannot overwrite reliable data, but a later decapsulated reliable packet can overwrite a low-priority packet.

■ In lost TS packets, the PID information in the packet header has been corrupted and the packet cannot be recognized as a part of the transport stream. The location of a lost TS packet can be found using the continuity counter in the header. If the counter indicates that a TS packet is lost, the byte positions of that packet are marked with 1 in the EIT. To facilitate the assignment of the following packets into the MPE-FEC frame, the location of a lost packet is filled with dummy data symbols.

After the decapsulation, hierarchical decoding is performed as described above. This decoding method is referred to as hierarchical transport stream (HTS) decoding.

20.5.2.5 Nonerasure Decoding

In pure error decoding, or nonerasure (NE) decoding, the allowed amount of byte errors is 32 per row for code rate 3/4, because the available erasure information is ignored. At first glance this approach may seem pointless, but the utilization of pure error decoding may be necessary if erasure information cannot be trusted, parsing of sections fails, or there exists a risk that received packets are decapsulated into the wrong place in the MPE-FEC frame.

20.5.3 *Comparison of Decoding Methods*

20.5.3.1 *Theoretical Analysis*

In the following, the results of the theoretical analysis on error correction capability at the DVB-H link layer are presented. The analysis based on decoding error probabilities for the five different decoding methods discussed above is presented in Paavola et al.[2] In the analysis a stationary memoryless channel for the bit stream arriving at the link layer is assumed. This starting point is justified by the interleaving procedures preceding the link layer decoding stage. The criterion for comparing different decoding methods is the MFER. A frame is considered erroneous whenever the decoding of the frame is not successful (i.e., the decoder was unable to decode at least one row). The payload of the sections belonging to the MPE-FEC frame is always assumed to cover the length N_s, coinciding with the number of rows.

The number of correctable errors and erasures was given in (20.3). Because only decoding within the code distance is considered and the MPE-FEC frame is considered erroneous anytime error correction fails, every violation of (20.3) is treated as a decoding error. Now under the assumption of independent symbol errors, for a code of length n there are $\binom{n}{t_e}$ equiprobable patterns of t_e erasures, and for each of them $\binom{n-t_e}{t_u}$ equiprobable placements of t_u undetected symbol errors on the $n - t_e$ positions left. We denote with p_e and p_u, respectively, the probability of erasure and undetected error. Because the probability of any fixed pattern of t_e erasures and t_u undetected errors is $p_e^{t_e} p_u^{t_u} (1 - p_e - p_u)^{n-t_e-t_u}$, the joint probability distribution of t_e, t_u is:[12]

$$p(t_e, t_u) = \binom{n}{t_e}\binom{n-t_e}{t_u} p_e^{t_e} p_u^{t_u} (1 - p_e - p_u)^{n-t_e-t_u}. \tag{20.6}$$

As a result, the probability P_c of correct decoding of one codeword for RS decoding is evaluated:[12]

$$P_c = P(t_e + 2t_u < d) = \sum_{t_e=0}^{d-1} \sum_{t_u=0}^{\frac{d-1-t_e}{2}} p(t_e, t_u).$$ (20.7)

Probabilities p_e and p_u are calculated for different decoding methods as a function of the byte error probability from the physical layer and substituted into (20.7). Then, decoding error probability P_f for the whole frame (MFER) can be evaluated with the help of (20.7) as

$$P_f = 1 - P_c^N,$$ (20.8)

where N is a number of codewords per a frame.

MFER as a function of byte error probability p_s observed at the physical layer output for different decoding methods are shown in figure 20.11. (The calculations are performed for MPE-FEC code rate 3/4. For simplicity $N_s = 535$, which corresponds to the payload of three complete TS packets for each column, was used.) The curve for SE decoding with $N_s = 512$ is included to show the negligible effect of deviation of $N_s = 535$ from the standard frame size. The ranking of the compared decoding methods follows from the order of frame error probabilities calculated with (20.8) (the decoding method is indicated in the subscript): $P_{f,HTS} < P_{f,NE} < P_{f,HSE} < P_{f,TSE} < P_{f,SE}$. As is seen, the best decoding strategies are those fully ignoring CRC as the source of erasure information. The reason behind the disadvantage of using erasure information over rather long sections in CRC-based decoding is that even one erroneous byte in the section is enough to erase all bytes carried in it. In general, TSE is more efficient than SE, when the average length of the MPE(-FEC) sections is more than the payload of a TS packet, 184 bytes.

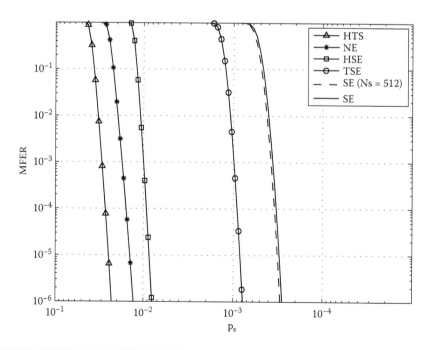

Figure 20.11 Comparison of different decoding methods.

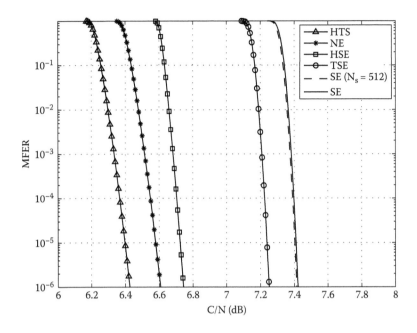

Figure 20.12 Comparison of different decoding methods in the AWGN channel.

20.5.3.2 Simulations for AWGN Channel

To compare experimentally the performance of different decoding methods in terms of MFER as a function of C/N, simulation of the physical layer was performed first in an AWGN channel. The comparison of the decoding methods in an AWGN channel is presented in figure 20.12 for the physical layer parameters: 16-QAM modulation, convolutional code rate 1/2, 8K OFDM mode, and guard interval duration 1/4 of pure OFDM symbol duration. It is evident that the HTS decoding outperforms SE decoding by approximately 1 dB in the AWGN channel. Also, NE decoding performs approximately 0.7 dB better than SE decoding, even though the number of errors that NE decoding can correct is only half of the number of erasures that SE decoding can correct.

20.5.3.3 Simulations for Mobile Fading Channels

The performance of HTS decoding, the theoretically best decoding strategy, was compared to the one of SE decoding and the uncoded case, using 16-QAM modulation, convolutional code rate ½, and MPE-FEC code rate 3/4 or 1 in the VU and MR channels (figures 20.13 and 20.14). Performance is compared with IP packet error ratio and byte or symbol error ratio (SER). For section erasure decoding the IP PER and SER will be the same, as all bytes of an erroneous IP packet are erased. This is a clear disadvantage of erasure decoding methods, when comparing the byte error ratios. Measured in SER, the gain of using HTS compared to SE is over 1 dB in the vehicular urban channel and over 2 dB in the motorway rural channel. Also, the gain of using SE over the uncoded case is not as significant, when measuring byte error ratio. It is predicted that future video codecs used in mobile TV applications will be able to utilize erroneous IP packets, instead of discarding them. Then byte errors remaining after HTS decoding will probably affect the video quality less than discarded IP packets after SE decoding.

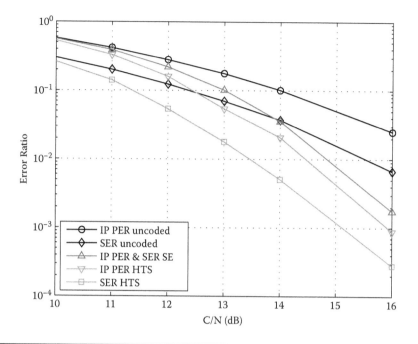

Figure 20.13 IP packet error ratios and byte error ratios in the vehicular urban channel.

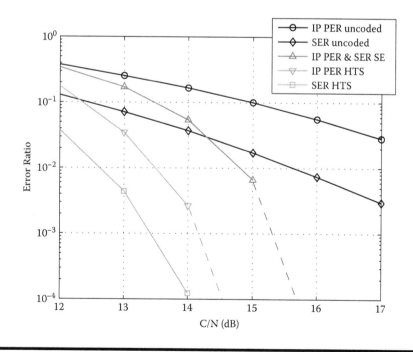

Figure 20.14 IP packet error ratios and byte error ratios in the motorway rural channel.

20.6 Summary

The main focus of this chapter was to inspect the effect of receiver speed on the error performance and how MPE-FEC can reduce the destroying effects of the Doppler shift. For those purposes, three channel models developed in the Wing TV project were utilized. One of the goals of simulation experiments described above was to work out an adequate strategy of choosing the most efficient combinations of physical layer modulation mode, convolutional code rate, and link layer MPE-FEC code rate. Simulation results thus collected may be looked at as establishing some initial guidelines on the selection of transmission mode.

It was seen that for slowly moving receivers MPE-FEC has negligible effect, but for high speeds the gain obtained from MPE-FEC is significant, even for the standard implementation of MPE-FEC. On the other hand, a detailed study, including other possible strategies of involving erasure information into the link layer decoding, has uncovered substantial resources of performance improvement versus just the standard-defined CRC-supported procedure. The principal idea of such advanced decoding methods rests on enabling the receiver to successfully retrieve more data from the MPE-FEC than would be possible when relying only on CRC.

References

1. European Telecommunication Standards Institute. 2005. *Digital video broadcasting (DVB); DVB-H implementation guidelines.* ETSI TR 102 377 V1.1.1.
2. J. Paavola, H. (Joki) Himmanen, T. Jokela, J. Poikonen, and V. Ipatov. 2007. The performance analysis of MPE-FEC decoding methods at the DVB-H link layer for efficient IP packet retrieval. *IEEE Trans. Broadcasting* 53:263–75.
3. ISO/IEC. 2002. *Information technology—Generic coding of moving pictures and associated audio information. Part 1. Systems.* 2nd ed. ISO/IEC 13818-1.
4. G. Faria, J. A. Henriksson, E. Stare, and P. Talmola. 2006. DVB-H: Digital broadcast services to handheld devices. *Proc. IEEE* 94:194–209.
5. Wing TV project (EUREKA/Celtic). 2006. Deliverable D15 "Simulation report." http://projects.celtic-initiative.org/WING-TV/.
6. Wing TV project (EUREKA/Celtic). 2006. Deliverable D6 "Common field trials report." http://projects.celtic-initiative.org/WING-TV/.
7. Wing TV project (EUREKA/Celtic). 2006. Deliverable D4 "Laboratory test results." http://projects.celtic-initiative.org/WING-TV/.
8. Wing TV project (EUREKA/Celtic). 2006. Deliverable D11 "Wing TV network issues." http://projects.celtic-initiative.org/WING-TV/.
9. R. E. Blahut. 2003. *Algebraic codes for data transmission.* Cambridge: Cambridge University Press.
10. K.-M. Cheung and R. J. McEliece. 1988. The undetected error probability for Reed-Solomon codes. Paper presented at Proceedings of MILCOM '88, San Diego.
11. H. Joki and J. Paavola. 2006. A novel algorithm for decapsulation and decoding of DVB-H link layer forward error correction. Paper presented at Proceedings of ICC 2006, Istanbul, Turkey.
12. A. Dür. 1998. On the computation of the performance probabilities for block codes with a bounded-distance decoding rule. *IEEE Trans. Inform. Theory* 34:70–78.

APPLICATIONS

IV

Chapter 21

DVB-H Systems and Receivers

F. Boronat, R. Llorente, J. Lloret, and D. Vicente

Contents

Keywords

DVB-H and DVB-SH standards, DVB-H distribution network, DVB-H receiver architecture, DVB-H receiver subsystem design, DVB-H performance trials, Mobile TV, IP datacast

21.1 Introduction

Several mobile TV standards are present today in the market, but they can be divided into two main groups: those using existing network equipment (notably third-generation (3G)/second-generation (2G) network streaming and to a lesser extent DAB-IP[1] and T-DMB[2]) and those requiring an ad hoc network infrastructure deployment. Digital TV broadcasting to mobile terminals has been introduced in a number of countries based on different video coding and transmission standards: ISDB-T used in Japan,[3] T-DMB in South Korea, Digital Video Broadcasting–Handheld (DVB-H) in Europe and North America,[4] and QUALCOMM'S MediaFLO[5] in North America. Also, the Advanced Television Systems Committee for Mobile and Handheld services (ATSC-M/H) standard[6] is being finished to target the North American market.

The 3G/2G cellular streaming approach has the advantage of using existing networks, but given the unicast nature of streaming, the number of mobile TV users per base station is very limited—typically 8 to 10.[7] On the other hand, broadcasting is an excellent way of reaching many users with a single service, that is, one-to-many communications, and this is the approach adopted by DVB-H, allowing dozens of channels. DVB-H combines broadcasting technology with several optimization techniques, allowing the receivers to operate from a battery and on the move, and can be deployed as a companion to 3G services, On the negative side, critics argue that DVB-H deployment costs are huge because it implies duplicating network infrastructure. The new S-band DVB-SH standard can help reduce these costs.[8]

21.2 DVB-H and DVB-SH Systems

21.2.1 DVB-H

DVB-H[9–11] is a technology standard for the transmission of digital TV and broadcast services to handheld devices such as mobile telephones and personal digital assistant (PDAs). The EN 302 304 standard[4] was published by ETSI in November 2004, and it has recently received the support of the European Commission as the standard of choice for mobile TV in Europe, although with controversy regarding the defendants of the DAB/Digital Multimedia Broadcasting (DMB) solution and the European Mobile Broadcasting Council (EMBC), who proposed technological neutrality in the regulation of the mobile TV market.[12]

DVB-H is a physical layer specification designed to enable the efficient delivery of IP-encapsulated data over terrestrial networks. The creation of DVB-H, closely related to the 1997 DVB-T standard, also implied modifications of some other DVB standards dealing with data broadcasting, service information, an so forth. It is designed to be used as a bearer in conjunction with the set of DVB-IPDC (Internet Protocol Data Casting) systems layer specifications. As an open standard, DVB-H has received broad support across the industry and has been the subject of a large number of technical and commercial trials around the world.

The DVB-H standard is built upon the principles of the DVB-T standard, and adds functional elements necessary for the requirements of the mobile handheld reception environment. Both DVB-H and DVB-T use the same physical layer, and DVB-H can be backwards compatible with DVB-T.

Like DVB-T, DVB-H can carry the same MPEG-2 transport stream and use the same transmitter and orthogonal frequency division multiplexing (OFDM) modulators for its signal. Up to 50 television programs can be transmitted in a single multiplex, or the capacity of a multiplex can be shared between DVB-T and DVB-H.

Although DVB-T presents an excellent mobile performance in terms of reliable, high-speed, high-data-rate reception, it was not specifically designed for mobile receivers. In this sense, the need for DVB-H is justified to address specific needs of mobile reception in handheld devices, namely, limited battery life and difficult reception conditions. DVB-H meets several important requirements: a significant power savings in the receiver compared to DVB-T, excellent performance and robustness in a cellular environment, and enhanced support for single-antenna reception in single-frequency networks (SFNs).

Typical DVB-T front ends consume too much power to support handheld receivers that are expected to last from one to several days on a single charge. Additional requirements for DVB-H are the ability to receive up to 15 Mbit/s in an 8MHz channel in a wide-area SFN while moving at high speed, and all this should be possible while maintaining maximum compatibility with existing DVB-T networks and systems.

To address the mentioned handheld constraints, DVB-H includes a series of new technical features that will be further explained in this chapter and are now just outlined:

1. Introduction of a new 4K carrier mode for network optimization, so that, with some 3,409 active carriers, it provides a compromise and greater flexibility in network design between the high-speed small-area SFN capability of 2K DVB-T and the lower-speed but larger-area SFN of 8K DVB-T.

2. Inclusion of a time-slicing mechanism to achieve power savings, in a manner that bursts of data are received at a time in an Internet Protocol Datacast (IPDC) carousel, instead of a continuous data transmission, as in DVB-T. This scheme allows each individual TV service in a DVB-H signal to be transmitted in bursts, and the receiver can be inactive (in a sleep mode), only waking up when the service to which it is tuned is transmitted, leading to a big reduction of power consumption.

3. To improve the robustness of DVB-H transmissions that have to cope with poor handheld antenna designs and hostile environments, the standard includes an optional, multiplexer-level, forward error correction (FEC) scheme named MPE-FEC, that is, a Multi-Protocol Encapsulation (MPE) mechanism, making it possible to transport data network protocols on top of MPEG-2 transport streams, and used in conjunction with a forward error correction (FEC) scheme to improve the robustness and thus the mobility of the signal.

4. An in-depth interleaver introduced for 2K and 4K modes leads to better tolerance against impulsive noise (helping to achieve a level of robustness similar to that of the 8K mode).

Specific mention needs to be made of IPDC as the higher layer to DVB-H. With IPDC, content is delivered in the form of data packets using the same technique as that for delivering digital content on the Internet. The use of IP packets to carry its data allows DVB-H to rely upon standard components and protocols for content manipulation, storage, and transmission. DVB-IPDC[13] is a set of specifications designed for use with the DVB-H physical layer, but which can be used as a higher layer for all DVB mobile TV systems, including DVB-SH. IPDC specifications define the system architecture, use cases, DVB PSI/SI signaling, electronic service guide (ESG), content delivery protocols (CDPs), and service purchase and protection (SPP).[14]

DVB-H is an extension to DVB-T (both use the same physical layer) with some backwards compatibility. It admits 6, 7, and 8 MHz channels and can share the same multiplex with DVB-T, so that an operator can choose to have two DVB-T services and one DVB-H service coexisting in the same DVB-T multiplex.

From a systems and business perspective, it is necessary to highlight that the usual approaches to build a complete end-to-end DVB-H system broadcasting TV services to handheld devices require compromises and a tight collaboration among different players such as broadcasters, broadcast network operators, and mobile UMTS carriers for the subscription billing capabilities.[15] Thus, DVB-H can be used alone or as an enhancement of mobile telecom networks, which many typical handheld terminals are able to access anyway.

21.2.2 DVB-SH

From the work of the DVB ad hoc group TM-SSP (Satellite Services to Portable Devices) emerged the DVB-SH draft specification (satellite services to Handhelds[16]) that introduces the option of using satellites operating in the S-band below 3 GHz as part of the mobile TV chain. This specification was approved in February 2007.[17]

Like DVB-H, DVB-SH is a transmission system standard designed to deliver video, audio, and data services to handheld devices such as mobile telephones and PDAs, but with the key feature that it is a hybrid satellite-terrestrial system that will allow the use of a satellite to achieve coverage of wide areas. DVB-SH considers the possibility of using a terrestrial gap filler to provide coverage in the areas where direct reception of the satellite signal is not possible.

DVB-H is primarily intended for use in the UHF bands, currently occupied by analogue TV and digital terrestrial television (DTT) services. DVB-SH seeks to exploit opportunities in the higher-frequency, adjacent to UMTS, S-band (2.17–2.2 GHz), where there is less congestion than in UHF.

At the time of this writing, the system and waveform specifications are to be sent to ETSI for formal standardization, but they have already been released in the form of DVB blue books.[18,19]

The S-band requires demanding signal coverage. The short wavelength requires a dense terrestrial repeater network in towns and cities, and to reduce the cost of this network, DVB-SH introduces new tools (partially based in DVB-S2) to enhance the signal robustness that allows the receiver to work with a low signal-to-noise ratio (SNR).

DVB-SH uses turbo coding (3GPP2) for FEC, and also a highly flexible channel interleaver designed to cope with the hybrid satellite-terrestrial network topology, offering time diversity from about 100 ms to several seconds, depending on the targeted service level and memory size of terminal.

DVB-SH is defined as a system capable of delivering IP-based media content and data to handheld terminals like mobile phones and PDAs via satellite. Whenever a line of sight between terminal and satellite does not exist, terrestrial gap fillers are employed to provide the missing coverage. The DVB-SH system has been designed for frequencies below 3 GHz in the S-band. It complements the existing DVB-H physical layer standard, and like its sister specification, it uses the DVB-IP Datacast (IPDC) as a higher layer.

DVB-SH includes features such as turbo coding for forward error correction and a highly flexible interleaver in an advanced system designed to cope with the hybrid satellite-terrestrial network topology. Satellite transmission ensures wide-area coverage, with a terrestrial component ensuring coverage where the satellite signal cannot be received, as may be the case in built-up areas. DVB-SH in fact specifies two operational modes: SH-A, which uses coded-OFDM (COFDM), i.e., forward error correction is included at the OFDM modulation stage, on both satellite and terrestrial links with the possibility of running both links in SFN mode, and SH-B, using a time division multiplex (TDM) on the satellite with COFDM on the terrestrial link.

DVB-SH terminals are in development by several manufacturers and are due to arrive to the market during 2008. At the European level, Alcatel-Lucent is the driving force for DVB-SH technology. This company, in collaboration with NXP Semiconductors,[20] is developing a receiver module for UHF and the L-band that will be extended to the S-band (2.2 GHz) based on the forthcoming DVB-SH standard. Chipsets are also reported to be under development by Dib-Com Semiconductor[21] for DVB-SH. First demonstration of DVB-SH reception at European level was done by Alcatel-Lucent demonstrating S-band (2.17 GHz to 2.20 Ghz) broadcasting employing prototype handsets from the French manufacturer SAGEM. Other companies, such as

Teamcast, are active in the development and commercialization of DVB-SH reference designs that are creating momentum in this new market.

Samsung Electronics also announced it would develop DVB-SH-compatible handsets. Satellite firms Eutelsat and SES Global have established a joint venture to provide satellites that should be in position in 2009,[22] as will be further explained in section 21.5.

21.2.3 System Elements

For a DVB-H receiver to be integrated in a mobile device, three key issues must be addressed: (1) reduction of the power consumption, (2) maintaining of performance in mobile environments, and (3) achievement of the minimum mechanical size.

The most important issue to be addressed by the handheld device is the battery life. DVB-H fosters battery saving, including a time-slicing technique. This technique increases the general robustness and improved error resilience compared to DVB-T using MPE-FEC.

Another major requirement for DVB-H was the ability to receive up to 15 Mbit/s in an 8 MHz channel and in a wide-area single-frequency network (SFN) at high speed.[9] These requirements were drawn up after much debate and with an eye on emerging convergence devices providing video services and other broadcast data services to 2.5G and 3G handheld devices. Furthermore, all this should be possible while maintaining maximum compatibility with existing DVB-T networks and systems.

The requirements above imply that a DVB-H system should incorporate several technological elements, already developed and included in DVB-T4 technology, in the physical and link layers.[23]

The DVB-H physical layer should include:

- Signaling in TPS-bits (transmitter parameter signaling) to enhance and speed up service discovery. A cell identifier is also carried in TPS-bits to support quicker signal scan and frequency handover on mobile receivers.
- 4K mode (although it is not mandatory for DVB-H) for balancing mobility and SFN cell size, allowing the use of single-antenna reception in medium SFN networks at very high speed, providing flexibility for the network design.
- In-depth symbol interleaver for the 2K and 4K modes (also not mandatory for DVB-H) to further improve the robustness in mobile environments and impulse noise conditions.

The DVB-H link layer should include:

- Time slicing (mandatory for DVB-H) to reduce the average power consumption of the receiving terminal and enable smooth and seamless frequency handover.
- MPE-FEC (not mandatory for DVB-H) to improve the contrast-to-noise ratio (C/N) performance, the Doppler performance in mobile channels, and the tolerance to impulse interference.

In brief, time slicing, cell identifier, and DVB-H signaling are mandatory in DVB-H. All other technical elements may be combined arbitrarily. Both time-slicing and MPE-FEC technology elements, both implemented on the link layer, do not affect the DVB-T physical layer in any way. It is also important to notice that the payload of DVB-H consists of IPDC datagrams. Other network layer packets can also be encapsulated into MPE sections.

21.2.3.1 Time Slicing

In any handheld device, battery life is a critical aspect. The DVB Project estimated the future power consumption of DVB-T implementations. The estimation for a mobile handheld terminal was that the power consumption of the RF and baseband processing may come down to 600 mW by the year 2007. However, the average power consumption of any additional receiver in a mobile handheld terminal should be less than 100 mW. This is required due to both the limited battery capacity and the extremely challenging heat dissipation in a miniaturized environment. In the future, the required reduction in power consumption may become as high as 90 percent.[23]

Services used in mobile handheld terminals require relatively low bit rates. The estimated maximum bit rate for streaming video using advanced compression technology like MPEG-4 is in the order of a few hundred kilobits per second (Kbps), one practical limit being 384 Kbps, arising from the 3G standard. Some other types of services, such as file downloading, may require significantly higher bit rates, though; therefore, there is a requirement for flexibility.

A DVB transmission system usually provides a bit rate of 10 Mbps or more. This circumstance allows the possibility to significantly reduce the average power consumption of a DVB receiver by introducing a scheme based on TDM. This scheme shall be called time slicing.

The objective of time slicing is to reduce the average power consumption of the terminal and enable smooth and seamless service handover.

The concept of time slicing is to send data in bursts using a significantly higher bit rate than the bit rate required if the data was transmitted continuously using traditional streaming mechanisms.

Within a burst, the time to the beginning of the next burst (Delta-t) is indicated. Between the bursts, the data of the elementary stream is not transmitted, allowing other elementary streams to use the bit rate otherwise allocated. This enables a receiver to stay active for only a fraction of the time, while receiving bursts of a requested service. If a constant lower bit rate is required by the mobile handheld terminal, this may be provided by buffering the received bursts.

Data representing a particular service is delivered to the handheld device in bursts at given intervals of time. Time slicing enables a receiver to stay active for only a fraction of the time when it is receiving bursts of a requested service. When the receiver is not receiving the wanted burst of data, the tuner contained in the handheld device is inactive and therefore uses less power. This is transparent to the user because the data bursts are stored in the receiver memory and played out continuously. Time slicing could allow for up to a 95 percent reduction in power consumption compared to conventional and continuously operating DVB-T tuners. Although the receiver is inactive for periods of time, the broadcasting transmitter remains active at all times, sending a series of time-slice bursts for each service in sequence. In addition, time-sliced and non-time-sliced services can be placed in the same multiplex.

21.2.3.2 Power Consumption

To get a reasonable power savings effect, the burst bit rate should be at least ten times the constant bit rate of the delivered service. In case of a 350 Kbps streaming service, this indicates a requirement of a 4 Mbps bit rate for the bursts. Note that if the burst bit rate is only twice the constant bit rate, this gives near to 50 percent power savings, which is still far from the required 90 percent mentioned above.

The power consumption depends on the duty cycle of the time-slicing scheme. We assume here a 10 percent duty cycle, which implies a 90 percent reduction in power consumption. The power consumption estimations took into account the duty cycle as well as the increase in power consumption due to the MPE-FEC. The results estimated about 2 mW additional power consumption with 0.13 μm technology, and about 1 mW using 0.18 μm technology for the MPE-FEC.

It should be emphasized that these power consumption estimations assume that all Reed–Solomon (RS) codewords are always decoded. However, for most of the time in normal receiving conditions (particularly low-speed reception) the RS decoding will not be used, because the MPEG-2 TS is already fully correct, and so no MPE-FEC decoding will be necessary. Even in situations when the MPE-FEC is used, it may only be for a subset of the received bursts. This leads to the conclusion that for a mixture of receiving conditions (probably typical to real user behavior), the MPE-FEC will consume the additional 2 mW estimated only occasionally. The effect on battery time will therefore be negligible.

21.2.3.3 Handover

For mobile reception in DVB-T multiple-frequency networks (MFNs), there is normally the need to hand over to another frequency when the reception quality of the present frequency becomes too low. Because DVB-T does not include seamless handover facilities, changing frequency normally results in a service interruption. In addition to this, the receiver will have to scan possible alternative frequencies to find out which of these provides the best or at least sufficient reception quality. Each time a frequency is scanned there will be an interruption, unless the receiver is equipped with an extra RF part dedicated for this purpose. The inclusion of such an extra RF part would increase the cost of receivers. There is therefore a requirement to allow for seamless handover and seamless scanning of alternative frequencies without having to include an additional RF part.

Time slicing supports the possibility of using the receiver to monitor neighboring cells during the off-times. By accomplishing the switching between transport streams during an off-period, the reception of a service is seemingly uninterrupted.

Although a detailed explanation is outside the scope of this chapter, with proper care, the bursts of a certain IP stream can be synchronized between neighboring cells in a way that the receiver can tune to the neighboring cell and continue receiving the IP stream without losing any data.

Using time slicing, within the burst currently received, the time (Delta-t) to the beginning of the next burst is given to indicate to the receiver when to expect the next burst. Between the bursts, data of the elementary stream is not transmitted, to allow other elementary streams to share the capacity otherwise allocated. The transmitter is constantly on (i.e., the transmission of the transport stream is never interrupted). Time slicing also supports the possibility of using the receiver to monitor neighboring cells during the off-times (between bursts). Then, switching the reception from one transport stream to another during an off-period, it is possible to accomplish a quasi-optimum handover decision as well as a seamless service handover.

In a SFN network, handover is only required when the terminal changes network, because all transmitters in the SFN form a single cell.

21.2.3.4 Delta-t Method

The functionality of the Delta-t method is to signal the time from the start of any MPE (or MPE-FEC) section—currently being received—to the start of the next burst within the elementary stream. Delta-t information in the MPE (or MPE-FEC) sections removes the need to synchronize clocks between transmitter and receiver. In this way, the receiver is flexible to parameters such as burst size, burst duration, burst bit rate, and the specific off-time of the system, which may vary

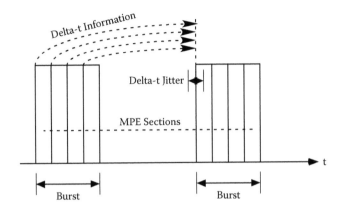

Figure 21.1 Each MPE section header contains Delta-t indicating time to the beginning of the next burst and Delta-t jitter.

with time. The receiver has to be very accurate regarding the off-time parameter, to avoid missing the next burst. Nevertheless, the accuracy requirement is relaxed as the clock is restarted by each, so only the maximum off-time should be considered.

To keep the Delta-t insensitive to any constant delays within the transmission path, Delta-t timing information is relative, as is depicted in figure 21.1.

In bad reception conditions, parts of a burst may be lost. In case the Delta-t information is lost, the receiver would not know the time to the next burst, and therefore is forced to continue waiting for the next burst. To avoid this situation, Delta-t (together with other real-time parameters) shall be delivered in the header of each MPE and MPE-FEC section within a burst. Even in very bad reception conditions, if only one MPE or MPE-FEC section is received, proper Delta-t information can be accessed and power savings achieved.

As Delta-t is relative rather than absolute, this method is insensitive to any constant delays within the transmission path. However, jitter does have an effect on the accuracy of Delta-t. This jitter is later referred as Delta-t jitter, as shown in figure 21.1. Note, however, that the accuracy of Delta-t has an effect on the achieved power savings. It is possible to perform a jitter estimation in the receiver, to ensure that the wakeup time for the next burst is not mistakenly too late because of the current burst being delayed. In time-slicing operation, a Delta-t jitter of 10 ms can be accepted, because 10 ms is the resolution of the Delta-t signaling. This should be easily achieved because typical transmission paths already support far better accuracy. On the other hand, virtually no gain is achieved by decreasing the value below 10 ms, as it is already less than a typical jitter in synchronization time.

Synchronization time is the extra time required by a receiver to reacquire lock onto the signal before the start of the reception of the next burst. In current DVB-T implementations the time is estimated to be at most in the order of 200 to 250 ms. Synchronization time is implementation dependent, and typically differs noticeably from time to time (i.e., has noticeable jitter).

One can see how Delta-t jitter has an effect similar to that of synchronization time. When the maximum Delta-t jitter is known accurately, we may assume that on average each burst starts $1/2 \times$ Delta-t jitter later than the time indicated by Delta-t. However, to be on the safe side, calculations later in this chapter add $3/4 \times$ Delta-t jitter to the synchronization time. This allows a network operator to use twice the accurate value of the Delta-t jitter.

21.2.3.5 MPE-FEC

Because the handheld devices have poorly designed antennas that require reception from many different locations and hostile environments, they need a robust transmission system with solid error protection. To better match the handheld environment, DVB-H offers, optionally, improved transmission robustness through the use of an additional level of forward error correction (FEC) at the Multi-Protocol Encapsulation (MPE) layer, and so independent of the DVB-T physical layer. MPE-FEC was added to the DVB-H specification to implement time interleaving and error correction.

The main purpose of the MPE-FEC is to improve the C/N and Doppler performance in mobile channels and to provide good immunity to impulse noise interference.

With the addition of FEC parity data in new sections, parallel to MPE sections carrying IP datagrams, it is possible to recreate (after MPE-FEC decoding) error-free IP datagrams even under bad reception conditions (with high packet loss ratio [PLR]) on the MPE level. Such high PLR may sometimes occur with DVB-T on mobile channels when the speed is too high or the C/N is too low. The amount of the transmission capacity allocated to parity overhead is flexible.

Performance estimations show that the proposed MPE-FEC should be able to output an error-free IP stream down to a PLR of about 10 percent. With the MPE-FEC, about 25 percent of TS data is allocated to parity overhead. For a given set of DVB-T parameters the MPE-FEC may require about the same C/N as if antenna diversity was used or if inner time interleaving was to be introduced in DVB-T, although with a 25 percent lower throughput, due to the parity overhead. This can, however, be compensated by choosing a slightly weaker code rate in DVB-T. For example, with 16-QAM, code rate 2/3, and MPE-FEC, the same throughput can be provided as with 16-QAM and code rate 1/2, but with a much better performance. This should allow high-speed, single-antenna DVB-T reception using 8K/16-QAM or even 8K/64-QAM signals.

The time interleaving that DVB-H offers is flexible and can be adapted to the service. Typical time-interleaving periods can be as high as 500 ms and as low as 50 ms. These periods can be even lower, but then it does not make sense at the system level.

After time interleaving, an RS encoder is applied to protect the data. The code rate is flexible and can also be adapted to the service. A typical code rate may be ¾, but rates as low as ½ or as high as 7/8 may be used.

With MPE-FEC, a flexible amount of the transmission capacity is allocated to parity overhead. For a given set of transmission parameters providing 25 percent of parity overhead, the receiver with MPE-FEC may require about the same C/N as a receiver with antenna diversity and without MPE-FEC.

The MPE-FEC overhead can be fully compensated by choosing a slightly weaker transmission code rate, while still providing far better performance than DVB-T (without MPE-FEC) for the same throughput. This MPE-FEC scheme should allow high-speed, single-antenna DVB-T reception using 8K/16-QAM or even 8K/64-QAM signals. In addition, MPE-FEC provides good immunity to impulse noise interference. The MPE-FEC, as standardized, works in such a way that MPE-FEC-ignorant (but MPE-capable) receivers will be able to receive the data stream, being fully backwards compatible, provided they do not reject the stream type information.

The MPE-FEC is defined on the MPE layer, that is, independent of the DVB-T physical layer. With the addition of FEC parity data in new sections, parallel to MPE sections carrying IP datagrams, it is possible to recreate error-free IP datagrams despite a very high PLR on the MPE level. Such high PLR may sometimes occur with DVB-T on mobile channels when the speed is too high or the C/N is too low. Performance estimations show that the proposed MPE-FEC should be able to output an error-free IP stream down to a PLR of about 10 percent. With the MPE-FEC, about 25 percent of TS data is allocated to parity overhead. For a given set of DVB-T parameters, the MPE-FEC may require about the same C/N as if antenna diversity was used or if inner time interleaving was to be introduced in DVB-T, although with a 25 percent lower throughput, due to the parity overhead. This can, however, be compensated for by choosing a slightly weaker code rate in DVB-T. For example, with 16-QAM, code rate 2/3, and MPE-FEC, the same throughput can be provided as with 16-QAM and code rate 1/2, but with a much better performance. This should allow high-speed, single-antenna DVB-T reception using 8K/16-QAM or even 8K/64-QAM signals.

21.2.3.6   Immunity to Impulsive Noise Interference

The MPE-FEC also provides good immunity to impulsive interference. With MPE-FEC, reception is fully immune to repetitive impulsive noise, causing a destruction of the OFDM symbols if the distance between the destroyed symbols is in the range of 6 to 24 ms. This depends on the chosen DVB-T mode.

21.2.3.7   Time Slicing and MPE-FEC Used Together

Time slicing and MPE-FEC constitute processes applied at the link layer and are fully compatible with the existing DVB physical layer. Moreover, the interface of the network layer supports bursty incoming of datagrams, and is therefore fully compatible with time slicing.

When time slicing and MPE-FEC are used together, one time-slice burst carries exactly one MPE-FEC frame. The first part of the burst is the MPE sections carrying the IP datagrams belonging to the MPE-FEC frame. Immediately following the last MPE section is the first MPE-FEC section carrying the parity bytes. All sections contain a table_boundary flag; this is set high in the last MPE section to indicate this is the last MPE section of the MPE-FEC frame. If all the MPE sections within the burst have been received correctly, the receiver can then neglect the MPE-FEC sections and go to sleep until the next burst. All sections contain a frame_boundary flag; this is set high in the last MPE-FEC section to indicate that this is the last MPE-FEC section, and hence the end of the MPE-FEC frame.[23]

21.2.3.8   4K Mode

The 2K transmission mode is known to provide significantly better mobile reception performance than the 8K mode, in DVB-T systems, due to the larger intercarrier spacing it implements. However, the duration of the 2K mode OFDM symbols, and consequently the associated guard interval, is very short, which makes this mode only suitable for small-size SFNs. It makes it very difficult for network designers to build spectrally efficient networks.

So, in DVB-H a new transmission mode is included in the DVB-T physical layer: the 4K mode. It is an intermediate mode between the 2K and 8K whose purpose is to improve the network planning flexibility by trading off mobile reception and SFN size. It intends to offer an additional trade-off between single-frequency network (SFN) cell size and mobile reception performance, providing an additional degree of flexibility for network planning. It is also, from an architectural standpoint, hardware compatible with existing DVB-T infrastructure, so the required changes in

Table 21.1 Terms of the Trade-Off between SFN Cell Size and Mobile Reception Performance

Mode	Use
DVB-T 8K mode	It can be used both for single-transmitter operation and for small, medium, and large SFNs. It provides a Doppler tolerance allowing high-speed reception.
DVB-T 4K mode	It can be used both for single-transmitter operation and for small and medium SFNs. It provides a Doppler tolerance allowing very high-speed reception.
DVB-T 2K mode	It is suitable for single-transmitter operation and for small SFNs with limited transmitter distances. It provides a Doppler tolerance allowing extremely high-speed reception

the modulator and the demodulator are minor. Terms of the trade-off can be expressed as shown in table 21.1.

With the addition of a 4K mode, DVB-H benefits from the compromise between the high-speed, small-area SFN capability of 2K DVB-T and the lower-speed, but larger-area SFN of 8K DVB-T.

With the aid of enhanced in-depth interleavers in the 2K and 4K modes, DVB-H has even better immunity to impulsive interference. For these modes, the in-depth interleavers increase the flexibility of the symbol interleaving, by decoupling the choice of the inner interleaver from the transmission mode used. This flexibility allows a 2K or 4K signal to benefit from the memory of the 8K symbol interleaver. This effectively quadruples (for 2K mode) or doubles (for 4K mode) the symbol interleaver depth to improve reception in fading channels.

The longer symbol duration of the 8K transmission mode makes it more resilient to impulsive interference. For a given amount of noise power occurring in a single impulsive noise event, the noise power is averaged over 8,192 subcarriers by the fast Fourier transform (FFT) in the demodulator. In the 4K and 2K transmission modes, the same amount of impulse noise power is averaged only over 4,096 and 2,048 carriers, respectively. The noise power per subcarrier is therefore doubled for 4K and quadrupled for 2K when compared with 8K. The use of the 8K symbol interleaver for 2K and 4K modes helps to spread impulse noise power across two symbols (for 4K mode) and four symbols (for 2K mode). If only one symbol suffers such an impulse noise event, then at the output of the interleaver, four consecutive symbols in the 2K mode would each have one carrier in every four with some noise, while in the 4K mode, one carrier in every two would have some noise over two symbols. This extended interleaving allows 2K and 4K modes to operate with impulse noise immunity quasi-similar to that of an equivalent 8K mode. This provides an extra level of protection against short noise impulses (for example, the ones caused by automobile ignition interference).

Note that when using in-depth interleavers in SFN configuration, it should be taken into account that, due to SFN synchronization, there is an additional delay of one OFDM symbol for the 4K mode and three OFDM symbols for the 2K mode, so the additional delay should be compensated.

Table 21.2 shows the guard interval lengths in time for all modes. It can be seen that a 4K OFDM symbol has a longer duration, and consequently a longer guard interval than a 2K OFDM symbol, allowing the building of medium-size SFNs. This gives to the network designers a better way to optimize SFNs, with regard to spectral efficiency.

Table 21.2   Guard Interval Lengths for All Modes[23]

	2K	*4K*	*8K*
1/4	56 μs	112 μs	224 μs
1/8	28 μs	56 μs	112 μs
1/16	14 μs	28 μs	56 μs
1/32	7 μs	14 μs	28 μs

Compared with the 8K mode, with a symbol duration shorter than in the that mode, channel estimation can be done more frequently in the demodulator, thereby providing a mobile reception performance that, although not as high as with the 2K transmission mode, is nevertheless adequate for the use of DVB-H scenarios. Furthermore, doubling the subcarrier spacing with respect to the 8K mode allows for mobile reception with reasonably low complexity channel estimators, thus minimizing both power consumption and cost of the DVB-H receiver.

The incorporation of a 4K mode provides a good trade-off for the two sides of the system: spectral efficiency for the DVB-H network designers and high mobility for the DVB-H users. The 4K mode also increases the options available to flexibly plan a transmission network while balancing spectral efficiency, mobile reception capabilities, and coverage.

Implementation-wise, compared to an existing 2K/8K DVB-T receiver, the addition of the 4K mode and the in-depth symbol interleaver does not require extra memory, significant amounts of logic, or extra power. Also, it could be envisaged that the future DVB-H demodulators will be designed to support only the subset of the standardized transmission modes that are most suitable for mobile applications. These savings would partially offset the increase in the silicon area, and then the power consumption, required for more advanced mobile receiver algorithms such as complex channel estimation. On the network side, it is expected that the changes in the transmitter will be marginal because they are only located in the modulator. In addition, the emitted spectrum is similar to existing 2K and 8K modes; thus, no changes in expensive RF transmitter filters are necessary.

There are some interesting aspects to take into account regarding the current DVB-T 2K and 4K physical layer specification and the proposed new DVB-H 4K mode with in-depth symbol interleaver. In principle, DVB-T receivers could not decode a DVB-H signal employing DVB-T transmission modes. However, there is some sort of compatibility with the current DVB-T specification in some aspects:

- **Spectrum coexistence:** Full compatibility exists: Current spectrum requirements of 2K and 8K DVB-T modes match DVB-H 4K requirements. The occupied bandwidth is the same, as well as the shape and interference characteristics.
- **System implementation:** The new 4K mode could be considered as an interpolation of the existing 2K and 8K modes, requiring only an additional parameter in the DVB-T demodulator and a little control logic in the receiver. This upgrade would not affect other blocks of the system (in the same way that some DVB-T 2K receivers cannot decode an 8K transmission, but the receiver subsystems are 100 percent DVB-T). As most of the current DVB-T equipment supports both 8K and 2K FFT modes, the additional complexity to support the 4K mode is minimal and consists mainly of additional control logic.

■ **Legacy receivers:** Current 2K or 8K receivers will be unable to receive 4K signals, but this is not a severe restriction, as any new DVB-H network using the 4K mode would be targeted toward new services and new types of hand-portable terminals. There is one restriction in this case when sharing the multiplex between traditional DVB-T and DVB-H services. The standard allows new 4K-capable receivers to receive both 2K and 8K transmissions. The actual implementation of all modes is a commercial decision. Another receiver-level compatibility consideration is the relative simplicity of adding the new 4K mode to the existing 8K/2K chip designs. This ensures low cost and fast time to market for the new hardware.

21.2.3.9 Frequency Bands

DVB-H is designed to work in the following bands (currently occupied by analogue TV and DTT services):

■ VHF-III (170–230 MHz, or a portion of it)
■ UHF-IV/V (470–862 MHz, or a portion of it)
■ L (1.452–1.492 GHz)

Like DVB-T, DVB-H can be used in 6, 7, and 8 MHz channel environments. However, a 5 MHz option is also specified for use in nonbroadcast environments. A key initial requirement, and a significant feature of DVB-H, is that it can coexist with DVB-T in the same multiplex. Thus, an operator can choose to have two DVB-T services and one DVB-H service in the same overall DVB-T multiplex.

The most problematic cellular radio from the interoperability point of view is the GSM 900 because of the very narrow guard band between the DVB-H band and GSM 900 uplink. The guard band is only 18 MHz wide. Therefore, the relative bandwidth of the guard band is very small. The problems are much less severe with GSM 1800, and even easier with WCDMA, because of the bigger guard band between Rx- and Tx-bands.

21.2.4 Network Architecture

A DVB-H network is a highly centralized, centrally managed infrastructure. Programs, applications and services are created and managed by the service providers. As it has been written before, a DVB-H network enables the transmission of several types of multimedia services for small devices, so there are two needs: to save power on the receiver side and to have a robust signal to provide service to receivers such as mobile cellular telephony. DVB-H uses different handheld reception modes, which are based on where the device is placed: indoor handheld reception (at very low or no speed) and outdoor handheld reception (from high to very high speed).

21.2.4.1 DVB-T Broadcasting Spectrum

In this type of network, DVB-H uses the same broadcasting spectrum as DVB-T.[24] Both have the same physical layer, so they have full-spectrum compatibility, and any DVB-T frequency allotment or assignment can be used also for DVB-H. DVB-H can be introduced sharing an existing DVB-T multiplex offering DVB-H and DVB-T services. One 8 MHz channel can deliver 30–50

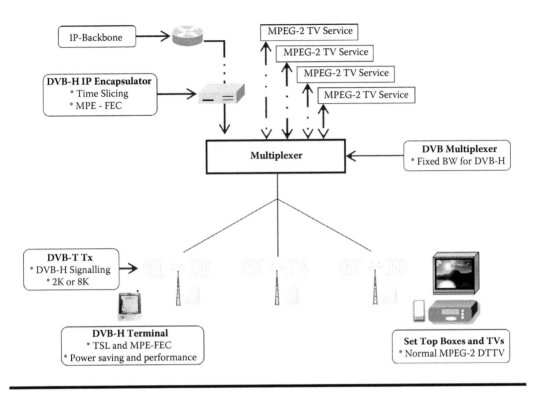

Figure 21.2   Shared network.

video-streaming services to the small screen terminals. It is 10 times more than SDTV or 20 times more than HDTV.[25] An example of a shared network could look like the one shown in figure 21.2. DVB-T transmitters are serving both DVB-H and DVB-T terminals. The current sharing is done at the multiplex level.

The existing DVB-T network has to be designed for portable indoor reception to provide high-enough field strength for the handheld terminals inside the wanted service area. DVB-H offers full flexibility to select the wanted portion of the multiplex to DVB-H services. The key DVB-H component in the network is the IP encapsulator, where the MPE of IP data, time slicing, and MPE-FEC are implemented.

One approach to constructing a total end-to-end DVB-H system making use of a mobile tele-communication network for subscription billing is shown in figure 21.3. Broadcast services can be delivered by DVB-H without the need for an interaction channel, or in the configuration shown, an interaction channel can easily be provided using a cellular network such as the GSM network. Methods for providing services payment can be built upon a proprietary encryption and payment solution or in conjunction with the telecoms network's inherent service statistics collection and billing functions.

While the DVB-T network is intended primarily for rooftop antenna reception, a DVB-H network will be designed for portable reception available even inside buildings. Hence, it will need a much higher signal power density. To reach the higher power density needed for mobile coverage levels, several network architectures can be used depending on available frequencies, allowed maximum transmitter powers, and antenna heights.

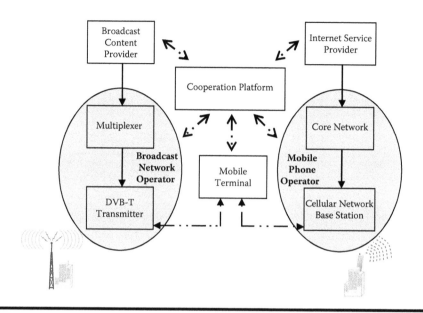

Figure 21.3 **System architecture for collaboration between mobile and broadcast operators.**

21.2.4.1.1 The Transmission Chain

DVB-T hierarchical modulation allows sharing the network too. In that case, MPEG-2 and DVB-H-IP services will have their own independent TS inputs in the DVB-T transmitters. Figure 21.4 shows how DVB-H is modulated to be transmitted over DVB-T. Figure 21.5 shows a blocks diagram taking into account the type DVB network where it is transmitted.

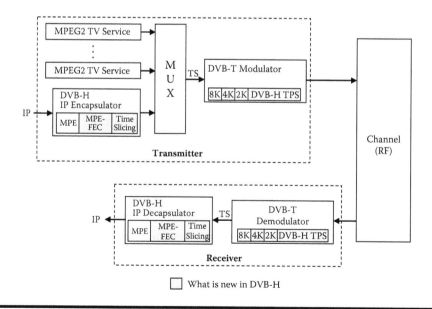

Figure 21.4 **DVB-H modulated to be transmitted over DVB-T.**

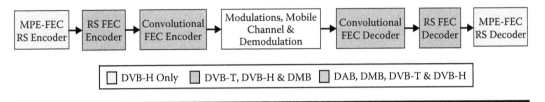

Figure 21.5 Blocks diagram taking into account the type of DVB network.

In a DVB network, RS coding is introduced as an extra precautionary measure for FEC, along with convolution encoding/Viterbi decoding, and MPE-FEC RS encoding/decoding. These are the FEC components. Modulation and demodulation, depending on the network, can be QPSK/16-QAM/64-QAM for transmission of the data over the network. Figure 21.5 shows the encoding/decoding flow. In addition to convolution coding with a constraint length of usually K = 7, puncturing logic is used with the help of various chips. The chips have built-in puncturing features, which provide these FEC coding provisions. The puncturing helps in attaining the required variable transmission rate depending on line capacity, weather conditions, transmission bandwidth, network parameters and load, MSF capability, and so on.

21.2.4.2 Dedicated DVB-H Networks

When the multiplex can be reserved fully for DVB-H, the planning of the network could be done freely as a dedicated DVB-H network. A typical network is composed of several SFN areas, each using its own frequency allotment. The maximum size of one SFN area depends on the FFT size (2K, 4K, or 8K mode), guard interval, and geographical properties in the network. Each SFN area could have several GPS-synchronized transmitters supported by a number of on-channel repeaters to cover some smaller holes. The on-channel DVB-H repeater receives a terrestrial DVB-H emission at a certain VHF/UHF frequency, amplifies the received channel, and retransmits it in the same frequency. The repeater is used to extend the coverage of an existing DVB-H network through emissions at a single frequency without the need for additional transmitters. One of the benefits of repeaters is that they are easy to deploy and have low cost, but a disadvantage is the instability of the devices. The key trade-off in their deployment is between the gain expected from the device and the isolation between the output of the repeater and its input. The process of reception, amplification, and transmission has to have a delay shorter than the guard interval of the used DVB-H mode, so that a receiver receiving both a signal from a transmitter and a signal from a repeater does not have to deal with interference.

Because the DVB-H network requires higher field strength, and the number of synchronized main transmitters is higher and the transmitter powers and antenna heights lower than in a traditional DVB-T network, the SFN is denser and the cost of the network is higher than in a conventional DVB-T network (the number of services in one multiplex is ten times higher).

21.2.4.3 Hierarchical Networks

The DVB-H standard allows the building of hierarchical networks. In ETSI 102 377,[26] a hierarchical network to either support a progressive quality-of-service (QoS) degradation or allow multiformat-multidevice transmissions is proposed. In hierarchical modulation, the possible digital states of the constellation (i.e., the number of states in a QAM) are interpreted differently than in the nonhierarchical case. Several data streams are available for transmission. The one defined by the number of the quadrant in which the state is located is named the first stream, and the one defined by the location of the state within its quadrant is named the second stream. The hierarchical modulation allows the transmission of two streams, having different bit rates and performance, in the same RF channel. The sum of the bit rates of the two streams is equal to the bit rate of a nonhierarchical stream using the same modulation.

21.2.4.4 DVB-SH System Architecture

DVB-SH is only a physical layer specification, using as a "higher layer" the DVB-IPDC specifications. The DVB-SH standard provides a very efficient transmission system in terms of reception threshold and resistance to mobile satellite channel impairments. The system relies on a hybrid satellite-terrestrial infrastructure to provide universal coverage. In a cooperative mode, the SC ensures geographical global coverage, while the complementary ground component (CGC) provides cellular-type coverage. The signals are broadcast to mobile terminals on two paths: a direct path from a broadcast station to the terminals via the satellite, and an indirect path from a broadcast station to terminals via terrestrial repeaters that form the CGC to the satellite. The CGC can be fed through satellite or terrestrial distribution networks.

A typical DVB-SH system is based on a hybrid architecture combining a satellite component and a CGC consisting of terrestrial repeaters fed by a broadcast distribution network (see figure 21.6). The repeaters could be used as broadcast infrastructure transmitters (to complement

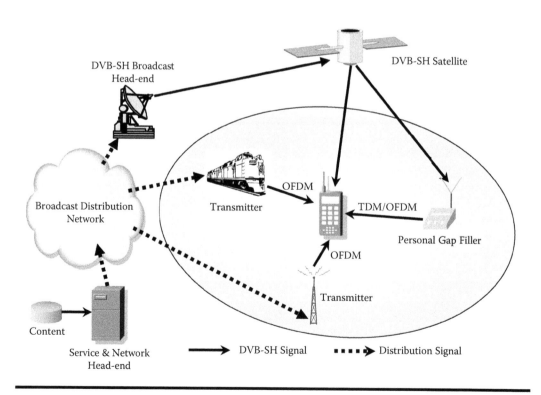

Figure 21.6 Overall DVB-SH system architecture.

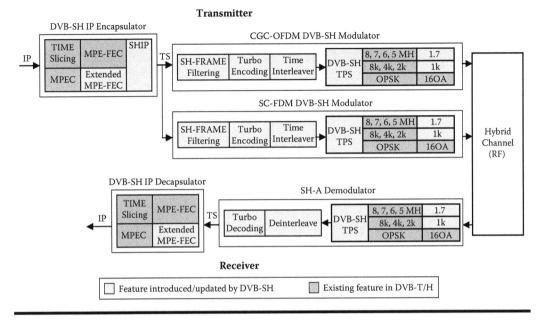

Figure 21.7   SH-A system architecture.

reception in areas where satellite reception is difficult), as personal gap fillers of limited coverage providing local on-frequency retransmission or frequency conversion, or as mobile broadcast infrastructure transmitters.

The system includes two transmission modes leading to two reference architectures:

■ An OFDM mode based on DVB-T standard with enhancements. This mode can be used on both the direct and indirect paths; the two signals are combined in the receiver to strengthen the reception in a SFN configuration. This mode leads to the SH-A reference architecture where OFDM is used on both the satellite and the terrestrial link (figure 21.7).

■ A time division multiplexing (TDM) mode partly derived from the DVB-S2 standard to optimize the transmission through satellite toward mobile terminals. This mode is used on the direct path only. The system supports code diversity recombination between satellite TDM and terrestrial OFDM modes so as to increase the robustness of the transmission in relevant areas (mainly suburban). The SH-B reference architecture uses TDM on the satellite link and OFDM for the terrestrial link (figure 21.8).

The terrestrial OFDM part of SH-B is identical to the OFDM part of SH-A, and terminals designed for SH-B architectures can also be used with SH-A architectures, their TDM processing branch being simply switched off.

The combination of a satellite footprint and a terrestrial complement in the S-band can deliver nationwide coverage to terminals that could implement the TDM and OFDM modes of SH, a combination of SH and DVB-H, or simply the OFDM mode of DVB-SH operating in SFN.

21.2.5 IP Datacasting (IPDC)

Transmission of IP packets over MPEG-2 is based on digital storage media command and control (DSMCC)[27] sections that are typically not repeated. The transparent transmission of IP

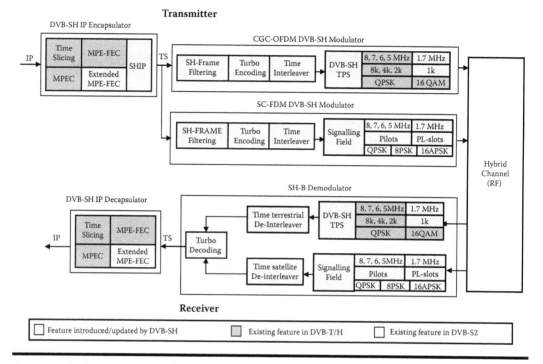

Figure 21.8   SH-B system architecture.

is accomplished by encapsulating the entire IP packet payload within the payload of a DSMCC section and by mapping the MAC address to the respective header and payload fields in the DSMCC section. The section format permits fragmenting datagrams into multiple sections. If the length of the datagram is less than or equal to 4,086 bytes, the datagram is sent in one section. In the case of IP, the maximum transmission unit (MTU) is set to 4,086 bytes or less so that the datagrams will never be fragmented. The MAC address has been divided into six bytes that are located in two groups.[28] Bytes 5 and 6 are mapped to the table_id_extension field of the DSMCC section, while bytes 1–4 are mapped to the payload area of the DSMCC section.

This mapping was done to utilize the limited capabilities of the first generation of demultiplexers.

21.2.5.1   IPDC over DVB-H System Architecture

IP Datacast (IPDC) over DVB-H is an end-to-end broadcast system for delivery of any type of digital content and services using IP-based mechanisms optimized for devices with limitations on computational resources and battery. Figure 21.9 shows the DVB-H network structure, and figure 21.10 shows an example with the sites that could be in the network. The set of specifications for IP Datacast (phase 1) was approved by DVB in October 2005. It was developed for handheld terminals like mobile phones.

IP Datacast uses DVB-MPE.[10] It was adopted as part of the specification for video streaming and data transfer for use in DVB-H systems. An inherent part of the IPDC system is that it comprises a unidirectional DVB broadcast path that may be combined with a bidirectional

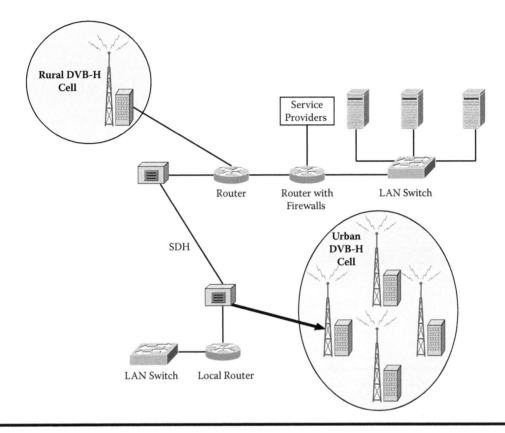

Figure 21.9   IPDC over DVB-H network structure.

mobile/cellular interactivity path. IP Datacast is thus a platform that can be used for enabling the convergence of services from broadcast/media and telecommunications domains such as mobile and cellular.

In figure 21.11, a full IPDC system with many components and elements is shown. First, there is a service system that is used to produce IP video streams to the network. They are then distributed over the multicast intranet to the IP encapsulators, which will output the DVB-H TS with time slicing and MPE-FEC included. TS is then distributed to the DVB-T/H transmitters of the broadcasting network. The IP Datacast system may include other functions via cellular networks, like General Packet Radio Service (GPRS) or Universal Mobile Telecommunications System (UMTS). IP Datacast allows integrating DVB-H in a hybrid network structure with a mobile communications network (such as GPRS or UMTS) and an additional DVB-H downstream.

In IP Datacast, content is delivered in the form of data packets (IP datagrams) using the same distribution technique as used for delivering digital content on the Internet. It allows DVB-H to rely upon standard components and protocols for content manipulation, storage, and transmission. In addition to video and audio stream broadcasting, IP Datacast over the DVB-H system also can be used for file delivery.

DVB-H-IP-encapsulated packets are multiplexed along with MPEG2 TV packets, producing a uniform transport stream (TS) that is encoded into DVB-T packets for transmission over the RF channel. Reverse processes are performed to recover both streams. The handheld terminal

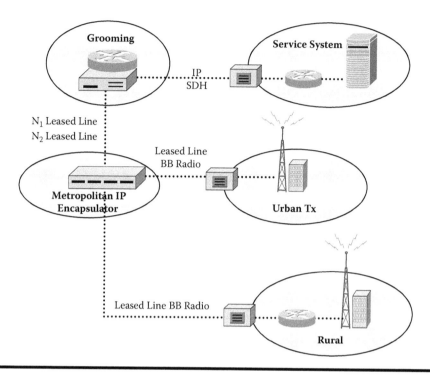

Figure 21.10   IPDC over DVB-H network with sites.

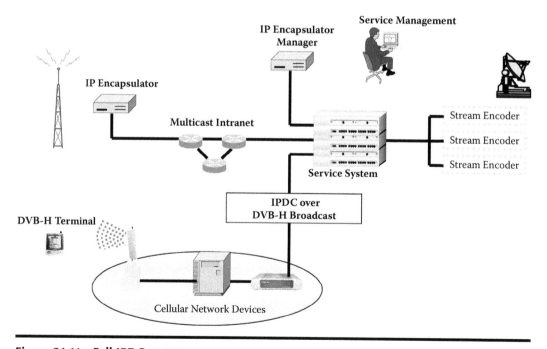

Figure 21.11   Full IPDC system.

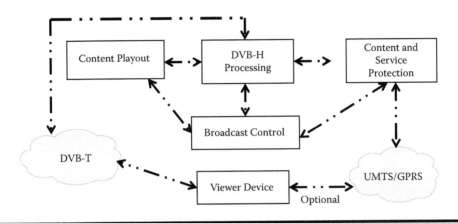

Figure 21.12   Overview of the DVB-IPDC over DVB-H architecture.

decodes/uses IP services only. The 4K mode and the in-depth interleavers are not available in the cases in which the multiplex is shared between services designed for fixed DVB-T receivers and services for DVB-H devices.

The flowchart of the DVB-IPDC over DVB-H architecture can be seen in figure 21.12. The "broadcast control" block controls what content is played at a given time, and using what access criteria. The "content playout segment DVB-H processing" block is responsible for the encoding, storage, and playout of content. The "content and service protection" block is responsible for the creation of the DVB-H stream and presenting this data to the DVB-T chart for transmission. Finally, the "viewer device" chart is responsible for scrambling content and maintaining containers used to carry entitlements and access rights.

21.3   DVB-H and DVB-SH Receivers

DVB-H receivers are based on DVB-T receivers.[29] A DVB-H receiver adds extra blocks to optimize power consumption and to improve robustness in difficult reception environments, as was discussed in section 21.2.1. Average power consumption is optimized including time slicing, which also facilitates frequency handover, and robust reception is facilitated by the introduction of the 4K transmission mode in addition to the 2K and 8K modes found in the DVB-T standard.

DVB-H receivers operate in the UHF bands IV and V (470 to 862 MHz), as was discussed in section 21.2.3.9. If the DVB-H receiver is combined with a GSM 900 mobile phone, the usable UHF band is then limited from 470 MHz to 702 MHz due to the interoperability requirements of GSM power amplifiers and the required spectral spacing by the GSM blocking filter. A summary of DVB-H receiver requirements is shown in table 21.3.

21.3.1   Receiver Architecture

Let us first consider a typical GSM/EDGE or WCDMA cellular device when a DVB-H receiver is attached, forming a converged terminal. The converged terminal architecture is shown in figure 21.13. In this figure, a DVB-H receiver and the cellular radio are shown, with two physically

Table 21.3 DVB-H Receiver Operational Parameters

Parameter	DVB-H
Frequency of operation	470 → 702 MHz
Adjacent channel protection ratio (ACPR)	−40 dB
Co-channel protection ratio (CCPR)	−22 dB
IIP3	≥ −12.4 dBm
Noise figure	≤6 dB
Input power range	−91.6 → −20 dBm
Channel bandwidths (MHz)	5, 6, 7, 8 (typical)
FFT sizes	8K, 4K, 2K
Guard intervals (U.S.)	224, 112, 56, 28, 14, 7
Inner modulations	QPSK, 16-QAM, 64-QAM
Error protection	Convolutional code + RS FEC + MPE-FEC
Convolutional code rates	1/2, 2/3, 3/4, 5/6, 7/8
Time interleaving	Practical: Up to 1000 ms depending on MPE-FEC selectionTypical: 200–500 ms
MPE-FEC code rate	Free selection (most likely 1/2 to 7/8)
Time slicing	Yes, power saving
Protocol stack	IP layer
Theoretical data rate range (Mbit/s)	2.49 31.67 (@ 8 MHz channel)
Practical data rate	3.32 13.8 (@ 8 MHz channel, 1/4 GI QPSK 1/2CR MPE-FEC 2/3 1/8 GI 16-QAM 3/4CR MPE-FEC 5/6)
Backward compatibility	DVB-S, DVB-T
Antenna gain	Low

Source: From Kampe and Olsson[30] and EICTA/TAC/MBRAI-02-16.[31]

separate antennas with frequency isolation between them. This is the most common case, but a properly engineered antenna could be also used.

As it can be seen in figure 21.13, the key difference between WCDMA and GSM/EDGE implementations is the duplex filter. WCDMA uses a duplex filter, but a majority of modern

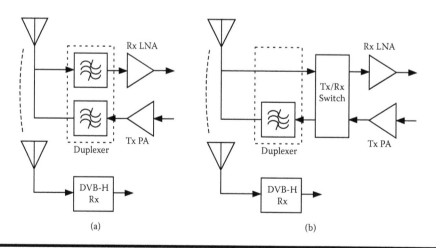

Figure 21.13 Terminal architectures: (A) WCDMA radio and DVB-H receiver. (B) GSM/EDGE radio and DVB-H receiver. LNA, low-noise amplifier; PA, power amplifier.

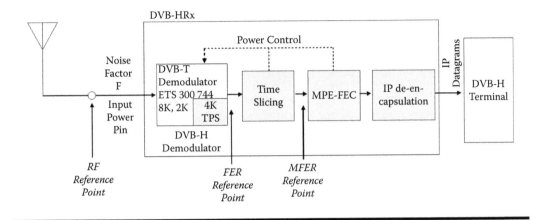

Figure 21.14   DVB-H typical receiver structure.

GSM/EDGE radios use a Tx/Rx switch. From the receiver point of view, this has a major implication: the cellular radio uplink unwanted signal interference to the DVB-H receiver will not be a problem in the WCDMA terminal; however, this problem will be severe in GSM/EDGE terminals employing the Tx/Rx switch.

21.3.2 Receiver Design

Figure 21.14 shows a conceptual DVB-H receiver structure.[29] This receiver comprises a DVB-T demodulator and MPE-FEC block. The DVB-H terminal shown in the figure can be seen as a player that demodulates the time-sliced streams' decode ATM/MPEG packets (note that other stream formats are also possible).

The DVB-H demodulator is based on a standard DVB-T demodulator, but including the 4K mode (optional) discussed in section 21.2.3.8 and a time-slicing management module. The MPE-FEC module, discussed in section 21.2.3.5, is also optional and may not appear in some manufacturer implementations.

The receiver comprises several blocks. First, the DVB-T demodulator. The demodulator recovers the MPEG-2 transport stream packets from the received DVB-T RF signal.[32] The RF signal can be configured in three transmission modes—8K, 4K, and 2K—ith the corresponding transmitter parameter signaling (TPS). Note that the 4K mode, the in-depth interleavers, and the TPS DVB-H signaling have been defined in the context of the DVB-H standard. The time-slicing module enables reduced power consumption and smooth frequency handover when required. Finally, the MPE-FEC module, as by DVB-H specification, offers in addition to the physical layer transmission a complementary forward error correction that allows the receiver to deal with difficult reception situations.

State-of-the-art DVB-H commercial receivers exhibit several implementation characteristics. These can be summarized as:

- Zero-IF (ZIF) silicon tuner. DVB-H receivers target lowest power consumption and minimum space. These are addressed by ZIF silicon tuners. These tuners downconvert the RF signal to baseband to feed the demodulator, whereas former tuner approaches (e.g., heterodyne) downconvert to intermediate frequency (IF). Zero-IF has the advantage that adjacent-channel filtering can be implemented in baseband and can be integrated into the silicon tuner. The ZIF concept is also called direct-downconversion (homodyne) architecture. This approach is attractive for its overall simplicity, though it is difficult to implement because the direct-current (DC) offset issue. If ZIF is used, the downconverted signal is centered at DC and must contend with DC leakage introduced by circuitry. The signal to be demodulated cannot exhibit a significant DC component because it would be impossible to distinguish which part of the DC belongs to the signal and which part is due to leakage. This issue can be solved by differential architectures with excellent matching and using modulations without significant DC content at the baseband. Direct-sequence spread spectrum (DSSS) that uses double-sideband suppressed-carrier phase-shift keying modulation, and orthogonal frequency division multiplexing (OFDM)—excluding the center carrier—are good choices. DC offset also has thermal and external dependence, which require matched, balanced automatic gain control (AGC) amplification to be employed in the receiver.

■ Doppler compensation. A primary target for the demodulator is to provide good reception in mobile environments compensating Doppler effect. Doppler compensation requirement in a DVB-H receiver is at least 100 Hz.[29] The receiver must be able to handle a channel with a central frequency variation and at the same time maintain an adequate signal-to-noise ratio. This point is critical for network planning, as network coverage must stay the same when the speed of the receiver varies.

■ Receiver desynchronization resistance. In addition to Doppler effect, the receiver will face channel profile variations. The different echoes received at the antenna will vary continuously. The receiver must compensate these variations to avoid desynchronization with the transmitter.

■ Received desensitization. Co-channel interference is an important transient disturbance that comes from different industrial sources that are generated in the operating environment, and sometimes even from other neighboring networks. The receiver must provide immunity from such disturbances and also be capable of maintaining good reception even if the interference exhibits power higher than the desired signal. In a GSM cell phone, for example, such interference typically comes from the upstream GSM channel.

■ High sensitivity. Sensitivity is also an important parameter to provide the best geographical coverage. The demodulator implementation typically exhibits very low degradation with respect to the theoretical DVB-H specification. An implementation margin of 0.5 to 1 dB is usually found.

■ Integrated link layer. In addition to the demodulation, the link layer or MAC is also integrated into the demodulator. This layer handles the time slicing, the IP data extraction, and the IP data error correction. Because of the high bit rates that DVB-H must support, the RS decoding is implemented in hardware.

■ Embedded application processing. The output of the DVB-H demodulator is an IP stream that is sent to the device application processor. The application processor is typically responsible for the decoding and rendering of video and audio and the integration of these services within the context of the device type, such as a cell phone or PDA.

■ Interoperatibility with cellular radio. Most of the services presented for convergence terminals require the coexistence and partly simultaneous operation of DVB-H receiver and cellular radios.[31] The cellular radio could be GMS/EDGE 900 or 1800, WCDMA, or a combination of these. The coexistence and simultaneous operation in a small-sized handheld terminal causes several challenges for the design. System-level interoperability issues for DVB-H reception can be divided into two main categories, described below. These issues must be solved by the proper terminal design.

- Cellular radio uplink wanted signal interference to DVB-H receiver. The transmitted cellular signal is very high power compared to the received DVB-H signals. Part of the cellular signal is coupled from the cellular transmitter antenna to the DVB-H receiver antenna. Without any filtering, the cellular TX interference signal level would cause severe blocking effects by two mechanisms: desensitization and cross-modulation. The practical solution is to insert a cellular-rejection filter in front of the DVB-H receiver.

- Cellular radio uplink unwanted signal interference to DVB-H receiver. This interference can be originated by (1) the transmitter power amplifier (PA) spurious responses, or (2) the Tx PA noise shown in figure 21.13.

- Undisturbed operation of a cellular radio must also be guaranteed. A possible impairment caused by the DVB-H receiver is the spurious responses that leak in the cellular downlink (Rx) band, or isolation (cross-talk) effects that arise from the cellular antenna pattern.

■ Delta-t method (time slicing). The Delta-t implementation must be highly flexible. Parameters such as burst size, burst duration, burst bit rate, and off-time may vary between elementary streams as well as between bursts within an elementary stream. The key requirement is the receiver must be sufficiently accurate inside one off-time period, as the clock is restarted in each burst.

■ MPE-FEC. As discussed in section 21.2.3.5, the MPE-FEC improves the C/N and Doppler performance in mobile channels and also improves the tolerance to impulse interference. The proposed additional MPE-FEC is introduced in such a way that MPE-FEC-ignorant (but IP/MPE-capable) DVB-T receivers will be able to receive the IP stream in a fully backwards-compatible way. This backwards compatibility holds when the MPE-FEC is used with and without time slicing. The use of MPE-FEC is not mandatory and is defined separately for each elementary stream in the transport stream. On each elementary stream it is possible to choose whether or not MPE-FEC is used in the first place, and if it is used, to choose the trade-off between FEC overhead and RF performance. Time-critical services, without MPE-FEC and therefore minimal delay, could therefore be used together with less time-critical services using the MPE-FEC, on the same transport stream but on different elementary streams.

■ Carrier-to-noise ratio. The required carrier-to-noise ratio (C/N) for reception of DVB-H signals is a very important parameter that highly affects network costs in general and in particular the possibilities to receive services carried over DVB-H at high reception speeds. Techniques like antenna diversity reception can improve performance but are not practically suited for small handheld devices, where single-antenna reception and low power consumption are required. From a spectrum efficiency, network cost, and coverage point of view, single-frequency networks (SFNs) are highly desirable. Such networks normally require the use of the DVB-T 8K mode, which also provides the highest bit rate. However, mobile single-antenna reception at high speeds using the 8K mode is very difficult, so there is a clear trade-off between low network costs and using higher bit rates in DVB-H.

21.3.2.1 Protocol Stack

Decoding high-bandwidth MPEG-2 encoded streaming video/audio is relatively difficult because of the processing capability required, often not available in converged terminal. Complete MPEG-2 decoding would also lead to excessive power consumption, which is also a nondesirable effect.

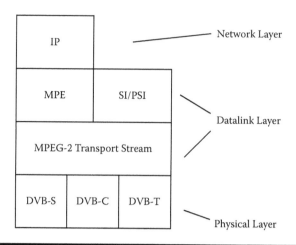

Figure 21.15   Protocol stack, OSI layers 1 to 3.

DVB-H uses plain IP packets embedded in the MPEG-2 stream. This approach decouples IP data from MPEG-2 transport, thus opening the door to a variety of encoding methods.

IP is relatively insensitive to any buffering or delays within the transmission (unlike MPEG-2). Therefore, IP is well suited for the time-sliced transmission present in convergence terminals. IP version 6 (IPv6) may be better suited in mobile environments (compared to IP version 4 [IPv4]). Therefore, IPv6 may be the preferred option on the broadcast interface. However, both time slicing and MPE-FEC may be used with IPv4 and IPv6. In this section IP indicates both IPv6 and IPv4 versions. No distinction in the time slicing or MPE-FEC implementations exists.

DVB specifies four data broadcasting methods: data piping, data streaming, MPE, and data carousel. Although they all can be used for IP delivering, data piping and data streaming are used rarely, and we will ignore them in this context. The data carousel method supports the delivery of files and other data objects, and it is not well suited for streaming. Also, implementing time slicing on data carousels may be difficult. We will focus on Multi-Protocol Encapsulation (MPE) broadcasting.

MPE is well suited to the delivery of streaming services as well as files and other data objects. Note that DVB has specified IP address resolution on MPE. In addition, MPE also supports delivery of other protocols, giving more flexibility. Finally, the time-slicing implementation on MPE is simple, which is a key advantage.

Figure 21.15 illustrates the protocol stack for delivering IP data on generic DVB receivers (DVB-S, DVB-C, or the DVB-T received found in DVB-H terminals). At the data link layer (OSI layer 2), time slicing could, in principle, be implemented either on the MPE level (Delta-t method, as described in section 21.2.3.4, delivered within the MPE section) or at the transport stream level (Delta-t delivered within transport packets). To enable MPE-FEC, time slicing has to be implemented at the MPE level due to the following reasons:

■ Simple and cost-efficient implementation on the receiver side. It can be implemented using existing hardware, because the handling of real-time parameters could be implemented in software. Depending on implementation, time slicing could be adopted even in existing integrated receivers/decoders (IRDs) by updating only the system software.

- Simple and cost-efficient implementation on the network side. All required functionality can be implemented within the IP encapsulator.
- Delivering real-time parameters has no effect on the bit rate. Parameters can be delivered within a field called MAC_address, which is of limited use in broadcast applications.
- Backward compatibility. The current MPE specification specifies a method to allocate a part of the MAC_address field for other uses. The minimum length of the MAC address is 1 byte, allowing up to 5 bytes to be used for real-time parameters. In the case of time slicing, the filtering function may use the MAC address or IP address.

21.3.2.2 Receiver Burst Size and Off-Time Specification

As was described in section 23.2.3.1, the objective of time slicing is to reduce the average power consumption of the terminal and enable smooth and seamless service handover. The time-slicing configuration is of special importance for proper operation of the DVB-H receiver. The time-slicing process generates bursts of data that must fit in the memory available in the receiver. The data buffered in the receiver must be processed during the time between bursts. In a typical configuration, it can be assumed that a receiver can include up to 2 Mb for buffering an incoming burst. A receiver supporting the reception of multiple time-sliced elementary streams simultaneously may typically need a 2 Mb buffer for each time-sliced elementary stream. This specification is typical. Elementary streams can use smaller burst sizes if desired.

Figure 21.16 shows the burst size configuration parameters. The burst size indicates the number of network layer bits within a burst. The network layer bits consist of section payload bits. Each MPE and MPE-FEC section contains 16 bytes overhead caused by the header and CRC-32 correction. Assuming an average IP datagram size of 1 kB, the header and CRC-32 include 1.5 percent overhead. In addition, the transport_packet header causes overhead, which depends on the length of a section. If the length of a section is 1 kB, the overhead is approximately 2.2 percent.

The burst bit rate parameter stands for the bit rate used by a time-sliced elementary stream while transmitting a burst. Constant bit rate is the average bit rate required by the elementary stream when not time sliced. Both burst and constant bit rate include transmission of transport_packets (188 bytes in length). For a burst size of 1 Mb and a burst bit rate of 1 Mbps, the burst duration (time from the beginning to the end of the burst) is 1.04 s (due to the 4 percent overhead).

The off-time parameter stands for the time between bursts. During off-time, no transport_packets are delivered on the relevant elementary stream. During the on-time (i.e., while a burst is

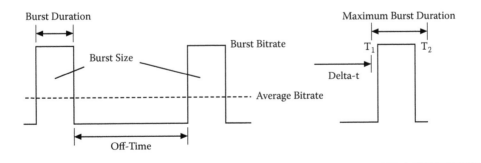

Figure 21.16 Burst parameters.

transmitted), transport_packets of other elementary streams may be also transmitted. This occurs when the burst bit rate is less than the bit rate of the transport stream (i.e., the burst uses only a part of the bit rate available on the transport stream).

In this case, the transport packets of the time-sliced and non-time-sliced elementary streams are multiplexed together on a packet-by-packet basis. This ensures that traditional DVB-T receivers, which receive non-time-sliced services, are not locked out from reception during a time-slice burst.

The maximum burst duration is the maximum duration of a burst, and shall be signaled for each time-sliced elementary stream. A burst shall not start before T_1 and shall end no later than T_2, where T_1 is the time indicated by Delta-t on the previous burst, and T_2 is T_1 + maximum burst duration. In poor reception conditions, the receiver may use this information to know when a burst has ended (timed out).

To enable a receiver to reliably distinguish bursts from each other, the next burst shall not start before T_2 of the current burst (i.e., Delta-t shall signal time beyond T_2). Distinction between bursts in a reliable way is required, especially when MPE-FEC is used.

Simplified formulas can be used to calculate the length of a burst and the length of the off-time to achieved a given power saving. These are given by equations (21.1) to (21.3), where B_d stands for burst duration (seconds), B_s for burst size (bits), B_b for burst bit rate (bits per second), O_t for off-time (seconds), S_t synchronization time (seconds), and C_b for the constant bit rate (bits per second). The correction factor 0.96 compensates for the overhead caused by transport_packet and section headers.

$$B_d = \frac{B_s}{B_b \cdot 0.96} \tag{21.1}$$

$$O_t = \frac{B_s}{C_b \cdot 0.96} \cdot B_d \tag{21.2}$$

$$P_s = \left[1 - \frac{\left(B_d + S_t + \frac{3}{4} D_j \right) \cdot C_b \cdot 0.96}{B_s} \right] \cdot 100\% \tag{21.3}$$

If the burst size is 2 Mb (over MPE and MPE-FEC section payloads) and the burst bit rate is 15 Mbps (over related transport packets), the maximum burst duration is 140 ms (from the beginning of the first transport packet to the end of the last one). If the elementary stream carries one streaming service at a constant bit rate of 350 Kbps, and MPE-FEC is not supported, the average off-time is 6.10 s. Assuming a synchronization time of 250 ms and a Delta-t jitter of 10 ms, a 93 percent savings on power consumption may be achieved. The Delta-t jitter has only a small effect on the power savings, as changing the value from 0 to 100 ms decreases the achieved power savings only from 94 to 92 percent.

Figure 21.17 shows how the burst bit rate increasing up to approximately 10 times the constant bit rate enhances the power savings. For a constant bit rate of 350 Kbps, increasing the burst bit rate from 1 to 2 Mbps increases the power savings from 60 to 78 percent (i.e., 30 percent). However, similar doubling on burst bit rate from 7 to 14 Mbps gives less than 3 percent benefit on power savings (91 to 93 percent).

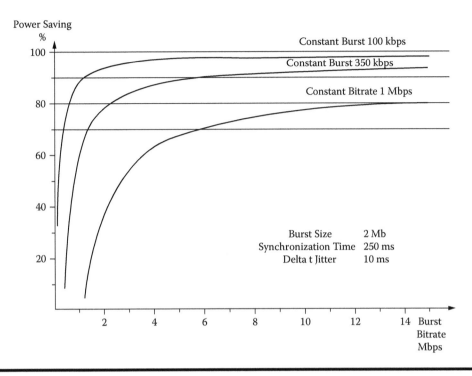

Figure 21.17   Relation between bust bit rate and power saving. From ETSI-ETR 102 377.[25] Copyright IEEE.

21.3.2.3   Complexity and Cost Considerations

From the cost and complexity point of view, the main component for the time slicing is 2 Mbit memory in the receiver.

When MPE-FEC is used, this 2 Mbit memory is reused and about 100 Kgates silicon implementation is needed for the MPE-FEC decoding. The complexity estimations assume pessimistically that full RS(255, 191) decoding is used. It should be pointed out that the MPE-FEC typically works with erasure-based RS decoding only, in which case the decoding can be significantly simplified with a consequent effect on the complexity, which can then be further reduced.

The additional complexity introduced by MPE-FEC is low, and the implement is straightforward. The additional complexity of combined time slicing and MPE-FEC should be well within the maximum 20 percent from the commercial requirements.

In addition, it should be noted that if a receiver does not have MPE-FEC or time-slicing functionality, it could be added later with full backwards compatibility. This is because both time slicing and MPE-FEC are backwards compatible with traditional IP delivery over MPE.

21.3.2.3.1   Time Slicing and Conditional Access

To support conditional access for DVB-H services, a fully IP-based conditional access system (IP-CAS) could be used. As all CAS-specific messaging would be on IP, the delivery of such messages could use time-sliced elementary streams, ensuring power savings for a receiver. Note, however, that the DVB-H environment does not necessarily support a bidirectional connection between the CAS and the receiver. The IP-CAS would need to support a broadcast environment, if a return channel is not supported by the DVB-H end-user equipment.

To support conditional access for DVB-H services, the DVB common scrambling algorithm on transport stream packets can also be used (DVB-CAS). A DVB-CAS uses entitlement control message (ECM) to deliver keys for descrambling. The delivery of ECMs is not time sliced; the receiver needs to get one ECM at wakeup to decipher the upcoming slice. Also, a typical DVB-CAS sends entitlement management messages (EMMs). These are time sliced.

In case DVB-CAS is used in the DVB-H receiver, the receiver must have the descrambling key before a burst of scrambled data is received. To do so, a receiver may switch on before the burst, to wait for an ECM.

The use of DVB-CAS has a slight effect on details described in section 21.3.3.2. Some modifications are required in equations (21.1) to (21.3), giving equivalent equations (21.4) to (21.6) when DVB_CAS is included in the system.

The new parameter ECM synchronization time (Ca) corresponds to the time required to receive an ECM before a data burst. Equations (21.4) to (21.6) give the burst length, off-time, and achieved savings on power consumption, where B_d stands for the burst duration (seconds), B_s for the burst size (bits), B_b for the burst bit rate (bits per second), O_t for the off-time (seconds), S_t for the synchronization time (seconds), C_b for the constant bit rate (bits per second), P_s for the power savings (percent) $(1 - B_s)*100$ percent, D_j for the Delta-t jitter (seconds), and C_a for the ECM synchronization time (seconds).

$$B_d = \frac{B_s}{B_b \cdot 0.96} \tag{21.4}$$

$$O_t = \frac{B_s}{C_b \cdot 0.96} - B_d \tag{21.5}$$

$$P_s = \left[1 - \frac{\left(B_d + S_t + C_a + \frac{3}{4}D_j\right) \cdot C_b \cdot 0.96}{B_s}\right] \cdot 100\% \tag{21.6}$$

Assuming a burst size of 2 Mb, the burst bit rate is 15 Mbps, the constant bit rate is 350 Kbps, the synchronization time is 250 ms, the Delta-t jitter is 10 ms, MPE-FEC is not supported, and the ECM synchronization time has the default value (100 ms), the achieved power savings would stay a little under 92 percent. For a 1 Mbps burst bit rate, the power savings would be 58 percent, for 2 Mbps it would be 76 percent, and for 7 and 14 Mbps it would be 89 and 91 percent, respectively. For simplicity, one could consider the effect of ECM synchronization time to be much the same as a slight increase of Delta-t jitter would cause.

21.4 Reference Receiver

DVB-H receivers share the same DVB-T physical layer. Differences can be summarized in the addition of the MPE-FEC in the link layer and by the handheld nature of the receiver. The payload in DVB-H is comprised by IP packets. In this section, we will report and point out the key operational parameters in a DVB-H receiver, highlighting relevant considerations.

First, we will focus on the carrier-to-noise ratio (C/N) gain parameter when the receiver is operating in a typical urban (TU) mobile channel. This gain depends on the Doppler frequency shift. At moderate Doppler between 10 and 90 Hz a DVB-H receiver exhibits a constant gain of 6 to 7 dB compared to a DVB-T receiver. At very low Doppler (in the order of a few hertz or less) the C/N requirement will raise as the virtual time interleaving of the MPE-FEC becomes shorter than the coherence time of the channel. The actual Doppler frequency where this happens is dependent on the length of the time-slice burst, which is roughly equal to the time-interleaving depth. Note that the C/N improvement is available even when MPE-FEC is applied to a nonmobile DVB-T demodulator.

C/N improvements in portable indoor and outdoor (pedestrian) cases are not as clear as in the TU channel profile. DVB has traditionally been using the DVB-Rayleigh channel to describe the portable reception conditions, and this chapter includes estimates for the theoretical DVB-H C/N performance in a 6-tap approximation of the DVB-Rayleigh channel. The effect of the MPE-FEC is, in this case, small, as no Doppler is present and the gain over the DVB-T figures is mainly coming from the fact that the error criterion is different from the quasi-error-free (QEF) used with DVB-T. It is, however, important to note that the experience from the field tests indicates that this static Rayleigh channel is not descriptive of the real portable indoor or outdoor channels, and the true benefit of the MPE-FEC is probably higher. Another approach to model the portable conditions is to use the TU channel with low Doppler frequencies. The experience from the field tests indicates that the TU channel is too pessimistic to describe the true portable conditions indoors or outdoors.

The MPE-FEC also improves tolerance to impulse interference. Laboratory tests have verified this effect, but it is very difficult to be quantified. However, it seems that the gain in C/I is higher in portable and mobile channels than in Gaussian channels, where only the impulses are present. This will probably emphasize the benefit of the MPE-FEC in portable and mobile reception conditions, as improvements are coming from both C/N and C/I.

It is expected that the noise figure of a handheld terminal with integrated antenna will be lower than in the current set top boxes. A noise figure of 5 dB has been specified for the reference receiver.

21.4.1 C/N Specification

The C/N is specified for two mobile DVB-H reference receivers. They have been derived from the measurement results of the DVB-H Verification Task Force and simulation results.[29] The first one is called typical receiver and should reflect a reasonable balance between performance and complexity expected from the first DVB-H receivers on the market. The second one, called possible receiver, reflects the already proven performance level, what can be achieved with the DVB-H specification, although with a higher level of complexity and cost. The performance for both reference receivers for the time-slicing parameters is given in table 21.4.

Table 21.4 DVB-H Reference Receiver Time-Slicing Parameters

DVB-H Burst Bit Rate	4 Mbit/s
MPE-FEC code rate	3/4
Number of rows in MPE-FEC	1,024
DVB-T/H bandwidth	7.61 MHz

The parameters in table 21.4 will result in approximately 0.5 s time-slicing burst duration. The degradation criterion has been 5 percent MFER. The MPE-FEC frame error rate (MFER) refers to the error rate of the time-sliced burst protected with the MPE-FEC. As an erroneous frame will destroy the service reception for the whole interval between the bursts, it is appropriate to fix the degradation point to the frequency of lost frames. Obviously the used burst and IP parameters will affect the final service quality obtained with certain fixed MFER, but experience has shown that the behavior is very steep and a very small change in C/N will result in a large change in MFER. MFER is the ratio of the number of erroneous frames (i.e., not recoverable) to the total number of received frames as given by equation (21.7). At least 100 frames shall be analyzed to obtain appropriate accuracy.

$$MFER[\%] = \frac{Erroneus\ Frames \times 100}{Total\ Frames} \qquad (21.7)$$

The service reception quality at the 5 percent MFER degradation point may not meet the QoS requirement in all cases. The criterion is nevertheless suitable for measurements, and a small 0.5 to 1 dB carrier power increase will improve the reception quality to less than 1 percent MFER.

The receiver should have the theoretical performance C/N given in table 21.5 for Gaussian channels, portable channels, and mobile channels, when noise (N) is applied together with the wanted carrier (C) in a signal bandwidth of 7.61 MHz. The values are calculated using the theoretical C/N figures given in EN 300 744[32] and the measured effect of the MPE-FEC. The difference between DVB-T QEF C/N and MFER 5 percent C/N is assumed to be 1.0 dB. Ideal transmitter and receiver are assumed.

The receivers should have the performance given in table 21.6 where a mobile channel is considered when noise (N) and Doppler shift (F_d) are applied together with the wanted carrier (C) in both typical receiver and possible receiver implementations.

All the receiver performance figures are specified at the RF reference point, which is the input of the receiver. Note that in some cases the GSM reject filter must be added in front of the receiver to prevent the high power from the GSM transmitter to enter the DVB-H receiver. Typically the insertion loss of the GSM rejection filter is in the order of 1 dB, raising the overall noise figure to 6 dB (see section 21.3.1).

The relation between field strength and input power is given by equations (21.8) and (21.9), where $\eta = 120\pi\ \Omega$:

$$E = \sqrt{4\pi\eta \frac{P_{in}}{G_a}} \cdot \frac{f}{c} \qquad (21.8)$$

$$E[dB\mu V/m] = P_{in}[dBm] - G_a[dB] + L_{GSM} + 77.2 + 20\log(f[MHz]) \qquad (21.9)$$

Table 21.5   C/N (dB) in Gaussian Channel and Portable Channel

Modulation	Code Rate	Gaussian Channel MPE-FEC CR = 3/4	Portable Channel MPE-FEC CR = 3/4
QPSK	1/2	2.5	3.9
QPSK	2/3	4.3	6.9
16-QAM	1/2	8.3	9.7
16-QAM	2/3	10.4	12.7

Table 21.6 C/N in Mobile TU Channel for the "Typical" and "Possible" Reference Receiver

Modulation	Code Rate	Bit Rate (Mbit/s)	C/Nmin (dB)	Fd3dB (Hz)	474 MHz	698 MHz
			"Typical" Reference Receiver			
QPSK	1/2	4.98	9.5	380	866	588
QPSK	2/3	6.64	12.5	360	820	557
16-QAM	1/2	9.95	15.5	340	775	526
16-QAM	2/3	13.27	18.5	320	729	495
QPSK	1/2	4.98	9.5	190	433	294
QPSK	2/3	6.64	12.5	180	410	279
16-QAM	1/2	9.95	15.5	170	387	263
16-QAM	2/3	13.27	18.5	160	365	248
QPSK	1/2	4.98	9.5	95	216	147
QPSK	2/3	6.64	12.5	90	205	139
16-QAM	1/2	9.95	15.5	85	194	132
16-QAM	2/3	13.27	18.5	80	182	124
			"Possible" Reference Receiver			
QPSK	1/2	4.98	8.5	520	1185	805
QPSK	2/3	6.64	11.5	520	1185	805
16-QAM	1/2	9.95	14.5	480	1094	743
16-QAM	2/3	13.27	17.5	480	1094	743
QPSK	1/2	4.98	8.5	260	592	402
QPSK	2/3	6.64	11.5	260	592	402
16-QAM	1/2	9.95	14.5	240	547	371
16-QAM	2/3	13.27	17.5	240	547	371
QPSK	1/2	4.98	8.5	130	296	201
QPSK	2/3	6.64	11.5	130	296	201
16-QAM	1/2	9.95	14.5	120	273	186
16-QAM	2/3	13.27	17.5	120	273	186

21.4.2 *Minimum Receiver Signal Input Levels for Planning*

Noise floor: The receiver should have a noise figure better than 5 dB at the reference point at the sensitivity level of each DVB-T mode. This corresponds to the noise floor power levels shown in table 21.7:

Table 21.7   DVB-H Reference Receiver Noise Floor

Channel	Noise Floor
8 MHz channels, BW = 7.61 MHz	Pn = −100.2 dBm
7 MHz channels, BW = 6.66 MHz	Pn = −100.7 dBm
6 MHz channels, BW = 5.71 MHz	Pn = −101.4 dBm
5 MHz channels, BW = 4.76 MHz	Pn = −102.2 dBm

Table 21.8   Minimum Input Levels

Channel BW	Pmin
8 MHz	−100.2 dBm + C/N (dB)
7 MHz	−100.7 dBm + C/N (dB)
6 MHz	−101.4 dBm + C/N (dB)
5 MHz	−102.2 dBm + C/N (dB)

21.4.3 Minimum Input Levels

The receiver should provide reference BER for the minimum signal levels (Pmin) stated below and higher, where C/N is specified in table 21.8 and depends on the channel conditions and DVB-H mode.

21.4.4 Antenna Design Issues for DVB-H Terminals

The antenna design in a small handheld terminal has to be an integral part of the terminal construction and will therefore be small when compared to the wavelength. If the antenna has to cover the whole wide tuning range of the UHF band, it probably has to be matched with a tunable matching circuit. The resistive part of antenna impedance (radiation resistance), which is to be matched to the receiver input impedance, will be rather small due to the small size of the antenna ($<1/10 \lambda$). This leads to rather high losses and to a low overall efficiency. Moreover, in this type of terminal the ground plane does not function anymore, but acts as a radiator. However, even the size of the radiating round plane is small compared to the wavelength, resulting in low radiation efficiency.

Another issue is the influence of the user on the radiation characteristic of the antenna. Depending on the relative position of the user to the handheld terminal, the human body could act as an absorber or a reflector.

Current understanding of the overall design problem indicates that the typical antenna gain at the lowest UHF band frequencies would be in the order of −10 dBi, increasing to −5 dBi at the end of the UHF band. Nominal antenna gain between these frequencies can be obtained by linear interpolation. In case a GSM 900 is used in a convergence terminal, the usable frequency range is limited to channel 49 (698 MHz) due to interoperability considerations. In case GSM 900 is not used, this limitation does not apply.

Generally, no polarization discrimination can be expected from this type of portable reception antenna, and the radiation pattern in the horizontal plane is omnidirectional. Typical gain of the antenna for planning purposes is presented in table 21.9.

Table 21.9   Typical Antenna Gain for Handheld Terminals

Frequency (MHz)	Gain (dBi)
474 (channel 21)	−10
698 (channel 49)	−7
858 (channel 69)	−5

21.4.5 Silicon Implementation

The traditional DVB receivers are based on a solution with multiple off-chip band selection filters,[10] and are not usable for handheld devices like mobile phones, where the physical size is of importance. The power consumption is of great concern due to the limited battery capacity for handheld receivers, and the use of low-cost CMOS is desirable. Several DVB-H receiver implementations have been reported in the literature: see, for example, DPS7040[33] and Drude and Klecha.[34]

In Drude and Klecha[34] the system architecture of an integrated circuit solution for TV-on-Mobile is described, and so the underlying design choices address the need for multistandard capability, ease of integration in the mobile phone architecture, ability to update in the field, small size and low power consumption, and ultimately low cost.

System-in-a-package technology (SiP) is used to combine a radio frequency tuner in BiCMOS technology with the digital channel decoder circuit in 90 nm CMOS technology in a single package. The decoding of the digital video stream is done in an application engine; its multistandard and multiformat handling capabilities are implemented by using programmable video and audio processor engines. The SiP solution for TV-on-Mobile is expected to consume only 40 mW when receiving DVB-H streams. It is targeted for a package with dimensions of 9×9 mm^2.

The system architecture for mobile digital television reception systems is based on three major blocks: the TV tuner, the channel decoder/demodulator, and the source decoder implemented on the application engine. Figure 21.18 presents an example of such a system for DVB-H.

The TV-on-Mobile module shown in figure 21.18 uses system-in-a-package (SiP) technology to integrate the TV tuner and the channel decoder/demodulator. SiP allows each functional block to be fabricated in the technology that serves it best, which is a clear advantage. Having different components on different dies permits a plug-and-play approach that would fit a range of markets. SiP implementation would give a more compact architecture because it provides high integration levels, including the antenna switch and power amplifier, altogether all passive components, multiple active dies, and specific RF components. The implementation is done in a single package: antenna signal going in, and digital data coming out.

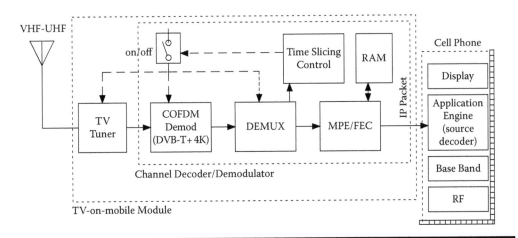

Figure 21.18 DVB-H system with the three main building blocks of a TV on mobile solution: TV tuner, the channel decoder, and the application engine needed for source decoding of the incoming stream.[33]

In some countries, the broadcasting of mobile TV operates in a frequency range close to the 900 MHz cellular network bands; therefore, a rejection filter has to be applied on the system level. Actual implementation requires an external passive component.

The channel decoder/demodulator is implemented in low-leakage CMOS 90 nm technology to reduce both active and standby power consumption.

To avoid duplication of multimedia functionality in a mobile phone, the source decoding is executed on the application engine. It interfaces with the module via a serial-parallel interface (SPI), which is compatible with a wide variety of application coprocessors and which makes it easy to integrate in the mobile phones. In case of DVB-H reception, the SPI data streams are entering the application engine as IP packages.

21.4.6 Silicon TV Tuner

The TV tuner is implemented in BiCMOS technology with a Zero-IF architecture to minimize the operating frequency for the on-chip channel-selectivity filters. This results in a reduced number of external components affecting both cost and size of the module. The silicon TV is designed for both DVB-T and DVB-H standards and consumes only 20 mW of power in DVB-H mode. It has a noise figure of 4 dB and in-band IP3 figure of –7 dBm. In the U.S. market, the tuner operates in the 1,670 MHz range, and for the European market in the UHF 470–862 MHz range.

21.4.7 Channel Decoder/Demodulator

An important function of the channel decoder in the context of suitability for low-power TV reception is the handling of time slicing. Time slicing is based on the time multiplex methodology of broadcasting in DVB-H. To take the most advantage of time slicing, a fast synchronization algorithm is implemented in the channel decoder.

Excellent Doppler performance (120 Hz) that guarantees stable reception under mobile conditions is achieved by applying Doppler compensation algorithms.

The Multi Protocol Encapsulation–Forward Error Correction (MPE-FEC) decoding algorithm,[29] specific for DVB-H, has been implemented for improving the performance during mobile reception.

The channel demodulation is based on an orthogonal frequency division multiplexing (OFDM) modulation scheme. In case of DVB-H, a 4K mode has been added to the standard to improve mobile reception. Based on decoding information, the channel decoder also provides information to the control processor about quality-of-service (QoS) parameters. Those parameters could be used to implement handover decisions. The channel decoder/demodulator function uses a digital signal processing (DSP) architecture, so it can be programmed to support multiple standards, including DVB-H/T and (in later versions) ISDB-T. Programming options support field upgradeability, so that the decoder can adapt to changing broadcast characteristics.

Other implementations reported[35,36] exhibit a low-power and area-efficient integrated CMOS architecture. In these cases, the receiver uses a low-IF quadrature downconversion for image rejection and is thereby insensitive to large flicker noise and DC offsets, normally associated with Zero-IF conversions. Another low-IF architecture, reported in Kampe and Olsson,[30] focuses on low power consumption. The total power consumption for the RF tuner is, in this case, estimated to be less than 20 mW, with a duty cycle of 10 percent. The receiver covers the receive bands from

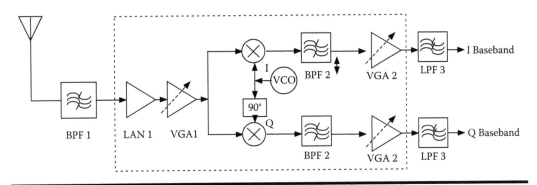

Figure 21.19 Low-IF receiver architecture.[30]

470 to 702 MHz, with an IF of 4.57 MHz. The proposed receiver meets the DVB-H requirements: the sensitivity is −88 dBm, the noise figure is 5.7 dB, and the adjacent channel protection ratio (ACPR) is −51 dB.

The noise figure and the linearity of a receiver are the two most important parameters to characterize the performance of the receiver in figure 21.19. The total noise figure is to a large extent determined by the first block in the receiver chain. The linearity, on the other hand, is usually dominated by the blocks at the end of the receiver.

The gain distribution of the receiver will seriously affect the overall performance. For a minimum noise figure, it is desirable to have a large gain in the beginning of the receiver. Unfortunately, this is in contrast to the linearity, for a trade-off between linearity and noise figure is required.

The high order of modulation combined with multicarrier makes the DVB-H receiver sensitive to intermodulation distortions, and the wide RF bandwidth makes the out-of-band intermodulation troublesome because the adjacent channels can be occupied by strong analogical signals. This results in strict requirements regarding linearity. Unfortunately, this also means high power consumption, as wide-bandwidth linear amplifiers are power hungry.

According to the DVB-H specification, the adjacent channels (N ± 1) can be up to 29 dB stronger, while the channels (N ± 2) can be up to 40 dB stronger. This is shown in figure 21.20.

In the worst case, the image channel is allowed to contain 39 dB more power. The required SNR for a handheld DVB-H receiver is 11.3 dB. The required image rejection ratio (IRR) is 60.3 dB. This means that a calibration of the amplitude and phase imbalance of the I and Q paths is needed to achieve the image rejection. Either a hardware[37] or a software[38] solution can be applied to solve this issue.

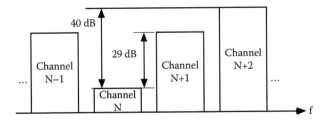

Figure 21.20 Channel power protection ratio.

21.4.8 Receiver Parameters and Network Design

Receiver characteristics have a major effect on the overall cost and performance of DVB-H networks. In the case of the DVB-H standard, a transmitter is required to provide a stated minimum signal quality. Receivers are expected to be able to decode that signal under certain stated environmental conditions. Both of these requirements are minima. In section 21.4 were introduced two receiver types: (1) typical reference receiver and (2) the proven performance receiver, also known as the possible reference receiver. Implicit in this specification is that different pieces of equipment—all fully compliant with the DVB-H standard—can have a very wide QoS disparity. The critical point is that seemingly small differences in equipment quality can have enormous consequences in network coverage and deployment cost. Homologation of higher-quality DVB-H receivers in a network has a direct effect in network planning by the:

1. Reduction of the number of base stations in a single-frequency network (SFN), that is, overall infrastructure cost is significantly reduced.
2. Improvement of the coverage area. The coverage area is extended without increasing the cost of infrastructure, just certifying high-quality receivers.
3. Bit rate/coverage trade-off. Network planning can be done from both points of view.

Let us further analyze these network planning considerations.

21.4.8.1 Reducing the Number of Base Stations

By reducing the number of base stations we can significantly reduce the total cost of the network. If the C/N of a given receiver is 1.5 dB better than the proven performance receiver and 3 dB better than the typical receiver C/N, the coverage area of each DVB-H transmitter is widened by approximately 50 percent. This is a key feature for the network operators because it reduces their capital expenditure dramatically. For example, an 80 percent indoor coverage simulation over Paris Indoor in UHF bandwidth with a cellular topography (several medium cells transmitting at 5 kW EIRP [equivalent isotropic radiated power] each) implies a minimum of 17 cells with a typical receiver and only 12 cells with the given receiver, that is, a 40 percent improvement.

21.4.8.2 Coverage Improvement

Another way to evaluate the benefit of a 3 dB gain is, for a given transmitted EIRP, to consider the related increase of covered population. A simulation with a single transmitter over Berlin (EIRP 100 kW) results in 940,000 covered people when using a reference receiver, instead of only 600,000 people watching TV on a basic typical receiver, that is, a potential end-user gain of 56 percent.

21.4.8.3 Bit Rate Increase

A 3 dB improvement on C/N can be used to increase the available bit rate. For example, instead of transmitting in a MUX only 6.65 Mbit/s using QPSK, CR2/3, GI1/4 modulation, a 3 dB gain in the receiver enables increasing the bit rate up to 9.95 Mbit/s through the use of 16-QAM, CR1/2, GI1/4 modulation. This 3.35 Mbit/s gain allows the transmission of approximately ten more video services.

21.4.8.4 Enhanced Doppler Compensation

Transmission impairments due to multipath interference are highly frequency and position specific. When a receiver moves even a few meters, multipath conditions can change radically. This is especially important in city areas. In DVB OFDM transmission, conditions on each subcarrier can change radically and rapidly when the receiver moves at even quite a slow speed. This tends to shift errors from subchannel to subchannel very quickly. The FEC algorithm of DVB-H corrects random bit errors very efficiently. Thus, movement of the receiver can significantly improve reception. (This effect is known by GSM users who have learned to walk briskly up and down when holding a conversation in a marginal reception area.) In city streets where multipath effects are severe, GSM works much better in a moving car than for a stationary user.

In DVB transmission subchannels are packed very close together and the receiver cannot reliably track Doppler frequency changes because changes can occur inside a single symbol period. This is reflected in interference by overlapping in adjacent channels. By example, a receiver able to compensate a Doppler frequency up to 130 Hz will guarantee watching DVB-H services at speeds up to:

- 185 km/h (114 km/h for typical receiver) in UHF in Europe (750 MHz, 8 MHz, 8k)
- 130 mph (80 mph for typical receiver) in L-band in the United States (1.67 GHz, 5 MHz, 2k)

When traveling in high-speed trains in Europe, the maximum speed of reception can reach up to 370 km/h (228 km/h for typical receiver) in UHF (750 MHz, 8 MHz, 4K).

This specification gives flexibility in the network design: Doppler effect is at its maximum when the receiver is moving directly toward or away from the transmitter. It is significantly reduced when the receiver is moving at an angle to the direct line of the transmitter. Thus, the network can be designed placing the transmitters carefully away from major roads or railway lines. This distance depends exclusively on the Doppler compensation specification of the receiver.

21.5 DVB-H Pilot Projects, Tests and Trials, and Market Outlook

DVB-H trials are an important element for system verification, frequency planning, and improving understanding of the interoperability with telecommunications networks and services. Since the early stages of DVB-H, extensive pilot projects and trials have been performed.[39] In fact, several pilot projects have already become commercial launches providing live television broadcast services that are currently under way or in development in many parts of the world. In fact, results from pilot tests of broadcast DVB-H mobile TV services among consumers in Finland, the United Kingdom, France, and Spain revealed clear consumer demand for these services and outlined future business models for commercial mobile TV services, so that it can be affirmed that television on mobiles has caught the attention of broadcasters and mobile network operators.

Just to provide the reader with an outlook, we highlight relevant trials at the time of this writing. These are summarized below.

21.5.1 DVB-H Trials

21.5.1.1 Australia

In Sydney, a year-long DVB-H trial in UHF began in July 2005. It was intended first to measure service coverage, and then expanded to include a commercial component with 375 users and

10 channels. Partners include the network operator Bridge Networks (Broadcast Australia), the mobile operator Telstra and Nokia, and others.

21.5.1.2 United States

Network operator Crown Castle has deployed a DVB-H pilot using a 5 MHz channel of L-band spectrum in Pittsburgh. This trial is evolving into a nationwide service provided by Modeo (owned by Crown Castle), beginning in New York City. A second service launch during 2007 was announced by SES Americom and Aloha Partners, named Hiwire Mobile TV. The service is set to begin trials in Las Vegas in 2007. Verizon announced a competing system based on QUALCOMM'S MediaFLO proprietary technology.

21.5.1.3 India

The Indian public broadcaster Prasar Bhati (Doodarshan) teamed with Nokia to start a DVB-H in Delhi. The trial is ongoing in various metropolitan areas to test the reception quality of the broadcast coverage. The full service is expected to be available from mid-2007.

21.5.1.4 China

In China two companies have been issued licenses by the government, Shanghai Media Group and China Central Television. Trials are currently under way, with service launch expected before the 2008 Beijing Olympic Games.

21.5.1.5 Germany

The Broadcast Mobile Convergence (BMCO) project conducted the first live broadcast of DVB-H services in May 2004 in Berlin within a public, digital terrestrial television network. T-Systems performed another trial in Berlin.

The media authority of the German state of Baden-Würtenberg announced the start of a new nationwide DVB-H pilot project at the beginning of 2007. The authority takes responsibility for the whole of the country. The commercial launch is expected in 2007.

Germany has also become the first country, apart from South Korea, to provide commercial T-DMB-based mobile TV services, which are available in 16 urban areas.

21.5.1.6 Holland

The first DVB-H trial took place during the IBC 2004 Exhibition. In July 2005, a second trial was launched in the Hague with several hundred "friendly" users. This trial includes the city center as well as some major motorways and railway lines leading to the Hague, so as to test the mobility of DVB-H. Trial partners include Digitenne, KPN, Nokia, and Nozema Services. At the CeBIT trade show in 2007, Samsung showed two mobile phones with DVB-H reception for introduction in the Dutch market when KPN will launch its DVB-H service in the country. The Dutch telecom operator has the sole DVB-H license after acquiring the Digitenne DVB-T platform and is now preparing a DVB-H launch for the second half of the year 2007.

21.5.1.7   Ireland

Mobile operator O2 and transmission company Arqiva are conducting the first consumer trial of mobile TV in Ireland in Dublin with 350 participants, selected from the O2 customer base. The users will have access to a package of 13 channels, including RTE1, RTE2, TV3, Sky Sports, Sky News, Setanta Sports, and the Discovery Channel. Interactive music and game channels will also be available. The DVB-H mobile TV standard will be used for the trial. Signals will be received on the Nokia N92 handset. The trial results will help determine how consumers responded to the next-generation TV services they were presented.

21.5.1.8   France

In Cannes, network operator TDF led two DVB-H trials in early 2005 in Metz and Paris. Nokia supplied the service system and receivers for these demos. Nationwide trial is expected to begin during 2007.

21.5.1.9   Hungary

The broadcast infrastructure provider Antenna Hungaria and the mobile operator T-Mobile Hungary have started testing DVB-H-based mobile TV. The test, focused in interoperability, includes versions of the four Hungarian public TV channels, M1, M2, Duna, and Duna 2. Commercial service launch is expected in one-year time frames if the test results are positive.

21.5.1.10   Sweden

In Sweden there have been three different trials in Stockholm and Gotenburg, each one of them led by a different operator (Viasat, Telia, and Teracom).

The Swedish Telco Telia set up a DVB-H-based trial for mobile TV covering the European Athletics Championship in Gotenburg. The test broadcast showed five TV channels on Nokia handsets.[41]

The DVB-H trial run by Teracom, the Swedish terrestrial broadcaster, was completed in early 2007, with nine of ten users saying they liked the service and had an overall positive opinion. Most viewing was done between 18h00 and 24h00 from Monday to Friday. Users also had a positive opinion of both the picture and sound quality. A large number also expressed an interest in being able to request the streaming of particular programs on demand. Fifty-seven percent of users would be prepared to pay up to approximately 4€ per month for the service. Teracom has to wait for a government decision on which frequencies would be made available for a future mobile TV network.

21.5.1.11   United Kingdom

Trials were conducted in both Oxford and Cambridge. In Oxford the commercial pilot supplied 350 test users with Nokia DVB-H receivers with telephone functionalities. Project partners include Arqiva, O2, Nokia, and Sony Semiconductors & Electronic Solutions. The main results showed that 83 percent of consumers expressed satisfaction with the service, which offered 16 branded TV channels. Seventy-six percent said they would subscribe for a year at €15 a month.

But consumers also made clear that 16 channels was the minimum number they would want, and the majority wanted many more.

The United Kingdom tested not only DVB-H but also the competing DAB-IP. Both trials of DVB-T and DAB-IP showed a demand for a full-scale offering of multichannel television to mobiles.

Both the Arqiva/O₂ DVB-H and BT Movio DAB-IP trials showed that consumers want full-simulcast television-to-mobile handsets rather than short-form TV snacks. The second most popular place to view mobile content turned out to be in the home.

BT Movio launched its UK commercial service with Virgin Mobile in October. The initial TV line-up includes BBC One, ITV1, Channel 4, and E4.[40]

In any case, BT Movio indicated that it chose DAB-IP because it was the only format in the United Kingdom with available spectrum. Spectrum for the DVB-H format has yet to be allocated in the United Kingdom by the regulator Ofcom.

21.5.1.12 Spain

In Spain, the government approved DVB-H pilot projects in Barcelona, Madrid, Seville, Valencia, and other cities. Partners include the network operator Abertis, Nokia, and Telefónica Móviles.

Besides, Telefónica Móviles has launched an interactive mobile TV trial in the city of Alcazar de San Juan. It is the first DVB-H-based experience in Spain to test interactive services through mobile TV. Thirty-five people, equipped with Motorola phones, will enjoy several interactive mobile TV services, which include live participation in quiz shows, chatting through the Net while watching mobile TV, or choosing mobile TV content. Telecom Castilla-La Mancha will provide the infrastructure to carry the signals, and Telefónica Móviles the service and interactive platform. Research among Telefonica Móviles clients revealed that more than 55 percent would be willing to pay for mobile TV services. Seventy-one percent said that they have already watched digital TV through their mobile phones for between 15 and 20 minutes, and 17 percent said they have done so for more than 25 minutes.

On the regulatory side, in Spain there is currently a process of public consultation relative to the regulatory approach that should be adopted for the development and deployment of mobile TV services. The Spanish Ministry of Industry opened public consultations with all key players involved in the market up to April 2007 aimed at laying the groundwork to speed up the commercial launch of the service by year's end. The industry has been called on to provide its opinion about how to develop the mobile TV market at a time when the government is drafting the new mobile TV regulation.[42] However, it does not seem likely that there will be a formal approval of the regulation before mid 2008.

Many other trials (Malaysia, Singapore, Switzerland, Qatar, Russia, etc.) are under way worldwide. A comprehensive list can be found at the DVB-H official Web site.[39]

21.5.2 From DVB-H Trials to Services

As it has been explained, a considerable number of DVB-H trials and pilot projects have been deployed and performed throughout Europe notably, but also elsewhere. Some of these trials have already evolved into the commercial service phase.

Results from pilot tests of broadcast DVB-H mobile TV services among consumers in Finland, the United Kingdom, France, and Spain revealed clear consumer demand for these services and pointed to future business models for commercial mobile TV services.

DVB-H networks have been commercially launched notably in Italy, Finland, Vietnam, and Albania, and South Africa, France, Germany, and Spain plan nationwide commercial services during 2007. The upcoming analogue switch-off processes across Europe will release spectrum in the UHF bands, thus enabling the widespread deployment of DVB-H networks.

In Albania, DigitAlb launched nationwide commercial service on December 20, 2006. In Vietnam, VTC launched the service on December 21, 2006.

21.5.2.1   The Italian Case

In Italy the trial was performed in Turin, and the first commercial service launch was in June 2006, with already hundreds of thousands of subscribers. Currently there are three commercial competing services in operation: 3 Italia since May 2006, Telecom Italia Mobile (TIM TV) since June 2006, and Vodafone since December 2006. Broadcasters RAI, Mediaset, and Sky are also involved.

The 3 Italia mobile TV platform states that the introduction of its DVB-H service in June 2006 is a commercial success, attracting over 250,000 subscribers within a couple of months. 3 Italia offers nine services for 29€ a month; in addition, there are three premium services. Most usage is outdoors (90 percent), and sports are the best-watched programming. Peak viewing time is during lunch and just before dinner.

Mobile TV take-up in Italy is already showing promise, and the regulation of mobile television in Italy is under review.

21.5.2.2   Finland

In Helsinki, Finland, Finnish Mobile TV launched the first commercial pilot for DVB-H services in 2005. As part of the pilot, 500 test users accessed television services on Nokia receivers. Project partners included Nokia for the technology, the main television broadcasters for the content, the two main telecom operators, and Digita, the broadcast network operator for the network. In August 2005, a similar trial was held during the World Championships in Athletics.

Finally, the license to operate a DVB-H network was awarded to Digita, who built the network, in March 2006. The network has a coverage area of 25 percent of the population in Helsinki, Oulu, and Turku. The service was supposed to be launched on December 2006; however, the commercial service is delayed until the end of 2007 according to information provided by YLE, the Finnish public broadcaster of information, who cited a combination of copyright issues of the broadcast material and the limited number of DVB-H-enabled handsets in the market.

For the mobile TV market to develop there are two main obstacles. The unavailability of spectrum is the largest barrier to the launch of more mobile TV services, particularly in Europe, the research firm says. Over the next ten years, as more spectrum is made available, in many cases when analogue TV signals are shut off, more mobile TV broadcast services will launch. Another issue limiting the market today is the small number of mobile TV broadcast-enabled handsets available in many markets.

The European Commission is to set a European standard for mobile TV, and DVB-H is the system of choice, and by making it mandatory, it will help grow the market further.

In Europe 17 countries have now chosen DVB-H as the standard; 5 countries also support DMB (including Germany and the United Kingdom).

21.5.3 DVB-SH Trials

Regarding DVB-SH, the first demonstration of the system was performed in the United Kingdom by Alcatel using Sagem myMobileTV terminals in December 2006, and with the participation of UK broadcasters Sky, ITV, and BBC.[44,45]

Alcatel also demonstrated two key technical features using the DVB-SH standard: antenna diversity in reception, using two antennas inside the same mobile device, thus enabling improvements in the signal quality under difficult conditions, and time interleaving that helps overcoming fading impairment in mobility conditions.

Alcatel gave also another demonstration of DVB-SH service at the 2007 3GSM Congress in Barcelona and, more recently, is demonstrating in Germany seamless access on a single device to a selection of mobile TV channels delivered via either 3G or broadcast networks, including German language channels RTL Mobile TV, National Geographic Channel, and Eurosport. The demonstration is based on the DVB-SH standard.

For DVB-SH technology, Alcatel is the company leading the creation of momentum, and so far an announcement has been made that both Eutelsat and SES Astra agreed on the launch of an S-band satellite (W2Aat 10 degrees East) covering Europe in 2009 through a 50/50 joint venture. Eutelsat has commissioned the W2A satellite from Alcatel Alenia Space, and the 2.2 GHz S-band payload is in addition to several Ku-band and C-band transponders on the spacecraft. The satellite will be designed for 15 years of useful life and is expected to provide sufficient power for indoor penetration of mobile TV across several countries, including France, Germany, Italy, Spain, and the United Kingdom.[43]

DVB-SH satellite services are expected to become operational with full-scale commercial deployment in 2009, but maybe DVB-SH operations will start earlier with terrestrial networks in certain regions of the world (in fact, the first DVB-SH terrestrial repeaters are expected to reach the market by late 2007). Other agents in this market include the chip maker DiBcom, who is designing a chipset that will be compatible with the DVB-SH standard; Sagem, who is developing DVB-H phones that support both UHF and S-band; and Samsung and ARCHOS, who are to develop mobile handsets and portable multimedia players, respectively, that will be compatible with the DVB-SH standard in the S-band.

Strategic for the right development of DVB-SH is the approval of a EU Commission decision harmonizing radio spectrum around the 2 GHz frequency use for mobile satellite service systems. This decision is expected to give industry the necessary confidence to invest in such services by establishing a stable regulatory framework, and will mean that DVB-SH will find a clear regulatory path to become a universal standard to provide a pan-European mobile TV service.[46]

Links

1. DVB, Digital Video Broadcasting: http://www.dvb.org/.
2. DVB-H, DVB–Handheld: http://www.dvb-h.org/
3. DVB-T, DVB–Terrestrial: http://www.dvb-t.org/.
4. DAB, Digital Audio Broadcasting (Eureka-147): http://www.worlddab.org/.
5. DMB, Digital Multimedia Broadcasting: http://www.t-dmb.org/.

References

1. ETSI. *Radio broadcasting systems; Digital audio broadcasting (DAB) to mobile, portable and fixed receivers.* ETSI ETS 300 401.
2. *Radio broadcasting systems; Digital audio broadcasting (DAB) to mobile, portable and fixed receivers.* ETSI EN 300 401 (DAB, Eureka-147). T-DMB (Korean DTTB standard, based upon DAB).
3. *Transmission system for digital terrestrial television broadcasting.* ARIB STD-B31 V1.5. ISDB-T.
4. *Digital video broadcasting (DVB); Transmission system for handheld terminals.* ETSI EN 302 304. DVB-H.
5. QUALCOMM Inc. 2007. FLO *technology overview. Revolutionizing multimedia.*
6. ATSC. http://www.atsc.org.
7. H. Holma and A. Toskala, eds. 2006. *HSDPA/HSUPA for UMTS.* New York: John Wiley & Sons, Ltd.
8. Mobile TV. *Euromedia*, January–February 2007, pp. 16–22.
9. DVB Project. 2004. *DVB-H handheld. IP broadcasting to handheld devices based on DVB-T.* White paper.
10. DVB Project. 2007. DVB-H. Broadcasting to handhelds. The global technology standard for mobile television. Fact sheet.
11. Wikipedia. DVB-H entry. http://en.wikipedia.org/wiki/DVB-H.
12. European commissioner backs DVB-H technology. Appeared on IBE Network News. March 30, 2007.
13. DVB-H guide IBC2006. http://www.dvb-h.org.
14. DVB Project. 2007. DVB-IPDC. Internet protocol datacast. Making mobile TV happen. Fact sheet.
15. *Digital Terrestrial Television Action Group. Television on a handheld receiver. Broadcasting with DVB-H.* DIGITAG 2005. http://www.digitag.org.
16. *Satellite services to handheld. Mobile digital TV in S-Band.* DVB-SH. Fact sheet published by the DVB Project office. 2007.
17. DVB approves DVB-SH specification. DVB Project press release. February 14, 2007.
18. *System specifications for satellite services to handheld devices (SH) below 3 GHz.* DVB document A110. 2007.
19. *Framing structure, channel coding and modulation for satellite services to handheld devices (SH) below 3GHz.* DVB document A111. 2007.
20. Alcatel-Lucent and NXP semiconductors strengthen their collaboration for broadcast mobile TV in the S-Band. Alcatel-Lucent press release. February 14, 2007.
21. Alcatel and DiBcom cooperate for unlimited mobile TV. Alcatel press release. April 25, 2006.
22. SES and Eutelsat announce joint investment to serve markets for mobile broadcasting and other communications services in Europe. SES and Eutelsat press release. October 30, 2006.
23. *DVB-H implementation guidelines.* DVB document A092. 2005.
24. H. Jenkač, T. Stockhammer, W. Xu, and W. Abdel Samad. 2006. Efficient video-on-demand services over mobile datacast channels. *Journal of Zhejiang University-Science A*, 7(5):873–884.
25. G. Faria, J. A. Henriksson, E. Stare, and P. Talmola. 2006. DVB-H: digital broadcast services to handheld devices. *Proceedings of the IEEE*, 94(1):194–209.
26. *Implementation guidelines for DVB-H services.* ETSI technical report 102 377 v1.2.1. 2005.
27. *Extension for digital storage media command and control.* ISO 13818-6.
28. *DVB specification for data broadcasting.* ETSI EN 301 197.
29. ETSI. 2001. *Digital video broadcasting (DVB); DVB-H implementation guidelines.* ETSI-ETR 102 377 v1.1.1.

30. A. Kampe and H. Olsson. 2005. A DVB-H receiver architecture. In *23rd NORCHIP Conference*, 265–68.

31. *Mobile and portable DVB-T radio access interface specification.* EICTA/TAC/MBRAI-02-16.

32. ETSI. *Digital video broadcasting (DVB); Framing structure, channel coding and modulation for digital terrestrial television.* EN 300 744.

33. *Single-chip RF tuner—Data sheet LSI logic.* 2004. DPS7040. http://www.lsilogic.com.

34. S. Drude and M. Klecha. *System aspects for broadcast TV reception in mobile phones.* Eindhoven, The Netherlands: Philips Semiconductors.

35. M. Dawkins, A. Payne Burdett, and N. Cowley. 2003. A single-chip tuner for DVB-T. *IEEE Journal of Solid-State Circuits* 38:1307–17.

36. S. Sanggyu, R. Kuhn, B. Pflaum, and C. Muschallik. 2002. A three-band-tuner for digital terrestrial and multistandard reception. *IEEE Transactions on Consumer Electronics* 48:709.

37. L. Der and B. Razavi. 2003. A 2-GHz CMOS image-reject receiver with LMS calibration. *Solid-State Circuits* 38:167–75.

38. A. Schuchert and R. Hasholzner. 2001. A novel IQ imbalance compensation scheme for the reception of OFDM signals. *IEEE Transactions on Consumer Electronics* 43:313–18.

39. Most of the information about the trials has been gathered from either the official DVB-H Web site (http://www.dvb-h.org/services.htm), where there is available a comprehensive list of trials and services launches, or else any other industry magazines, sources, and e-mailing lists:

    Euromedia: http://www.advanced-television.com
    TVB Television Broadcast: http://digitaltelevision.com
    Broadband TV News: http://www.broadbandtvnews.com
    Broadcast Engineering: http://broadcastengineering.com
    CabSat Daily: http://www.informamedia.com
    European Communications: http://www.eurocomms.co.uk

40. N. Radlo. 2006. Mobile television. *Broadcast Engineering World*, pp. 10–13.

41. European mobile TV market needs room to breathe. *Euromedia*, September 2006, p. 10. Championship trial for Telia mobile TV. *Euromedia*, October 2006, p. 13.

42. http://www.mityc.es/Telecomunicaciones/Herramientas/Novedades/ConsultaTDTmovil.htm (Spain's Ministry of Industry Web site).

43. http://www.newvideobusiness.com (New Video technology magazine news Web site).

44. http://www.umtsforum.net.

45. SatNews Daily. http://www.satnews.com.

46. Commission Decision 2007/98/EC on the harmonized use of radio spectrum in the 2 GHz frequency bands for the implementation of systems providing mobile satellite services. EU Radio Spectrum Policy site: http://ec.europa.eu/information_society/policy/radio_spectrum/index_en.htm; http://ec.europa.eu/information_society/policy/radio_spectrum/ref_documents/index_en.htm. February 14, 2007.

Chapter 22

Video Streaming and Related Applications in the Mobile Environment

P. Fouliras

Contents

Keywords

video streaming, broadcasting, T-DMB, DVB-H, IPDC, MediaFLO, applications, services, economics, regulatory

Mobile devices such as PDAs and mobile phones in particular have witnessed an explosion in their popularity among consumers during the past decade. Many factors have contributed to this: small device size and weight, advances in mobile telecommunication technology, and longer battery time. Although many more features and services are constantly added for the users of such devices, the most popular have been voice communication and SMS; it is only relatively recently that the inclusion of cameras and the availability of ample bandwidth have triggered the popularity of other, multimedia-based services such as MMS and live video broadcasting.

Experience shows that some of the principal factors that may limit the popularity of the various services are cost, the small size of a typical mobile phone screen, and battery life. Advances in energy savings, for example, have been counterbalanced by the increase in processor and memory cycles due to the new features: games with three-dimensional (3D) graphics, picture, audio and video encoding/decoding employing complex compression algorithms—all of which are power hungry.

Nevertheless, the new services are here to stay, and video streaming is one of the basic building blocks to achieve this goal. Many recent studies appearing in daily press show that video on demand and satellite video on demand increased in EU households between 2001 and 2006 by an annual rate of 109 percent, reaching 10 million households at the end of 2006. In addition, the respective expenses by those households have increased from 5 million Euros (2001) to 541.67 million Euros (2006). It is also estimated that by the end of 2009 this sum will rise to more than 1.39 billion Euros under "conservative assumptions." Video streaming, in particular, can be combined with other elements to provide more sophisticated services, such as video conferencing, mobile TV, or even social networking.

In this chapter we start with the presentation of the basic aspects of video streaming, first in traditional wired networks (IP based) and then in wireless networks, focusing on the standards examined in this book, namely, ISDB-T, DVB-H, DMB, and MediaFLO. Some of the most important performance parameters are presented next, affecting the design of end users' devices, which in itself represents a large market. Additional video-streaming terms and basic services are presented next, followed by examples of more advanced applications. To make the presentation complete, we then look at economic considerations and the challenges that arise from the regulatory environment.

22.1   Video-Streaming Basics

The term *multimedia streaming* typically represents multimedia displayed or processed in any way on the end user's system while being received from the sender. Video streaming is the most common manifestation.

The benefits of this approach for a client are minimal waiting time as well as memory/disk space requirements. Furthermore, it is possible to broadcast live video, because both sending and receiving ends introduce minimal delays. On the other hand, the workload in terms of processor time may be increased, because encoding and decoding must be performed on the fly. Even more important, the jitter introduced by the intermittent network nodes must not exceed a predetermined amount or the result will be unacceptable. These problems were first studied in wired, packet-switched networks, the experience of which proved useful in the wireless environment.

The basic elements of a video-streaming system are the video server(s) and the clients. Although it is possible, as for other cases, to have video streaming over communication networks without packets (e.g., circuit-switching networks), packet switching is more efficient, because it allows better overall use of network resources. From the server viewpoint, broadcast scheduling methods may be broadly classified into two classes: push and pull. Under push-based scheduling, the clients continuously monitor the broadcast process from the server and retrieve the required data without explicit requests, whereas under the second class the clients make explicit requests, used by the server to make a schedule that satisfies them. Typically, a hybrid approach is employed with push-based scheduling for popular videos and pull-based for less popular.

In the case of wired IP networks, video is typically streamed over Real-Time Transport Protocol (RTP)/User Datagram Protocol (UDP) as the session/transport protocols, together with Real-Time Transport Control Protocol (RTCP), primarily for measuring and reporting stream statistics, and Session Description Protocol (SDP), for media description and announcement. Nevertheless, IP is the core protocol, which was not designed for carrying time-critical data. Hence, in the case of congestion, very little can be done in terms of quality of service (QoS) unless the intermittent network nodes (border routers of autonomous systems in particular) are configured to take appropriate action. Such action typically involves the assignment and policing of priorities to IP packets depending upon some predetermined mechanism, resulting in some packets being dropped more often than others.

One such general mechanism is integrated services (IntServ), which resembles telephone communication networks: routers along a specific path first make certain that there are available network resources, reserved prior to the actual data transfer. This approach, however, suffers from scalability because it is per-flow based.[35] Differentiated services (DiffServ), on the other hand, is a mechanism under which designated routers (typical at the borders of an autonomous system) forward ingress packets according to their priority tag value, regardless of their origin. This approach is per-packet tag based, and hence scalable, although still not a complete solution to the problem.

Live video streaming (in the sense that the video content is generated in real-time at the source) is more demanding, because the respective codecs must complete encoding on the fly. Even if this may be relatively easy at the server side, it must be carefully planned at the client side, taking into account all the complexities involved (e.g., minimum available processor power, network heterogeneity, battery power consumption, etc.).

It is also possible to use multicast transmissions, given that IP inherently supports them. Nevertheless, for any approach based on multicast to work, all intermediate routers must support it; moreover, multicast does not always work in the case of heterogeneous networks and client types. To address this challenge, the normal solution is to allow for multiple multicasts, but this approach rules out the case for packet retransmission and is very difficult to coordinate, especially in wide-area networks.

The peer-to-peer (P2P) model of communication has been successfully used for file-sharing applications. Hence, it is of no surprise that considerable research effort has been directed to its adoption for video streaming. Due to its nature, however, P2P is more applicable for popular, stored videos and is not scalable for real-time events. This is because for P2P to be effective, many participating nodes must exist at the required time with a copy of the required video portion in their buffers. In addition, each packet hop at successive nodes incurs a time delay; hence, long P2P "chains" are not suitable for real-time events because the aggregate delay from the source to the last destination may prove prohibitive, especially in the case of an interactive service.

22.2 Video Streaming in the Wireless Environment

There are more challenges in the wireless environment due to the diverse nature of the underlying technologies and device capabilities involved. By just extending IP in the wireless environment, additional problems emerge. Hence, direct approaches such as mobile IP are not an answer for video streaming.

Furthermore, wireless communication networks impose additional complexity, because the probability of lost or damaged packets is significantly higher than that over wired networks. Moreover, there is a large variety of mobile devices that may be used by the end users with diverse capabilities: set-top boxes or laptops can receive video streams of high quality and detail without power consumption considerations (e.g., when in commuter trains with power outlets available), but so can battery-powered devices like PDAs and cell phones with small displays. In general, the requirements for the mobile environment can be summarized as follows:

- **Minimum power consumption** to support battery-powered devices effectively
- **Adequate QoS level** even in the case of handover from one cell area to another, and when the mobile client device moves at high speed
- **Support for different transmission rates and respective codecs**
- **Advanced error correction mechanism**, which does not depend on packet retransmission

The first requirement can be addressed in several ways, one of which is time division multiplexing: multiple broadcasting channels are multiplexed into one physical channel, so that the receiver is actively receiving packets for only a fraction of the multiplexing period. The trick here is that these packets correspond to video content spanning the whole of the multiplexing period. In this way, most of the time the receiver circuit does not consume power.

The second requirement is of obvious importance if appealing media streaming services are to be offered. Examples where such conditions may appear are train commuters moving at high speed or users who simply pass from one cell area to another.

The third requirement is a direct consequence of the diversity of mobile device capabilities (e.g., different screen sizes or processing power, decoding complexity, etc.) and reception quality, with MPEG-4 and H.263 (or the most recent H.264) as the typical encoding formats. This diversity is presented more clearly in table 22.1.[3]

Finally, the last requirement increases reception reliability by avoiding delays due to packet retransmission.

Table 22.1 Mobile Consumer Devices

	Device	Screen Size	Power Consumption	Input Method
Portable receivers	Receivers in cars	800 × 600	Not critical	Touch screen
	Laptops	1024 × 768	Not critical	Keyboard
Handheld receivers	PDA	320 × 240	Critical	Thumb keyboard
	Mobile phones	192 × 176	Critical	Keypad

According to Hartung et al.,[32] broadcasting enforces fixed schedules that are acceptable by traditional TV viewers, something that is not valid for mobile TV; such viewers prefer to view programs only a few minutes per day (5 min on average) as known from trials, with a realistic upper limit of 30 minutes. Personalized content and limited viewing time lead to a more on-demand access model, which in turn dictates that such users are not likely to be satisfied with traditional-type broadcasts because they impose fixed schedules.

Therefore, unicast streaming approaches like using packet-switched streaming (PSS) in third-generation (3G) networks are far better suited for mobile TV users, except for cases or real-time events of particular interest from a single provider. PSS also defines a Synchronized Multimedia Integration Language (SMIL)–based application for multimedia presentations, combining video streams with downloaded images, graphics, and so on. Channel switching in PSS, however, is inadequate, because it takes 8–10 s to switch from one channel to another.

Again, as for the wired network environment, a hybrid approach of broadcasting for popular content and unicasting for less popular seems to be the optimal approach in terms of network resources, simultaneous content availability, and user satisfaction.

Despite the above and other problems, several novel proposals have been standardized for use in the wireless environment, most of them offering two-way communication, so that some form of interaction may become available to the user. Data broadcasting has also been added to multimedia streaming, making added-value services possible and customer subscription more attractive. We examine some of them in the following section.

22.3 Digital Wireless Broadcasting Standards

ISDB-T (Integrated Services Digital Broadcasting–Terrestrial) is a standard for digital terrestrial TV and radio developed in Japan. It was launched commercially in Japan in 2003. Outside Japan, extensive tests were conducted in Brazil resulting in the adoption of this technology.[26] ISDB-T uses MPEG-2 for video and audio, as well as data broadcasting. It also provides interaction indirectly, realizing the upstream channel using other media (e.g., WiFi). One distinctive characteristic is that the signal is always encrypted, although no charge is made. It also supports rights protection and management as other broadcasting standards in various ways. This is important in the case that a content provider expects to generate income directly from its subscribers rather than from other means (e.g., advertisements).

No other country apart from Japan and Brazil has officially adopted ISDB-T at present. A special decoder card is required for content decoding, which is a matter for concern. Another issue for ISB-T, overall, is that its version for low-powered mobile devices (e.g., cell phones and PDAs), called ISDB-T 1-seg, has not been successful. Its protocol stack is shown in figure 22.1. Note the PSI (program standard information), SI (service information), and ESG (electronic service guide).

DVB (Digital Video Broadcasting) is a family of standards published by the European Telecommunications Standards Institute (ETSI) to implement digital television networks. There are three main variations depending on the media used for broadcasting: satellite (DVB-S), terrestrial (DVB-T), and handheld (DVB-H). Here we must note that although DVB-T supports mobile reception, the specific requirements of mobile devices, especially in terms of power consumption, have led to DVB-H. The protocol stack for DVB is shown in figure 22.2 and for DVB-H in figure 22.3. For the latest DVB publications, the interested reader is referred to the respective ETSI portal.[39]

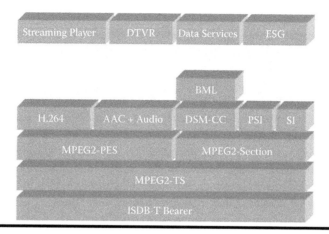

Figure 22.1   ISDB-T 1-seg protocol stack.

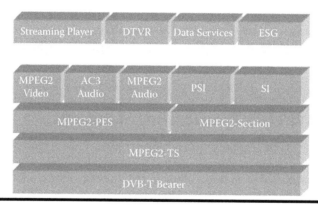

Figure 22.2   DVB protocol stack.

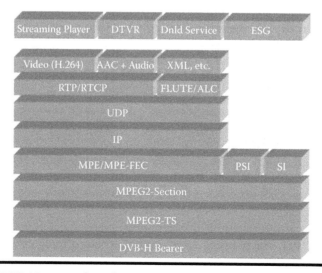

Figure 22.3   The DVB-H protocol stack.

DVB-MHP (DVB Multimedia Home Platform)[10] is an open middleware system designed for interactive digital television by the DVB Project. Its purpose is to facilitate the development of interactive Java applications for reception and execution on a TV over DVB. Applications can be of any type, ranging from e-mail to secure transactions for e-commerce and games. The main appeal of DVB-MHP is that it does not need extensive support from manufactures, as in the case of vertical set-top boxes; a single middleware platform can be deployed and supported at a shorter time and lower cost. As such, it will be interesting to see whether it will expand to cover handheld devices. As of September 2006 MHP has been "on-air" mainly in several European countries.

DVB-H achieves its goals by the introduction of several novel characteristics, such as:[3,30]

- **Time slicing.** Reduces the average power consumption of the mobile device and makes smooth handovers possible.
- **MPE-FEC** (Multi-Protocol Encapsulation–Forward Error Correction). For improving C/N Doppler performance in mobile channels and tolerance to impulse interferences.
- **TPS** (extended Transmission Parameter Signaling). Allows transmission of enough information about the services carried by the multiplex, thus enhancing and speeding up service discovery.
- **5 MHz channel bandwidth operation**, outside the traditional broadcasting bands.
- **Use of H.264** (MPEG-4 Part 10) and CIF resolution for video streaming, resulting in a maximum bit rate of 384 kbps, while allowing for an aggregate downstream rate of 10 Mbps. Consequently, many audio and video channels may be multiplexed, and other services such as file downloading can be offered.

IPDC (IP Datacast over DVB-H) is an end-to-end broadcast system for delivery of any type of digital content and services using IP-based mechanisms optimized for devices with limitations on computational resources and battery.[9] An inherent part of the IPDC system is that it comprises a unidirectional DVB broadcast path that may be combined with a bidirectional mobile/cellular interactivity path. Hence, encapsulated IP datagrams over MPEG transport streams are used. A simplified view of the protocol stack of DVB-H IPDC is shown in figure 22.4 and a typical system in figure 22.5.[40]

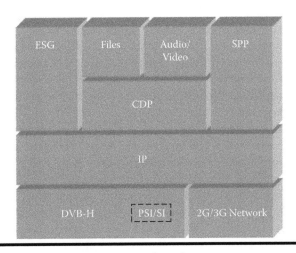

Figure 22.4　A simplified DVB-H IPDC protocol stack.

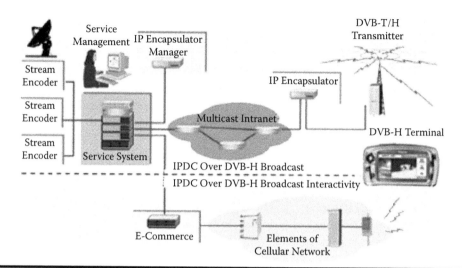

Figure 22.5   A typical DVB-H IPDC system.

An issue involving considerable research for DVB-H as well as for other standards is the handover speed. This is important not only for the video streaming itself, but also for datacasting (e.g., IP streams). Time slicing, employed in DVB-H, has many advantages, as stated above, one of which is the possibility for smooth and seamless service handover by switching from one transport system to another during the period that is the off-period of the receiver. Nevertheless, when a user wishes to change channels quickly (zapping), the necessary time increases, having a negative effect on user satisfaction.

The zapping time is the aggregate result of the time to tune into a new frequency, synchronize to the respective broadcast stream and receive all its metadata, and wait for the beginning of the next group of pictures of the elementary stream. This problem should not manifest itself to a delay of more than 1.5 seconds for video channel zapping. One recent proposal[20] is to use a combination of DVB-H and UMTS, so that the same IP datagram stream is sent to both networks simultaneously. When a problem arises with the DVB-H network, the receiver can automatically switch to the UMTS network, so that the user experiences no interruption.

Another important issue is the optimal IP packet size for efficient data transmission over DVB-H, given that the latter is prone to errors due to both the physical characteristics of the radio channel and the small size of the antenna. Obviously, the packet error probability is lower for smaller packets, but the header overhead becomes significant. One such experimental study is reported in Vadakital et al.,[22] where the typical encapsulation was of IP/UDP/RTP protocols into MPE (Multi-Protocol Encapsulated) sections before transmission. Even though the maximum transfer unit (MTU) of an MPE section is 4,096 bytes, it was found that under error conditions optimal efficiency was achieved for IP datagram sizes between 1,024 and 2,048 bytes.

At this point it is interesting to note that the only transport protocol used in the discussion above IP is UDP[30] (see figure 22.3). The reason for this decision is that DVB-H does not provide by itself a return channel. Hence, it was impossible to support TCP that requires a two-way communication channel. Consequently, FTP cannot be used for the purpose of file transfer; FLUTE (File Delivery over Unidirectional Transport) is used instead.[31]

Overall, DVB-H has been tested in various countries at both the laboratory and field levels with positive results. One such extensive test and evaluation was performed at the Braunschweig Technical University, some results of which are summarized in Kornfeld.[8] According to these results, DVB-H can recover damaged frames completely at vehicle speeds of approximately 100 km/h, which is adequate for urban environments.

The first place to launch a commercial DVB-H-IP Datacast service was Italy in June 2006.[30] Within the first six weeks, more than 100,000 DVB-H handsets were sold. According to official site of DVB-H,[40] a commercial service is running in nine countries at the time of writing (mainly European, but also in the United States, Russia, Vietnam, and South Africa), and trial services have been completed in a dozen others. Nearly all major cell phone manufacturers have shown prototype DVB-H terminals, and market prospect sales figures of DVB-H devices are predicted to top tens of millions by 2008.

DMB (Digital Multimedia Broadcasting) is a digital radio transmission system for sending multimedia to mobile devices. It is based on the Eureka 147 DAB (Digital Audio Broadcasting), standard and there are two variations: S-DMB (Satellite) and T-DMB (Terrestrial). It is the main competitor for DVB-H in the sense that it has many similarities to it (e.g., it uses H.264 for video). It adds various coding, networking, and error-correcting tools for multimedia content processing to Eureka 147. It boasts an overall bit error rate (BER) of 10^{-9}. This characteristic corresponds to trouble-free CD-quality audio and real-time video-streaming reception for portable receivers at 200 km/h.[18] The leading country in terms of deployment is South Korea, with other countries of the Far East and Europe currently either running a pilot service or offering a fully commercial service. However, in Europe, the L-band used by T-DMB is not available in all countries; moreover, DVB-H is far more standardized, and the required infrastructures installed to a large extend already. The protocol stack for T-DMB is shown in figure 22.6.

Commercial T-DMB service was launched on December 1, 2005, in the Seoul metropolitan area, South Korea.[23] As in other video broadcasting systems, a middleware platform for T-DMB has been set up called T-DMB MATE (Mobile Application Terminal Equipment). Its structure is depicted in figure 22.7. T-DMB MATE is designed to provide an interactive data service by using a broadcasting and telecommunication network. The data channel bandwidth is low (approximately

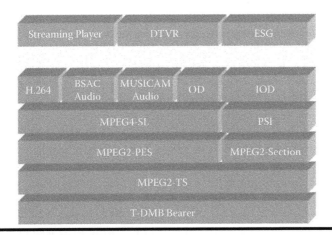

Figure 22.6   The T-DMB protocol stack.

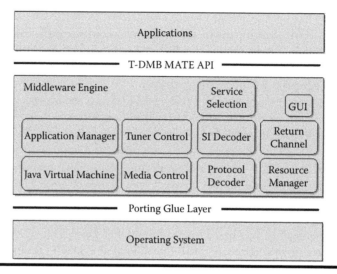

Figure 22.7   The structure of T-DMB MATE.

256 kbps) compared to the video-streaming channel. For this reason, T-DMB MATE typically schedules methods to be downloaded (essentially application chunks and static data), and after storing them on terminal memory, it transmits real-time data with low capacity. The respective application programming interface (API) results in easier application development, because the application code is independent from the underlying hardware.

When T-DMB services were launched in South Korea (December 2005) for the first time, there were no mobile-compatible terminals that could support all of the available data services. This was a gap recently filled with research efforts for suitable low-powered terminals (e.g., Lee et al.[24]). The popularity of T-DMB has been high in South Korea, because it is free and the quality of service quite good as perceived by users. However, unless advertisements or other sources of income grow significantly, service providers are not likely to invest more in expanding the range of services offered.

MediaFLO is another technology, by Qualcomm, for broadcasting audio, video, and data to PDAs and cell phones, specifically designed for multicasting. FLO stands for Forward Link Only. It allows mobile operators to provide live streaming video channels, in addition to supporting 50–100 national local contents channels. MediaFLO requires only two or three broadcast towers per metropolitan area, resulting in 30–50 times less than those required by cellular network systems.[28] In March 2007, Verizon Wireless launched a commercial mobile TV service based on MediaFLO, available in 20 states in the United States. A recent article (March 29, 2007) in *USA Today* tested this service together with a pilot DVB-H in New York and found that channel switching with the former was faster (2 s) compared to the latter (8 s).[37]

The main idea is that a video stream in MPEG-2 format (704 or 720 × 480 or 576 pixels) is received, which is then transcoded to H.264 QVGA resolution, supported by the FLO network. Contrary to other technologies presented earlier, MediaFLO is new in the sense that it has the least features in common with others. The MediaFLO protocol stack is shown in figure 22.8.

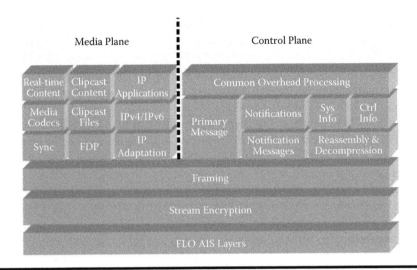

Figure 22.8   The MediaFLO protocol stack.

22.4   Important Performance Parameters and Design of End-User Devices

The multitude of standards, the rapid rate at which technology evolves, and strong competition push not only service providers but also mobile devices manufacturers. The development of an appealing consumer product in this context may cost considerably, in terms of both time and money. Consumer devices these days are a mixture of hardware and software. The latter can be very complex, supporting both old and new standards, as well as a multitude of services.

All this requires millions of lines of code and hundreds of man-years to build a product. Moreover, a large part of the software concerns common services any user should have, with the additional effort to make, maintain, and improve it, without any differentiating advantage gained from competitors. For example, a DVD recorder interface is not very different from a VCR.

A solution adopted by many manufacturers (e.g., the Nexperia Home Platform by Philips) is to follow a layered approach, under which a generic middleware is placed between the end applications/services offered to the consumers and the underlying hardware.[7] In this way, a single streaming platform may be reused for a family of products, reducing overall time and cost. It is therefore not surprising to see this solution applied in different forms in the new broadcasting standards described in the previous section, with the introduction of a middleware layer.

Nevertheless, for device manufacturers to evaluate the performance of any device under a new technology, it is first necessary to determine the most important parameters that affect it. Some of these are outlined below:[2]

- **Sensitivity.** As little as 2 or 3 dB of performance improvement in this area can increase range by as much as 60 percent, which gives consumers much better reception in the fringe coverage areas between transmission towers. Efforts to improve sensitivity in the mobile TV solution need to be accompanied by equal efforts in the area of antenna design. Neglecting this part of the system solution will have a negative impact on overall performance.

- **Interference.** There are many potential sources, either analog (old TV sets, radios, power transformers) or digital (PCs, cell phones, etc.). It is important that specification interference levels are set higher than current standards for new technologies, because experience has shown that such standards may need to be updated very quickly.
- **Doppler performance.** This is a measure of the quality of reception at higher speeds.
- **Channel-switching time.** For example, while MediaFLO supports channel change times of less than 2 s, in power-saving sleep mode, it can take 5 s or more from the time a command is issued to when the channel is switched in DVB-H. This can be extremely annoying to the consumer.
- **Power consumption.** Using a DVB-H solution as an example, if the user is only watching one channel, the dominant power contributors are the screen, backlight, and application processor. However, some more power-hungry features allow consumers to do things like record two to three programs in the background.
- **User interface.** This must be extremely friendly, including the provision of program and service guides and the ability to download files, receive alerts, and so on. The program guide, in particular, is a key feature, which can be implemented in a visual manner that enables users to actually see small clips of programs, instead of reading text, before selecting one to watch.
- **Compatibility with multiple broadcasting standards.** As for traditional cell phones, consumers prefer mobile TV devices that allow them to roam freely among the wide range of related services. With a plethora of such standards there is great pressure on manufacturers to provide such devices at reasonable cost that adhere to the parameters discussed so far. This task becomes even more difficult, because devices of this type must not provide adequate performance for a single standard, but all supported standards.

Despite manufacturers' concerns and end-user devices' performance, we must also turn our attention to the content creators, because the result depends on the new features that they can provide. A significant burden is placed on them simply by the heterogeneous nature of receiving devices. For example, it costs more to develop multiple versions of a Web page with different formats.

Obviously, it is necessary for device capabilities to be automatically detected by the network system, so that the appropriate transformation may take place automatically. As an example, Broadcast Markup Language (BML) is an eXtensible Markup Language (XML)–based derivative used in HDTV in Japan. In Matsumura et al.[14] a predetermined template corresponding to a pattern existing in BML is used to extract data and identify the semantic contents; these, in turn, can then be transformed to fit different user interfaces.

A similar case is presented in Ferreti et al.[15] in the DVB-MHP environment outlined above: a suitable DVB-J browser is broadcasted and locally executed at the user's terminal. In this way, "normal" HTML pages are transformed by the DVB-J browser so that they are appropriately viewable at different terminals (e.g., large tables are fragmented into subtables, font size is adjusted, etc.).

22.5 Video Streaming and Basic Services

Video on demand (VoD) is a service directly built over video streaming. Typically, it is associated with RTSP in order for the end user to have some form of control over the video stream in a way similar to the control of a DVD player. Nevertheless, there are many other services.

Video streaming is usually associated with broadcasting, because this type of media delivery can accommodate many end users simultaneously, with minimal network resource reservation. Moreover, it is modeled along classical TV broadcasts, something that both users and service providers are accustomed to. This is the reason that all of the standards discussed earlier follow this model. Even cellular network operators followed this approach in the effort to provide mobile TV service over 2.5G and 3G networks.

Such direct broadcasting technology did not exist in the 3G environment. For this reason, as early as 2002, the Third Generation Partnership Project (3GPP) and later 3GPP2 (Third Generation Partnership Project 2) provided suitable work items, namely, MBMS by the former (discussed below) and BCMCS by the latter. Despite this, cellular network operators could not wait for 3G broadcast technologies to emerge and started deploying mobile TV services over unicast 3G networks using packet-switched streaming (PSS), later enhanced by the introduction of the advanced H.264 video codec.

MBMS (Multimedia Broadcast Multicast Service) is a service that can be offered over Global System for Mobile (GSM) and Universal Mobile Telecommunications System (UMTS) networks. A distinctive advantage of MBMS is that it uses IP multicast for IP flows (hence IP multicast addresses) at the network core, which allows resource sharing, and hence better network bandwidth utilization. The video resolution is QCIF (176 × 144), adequate for mobile phones; therefore, the bit rate required is only 128 kbps. MBMS also offers an additional—uplink—channel for each user, which allows user interaction.

MBMS was primarily designed for mobile television services, and as such, it is a competitor to DVB-H and DMB. Its major disadvantages are economics (the cost of building new broadcast infrastructure to support it) and the lack of extensive field tests.

IPTV uses IP for delivering digital television to subscribers. However, it is not just another concept for passively watching television programs; in IPTV, the customer is able to view only those channels that he wishes at any given time. Moreover, he can perform additional operations, such as pause, rewind, and replay. What is more important is that the user can have a truly interactive and personalized service through a two-way communication with the server.

IPTV is often offered as part of a triple-play service, that is, Voice-over-IP (VoIP; telephone), data, and video on demand. As such, it is in competition with the other wireless broadcasting standards examined earlier. Unfortunately, IPTV remains an immature technology, and recent experience has shown that loyalty is hard to obtain but easy to lose if it fails to ensure quality for its customers.[4]

On the other hand, conservative forecasts predict broadband subscribers availing themselves specifically of IPTV to reach 37 million with about 4.5 million in Europe alone by 2008.[5] Therefore, the predicted market size is not negligible. However, several problems have to be overcome for this market to evolve into a substantial source of income for the service providers.

Datacasting is the term often used for the broadcasting of data over wireless networks, often used in conjunction with digital television for the provision of news, broadcasting programs, and other information. It can also appear in an interactive form, so that the end user can play games and perform shopping or other transactions.

IP Datacasting (IPDC) is the broadcasting of IP datagrams using some digital video broadcasting system as the carrier for the downstream channel. Again, the upstream channel is carried over typical wireless networks such as UMTS, so that end users can be offered some means of interaction. A simpler variant of this concept has been implemented under Eureka 147 (or Digital Audio Broadcasting [DAB]), which started as a European Union research project for the digital broadcasting of audio only, simultaneously multiplexing multiple audio channels.[1] DVB-H also includes its own version (DVB-H IPDC).

One last point that needs to be discussed is datacasting real-time data. More specifically, up until now we have examined datacasting in conjunction with video streaming using some form of wireless broadcasting. Therefore, the latter is supposed to handle all time-critical data (which does not have to be reliable, as in the case of video), with datacasting for non-time-critical tasks (e.g., file downloading).

If this assumption is not true, and there is a demand for reliable transfer of data, probably combined with interaction, the situation is altogether different. This occurs because wireless or hybrid networks can experience wide delay variations that cannot be safely attributed to congestion as in the case of wired networks. Packets may be lost and retransmitted by intermediate nodes in a path between the end communication points (e.g., because the underlying data link layer retransmits lost or damaged packets).[33]

Some of these problems (reliability, availability, cost, flexibility, and ease of use) may be common to other technologies. However, the key point for a new technology is that the content and type of services must be differentiated and the public be "trained" to appreciate the merits of the new services. If a new technology creates services that potential customers do not think they need, this top-down approach has a high probability of failure. Therefore, it is imperative to find (or convince potential customers of) needs that the new technology offers to satisfy.

22.6 Advanced Applications and Services

22.6.1 3D Audiovisual Service

The audiovisual (AV) service is at the first line of services built upon video streaming. Increased bandwidth and the new 3D flat-panel displays combined with mobility are promising derivatives. Such a 3D-AV service is a novel approach.

A pilot effort is presented in Cho et al.[21] using the T-DMB system in Korea. Although wide flat panels with 3D capability for multiple viewers without special glasses do not exist at the time, such a technology is relatively mature for a small display and a single user. Furthermore, such displays are available commercially. In order for this to work, however, the authors had to implement a special form of T-DMB (called 3D DMB), whose 2D content could be extracted from ordinary T-DMB receivers. Experimental results showed that bit rates between 512 and 768 kbps were sufficient to produce adequate results. Of course, further technological advances are necessary in terms of display (e.g., special 3D glasses to provide a stereoscopic vision or displays that can produce holographic representation of the broadcasted content).

22.6.2 Traveler Information

This is one of the most common but extremely useful services for the mobile user. For this to be effective, each particular user must spend considerable time selecting his personal points of interest from a general categorization. Cho et al.[19] present a recent implementation where such a service is enhanced by the use of T-DMB and a common cellular network. T-DBM has the functionality to deliver the Transport Protocol Expert Group (TPEG) application services. TPEG, in turn, has a well-organized service message structure.

22.6.3 Social Networking

Simply put, social network means a graph with nodes (agents) often representing individuals and edges (ties) to one or more specific type of relations. Such relations may include friendship, certain habit, or even a commercial link. Consequently, social networking is a service that allows the formation and maintenance of social networks. This type of service is not new; classmates.com was perhaps one of the first Web sites (1995) to offer a school social networking service. ICQ is another early and highly successful example that allowed not only friends to communicate, but also users totally unknown to each other to communicate for chatting or even dating. It should not come as a surprise that similar services have been researched for the mobile arena.

One characteristic example of mobile social networking is Socialight,[11] from which a spin-off product has evolved. The Socialight server tracks the user's current location as well as those of the other users, notifying one another via their cell phones when friends are within a certain distance.

Sticky Shadow is more novel: The user can leave a message (in text, audio, or even video form) that is attached for a predetermined time with a particular location and only available to his friends. Whenever a friend logs into the system and passes within a certain distance from the particular place, he is notified so that he can see the message. Some examples are the grocery shop (pick up milk), restaurant (personal review), and so on. Obviously, any location-based system can be used, such as GPS or a Bluetooth hybrid.

22.6.4 Mobile Users' Behavioral Patterns

In a recent work,[12] researchers have implemented and field-tested two particular ambient mobile phone applications: music and motion presence. The former allows a user to see the music title and artist his friends have played at home; the latter allows a user to see when a friend was moving between particular places.

The important results of this study are the moments that a user might use these applications. These were micromoments (~10 s), when bored and when seeking interaction. In some cases, these applications were purposeful (e.g., when trying to coordinate home arrival) and were mostly used during weekdays when people were busy and tried to coordinate their schedules. During weekends, though, the users did not use the applications much, because they had arranged in advance to be with their family or friends. Therefore, such a service is useful when an individual is bored or generally looking for a distraction to keep him occupied.

22.6.5 Mobile Multimedia Portals

Given the diversity of services built around video streaming, it is not surprising that there have been proposals for multimedia event portals. One such complete example is outlined in Baldzer et al.[17] Night Scene Live is targeted toward teenagers, who tend to go out without planning anything in advance. Instead of using phone calls or a series of SMS messages so that they can make up their minds on where to go, they connect to the aforementioned portal. This provides them with live videos from parties and other events in the vicinity, using DVB-H. The resulting network is a hybrid one, because UMTS/GPRS is used for the uplink, so that the end result can be an interactive service. To make better use of the network resources, videos in high demand are transmitted using DVB-H, so that more users can benefit from a single transmission. The rest are not transmitted, but a picture preview is presented instead in the interactive point-to-point network.

Geographica is another system for spatial socializing through mobile devices and DMB in particular, proposed by researchers at Samsung.[25] This system is composed of a client application running on the user's mobile device and a server application running on a gateway server. The former requires that the user's device is equipped with GPS for location tracking, J2ME, and some wireless local area network (LAN) connectivity (e.g., WiFi). The server follows the J2EE specification, delivering contents in either HTML or XML format to the mobile phone. The main idea is that when a user selects a particular media to be streamed to his mobile phone (e.g., song or video clip), the accompanying metadata is also downloaded, which can in turn be used for real-time searches of users in his vicinity with similar interests. In this way, socializing is performed focusing on users with similar interests at the same time that a particular user has these interests. This is important in the sense that some or most user attributes or characteristics (e.g., what sort of music one likes to listen to) can change with time or mood. Moreover, for devices with larger screens, a visual result of any search may be displayed as a set of icons on a map.

22.6.6 Military Applications

As expected, video-streaming applications have been extensively developed and used by the military. In Bennett and Hemmings[27] one such application framework is outlined using DVB-H, due to its widespread acceptance and consequent low cost. DVB-H is proposed as an extension to the existing satellite systems such as Global Broadcast Services (GBS), which provides a high-bandwidth, one-way digital broadcast of tactical video, large files, and IP services over a digital video broadcast (DVB) architecture to warfighters deployed worldwide.

The problem here is that GBS satellite receive suites (RSs) are bulky and heavy. Nevertheless, appropriate receivers are critical for the support of real-time tactical engagements. DVB-H is considered a fitting alternative for last-mile data broadcasting to relay broadband services from the GBS RS to the forward-deployed, dismounted warfighter. Several one-way key services can be offered in this way:

- **Streaming of IP video.** This is an obvious one, rebroadcasting live streaming video from various reconnaissance sources, command posts, intelligence sources, and forward-deployed troops. A software application such as electronic service guide (ESG) can advertise existing services so that the user can select the appropriate one. This can significantly increase situation awareness on the battlefield.
- **Reliable file transfer.** This service is important because valuable information such as high-resolution images, maps, and other large files can be downloaded reliably.
- **Hierarchical modulation.** This makes use of the hierarchical modulation supported by DVB-H, allowing two streams with differing service requirements to be transmitted over the same RF channel. In this way, one more robust stream service could always be available at a higher priority, whereas a second, lower-priority stream could convey a high-definition version of the same video over a smaller area.

Key to the success of such an approach is, of course, security. Typically, this is achieved with software clients using a special smart card. Given that external card readers are small, cheap, and easily connected via a USB port, this problem is addressed adequately.

22.6.7 Medical Applications

Some of the applications of video streaming in the mobile environment as envisioned by manufacturers are outlined in Franz.[13] At the hospital, doctors and nurses can have wireless, intelligent monitoring devices. In addition, pocket medical imaging devices can provide high-definition, real-time body scanning. Emergency personnel can then convey this information to the hospital in real-time so that proper treatment can be administered en route, saving precious time. Digestible cameras can provide detailed information as they travel through the body, and of course, medical education can benefit from video streaming (e.g., see Dufour et al.[16]).

For all this to happen, the existing complexity of video streaming has to be reduced. As stated by other researchers (and mentioned elsewhere in this chapter), standardization and a generic programmable platform will allow application developers to use off-the-shelf code with support for a wide variety of codecs without any change to their code at the application level. This will also increase modularity in design and allow application developers to focus on their task without the need to be digital signal processing (DSP) experts so that they can deal with hardware and other problems at lower levels.

22.7 Economic Considerations

As in all new products and services, economics is a crucial parameter for the success or failure of a new venture: new investment is required and all investors must be ensured that substantial revenues will be generated within a reasonable time, not only to account for the original investment and maintenance costs, but also to generate adequate profit. In this section, we look at two such analyses: the first for DMB and the second for IPDC over DVB-H. Finally, we look at the remaining important elements of a business model.

DMB has been launched commercially in South Korea, as stated earlier. The Korean DMB market is expected to surpass $800 million in annual revenues by the end of the decade according to the Korean Ministry of Information and Communication. Despite all this, uncertainty still remains about where DMB will evolve. In a recent study, the prospectus, limitation, and uncertainty in the development of DMB are identified, using tools drawn from social constructionism.[36] A large number of reports, forecasts, and targeted interviews were performed or collected and then thoroughly analyzed.

The author questions the hype generated by the industry and other bodies, pointing out that the various revenue-generating services did not evolve from market needs, but rather from the industry itself; the technology was relatively stable in terms of research, and the industry wanted to provide mobile TV and other services commercially at that time. Several market surveys have identified that mobile users' main interest lies in multimedia content related to handset personalization, news, sports, and music. This typically corresponds to short-format content, because customers use their devices in short intervals. Nevertheless, DMB (and other technologies of this type) still faces a challenge when broadcasting content over such small screens compared to large LCD TV screens in certain cases. For example, no one would claim that the experience of watching a film like *Lord of the Rings* is equally rewarding on both media types. In fact, watching anything on a small display for a long period may cause eyestrain. Regulatory obstacles have also appeared. Does DMB belong to the telecommunication industry, or is it a functional extension of broadcasting? The latter is the current practice adopted in South Korea.

Despite all the above reservations, the author concludes that although DMB will not turn into a bubble, due to high financial investments and government intervention, it will not evolve as a killer application. DMB will most likely draw critical mass slowly, with new innovative services, destructing current markets and gradually introducing a new market structure.

A simplistic but highly instructive analysis for the case of IPDC over DVB-H is described by a Nokia research engineer in Hoikkanen.[29] It is assumed that the IPDC over DVB-H network will be installed in a country like the United States in a period of seven years (2006–2012), excluding only the most remote of areas. Under this area model, the total area is 8.9 million km² and the population 280 million; 60 percent of the population lives in or close to large- or medium-size cities (total of 100,000 km²), about 20 percent in smaller regional centers (200,000 km²), and the final 2 percent in rural areas (8.6 km²). Furthermore, this scenario assumes 95 percent indoor coverage for all cities and 70 percent outdoor coverage for rural areas, with mast heights of 70–100 m and respective output power of 2–10 kW at each site. Finally, it is assumed that the network deployment will take place gradually, starting from major cities.

With simple calculations, the total number of transmitters would be 7,000–8,000. Repeaters (gap fillers) for urban areas would be needed to cover all gaps at approximately twice the normal transmitter number—roughly 15,000. With some necessary additional hardware, the total CAPEX (capital expenditures) would range from 450 million Euros (2006) down to 150 million Euros (2012).

This figure can be significantly reduced if only the urban and suburban population is to be covered, because this is the vast majority of clients who would in the end produce revenues. The latter scenario would lead to a CAPEX of only 500 million Euros, or approximately 3 Euros per person (although not included in this analysis). In addition, if high reuse of existing cellular sites is achieved, then the total cost can remain low. Under this assumption (100 percent reuse of urban cellular sites and 75 percent of those in rural areas), the author ends up with OPEX (operating expenses) ranging from 40 million Euros (2006) to 120 million Euros (2012). With all the above in mind and other costs, such as the use of transport networks (the backbone of the whole scheme), be it satellite or leased lines, the total CAPEX is approximately 1.8 billion Euros for the seven-year period. This translates to approximately 6.50 Euros per person.

The respective figure for most large European countries is 5–10 Euros per person. Similarly, if only cities are to be covered, then the cost is reduced to 3 Euros per person, which is in line with large European countries where the cost is 1–3 Euros per person.

In the case of interactive audiovisual services, three major factors drive their development in strict technical terms:[34]

- Encoding and compression techniques
- Development of IP-based technologies
- Availability of appropriate user terminals and platforms

In the case of wired networks, despite the fact that multicast technology has been around for considerable time, lack of support by most of the routers had hampered its development, until QoS with the prioritizing of video and audio appeared.

Similarly, the lack of standardization among the plethora of terminal types for wireless broadcasting, with a wide range of features, and the multitude of standards affect the penetration of these systems. From a business perspective, there are some attractive elements in the new market.

First, there is greater interaction with the user, leading to a significant probability that many individuals will be enticed into joining it. Second, the initial investment can be quite low, especially if only a small percentage of a country (but a large percentage of the population) can be covered and existing communication infrastructure and technology reused. The third element is standardization, which allows the potential market to be as large as possible. Finally, it is important to prove to the advertising companies and potential sponsors that this new market has a new type (and large enough number) of users. This can generate substantial revenues.

In the end, there are two basic business models: subscription based and free for the user, where advertisements yield the income necessary to pay for the service. Both models have existed in traditional TV. However, the subscription-based model seems at first to be the best, because advertisers have to be persuaded that it is worth paying for the new broadcasting services. Competition, however, dictates that partnerships combining content providers and network operators are probably the best way for a profitable investment in this new environment.

For a thorough discussion on this subject as well as for accounting models regarding the (similar) 3G environment, the interested reader is referred to Ghys et al.[38]

22.8 The Regulatory Environment Challenge

The regulatory approach in a market may significantly affect the "new" audio/visual market, just as any other. We therefore have to examine the issues involved, and especially the strongly related competition legislation.

Competition legislation essentially has to clarify what is not permissible practice by the competing content operators. However, this works both ways in the sense that if public television channels are allowed (hence public operators in the new market), one has to define precisely their function and actual practices in a liberalized environment. As in the case of television, this is important because television channels tend to have and provide subjective opinions that can affect many citizens over a series of issues, including voting in national elections. Therefore, in principle, each such regulation deals not only with fair competition, but also with rules (or the lack thereof) forcing a more or less objective set of guidelines on an otherwise subjective collection of practices. This is more important for public television because it belongs to the whole society. The practical steps necessary toward this goal are:[34]

- **Defining public service obligations.** These are typically linked to the protection of the basic rights of the citizens (e.g., social and cultural pluralism, protection of expression and communication). Furthermore, there is a strict obligation of free access to a balanced set of informative, cultural, educational, and entertainment content.
- **Deciding on the most appropriate model for the provision of services.** Typically, this involves a broadcast service owned and managed directly by the public authorities, but other forms are also possible (e.g., indirect management or even contracting with an agent already established in the market).
- **Financing method.** The traditional method for financing has been public subsidy. This may be considered a state aid, unless a series of criteria, clearly defined in the aforementioned legislation, are followed. As expected, this has been a hot topic for controversy between the public sector and private operators in many types of markets (e.g., telecommunications, aviation, etc.).

Regulation is currently a hot topic, particularly in Europe, with the latest proposals from the European Commission for the revision of the "Television without Frontiers" directive,[5,34] which we are examining here in detail, due to the respective size of the market and the diverse countries, languages, and cultural elements that constitute it.

More specifically, regulatory provisions were to be extended to Web sites and other online services streaming audiovisual digital content to customers, including the advertisement therein. This sparked a controversy over possible regulation proposed by the European Commission regarding the new broadcasting services, which has led to a recent amended proposal for a "Directive of the European Parliament and the Council Amending Council Directive 89/552/EEC."[6] Some of the most important points are:

- The original directive (89/552/EEC) is extended to new technologies that offer audiovisual services in both linear and nonlinear form. Any audiovisual media service is considered as much a cultural service as it is an economic service. Therefore, the effect of such services in a democratic environment justifies the application of specific rules. However, the member states (Amendment 17) should impose no new systems of licensing or administrative authorization.

- Any audiovisual service generating commercial activity directly (e.g., advertising or teleshopping) or indirectly (e.g., promoting sponsors) is covered by the proposed directive; public service announcements and, charity appeals broadcast free of charge are excluded (Amendment 28). This is also the case for services or activities not in competition with television broadcasting, such as private Web sites/services consisting among others, of audiovisual content generated by private users for sharing within communities of interest. This is also valid for services where any audiovisual content is merely incidental (Amendment 18).

- Member states should set up one or more independent regulatory authorities to enforce this directive; apart from this, no obstacle is placed, so that member states can, in addition, apply their constitutional rules relating to freedom of press and expression in the media (Amendments 13 and 110).

- The transmitting state principle and common minimum standards are retained. In case of legal uncertainty, a member state should argue a case against a broadcast from another origin, by providing evidence that advertisements and other revenues or languages used are directed toward its population (Amendment 34).

- Broadcasters with exclusive rights should grant others the right to select an excerpt of up to 90 seconds of an event (e.g., news or sports) before it concludes for better and more objective information service to the public (Amendment 218).

- There is more freedom to broadcasters regarding advertisements, though. In Amendments 58 and 219, it is stated that given the widespread existence of personal DVD recorders and increased choice of channels, detailed regulation for the insertion of spot advertising for protecting viewers is no longer justified. Sponsorship is also allowed by permitting advertising in a program, if it is not part of the plot (in which case it is considered product placement and is prohibited).

Whether this recent proposal will be approved in the end or more amendments made prior to its introduction to the European Parliament is unknown. Nevertheless, it is basically a legal document on a new and highly technical environment that combines a multitude of factors and players from many fields. As such, it is questionable whether it can clearly set effective rules that will help instead of inhibit through their rigidness the development of the new market.

22.9 Summary

In this chapter, we presented an overview of the basic terms and elements that constitute video streaming and related services, focusing on the wireless/mobile environment. The most important of the new standards were outlined, together with their current status. We then turned our attention to the basic services that can be built over or offered with them. These represent the foundation upon which more complex and advanced services may be designed on a diverse set of applications, some of which were examined.

We also looked at the socioeconomics issues involved in these new technologies that could turn such an enterprise into a big success or another big bubble. Finally, we examined them from the regulatory point of view, focusing on the current controversy in the European Union that may prove a raw model in an ever-more global market.

Links

1. ETSI DVB portal: http://portal.etsi.org/radio/digitalvideobroadcasting/dvb.asp/.
2. DVB-H: http://www.dvb-h.org/.
3. T-DMB: http://eng.t-dmb.org/.
4. Qualcomm MediaFLO: http://www.qualcomm.com/mediaflo/index.shtml.

References

1. DAB FAQ. http://www.worlddab.org/technology_faq.php.
2. *FLOFocus*, Issue 03, April 2007. http://www.floforum.org/.
3. C. Herrero and P. Vuorimaa. 2004. Delivery of digital television to handheld devices. In *Proceedings of the 1st IEEE International Symposium on Wireless Communication Systems*, Mauritius, 240–44.
4. B. Wagner. 2005. *Driving IPTV growth: The challenges and perspectives*. White paper on DaVinci Technology, Texas Instruments. http://www.ti.com/litv/pdf/sphy004.
5. C. Burbridge. 2006. IPTV: The dependencies for success. *Journal of Computer Law & Security Report* 22:409–12.
6. European Commission. 2007. Amended proposal for directive of the European parliament and of the council amending council directive 89/552/EEC. Brussels. http://ec.europa.eu/avpolicy/docs/reg/modernisation/proposal_2005/com_2007_170_en.pdf.
7. G. van Doren and B. Engel. 2005. Streaming in consumer products. Beyond processing data. In *Dynamic and Robust Streaming in and between Connected Consumer-Electronic Devices*, chap. 6, 139–65. Philips Research Book Series, Vol. 3. Dordrecht: Springer.
8. M. Kornfeld. 2006. DVB-H for wireless broadband terminal access—A performance evaluation. In *Proceedings of the IEEE International Conference on Consumer Electronics*, Digest of Technical Papers, 413–14.
9. ETSI. 2006. *Digital video broadcasting (DVB); IP datacast over DVB-H: Content delivery protocols*. TS 102 472, V1.2.1.
10. MHP. http://www.mhp.org/index.xml.
11. D. Melinger, K. Bonna, M. Sharon, and M. SantRam. 2004. Socialight: A mobile social networking system. Poster paper presented at Proceedings of the Sixth International Conference on Ubiquitous Computing (Ubicomp), Nottingham, England.
12. F. Bentley, J. Tullio, C. Metcalf, D. Harry, and N. Massey. 2007. A time to glance: Studying the use of mobile ambient information. Paper presented in Proceedings of the Workshop at Pervasive 2007, Toronto, Canada.
13. G. Franz. 2005. *The future of digital video*. White paper, Texas Instruments. http://www.techonline.com/learning/techpaper/193103047.
14. K. Matsumura, K. Kai, H. Hamada, and N. Yagi. 2005. Transforming data broadcast contents to fit different user interfaces—Generating a readout service for mobile DTV receiver. In *Proceedings of the 7th ACM International Conference on Human Computer Interaction with Mobile Devices & Services*, Salzburg, 323–24.
15. S. Ferreti, M. Roccetti, and J. Andrich. 2006. Living the TV revolution: Unite MHP to the web or face IDTV irrelevance! In *Proceedings of the 15th International ACM WWW Conference*, Chiba, Japan, 899–900.
16. J. C. Dufour, M. Cuggia, G. Soula, M. Spector, and F. Kohler. 2007. An integrated approach to distance learning with digital video in the French-speaking virtual medical university. *International Journal of Medical Informatics* 76:369–76.

17. J. Baldzer, S. Thieme, S. Boll, H.-J. Appeltrath, and N. Rosenhaeger. 2005. Night scene live—A multimedia application for mobile revellers on the basis of a hybrid network, using DVB-H and IP datacast. In *Proceedings of the 2005 IEEE International Conference on Multimedia & Expo (ICME)*, Amsterdam, 1567–70.

18. V. H. S. Ha, S. Choi, J. Jeon, G. Lee, W. Jang, and W. Shim. 2004. Real-time audio/video decoders for digital multimedia broadcasting. In *Proceedings of the 4th IEEE International Workshop on System-on-Chip for Real-Time Applications (IWSOC'04)*, Banff, 162–67.

19. S. Cho, Y. Jeong, K. Geon, K. Soon-Choul, and A. Hyun. 2006. A traveler information service structure in hybrid T-DMB and cellular communication network. In *Proceedings of the IEEE Vehicular Technology Conference*, Vancouver, 747–50.

20. V. Ollikainen and C. Peng. 2006. A handover approach to DVB-H services. In *Proceedings of the 2006 IEEE International Conference on Multimedia & Expo (ICME 2006)*, Toronto, 629–32.

21. S. Cho, N. Hur, J. Kim, K. Yun, and S. Lee. 2006. Carriage of 3D audio-visual services by T-DMB. In *Proceedings of the 2006 IEEE International Conference on Multimedia & Expo (ICME 2006)*, Toronto, 2165–68.

22. V. Vadakital, M. Hannuksela, M. Razaei, and M. Gabbouj. 2006. Optimal IP packet size for efficient data transmission in DVB-H. In *Proceedings of the 7th IEEE Nordic Signal Processing Symposium (NORSIG 2006)*, Iceland, 82–85.

23. S. Cho, G. Lee, B. Bae, K. Yang, C. Ahn, S. Lee, and C. Ahn. 2007. System and services of terrestrial digital multimedia broadcasting (T-DMB). *IEEE Transactions on Broadcasting* 53:171–78.

24. K. Lee, Y. Park, S. Park, J. Paik, and J. Seo. 2007. Development of portable T-DMB receiver for data services. *IEEE Transactions on Consumer Electronics* 53:17–22.

25. K. Sagoo and Y. Rhee. 2006. Real-time spatial socializing through mobile device. In *Proceedings of the 8th ACM Conference on Human-Computer Interaction with Mobile Devices and Services (Mobile-HCI'06)*, Helsinki, 267–68.

26. G. Bedicks Jr., F. Yamada, F. Sukys, C. Dantas, L. Raunheitte, and C. Akamine. 2006. Results of the ISDB-T system test, as part of digital TV study carried out in Brazil. *IEEE Transactions on Broadcasting* 52:38–44.

27. B. Bennett and P. Hemmings. 2005. Digital video broadcast–handheld (DVB-H)—A mobile last-mile tactical broadcast solution. In *Proceedings of the 2005 Military Communications Conference (MILCOM 2005)*, Atlantic City, NJ, 1–7.

28. K. Kim. 2006. Key technologies for the next generation wireless communications. In *Proceedings of the 4th International Conference on Hardware/Software Codesign and System Synthesis*, Seoul, 266–69.

29. A. Hoikkanen. 2006. Economics of wireless broadcasting over DVB-H networks. In *Proceedings of the 2006 IEEE Wireless Telecommunications Symposium (WTS'06)*, 1–5.

30. M. Kornfeld and G. May. 2007. DVB-H and IP datacast broadcast to handheld devices. *IEEE Transactions on Broadcasting* 53:161–70.

31. FLUTE. IETF RFC 3296.

32. F. Hartung, U. Horn, J. Huschke, M. Kampmann, T. Lohmar, and M. Lundevall. 2007. Delivery of broadcast services in 3G networks. *IEEE Transactions on Broadcasting* 53:188–99.

33. G. Cheung, W. Tan, and T. Yoshimura. 2005. Real-time video transport optimization using streaming agent over 3G wireless networks. *IEEE Transactions on Multimedia* 7:777–85.

34. C. Feijóo, J. Fernández-Beaumont, J. Gómez-Barroso, A. Marín, and D. Rojo-Alonso. 2007. The emergence of IP interactive multimedia services and the evolution of the traditional audiovisual public service regulatory approach. *Journal of Telematics and Informatics*, 24(4):272–284.

35. J. F. Kurose and K. W. Ross. 2007. *Computer networking: A top-down approach featuring the Internet.* 4th ed. Boston: Addison Wesley.

36. D. Shin. 2006. Socio-technical challenges in the development of digital multimedia broadcasting: A survey of Korean mobile television development. *Technological Forecasting & Social Change* 73:1144–60.

37. E. Baig. Mobile TV has a future. *USA Today*, March 29, 2007. http://www.usatoday.com/tech/columnist/edwardbaig/2007-03-28-mobile-tv_N.htm.

38. F. Ghys, M. Mampaey, M. Smouts, and A. Vaaraniemi. 2003. *3G multimedia: Network services, accounting, and user profiles*. Norwood, MA: Artech House.

39. ETSI DVB portal. http://portal.etsi.org/radio/digitalvideobroadcasting/dvb.asp/.

40. DVB-H. http://www.dvb-h.org/.

Chapter 23

Live Video and On-Demand Streaming

I. S. Venieris, E. Kosmatos, C. Papagianni, and G. N. Prezerakos

Contents

Keywords

live video, streaming, mobile TV, video compression, MPEG-2, MPEG-4, VC-1, RTP/RTCP, FLUTE, DSM-CC, DVB-H, IPDC, MBMS, OMA BCAST, FLO, DMB, ISDB-T, broadcast-mobile interworking

23.1   Introduction

Video has been an important medium for communications and entertainment for many decades. Video technology describes the means for electronically reconstructing a sequence of images. The creation and composition of the sequence of still images involve basic functions such as capturing, recording, and processing. Video technology has evolved rapidly over the last 30 years. Initially video was captured and transmitted in analog form. The success of the analog videotape format left its mark on the entire 1980s decade. However, video technology, from the 1980s and on, followed the general trend toward "everything digital" due to the advent of digital integrated circuits and computers. Nowadays video can be delivered via communication networks and appropriate storage discs (e.g., digital video discs).

The major advantages of digital video format are that it is not susceptible to degradation of quality over the years, it is editable, and it can be easily integrated with audio, text, and graphics. In addition, it can be transmitted without loss in quality, depending on the physical medium (cable, air, etc.). The quality of the digital video is based on three key factors: frame rate, color depth, and frame resolution. Frame rate is the number of images displayed per second that provokes the perception of motion, for example, the National Television Standards Committee (NTSC) standard for full motion is 30 frames per second. Color depth is the number of bits per pixel for color information representation, for example, 24 bits can represent 16,777,216 colors ($256 \times 256 \times 256$). Frame resolution is the display dimension measured in pixels, represented by the number of horizontal pixels times the number of vertical pixels, for example, 800×600. Typical digital video formats, analogous to the component analog video, are RGB and YCbCr, where Y represents a pixel brightness and CbCr pure color. Because the eye is less sensitive to color than brightness, chrominance is usually subsampled, and thus represented by fewer bits (e.g., 4:2:2) to reduce the size of the digital output. However, subsampling is not enough. Further techniques that cut down the size of digital output are scaling of the aforementioned key factors (e.g., reduce frame rate to 15 frames per second) and compression. Video compression standards will be analyzed later in this chapter.

Once digital video has been produced, it can be either presented asynchronously through the progressive (or not) download-store-present method or streamed across a communication network. Video streaming takes advantage of advances in video scaling, compression techniques, and protocols that help restore the temporal relationships within each stream containing the media information, or synchronize various streams of multimedia information. The major advantage of video streaming is that the video is reproduced at the media client while it is being downloaded; therefore, it is not stored at the client. In addition, because the file is not downloaded, copyright is protected.

In this chapter, the basic architecture and components of mobile broadcast multimedia delivery systems will be presented, including corresponding protocols for video delivery and video compression standards. In addition, the basic video services and video delivery schemes in digital broadcast networks will be presented, as well as a digest of the application that brought forth the need for massive deployment of mobile broadcast technologies: mobile television.

23.2 Mobile Video Delivery over Digital Broadcast Networks

Wireless carriers are incorporating mobile broadcast/multicast technologies such as Digital Video Broadcast transmission to handheld terminals (DVB-H), Mobile Broadcast and Multicast Service (MBMS), Media Forward Link Only (MediaFLO), and Terrestrial Digital Multimedia Broadcasting (T-DMB) based on Digital Audio Broadcasting (DAB) standards, and Integrated Services Digital Broadcasting Terrestrial (ISDB-T) to provide advanced video-centric mobile services. To gain an understanding of the factors that led to the development and deployment of video delivery over mobile broadcast networks, the basic video transport mechanisms and applications must be explained.

23.2.1 Video Services

Streaming video is a trend that has emerged in the last decade. However, it is only recent widespread use of broadband downlinks and mobile wireless technology that has enabled its massive user adoption. There are two main types of streaming: progressive streaming and real-time streaming. Progressive streaming enables a file to be watched as it is being downloaded. Real-time video streaming diversifies in the time frame between the video production and the video delivery. Concurrent production/encoding and delivery of media content constitute live streaming. Streaming of preencoded, stored video content defines on-demand streaming service. In other words, live streaming is when the event is actually happening at that moment you are watching it, while on the other hand, the video has been produced at a previous point in time and has been stored to become available for streaming. Sports games and news coverage are usually coming live over various communication networks to the users.

There is a tough challenge concerning the live streaming of events. The determining factors in performance and cost are the quality of the necessary equipment; appropriate video content enhancement with text, graphics, and so on; encoding of the material; and latency demands, as a live coverage has to be as "live" as possible. Latency is comprised of the delay introduced by the content production process as well as the encoding, transmission, and decoding delay. What is more important is that live content producers have got just one content. Therefore, to maximize their investment, it must be delivered accordingly to any possible terminal. As far as mobile users are concerned, third-generation (3G) cellular networks are currently not optimized to deliver multimedia content to a multiplicity of receivers, from both a cost and technical viewpoint. On the other hand, video delivery over broadcast systems can easily support the demand for an unlimited number of users. Live video delivery over broadcast channels is a direct parallel to viewing a traditional network television broadcast. The content is a live feed received usually from a traditional broadcaster in appropriate broadcast format and delivered immediately to users as soon as it is ingested and transcoded.

In general, video-on-demand systems either stream content to enable viewing in real time, or allow users to download it, in which case it is available on the terminal (set top box, mobile handset, etc.) before viewing starts. When a user utilizes the particular service, he or she expects that content is reproduced as a whole with a low start-up latency. In traditional IP networks (fixed, 3G) employing unicast sessions, this is easily realized by issuing a request to the streaming server, and consequently starting a point-to-point streaming session to the client. However, bandwidth in wireless environments is an extremely costly and valuable resource; therefore, employing unicast delivery is not appropriate.

In the case of pure broadcast delivery, users cannot issue a request to the streaming server, as there is no return channel. To accomplish the VoD service, in many cases conventional p-t-p connections (e.g., GPRS or UMTS) are utilized for the return channel. However, with an increasing number of subscribers, this solution does not scale appropriately. For that purpose, VoD broadcasting schemes have been proposed and are being evaluated on mobile broadcast systems. The idea behind VoD broadcasting schemes is to divide the video into a series of segments and broadcast each segment periodically on dedicated channels. The most trivial solution for enabling asynchronous access to broadcasted content is the periodic transmission (carousel delivery) of the video stream. A receiver subscribing to a carousel may have to wait in the worst-case scenario for T seconds, with T denoting the duration of the media stream. The carousel broadcasting scheme is presented in figure 23.1.

An alternative solution would be to transmit the same media stream into K parallel channels in a time-shifted manner. A receiver subscribing to the service may have to wait in the worst-case scenario for T/K seconds, but required bandwidth increases by a factor of K. More sophisticated VoD broadcasting schemes can be subdivided into three groups. Protocols in the first group partition the video into increasing size of segments and transmit them in logical channels of the same bandwidth. They are based on pyramid broadcasting. Broadcasting protocols based on the harmonic broadcasting protocol divide the video into equal-size segments and transmit them in logical channels of decreasing bandwidth. The last family of broadcasting protocols includes pagoda

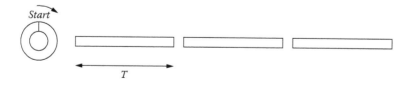

Figure 23.1 Video carousel.

broadcasting schemes. These protocols are a hybrid of pyramid and harmonic-based protocols. They partition each video into fixed-size segments and map them into a small number of data streams of equal bandwidth and use time division multiplexing to ensure that successive segments of a given video are broadcast at the proper decreasing frequencies.

23.2.2 Video Delivery

Unicast delivery of content incorporates point-to-point stream delivery to each client. Unicast streaming services such as video on demand include streaming media content over fixed/mobile and wired/wireless networks to the subscriber/user. Such mobile video services are already provided to users on their mobile handsets by various service providers over cellular networks. Due to the nature of unicast delivery and bandwidth limitations, massive user demand is not supported in this mode, although a wide range of services and content can be provided. Multicasting, by contrast, delivers one stream simultaneously to multiple clients, assimilating point-to-multipoint transmission of content on communication networks where a subset of users must view the same content. Broadcast transport scheme is a special case of one-to-many delivery of multimedia content, because a single stream is delivered simultaneously to all communication network clients. Broadcast is a very efficient form of communication for popular content, as it can deliver this content to all receivers at the same time.

23.2.3 Mobile TV

To deliver the aforementioned applications to handheld terminals that support the mobility of users, a new service has emerged combining broadcast technologies and advanced video compression and delivery schemes called *mobile television*. Mobile TV usually refers to live simulcast TV on mobile handheld devices that provides the same content as seen on regular digital TV at home. The service usually encompasses on-demand video or short clips that a user could download or that could be broadcasted to a large number of users. Mobile television integrates two of the most successful products ever, television and the mobile phone, *and* therefore it is expected to be the key factor driving demand for next-generation mobile services.

Mobile video is already a reality on existing 2.5G and 3G infrastructures in point-to-point streaming and downloading modes. However, spectrum is an expensive and limited resource and unicast transmission does not scale well for a large number of subscribers. In a cellular network like UMTS, a cell can serve only a very limited number of point-to-point multimedia sessions. Therefore, an important bandwidth limitation factor is imposed concerning cellular transmission of multimedia content. UMTS capacity was improved with the emergence of High Speed Downlink Packet Access (HSDPA) and 3GPP Long Term Evolution (3G LTE) technologies. 3GPP and 3GPP2 have addressed the matter of efficient multicast/broadcast transmission over cellular networks, and the outcome of the effort is called Mobile Broadcast and Multicast Service (MBMS) for 3GPP and Broadcast and Multicast Service (BCMCS) for 3GPP2. MBMS and BCMCS introduce efficient multicast/broadcast transport in mobile cellular networks. The OMA BCAST working group specified the multicast/broadcast–related service layer functionalities, such as content protection, service and program guides, and so on.

An alternative approach is to use a dedicated broadcast network to offer mobile TV services. Broadcast video services employ a separate radio network where the same content is received by broadcast-capable mobile handsets without limiting the capacity of the operator's network.

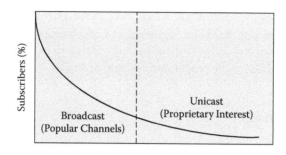

Figure 23.2   Long tail–power law distribution of digital media.

This exact service is usually referred to as mobile TV, as it resembles the functionality of traditional television. Nowadays digital broadcast networks and their extensions, such as the European DVB-H, the Korean-led T-DMB, QUALCOMM's MediaFLO, and Japanese ISDB-T, are being deployed worldwide to provide efficient mobile TV services. However, mobile broadcast technologies are limited regarding the number of channels they can offer as well as nonbroadcast services. For that reason, the available content that they can deliver is also limited, as mobile broadcast is not designed to address the "long tail" (figure 23.2) distribution pattern of digital media consumption.

Combining unicast delivery with mobile broadcast provides an extended selection of content and services to end users and increases revenues for the mobile operator. The most popular channels can be broadcast and additional content can be selectively sent to particular users via unicast channels. In such an integrated system, radio resources are preserved through the use of broadcasting delivery, and yet users can receive the desirable video content through unicast delivery.

23.3 Video Compression Standards Adopted in Mobile Broadcasting

Compression reduces the number of bits used to represent each pixel in the image. The algorithms used to compress the images inevitably introduce, to a lesser or greater extent, visible artifacts. In general, the complexity of a compression algorithm is conversely proportional to the number of artifacts. The processing power of an encoding/decoding system increases with time, so more sophisticated algorithms are always being introduced into the codec products. The selection of the appropriate algorithm should be done in a way that uses realizable computation, and needs short processing time. The processing should not distort the image so that visible artifacts are generated.

Compression systems take advantage of the mechanisms of human visual perception to remove redundant information, but still produce a compelling viewing experience. A typical example of the process is the persistence of vision, where an instantaneous view of a scene fades over about one-tenth of a second. This allows the encoding mechanism to portray the continuum of time by using a series of discrete pictures or frames. In general, viewing a motion picture shot at 25 frames per second (fps) gives a good approximation of motion in the original scene.

The most common video compression standards used in digital broadcasting systems are presented below.

23.3.1 MPEG-2

MPEG is an encoding and compression system for digital multimedia content defined by the Motion Pictures Expert Group (MPEG). MPEG-2 extends the basic MPEG system to provide compression support for transmitting digital TV and digital video quality.

On understanding the importance of video compression, just consider the vast bandwidth required to transmit uncompressed digital TV/video pictures. The phase alternate line (PAL) is an analog TV transmission standard used in many parts of the world. An uncompressed PAL TV signal requires a massive 216 Mbps, which is far beyond the capacity of most radio frequency links. The U.S. standard used in analog TV is called NTSC. The NTSC system provides less precise color information and a different frame rate. An uncompressed NTSC signal requires slightly less transmission capacity, approximately 168 Mbps. This situation becomes much more intense in case of high-definition TV (HDTV), which is becoming very popular nowadays. An HDTV signal requires a raw bandwidth exceeding 1 Gbps.

The MPEG-2 standard provides good compression using standard algorithms, so it has become the standard for digital television. MPEG-2 has the following features:

- Achieves high video compression and is backwards compatible with MPEG-1 standard
- Supports both full-screen interlaced and progressive video capable for TV screens and computer displays
- Succeeds high audio coding (high quality, mono, stereo)
- Supports transport multiplexing, therefore allowing the combining of different MPEG streams in a single-transmission stream
- Provides a variation of services: graphical user interface (GUI), interaction, encryption, an so on
- Supports interaction with the user

The list of systems that use MPEG-2 is extensive and continuously growing: digital TV (cable, satellite, and terrestrial broadcast), video-on-demand systems, digital versatile disc (DVD), personal computing, an so on.

23.3.1.1 Technical Overview

The MPEG-2 video compression algorithm achieves high rates of compression by exploiting the redundancy in video information. MPEG-2 takes advantage of both the temporal redundancy and spatial redundancy present in motion video.

Temporal redundancy appears when successive frames of the video display images of the same scene. The temporal redundancy is based on the fact that the content of the scene usually remains fixed or changes only slightly between successive frames.

Spatial redundancy occurs because small pieces of the picture, which are called pels, are often replicated (with minor changes) within a single frame of video.

The basic operation of the MPEG-2 encoder is depicted in figure 23.3.

MPEG-2 includes a wide range of compression mechanisms. The MPEG-2 encoders must therefore have the ability to choose the proper compression mechanism best suited to a particular scene/sequence of scenes. In general, the more sophisticated the encoder, the more efficient it is at selecting the most appropriate compression mechanism, and therefore the higher the picture quality for a given transmission bit rate. On the other hand, the MPEG-2 decoders also come in

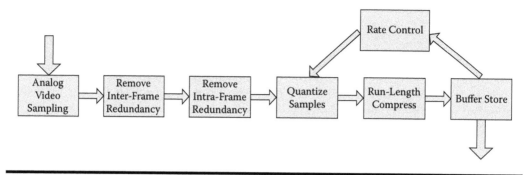

Figure 23.3 MPEG-2 encoder basic operations.

various types and have varying capabilities, such as ability to handle high-quality video and cope with errors, jitters, and delays.

23.3.1.2   Interaction

An extremely important feature that led to the adoption of MPEG-2 in digital TV is the ability to use a return channel to allow the user to control the content or scheduling of the transmitted video/audio/data. This feature is known as interaction and has become the key discriminator between traditional video and MPEG-2. MPEG-2 defines an interaction channel using the digital storage media command and control (DSM-CC) toolkit.

Interaction channels may be used for diverse services, including:

- Display and control of small video clips
- Ability to select and pay for video on demand
- Access to remote information servers
- Access to remote databases, therefore having access to systems providing online shopping, banking, and so on
- Internet access

23.3.1.3   Profiles

A number of levels and profiles have been defined for MPEG-2 video compression. Each of these describes a useful subset of the total functionality offered by the MPEG-2 standards. An MPEG-2 system is usually developed for a certain set of profiles at a certain level. The profile is expressing the quality of the video, while the level is expressing the resolution of the video.

The basic functionality of the standard is known as main profile main level (MP@ML), which covers video compression concluding to a bit rate of 1 to 15 Mbps. There are other levels, such as high level, high level-1440, and low level. There are also other profiles: simple, signal-to-noise ratio (SNR), spatial, 4:2:2, and high.

Regarding decoders, the typical decoder specifications are:

- $720 \times 576 \times 25$ fps (PAL CCIR 601)
- $352 \times 576 \times 25$ fps (PAL Half-D1)

- 720 × 480 × 30 fps (NTSC CCIR 601)
- 352 × 480 × 30 fps (NTSC Half-D1)

Most decoders have also backward compatibility supporting MPEG-1 specifications:

- 352 × 288 × 25 fps (PAL SIF)
- 352 × 240 × 30 fps (NTSC SIF)

23.3.2 *MPEG-4*

MPEG-4 is a standard used primarily to compress audio and visual (AV) digital data. Introduced in late 1998, it is the designation for a group of audio and video coding standards and related technology agreed upon by the ISO/IEC Moving Picture Experts Group (MPEG) under the formal standard ISO/IEC 14496. The uses for the MPEG-4 standard vary from Internet streaming media and CD distribution to videoconferencing and broadcast television, all of which benefit from compressing the audiovisual stream.

MPEG-4 supports many of the features of MPEG-1 and MPEG-2 and other related standards, while adding new features such as:

- Virtual Reality Modeling Language (VRML) support for three-dimensional rendering
- Object-oriented composite files (including audio, video, and VRML objects)
- Externally specified digital rights management
- Many types of interactivity

MPEG-4 is still a developing standard and is divided into a number of parts. Unfortunately, the companies promoting MPEG-4 compatibility do not always clearly state which par" of the standard they are supporting. The commonly used parts are MPEG-4 Part 2 and MPEG-4 Part 10.MPEG-4 Part 2 (MPEG-4 SP/ASP) is used by codecs such as DivX, XviD, and Quicktime 6. MPEG-4 Part 10 (MPEG-4 AVC/H.264) is used by the x264 codec, Quicktime 7, and next-generation DVD formats like high-definition DVD and Blu-ray disc.

Most of the features included in MPEG-4 are left to individual developers to decide whether to implement them. This means that there are probably no complete implementations of the entire MPEG-4 set of standards. The standards, to deal with this, include the concept of profiles and levels, allowing a specific set of capabilities to be implemented in a manner appropriate for a subset of supported applications.

MPEG-4 Part 2 and MPEG-4 Part 10 are described in detail in the following paragraphs.

23.3.2.1 *MPEG-4 Part 2*

MPEG-4 Part 2 is a video compression technology developed by MPEG. It belongs to the MPEG-4 ISO/IEC standard (ISO/IEC 14496-2). It is a discrete cosine transform (DCT) compression standard, similar to previous standards such as MPEG-1 and MPEG-2.

23.3.2.1.1 Profiles

The MPEG-4 Part 2 standard, to cover various applications ranging from low-quality, low-resolution surveillance cameras to high-definition TV broadcasting and DVDs, has defined a set of profiles and levels covering subsets of the features. MPEG-4 Part 2 has approximately

21 profiles, including profiles called Simple, Advanced Simple, Main, Core, Advanced Coding Efficiency, Advanced Real Time Simple, and so on. The most commonly deployed profiles are Advanced Simple and Simple, a subset of Advanced Simple.

Most of the video compression schemes standardize the bit stream and thus the decoder side, leaving the encoder side to individual implementations. Therefore, implementations for a particular profile (such as Xvid or DivX, which are implementations of Advanced Simple Profile) are all technically identical on the decoder side, whether it was created through Real Media, Windows Media Player, or another common encoder.

23.3.2.1.1.1 Simple Profile (SP)—Simple Profile is mostly aimed for use in situations where low bit rate and low resolution are imposed by external factors such as network bandwidth, device size, and device processing power. Examples are cell phones, low-end videoconferencing systems, and surveillance systems.

23.3.2.1.1.2 Advanced Simple Profile (ASP)—Advanced Simple Profile technical features are a superset of Simple Profile features, and it is roughly similar to H.263. The ASP includes:

■ B-frames/B-VOPS/bidirectional encoding/prediction: Contrary to I-frames/key frames (which include the entire image and do not depend on other frames), and P-frames (which include only the changed parts of the image from the previous I- or P-frame), B-frames are constructed using data from the previous and next I- or P-frame. B-frames can be compressed much more than other frame types, which overall should help quality and compressibility.

■ Quarter-Pixel Motion Search Precision (QPEL): Basically, most MPEG-4 codecs by default detect motion between two frames down to a half pixel (HalfPel). QuarterPel feature has the ability to detect motion that is only a quarter of a pixel per frame, therefore effectively doubling precision. Practically, this means that using QPEL results in a much sharper image.

■ Global motion compensation (GMC): GMC detects if there is an amount of motion that big parts of the frame have in common. If that is the case, GMC becomes active, using a single motion vector for all similar parts of the frame instead of multiple ones. Practically, this helps saving bits when panning, zoom, or rotation occurs. This feature also gives more sharpness to the outcome.

■ MPEG/custom quantization: In MPEG-4 Simple Profile the H.263 quantization type is used, while the ASP allows the use of custom quantization types also. The H.263 quantization type using the default MPEG matrix results in a softer image and is suitable for higher bit rates. On the other hand, a custom quantization type using the popular custom matrix hvs_good stands well in lower bit rates, while preserving more details.

■ Adaptive quantization: During encoding of a variable-bit-rate video each frame can get compressed with a different quantization mechanism (the higher the quantization the smaller the size/bit rate of the frame). The "rate control" is responsible for deciding the quantization mechanism of each frame. If adaptive quantization is supported, the different quantization mechanism is applied inside each frame; for example, applying higher quantization/higher compression to high motion/dark parts of the frame, while applying lower quantization to faces.

■ Support for interlaced video.

The MPEG quantization and interlace support follow the MPEG-2 Part 2 application, while the B-frame support is designed in basically a way similar to that of MPEG-2 Part 2 and H.263 v2. The quarter-pixel motion compensation feature of ASP was innovative, and was also implemented in MPEG-4 Part 10 and VC-1. On the contrary, some implementations omit supporting this feature, because it has a significantly harmful effect on speed and is not always beneficial for quality. The global motion compensation feature is not actually supported in most implementations of MPEG-4 Part 2, although the standard officially requires decoders to support it. Some experts agree that it does not ordinarily provide any benefit in compression because when used it has a large unfavorable impact on speed while adding considerable complexity to the implementation.

23.3.2.2 MPEG-4 Part 10 AVC/H.264

The H.264, MPEG-4 Part 10, or AVC (for Advanced Video Coding), is a digital video codec standard that is prominent for achieving very high data compression. The standard was proposed by the ITU-T Video Coding Experts Group (VCEG) together with the ISO/IEC MPEG (Moving Picture Experts Group) as the result of a collective partnership effort known as the Joint Video Team (JVT). The ITU-T H.264 standard and the ISO/IEC MPEG-4 Part 10 standard are jointly maintained so that they have identical technical content.

The intent of the H.264/AVC project was to create a standard that would provide high video at substantially lower bit rates than previous standards would require (e.g., relative to MPEG-2, H.263, or MPEG-4 Part 2). The target was to create this standard in a way that would not increase the complexity of the design so much that it would be impractical and therefore excessively expensive to implement. An additional goal was to provide a standard flexible enough to be applied to a wide variety of applications (for both low and high bit rates and for both low- and high-resolution video) and effective enough on a wide variety of networks and systems (broadcast, DVD storage, packet networks, and multimedia telephony systems).

23.3.2.2.1 Technical Overview

The H.264/AVC design supports the coding of video in 4:2:0 chroma format, which contains either progressive or interlaced frames, which may be mixed together in the same sequence. Generally, a frame of video contains two interleaved fields, the top and the bottom field. These two fields of an interlaced frame may be coded separately as two field pictures or together as a frame picture. A progressive frame should always be coded as a single frame picture.

The H.264/AVC standard specifies a Video Coding Layer (VCL), which efficiently represents the video content, and a Network Abstraction Layer (NAL), which formats that representation and provides header information in a manner appropriate for conveyance by particular transport layers or storage media.

23.3.2.2.1.1 Network Abstraction Layer—While the VCL is specified to represent, efficiently, the content of the video data, the NAL is specified to format that data and provide header information in a manner appropriate for delivery through transport layers or storage to media. Each NAL unit contains an integer number of bytes, while the total NAL units contain all data. A NAL unit specifies a generic format for use in both packet-oriented and bit stream systems. The format of NAL units for both packet-oriented transport and bit stream delivery is identical, with the only alteration that in a bit-stream-oriented transport layer, each NAL unit can be preceded by a start code prefix.

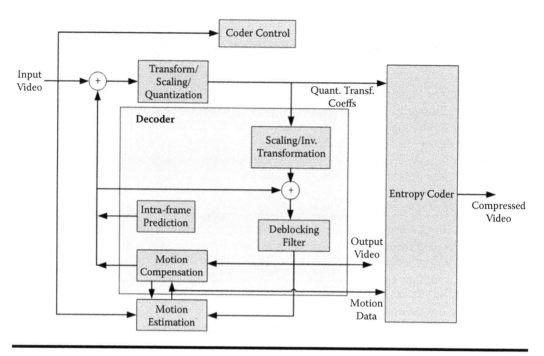

Figure 23.4   Block diagram of the video coding layer MPEG-4 AVC.

23.3.2.2.1.2   Video Coding Layer—The video coding layer of H.264/AVC is similar in design to previous standards such as MPEG-2. It consists of a hybrid mechanism of temporal and spatial prediction, in conjunction with transform coding. The block diagram of the video coding layer for a macroblock is depicted in figure 23.4.

In summary, the initial picture is split into blocks. The first picture of a sequence or a random access point is typically "intra" coded using information only available in the picture itself. Each sample of a block in an intraframe is predicted using spatial compensation of previously coded blocks. The encoding process chooses which and how the neighboring already-coded blocks are used for intraprediction, which is simultaneously conducted at the encoder and decoder using the transmitted intraprediction side information.

For all remaining pictures of a sequence or between random access points, typically "inter" coding is used. Intercoding uses prediction based on motion compensation from other previously decoded pictures. The encoding process for interprediction consists of choosing motion data, comprising the reference picture, and applying a spatial displacement to all samples of the block. The motion data that are transmitted as side information are used by the encoder and decoder to simultaneously provide the interprediction signal.

The residual of the prediction (either intra or inter), which is the difference between the original and the predicted block, is transformed. The transform coefficients are scaled and quantized. The quantized transform coefficients are entropy coded and transmitted together with the side information for either intra- or interframe prediction.

The encoder contains the decoder to conduct prediction for the next blocks or the next picture. Therefore, the quantized transform coefficients are inverse scaled and inverse transformed in the same way as at the decoder side, resulting in the decoded prediction residual. The decoded prediction residual is added to the prediction. The result of that addition is fed into a deblocking filter that provides the decoded video as its output.

23.3.2.2.2 Profiles and Levels

A number of profiles and levels have been defined for the H.264/AVC video compression standard. Each of these describes a useful subset of the total functionality, while facilitating interoperability between various applications of the H.262/AVC standard that have similar functional requirements. A profile defines a set of coding tools or algorithms that can be used in generating a compliant bit stream, whereas a level places constraints on certain key parameters of the bit stream. All decoders conforming to a specific profile have to support all features in that profile. Encoders are not required to make use of any particular set of features supported in a profile, but have to provide conforming bit streams.

In the H.264/AVC standard, three profiles are defined: baseline, main, and X:

- The baseline profile supports all features proposed in H.264/AVC except the following two feature sets:
  - B slices, weighted prediction, context-adaptive binary arithmetic coding (CABAC), field coding, and macroblock adaptive switching between frame and field coding
  - SP and SI slices
- The main profile supports all features of baseline profile except for the feature map output (FMO) aspect. The main profile also supports the first set of features not supported by baseline profile.
- Profile X supports both sets of features on top of the baseline profile, except for CABAC and macroblock-adaptive switching between frame and field coding.

In H.264/AVC, the same set of level definitions is used with all profiles, but individual implementations may support a different level for each supported profile. Eleven levels are defined, specifying upper limits for the picture size in macroblocks, the decoder processing rate in macroblocks per second, the size of the multipicture buffers, the video bit rate, and the video buffer size.

23.3.2.2.3 Application Areas

The high compression efficiency of H.264/AVC offers new application areas and business opportunities. The standard makes possible video signals of TV (PAL) quality to be transmitted at about 1 Mbps, which makes them capable of streaming over xDSL connections. In the area of TV transmission over satellite, by choosing 8-PSK and turbo coding in combination with the usage of H.264/AVC, one can triple the number of TV channels per satellite, in comparison to the current DVB-S systems using MPEG-2. Given this huge amount of additional transmission capacity, even the exchange of existing set-top boxes might become an interesting option.

Regarding DVB-T, H.264/AVC is a very attractive option. If we assume common transmission parameters (8k mode, 16-QAM, code rate 2/3, and ¼ guard interval), we end up with a bit rate of 13.27 Mbit/s available in each 8 MHz physical channel. By using MPEG-2 coding, the number of TV channels per physical channel is restricted to four, whereas by using H.264/AVC, the number of TV channels could be raised to ten or even more. This happens because both the coding efficiency and the statistical multiplex gain for variable bit rates are higher due to the higher number of different TV channels. Another interesting alternative is to use QPSK, code rate ½, in conjunction with H.264/AVC to reduce "electrosmog" (electric pollution). This combination would allow a system to retain four TV channels per physical channel, but would decrease the transmitted power by 15 percent in comparison to the transmission mode mentioned above.

Another interesting area is HD transmission and storage. It now becomes possible to encode HD signals at about 8 Mbit/s, which fits onto a conventional DVD. This fact will accelerate the DVD cinema market, because it is no longer necessary to wait for the more expensive and unreliable blue DVD laser. It is also possible to transmit four HD TV channels per satellite or cable channel, which makes this service much more attractive to broadcasters, as the transmission costs are much lower than with MPEG-2.

Also in the field of mobile communication, where the bandwidth is precious, H.264/AVC will surely play an important role because of the high compression rate. The compression efficiency will be doubled in comparison to the coding schemes currently specified by 3GPP standards for streaming, that is, H.263 Baseline, H.263+, and MPEG-4 Simple Profile.

23.3.3 VC-1

The VC-1 video codec was initially developed by Microsoft under the formal name SMTPTE 421M. The VC-1 codec specification was based on the Windows Media Video 9 codec, also called WMV3. On April 2006, the formal release of the VC-1 standard was announced, while its most popular implementation is Windows Media Video 9.

The design of the VC-1 codec is based on the conventional DCT-based video codec design incorporated in previous video codecs, like H.261, H.263, MPEG-1, MPEG-2, and MPEG-4. It is an alternative to the latest ITU-T and MPEG video codec standard known as H.264/MPEG-4 AVC. Coding tools for interlaced video sequences as well as progressive encoding are incorporated in the VC-1 standard. The main advantage of VC-1 development is the fact that it supports compression of interlaced content without first converting it to non-interlaced, something very convenient for broadcast and *the* video delivery market.

Although there is a Microsoft product, a variety of companies support the development of VC-1. As an SMPTE standard, VC-1 is open to implementation by anyone, with the prerequisite to pay licensing fees to the MPEG LA, LLC licensing body or directly to its members who are participating in the development of format.

An essential step toward the proliferation of VC-1 *was* accomplished with the adaptation of VC-1 as a mandatory video standard in HD DVD and Blu-ray discs. This means, that their video playback devices will be capable of decoding and playing video content compressed using VC-1. Windows Vista *also* supports the VC-1 compression scheme.

23.4 Protocols for Mobile Delivery

With the emergence of new video compression standards that allow for multimedia content delivery over communication networks bearing bandwidth limitations and the evolution of associated network technologies, streaming has become increasingly important in 2.5/3G and mobile broadcast networks as in the Internet. Whereas the Internet streaming market is dominated by proprietary solutions, the Third Generation Partnership Project (3GPP) developed a commonly agreed upon streaming standard for 2.5/3G networks called Packet-Switched Streaming Service (PSS). PSS provides an entire end-to-end streaming and download framework for mobile networks describing all

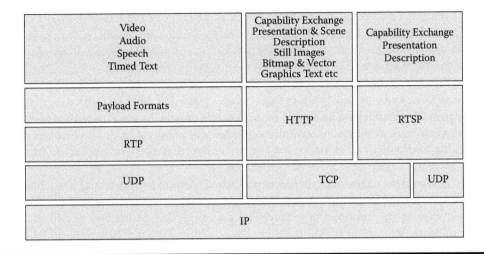

Figure 23.5   PSS protocol stack.

functional modules and protocols of the reference architecture. The main specification defines associated protocols and media codecs, while the 3GP file format defines a storage format for servers. In addition, the standard specifies the timed text format for subtitling and the 3GPP Synchronized Multimedia Integration Language (SMIL) Language Profile for scene descriptions. As far as audio and video *are* concerned, AMR, AMR-WB (Wide Band), Extended AMR-WB (AMR-WB+), MPEG-4 High Efficiency AAC v2, and H.263, H.264/AVC are the defined codecs for PSS clients. More information on the relevant media codecs and other mechanisms such as digital rights management and media selection is available in the relevant specifications.

PSS is based on protocols developed by the Internet Engineering Task Force (IETF), as presented in figure 23.5. The main protocols adopted include the Real-Time Streaming Protocol (RTSP) for session control, the Session Description Protocol (SDP) for presentation description, the Real-Time Transport Protocol (RTP) for media transport, and the Hypertext Transfer Protocol (HTTP) for download of scene and presentation description.

Emerge of mobile broadcast *networks* and the diversity of adopted technologies and systems have led to different protocol stack specifications per technology. However, several analogies are observed among them; for example, the MPEG-2 transport stream specification is adopted by many mobile broadcast technologies for multiplexing digital video and audio synchronizing the output. Therefore, a basic description of the most commonly used protocols for video delivery is presented subsequently.

23.4.1 MPEG-2 Transport Stream

The systems part of the MPEG-2 recommendation (ISO/IEC 13818-1) addresses the combining of video and audio, as well as other data, into single or multiple streams that are suitable for storage or transmission. System coding was specified in two forms: the *transport stream* and the *program stream*, each optimized for a different set of applications. Multiplexing of video, audio, and data portions of a single program for transmission in relatively error-free environments such as digital storage recorders (e.g., DVD) is described by the program stream, while the transport stream (TS) can be used for applications such as broadcast of digital multimedia.

The MPEG-2 TS is so called, to signify that it is the input to the transport layer in the ISO Open System Interconnection (OSI) seven-layer network reference model. It is not, in itself, a transport layer protocol, and no mechanism is provided to ensure the reliable delivery of the transported data. MPEG-2 relies on underlying layers for such services.

The MPEG-2 TS defines a protocol for packet-oriented multiplexing of multiple MPEG compressed programs, and program directory (e.g., channel guide) information into a packetized fixed-length format for transmission on digital networks. It is designed for use in environments where errors are likely, such as storage or transmission in lossy or noisy media. It also includes sophisticated timing distribution, synchronization, and jitter correction mechanisms that are essential for the transmission of video signals over long distances.

Basic multiplexing takes place between packetized elementary streams (PESs). Elementary streams are diversified in control, audio, video, and data elementary streams. These compressed elementary streams are packetized to produce PES packets. A PES packet may be a fixed- (or variable-) sized block, with up to 65,536 bytes per block, and includes a 6-byte protocol header. The PES header starts with a 3-byte start code, followed by a 1-byte stream ID conveying the type of media and a 2-byte length field. The next field contains the PES indicators. These provide additional information about the stream to assist decoding at the receiver. The indicators that are defined are the *PES scrambling control*, which defines whether scrambling is used as well as the chosen scrambling method; *PES priority*, which indicates the priority of the current PES packet; data alignment indicator, which specifices if the payload starts with a video or audio start code; copyright information, which determines whether the payload is copyright protected; and original or copy, which defines if this is the original ES. A 1-byte flag field completes the PES header. This defines whether the following fields are present: presentation time stamp (PTS) and possibly a decode time stamp (DTS) used for synchronization; elementary stream clock reference (ESCR); elementary stream rate, the rate at which the ES was encoded; trick mode, which indicates the video/audio is not the normal ES; copyright information, which indicates a copyright ES; CRC for monitoring errors in the previous PES packet; and PES extension information. The PES packet payload includes the ES data. The information in the PES header is independent of the transmission method used.

For the MPEG-2 TS each PES packet is broken into fixed-sized transport packets of 188 to 184 bytes of payload and a 4-bytes header. The header starts with a synchronization byte followed by a set of three flags. Bits 12–25 of the packet header define the packet identifier (PID). Each packet is associated with a PES through the setting of this PID value. Two scrambling control bits are used for access control to the payload of the TS packet. Finally, two adaption field control bits are defined for payload specification and a four-bit field called continuity counter incremented with each transport stream packet of the same PID.

If no packets are available, null packets with a specific PID are inserted to retain the specified TS bit rate. To receive a particular transport stream, packets are filtered according to their PID value. Alignment of PID to their corresponding program is performed through the use of signaling tables, which are also transmitted within the MPEG-2 transport stream. Signaling tables are sent separately to PES and are not synchronized with the elementary streams. These tables are called program-specific information (PSI) and define the set of elementary streams that belong to a program and its description.

A transport stream may correspond to a single TV program, *for example,* containing a video and an audio PES, which is called as *single-program transport stream* (SPTS). SPTSs are combined and form *multiple-program transport streams* (MPTSs), which also contain the corresponding PSI.

The transport protocol also provides features for broadcast scrambling and transmission of encryption keys. It has a special message format, called entitlement management message (EMM), for individually addressing decoders in a broadcast system. EMMs carry information such as encryption keys and the addresses of the devices that have paid for this program.

Adoption of the MPEG-2 TS format has provided broadcasters and other content delivery companies a comprehensive means to interchange and deliver multiple streams of video, audio, and data in one transport stream.

23.4.2   Real-Time Transport Protocol

The real-time transport protocol (RTP) provides end-to-end delivery services for data with real-time characteristics, such as video and audio. In general, multimedia applications require appropriate timing in data transmission and playing back. RTP provides time-stamping, sequence numbering, and other mechanisms to take care of the timing issues. Through these mechanisms, RTP provides end-to-end transport for real-time data over communication networks.

Time-stamping is extremely important for real-time applications. The sender sets the time stamp according to the instant the first octet in the packet was sampled. Time stamps increase by the time covered by a packet. On receiving the data packets, the receiver uses the time stamp to reconstruct the original timing order so as to play out the data at the correct rate. Time stamps are also used to synchronize different streams with timing properties, such as audio and video data. However, synchronization is performed in the application level. Because UDP does not deliver packets in a timely order, sequence numbers are used to place the incoming data packets in the correct order. Sequence numbers are also used for packet loss detection. Due to video frame fragmentation in several video formats, many RTP packets bear the same time stamp. Therefore, time-stamping is not sufficient for correct reordering of packets at the receiver. The payload type identifier specifies the payload format as well as the encoding/compression schemes. It provides a means for the receiving application to identify, interpret, and play out the payload data. Default payload types are defined in RFC 1890. Example specifications include MPEG4 Part 10, and so forth. Additional payload types can be added by providing a profile and a payload format specification. At any given time of transmission an RTP stream can include only one type of payload. Source identification discloses information about the sender of the media stream. For example, in an audio conference the particular field identifies the speaker. All the aforementioned mechanisms are implemented via the RTP header.

RTCP is the control protocol designed to work in conjunction with RTP. In an RTP session, participants periodically send RTCP packets to convey feedback on the quality of data delivery and information of membership. RTCP provides feedback to the application concerning the quality of the data distribution. Control information is available to senders, receivers, and third parties monitoring the RTP sessions. Transmission may be adjusted based on the receiver report feedback. Moreover, the receivers can determine whether congestion is local. Network administrators may evaluate network performance for multicast distribution. In RTP data packets sources are identified by randomly generated 32-bit identifiers. Identifiers as such are not suitable for end users. RTCP SDES (source description) packets contain textual information called canonical names as globally unique identifiers of a session's participants (e.g., username, telephone number, e-mail address, etc.). RTCP sender reports contain an indication of real-time and the corresponding RTP time stamp, which can be used in intermedia synchronization such as lip synchronization in video. RTCP packets are sent periodically among participants. When the number of participants increases, it is necessary to balance reception of up-to-date control information and limiting redundant control traffic. Moreover, to scale up to large multicast groups, RTCP has to prevent overwhelming network resources with control traffic. RTCP limits control traffic to 5 percent of the overall session traffic.

RTP/RTCP typically runs on top of UDP/IP to make use of its multiplexing and checksum functions. However, there has been substantial effort to become transport independent. UDP was chosen as the transport protocol of RTP/RTCP basically for two reasons. RTP is primarily designed for multicast; connection-oriented TCP does not scale well and therefore is unsuitable. In addition for real-time data, reliability is not as important as timely delivery. Moreover, reliable transmission provided by retransmission (TCP) is not desirable.

In practice, RTP is usually implemented within the application. To set up an RTP session, the application defines a particular pair of destination transport addresses (one network address plus a pair of ports for RTP and RTCP). In a multimedia session, each stream is carried in a separate RTP session having its own RTCP packets reporting the reception quality of the session. For example, audio and video streams would be transmitted on different RTP sessions.

23.4.3 File Delivery over Unidirectional Transport/ Asynchronous Layered Coding

File Delivery over Unidirectional Transport (FLUTE) is a protocol for the delivery of files, for example, images, documents, software, video, audio, and so on, enabling progressive as well as background download. Initially FLUTE was designed for transmission over the Internet on top of UDP/IP. FLUTE is compatible with both IPv4 and IPv6, as no part of the packet is IP version specific. However, it has been adopted as the protocol for download delivery services in many broadcast technologies, as it is suited for one-to-many delivery over wireless broadcast radio systems.

The specification builds on Asynchronous Layered Coding (ALC), the base protocol designed for massively scalable multicast distribution. ALC combines the layered coding transport (LCT) building block, a congestion control (CC) building block, and the forward error correction (FEC) building block to provide congestion-controlled reliable asynchronous delivery of content to an unlimited number of concurrent receivers from a single sender. ALC uses the LCT building block to provide in-band session management functionality, and the FEC building block to provide reliability. The use of CC and FEC building blocks with FLUTE is optional. By default, no CC is used and the FEC code is compact no-code FEC, which means that there is no actual FEC encoding or decoding, and encoding symbols contain only the source symbols. ALC is underspecified and used for transporting binary objects. FLUTE is a fully specified protocol for transporting files and utilizes the File Delivery Table (FDT) to provide an index of files and the associated reception parameters in-band.

A FLUTE session (i.e., an ALC/LCT session) consists of one or more ALC/LCT channels defined by the combination of a sender and an address associated with the channel by the sender. A receiver joins a channel to start receiving the data packets sent to the channel by the sender, and the receiver leaves the channel to stop receiving data packets from the channel. One of the fields carried in the ALC/LCT header is the transport session identifier (TSI). The TSI is scoped by the source IP address, and the source IP address and TSI pair uniquely identifies a session. In case multiple objects are carried within a session, the transport object identifier (TOI) field within the ALC/LCT header identifies from which object the data in the packet was generated. Each object is associated with a unique TOI within the scope of a session.

Usually the size of a file (transport object) to be transported is larger than the maximum transmission unit of the underlying network. For that reason, FLUTE supports an appropriate packetization mechanism (figure 23.6) where the transport object is initially segmented into a number of

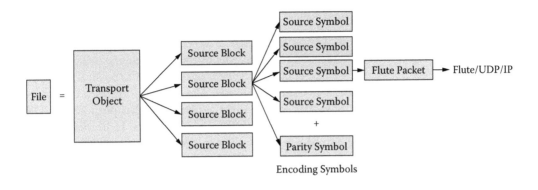

Figure 23.6   FLUTE packet creation.

smaller blocks, referred to as source blocks, by a blocking algorithm. Blocking algorithms define the structure of the source block in terms of source size K. Thereupon source blocks are fragmented into K equal-sized source symbols, where K denotes the source block size.

Following the segmentation process, FLUTE supports the usage of a symbol encoding algorithm (FEC) to provide reliability. FEC is applied on each source block, and encoding symbols are equally sized to source symbols. In the case of a systematic FEC encoding, the first K encoding symbols are identical to source symbols and the remaining symbols are parity symbols. The selection of source symbol length and source block size is determined by the transmitter. The encoding symbols are encapsulated in FLUTE packets. Correspondingly, the FLUTE header precedes each encoding symbol/group of encoding symbols/FDT. The FLUTE header provides information concerning the position of each encoding symbol within the encoding block, the relevant source block number, and transport object identifier.

23.4.4 *Digital Storage Media Command and Control Carousel*

DSM-CC covers a number of distinct protocol areas for the delivery of multimedia services over broadband networks. DSM-CC defines a user-to-network (U-N) protocol for session and resource management as well as protocols for client configuration and download. DSM-CC includes an interface for VCR-like control of video streams, and interfaces for a multitude of other generic interactive application services. Also, through the definition of a U-N data carousel and a user-to-user (U-U) object carousel, DSM-CC provides protocols for generic broadcast application services (data and object carousel). For channel changing in broadcast networks, DSM-CC includes a protocol for switched digital channel change. DSM-CC is transport layer independent; the same application can be delivered over a multitude of broadband networks (e.g., MPEG-2 TS).

The data carousel protocol area provides periodic broadcast of data to a set of clients through the use of non-flow-controlled download messages. A download control message provides a list of available modules from a particular data carousel so that each client can identify which modules it wishes to capture. However, for the delivery of more complex data structures, DSM-CC specifies a U-U object carousel and a Broadcast Inter-ORB Protocol (BIOP). BIOP provides a standard way of embedding in broadcast carousels object references that describe actual locations of object representations within the same broadcast channel.

23.4.4.1 DSM-CC Data Carousel

The data carousel is a transport mechanism that allows a server to present a set of distinct data modules to a decoder by cyclically repeating the contents of the carousel, one or more times. A module is the highest-level structure that contains one item of data, such as a file. If the decoder wants to access a particular module from the data carousel, it may wait for the next time that the data for the requested module is broadcast (e.g., teletext system). For that purpose, the DSM-CC data carousel defines Download Server Initiate (DSI), Download Info Indication (DII), and Download Data Block (DDB). The fixed-size DDB message corresponds to a block of data that is broadcasted as a single unit and contains the module ID and version that it is part of, the block number within that module, and the corresponding data. The DSI and DII are the control messages used for transporting DSM-CC modules. The DII contains delivery parameters such as module ID, module size, module version, as well as the size of the DDB messages that are used to transmit the module. Modules that are listed in the same DII message are said to be members of the same group. The DSI message acts as a top-level control message for those carousels that have several DII messages. It groups together a number of DII messages, and the groups of modules associated with them, into a single supergroup. Carousels where a number of DII messages are linked to a supergroup by a DSI message are called two-layer carousels, while carousels that only use a single DII message are called one-layer carousels.

23.4.4.2 DSM-CC Object Carousel

A DSM-CC object carousel facilitates the transmission of a structured group of objects from a broadcast server to broadcast receivers using directory objects, file objects, and stream objects. Object carousels are built on top of the data carousel model, but they extend it to add the concept of files, directories, and streams. The directory objects convey the file system's structural information, such as the names of the files and the directory structure. The service gateway object that is a type of directory object identifies the root directory of the object carousel. The stream objects allow the broadcaster to control the audiovisual session with commands such as pause, play, and stop. The stream event object is a special object related to the stream. The stream event objects are used to carry the information of the stream event descriptors in the stream, which are used to provide a mechanism whereby the interactive application can respond to events in the audiovisual content. File objects represent real files. They contain the actual data that makes up the file.

The data and attributes of one U-U *object* in an object carousel are transmitted in one message. The message format is specified by the BIOP and is referred to as the BIOP *generic object message* format. A BIOP *message* consists of a MessageHeader, MessageSubHeader, and MessageBody. The MessageHeader provides information about the version of the BIOP protocol and the length of the BIOP message. The MessageSubHeader contains information about the conveyed *object*, such as objectType (*file, stream, directory*) and objectKey (the unique identifier within a *module*). The MessageBody depends on the objectType and contains the actual U-U *object's* data. The size of a BIOP message is variable.

23.4.5 Real-Time Streaming Protocol

The Real-Time Streaming Protocol (RTSP) is an application-level protocol for out-of-band control over the delivery of data with real-time properties. It provides an extensible framework that enables controlled, on-demand delivery of real-time data, such as audio and video. Sources of data can include both live data feeds and stored clips. This protocol is intended to control multiple data delivery sessions *and* provide a means for choosing delivery channels such as UDP, multimedia UDP and TCP, and delivery mechanisms based upon RTP.

Its functionality is similar to HTTP. However while HTTP is stateless, RTSP is not. A session ID is used to keep track of sessions when needed. This way, no permanent TCP connection is needed. The operation of RTSP is based in the client/server model, but both the server and the client can issue request for connection. The most basic commands that are defined by RTSP to control the multimedia session are SETUP, PLAY, RECORD, PAUSE, and TEARDOWN. In a typical session where the client retrieves media content from the server, as soon as the client has collected information about the file it sends a SETUP message to the server to request a connection. The server authenticates the user and establishes the connection. The client sends a PLAY message indicating the wish to stream the file. The server establishes two RTP sessions (audio and video) and an RTCP session for receiving reports from the client. A transmission of a TEARDOWN message informs the server to close corresponding sessions.

23.4.6 Session Description Protocol

The Session Description Protocol (SDP) is used to describe multimedia sessions, for the purpose of session announcement, session invitation, and other forms of multimedia session initiation. SDP is a format for session description and is used in conjunction with transport protocols *such* as the Session Announcement Protocol (SAP), Session Initiation Protocol (SIP), Real-Time Streaming Protocol, and Hypertext Transport Protocol. The SDP carries information concerning the multimedia session, such as session name and purpose, time the session is active, media comprising the session, and information to receive the media (address, etc.).

23.5 Transport Network Architectures

Mobile phones prove to be versatile devices. They can be simultaneously connected to cellular networks, receive FM broadcasts, or connect to wireless LANs using Wi-Fi. The delivery of broadcast services, and especially mobile TV, can similarly be multimodal through the 3G networks themselves, 3G network broadcast extensions (MBMS or MCBS), or satellite or terrestrial broadcast networks. In all these expressions of delivery mechanisms, the common resource that is used is the frequency spectrum. The rapid growth of mobile broadcasting was indeed an event that was not foreseen by the majority of the industry. The result has been that the mobile broadcasting industry has been left to search for ways to find available spectrum and deliver broadcast services, especially mobile TV.

In the United Kingdom and United States, the traditional TV broadcast spectrum in UHF and VHF stands occupied by the transition to digital and the need to simulcast content in both modes. In the United Kingdom, BT Movio has fallen back on the use of the digital audio broadcast (DAB) spectrum to deliver mobile TV using a standard called DAB-IP. In Korea, the DAB spectrum for satellite services was used to deliver services in a format named DMB-S (Digital Multimedia Broadcast–Satellite). DVB-H is a standard largely designed to use the existing DVB-T networks to also carry DVB-H services and ideally use the same spectrum. This is indeed the case in many countries, with the UHF spectrum being designated for such services. In the United States, where the ATSC systems do not permit "ride on" of mobile transmissions, the UHF spectrum remains occupied with digital transitions and spectrum is auctioned. Modeo, a DVB-H operator, has ventured to lay out an entirely new network based on DVB-H using the L-band at 1670 MHz. Another operator, HiWire, having spectrum in the 700 MHz band, is launching its DVB-H services using this spectrum slot. The United States (along with Korea and India) is also the stronghold of the code division multiple access technologies originated by QUALCOMM. A broadcast technology for mobile TV called MediaFLO was developed in these countries, which is used by many operators to provide mobile TV in a broadcast mode. Many other countries are set to use the same technology. In Korea, the government has also allowed the use of the VHF spectrum for mobile TV services, and this has led to the launching of the terrestrial version of the DMB services, called DMB-T. In Japan, which uses ISDB-T broadcasting, the industry chose to allow the same to be used for mobile transmission.

The scramble to provide broadcast services by using the available networks and resources partly explains the multiple standards that now characterize this industry. Serious efforts are now on to find spectrum and resources for mobile broadcasting on a regional or global basis that will in the future lead to convergence of the standards. The standards mentioned above are described in detail in the next paragraphs.

23.5.1   Digital Video Broadcasting–Handheld (DVB-H)

Digital Video Broadcasting–Handheld is a digital broadcast standard for the transmission of broadcast content to handheld terminal devices, developed by the international Digital Video Broadcasting (DVB) Project and recently published by the European Telecommunications Standards Institute (ETSI). DVB-H is based on the DVB-T standard for digital terrestrial television but tailored to the special requirements of the pocket-size class of receivers.

The DVB-H technology is a spin-off of the DVB-T standard. It is to a large extent compatible with DVB-T, but takes into account the specific properties of typical terminals, which are expected to be small, lightweight, portable, and battery powered. DVB-H can offer a downstream channel at a high data rate, which will be an enhancement to the mobile telecommunications network, accessible by most of the typical terminals. Therefore, DVB-H creates a bridge between the classical broadcast systems and the world of cellular radio networks. The broadband, high-capacity downstream channel provided by DVB-H will feature a total data rate of several Mbit/s and may be used for audio and video streaming applications, file downloads, and many other kinds of services. The system thereby introduces new ways of distributing services to handheld terminals, offering greatly extended possibilities for content providers and network operators.

23.5.1.1   DVB-H Architecture and Protocols

DVB-H, as a transmission standard, specifies the physical layer as well as the elements of the lowest protocol layers. In contrast to other broadcast transmission systems that have adopted the MPEG-2 transport stream for content delivery, DVB-H provides for the use of services based on Internet Protocol (IP) Datacastingm, which enables the delivery of lower-resolution video. Nevertheless, the MPEG-2 transport stream is still used as the baseband layer. IP Datacast (IPDC) is the name of a set of specifications that were developed to bridge the gap between the independent broadcast and mobile communications networks developed by the DVB Project. DVB-H is one of the very first standards that have been developed clearly keeping the idea of convergent networks in mind. According to the relevant specification:

IP Datacast over DVB-H is an end-to-end broadcast system for delivery of any types of digital content and services using IP-based mechanisms optimized for devices with limitations on computational resources and battery. An inherent part of the IPDC system is that it comprises of a unidirectional DVB broadcast path that may be combined with a bi-directional mobile/cellular interactivity path. IPDC is thus a platform that can be used for enabling the convergence of services from broadcast/media and telecommunications domains (e.g., mobile/cellular).

The IP Datacast system is designed to transport files such as video, audio, text, and so on. Moving up in the IPDC delivery platform, bearers provide the mechanism by which IP data is transported. The DVB-H bearer, which is used to transport multicast and broadcast traffic, may be complemented by point–to-point bearers to offer complete video delivery services. The content delivery services offered are classified to streaming and file delivery services. Different applications impose different requirements to deliver content to end users; therefore, they use different delivery methods. For example, a TV broadcast application would use the streaming delivery, while distribution of software updates requires an efficient downloading method. The delivery layer provides functionality such as security and key distribution, reliability control by means of FEC techniques, and associated procedures such as file repair and reception reporting. As far as video delivery is concerned, it is obvious that the broadcast network is used for video-streaming services of general interest to a large number of subscribers as well as pushed video-on-demand services where content is pushed (downloaded) into the terminal at the time of low network usage, ready to be consumed on demand asynchronously.

Figure 23.7 presents the protocol stack of the IP Datacast transmission system. The center part of the figure presents the protocols associated with video delivery (streaming or downloading). Services delivering real-time content as video audio and subtitling utilize the Real Time Protocol. Services delivering non-real-time data employ FLUTE/ALC object/data carousel. For navigation between the services, an eXtensible Markup Language (XML)–based ESG is used. It contains metadata and access information about the available services and is also broadcast with FLUTE/ALC. The DVB-H layer uses Multi-Protocol Encapsulation (MPE) to encapsulate the IP packets into an MPEG-2 transport stream, which is used by the DVB-T transmission system.

Focusing on the streaming service that provides real-time video delivery due to the unidirectional nature of the IPDC system, control information is not provided by the receivers in the form

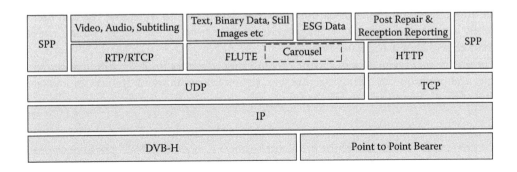

Figure 23.7   IPDC protocol layer architecture.

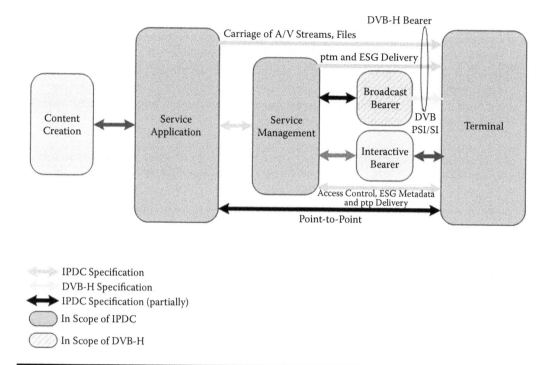

Figure 23.8 IPDC reference architecture.

of RTCP receiver reports. SDP is provided to the IPDC terminal to describe the streaming delivery session. Session description of an IPDC streaming session shall include parameters such as the IP address of the sender, the list of the session's media components, the destination IP address and port number for all media components, the start/end time of the session, the transport protocol, the media type and formats, the data rate, and service languages per media. The relevant specifications propose that encoded video bit streams for DVB-IP Datacast applications shall conform to either H.264/AVC or Windows media 9 video encoding (VC-1). As far as audio is concerned, MPEG-4 High Efficiency AAC v2 and AMR-WB+ are preferred.

According to the IP Datacast reference architecture as it is depicted in figure 23.8, the content to be delivered to the terminal is created on the content creation module, which is outside the scope of IPDC. Services are delivered over different communication networks; therefore, the service application provides a logical link between the content provider and the end user independently of the bearer technology. It aggregates content from multiple sources and their related metadata to provide a particular service.

The service management is in charge of allocating resources from the different bearer technologies and performs the billing together with the service application. The service management consists of four subentities:

■ Service configuration and resource allocation: Registration of service applications that contend for bandwidth of the broadcast bearer (i.e., one DVB-H-IP platform in one DVB transport stream). Assignment of services to location (with respect to broadcast network topology), to bandwidth, and scheduling services over time. There is one instance of this subentity associated with a broadcast bandwidth contention domain.

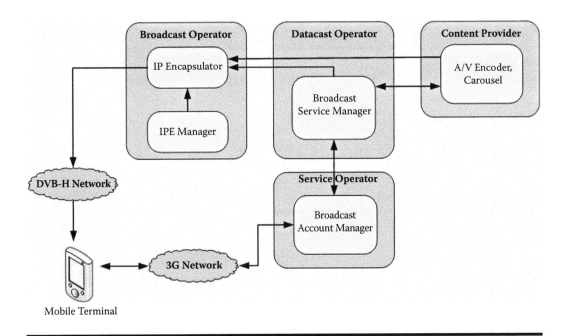

Figure 23.9   DVB-H system architecture.

■ Service guide provisioning application: Aggregation of ESG (metadata information) pieces from the service applications. There may be multiple instances of this subentity.
■ Security/service protection provision: Management of user access to service applications.
■ Location services: The service management entity may provide location services to service application(s) in a manner that is independent of the way they are actually obtained (such as interaction bearer network functionality or GPS).

The broadcast network multiplexes service applications at the IP level. It also performs the assignment of IP flows on DVB-H time slices (IP encapsulation), the transmission over DVB-H, and the security/service protection. The terminal represents the user device as a point of acquisition and consumption for the content and client of network and service resources. The terminal may or may not implement the support of an interaction channel.

Recently Nokia has announced a commercial service management solution for DVB-H services, the Nokia Mobile Broadcast Solution. The solution supports the broadcasting of different types of digital content such as live TV, radio, and video clips over DVB-H networks to mobile devices. The proposed architecture is depicted in figure 23.9, while the main modules are the electronic service guide, the broadcast service manager, the broadcast account manager, and the DVB-H broadcast encapsulator, described in detail below.

23.5.1.1.1   Electronic Service Guide

Electronic service guide (ESG) is a service discovery tool for both user and client applications on the mobile terminal. The ESG provides the user with information about the available services. ESG also serves the mobile terminal middleware with signaling data to enable service lookup from the DVB-H stream and playback with the correct client software and codecs. The ESG enables the user of a mobile TV–capable device to automatically discover all the service platforms and services available in the usage area. In addition to the multiple audio and video streams, a service in the broadcast ESG can include dynamic links and a whole dedicated data stream that populates the mobile terminal memory with files supporting the experience with the live stream.

23.5.1.1.2 Broadcast Service Manager

The broadcast service manager controls the encapsulation, multicast routing, encryption, ESG generation, and digital rights management (DRM) aspects of the mobile broadcast services. The broadcast service manager allows each content provider to manage its accessible capacity as if it were its own, via either user or system-to-system interfaces. The broadcast plans of all the different content providers are stored in a central repository. The broadcast service manager constantly controls the IP encapsulation process to execute these stored plans in all parts of the network, so that correct streams are broadcasted.

The broadcast service manager also generates and multicasts the ESG for each broadcast area. The mobile terminals in that area pick up the ESG transmission and update their local ESG databases accordingly.

Service protection is controlled by the service operators, via an interface similar to the one content providers have at their disposal. Service operators can configure their own service packages and prices, which are also stored in the central repository. The stored packaging data continuously drives the encryption process for pay services and the ESG generation, so that the consumers are presented with the correct purchase options. Each service operator can also offer its interactive services for the consumers independently of each other.

23.5.1.1.3 Broadcast Account Manager

The broadcast account manager is an online service fulfillment and charging solution for paid mobile broadcast services. A consumer who selects a pay service from the ESG is prompted to try it or buy it. If the user agrees to make the purchase, a request is routed via the cellular packet IP services to the broadcast account manager, which processes the request and responds with the appropriate rights. The consumer may then proceed to enjoy the service while the service operator is entitled to charge the consumer.

The broadcast account manager is able to fulfill customer orders in an automatic way. There is nothing in the network that needs to be activated or provisioned for a subscriber to make a subscription. To maximize the applicable user base, the broadcast account manager supports both postpaid and prepaid subscriber types.

From the service operator point of view, the broadcast account manager is a mediator between the DVB-H broadcast services and the existing authentication and authorization as well as charging solutions. The interfaces offered make it possible to integrate the broadcast service offering quickly into the existing infrastructure.

23.5.1.1.4 DVB-H Broadcast Encapsulator

The IP encapsulator is a gateway between the multicast IP network and the DVB-H transmitters. Each IP encapsulator routes the applicable multicast groups from its ingress Ethernet interface to the egress DVB ASI port, optionally encrypting the content. The broadcast service manager controls both the routing and the encryption of each IP encapsulator element.

The IPE manager mediates the control messages from the broadcast service manager to each IP encapsulator element. The mediation layer enables control of very large DVB-H deployments by a single broadcast service manager cluster.

The various contents, whether audio, video, or data, are transmitted as multicast streams with the depicted infrastructure. The signaling for the mobile terminal to be able to select the correct application is achieved with the SDP files in the ESG, which are generated by the broadcast service manager based on administrator-generated profiles.

The network is relying on IP multicast in the end, something that have many advantages. First, while the nature of DVB-H subchanneling is quite static, the multicast routing of streams to these channels is not. Thus, the bandwidth allocated for a service, such as a TV channel, may vary based on the type of content scheduled for broadcast. Additionally, the set of offered services can vary from one network area to another, so a generation of ESG takes place to match the content offering. Finally, using multicast optimizes the capacity of each link in the transport network between the head end and broadcast cells provided that IP transport is used. While the core multicast network connecting the broadcast areas can be IPv4 based, the air interface is based on IPv6.

23.5.2 *Multimedia Broadcast/Multicast Service (MBMS)*

The Multimedia Broadcast/Multicast Service was proposed by 3GPP trying to address broadcast/multicast services in GSM/WCDMA. The Broadcast and Multicast Service (BCMCS) was proposed by 3GPP2 with the same intention for CDMA2000, respectively. 3GPP MBMS and 3GPP2 BCMCS have many commonalities, and in the rest of this chapter we will only distinguish between them when needed.

MBMS and BCMCS succeeded in introducing only minor changes to the existing radio and core network protocols. This has as a consequence of low implementation costs in terminals and the network, while it makes mobile broadcast a relatively inexpensive technology compared to nonmobile broadcast technologies, which require new receiver hardware in the terminal and additional investments in the network infrastructure. Mobile operators, by using MBMS technology, can maintain already established business models while new bandwidth-demanding applications like mobile TV are boosting.

23.5.2.1 *MBMS Architecture and Protocols*

MBMS can be used for any kind of data transmission. However, constraints on data transmission imposed by 2.5G and 3G systems as low data rate have to be taken into consideration because they affect the quality of the service. Therefore, MBMS is a suitable solution for transferring light multimedia clips. For streaming of large video and audio clips, however, pure broadcast systems (e.g., DVB-H) are more suitable.

MBMS is split into MBMS bearer and user services. The MBMS bearer service addresses MBMS transmission procedures below the IP layer, whereas the MBMS user services address service layer protocols and procedures. The MBMS bearer service is supported by both UMTS Terrestrial Radio Access Network (UTRAN) and GSM/EDGE Radio Access Network (GERAN). In addition, point-to-point as well as point-to-multipoint content delivery is enabled, through the *broadcast, enhanced broadcast, and multicast* modes supported for data delivery.

The MBMS user services include a streaming and download delivery method. In addition, carousel delivery is supported. Carousel transmission combines streaming and downloading,

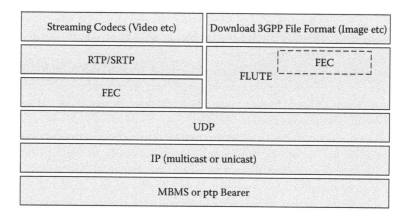

Figure 23.10 MBMS user-service protocol layer architecture.

because content is delivered with cyclical transmission of data. The target media of this service is only static media (e.g., text or still images). The streaming delivery method is intended for continuous receptions (e.g., mobile TV) and is in line with the packet-switched streaming (PSS) service defined by 3GPP. A part of the MBMS protocol stack for streaming and download delivery is shown in figure 23.10. Streaming data such as video streams are encapsulated in RTP and transported over the streaming delivery network. In this case, application layer FEC is applied on UDP flows, either individually or on bundles of streams. Discrete objects such as multimedia streams encapsulated in file formats or other binary data are transported using the FLUTE protocol. MBMS also specifies the media codecs supported for the delivery of multimedia data. For the MBMS streaming delivery method, H.264 (AVC) is the recommended video codec. However, because H.263 is supported by PSS, it may be used also for the MBMS user service. As far as audio is concerned, AMR narrowband (NB), AMR wideband (WB), enhanced AMR-WB, and high-efficiency AAC v2 are supported.

The new features added by MBMS and BCMCS to a mobile infrastructure are:

■ A set of functions that control the broadcast/multicast delivery service. MBMS uses the term *BM-SC* (Broadcast/Multicast Service Center), while BCMCS calls it *BCMCS controller.*
■ Core network supports broadcast/multicast routing of data flows.
■ Efficient radio bearers are used for point-to-multipoint radio transmission within a cell.

The MBMS architecture is depicted in figure 23.11. The dark gray color indicates the modules/function affected by MBMS, while the boxes with stripes indicate the new module BM-SC, which is proposed by MBMS. The BM-SC is responsible for providing and delivering mobile broadcast services. It serves as an access point for content delivery services that want to use MBMS. It is responsible of setting up and controlling MBMS transport bearers to the mobile core network and can be used to schedule and deliver MBMS transmissions. The BM-SC also provides service announcements to end devices. These announcements are necessary for a terminal to join an MBMS service and contain information regarding multicast service identifier, IP multicast addresses, time of transmission, media descriptions, and so on. The BM-SC also manages the

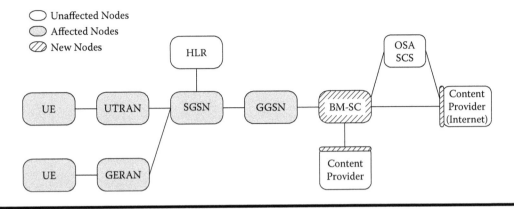

Figure 23.11 MBMS network architecture.

security functions and is capable of generating charging records for the data already transmitted from the content provider. The MBMS standard does not constrain the methods by which the BM-SC functions are to be implemented. So, it can be offered in a separate node or even integrated into existing core and service network nodes.

The broadcast and multicast data distribution trees are created and managed in the core network using functions and protocol messages defined in MBMS. Network operators are able to define broadcast and multicast services for specific geographical areas at very fine granularity, even down to the size of individual radio cells. The necessary configuration is done via the MBMS service area. Each node in the core network uses the list of downstream nodes to determine to which nodes it should forward MBMS data. At the gateway GPRS support node (GGSN) level, the list contains every serving GSN (SGSN) to which the data should be forwarded. At the SGSN level, the list contains every radio network controller (RNC) node of the WCDMA terrestrial radio access network, or in the case of the GSM radio access network, every base station controller (BSC) node, that needs to receive the data. The MBMS system also supports multicast by managing a dynamic data distribution tree and keeping track of users currently registered to the service. Similar to traditional IP multicasting, each core network node forwards MBMS data to the downstream nodes that are serving registered users.

For example, suppose that a network is supporting delivery of mobile TV. If the MBMS system is absent, each user holds a separate streaming connection to the server. Therefore, server and network traffic load are thus directly linked to the number of users. On the other hand, if MBMS is present, the server delivers just one stream per channel to the MBMS BM-SC. The data flow for each channel in the core and radio network is solely replicated when necessary, resulting in reducing the number of simultaneous streams in the network. Furthermore, radio resources in the bottom-most cell need only be allocated for fewer parallel broadcast transmissions, instead of many separate unicast transmissions.

23.5.2.2 *Open Mobile Alliance Broadcast Specification*

Open Mobile Alliance (OMA) is an international organization that works to facilitate the global adoption of mobile data services by specifying market-driven standardized service enablers. The Open Mobile Alliance issued the Mobile Broadcast Services Enabler Suite 1.0 (OMA BCAST 1.0),

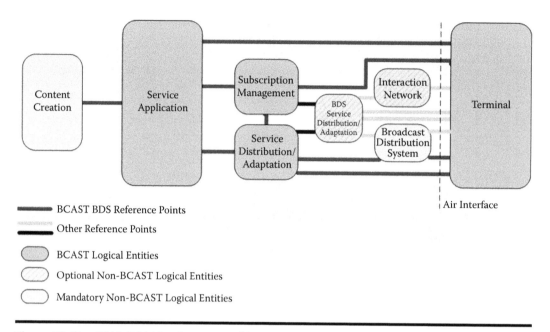

Figure 23.12   OMA BCAST reference architecture.

which supports three underlying broadcast bearers. These are DVB-H, 3GPP MBMS, and 3GPP2 BCMCS. For each underlying bearer, OMA BCAST has created two types of adaptation specifications. The first specification describes how pure BCAST functionality can be deployed over the underlying bearer. The second type of adaptation ensures interoperability as it describes how BCAST functionality should be adapted so the service layer of the underlying bearer and the OMA BCAST service layer can coexist with reuse of overlapping service functionality.

The OMA BCAST reference architecture as it is depicted in figure 23.12 and the BCAST modules will be explained further on. The BCAST service application represents the service application of the BCAST service (e.g., streaming video or movie file download). The BCAST service distribution/adaptation is responsible for the aggregation and delivery of BCAST services and performs the adaptation of the BCAST enabler to underlying broadcast distribution systems. The BCAST subscription management is responsible for service provisioning, such as subscription- and payment-related functions, the provision of information used for BCAST service reception, and BCAST terminal management. The terminal represents the user device as a point of acquisition and consumption for content and service resources.

23.5.3   FLO

The FLO air interface is the bearer technology of the MediaFLO system developed by QUAL-COMM as an alternative mobile broadcast technology for the transmission of multiple multimedia streams to mobile devices. FLO technology allows users to search for channels of content using the same mobile handsets they use for traditional cellular voice and data services, because it has been designed to enable deployment of the broadcast network as an overlay to the cellular network.

The Forward Link Only specification for terrestrial mobile multimedia multicast is standardized within the Telecommunications Industry Association (TIA). The FLO *Air Interface Specification* is the published standard TIA-1099 that specifies the lowest or physical layer of the FLO communications protocol. Associated with the air interface specification is the set that is comprised of the *Minimum Performance Specification for FLO Devices* (TIA-1102), *Minimum Performance Specification FLO Transmitters* (TIA-1103), and the *FLO Test Application Protocol* (TIA-1104) that ties them together. The upper layers that specify the functions and protocols for delivering services over the FLO air interface to FLO-enabled devices are currently the focus of technical discussion within the FLO Forum, which is a multicompany initiative led by QUALCOMM and dedicated to the global standardization of FLO technology. The FLO transport specification was published in 2007 as TIA-1120, following approval by the Telecommunications Industry Association's TR-47.1 Engineering Subcommittee.

23.5.3.1 FLO Architecture and Protocols

FLO technology enables the efficient multicasting of multiple, multimedia services, including real-time video and audio delivery, non-real-time through the clipcast service, and IP Datacast to mobile (FLO) devices. Therefore, among the main tasks that are specified for a FLO network are the transcoding of real-time content to enable display on portable mobile devices and compression of video (and audio) content for efficient transmission in terms of bandwidth, the delivery of IP Datacast content, and the delivery of content to the stream layer of the FLO air interface.

The upper-layer protocols provide multiple functions, including compression of multimedia content, controlling access to the multimedia content, formatting of control information, application of forward error correction for file-based applications, and so on. The FLO protocol stack is shown in figure 23.13, where the FLO media transport and FLO transport layers are presented in light gray.

In a FLO network each data flow is mapped to a stream and multiple streams are combined into multicast logical channels (MLCs). The data flow is finally delivered to the FLO-enabled devices in stream packets. Each service is carried over one or more MLCs. For example, the video component of a service is transmitted over an MLC. The core function of the FLO transport framing layer is to deliver variable-sized service packets over the FLO air interface stream layer as a set of fixed-size frames. In addition, the framing layer provides an optional cyclic redundancy check to verify data integrity. The stream encryption layer provides optional encryption. The FLO media transport layer is a collective label for the layers above the FLO transport layer, and its responsibility is to supply protocol adaptations specific to the class of content being transported. To enable synchronization among real-time video, audio, and data flows, the Sync Protocol is used, while the File Distribution Protocol is used to deliver files efficiently over a FLO network. IP Datacasting is possible through the use of the IP Adaptation Protocol, which adapts IP packets to the FLO framing layer and maps IP addresses to flows.

FLO technology, to provide the best possible quality of service, supports the technique of layered modulation. Therefore, each FLO data stream is divided into a base layer that all users can decode, and an enhancement layer that is decoded in areas where a higher signal-to-noise ratio (SNR) is available. The base layer can deliver 15 fps video quality and has superior coverage compared to an unlayered mode of similar total capacity. Both layers (basic and enhancement layers) can deliver 30 fps video quality. The combined use of layered modulation and source coding allows for graceful degradation of service and the ability to receive in locations or at speeds that could not otherwise have reception.

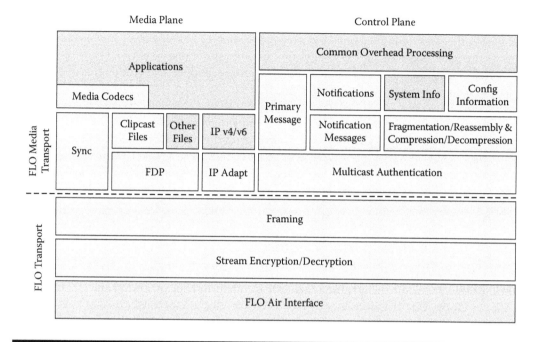

Figure 23.13   FLO MDNI protocol layer architecture.

The FLO system use H.264 for real-time media. The H.264 encoding is extended H.264 compliant for nonlayered applications, and the base layer is H.264 extended compliant in applications in which a layered codec is applied. The picture size supported is QQVGA, QVGA, CIF, and QCIF. The audio codec supported is high-efficiency AAC v2.

A FLO system is comprised of four subsystems: the network operation center, the FLO transmitters, the 3G network, and FLO-enabled devices. The FLO network architecture is depicted in figure 23.14.

The network operation center is the major subsystem of the FLO network. It is comprised of one national operations center (NOC) and one or more local operation centers (LOC). The NOC is responsible for the billing, distribution, and content management of the network. The NOC can manage the various elements of the network, the user-service subscriptions, and the delivery of access and encryption keys, while providing billing information to cellular operators. The NOC plays the role of an access point for national and local content providers, to distribute wide-area content and program guide information to mobile devices. The network operation center may include one or more LOCs that serve as an access point for the local content providers in distributing their contents. The FLO transmitters are responsible for the delivery of multimedia data through FLO waveforms to mobile devices. The 3G network allows mobile devices to communicate with the NOC and LOCs to perform service subscriptions and access key distribution. The 3G network belongs to the wireless operators. FLO-enabled devices are capable of receiving FLO waveforms containing multimedia content and program guide information. In practice, FLO-enabled devices may be cell phones, enhanced to receive FLO waveforms.

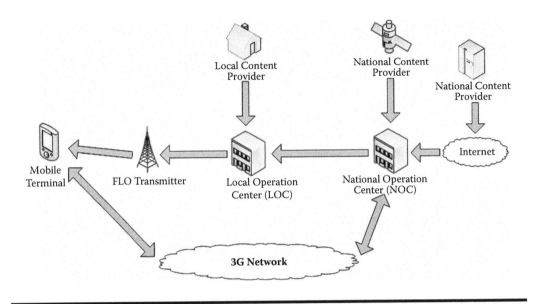

Figure 23.14   FLO system architecture.

In a FLO network, real-time multimedia content is received directly from content providers in MPEG-2 format, typically via a C-band. The MPEG-2 format (704 or 720 × 480 or 576 pixels) is the most common format utilized to enable simple cooperation among content and service providers. The FLO system receives the multimedia content and transcodes it to H.264 QVGA resolution, supported by the rest of the FLO infrastructure. Non-real-time content is received by a content server, typically through a wired link, and then reformatted into FLO packet streams and redistributed. This non-real-time content is delivered according to a prearranged schedule.

The distribution of the content to the FLO transmitters can be done using either wireless or wired links. The content is received by the FLO transmitters, and then the FLO packets are converted to FLO waveforms and radiated out to the mobile devices. If any local content is provided, it will be combined with the wide-area content and delivered to the mobile users. The content is received by only those devices that have subscribed to the service. Then, the content can be stored on the mobile device for future viewing, in accordance with a service program guide, or as a linear feed of content, delivered in real-time to the device. The typical content may consist of high-quality video (QVGA) and audio (MPEG-4 HE-AAC3) as well as IP data streams. Control functions including interactivity and user authorization are provided by a 3G cellular network, such as UMTS or HSDPA.

23.5.4   *Digital Multimedia Broadcasting (DMB)*

DMB constitutes an extension of Digital Audio Broadcasting (DAB), which was proposed and developed in the late 1980s with the intention of transmitting digital radio programs DAB has been adopted by many countries of the world since the 1990s. The Eureka, a pan-European consortium for the funding and coordination of research and development activities, originally initiated the development of DAB. Later, DAB was adopted as a European standard by ETSI, which is also responsible for the standardization of DMB since 2005.

23.5.4.1 DMB Architecture and Protocols

The DMB approach is based on DAB transmission technology, but also proposes several features like additional coding schemes for video and audiovisual content. DMB is enhanced with efficient methods for error correction, enabling high-quality reception of mobile TV programs even for viewers traveling at high speed (up to 200 km/h).

DAB provides two variants for video delivery on mobile handheld devices, DMB and DAB-IP. DMB is merely an addition of video services to the DAB standard to facilitate TV service. DAB has the ability to deliver DMB or DAB-IP mobile TV together with DAB radio within a single system. DAB, DMB, and DAB-IP all share the same infrastructure, spectrum, transmission protocol, and receiver components, which provides flexibility, speed, and economy of scale to any deployment. The video, audio, and interactive services of DMB as well as DAB-IP are explained in detail in the following paragraphs.

The DMB video service enables the digital broadcast transmission of mobile TV programs that are encoded for reception on handheld mobile devices. The video service is delivered through the main service channel stream mode data channel of DAB. The video service is composed of three layers: content compression, synchronization, and transport layers. In the content compression layer, specific video and audio compression methods are employed, as well as MPEG-4 Binary Format for Scenes (BIFS) for the auxiliary interactive data services. To synchronize the audio and video content, both temporally and spatially, MPEG-4 synchronization layer (SL) is employed. In the transport layer, MPEG-2 TS with some appropriate restrictions is employed for the multiplexing of the compressed audio and video content. Figure 23.15 presents the protocol stack for the delivery of DMB video service.

The DMB radio channel has a net data rate of 1.5 Mbps. Therefore, for the compression of video streams, DMB uses H.264/AVC (Advanced Video Coding). Different video resolutions can be utilized for handheld devices such as CIF (352 × 288 pixels), QCIF (176 × 144 pixels), QVGA (320 × 240 pixels), and WDF (384 × 224 pixels) with a maximum refresh rate of 30 fps. For encoding audio, DMB supports MPEG-4 Bit Sliced Arithmetic Coding (BSAC) and MPEG-4 High Efficiency AAC v2. Synchronization of independent video- and audio-associated elementary streams is performed at the MPEG-4 SL, where they are packetized and time-stamped. The synchronized data streams are multiplexed according to the MPEG-2 TS standard.

DAB-IP has been developed for the delivery of packet mode audio, video, and data services to handheld portable devices over a Eureka 147 bearer platform. According to the IP tunneling data

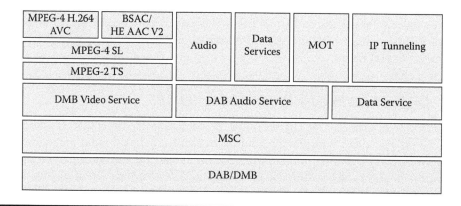

Figure 23.15 DMB video service and DAB-IP protocol layer architecture.

service, IP packets are encapsulated in a main service channel data group on the packet mode transport level. The encapsulation is done by carrying the IP datagram as payload in the MSC data group data field. From the IP point of view, the packet mode SC will act as the Internet data link layer.

IP tunneling may be used apart from the efficient distribution of Internet traffic over the air interface, and also for the transmission of video as an alternative to the DMB video service. However, video generated and encoded for transmission over DMB over the video service is more efficient, as encapsulation and fragmentation of IP (/UDP/RTP) packets as well as the overhead of the IP protocol is avoided. A compromise leading to low overhead, but retaining the advantages of IP Datacast is the adoption of robust header compression according to RFC 3095. Apart from MPEG-4 AVC, Windows media 9 video encoding (vc-1) is also used to encode the source content. RFC 4425 specifies the associated RTP payload format for RTP/UDP/IP delivery.

Multimedia object transfer allows the repeating transmission of frequently demanded objects as video sequences, and so on. Objects are fragmented into smaller segments and their transmission is periodically repeated (carousel). Utilizing this service, the user does not have to make an explicit request to the remote server. Instead, the user tunes in and receives the segments of the desired content on the carousel broadcast.

However, the initial objective of DAB was the distribution of digital radio programs, and therefore the substitution of analog radio. DAB has been designed for transmitting error-free CD quality audio content in contrast to VHF quality analog audio. The DAB audio service uses the MPEG-1 layer 2 source coding scheme. The required data rates depend on the desired sound quality and may vary between 8 and 384 kbps. The DAB audio service supports the transmission of radio programs in mono and stereo mode. Besides radio programs, the DAB service also enables the transmission of short messages, which carry program-associated data (PAD). These messages are transmitted in parallel to ongoing programs and are shown on the terminal's display.

The combination of DMB with mobile cellular networks like GSM and UMTS (figure 23.16) enables the distribution of interactive mobile TV programs, where the viewer can make choices

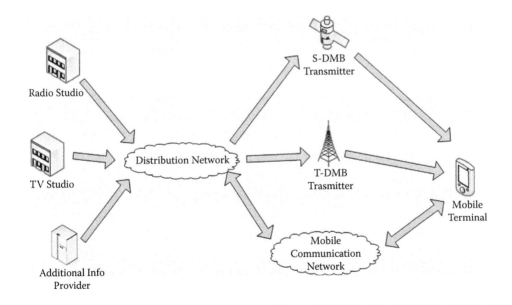

Figure 23.16 **DMB system architecture.**

and take actions. Data services like Short Message Service (SMS) and the General Packet Radio Service (GPRS) of the mobile network can serve as feedback channels for making transactions and returning data of the viewers back to the providers of such programs. However, the underlying protocols needed for interactive TV are not fixed in the DMB standards, but are based in most cases on proprietary solutions.

DAB/DMB uses for transmitting broadcast data frequency channels of 1.536 MHz bandwidth and network data rates of between 1 and 1.5 Mbps. The DMB systems can operate in a very flexible manner between 30 MHz and 3 GHz in the electromagnetic spectrum. This is based on the fact that DMB supports several transmission modes that have been adapted to special propagation phenomena of radio signals in the different frequency ranges. As a consequence, DMB transmission is not limited to terrestrial networks (Terrestrial DMB [T-DMB]), but can also be used by satellites networks (Satellite DMB [S-DMB]). For transmission of data in T-DMB, common frequency bands are in the ranges 174–240 MHz and 1452–1492 MHz, while in S-DMB they are in the range 2605–2655 MHz. In practice, the real utilization of the bands depends on the regulatory constrains of each country in which DMB is developed.

The T-DMB infrastructure is comprised of a network of several transmitters, which is operated as either a single-frequency network (SFN) or multifrequency network (MFN). A SFN network is characterized by the fact that one frequency channel is occupied by all transmitters. To avoid co-channel interferences at the receivers, all transmitters must simultaneously emit the same stream of data and must thus be synchronized with each other. Most SFN networks in DMB occupy frequency channels in band 174–240 MHz, while a coverage range of up to 100 is succeeded by each transmitter. In MFN networks, on the other hand, different frequency channels are assigned to neighboring transmitters. The coverage range of a transmitter is up to 25 km. Therefore, MFNs are a more expensive solution because they need more roll-out and operating costs. In addition, MFNs require the receivers to have installed a handover operation to avoid interruptions during reception when the terminal is crossing the boundary of two neighboring coverage areas.

23.5.5 *Integrated Services Digital Broadcasting–Terrestrial (ISDB-T)*

Integrated Services Digital Broadcasting services were first implemented in Japan in 2000 in the form of ISDB–Satellite systems. ISDB–T is the associated digital radio system developed by the Japanese organization ARIB (Association of Radio Industries and Businesses) for massive delivery of radio and television stations over the air interface. The objective of efforts behind ISDB-T was the design of a flexible system capable of not only sending television or sound programs as digital signals, but also offering multimedia services in which a variety of digital information, such as video, sound, text, and computer programs, will be integrated. ISDB-T is trying to exploit the advantages in terrestrial radio waves so that stable reception can be provided by compact, light, and inexpensive mobile receivers in addition to integrated terminal receivers based on the BST (Band Segmented Transmission)-OFDM scheme. ISDB-T services were initiated in December 2003, while corresponding services for mobile users have been provisioned since 2006.

23.5.5.1 *ISDB-T Architecture and Protocols*

One-segment is the name of the associated terrestrial digital audio/video and data broadcasting service over ISDB-T for mobile devices. The one-seg (segment) name comes from the fact that the service uses one segment of the frequency band allocated to digital terrestrial broadcasts.

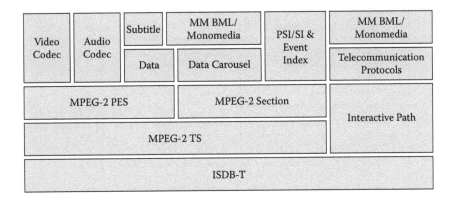

Figure 23.17   ISDB-T protocol layer architecture.

The protocol stack specified in the technical specification standards for digital data broadcasting in Japan set by ARIB is depicted in figure 23.17. Multimedia-encoded data is transmitted in the form of a data carousel in the MPEG-2 TS along with packetized elementary stream–formatted video and audio streams. These multimedia data consist of Broadcast Markup Language (BML) document files and monomedia files such as still images, sound, binary-coded data, and so on. BML is an XML-based multimedia coding scheme, which is defined in ARIB STD-B24. Regarding the one-seg service, ISDB has adopted the H.264 AVC standard (QVGA—4:3 and 16:9) for video delivery and MPEG-2 Advanced Audio Coding (AAC) or MPEG-2 Advanced Audio Coding + Spectral Band Replication (AAC+SBR) for audio delivery. The MPEG-2 video and audio compression system is adopted for high-definition television (HDTV).

The ISDB-T occupies two transmission bands of 5.6 MHz and 432 kHz, each oriented to particular types of broadcasting services. The 5.6 MHz bandwidth is designated for digital broadcasting of television programs, while the 432 kHz bandwidth is designated for audio programs. All other system parameters, such as encoding format, multiplexing format, and OFDM carrier interval and frame configuration, are shared among the above two modes. In a T-ISDB system, the 5.6 MHz wideband can include the 432 kHz narrowband directly because both bands share the same OFDM parameters. Therefore, a 432 kHz receiver can receive some 5.6 MHz services, and a 5.6 MHz receiver can receive all services at 432 kHz.

STD-B24 consists of three parts. Volume 1 is for data coding, volume 2 is for the XML-based multimedia coding scheme, and volume 3 is for data transmission specification. The datacasting system structure is depicted in figure 23.18. STD-B24 specifies all the interface points of 1–5 in detail (Monomedia Coding-1, Coding of Subtitle/Superimpose-2, Multimedia Coding-3, Content Transmission Format-4, Subtitle/Superimpose Transmission Format-5).

23.5.6   3G Broadcast: Mobile Interworking

New services will emerge from a successful interworking between broadcast and mobile systems. These services, to be successful, must fulfill two important criteria: meet the needs of the real user and be easy to use. Successful new services are difficult to predict. In recent years, a lot of effort has been spent in mobile systems on searching for a "killer application." Many researchers believe today that the killer application will in fact comprise a multitude of familiar applications, servicing the many

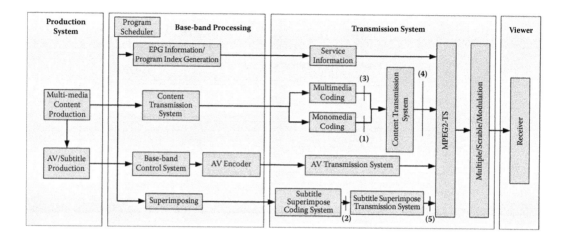

Figure 23.18 ISDB-T system architecture.

market segments that will continue to emerge. In fact, the new emerging applications will allow us to do something we were already doing before, but in a more convenient or effective manner. The wireless technology is based on two major applications: exchange of information and communication, both of which are quite mature. This means that what 3G is actually likely to represent is an increase in convenience and depth over a number of applications that are based on these two simple applications. The profits for the telecom companies will come from increased use of these services as their costs fall and their usefulness rises, rather than from a small number of killer applications.

Future services must be tailored to their application. When TV had emerged, many predicted that it would cause the downfall of cinema. In reality, while TV prevails today, people continue to prefer cinema for watching films. A similar situation applies between newspapers and broadcast news, where newspapers can give more in-depth coverage. This suggests a potential for services that are much more than just the sending of existing broadcast content to mobile handsets. These services, to be applied, need a network comprised of broadcast and mobile infrastructure harmonically coupled.

The important issue is what mobile networks can offer to the broadcast networks. This breaks down into several key attributes, including interactivity, mobility, personalized subscriber information, and real-time billing systems.

23.5.6.1 Interactivity

Interactivity has begun to be implemented over the last years, within the context of digital television by using standard, wired telephone lines. Some years ago, mobile and fixed phones were very effectively, and profitably, used as a back channel for mobile and classical TV, respectively.

23.5.6.2 Mobility

Mobility is essential when people are on the move and want specific, time-sensitive, and often location-sensitive information. Mobile networks can provide location information to broadcast networks, or can allow location-specific information to be provided outside that area. For example, users can watch their own local news program inside and outside of their local territory.

23.5.6.3 Personalization

Personalization is a very sensitive issue in a broadcasting network. If data must be sent individually to a subscriber, then that part of the radio spectrum is serving only one user, which is expensive and imposes severe capacity and cost restraints. However, in many cases, while some of the information is personalized, much may be common to a number of users, making broadcast a more efficient delivery mechanism for such services.

Mobile users will be served by a hierarchy of networks, including cellular, broadcast, local WLAN, and personal area networks (PANs) interconnecting a variety of user terminals. Wireless devices will have reduced in cost to the point where users will have many small specialized devices, most of them operating entirely autonomously.

Machine-to-machine communication has become more important than the traditional person-to-person communication of today's mobile networks. Emerging systems have five key elements:

- Converged services: The users access a seamless pool of personal communication and broadcast services.
- Ubiquitous mobile access: Mobile access to services becomes the most convenient means of accessing communications and information services, and will also take a more significant role in broadcast and entertainment services.
- Diversity of end-user devices: The end users have a large number of small devices tailored to specific tasks.
- Autonomous networks: The networks capable of supporting these devices will be highly adaptive, efficient, and self-managing.
- Software dependency

The interworking of broadcast and 3G networks takes into consideration solutions that use a set of functionality in one network to assist another network in delivering services to users, perhaps services that could not be supported by a single network on its own. In fact, the networks do not converge, but they remain autonomous independent entities.

The interworking network must fulfill the requirements of the below research areas to succeed in delivering new services to users.

- Mobility management: The interworking network should support both IP- and non-IP-based transportation of multimedia and data content, fulfilling the requirements of broadcast and cellular networks.
- Resource management: An effective resource management presupposes exchange of information between networks. Given that such information is unlikely to be freely available, the trade-offs must be examined between the operational efficiency and the information flow and level of trust.
- End-to-end security: The reservation of end-to-end security must be taken into account, focalizing on the point of interworking between the networks.
- Interworked Network Architecture: An overall framework and architecture must be provided to test and verify solutions, including the construction of a tested, demonstrated handover of flows, based on IP.

References

1. ETSI. *Radio broadcasting systems; Digital audio broadcasting (DAB) to mobile, portable and fixed receivers.* ETSI EN 300 401.
2. ETSI. *Digital audio broadcasting (DAB); DMB video service; User application specification.* ETSI TS 102 428.
3. ETSI. *Digital audio broadcasting (DAB); Data broadcasting—MPEG-2 TS streaming.* ETSI TS 102 427.
4. ETSI. *Digital audio broadcasting (DAB); Multimedia object transfer (MOT) protocol.* ETSI EN 301 234.
5. ETSI. *Digital audio broadcasting (DAB); Internet protocol (IP) datagram tunnelling.* ETSI ES 201 735.
6. Samsung. *Digital multimedia broadcasting.* White paper.
7. K. Maalej and S. Pekowsky. 2005. *DVB-H architecture for mobile communications systems.* Broadcast Satellite Communications.
8. ETSI. *DVB specification for data broadcasting.* ETSI EN 301 197.
9. ETSI. *Digital video broadcasting (DVB); IP datacast over DVB-H: Content delivery protocols.* ETSI TS 102 472.
10. ETSI. *Digital video broadcasting (DVB); Specification for the use of video and audio coding in DVB services delivered directly over IP protocols.* ETSI TS 102 005.
11. ETSI. *Digital video broadcasting (DVB); IP datacast over DVB-H: Use cases and services.* ETSI TR 102 473.
12. Nokia. *IP datacasting– Bringing TV to the mobile phone.* White paper.
13. H Bürklin, R. Schäfer, and D. Westerkamp. 2007. DVB: From broadcasting to IP delivery. *SIGCOMM Comput. Commun. Rev.* 37:65–67.
14. Texas Instruments. *DVB-H mobile digital TV.* White paper.
15. M. Kornfeld and U. Reimers. 2005. *DVB-H—The emerging standard for mobile data communication*, 1–10. EBU technical review 301.
16. G. May. 2004. The IP datacast system—Overview and mobility aspects. In *Proceedings of the IEEE International Conference on Consumer Electronics*, 509–514.
17. G. Faria, J. Henriksson, E. Stare, and P. Talmola. 2006. DVB-H digital broadcast services to handheld devices. *Proceedings of IEEE* 94:194–209.
18. UDCAST. *DVB-H mobile TV flexible satellite distribution.* White paper.
19. G. May. 2005. IP datacast over DVB-H. Technischer Bericht, Institut fur Nachrichtentechnik, Technische Universitat Braunschweig.
20. Digitag. 2005. *Television on a handheld receiver—Broadcasting with DVB-H.* White paper.
21. bmco-forum e.V. 2007. *Mobile broadcast bearer technologies—A comparison.*
22. QUALCOMM. 2005. *MediaFLO—FLO technology overview.*
23. QUALCOMM. 2005. *MediaFLO—Media distribution system—Product overview.*
24. QUALCOMM. 2007. *MediaFLO—Technology comparison: MediaFLO and DVB-H.*
25. QUALCOMM. 2005. *MediaFLO—FLO technology brief.*
26. TIA. *Forward link only air interface specification for terrestrial mobile multimedia multicast.* TIA-1099.
27. TIA. *Minimum performance specification for terrestrial mobile multimedia multicast forward link only devices.* TIA-1102.
28. TIA. *Minimum performance specification for terrestrial mobile multimedia multicast forward link only transmitters.* TIA-1103.
29. TIA. *Test application protocol for terrestrial mobile multimedia multicast forward link only transmitters and devices.* TIA-1104.
30. M. Selby. Progress on FLO protocol stack standardization. *FLOFocus*, Issue 3, April 2007.
31. ARIB STD-B24 Version 3.2. 2001.
32. ARIB STD-B24 Version 5.0. 2003.
33. H. Asami and M. Sasaki. 2006. Outline of ISDB systems. *Proceedings of the IEEE* 94.
34. IETF. 2003. *RTP: A transport protocol for real-time application.* RFC 3550.

35. T. Paila, M. Luby, R. Lehtonen, V. Roca, and R. Walsh. 2004. *FLUTE–File delivery over unidirectional transport*. RFC 3926.

36. M. Luby, J. Gemmell, L. Vicisano, L. Rizzo, and J. Crowcroft. 2002. *Asynchronous layered coding (ALC) protocol instantiation*. RFC 3450.

37. J. Peltotalo, S. Peltotalo, and J. Harju. 2000. Analysis of the FLUTE data carousel. Paper presented at Proceedings of the 10th EUNICE Open European Summer School, Colmenarejo, Spain.

38. Real. 2006. *Delivering total mobile TV. Combining streaming and broadcast for a complete mobile television solution*. White paper.

39. Alcatel. 2006. *Unlimited mobile TV for the mass market*. White paper.

40. Texas Instruments. 2005. *Digital broadcast TV—Coming soon to a mobile phone near you*. White paper.

41. ISO/IEC. *Information technology—Generic coding of moving pictures and associated audio information: Systems*. ISO/IEC 13818.

42. 3GPP. *Transparent end-to-end packet-switched streaming service (PSS) protocols and codecs*. 3GPP TS 26.134.

43. 3GPP. *Multimedia broadcast/multicast service; Stage 1*. Release 6. 3GPP TS 22.146.

44. 3GPP. *Multimedia broadcast/multicast service (MBMS) user services; Stage 1*. Release 6. 3GPP TS 22.246.

45. 3GPP. *Multimedia broadcast/multicast service (MBMS); Architecture and functional description*. Release 6. 3GPP TS 23.246.

46. 3GPP. *Multimedia broadcast/multicast service (MBMS); Architecture and functional description*. Release 6. 3GPP TS 23.846.

47. 3GPP. *Introduction of the multimedia broadcast multicast service (MBMS) in the radio access network (RAN); Stage 2*. Release 6. 3GPP TS 25.346.

48. 3GPP. *Multimedia broadcast multicast service (MBMS); UTRAN/GERAN requirements*. Release 6. 3GPP TR 25.992.

49. 3GPP. *Multimedia broadcast/multicast service (MBMS); Protocols and codecs*. Release 6. 3GPP TS 26.346.

50. R. Schafer and T. Twiegand. *The emerging H.264/AVC standard*.

51. Nokia. *Nokia mobile broadcast solution*. White paper.

52. M. Bakhuizen and U. Horn. *Mobile broadcast/multicast in mobile networks*.

53. H. Jenkac, T. Stockhammer, W. Xu, and W. Abdel Samad. 2006. Efficient video-on-demand services over mobile datacast channels. *Journal of Zhejiang University Science A* 7:873–84.

54. T. Stockhammer, T. Gasiba, W. Abdel Samad, T. Schierl, H. Jenkac, T. Wiegand, and W. Xu. 2007. Nested harmonic broadcasting for scalable video over mobile datacast channels. *Wireless Communications and Mobile Computing* 7:235–56.

55. K. Matsumura, K. Kai, H. Hamada, and N. Yagi. 2005. Transforming data broadcast contents to fit different user interfaces: Generating a readout service for mobile DTV receiver. In *Proceedings of the 7th International Conference on Human Computer Interaction with Mobile Devices and Services*, 323–24. Vol. 111. New York: ACM Press. http://doi.acm.org/10.1145/1085777.1085847.

56. V. Balabanian, L. Casey, N. Greene, and C. Adams. 1996. An introduction to digital storage media-command and control. *IEEE Communications Magazine* 34:122–27.

57. D.-H. Park, T.-Y. Ku, and K.-D. Moon. 2006. Real-time carousel caching and monitoring in data broadcasting. *IEEE Transactions on Consumer Electronics* 52:144–49.

58. ETSI. *DVB implementation guidelines for data broadcasting*. ETSI TR 101 202.

59. ISO/IEC. *MPEG extensions for DSM-CC*. ISO/IEC 13818-6.

60. A. Hu. 2001. *Video-on-demand broadcasting protocols: A comprehensive study*. INFOCOM.

61. A. Hori and Y. Dewa. 2006. Japanese datacasting coding scheme BML. *Proceedings of the IEEE* 94:312–17.

Chapter 24

Broadcasting Techniques for Video on Demand in Wireless Networks

Duc A. Tran and Thinh Nguyen

Contents

Keywords

video on demand, video streaming, ad hoc networks, wireless networks, mobile networks, periodic broadcast

24.1 Introduction

Today's wireless technologies such as IEEE 802.16 (a.k.a. WiMAX)[3] for long-haul communications and IEEE 802.11 (e.g., WiFi)[15] and Bluetooth[5] for short distances are widely deployed. While WiFi is suitable for a small local wireless network, WiMAX (Worldwide Interoperability for Microwave Access) allows for communications over tens of kilometers and thus can be used to provide a high-speed wireless "last mile" service to local networks. As beneficiaries, users can move freely without disconnection from the network. Wireless networks are therefore an excellent infrastructure for important applications such as disaster relief efforts and military collaborative networks.

Being wireless and mobile also means that users may enjoy ubiquitous entertainment services. For example, they may play games or watch videos of their interest online wherever they are. This chapter focuses on technologies that enable mobile video-on-demand (VOD) services. A VOD system is an interactive multimedia system working like cable television, the difference being that the client can select a movie from a large video database stored at a distant video server. Unlike other video services such as pay-per-view (PPV) and video in demand (VID), individual VOD clients in an area are able to watch different programs *whenever* they wish to, not just in time as in VID or prescheduled as in PPV. A VOD system is therefore a realization of the video rental shop brought into the home.

A mobile wireless VOD system has many practical applications. For instance, airlines could provide VOD services in airport lounges to entertain passengers on their PDAs (personal digital assistants) while they are waiting for a flight; a museum could provide video information on the exhibits on demand over the wireless network; in education, a university could also install such a system on campus to allow students to watch video recorded earlier from lectures they were not able to attend.

24.1.1 VOD on the Internet

VOD services are already available on the Internet. News video clips can be rendered on demand on most Web media outlets (e.g., cnn.com, espn.com). Video commercials in business areas ranging from automotive to real estate to health and travel can also be played on demand via a product of Comcast called ComcastSpotlight.[7] The major server providers for VOD deployment include Motorola On-Demand Solutions,[19] SeaChange International,[23] and Concurrent Corp.[8] Informa Telecom* predicted a revenue of more than U.S. $10.7 billion from VOD services offered to more than 350 million households by 2010. The future of VOD business is very bright.

The designs for current VOD systems can be categorized into three main approaches: client/server, peer-to-peer, and periodic broadcast. They are described below:

Client/server:[14,10,11,31] The video content is stored at the server. Each client connects to the server and plays the requested video from the server, independently from other clients' transactions. The server workload may be heavy if there are many client requests. Proxy servers

---

* http:/www.information.com

can be deployed to reduce this load. Alternatively, clients requesting the same video can be grouped into a single multicast sent by the server. The client/server approach is used in most commercial VOD products nowadays because of its simplicity and central management.

Peer-to-peer (P2P):[25,26,20,32] The motivation for P2P is due to the bottleneck problem at the server side. In this approach, not only the video server but also clients can provide the video content to each other, thus saving server bandwidth. For example, using P2P, a TiVo subscriber could not only replay a video previously recorded in his or her TiVo box, but potentially play a video recorded in some other subscriber's box without requesting the TV central server. Most P2P VOD systems are implemented in laboratory settings, but we expect them soon to be on the market.

Periodic broadcast:[1,12,22,28,24] For a popular video, it is better for the server to proactively and periodically broadcast it to all the clients, rather than send it out reactively upon a client request. The advantage of the broadcast approach is that the server bandwidth required is constant and the system can satisfy any number of clients. It avoids the bottleneck problem of the client/server approach and the service vulnerability of the P2P approach.

These three approaches are like "apples and oranges" when it comes to comparison because each approach is effective for some subset of VOD applications. Video popularity is known to practically follow a 80/20-like rule of thumb;[9] that is, most clients would be interested in only a few popular videos. As such, periodic broadcast should be the best design for transmitting popular videos to a large number of clients, while client/server and P2P techniques are better suited for nonpopular videos or for videos requested by a small client population.

24.1.2   VOD for Mobile Wireless Users

When deployed to a wireless environment, the success of existing VOD designs may not remain. The first reason is due to the limitation of wireless bandwidth. Because the most widely used form of wireless communications in a local area is using IEEE 802.11 technologies (a, b, or g), the network bandwidth shared by all the users covered by an access point is typically no more than 54 Mbps. Therefore, no more than 36 MPEG-1 video streams can be delivered simultaneously to this local area. Taking into account signal inference and distance factors, the effective number of such streams would be much fewer.

The second reason is due to the limited coverage of wireless transmission. An 802.11-enabled host can only reach other devices within 100 m of its radius, while that radius is 10 m for Bluetooth. If a user is too far away from any access point, how can it get the video service? Fortunately, wireless hosts are nowadays able to concurrently participate in multiple connections: with the access point in the infrastructure-based mode and with a nearby host in the ad hoc mode. Therefore, it is possible that a distant user could get the video service from the access point via several intermediate hosts. The problem is that significant amounts of bandwidth and energy of the intermediate mobile hosts are consumed. We do not have this problem with the typical Internet.

So, we have the following two questions:

1. What should be the architecture for a mobile wireless VOD system?
2. What should be the communication protocol for a client to download a video from the video server?

To cope with the limited wireless bandwidth, the client/server approach should not be the first design option for VOD in a local wireless area. Also, as each wireless transmission is a broadcast where every host can hear, it is natural to think that it would be more efficient if we adopt the broadcast approach. It would be best if we apply the broadcast approach to the most popular videos and the client/server approach only for ad hoc unpopular video requests.

To cope with the limited wireless coverage, we should allow sharing of video contents among the users. For instance, instead of playing a video through multiple hops from an access point, we hope to play the video or part of it in some existing users nearby. In other words, we should adopt the P2P approach for this kind of users.

In the next section, we review existing periodic broadcasting techniques for VOD on the Internet and their potential for deployment in a mobile wireless network. We then introduce MobiVOD, a mobile VOD technique that combines the strength of periodic broadcast and P2P approaches.

24.2 VOD Periodic Broadcast Techniques

Cable TV and satellite TV follow the broadcast approach, in which TV programs are broadcast at prescheduled time slots. This simple broadcast schedule is not designed for VOD because if a TV client wants to watch a movie instantly, the movie will not be shown until the scheduled time. A TiVo box can offer VOD, but only *locally*. What if the client wants a movie not prerecorded in the TiVo box? Periodic broadcast (PB) techniques are aiming for that service.

In the PB approach,[1,12,22,24,28] a video is divided into several segments, each repeatedly broadcast on a separate communication channel from the server. A client receives a video by tuning to one or more channels at a time to download the data. The broadcast schedule at the server and playback synchronization protocol at the client ensure that the broadcast of the next video segment is available to the client before the playback of the current segment runs out. Many other PB techniques have been proposed, but the most popular ones are staggered broadcast,[18] skyscraper broadcast,[12] pyramid broadcast,[28] pagoda broadcast,[22] harmonic broadcast,[16] permutation-based broadcast,[1] and fast broadcast.[17] We review these techniques in this section. For other PB techniques, a comprehensive survey can be found in Hua et al.[13]

24.2.1 Staggered Broadcasting

Staggered broadcast (SB)[18] is a simple periodic broadcast technique. It works as follows. Without loss of generality, we focus on a single video. This video is partitioned into K equally sized segments $\{s_1, s_2, \ldots, s_K\}$. The duration of each segment is $s = V/K$, where V is the duration of the entire video. Given the video consumption rate r (bps), we allocate a server bandwidth of $r \times K$ for the video.* This bandwidth is divided into K equal channels, each repeatedly broadcasting the video with a transmission rate equal to the consumption rate. The scheduling of these broadcasts is illustrated in figure 24.1.

The playback procedure at a client follows the simple algorithm below:

Algorithm 24.2.1   (Staggered Broadcasting) *Client Playback*

1. *Tune in a random channel i. Suppose that channel i is currently broadcasting segment s_h.*
   (a) *If $h = K - 1$, let $j = i$ and go to step 2;*

---

* We assume that the server has enough bandwidth for this allocation.

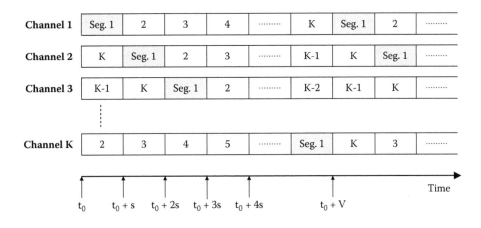

Figure 24.1 **Broadcasting video segments at the server: *s* is the duration of a segment, t_0 the start time of server broadcast, and *V* the duration of the video.**

(b) *Else, compute*

$$j = \begin{cases} i+h-K, & \text{if } i+h-K > 0 \\ i+h, & \text{otherwise} \end{cases} \tag{24.1}$$

■ *Channel j must be currently broadcasting segment s_K and about to broadcast segment s_1.*
2. *Wait until channel j starts broadcasting segment s_1 and join this channel when that time comes.*
3. *Play the video data received from this channel and quit when the video is finished playing.*

A client joins only one broadcast channel at any time; thus, the client bandwidth required is no more than the playback rate. In addition, the rendering at the client only consists of receiving a video packet, decoding, and displaying it. Therefore, the computational complexity required for playback is minimal.

A disadvantage of SB is a high service delay. If a client requests the video during the broadcast of segment s_1, this client has already missed the already-broadcast packets belonging to s_1 and must wait until the next broadcast of this segment. For instance, in figure 24.1, if a client requests at time $t_0 + s + \delta(0 < \delta < s)$, it must wait until time $t_0 + 2s$ to start downloading segment s_1. Hence, the service delay is $s - \delta$; in the worst case, it is s. Supposing $K = 5$ channels, which is likely to be the case with a IEEE 802.11g video server and MPEG-1, broadcasting a 60-minute video results in a worst-case delay of $s = V/K = 60/5 = 12$ minutes.

24.2.2 Harmonic Broadcasting

Similar to SB, harmonic broadcasting (HB)[16] divides each video into *K* equally sized segments {s_1, s_2, … s_K}. The server bandwidth for a video is also divided into *K* channels, however with different bandwidth allocation. Specifically, each segment s_i is broadcast repeatedly on channel *i* at rate

r/i (bps), hence the name *harmonic*. As such, it takes longer to download a later segment than a previous segment.

At the client side, it has to tune in all K channels and receive the data for all the K segments. The client starts the playback as soon as it can download the first segment. Before that, all the received segments are stored in the client cache. The cache space required can be up to 40 percent of the video.[16] The good thing is that, because of the harmonic rate allocation, it is guaranteed that by the time a segment is finished playing, the next segment is either in the cache or readily available from its broadcast channel. Therefore, the playback is smooth.

An advantage of SB is that it requires reasonable server bandwidth, which is equal to $\sum_{i=1}^{K} \frac{r}{i} = r \sum_{i=1}^{K} \frac{1}{i}$. This is much less than K_r—the server bandwidth required by SB. On the other hand, the client bandwidth requirement is very high, same as the server bandwidth. For example, the client would need 240 channel tuners if desiring a service delay of 30 seconds for a 2-hour video. This is practically not desirable. Also, the service delay of HB is as long as that of SB and equal to V/K—the duration of a segment. Later techniques such as cautious harmonic broadcasting and quasi-harmonic broadcasting[21] attempt successfully to erase the long service delay of HB, but the caching and bandwidth cost at the client side remains substantial.

24.2.3 Fast Broadcasting

SB broadcasts all segments of each video on every channel. HB broadcasts each segment on a different channel. Unlike these two strategies, fast broadcasting (FB)[17] allows the broadcasting of more than one segment on each channel.

FB divides each video into n equally sized segments. The server bandwidth for a video is also divided into K equal-rate channels as in SB. The broadcast schedule at the server is as follows:

- Channel 1 broadcasts segment s_1 repeatedly.
- Channel 2 broadcasts two segments, s_2 and s_3, repeatedly one after another.
- Channel 3 broadcasts 4 segments, s_4, s_5, s_6, s_7, repeatedly one after another.
- Channel i ($i > 3$) broadcasts 2^{i-1} segments, s_{2i-1}, $s_{2i-1}+1$, ... , s_{2i-1}, repeatedly one after another.

The number of segments n therefore must be $n = 1 + 2 + 2^2 \ldots + 2^{K-1} = 2^{K-1}$.

For playback, the client listens to all the channels and stores the segment downloaded but not yet played into the client cache. Therefore, the client bandwidth required is as large as the server bandwidth (K channels) and the client cache is also significant. The service delay is the wait time for the first segment. In the worst case, this delay is $V/n = V/(2^K - 1)$. An advantage over SB and HB is that this delay decreases exponentially as K increases.

24.2.4 Pagoda Broadcasting

Similar to FB, pagoda broadcasting (PaB)[22] allows the broadcasting of more than one segment on each channel. PaB differs from FB in the selection of segments to broadcast on each channel.

PaB divides each video into n equally sized segments. The server bandwidth for a video is also divided into K equal-rate channels. Each channel is logically divided into time slots, each for a duration of a segment. The time slots are indexed as $slot_0$, $slot_1$, $slot_2$, ...

At the server side, channel 1 broadcasts s_1 repeatedly $\{s_1, s_2, \ldots\}$. We consider channel i. There are two cases:

i is even: Suppose that s_z is the earliest segment not broadcast on channels 1, 2, ..., $i-1$. It will be broadcast in every slot $slot_{jz}$ ($j = 0, 1, ...$). All the other even-indexed slots will be equally allocated to segments s_{z+1}, s_{z+2}, ..., $s_{3z/2-1}$. All the other slots will be equally allocated to segments s_{2z}, s_{2z+2}, ..., s_{3z-1}.

i is odd: Supposing that s_z is the first segment broadcast on channel $i-1$, the earliest segment not yet broadcast is segment $s_{3z/2}$. $s_{3z/2}$ is broadcast on channel i in every slot $slot_{j(3z/2)}$ ($j = 0, 1, ...$). Every third slot is equally allocated for segments $s_{3z/2}$ to s_{2z-1}. The remaining slots are equally allocated to segments s_{3z} to s_{5z-1} in such a way that each pair of consecutive sets of $3z/2$ slots contains exactly one instance of each of these $2z$ segments.

Therefore, channel 2 will be broadcasting in the order s_2, s_4, s_2, s_5, s_2, s_4, s_2, s_5, ..., and channel 3 will be broadcast in the order s_3, s_6, s_8, s_3, s_7, s_9, s_3, s_6, s_8, s_3, s_7, s_9, ...

Similar to HB and FB, PaB requires the client to tune in all channels to download data and store yet-to-be-played data into the client cache. The playback starts as soon as the client receives the first segment. Paris et al.[22] proved the following relationship between the number of segments *n* and the number of channel *K*:

$$n = \begin{cases} 4(5^{k-1})-1, & \text{if } K = 2k \\ 2(5^k)-1 & \text{if } K = 2k+1 \end{cases} \tag{24.2}$$

The service delay is due to the wait time for the first segment on channel 1. Therefore, it decreases exponentially as *K* increases.

24.2.5 Pyramid Broadcasting

All the aforementioned broadcasting techniques partition each video into segments of equal size. On the contrary, pyramid broadcast (PyB)[28] partitions each video into *K* segments of increasing size (hence the name *pyramid*):

$$S_i = \begin{cases} V(\alpha-1)/(\alpha^K-1), & \text{if } i=1 \\ s_1 \times \alpha^{i-1}, & \text{otherwise} \end{cases} \tag{24.3}$$

Suppose that the server bandwidth allocated for each video is B, which is divided into K channels. Thus, the broadcast rate at the server is B/K on each channel. At this rate, channel 1 broadcasts repeatedly segment s_1, channel 2 broadcasts repeatedly segment s_2, and so on. The client has two video loaders that can download data from two channels concurrently. The video playback is done as follows:

Algorithm 24.2.2   (Pyramid Broadcasting) Client Playback

1. *Tune in channel 1 and start to download the first segment s_1 at the first occurrence and play it concurrently.*
2. *Set i = 1.*
3. *While ($i \leq K$):*
   (a) *As soon as segment s_i starts playing, also tune in channel $i+1$ to download segment s_{i+1} at the earliest possible time and store it in a buffer.*

(b) *Once segment s_i finishes playing, switch to play segment s_2 from the buffer.*
(c) *Set i: i + 1.*
4. *End While.*

The α value is chosen in such a way to ensure that the playback duration of the current segment must be longer than the worst delay in downloading the next segment. This is equivalent to

$$\frac{s_i}{r} \geq \frac{s_{i+1}}{B/K} \Leftrightarrow \alpha \leq \frac{B}{rK} \tag{24.4}$$

Based on this inequality, Viswanathan and Imielinski[28] suggested two options for the α value:

$$\alpha_1 = \frac{B}{r \times \lfloor \frac{B}{re} \rfloor} \text{ or } \alpha_2 = \frac{B}{r \times \lceil \frac{B}{re} \rceil} \tag{24.5}$$

where $e \approx 2.72$ is Euler's constant. Therefore, α is close to this constant. The number of channels K is chosen to be $K_1 = \lfloor \frac{B}{re} \rfloor$ (case α_1) or $K_2 = \lceil \frac{B}{re} \rceil$ (case α_2).

The wait time before the video starts playing equals the wait time until the first segment is first downloaded from channel 1. In the worst case, this delay is $\frac{VKr(\alpha-1)}{B(\alpha^K-1)}$. Assuming that $r = B/K$, this delay equals $V(\alpha-1)/(\alpha^K-1) = V/(1+\alpha+\alpha^2+\cdots+\alpha^{K-1}) \leq V/K$ because $\alpha \approx 2.72$. Therefore, PyB's delay is much better than both SB and HB's. In addition, the former is improved exponentially when K increases, while the latter can only improve linearly.

On the other hand, PyB requires caching at the client side. The memory space needed for this caching is as large as $r(s_K + s_{K-1} - rKs_K/B)$. This is approximately the size of the second-last segment s_{K-1}. Since segment size increases exponentially, size s_{K-1} is significant. If α is kept around e, each client must have a disk space large enough to buffer more than 70 percent of the video file.

24.2.6 Permutation-Based Broadcasting

Permutation-based broadcasting (PbP)[1] is an extension to PyB in an attempt to reduce the client cache space requirement, however at a cost of more complexity. In PyB, each segment is broadcast sequentially on its corresponding channel. In PbP, each channel is further divided into P subchannels with $\frac{B}{Kp}$ (bps) each. A replica of segment s_i is repeatedly broadcast on each of the P subchannels of channel i with a phase delay of s_i/P seconds. Because the subchannels are time multiplxed on their parent channel, the client cache space required is less. The new requirement at each client is

$$\frac{rVK(\alpha^K - \alpha^{K-2})}{B(\alpha^K - 1)} \tag{24.6}$$

The service delay is simply the access time of the first segment; hence, $\frac{s_1}{P+\alpha}$. In PbP, we choose $K = \lfloor \frac{B}{Kr} \rfloor$, but its value must be in the range [2, 7]. We can choose the following values for α and P:

$$P = \left\lfloor \frac{B}{Kr} - 2 \right\rfloor \quad \text{and} \quad \alpha = \frac{B}{Kr} - P \tag{24.7}$$

Typically, PbP requires a client to cache about 50 percent of the video size, much less than that required by PyB. However, when the server bandwidth is more generous, the delay of PbP can only increase linearly as opposed to exponentially, as in the case of pyramid broadcasting. Hua and Sheu[12] showed that PbP actually performs worse than PyB.

24.2.7 Skyscraper Broadcasting

Similar to PyB, skyscraper broadcasting (SkB)[12] also divides the server bandwidth for each video into K channels, each of rate B/K and broadcasting repeatedly a segment of the video. However, unlike PyB, SkB decomposes the video into K segments of the following sizes:

$$s_i = \begin{cases} s_1, & i=1 \\ 2s_1, & i=2,3 \\ 2s_{i-1}+1, & i \bmod 4 = 0 \\ s_{i-1}, & i \bmod 4 = 1 \\ s_{i-1}+2, & i \bmod 4 = 2 \\ s_{i-1}, & i \bmod 4 = 3 \end{cases} \tag{24.8}$$

That is, the sizes of the segments are

$$[1, 2, 2, 5, 5, 12, 12, 25, 25, 52, 52, \ldots] \times s_1.$$

The technique is called skyscraper because of this segment sizing.[*] SkB uses a control value W to restrict the segments from being too large. If a segment s_i becomes larger than W_{s_1}, its size is rounded to W_{s_1}. This is because if we allow a segment to grow too large, a large caching space would be required for each client. The size of the first segment s_1 is chosen so that $\sum_{i=1}^{K} s_i = V$. Thus, we can control the size s_1 by changing W. A nice thing about SkB is the following relationship between the serviced delay and the control factor W:

$$delay = s_1 = \frac{V}{\sum_{i=1}^{K} \min(s_i, W)}, \tag{24.9}$$

which can be used to determine W given the desired service delay.

The video segments are classified into groups, each containing consecutive segments of the same size. For example, $[s_1]$, $[s_2, s_3]$, and $[s_4, s_5]$ are different groups of sizes 1, 2, and 5, respectively. A group is called an odd (or even) group if its segment size is odd (or even). The client uses two software modules: an even loader to download even-group segments and an odd loader to download odd-group segments. Each loader tunes into the appropriate channels to download its groups one at a time and in the order they appear in the video. The downloaded segments are forwarded into a buffer that can be played by the video player module at the playback rate.

---

[*] If we stack boxes of these sizes in decreasing order from bottom up, the shape looks like a skyscraper. The shape for pyramid broadcasting looks like a pyramid.

The client bandwidth and cache requirement is as follows:

$$cache\ space = (W - 1)_{s_1} B/K \tag{24.10}$$

$$bandwidth\ required = \begin{cases} 0, & W = 1\ or\ K = 1 \\ 2B/K & W = 2\ or\ K = 2,3 \\ 3B/K & otherwise. \end{cases} \tag{24.11}$$

Hua and Sheu[12] showed that SkB is able to retain the low latency of PyB while using significantly less buffer space (20 percent) than that required by PbP.

24.3 MobiVoD: A Mobile Wireless VOD Solution

Despite many periodic broadcast designs for the typical Internet-based networks, when applied to a wireless network, they may not be directly applicable. Because mobile wireless clients are usually of limited resources, some of these techniques (HB, FB, PaB) are not well suited because they require significant client bandwidth and caching space. Such is one case with PbP because it incurs complex client playback. PyB is better in terms of client bandwidth, but its client caching requirement remains very high. The two techniques with potential for efficient deployment in a large-scale wireless environment are SB and SkB. We summarize the client resource requirement of these periodic broadcast techniques in table 24.1.

None of these periodic broadcast techniques can provide true VOD because their service delay is nonzero. Between SkB and SB, the former provides better service delay. On the other hand, SkB is more complex and requires that the client be capable of downloading at a rate twice as large as the playback rate and have caching space enough for approximately 10 percent of the video length. For current wireless architectures, SB seems a better choice because of its simplicity. In this section, we discuss how SB can be adapted to work for large-scale wireless networks. As wireless and memory technologies continue to advance, SkB could be the better choice in the future. The technique discussed here, called MobiVOD and proposed in Tran et al.,[27] is also applicable to SkB (with a slight change).

The main problem with SB is its service delay. MobiVOD is an adaptation of SB that erases this delay by leveraging video content sharing between the wireless clients.

Table 24.1. Typical Client Requirement in Periodic Broadcasting Solutions (*r* is Consumption Rate)

Solution	Caching Space	Bandwidth
Staggered[18]	0% of video	$1 \times r$
Skyscraper[12]	10% of video	$2 \times r$
Pyramid[28]	75% of video	$\geq 4 \times r$
Permutation-based[1]	20% of video	$\geq 2 \times r$
Pagoda[22]	45% of video	$\geq 5 \times r$
Harmonic[16]	40% of video	$\geq 5 \times r$
Fast[17]	50% of video	$\geq 6 \times r$

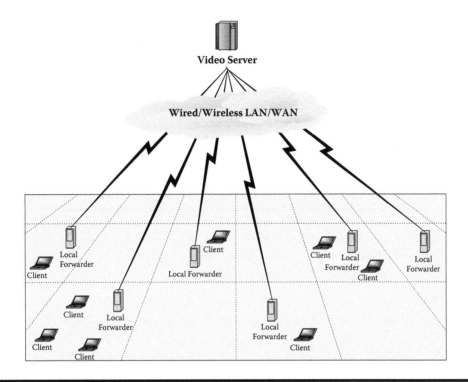

Figure 24.2 MobiVoD system architecture: server, local forwarders, and clients.

24.3.1 Network Architecture

Illustrated in figure 24.2, the system architecture for MobiVOD consists of three components: *video server*, *clients*, and *local forwarders*. The video server stores video files. Clients are the mobile users (devices or the people who use them), who subscribe for the VOD service provided by our system. Because the only way to communicate with the clients is via wireless transmissions, it is not possible for the video server to transmit a video to clients located in a too-wide geographic area. Therefore, we may deploy a scatter of local forwarders $\{LF_1, LF_2, ..., LF_k\}$. A local forwarder LF_j is a stationary and dedicated computer and used to relay the service to LF_j's transmission coverage range. This area is called a local service area.

The local forwarders and clients are referred to as nodes. Each of the nodes is equipped with a wireless network interface card or two, so that they can receive data from their local forwarder and, at the same time, form among themselves an ad hoc network. The rationale for this multiconnectivity is that MobiVOD allows clients to share and exchange video information with each other, directly without going through a broker node like the server or any local forwarder.

Every local forwarder receives the video packets from the server, via either a wired broadband connection or a wireless broadband connection like WiMAX. This local forwarder then broadcasts the packets to its local coverage, for example, using IEEE 802.11. If a client is within the service area of a local forwarder, the former can receive the video packets.

The topology for disseminating the video packets from the server to every local forwarder can be a star rooted at the server or any overlay topology connecting the server and all the local forwarders. The working environment the system is running in determines the locations of the local forwarders. For instance, on a campus or at an airport terminal, the local forwarders should be geographically uniformly distributed. In a big building with closed-door halls, however, we should install a local forwarder in each hall.

24.3.2 Client Playback

Each local forwarder (or the server in the case without local forwarders) uses staggered broadcasting to broadcast each video. As described in algorithm 24.2.1, the client tunes in the channel that is to broadcast the first segment the soonest and starts the playback as soon as the first segment arrives. Therefore, the service delay can be as large as the duration of the first segment.

The idea of MobiVOD is that when a new client starts the video request and the first segment is not yet available on any broadcast channel, the new client can get this segment instantly from a nearby client who has a cache of it. Thus, we need to determine who should cache or who should not. Obviously, if a cache is very far from the new client, it is helpless because there is no efficient way for the client to download the segment in a multihop manner. Thus, a key design component of MobiVOD is client caching.

A MobiVOD client has the following two buffers:

- *Reusable buffer*: This buffer is used to cache the first segment of the video. Therefore, the size of this buffer is that of the first segment. A client needs a reusable buffer if it is selected to cache the first segment.
- *Prefetched buffer*: This buffer is used for video playback. The size of this buffer is that of the already-broadcast portion that the client misses. A client needs this buffer if it opts to make use of the first segment cached at a nearby client's reusable buffer.

Let us consider a new arriving client X who detects that it already missed the current broadcast of the first segment. For example, in figure 24.1, the new client requests the video at some time between t and $t + s$; thus, the next broadcast of the first segment is at time $t + s$. Instead of waiting for the next broadcast of the first segment, this client looks for an existing client Y in its transmission range, who has a cache of the first segment in the reusable buffer. If such Y exists, X can download and play the missing portion from Y, and, at the same time, store the packets broadcast from X's local forwarder into the prefetched buffer. Once X finishes playing the missing portion, it switches to play the data in the prefetched buffer. In this case, though X misses the current broadcast of the first segment, X still manages to watch the entire video immediately.

There are three caching strategies:

- ALL-CACHE: Every client caches the first video segment.
- RANDOM-CACHE: A client randomly decides whether it should cache the first video segment.
- DOMINATING-SET-CACHE: Only the clients that belong to a dominating set of the clients cache the first segment.

24.3.2.1 ALL-CACHE

ALL-CACHE requires every existing client to cache the first segment. Since the clients are bandwidth limited, we should not let an existing client forward the cached data to more than one other client at the same time. Therefore, in choosing Y, we skip clients that have been forwarding their cache to other clients.

24.3.2.2 RANDOM-CACHE

Clearly, by caching the first segment at every existing client, we significantly increase the chance subsequent clients can join the service instantly. The trade-off of this strategy, however, is the amount of storage space needed for caching. Alternatively, we can use selective caching, in which only a number of selected clients need to cache. The advantage of selective caching is the saving on caching space.

The simplest selective-caching algorithm is RANDOM-CACHE, in which a client decides to cache the first segment based on a probability. When a new client starts a video request, if it finds an existing client in its neighborhood who has the cache, the playback is the same as that in the case of ALL-CACHE. If no cache exists in the neighborhood, the new client has to wait for the first segment to arrive from the broadcast channels as usual. Obviously, the average service delay of RANDOM-CACHE is longer than that for ALL-CACHE.

24.3.2.3 DOMINATING-SET-CACHE

Is there any way that a new client can always find a cache of the first segment in its neighborhood? The answer is yes. The key idea of DOMINATING-SET-CACHE (DSC) is to maintain a dominating set of all the clients. A dominating set of the set of clients is $\{Y_1, Y_2, \ldots, Y_k\}$, such that for any other client X there exists Y_j, which is a neighbor of X. We denote this dominating set by *Dset*. DSC requires that all clients in *Dset* cache the first segment.

The use of a dominating set of mobile hosts was proposed in many MANET works, mostly in wireless broadcasting/routing protocols.[2,29,30] The playback of a new client X's arrival is as follows (illustrated in figure 24.3):

- *Case 1: X* is in the transmission range of a client $Y \in Dset$ and Y is not currently forwarding cache to any other client.
  1. X makes use of the cache at Y and immediately plays the video as explained earlier.
- *Case 2: X* is in the transmission range of a client Y outside *Dset* and not in the transmission range of any client in *Dset*.
  1. Y finds a neighbor $Z \in Dset$ such that Z currently is not forwarding cache to any other client.

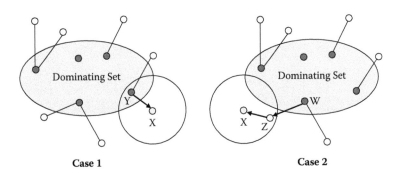

Case 1   Case 2

Figure 24.3   If a new client can reach an existing client, the former always finds a cache within two hops.

> 2. *Y* downloads the broadcast portion that *X* misses from *Z* and forwards it to *X*. *X* can play the video immediately as explained earlier.

■ *Case 3*: Neither case 1 nor case 2 holds. In this case, *X* waits until the next broadcast of segment 1. As soon as the first segment arrives, *X* starts playing it and remains on the same channel to play the rest of the video.

DSC guarantees that if a new client can reach an existing client, there is always a cache of the missing portion within two hops. Although streaming video is a challenging problem in multihop wireless networks, DSC should work well because two hops is short and the size of the missing portion is small; thus the cache downloading is quick.

A failure may occur while a new client is downloading its missing portion from an existing node; for example, when the existing node moves far away or quits the system. The new client detects this failure by observing that it has been waiting for the next packet for long enough. In this case, the new client can repeat the cache search above. However, if a new cache is found, the new client just needs to download part of the missing portion, which has not been downloaded from the previous cache. Again, since the cache is less than two hops away and the missing portion is short, we expect a small probability that a client needs to switch to a new cache.

Many distributed algorithms were proposed to solve the dominating set problems in wireless ad hoc networks (e.g., references 2, 6, and 30). Because mobile hosts may move or fail, these algorithms allow a mobile host to change status from "not in dominating set" to "belong to dominating set." Our situation is different. We need to decide if a client belongs to *DSet* as soon as it joins the system. If it is in *DSet*, it will cache the first segment. If DSC decides a client is not in *DSet*, this client will not hold any cache and therefore will never belong to this set in the future. Therefore, we just use the following policy to decide whether a new client is going to cache: *Initially, there is no client and DSet is empty. A new client belongs to DSet if and only if no client in DSet is within the transmission range of the new client.* To implement this policy, the new client *X* broadcasts a request and any client *Y* who intercepts the request will send a reply back to *X* if *Y* holds a cache. If *X* receives at least a reply, the new client decides that it will not cache (which also means *X* is not in *DSet*). If *X* does not receive any reply, *X* decides that it will cache (which also means *X* belongs to *DSet*).

We may find a case where *DSet* becomes not a dominating set of clients, illustrated in figure 24.4. The simulation results reported in Tran et al.[27] showed that the effectiveness of MobiVoD remained even in such a case.

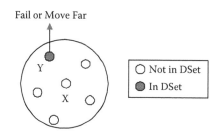

Figure 24.4 *X* **is currently not in** *DSet*. **After** *Y* **fails or moves far away,** *X* **cannot be reached by any caching node. If a new client arrives close to** *X*, **the new client cannot make use of any cache and therefore has to wait until the next broadcast of the first segment from its local forwarder.**

24.3.2.4 Comparison

Simulation results are provided in Tan et al.[27] for comparison of ALL-CACHE, RANDOM-CACHE, and DSC. In its study, ALL-CACHE provides almost true VOD services; however, storage is required for caching 20 percent of the video. This is a drawback that makes ALL-CACHE least desirable by current mobile clients. DSC and RANDOM-CACHE, with much less caching space occupancies, perform similarly to ALL-CACHE in terms of client bandwidth requirement, cache distance, and start-up overhead. Inaddition, DSC and RANDOM-CACHE offer service delays much better than without caching. Indeed, in most scenarios, they are more than nine times better than without caching. Between DSC and RANDOM-CACHE, DSC is more preferable in terms of service delay. On the other hand, the advantage of RANDOM-CACHE is its flexibility in choosing the number of clients who will cache.

24.4 Summary and Future Work

This chapter explored periodic broadcast techniques for video-on-demand (VOD) services in a wireless networking environment. Although broadcasting is the nature of wireless communications, VOD broadcasting is not to be trivially implemented in a wireless network. This is because, unlike the Internet where one-to-one communication does not affect nodes not involved in the communication, transmission in a wireless environment between two nodes may interfere with transmissions by other nodes. Therefore, video broadcasts, if not scheduled properly, may result in significant signal loss and expensive resource consumption.

We reviewed VOD periodic broadcast techniques already designed for the Internet. We also discussed their pros and cons if adapted to work for wireless networks. Most of these techniques require significant client bandwidth and caching space, which is not suitable for mobile wireless clients. On the other hand, none of them can offer true VOD. We presented MobiVoD—a novel technique aimed to be simple so that it is feasible for wireless environment, yet providing much shorter service delay than the traditional periodic broadcasting techniques. Using today's technologies, clients can communicate wirelessly through access points, base stations, or one-to-one in an ad hoc manner with each other. The key idea of MobiVoD is to utilize the collaboration among the clients.

We implicitly assumed in this chapter that clients are homogeneous, meaning they have the same capabilities (ratio transmission radius, wireless bandwidth, etc.). It is better to relax this assumption so the system is more accessible to different types of clients, especially those having bandwidth less than the video consumption rate. For this purpose, MobiVoD can be extended by employing a multiresolution or layered video coding approach.[4,20] In a heterogeneous system, two clients are considered neighbors if and only if they are in the transmission range of each other. Each video is encoded into several "layers," including a base layer and one or more enhancement layers. The base layer provides the version of least quality, while its combination enhancement layers provide better quality. Instead of broadcasting the entire video as in pure periodic broadcasting, the server, and thus every local forwarder, broadcasts all layers on separate channels. A new client selects a combination of layers that best match its resource constraints and only tunes in the corresponding channels to download such layers. As for the initial portion that the client misses from the current broadcasts, it searches for a nearby client who caches a version of the first segment (a version is a combination of the base layer and one or more enhancement layers). If more than one such client is found, the client with the highest-quality version is selected.

Another extension of MobiVoD is to allow a client to download a cache more than two hops away. This enhancement would increase the chance of providing true on-demand services to clients, which is, however, suggested only when wireless bandwidth is more advanced than the current.

References

1. C. C. Aggarwal, J. L. Wolf, and P. S. Yu. 1996. A permutation-based pyramid broadcasting scheme for video-on-demand systems. In *Proceedings of the IEEE International Confereence on Multimedia Systems '96*, 118–126.
2. K. M. Alzoubi, P.-J. Wan, and O. Frieder. 2002. Message-optimal connected-dominating-set construction for routing in mobile ad hoc networks. In *ACM International Symposium on Mobile Ad Hoc Networking and Computing (MobiHoc)*.
3. J. G. Andrews, A. Ghosh, and R. Muhamed. 2007. *Fundamentals of WiMAX: Understanding broadband wireless networking—The definitive guide to WiMAX technology*. Englewood Cliffs, NJ: Prentice Hall Professional.
4. S. Bajaj, L. Breslau, and S. Shenker. 1998. Uniform versus priority dropping for layered video. In *ACM SIGCOMM*, 131–143.
5. Bluetooth. http://www.bluetooth.com/.
6. Y. P. Chen and A. L. Liestman. 2002. Approximating minimum size weakly-connected dominating sets for clustering mobile ad hoc networks. In *ACM International Symposium on Mobile Ad Hoc Networking and Computing (MobiHoc)*.
7. Comcast. Comcast spotlight: Video advertising on demand. http://www.comcastspotlight.com.
8. Concurrent. On-demand and real-time linux solutions. http://www.ccur.com/.
9. A. Dan, D. Sitaram, and P. Shahabuddin. 1994. Scheduling policies for an on-demand video server with batching. In *Proceedings of ACM MULTIMEDIA*, 15–23.
10. S. Gruber, J. Rexford, and A. Basso. 2000. Protocol considerations for a prefix-caching proxy for multimedia streams. In *Proceedings of the 9th International WWW Conference*, 657–68.
11. K. A. Hua, Y. Cai, and S. Sheu. 1998. Patching: A multicast technique for true video-on-demand services. In *Proceedings of ACM MULTIMEDIA*, 191–200.
12. K. A. Hua and S. Sheu. 1997. Skyscraper broadcasting: A new broadcasting scheme for metropolitan video-on-demand systems. In *Proceedings of the ACM SIGCOMM' 97*, 89–100.
13. K. A. Hua, M. A. Tantaoui, and W. Tavanapong. 2004. Video delivery technologies for large-scale deployment of multimedia applications. *Proceedings of the IEEE* 92:1439–51.
14. K. A. Hua, D. A. Tran, and R. Villafane. 2000. Caching multicast protocol for on-demand video delivery. In *Proceedings of the ACM/SPIE Conference on Multimedia Computing and Networking*, 2–13.
15. IEEE. 1999. Wireless LAN medium access control (mac) and physical layer (phy) specification.
16. L. Juhn and L. Tseng. 1997. Harmonic broadcasting for video-on-demand service. *IEEE Transactions on Broadcasting* 43:268–71.
17. L. Juhn and L. Tseng. 1998. Fast data broadcasting and receiving scheme for popular video service. *IEEE Transactions on Broadcasting* 44:100–5.
18. J. B. Kwon and H. Y. Heom. 2002. Providing VCR functionality in staggered video broadcasting. *IEEE Transactions on Consumer Electronics* 48:41–48.
19. Motorola. Motorola on-demand solutions. http://broadband.motorola.com/business/ondemand/VideoOnDemand.html.
20. V. N. Padmanabhan, H. J. Wang, P. A. Chou, and K. Sripanidkulchai. 2002. Distributing streaming media content using cooperative networking. In *ACM/IEEE NOSSDAV*, 177–86.
21. J. F. Paris, S. W. Carter, and D. D. E. Long. 1998. Efficient broadcasting protocols for video on demand. In *Proceedings of the 6th International Symposium on Modeling, Analysis and Simulation of Computer and Telecommunication Systems (MASCOTS '98)*, 127–32.

22. J. F. Paris, S. W. Carter, and D. D. E. Long. 1999. A hybrid broadcasting protocol for video on demand. In *ACM/SPIE Conference on Multimedia Computing and Networking.*

23. SeaChange. Seachange international: The global leader in video on demand. http://www.schange.com/.

24. S. Sen, L. Gao, and D. Towsley. 2001. Frame-based periodic broadcast and fundamental resource tradeoffs. In *IEEE Performance, Computing and Communications Conference,* 77–83.

25. D. A. Tran, K. Hua, and T. Do. 2004. A peer-to-peer architecture for media streaming. *IEEE JSAC,* special issue on advances in service overlay networks, 22.

26. D. A. Tran, K. A. Hua, and T. T. Do. 2003. Zigzag: An efficient peer-to-peer scheme for media streaming. In *IEEE INFOCOM.*

27. D. A. Tran, K. A. Hua, and M. Le. MobiVOD. 2004. A video-on-demand system design for mobile ad hoc networks. In *Proceedings of IEEE International Conference on Mobile Data Management (MDM 2004).*

28. S. Viswanathan and T. Imielinski. 1996. Metropolitan area video-on-demand service using pyramid broadcasting. *ACM Multimedia Systems Journal* 4:179–208.

29. B. Williams and T. Camp. 2002. Comparison of broadcasting techniques for mobile ad hoc networks. In *ACM Symposium on Mobile Adhoc Networking and Computing (MOBIHOC 2002).*

30. J. Wu. 2002. Extended dominating-set-based routing in ad hoc wireless networks with unidirectional links. *IEEE Transactions on Parallel and Distributed Systems* 13:866–81.

31. K.-L. Wu, P. S. Yu, and J. L. Wolf. 2001. Segment-based proxy caching of multimedia streams. In *Proceedings of the 10th International WWW Conference,* 36–44.

32. D. Xu, M. Hefeeda, S. Hambrusch, and B. Bhargava. 2002. On peer-to-peer media streaming. In *IEEE Conference on Distributed Computing and Systems,* 363–71.

Index